Waste Treatment in the Process Industries

Waste Treatment in the Process Industries

edited by

Lawrence K. Wang
Yung-Tse Hung
Howard H. Lo
Constantine Yapijakis

A CRC title, part of the Taylor & Francis imprint, a member of the Taylor & Francis Group, the academic division of T&F Informa plc.

This material was previously published in the *Handbook of Industrial and Hazardous Wastes Treatment, Second Edition*

Published in 2006 by
CRC Press
Taylor & Francis Group
6000 Broken Sound Parkway NW, Suite 300
Boca Raton, FL 33487-2742

CRC Press is an imprint of Taylor & Francis Group

Printed in the United States of America on acid-free paper
10 9 8 7 6 5 4 3 2 1

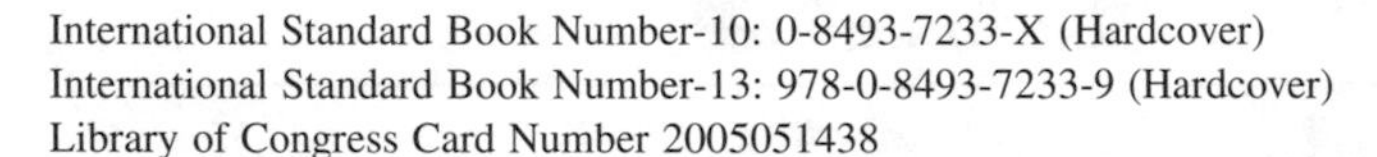
International Standard Book Number-10: 0-8493-7233-X (Hardcover)
International Standard Book Number-13: 978-0-8493-7233-9 (Hardcover)
Library of Congress Card Number 2005051438

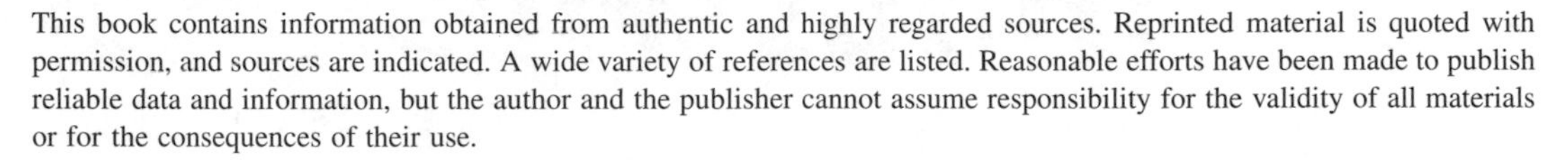

Library of Congress Cataloging-in-Publication Data

Waste treatment in the process industries / editors, Lawrence K. Wang ... [et al.].
p. cm.
Includes bibliographical references and index.
ISBN 0-8493-7233-X (alk. paper)
1. Factory and trade waste--Management. 2. Hazardous wastes--Management. 3. Manufacturing processes--Environmental aspects. 4. Industries--Environmental aspects. I. Wang, Lawrence K.

TD897W37 2005
628.4--dc22 2005051438

Visit the Taylor & Francis Web site at
http://www.taylorandfrancis.com

and the CRC Press Web site at
http://www.crcpress.com

Preface

Environmental managers, engineers, and scientists who have had experience with process industry waste management problems have noted the need for a book that is comprehensive in its scope, directly applicable to daily waste management problems of the industry, and widely acceptable by practicing environmental professionals and educators.

Many standard industrial waste treatment texts adequately cover a few major technologies for conventional in-plant environmental control strategies in the process industry, but no one book, or series of books, focuses on new developments in innovative and alternative technology, design criteria, effluent standards, managerial decision methodology, and regional and global environmental conservation.

This book emphasizes in-depth presentation of environmental pollution sources, waste characteristics, control technologies, management strategies, facility innovations, process alternatives, costs, case histories, effluent standards, and future trends for the process industry, and in-depth presentation of methodologies, technologies, alternatives, regional effects, and global effects of important pollution control practices that may be applied to the industry. This book covers new subjects as much as possible.

Special efforts were made to invite experts to contribute chapters in their own areas of expertise. Since the area of process industry waste treatment is very broad, no one can claim to be an expert in all areas; collective contributions are better than a single author's presentation for a book of this nature.

This book is one of the derivative books of the *Handbook of Industrial and Hazardous Wastes Treatment*, and is to be used as a college textbook as well as a reference book for the process industry professional. It features the major industrial process plants or installations that have significant effects on the environment. Specifically this book includes the following process industry topics: industrial ecology, bioassay, biotechnology, in-plant management, pharmaceutical industry, oil fields, refineries, soap and detergent industry, textile mills, phosphate industry, pulp mills, paper mills, pesticide industry, rubber industry, and power industry. Professors, students, and researchers in environmental, civil, chemical, sanitary, mechanical, and public health engineering and science will find valuable educational materials here. The extensive bibliographies for each type of industrial process waste treatment or practice should be invaluable to environmental managers or researchers who need to trace, follow, duplicate, or improve on a specific process waste treatment practice.

The intention of this book is to provide technical and economical information on the development of the most feasible total environmental control program that can benefit both process industry and local municipalities. Frequently, the most economically feasible methodology is combined industrial-municipal waste treatment.

We are indebted to Dr. Mu Hao Sung Wang at the New York State Department of Environmental Conservation, Albany, New York, who co-edited the first edition of the

Handbook of Industrial and Hazardous Wastes Treatment, and to Ms. Kathleen Hung Li at NEC Business Network Solutions, Irving, Texas, who is the consulting editor for this new book.

Lawrence K. Wang
Yung-Tse Hung
Howard H. Lo
Constantine Yapijakis

Contents

Contributors

Donald B. Aulenbach Rensselaer Polytechnic Institute, Troy, New York, U.S.A.

Thomas Bechtold Leopold Franzens University, Innsbruck, Austria

Eduard Burtscher Leopold Franzens University, Innsbruck, Austria

Sudhir Kumar Gupta Indian Institute of Technology, Bombay, India

Sunil Kumar Gupta Indian Institute of Technology, Bombay, India

Yung-Tse Hung Cleveland State University, Cleveland, Ohio, U.S.A.

Volodymyr Ivanov Nanyang Technological University, Singapore

Venera Z. Latypova Kazan State University, Kazan, Russia

Kathleen Hung Li NEC Business Network Solutions, Irving, Texas, U.S.A.

Howard H. Lo Cleveland State University, Cleveland, Ohio, U.S.A.

Svetlana Yu. Selivanovskaya Kazan State University, Kazan, Russia

Nadezda Yu. Stepanova Kazan Technical University, Kazan, Russia

Suresh Sumathi Indian Institute of Technology, Bombay, India

Jerry R. Taricska Hole Montes, Inc., Naples, Florida, U.S.A.

Joo-Hwa Tay Nanyang Technological University, Singapore

Stephen Tiong-Lee Tay Nanyang Technological University, Singapore

Lawrence K. Wang Lenox Institute of Water Technology and Krofta Engineering Corporation, Lenox, Massachusetts and Zorex Corporation, Newtonville, New York, U.S.A.

Joseph M. Wong Black & Veatch, Concord, California, U.S.A.

Constantine Yapijakis The Cooper Union, New York, New York, U.S.A.

1

Implementation of Industrial Ecology for Industrial Hazardous Waste Management

Lawrence K. Wang
Lenox Institute of Water Technology and Krofta Engineering Corporation, Lenox, Massachusetts and Zorex Corporation, Newtonville, New York, U.S.A.

Donald B. Aulenbach
Rensselaer Polytechnic Institute, Troy, New York, U.S.A.

1.1 INTRODUCTION

Industrial ecology (IE) is critically reviewed, discussed, analyzed, and summarized in this chapter. Topics covered include: IE definitions, goals, roles, objectives, approach, applications, implementation framework, implementation levels, industrial ecologists' qualifications, and ways and means for analysis and design. The benefits of IE are shown as they relate to sustainable agriculture, industry, and environment, zero emission and zero discharge, hazardous wastes, cleaner production, waste minimization, pollution prevention, design for environment, material substitution, dematerialization, decarbonation, greenhouse gas, process substitution, environmental restoration, and site remediation [1–46]. Case histories using the IE concept have been gathered by the United Nations Industrial Development Organization (UNIDO), Vienna, Austria [39–41]. This chapter presents these case histories to illustrate cleaner production, zero discharge, waste minimization, material substitution, process substitution, and decarbonization.

1.2 DEFINITIONS OF INDUSTRIAL ECOLOGY

Industry, according to the Oxford English Dictionary, is "intelligent or clever working" as well as the particular branches of productive labor. Ecology is the branch of biology that deals with the mutual relations between organisms and their environment. Ecology implies more the webs of natural forces and organisms, their competition and cooperation, and how they live off one another [2–4].

The recent introduction of the term "industrial ecology" stems from its use by Frosch and Gallopoulos [10] in a paper on environmentally favorable strategies for manufacturing. Industrial ecology (IE) is now a branch of systems science for sustainability, or a framework for designing and operating industrial systems as sustainable and interdependent with natural systems. It seeks to balance industrial production and economic performance with an emerging understanding of local and global ecological constraints [10,13,20].

A system is a set of elements inter-relating in a structured way. The elements are perceived as a whole with a common purpose. A system's behavior cannot be predicted simply by analysis of its individual elements. The properties of a system emerge from the interaction of its elements and are distinct from their properties as separate pieces. The behavior of the system results from the interaction of the elements and between the system and its environment (system + environment = a larger system). The definition of the elements and the setting of the system boundaries are "subjective" actions.

In this context, industrial systems apply not only to private sector manufacturing and service, but also to government operations, including provision of infrastructure. A full definition of industrial systems will include service, agricultural, manufacturing, military and civil operations, as well as infrastructure such as landfills, recycling facilities, energy utility plants, water transmission facilities, water treatment plants, sewer systems, wastewater treatment facilities, incinerators, nuclear waste storage facilities, and transportation systems.

An industrial ecologist is an expert who takes a systems view, seeking to integrate and balance the environmental, business, and economic development interests of the industrial systems, and who will treat "sustainability" as a complex, whole systems challenge. The industrial ecologist will work to create comprehensive solutions, often simply integrating separate proven components into holistic design concepts for possible implementation by the clients.

A typical industrial ecology team includes IE partners, associates, and strategic allies qualified in the areas of industrial ecology, eco-industrial parks, economic development, real estate development, finance, urban planning, architecture, engineering, ecology, sustainable agriculture, sustainable industry systems, organizational design, and so on. The core capability of the IE team is the ability to integrate the contributions of these diverse fields into whole systems solutions for business, government agencies, communities, and nations.

1.3 GOAL, ROLE, AND OBJECTIVES

An industrial ecologist's tasks are to interpret and adapt an understanding of the natural system and apply it to the design of man-made systems, in order to achieve a pattern of industrialization that is not only more efficient, but also intrinsically adjusted to the tolerances and characteristics of the natural system. In this way, it will have a built-in insurance against further environmental surprises, because their essential causes will have been designed out [29].

A practical goal of industrial ecology is to lighten the environmental impact per person and per dollar of economic activity, and the role of the industrial ecologist is to find leverage, or opportunities for considerable improvement using practical effort. Industrial ecology can search for leverage wherever it may lie in the chain, from extraction and primary production through final consumption, that is, from cradle to rebirth. In this regard, a performing industrial ecologist may become a preserver when achieving endless reincarnations of materials [3].

An overarching goal of IE is the establishment of an industrial system that recycles virtually all of the materials. It uses and releases a minimal amount of waste to the environment. The industrial systems' developmental path follows an orderly progression from Type I, to Type II, and finally to Type III industrial systems, as follows:

1. Type I industrial systems represent an initial stage requiring a high throughput of energy and materials to function, and exhibit little or no resource recovery. It is a once flow-through system with rudimentary end-of-pipe pollution controls.

2. Type II industrial systems represent a transitional stage where resource recovery becomes more integral to the workings of the industrial systems, but does not satisfy its requirements for resources. Manufacturing processes and environmental processes are integrated at least partially. Whole facility planning is at least partially implemented.
3. Type III industrial systems represent the final ideal stage in which the industrial systems recycle all of the material outputs of production, although still relying on external energy inputs.

A Type III industrial ecosystem can become almost self-sustaining, requiring little input to maintain basic functions and to provide a habitat for thousands of different species. Therefore, reaching Type III as a final stage is the goal of IE [11]. Eventually communities, cities, regions, and nations will become sustainable in terms of natural resources and the environment.

According to Frosch [9]:

> "The idea of industrial ecology is that former waste materials, rather than being automatically sent for disposal, should be regarded as raw materials – useful sources of materials and energy for other processes and products. The overall idea is to consider how the industrial system might evolve in the direction of an interconnected food web, analogous to the natural system, so that waste minimization becomes a property of the industrial system even when it is not completely a property of a individual process, plant, or industry."

IE provides a foundation for sustainable industrialization, not just incremental improvement in environmental management. The objectives of IE suggest a potential for reindustrialization in economies that have lost major components of their industrial base. Specifically, the objective of industrial ecology is not merely to reduce pollution and waste as traditionally conceived, it is to reduce throughput of all kinds of materials and fuels, whether they leave a site as products, emissions, or waste.

The above objectives of IE have shown a new path for both industrial and developing countries. Central objectives of an industrial-ecology-based development strategy are making economies profoundly more efficient in resource use, less dependent upon nonrenewable resources, and less polluting. A corollary objective is repair of past environmental damage and restoration of ecosystems. Developing countries that recognize the enormous opportunity opened by this transformation can leapfrog over the errors of past industrialization. They will have more competitive and less polluting businesses [21].

1.4 APPROACH AND APPLICATIONS

The IE approach involves (a) application of systems science to industrial systems, (b) defining the system boundary to incorporate the natural world, and (c) seeking to optimize that system.

Industrial ecology is applied to the management of human activity on a sustainable basis by: (a) minimizing energy and materials usage; (b) ensuring acceptable quality of life for people; (c) minimizing the ecological impact of human activity to levels natural systems can sustain; (d) conserving and restoring ecosystem health and maintaining biodiversity; (e) maintaining the economic viability of systems for industry, trade, and commerce; (f) coordinating design over the life cycle of products and processes; and (g) enabling creation of short-term innovations with awareness of their long-term impacts.

Application of IE will improve the planning and performance of industrial systems of all sizes, and will help design local and community solutions that contribute to national and global solutions. For small industrial systems applications, IE helps companies become more competitive by improving their environmental performance and strategic planning. For medium-sized industrial systems, IE helps communities develop and maintain a sound industrial base and infrastructure, without sacrificing the quality of their environments. For large industrial systems, IE helps government agencies design policies and regulations that improve environmental protection while building business competitiveness.

Several scenarios [20] offer visions of full-blown application of IE at company, city, and developing country levels. Lists of organizations, on-line information sources, and bibliographies in the book provide access to sources of IE information.

1.5 TASKS, STEPS, AND FRAMEWORK FOR IMPLEMENTATION

Pratt and Shireman [25] propose three simple but extraordinarily powerful tasks, over and over again, for practicing industrial ecological management:

1. *Task 1, Eco-management*: Brainstorm, test, and implement ways to reduce or eliminate pollution;
2. *Task 2, Eco-auditing*: Identify specific examples of materials use, energy use, and pollution and waste reduction (any form of throughput);
3. *Task 3, Eco-accounting*: Count the money. Count how much was saved, then count how much is still being spent creating waste and pollution, and start the cycle over.

The above three tasks are essentially eco-management, eco-auditing, and activity-based eco-accounting, which are part of an inter-related ecological management framework. Pratt and Shireman [25] further suggest a way to implement the three tasks by going through a series of perhaps 14 specific steps, spiraling outward from the initial Step 1, "provide overall corporate commitment," to the final Step 14, "continue the process," which flows back into the cycle of continuous improvement:

Step 1: Provide overall corporate commitment.
Step 2: Organize the management efforts.
Step 3: Organize the audit.
Step 4: Gather background information.
Step 5: Conduct detailed assessment.
Step 6: Review and organize data.
Step 7: Identify improvement options.
Step 8: Prioritize options.
Step 9: Implement fast-track options.
Step 10: Analyze options.
Step 11: Implement best options.
Step 12: Measure results.
Step 13: Standardize improvement.
Step 14: Continue the process.

Each of the components within the "three tasks" does not necessarily fall into discrete categories. For clarity of presentation, each of the tasks is divided into steps. Table 1 shows that these steps overlap and are repeated within this systematic approach. The names of tasks and steps have been slightly modified by the current author for ease of presentation and explanation.

Table 1 Implementation Process for Applying Industrial Ecology at Corporate Level

Task 1: Eco-management		Task 2: Eco-auditing		Task 3: Eco-accounting	
Step 1	Overall corporate commitment	*Step 3*	Organize the audit	*Step 5*	Conduct detailed assessment
Step 2	Organize management efforts	*Step 4*	Gather background information	*Step 12*	Measure results
Step 7	Identify improvement options	*Step 5*	Conduct detailed assessment		
Step 8	Prioritize options	*Step 6*	Review and organize data		
Step 9	Implement fast-track options	*Step 7*	Identify improvement options		
Step 10	Analyze options	*Step 12*	Measure results		
Step 11	Implement best options				
Step 13	Standardize improvements				
Step 14	Continue the process				

As shown in Table 1, the company must initially provide the overall corporate commitment (Step 1) and organize the management efforts (Step 2) in Task 1 that will drive this implementation process forward (and around). Once the industrial ecological implementation process is initiated by the eco-management team in Task 1 (Steps 1 and 2), the eco-auditing team begins its Task 2 (Steps 3–7) with background and theory that support an industrial ecology approach, and the eco-accounting team begins its Task 3 (Step 5) to conduct detailed assessment. The eco-management team must then provide step-by-step guidance and directions in Task 1 (Steps 7–11) to identify, prioritize, implement, analyze, and again implement the best options. Subsequently, both the eco-auditing team (Task 2, Step 12) and the eco-accounting team (Task 3, Step 12) should measure the results of the implemented best options (Task 1, Step 11). The overall responsibility finally to standardize the improvements, and to continue the process until optimum results are achieved (Task 1, Steps 13, 14), will still be carried out by the eco-management team.

1.6 QUALIFICATIONS OF INDUSTRIAL ECOLOGISTS

The implementation process for applying industrial ecology at the corporate level (as shown in Table 1) may sound modest in its concept. In reality, each step in each task will face technical, economical, social, legal, and ecological complexity, and can be accomplished only by qualified industrial ecologists.

Accordingly, the most important element for industrial ecology implementation will be drawing on in-company expertise and enthusiasm as well as outside professional assistance. The qualified industrial ecologists retained for their service must have their respective knowledge in understanding the rules and regulations, assessing manufacturing processes and wastes, identifying various options, and measuring results. Because it is difficult to find a single industrial ecologist who has all the required knowledge, several experts in different areas are usually assembled together to accomplish the required IE tasks.

The team of qualified industrial ecologists assembled should have a clear sense of the possibilities and methodologies in the following professional areas specifically related to the problem:

1. Industrial or manufacturing engineering of the target industrial system;
2. Energy consumption and material balances for environmental auditing;
3. Cleaner production, materials substitution, and dematerialization;
4. Zero emission, decarbonization, waste minimization, and pollution prevention;
5. Sustainable agriculture and sustainable industry;
6. Industrial metabolism and life-cycle analyses of products;
7. Site remediation and environmental restoration;
8. Ecological and global environmental analyses;
9. Accounting and economical analyses;
10. Legal, political affairs, and IE leverage analyses.

An IE team may not be required to have all of the above expertise. For example, the expertise of site remediation may not be required if the industrial system in question is not contaminated by hazardous substances. The expertise of global environmental analyses may not be needed if the IE level is at the company level, instead of at the regional or national level.

1.7 WAYS AND MEANS FOR ANALYSIS AND DESIGN

Each task and each step outlined in Table 1 for implementation of an industrial ecology project cannot be accomplished without understanding the ways and means for IE analysis and design. Indigo Development, a Center in the Sustainable Development Division of RPP International [13] has identified seven IE methods and tools for analysis and design: (a) industrial metabolism; (b) urban footprint; (c) input–output models; (d) life-cycle assessment; (e) design for environment; (f) pollution prevention; and (g) product life extension. Ausubel [2] and Wernick et al. [45] suggest that searching for leverage will be an important tool for IE implementation.

The United Nations Industrial Development Organization [39–41] and Ausubel and Sladovich [4] emphasize the importance of cleaner production, pollution prevention, waste minimization, sustainable development, zero emission, materials substitution, dematerialization, decarbonization, functional economic analysis, and IE indicators. These ways and means for analysis and design of industrial ecology are described separately herein.

1.8 SUSTAINABLE AGRICULTURE, INDUSTRY, AND ENVIRONMENT

Because IE is a branch of systems science of sustainability or a framework for designing and operating industrial systems as sustainable living systems interdependent with natural systems, understanding and achieving sustainable agriculture and industry will be the most important key to the success of sustainable environment.

An industrial ecologist may perceive the whole system required to feed planet Earth, preserve and restore its farmlands, preserve ecosystems and biodiversity, and still provide water, land, energy, and other resources for a growing population. The following is only one of many possibilities for achieving sustainable agriculture and industry: utilization of large volumes of carbon dioxide gases discharged from industrial and commercial stacks as a resource for decarbonation, pollution control, resource development, and cost saving [22,24,39–42].

Meeting the challenges involved in sustainable systems development, which can be either technical or managerial, will require interdisciplinary coordination among many technical, economic, social, political, and ecological research disciplines.

1.9 ZERO EMISSION, ZERO DISCHARGE, CLEANER PRODUCTION, WASTE MINIMIZATION, POLLUTION PREVENTION, DESIGN FOR ENVIRONMENT, MATERIAL SUBSTITUTION, DEMATERIALIZATION, AND PROCESS SUBSTITUTION

1.9.1 Terminologies and Policy Promotion

The terms of zero emission, zero discharge, cleaner production, waste minimization, pollution prevention, design for environment, material substitution, and dematerialization are all closely related, and each is self-explanatory. The U.S. Environmental Protection Agency (USEPA), the United Nations Industrial Development Organization (UNIDO), and other national and international organizations at different periods of time have promoted each [8,19,23,30–34,39–46].

Design for environment (DFE) is a systematic approach to decision support for industrial ecologists, developed within the industrial ecology framework. Design for environment teams apply this systematic approach to all potential environmental implications of a product or process being designed: energy and materials used; manufacture and packaging; transportation; consumer use, reuse, or recycling; and disposal. Design for environment tools enable consideration of these implications at every step of the production process from chemical design, process engineering, procurement practices, and end-product specification to postuse recycling or disposal. It also enables designers to consider traditional design issues of cost, quality, manufacturing process, and efficiency as part of the same decision system.

1.9.2 Zero Emission

Zero emission has been promoted by governments and the automobile industry in the context of energy systems, particularly in relation to the use of hydrogen as an energy source. Recent attention has focused on electric cars as zero-emission vehicles and the larger question of the energy and material system in which the vehicles are embedded. Classic studies about hydrogen energy may be found in a technical article by Hafele et al. [12]. The term "zero emission" is mainly used in the field of air emission control.

1.9.3 Zero Discharge

Zero discharge is aimed at total recycling of water and wastewater within an industrial system, and elimination of any discharge of toxic substances. Therefore, the term "zero discharge" is mainly used in water and wastewater treatment plants, meaning total water recycle. In rare cases, total recycling of air effluent within a plant is also called "zero discharge." Wastewater recycling is important, not only for environmental protection, but also for water conservation in water shortage areas, such as California, United States. Several successful IE case histories are presented to show the advantages of zero discharge:

Total Wastewater Recycle in Potable Water Treatment Plants

The volume of wastewater produced from a potable water treatment plant (either a conventional sedimentation filtration plant or an innovative flotation filtration plant) amounts to about 15% of a plant's total flow. Total wastewater recycle for production of potable water may save water and cost, and solve wastewater discharge problems [15,35–38].

Total Water and Fiber Recycle in Paper Mills

The use of flotation clarifiers and fiber recovery facilities in paper mills may achieve near total water and fiber recycle and, in turn, accomplish the task of zero discharge [16].

Total Water and Protein Recycle in Starch Manufacturing Plants

The use of membrane filtration and protein recovery facilities in starch manufacturing plants may achieve near total water and protein recycle and, in turn, accomplish the task of zero discharge [39–41].

Cleaner production, waste minimization, pollution prevention, designs for benign environmental impacts, material substitution, and dematerialization are all inter-related terms. Cleaner production is formally used and promoted by UNIDO (Vienna, Austria) [39–40], while waste minimization and pollution prevention are formally used and promoted by USEPA and U.S. state government agencies. Design for minimal environmental impact is very similar to cleaner production, and is mainly used in the academic field by researchers. Cleaner production emphasizes the integration of manufacturing processes and pollution control processes for the purposes of cost saving, waste minimization, pollution prevention, sustainable agriculture, sustainable industry, and sustainable environment, using the methodologies of material substitution, dematerialization, and sometimes even process substitution. Accordingly, cleaner production is a much broader term than waste minimization, pollution prevention, sustainability, material substitution, process substitution, and so on, and is similar to design for benign environmental impact. Furthermore, cleaner production implementation in an industrial system always saves money for the plant in the long run. Considering that wastes are resources to be recovered is the key for the success of an IE project using a cleaner production technology.

1.10 CASE HISTORIES OF SUCCESSFUL HAZARDOUS WASTE MANAGEMENT THROUGH INDUSTRIAL ECOLOGY IMPLEMENTATION

Several successful IE case histories are presented here to demonstrate the advantages of cleaner production for hazardous wastes management [40].

1.10.1 New Galvanizing Steel Technology Used at Delot Process SA Steel Factory, Paris, France

Galvanizing is an antirust treatment for steel. The traditional technique consisted of chemically pretreating the steel surface, then immersing it in long baths of molten zinc at 450°C. The old process involved large quantities of expensive materials, and highly polluting hazardous wastes. The cleaner production technologies include: (a) induction heating to melt the zinc, (b) electromagnetic field to control the molten zinc distribution, and (c) modern computer control of the process. The advantages include total suppression of conventional plating waste, smaller inventory of zinc, better process control of the quality and thickness of the zinc coating, reduced labor requirements, reduced maintenance, and safer working conditions. With the cleaner production technologies in place, capital cost is reduced by two-thirds compared to the traditional dip-coating process. The payback period was three years when replacing existing plant facilities.

1.10.2 Reduction of Hazardous Sulfide in Effluent from Sulfur Black Dyeing at Century Textiles, Bombay, India

Sulfur dyes are important dyes yielding a range of deep colors, but they cause a serious pollution problem due to the traditional reducing agent used with them. The old dyeing process involved four steps: (a) a water soluble dye was dissolved in an alkaline solution of caustic soda or sodium carbonate; (b) the dye was then reduced to the affinity form; (c) the fabric was dyed; and (d) the dye was converted back into the insoluble form by an oxidation process, thus preventing washing out of the dye from the fabric. The cleaner production technology involves the use of 65 parts of starch chemical Hydrol™ plus 25 parts of caustic soda to replace 100 parts of original sodium sulfide. The advantages include: reduction of sulfide in the effluent, improved settling characteristics in the secondary settling tank of the activated sludge plant, less corrosion in the treatment plant, and elimination of the foul smell of sulfide in the work place. The substitute chemical used was essentially a waste stream from the maize starch industry, which saved them an estimated US$12,000 in capital expenses with running costs at about US$1800 per year (1995 costs).

1.10.3 Replacing Toxic Solvent-Based Adhesives with Nontoxic Water-Based Adhesives at Blueminster Packaging Plant, Kent, UK

When solvent-based adhesives were used at Blueminster, UK, the components of the adhesive, normally a polymer and a resin (capable of becoming tacky), were dissolved in a suitable organic solvent. The adhesive film was obtained by laying down the solution and then removing the solvent by evaporation. In many adhesives, the solvent was a volatile organic compound (VOC) that evaporated to the atmosphere, thus contributing to atmospheric pollution. The cleaner production process here involves the use of water-based adhesives to replace the solvent-based adhesives. In comparison with the solvent-based adhesives, the water-based adhesives are nontoxic, nonpolluting, nonexplosive, nonhazardous, require only 20–33% of the drying energy, require no special solvent recovery systems nor explosion-proof process equipment, and are particularly suitable for food packaging. The economic benefits are derived mainly from the lack of use of solvents and can amount to significant cost savings on equipment, raw materials, safety precautions, and overheads.

1.10.4 Recovery and Recycling of Toxic Chrome at Germanakos SA Tannery Near Athens, Greece

Tanning is a chemical process that converts hides and skins into a stable material. Tanning agents are used to produce leather of different qualities and properties. Trivalent chromium is the major tanning agent, because it produces modern, thin, light leather suitable for shoe uppers, clothing, and upholstery. However, the residual chromium in the plant effluent is extremely toxic, and its effluent concentration is limited to 2 mg/L. A cleaner production technology has been developed to recover and reuse the trivalent chromium from the spent tannery liquors for both cost saving and pollution control. Tanning of hides is carried out with chromium sulfate at pH 3.5–4.0. After tanning, the solution is discharged by gravity to a collection pit. In the recovery process, the liquor is sieved during this transfer to remove particles and fibers originating from the hides. The liquor is then pumped to a treatment tank where magnesium oxide is added, with stirring, until the pH reaches at least 8. The stirrer is switched off and the chromium precipitates as a compact sludge of chromium hydroxide. After settling, the clear liquid is decanted off. The remaining sludge is dissolved by adding concentrated sulfuric acid until a pH of 2.5 is reached. The liquor now contains chromium sulfate and is pumped back to a storage tank for reuse. In the conventional chrome tanning processes, 20–40% of the chrome used was discharged into wastewaters as hazardous substances. In the new cleaner production process, 95–98% of the spent trivalent chromium can be recycled for reuse. The required capital investment for the Germanakos SA plant was US$40,000. Annual saving in tanning agents and pollution control was $73,750. The annual operating cost of the cleaner production process was $30,200. The total net annual savings is $43,550. The payback period for the capital investment ($40,000) was only 11 months.

1.10.5 Recovery of Toxic Copper from Printed Circuit Board Etchant for Reuse at Praegitzer Industries, Inc., Dallas, Oregon, United States

In the manufacture of printed circuit boards, the unwanted copper is etched away by acid solutions as cupric chloride. As the copper dissolves, the effectiveness of the solution falls and it must be regenerated, otherwise it becomes a hazardous waste. The traditional way of doing this was to oxidize the copper ion produced with acidified hydrogen peroxide. During the process the volume of solution increased steadily and the copper in the surplus liquor was precipitated as copper oxide and usually landfilled. The cleaner production process technology uses an electrolytic divided cell, simultaneously regenerating the etching solution and recovering the unwanted copper. A special membrane allows hydrogen and chloride ions through, but not the copper. The copper is transferred via a bleed valve and recovered at the cathode as pure flakes of copper. The advantages of this cleaner production process are: improvement of the quality of the circuit boards, elimination of the disposal costs for the hazardous copper effluent, maintenance of the etching solution at optimum composition, recovery of pure copper for reuse, and zero discharge of hazardous effluent. The annual cost saving in materials and disposal was US$155,000. The capital investment cost was $220,000. So the payback period for installation of this cleaner production technology was only 18 months.

1.10.6 Recycling of Hazardous Wastes as Waste-Derived Fuels at Southdown, Inc., Houston, Texas, United States

Southdown, Inc., engages in the cement, ready-mixed concrete, concrete products, construction aggregates, and hazardous waste management industries throughout the United States. According to Southdown, they are making a significant contribution to both the environment and energy conservation through the utilization of waste-derived fuels as a supplemental fuel source. Cement kiln energy recovery is an ideal process for managing certain organic hazardous wastes. The burning of organic hazardous wastes as supplemental fuel in the cement and other industries is their engineering approach. By substituting only 15% of its fossil fuel needs with solid hazardous waste fuel, a modern dry-process cement plant with an annual production capacity of 650,000 tons of clinker can save the energy equivalent of 50,000 barrels of oil (or 12,500 tons of coal) a year. Southdown typically replaces 10–20% of the fossil fuels it needs to make cement with hazardous waste fuels.

Of course, by using hazardous waste fuels, the nation's hazardous waste (including infectious waste) problem is at least partially solved with an economic advantage.

1.10.7 Utilization and Reduction of Carbon Dioxide Emissions at Industrial Plants

Decarbonization has been extensively studied by Dr. L. K. Wang and his associates at the Lenox Institute of Water Technology, MA, United States, and has been concluded to be technically and economically feasible, in particular when the carbon dioxide gases from industrial stacks are collected for in-plant reuse as chemicals for tanneries, dairies, water treatment plants, and municipal wastewater treatment plants [22,23,42]. Greenhouse gases, such as carbon dioxide, methane, and so on, have caused global warming over the last 50 years. Average temperatures across the world could climb between 1.4 and 5.8°C over the coming century. Carbon dioxide emissions from industry and automobiles are the major causes of global warming. According to the UN Environment Program Report released in February 2001, the long-term effects may cost the world about 304 billion U.S. dollars a year in the future. This is due to the following projected losses: (a) human life loss and property damages as a result of more frequent tropical cyclones; (b) land loss as a result of rising sea levels; (c) damages to fishing stocks, agriculture, and water supplies; and (d) disappearance of many endangered species. Technologically, carbon dioxide is a gas that can easily be removed from industrial stacks by a scrubbing process using any alkaline substances. However, the technology for carbon dioxide removal is not considered to be cost-effective. Only reuse is the solution. About 20% of organic pollutants in a tannery wastewater are dissolved proteins that can be recovered using the tannery's own stack gas (containing mainly carbon dioxide). Similarly, 78% of dissolved proteins in a dairy factory can be recovered by bubbling its stack gas (containing mainly carbon dioxide) through its waste stream. The recovered proteins from both tanneries and dairies can be reused as animal feeds. In water softening plants using chemical precipitation processes, the stack gas can be reused as precipitation agents for hardness removal. In municipal wastewater treatment plants, the stack gas containing carbon dioxide can be reused as neutralization and warming agents. Because a large volume of carbon dioxide gases can be immediately reused as chemicals in various in-plant applications, the plants producing carbon dioxide gas actually may save chemical costs, produce valuable byproducts, conserve heat energy, and reduce the global warming problem [47].

By reviewing these case histories, one will realize that materials substitution is an important tool for cleaner production and, in turn, for industrial ecology. Furthermore, materials substitution is considered a principal factor in the theory of dematerialization. The theory asserts that as a nation becomes more affluent, the mass of materials required to satisfy new or growing economic functions diminishes over time. The complementary concept of decarbonization, or the diminishing mass of carbon released per unit of energy production over time, is both more readily examined and has been amply studied by many scientists. Dematerialization is advantageous only if using fewer resources accompanies, or at least leaves unchanged, lifetime waste in processing, and wastes in production [43].

It is hoped that through industrial ecology investigations, strategies may be developed to facilitate more efficient use of material and energy resources and to reduce the release of hazardous as well as nonhazardous wastes to our precious environment. Hopefully, we will be able to balance industrial systems and the ecosystem, so our agriculture and industry can be sustained for very long periods of time, even indefinitely, without significant depletion or environmental harm. Integrating industrial ecology within our economy will bring significant benefits to everyone.

REFERENCES

1. Allen, D.T.; Butner, R.S. Industrial ecology: a chemical engineering challenge. Chem. Engng. Prog. **2002**, *98* (11), 40–45.
2. Ausubel, J.H. The virtual ecology of industry. J. Ind. Ecol. **1997**, *1* (1), 10–11.
3. Ausubel, J.H. Industrial ecology: a coming of age story. Resources **1998**, *130* (14) 28–31.
4. Ausubel, J.H.; Sladovich, H.E. *Technology and Environment*; National Academy of Science: Washington, DC, 1989.
5. Ayres, R.U.; Ayres, L.W. *Industrial Ecology: Towards Closing the Materials Cycle*; Edward Elgar Publishing: Cheltenham, UK, 1996.
6. Cox, B. High-mileage precept still just a high-priced concept. Times Union, Automotive Weekly, February 22, 2001; 16 pp.
7. AIChe. Society merges technology and ecology. Chem. Engng. Prog. **2001**, *97* (4), 13–14.
8. Evers, D.P. Facility pollution prevention. In *Industrial Pollution Prevention Handbook*; Freeman, H.M., Ed.; McGraw-Hill: New York, 1995; 155–179.
9. Frosch, R.A. Toward the end of waste: reflections on a new ecology for industry. Daedalus **1996**, *125* (3), 199–212.
10. Frosch, R.A.; Gallopoulos, N.E. Strategies for manufacturing. Scientific American **1989**, 144–152.
11. Graedel, T.E.; Allenby, B.R.; Comrie, P.R. Matrix approaches to abridged life cycle assessment. Environ. Sci. Technol. **1995**, *29*, 134A–139A.
12. Hefele, W.; Barner, H.; Messner, S.; Strubegger, M.; Anderer, J. Novel integrated energy systems: the case of zero emissions. In *Sustainable Development of the Biosphere*; Clark, W.C., Munns, R.E., Eds.; Cambridge University Press: Cambridge, UK, 171–193.
13. Indigo Development. *Creating Systems Solution for Sustainable Development through Industrial Ecology*; RPP International: Oakland, California, elowe@indigodev.com, June 5, 2000.
14. Klimisch, R.L. *Designing the Modern Automobile for Recycling. Greening Industrial Ecosystems*; Allenby, B.R., Richards, D., Eds.; National Academy Press: Washington, DC.
15. Krofta, M.; Wang, L.K. *Development of Innovative Floatation Processes for Water Treatment and Wastewater Reclamation*, National Water Supply Improvement Association Conference, San Diego, August 1988, 42 pp.
16. Krofta, M.; Wang, L.K. Total closing of paper mills with reclamation and deinking installations. Proceedings of the 43rd Industrial Waste Conference, Purdue University: W. Lafayette, IN, 1989; 673 pp.
17. Lovins, A.B.; Lovins, L.H. *Supercars: The Coming Light-Vehicle Revolution*, Technical report, Rocky Mountain Institute: Snowmass, CO, 1993.
18. Lovins, A.B.; Lovins, L.H. Reinventing the wheels. *Atlantic Monthly* **1995**, *January*.
19. Lowe, E.; Evans, L. Industrial ecology and industrial ecosystems. J. Cleaner Prod. **1995**, *3*, 1–2.
20. Lowe, E.A.; Warren, J.L.; Moran, S.R. *Discovering Industrial Ecology: An Executive Briefing and Sourcebook*; Battelle Press: Columbus, OH, 1997. ISBN 1-57477-034-9.
21. Lowe, E.A. *Creating Systems Solutions for Sustainable Development through Industrial Ecology: Thoughts on an Industrial Ecology-Based Industrialization Strategy*, Indigo Development Technical Report, RPP International: 26 Blachford Court, Oakland, California, USA, 2001.
22. Nagghappan, L. *Leather Tanning Effluent Treatment*; Lenox Institute of Water Technology: Lenox, MA. Master Thesis (Wang, L.K., Krofta, M., advisors), 2000; 167 pp.
23. NYSDEC. *New York State Waste Reduction Guidance Manual*; NYS Department of Environmental Conservation: Albany, NY, 1989.
24. Ohrt, J.A. *Physicochemical Pretreatment of a Synthetic Industrial Dairy Waste*. Lenox Institute of Water Technology: Lenox, MA. Masters Thesis (Wang, L.K.; Aulenbach, D.B., advisors), 2001; 62 pp.
25. Pratt, W.B.; Shireman, W.K. *Industrial Ecology: A How-to Manual: The Only 3 Things Business Needs to Do to Save the Earth*, Technical Manual. Global Futures Foundation: Sacramento, CA, 1996, www.globalff.org.
26. Renner, M. *Rethinking the Role of the Automobile*. Worldwatch Institute: Worldwatch Paper 84: Washington, DC, 1988.

27. Rittenhouse, D.G. Piecing together a sustainable development strategy. Chem. Engng. Prog. **2003**, *99* (3), 32–38.
28. Swan, C. *Suntrain Inc. Business Plan*. Suntrain Inc.: San Francisco, CA, 1998.
29. Tibbs, H. Industrial ecology: an environmental agenda for industry. Whole Earth Rev. **1992**, *Winter*, 4–19.
30. U.S. Congress. *From Pollution to Prevention: A Progress Report on Waste Reduction*. U.S. Congress, Office of Technology Assessment, U.S. Government Printing Office: Washington, DC, 1992; OTA-ITE-347.
31. USEPA. *Waste Minimization Issues and Options*. U.S. Environmental Protection Agency: Washington, DC, 1986; 530-SW-86-04.
32. USEPA. *Waste Minimization Benefits Manual, Phase I*. U.S. Environmental Protection Agency: Washington, DC, 1988.
33. USEPA. *Pollution Prevention Benefits Manual, Phase II*. U.S. Environmental Protection Agency: Washington, DC, 1989.
34. USEPA. *Facility Pollution Prevention Guide*. U.S. Environmental Protection Agency, Office of Solid Waste: Washington, DC, 1992; EPA/600/R-92/083.
35. Wang, L.K. Recycling and reuse of filter backwash water containing alum sludge. Water Sewage Works **1972**, *119* (5), 123–125.
36. Wang, L.K. Continuous pilot plant study of direct recycling of filter backwash water. J. Am. Water Works Assoc. **1973**, *65* (5), 355–358.
37. Wang, L.K. Design and specifications of Pittsfield water treatment system consisting of air flotation and sand filtration. Water Treatment **1991**, *6*, 127–146.
38. Wang, L.K.; Wang, M.H.S.; Kolodzig, P. Innovative and cost-effective Lenox water treatment plant. Water Treatment **1992**, *7*, 387–406.
39. Wang, L.K.; Cheryan, M. *Application of Membrane Technology in Food Industry for Cleaner Production*. The Second International Conference on Waste Minimization and Cleaner Production. United Nations Industrial Development Organization: Vienna, Austria, 1995; Technical Report No. DTT-8-6-95, 42 pp.
40. Wang, L.K.; Krouzek, J.V.; Kounitson, U. *Case Studies of Cleaner Production and Site Remediation*. United Nations Industrial Development Organization: Vienna, Austria, 1995; Training Manual No. DTT-5-4-95, 136 pp.
41. Wang, L.K.; Wang, M.H.S.; Wang, P. *Management of Hazardous Substances at Industrial Sites*. United Nations Industrial Development Organization: Vienna, Austria, 1995; Technical Report No. DTT-4-4-95, 105 pp.
42. Wang, L.K.; Lee, S.L. *Utilization and Reduction of Carbon Dioxide Emissions: An Industrial Ecology Approach*. The 2001 Annual Conference of Chinese American Academic and Professional Society (CAAPS), St. Johns University, New York, NY, USA, April 25, 2001.
43. Wernick, I.K.; Herman, R.; Govind, S.; Ausubel, J.H. Materialization and dematerialization measures and trends. Daedalus **1993**, *125* (3), 171–198.
44. Wernick, I.K.; Ausubel, J.H. *Industrial Ecology: Some Directions for Research*; The Rockefeller University: New York, 1997. ISBN 0-9646419-0-7.
45. Wernick, I.K.; Waggoner, P.E.; Ausubel, J.H. Searching for leverage to conserve forests: the industrial ecology of wood products in the U.S. Journal of Industrial Ecology **1997**, *1* (3), 125–145.
46. Wernick, I.K.; Ausubel, J.H. National Material Metrics for Industrial Ecology. In *Measures of Environmental Performance and Ecosystem Condition*; Schuize, P., Ed.; National Academy Press: Washington, DC, 1999; 157–174.
47. Wang, L.K.; Pereira, N.C.; Hung, Y. *Air Polution Control Engineering*; Human Press, Totowa, NJ, 2004.

2
Bioassay of Industrial Waste Pollutants

Svetlana Yu. Selivanovskaya and Venera Z. Latypova
Kazan State University, Kazan, Russia

Nadezda Yu. Stepanova
Kazan Technical University, Kazan, Russia

Yung-Tse Hung
Cleveland State University, Cleveland, Ohio, U.S.A.

2.1 INTRODUCTION

Persistent contaminants in the environment affect human health and ecosystems. It is important to assess the risks of these pollutants for environmental policy. Ecological risk assessment (ERA) is a tool to estimate adverse effects on the environment from chemical or physical stressors. It is anticipated that ERA will be the main tool used by the U.S. Department of Energy (USDOE) to accomplish waste management [1]. Toxicity bioassays are the important line of evidence in an ERA. Recent environmental legislation and increased awareness of the risk of soil and water pollution have stimulated a demand for sensitive and rapid bioassays that use indigenous and ecologically relevant organisms to detect the early stages of pollution and monitor subsequent ecosystem change.

Aquatic ecotoxicology has rapidly matured into a practical discipline since its official beginnings in the 1970s [2–4]. Integrated biological/chemical ecotoxicological strategies and assessment schemes have been generally favored since the 1980s to better comprehend the acute and chronic insults that chemical agents can have on biological integrity [5–8]. However, the experience gained with the bioassay of solid or slimelike wastes is as yet inadequate.

At present the risk assessment of contaminated objects is mainly based on the chemical analyses of a priority list of toxic substances. This analytical approach does not allow for mixture toxicity, nor does it take into account the bioavailability of the pollutants present. In this respect, bioassays provide an alternative because they constitute a measure for environmentally relevant toxicity, that is, the effects of a bioavailable fraction of an interacting set of pollutants in a complex environmental matrix [9–12].

The use of bioasssay in the control strategies for chemical pollution has several advantages over chemical monitoring. First, these methods measure effects in which the bioavailability of the compounds of interest is integrated with the concentration of the compounds and their intrinsic toxicity. Secondly, most biological measurements form the only way of integrating the effects on a large number of individual and interactive processes. Biomonitoring methods are often cheaper, more precise, and more sensitive than chemical analysis in detecting adverse conditions in the environment. This is due to the fact that the biological response is very integrative and accumulative in nature, especially at the higher levels of biological organization. This may lead to a reduction in the number of measurements both in space and time [12].

A disadvantage of biological effect measurements is that sometimes it is very difficult to relate the observed effects to specific aspects of pollution. In view of the present chemical-oriented pollution abatement policies and to reveal chemical specific problems, it is clear that biological effect analysis will never totally replace chemical analysis. However, in some situations the number of standard chemical analyses can be reduced, by allowing bioeffects to trigger chemical analysis (integrated monitoring), thus buying time for more elaborate analytical procedures [12].

2.2 GENERAL CONSIDERATIONS

According to USEPA, the key aspect of the ERA is the problem formulation phase. This phase is characterized by USEPA as the identification of ecosystem components at risk and specification of the endpoints used to assess and measure that risk [13]. Assessment endpoints are an expression of the valued resources to be considered in an ERA, whereas measurement endpoints are the actual measures of data used to evaluate the assessment endpoint.

Toxicity tests can be divided according to their exposure time (acute or chronic), mode of effect (death, growth, reproduction), or the effective response (lethal or sublethal) (Fig. 1) [11]. Other approaches to the classifications of toxicity tests can include acute toxicity, chronic toxicity, and specific toxicity (carcinogenicity, genotoxicity, reproduction, immunotoxicity, neurotoxicity, specific exposure to skin and other organs). For instance, genotoxicity reveals the risks for interference with the ecological gene pool leading to increased mutagenicity and/or carcinogenicity in biota and man. Unlike normal toxicity, the incidence of genotoxic effect is thought to be only partially related to concentration (one-hit model).

A toxicity test may measure either acute or chronic toxicity. Acute toxicity is indicative for acute effects possibly occurring in the immediate vicinity of the discharge. An acute toxicity test

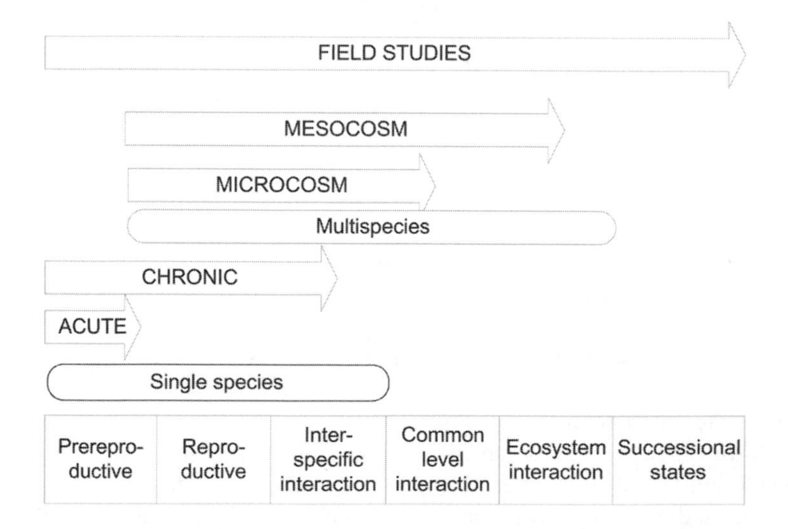

Figure 1 Classification of toxicity tests in environmental toxicology.

is defined as a test of 96 hours or less in duration, in which lethality is the measured endpoint. Acute responses are expressed as LC_{50} (lethal concentration) or EC_{50} (effective concentration) values, which means that half of the organisms die or a specific change occurs in their normal behavior. Sometimes in toxicity bioassays the NOEC (no observed effect concentration) can be used as the highest toxicant concentration that does not show a statistically significant difference with controls. The EC_{10} can replace the NOEC. This is a commonly used effect parameter in microbial tests [14–17]. At the EC_{10} concentration there is a 10% inhibition, which might not be very different from the NOEC concentration, but the EC_{10} does not depend on the accuracy of the test.

Acute toxicity covers only a relatively short period of the life-cycle of the test organisms. Chronic toxicity tests are used to assess long-lasting effects that do not result in death. Chronic toxicity reflects the extent of possible sublethal ecological effects. The chronic test is defined as a long-term test in which sublethal effects, such as fertilization, growth, and reproduction, are usually measured in addition to lethality. Traditionally, chronic tests are full life-cycle tests or a shortened test of about 30 days known as an "early-stage test." However, the duration of most EPA tests have been shortened to 7 days by focusing on the most sensitive early life-cycle stages. The chronic tests produce the highest concentration percentage tested that caused no significant adverse impact on the most sensitive of the criteria for that test (NOEC) as the result. Alternative results are the lowest concentration tested that causes a significant effect (lowest observed effect concentration; LOEC), or the effluent concentration that would produce an observed effect in a certain percentage of test organisms (e.g., EC_{10} or EC_{50}). The advantage of using the LC or EC over the NOEC and LOEC values is that the coefficient of variation (CV) can be calculated. In some cases, since toxicity involves a relationship with the effect concentration (test result; the lower the EC, the higher the toxicity), all test results are converted into toxic units (TU). The number of toxic units in an effluent is defined as 100 divided by the EC measured (expressed as a dilution percentage). Two distinct types of TUs are recognized by the EPA, depending on the types of tests involved (acute: $TU_a = 100/LC_{50}$; chronic $TU_c = 100/NOEC$). Acute and chronic TUs make it easy to quantify the toxicity of an effluent, and to specify toxicity-based effluent quality criteria.

However, the effect of a harmful compound should be studied with respect to the community level, not only for the organism tested. Tests with several species are realized in microcosm and mesocosm studies. Mesocosms are larger with respect to both the species number and the species diversity and are often performed outdoors and under natural conditions.

Choice of method is the most important phase if reliable data are to be obtained successfully. A good toxicity test should measure the right parameters and respond to the environmental requirements. When selecting from among available test organisms, the investigator should choose species that are relevant to the overall assessment endpoints, representative of functional roles played by resident organisms, and sensitive to contaminants. In addition, the test should be fast, simple, and repetitive [1,11,18]. The selection of ecotoxicological test methods also depends on the intended use of the waste and the entities to be protected. Usually a single test cannot be used to detect all biological effects, and several biotests should therefore be used to reveal different responses. The ecological relevance of the single species tests has been criticized, and the limits associated with these tests representing only one trophic level have to be acknowledged.

Biological toxicity tests are widely used for evaluating the toxicants contained in the waste. Most toxicity bioassays have been developed for liquid waste. Applications of bioassays in wastewater treatment plants fall into four categories [19]. The first category involves the use of bioassays to monitor the toxicity of wastewaters at various points in the collection system, the major goal being the protection of biological treatment processes from toxicant action.

These screening tests should be useful for pinpointing the source of toxicants entering the wastewater treatment plant. The second category involves the use of these toxicity assays in process control to evaluate pretreatment options for detoxifying incoming industrial wastes. The third category concerns the application of short-term microbial and enzymatic assays to detect inhibition of biological processes used in the treatment of wastewaters and sludges. The last category deals with the use of these rapid assays in toxicity reduction evaluation (TRE) to characterize the problem toxic chemicals. In addition to the abovementioned categories, we could point out another one: whole effluent testing (WET) in accordance with International (National) Environmental Policy.

Ecotoxicological testing of the pollutants in solid wastes should be considered in the following cases: supplementary risk assessment of contaminated waste; assessment of the extractability of contaminants with biological effects in cases where the waste can affect the groundwater; ecotoxicological assessment of the waste intended for future utilization as soil fertilizer, conditioner, or amendment (for example, compost from organic fraction of municipal solid waste, sewage sludge, etc.); and control of the progress in biological waste treatment.

All the tests used for estimation of solid waste toxicity can be divided into two groups: tests with water extracts (elutriate toxicity tests) and "contact" toxicity tests. The majority of the assays (e.g., with bacteria, algae, Daphnia) for testing toxicity have been performed on water extract. The water path plays a dominant role in risk assessment. Water may mobilize contaminants, and water-soluble components of waste contaminants have a potentially severe effect on microorganisms and plants, as well as fauna. Owing to their low bioavailability, adsorbed or bound species of residual contaminants in waste represent only a low risk potential. However, mobilized substances may be modified and diluted along the water path. Therefore investigations of water extracts may serve as early indicators [9]. Meanwhile, owing to the different solubility of each contaminant in the water, water extracts represent only a part of contamination. Water elutriation could underestimate the types and concentrations of bioavailable organic contaminants present [20,21]. Evaluation of results requiring sample extraction appears extremely difficult. The evaluation of toxicity with extracts sometimes ignores the interactions that may occur in contacts with substances in a solid phase. Therefore "contact" tests involve the use of organisms in contact with the contaminated solids. Such tests have been standardized and used for soils, for example, using higher plants [9,22,23]. During the past few years some applications of bacterial contact assays have been suggested [17,21,24–27]. We also present the bioassays that have been used for estimation of toxicity of liquid and solid wastes.

2.3 MICROBIAL TESTS

Microbial toxicity tests are known to be fast, simple, and inexpensive. These properties of the tests have resulted in their ever-increasing use in environmental control, assessment of pollutants in waste, and so on. Toxicity test methods based on the reaction of microbes are useful in toxicity. In particular they can be a very valuable tool for the toxicity classification of samples from the same origin. Microbial tests can be performed using a pure culture of well-defined single species or a mixture of microbes. The variables measured in toxicity tests may be lethality, growth rate, change in species diversity, decrease in degradation activity, and energy metabolism or activity of specific enzymes. The results are generally expressed as the dose–response concentration and the EC_{50} or EC_{10} value [11,15,17,28,29].

2.3.1 Tests Based on Bioluminescence

One of the commonly used tests is the bioluminescence-measuring test. It is based on the change of light emission by *Vibrio fischeri* (*Photobacterium phosphoreum*) when exposed to toxic chemicals. The bioluminescence is directly linked to the vitality and metabolic state of the cells, therefore a toxic substance causing changes in the cellular state can lead to a rapid reduction of bioluminescence. Thus a decrease in the light emission is the response to serious damage to metabolism in the bacterial cells. This test is a fast and reliable preliminary toxicity test and is comparable with other toxicity tests [11,29–31]. The procedure has been developed for the investigation of water, for example, wastewater, but can be applied without problems to the investigations of soil and waste extracts. Toxicity extracts can be determined using standard test methods such as the BioTox or Microtox methods [32]. The test criterion is the inhibition of light emission. The result is expressed as the G_L value (or lowest inhibitory dilution, LID, value). This is the lowest value for dilution factor of the extract which exhibits less than 20% inhibition of light emission under test conditions. In the case of individual toxicants the result is presented as EC_{50} or EC_{20}. This test is probably the most popular commercial test for assessing toxicity in wastewater treatment plants [19,33] and whole effluence testing. However, an expensive luminometer is required for the scoring of results. One of the reasons for the widespread application of this assay is the (commercial) availability of the bacteria in freeze-dried form, which eliminates the need for culturing of the test organisms [34–37].

A "direct contact test" has been developed for solid samples. A solid-phase assay eliminates the need for soil extracts and utilizes whole sediments and soils. In the current procedure the solid sample is suspended in 2% NaCl. Dilutions of the stock suspension are measured to determine the EC_{50} and EC_{10} at 5 and 15 minute contact times. For this the homogenized sample and photobacterial suspension mixture are incubated. The suspended solid material is then centrifuged out and light emission of the supernatant determined [24–26,32].

The bioluminescent direct contact flash test has been proposed as a modification of the direct contact luminescent bacterial test [24,38]. This method was developed for measuring the toxicity of solid and color samples, and involves kinetic measurements of luminescence started at the same time that the *V. fischeri* suspension is added to the sample. The luminiscence signal is measured 20 times per second during the 30 second exposure period.

2.3.2 Tests Based on Enzyme Activity

Enzyme activity tests can be used to describe the functional effects of toxic compounds on microbial populations. Many enzymes are used for toxicity estimation. The enzymes used to assess the toxicity of solid-associated contaminants (soils, composts, wastes) are phosphatase, urease, oxidoreductase, dehydrogenase, peroxidase, cellulase, protease, amidase, etc. Determining dehydrogenase activity is the most common method used in enzyme toxicity tests [11,29]. The method measures a broad oxidizing spectrum and does not necessarily correlate with the number of microbes, production of carbon dioxide, or oxygen demand. In ecological studies, correlations have been determined between dehydrogenase activity and the concentration of harmful compounds. Substrates for dehydrogenase activity are triphenil tetrazoliumchloride (TTC), nitroblue tetrazolium (NBT), 2-(*p*-iodophenyl)-3-(*p*-nitrophenyl)-5-phenyl tetrazoliumchloride (INT), and resasurine [21,29].

Toxi-Chromotest™ is a commercial toxicity assay that is based on the assessment of the inhibition of β-galactosidase activity, measured using a chromogenic substrate and a colorimeter. A mutant strain of *Escherichia coli* is revitalized from a lyophilized state prior to the test [39]. The principle of the MetSoil™ test is similar to that of the Toxi-Chromotest™.

The bacterial mutant is mainly sensitive to metals and should therefore be used in conjunction with another bacterial test. This microbiotest is commercially available and is designed specifically for testing soils, sediments, and sludges. Semiquantitative results are obtained after three hours [40].

The MetPAD™ test kit (Group 206 Technologies, Gainesville, Florida) has been developed for the detection of heavy metal toxicity. It has been used to determine the toxicity of sewage water and sludge, sediments, and soil [41]. The test is based on the inhibition of β-galactosidase activity in an *Escherichia coli* mutant strain. Performance of the test does not require expensive equipment and it is therefore easily applied as a field test.

The MetPLATE™ test (Group 206 Technologies, Gainesville, Florida) is a fast β-galactosidase activity microtiter plate test [40]. The test is specific for heavy metal toxicity. MetPLATE is in a 96-well microtitration plate format and is suitable for determination of toxicity characteristics such as median inhibitory concentrations. MetPLATE is based on the activity of β-galactosidase from a mutant strain of *E. coli* and uses chlorphenol red galactopyranoside as the enzyme substrate. The test is suitable for sewage water as well as for sewage sludge, sediments, and soil. The MetPLATE test is more sensitive to heavy metals than the Microtox™ test, which is based on bioluminescence inhibition. However, this test does not react sensitively to organic pollutants. The MetPAD and the MetPLATE tests are available in kit form.

The ECHA (Cardiff, England) Biocide Monitor™ is a qualitative test developed for environmental samples and is based on measurement of dehydrogenase activity [41,42]. This test is performed with a small plastic strip carrying an absorbent pad impregnated with a sensitive microorganism, nutrients, and an indicator of metabolic activity and growth. Solid samples are tested directly without extraction. Semiquantitative results are evaluated after 5–24 hours with this assay, which is available as a commercial kit.

A toxicity testing procedure using the inhibition of dehydrogenase enzyme activity of *Bacillus cereus* as test parameter has been developed [21]. This microbial assay includes direct contact of bacteria with solids over 2 hours and the following measurement of dehydrogenase enzyme activity on the base of resazurine reduction. It is the authors' opinion that this method can integrate the real situation in a more complex system much better than extracts. There are numerous results from different solid phases assayed with *B. cereus*. Experiments were conducted with several contaminants, which show differences in environmental behavior: Tenside and heavy metals (high adsorption, good solubility in water), para-nitrophenol (low adsorption, good solubility in water), polycyclic aromatic hydrocarbons (high adsorption, low solubility in water). For most of the substances, the contact assay shows higher sensitivity than elutriate testing; that is, the EC_{50} is lower (Table 1). Studies with soil samples spiked with organic compounds and copper indicate the higher sensitivity of solid-phase bioassay compared to water extract testing [17]. A comparison of the sensitivity of the *B. cereus* contact test and the *Photobacterium phosphoreum* solid-phase test demonstrates that the *B. cereus* test is more sensitive for copper. The test is the scientific tool to elucidate the importance of exposure routes for compounds in soils and solid wastes. However, the authors note that the problems in predicting ecological effects of contaminants (e.g., soil contaminants) exist.

Toxi-ChromoPad™ (EBPI, Ontario, Canada) is a simple method for evaluation of the toxicity of solid particles [25,26,32,39]. The test is based on the inhibition of the synthesis of β-galactosidase in *E. coli* after exposure to pollutants. The method has been used to measure acute toxicity of sediment and soil and other solid samples. The test bacterial suspension is mixed with homogenized samples and incubated for 2 hours. A drop of the test solution is pipetted onto a fiberglass filter containing an adsorbed substrate. A color reaction indicates the synthesis of enzyme, while a colorless reaction indicates toxicity. It has previously been shown

Table 1 Comparison of the Results of *Bacillus cereus* Contact Assay and Elutriate Toxicity for Some Spiked Soils

Substance	EC_{50} for contact assay	EC_{50} for elutriate assay
Benzalkonium-chloride	500 mg/kg	Up to 2000 mg/kg no effect
Alkylphenolpolyethylene-glycolether	3700 mg/kg (EC_{30})	Up to 4200 mg/kg no effect
Sodium alkylbenzenesulfonate	130 mg/kg	450 mg/kg
p-Nitrophenol	250/750/1000 mg/kg: 40.7/86.3/95.0% inhibition	250/750/1000 mg/kg: 13.2/65.0/82.3% inhibition
bis-tri-*n*-butyltinoxide	250/500 mg/kg: 94.2/95.1% inhibition	250/500 mg/kg: 34.0/80.5% inhibition
Naphthol	450 mg/kg	1000 mg/kg
Catechol	20 mg/kg	400 mg/kg
Lubricant oil	1.15 Gew%	3.40 Gew%
Copper	200 mg/kg	Up to 500 mg/kg no effect

that inducible enzyme metabolism can be considered a sensitive indicator for detecting the effects of harmful compounds [43]. Moreover Dutton *et al.* [44] found that β-galactosidase *de novo* biosynthesis in *E. coli* was a more sensitive reaction to harmful compounds than enzymatic activity.

2.3.3 Tests Based on Growth Inhibition

Growth inhibition tests are available for determination of the toxicity of harmful compounds. *Pseudomonas putida* is a common heterotrophic bacteria in soil and water and the test is therefore suited for evaluation of the toxicity of sewage sludge, soil extracts, and chemicals [45]. The test criterion is the reduction in cell multiplication determined as the reduction in growth of the culture. According to the standard test ISO 10712 [46] *P. putida* is grown in liquid culture to give a highly turbid culture, which is then diluted by mixing with the sample solution. After incubation of the culture for 16 hours, growth is measured as turbidity during this period. Inhibition of an increase in turbidity in the samples is compared with that of the control using the following equation:

$$I = \frac{B_c - B_n}{B_c - B_o} \times 100$$

where I is the cell multiplication inhibition, expressed as a percentage, B_n is the measured turbidity of biomass at the end of the test period, for the nth concentration of test sample, B_c is the measured turbidity of biomass at the end of the test period in the control, and B_o is the initial turbidity measurement of biomass at time t_0 in the control.

The inhibition values (I) for each dilution should then be plotted against the corresponding dilution factor. The desired values of EC_{50}, EC_{20}, and EC_{10} are located at the intersection of the straight lines with lines parallel to the abscissa at ordinate values of 10, 20, and 50%. The evaluation may also be performed using an appropriate regression model on a computer.

Another growth inhibition test of *B. cereus* is used to determine the toxicity of chemicals and sediments [41]. This test is based on the measurement of an inhibition zone.

An agar plate method is presented by Liu *et al.* [47]. On an agar plate covered by a bacterial suspension, an inhibition zone is formed and measured around the spot where the toxic sample has been placed. The duration of the test depends on the growth of the bacterial species (from 3 to 24 hours). This assay is not available in a commercial kit but it is simple to perform as part of routine testing. Any bacterial strain can be used, but solid samples can only be tested as extracts.

2.3.4 Test Based on the Inhibition of Motility

The test based on motility inhibition of the bacterium *Spirillum volutants* is a very simple and rapid test for the qualitative screening of wastewater samples or extracts [48]. The organisms are observed under the microscope immediately after the addition of the test solution. The maintenance of a bacterial culture is necessary as in the previous type of assay.

2.3.5 Tests Based on Respiration Measurements

The assay microorganisms in Polytox are a blend of bacterial strains originally isolated from wastewater [48]. The Polytox kit (Microbiotest Inc., Nazareth, Belgium), specifically designed to assess the effect of toxic chemicals on biological waste treatment, is based on the reduction of respiratory activity of rehydrated cultures in the presence of toxicants. The commercially available kit is specifically designed for testing wastewaters. Quantative results can be obtained in just 30 minutes.

Respiration inhibition kinetics analysis (RIKA) involves the measurement of the effect of toxicants on the kinetics of biogenic substrate (e.g., butyric acid) removal by activated sludge microorganisms. The kinetic parameters studied are q_{max}, the maximum specific substrate removal rate (determined indirectly by measuring V_{max}, the maximum respiration rate), and K_S, the half-saturation coefficient [19]. The procedure consists of measuring with a respirometer the Monod kinetic parameters, V_{max} and K_S, in the absence and in the presence of various concentrations of the inhibitory compound.

2.3.6 Genotoxicity

Genotoxicity is one of the most important characteristics of toxic compounds in waste. The Ames test with *Salmonella* is the most widely used test for studying genotoxicity [49]. The test has been applied in genotoxic studies on waste, contaminated soil, sewage sludge, and sediments [11,19,50–52]. Specific *Salmonella typhimurium* strains with obligatory requirements for histidine are used to test mutagenicity. On a histidine-free medium, colonies are formed only by those bacteria that have reverted to the "wild" form and can produce histidine. Addition of a mutagenic agents increases the reversion rate.

The SOS Chromotest™ (Labsystems, Helsinki, Finland) is a test based on *E. coli* with an additional *lacZ* gene with SOS gene promoter *sfiA*. Under the influence of mutagenic agents, the DNA of the bacterial cells is damaged and an enzymatic SOS-recovering program and *stifA* gene promoter induce *de novo* transcription and synthesis of β-galactosidase. Commercial SOS Chromotests™ are used for estimation of soil and sediment contaminants [41,42,53].

Genotoxicity may also be tested with a Mutatox™ test (Azur Environmental Ltd., Berkshire, England), using a dark mutant strain of bioluminescent bacterium *V. fischeri* [54]. DNA-damaging substances are recognized by measuring the ability of a test sample to restore the luminescent state in the bacterial cells. The authors pointed to the sensitivity of the test to chemicals that damage DNA, bind DNA, or inhibit DNA synthesis.

Muta-Chromoplate is a modified version of the classical Ames test for the evaluation of mutagenicity. The bioassay uses a mutant strain of *S. typhimurium*. The reverse mutation is recorded as absence of bacterial growth after 5 days incubation [55].

2.3.7 Tests Based on Nutrient Cycling

Sometimes the risk of waste is estimated on the basis of nutrient cycling tests. As a rule such investigation is carried out for surface waste disposal or its land application. The carbon cycle is very sensitive to harmful compounds. Soil respiration is considered a useful indicator of the contaminants' effects on soil microbial activity [56–59]. The production of carbon dioxide can be followed as short-term and long-term respiration tests.

Many organisms take part in processes that release inorganic nitrogen as a result of the mineralization of organic matter, leading initially to the formation of NH_4^+ ions. In contrast, relatively few genera of autothrophic bacteria, such as *Nitrosomonas* and *Nitrobacter* acting in sequence, take part in the transformation of ammonium to nitrite and nitrate. Toxicity assays based on the inhibition of both *Nitrosomonas* and *Nitrobacter* have been developed for determining the toxicity of wastewater samples [19]. However, *Nitrosomonas* appears to be much more sensitive to toxicants than *Nitrobacter*. A rapid method for testing potential nitrification on the basis of ammonium oxidation in soil is under development at ISO [11]. This method is used to estimate the effects of toxicants contained in soil or sewage sludge [60,61].

Soil microbial processes, like mineralization of organic matter or soil respiration, can be relatively little affected by moderate levels of heavy metals, while the processes carried out by a few specialized organisms, that is, nitrogen fixation, are more sensitive [56–60,62]. Toxicity tests exist for both symbiotic and free-living nitrogen-fixing microorganisms. It is generally agreed that N_2 fixation is more sensitive than soil respiration to toxicants such as metals.

One of the most commonly used parameters in soil biology is microbial biomass. The level of microbial biomass is used for assessment of the effects of contaminants in sewage sludge or compost of municipal solid waste in short-term or long-term experiments [56–59,63–69].

2.4 TESTS WITH FAUNA SPECIES

2.4.1 Tests with Crustaceans

Throughout the last three decades, only one taxon has emerged (for reasons of practicality as well as of sensitivity) as the key group for standard ecotoxicological tests with invertebrates, namely the cladoceran crustaceans, and more particularly the daphnids. *Daphnia* tests are currently the only type of freshwater invertebrate bioassay that are formally endorsed by international organizations such as the USEPA, the EEC, and the OECD, and that are required by virtually every country for regulatory testing [70]. The reasons for the selection of daphnids for routine use in toxicity testing are both scientific and practical. Daphnids are widely distributed in freshwater bodies and are present throughout a wide range of habitats. They are an important link in many aquatic food chains (they graze on primary producers and are food for many fish species). They have a relatively short life-cycle (important for reproduction tests) and are relatively easy to culture in the laboratory. They are sensitive to a broad range of aquatic contaminants. Their small size means that only small volumes of test water and little benchspace are required. *Daphnia magna* and *D. pulex* are the most frequently used invertebrates in standard acute and chronic bioassays. *Ceriodaphnia* species are used extensively in the United States, mainly in short-term chronic bioassays [71].

A large number of papers have been published on the use of acute *Daphnia* toxicity tests, on a whole range of fundamental and applied toxicological problems. Excellent reviews of ecotoxicological testing with *Daphnia* have been written by Buikema *et al.* [72] and Baudo [73]. Standard protocols are introduced in Refs. 74–83. Acute bioassays with *Daphnia* sp. are among the most frequently used toxicity tests because, once a good laboratory culture is established, the tests are relatively easy to perform on a routine basis and do not require highly skilled personnel. Moreover, compared to acute toxicity tests with fish, acute *Daphnia* tests are cost-effective because they are shorter (48 vs. 96 hours) and the culture and maintenance of the daphnids requires much less space, effort, and equipment.

The acute *Daphnia* bioassay is recognized to be one of the most "standardized" aquatic toxicity tests presently available and several intercalibration exercises report a reasonable degree of intra- and interlaboratory reproducibility [84–87].

In addition to acute toxicity tests, two standard chronic toxicity test methods are widely accepted by various regulatory agencies: the seven-day *Ceriodaphnia* survival and reproduction test and the 21-day *Daphnia* reproduction test.

Cereodaphnia dubia was first identified in toxicity testing as *Cereodaphnia reticulata* [88] and subsequently as *Cereodaphnia affinis* [89]. The *Ceriodaphnia* survival and reproduction test is a cost-effective chronic bioassay for on-site effluent testing and is now one of the most used invertebrate chronic freshwater toxicity tests in the United States. The major arguments for introducing this method are that it is a more ecologically relevant test species in the United States (than *D. magna*), is easier to culture, and has an exposure period that is only one-third of that of the *D. magna* chronic test [88]. Owing to its ease of culturing, short test duration, low technical requirements, and high sensitivity, the seven-day *Ceriodaphnia* chronic test is a very attractive and relatively cost-effective bioassay, which can be performed by moderately skilled personnel. Key documents and standard protocols may be found in Refs. 71, 88, and 90. Different standard bioassays (Toxkit tests) are now available. In Daphtoxkit F^{TM} magna (Microbiotest Inc., Nazareth, Belgium) and pulex inhibition of mobility of *D. magna* and *D. pulex* is recorded after 24 and 48 hours exposure [91]. The test organisms are incorporated into commercial kits Daphtoxkit F^{TM} magna and Daphtoxkit F^{TM} pulex as dormant eggs and can be hatched on demand from the dormant eggs 3 to 4 days before testing [92,93]. IQ^{TM} Fluotox-test is presented by Janssen and Persoone [94]. The damaged enzyme systems (β-galactosidase) of the crustacean *D. magna* after exposure to toxic substances can be detected by their inability to metabolize a fluorescently marked sugar. Healthy organisms with unimpaired enzyme systems will "glow" under long-wave ultraviolet light, while damaged organisms will not. This microbiotest is commercially available and only takes a one-hour exposure. CerioFast^{TM} is a rapid assay based on the suppression of the feeding activity of *C. dubia* in the presence of toxicants [93,95,96]. After a one-hour exposure to the toxicant, the *C. dubia* is fed on fluorescently marked yeast and the fluorescence is observed under an epifluorescent microscope or long-wave ultraviolet light. The presence or absence of fluorescence in the daphnid's gut is used as a measure of toxic stress. This microbiotest is commercially available and only takes a few hours to complete.

The test organisms are exposed for 24, 48, and 96 hours to different concentrations of testing water. After the exposure period the number of dead organisms is counted. Each test sample container is examined and the number of dead organisms counted (looking for the absence of swimming movements). A test is regarded as valid if the mortality in the control is $<10\%$. Toxicity is calculated as:

$$T = \frac{N_0 - N_t}{N_0} \times 100\%$$

where T is toxicity in %, N_0 is the average quantity of test organisms at time 0, and N_t is the average quantity of test organisms at time t.

There are many procedures for calculating LC_{50}s. LC_{50} or EC_{50} values are calculated using the probit-derived method. A very simple procedure consists of plotting the calculated percent mortalities on a log concentration/% mortality sheet. The procedure for estimation of the LC_{50} is as follows:

1. Indicate the concentrations or dilutions used in the dilution series on the Y-axis.
2. Plot the calculated percent mortality on the horizontal line at the height of each concentration or dilution.
3. Connect the plotted mortality points on the graph with a straight line.
4. Locate the two points on the graph that are separated by the vertical 50% mortality line and read the LC_{50} at the intersect of the two lines. Expression and interpretation of the toxicity data of wastewaters: all median toxicity values are converted into toxic units (TU), that is, the inverse of the LC/EC_{50} expressed in %, according to the formula $TU = [1/L(E)C_{50}] \times 100$.

This expression is the dilution factor, which must be applied to the effluent so as to obtain a 50% effect, and is directly proportional to toxicity. The result of several toxicity tests is applied on the base of the most sensitive test species.

2.4.2 Tests with Protozoa

Dive and Persoone [97] advanced a number of arguments in favor of tests with protozoa: unicellular organisms combine all biological mechanisms and functions in one single cell; the generation time of protozoa is very short in comparison to metazoa; large numbers of organisms can be produced in a small volume; and unicellular organisms play a significant role in aquatic ecosystems, especially in the transformation and degradation of organic matter.

The standard *Colpodium campylum* toxicity test developed by Dive and colleagues [98,99] measures the inhibition of growth of this ciliate, cultured monoxenically on *E. coli*. The reduction of the number of generations is measured in increasing concentrations of the toxicant, and the effects are expressed as 24 hour IC_{50} values. This bioassay is relatively easy to learn, to carry out, and to interpret.

The microbiotest with ciliate protozoan *Tetrahymena thermophila* (Protoxkit F^{TM}, which only became available commercially recently) evaluates the growth inhibition of the unicellulars submitted for 20 hours to a toxicant [100]. The decreased multiplication of the ciliates is determined indirectly via the reduction in their food uptake, by optical density measurement in 1 cm spectrophotometric cells.

A test with *Paramecium caudatum* was suggested for estimation of the toxicity of inflowing municipal wastewater entering the treatment plant as well as of local wastewater during the process of channeling [18,101,102]. Use of *P. caudatum*, a typical representative of the organisms of activated sludge, permits us to foresee the impact of toxicants on the processing of the wastewater treatment plant. The test reaction is the death of the test organism when exposed to tested wastewater or waste extract for 1 hour. The toxicity is calculated as:

$$T\ (\%) = (N_f : N_i) \times 100$$

where N_f is the number of dead *P. caudatum* (the average from five replications), and N_i is the initial number of *P. caudatum.*

Another test organism suggested for the estimation of wastewater entering the treatment plant is *Euplotes patella* [103].

2.4.3 Tests with Cnidaria

The freshwater cnidarian *Hydra attenuata* was only recently exploited to assess the acute lethal toxicity of wastewaters [37,104]. The advantages of using *Hydra* for bioassay include its wide distribution in freshwater environments, thereby making it a representative animal for conducting environmental hazard assessment, as well as its robustness, which makes it easily manipulable, and easily reared and maintained in the laboratory. Upon exposure to bioavailable toxicants, *Hydra* undergoes profound morphological changes, which are first manifested by sublethal and then lethal effects. From their normal appearance, the animals progressively exhibit bulbed (clubbed) tentacles as an initial sign of toxicity, followed by shortened tentacles and body. After these sublethal manifestations, and if toxicity continues to prevail, *Hydra* reaches the tulip phase, where death then becomes an irreversible event. The postmortem stage is finally indicated by disintegration of the organism. Noting *Hydra* morphology during exposure allows for simple recording of (sub)lethal toxicity effects. *Hydra* assay demonstrates good sensitivity in detecting effluent toxicity [105].

2.4.4 Tests with Fish

Toxic characteristics of industrial wastewater in many countries are still assessed using fish [106–108]. The standardized procedure describes testing with different species in different life stages. For ethical reasons, as well as those linked to cost- and time-effectiveness, labor-intensiveness, analytical output, and effluent sample volume requirements, there is unquestionable value in searching for alternative procedures that would eliminate the drawbacks associated with fish testing. Investigators therefore use an *in vitro* cell system, which can greatly decrease the need for the *in vivo* fish model [37].

2.4.5 Tests with Invertebrates

Soil invertebrates are also good subjects for evaluating the possible harmful effects of toxic substances. There is a wide range of methods that involve soil invertebrates in toxicity testing. There are standard methods for earthworms (*Eisenia fetida*), collembola (*Folsomia candida*), and enchytraeide (*Enchytraeidae* sp.) [11,19,110,111]. When considering the use of invertebrates for ecological testing, the species should be selected with respect to how well it represents the community of organisms in question and how feasible is the culture of the species in the laboratory throughout the year.

As protozoa and nematodes live in pore water in the soil, most of the methods are adapted from toxicity tests designed for aquatic samples. Among the protozoa the tests with ciliates *Tetrahymena pyriformis*, *Tetrahymena thermophiia*, *Colpoda cucullus*, *Colpoda inflata*, *Colpoda steinii*, *Paramecium caudatum*, and *Paramecium aurelia* have been developed [102,112–117]. It is the opinion of some authors that the sensitivity of infusorians is higher than that of microorganisms [115,116].

Bacteriovorus nematodes offer possibilities for toxicity testing because a large number of different species can be extracted from the soil and reared in the laboratory. Among the nematodes used are *Caenorhabditis elegans*, *Panagrellus redivivus*, and *Plectus acuminatus* [118–120]. The endpoint most often used has been mortality of the test organisms, expressed as the LC_{50}. Furthermore, fecundity, development, morphology, growth, population growth rate, and behavior have been used to assess toxic effects. Recently, assays for *C. elegans* that measure the induction of stress reporter genes have been developed [119]. The major problem in tests with nematodes and protozoans is extrapolation of the results for environmental risk assessment of hazardous compounds. Usually the tests are performed with artificial media; the composition of the media thus has a bearing on the results [11]. The survival, growth, and maturation of the nematode *P. redivivus* is evaluated such that three endpoints can be measured from this toxicity test: acute, chronic, and genotoxic [121]. This microbiotest is not available in commercial form, but the maintenance of these organisms is rather simple. Extracts from solid samples are prepared by a simple procedure, directly in the test media. A disadvantage of this 96 hour test is that qualified staff is needed to evaluate the results under the microscope.

Earthworms are often used for the assessment of toxicant effects due to their sensitivity to most of the factors affecting soil ecosystems, especially those associated with the application of agriculture chemicals. Earthworms respond to chemicals in several ways, for example, increase in body burdens, increase in mortality, and overall decrease in activities normally associated with viable earthworm populations [122]. Species recommended by standards ASTM (American Society for Testing and Materials) and OECD (Organization for Economic Cooperation and Development) are *Eisenia fetida* and *Eisenia andrei*, which commonly occur in compost and dung heaps, and can be easily cultured in the laboratory [11,123]. Another recommended species is *Limbricus terrestris* [124,125]. The ASTM standard test for soil toxicity with *E. fetida* is designed to assess lethal or sublethal toxic effects on earthworms in short-term tests. The sublethal effects examined can be growth, behavior, reproduction, and physiological processes, as well as observations of external pathological changes, for example, segmental constrictions, lesions, or stiffness. Callahan [122] has presented three different earthworm bioassays: the 48 hour contact test, 14 day soil test, and a neurological assay. The contact test is effective in detecting toxicity when the toxicant is water-soluble, and the soil test is effective in indicating the toxicity of a range of toxicants, both water-soluble and water-insoluble. Nerve transmission rate measurements have been found to be very efficient in picking up toxicity at lower concentrations and shorter exposure times. The contact test and the soil test appear to be adequate for toxicity assessment of pollutants in hazardous wastes.

In the past few years the use of rotifers in ecotoxicological studies has substantially increased. The main endpoints used are mortality, reproduction, behavior, cellular biomarkers, mesocosms, and species diversity in natural populations [126]. Several workers have used *Brachionus calyciflorus* for various types of toxicity assessments. Thus, comprehensive evaluation of approximately 400 environmental samples for the toxicity assessment of solid waste elutriates, monitoring wells, effluents, sediment pore water, and sewage sludge was carried out by Persoone and Janssen [127]. The mortality of rotifers hatched from cysts is evaluated after 24 hours exposure. This microbiotest has been commercialized in a Rotoxkit F™ [128,129].

2.5 ALGAE TESTS

Algae may also serve as test organisms in toxicity testing. In standard algal toxicity test methods published by various organizations such as APHA, ASTM, ISO, and OECD [130–133], a rapidly growing algal population in a nutrient-enriched medium is exposed to the toxicant for 3 or 4 days. *Selenastrum capricornutum* (renamed *Raphidocelis subcapitata*) and *Scenedesmus subspicatus* are the most frequently used, although others have also been used or recommended.

Increasing the simplicity and cost efficiency of algal tests has been an important research activity in recent years. [134,135] New tests procedures involve the application of flow cytometry, microplate techniques, and immobilized algae [135–140].

A miniaturized version of the conventional flask method with *S. capricornutum* has been developed by Blaise *et al.* [136]. In this assay the algae are exposed to the toxicant in 96-well microplates for a period of 96 hours, after which the cell density is determined using a hemocytometer or electronic particle counter. ATP content measurements [136] or chlorophyll fluorescence [141,142] have also been proposed as test criteria. Compared to the flask method, the main advantages of the microplate assay are: (a) the small sample volumes and reduced bench space requirements, (b) the use of disposable materials, (c) the large number of replicates, and (d) the potential for automation of the test set-up and scoring [136,143,144].

Another alternative algal assay with *S. capricornutum* that has recently been developed is the Algaltoxkit F^{TM} (Microbiotest Inc., Nazareth, Belgium) [135,140,145]. One of the main features of this kit test is that no pretest culturing of algae is required as the algae are supplied in the form of algal beads that can be stored for several months. The algae are de-immobilized from the beads in order to test for growth inhibition by optical density measurement in "long-cell" test cuvettes.

2.5.1 Calculation of Percent Growth Inhibiton in Algae Tests

A growth curve for the algae test is drawn up by assessing the cell concentration (number of cells/mL) or optical density for each concentration of the sample being investigated and plotting against time. In order to evaluate the relationship between growth and concentration the EC_{50} is calculated for every period of time at which the biomass was measured during the test (24, 48, and 72 hours) according to OECD [133]. The effect is estimated by using the area under the growth curves as a measure of the growth (EC_{50}: measure of the effect on biomass = concentration at which the area under the growth curve comes to half of the area under the growth curve of the control). The curves are constructed using the average values of the replicates.

The area under the growth curve is calculated for each of the points in time as:

$$A = \frac{N_1 - N_0}{2} t_1 + \frac{N_1 + N_2 - 2N_0}{2}(t_2 - t_1) + \frac{N_{n-1} + N_n - 2N_0}{2} \times (t_n - t_{n-1})$$

where A is area, N_0 is the nominal number of cells or absorption measured at time t_0, N_1 is the number of cells measured or absorption measured at time t_1, N_n is the number of cells measured or absorption measured at time t_n, t_1 is the point in time at which the first measurement was made after the start of the test, and t_n is the time of the nth measurement after the start of the test.

The percent growth inhibition for each test concentration is calculated by

$$I_a = \frac{(A_c - A_a) \times 100}{A_c}$$

where I_a is the percent inhibition of concentration a, A_c is the area of the control growth curve, and A_a is the area of the growth curve of concentration a. The concentration–effect curves and the EC_{50} are determined by means of linear regression analysis.

2.6 PLANT TESTS

Plants constitute the most important components of ecosystems because of their ability to capture solar energy and transform it into chemical energy. Oxygen and the sugars produced by plants from solar energy and carbon dioxide are essential to all living organisms. The sensitivity of plants to chemicals in the environment varies considerably. Plants sensitive to harmful substances can be used as bioindicators. The plant tests used in environmental analysis can be classified into five groups: (a) biotransformation (detecting changes in the amounts of chemicals caused by plants); (b) food chain uptake (determining the amounts and concentrations of toxic chemicals that enter the food chains via plant uptake); (c) phytotoxicity (determining the toxicity and hazard posed by pollutants to the growth and survival of plants); (d) sentinel (monitoring the pollutants by observing toxicity symptoms displayed by plants); and (e) surrogate (instead of animal or human assay).

Most attention has been devoted to phytotoxicity tests. Many plant species and numerous phytotoxic assessments endpoints have been used to characterize toxicant impacts on vegetation. Phytotoxicity can be determined as seed germination, root elongation, and seedling growth [22,23,57,58,146–151]. The tests can be carried out in pots or in petri dishes. The majority of plants commonly used in phytotoxicity tests have been limited to species of agricultural importance. A recent update of ASTM methodology for terrestrial plant toxicity testing lists nearly 100 plant taxa [152]. OECD has developed a plant bioassay [11]. This test is a simple test and includes at least one monocotyledon and one dicotyledon plant. The plant species recommended for growth experiments by OECD are listed in Table 2. Some test species are also recommended in ISO documents (Table 3). The test examines the reaction of growth of a plant species in the early stages of development. The efficiency of plant growth within 14 days is determined by establishing the average fresh mass after cutting the shoots above the soil surface. The calculation of the reduction in growth as a percentage of the average mass of the plants from the test samples compared to that of controls is then carried out, such that

$$\text{Percentage growth reduction} = \frac{C - T}{C} \times 100$$

where C is the average fresh mass in the control, and T is the average fresh mass of the plants from the diluted test waste or soil. The level of significance of any growth inhibition observed is computed using Student's t-test or Dunnett's t-test.

The other parameter for phytotoxicity assessment is emergence, calculated as

$$\text{Percentage emergence} = \frac{C_e - T_e}{C_e} \times 100$$

Table 2 Plant Species Recommended for Assessment of Toxicity by OECD

Common name	Latin name
Ryegrass	*Lolium perenne*
Rice	*Oryza sativa*
Oat	*Avena sativa*
Wheat	*Triticum aestivum*
Sorghum	*Sorghum bicolor* (L.) *Moench*
Mustard	*Sinapis alba*
Rape	*Brassica napus*
Radish	*Raphanus sativus*
Turnip	*Brassica rapa*
Chinese cabbage	*Brassica campestris*
Vetch	*Vicia sativa*
Mungbean	*Phaseoli aureus*
Red clover	*Trifolium opratense*
Lettuce	*Lactuca sativa*
Cress	*Lepidium sativum*

Table 3 Test Species of the Plants Recommended for Phytotoxicity Assessment by ISO

Monocotyledonous		Dicotyledonous	
Common name	Latin name	Common name	Latin name
Rye	*Secale cereale* L.	Mustard	*Sinapis alba*
Ryegrass	*Lolium perenne* L.	Rape	*Brassica napus* (L.)
Rice	*Oryza sativa* L.	Radish	*Raphanus sativus* L.
Wheat, soft	*Triticum aestivum* L.	Chinese cabbage	*Brassica campestris* L.
Oat	*Avena sativa* L.	Birdsfoot fenugreek	*Trifolium ornithopodioides* L.
Spring or winter barley	*Hordeum vulgare* L.	Lettuce	*Lactuca sativa* L.
Sorghum	*Sorghum bicolor* (L.) *Moench*	Cress	*Lepidium sativum* L.
Sweetcorn	*Zea mays* L.	Tomato	*Lycopersicon esculentum* Miller
		Bean	*Phaseolus aureus* Roxb.

where C_e is the average number of emerged seeds in the control, and T_e is the average number of emerged seeds in the diluted test waste or soil.

Plant growth and germination tests are the most common techniques used to determine compost maturity and toxicity. A large number of studies have been carried out with different plant species such as ryegrass [153], barley [57,58,149], barley and radish [154], poplar [150], red maple, white pine, pin oak [155], and lettuce [151]. Furthermore, phytotoxicity parameters are used to ascertain whether the different kinds of waste (sewage sludge and municipal solid waste) are suitable for agricultural use or soil rehabilitation [57,58]. However, whether sewage sludge or municipal solid waste is used, it is convenient to submit it first to a process of composting to avoid risks associated with the presence of the phytotoxic substances. On the other hand, the fresh products are suitable for addition to soils with a view to their rehabilitation.

2.7 COMMERCIAL STANDARD BIOASSAYS: TOXKIT TESTS

The impact of xenobiotics on aquatic environments, including wastewaters, is generally determined by acute and chronic toxicity tests. However, because of the large inventory of chemicals, short-term bioassays are now being considered for handling this task. The major attraction of the new bioassays is that they bypass one of the major handicaps of toxicological testing, namely the necessity of continuous recruitment and/or culturing of live stock of test species in good health and in sufficient numbers [156]. On the basis of the information supplied by 35 ecotoxicological laboratories in Europe, Persoone and Van de Vel [157] performed a cost analysis of the three acute aquatic toxicity tests recommended by the OECD, and came to the conclusion that maintenance and culturing of live stocks makes up at least half of the expense of any of those bioassays. Maintenance and culturing of live organisms furthermore requires highly skilled personnel and the availability of temperature-controlled rooms provided with specific equipment.

In a review on "Microbiotests in aquatic ecotoxicology," Blaise [28] comments on 25 different test procedures with bacteria, protozoa, microalgae, invertebrates, and fish cell lines, worked out and used to date by different research laboratories. When examining each of these tests from the point of view of practical features according to five criteria (availability in kit format, portability, maintenance-free bioindicator, performance in microplates, minimal training, and equipment requirement), Blaise comes to the conclusion that the Toxkit tests are the only types of bioassays that abide by all five features.

The first steps in bypassing of the biological, technological, and financial burden of live stock culturing or maintenance were made more than 20 years ago through the development of a "bacterial luminescence inhibition test" [34,35]; this bioassay is presently known and used worldwide as the Microtox® test. The revolutionary principle of this test is that it uses a "lyophilized" strain of a (marine) bacterium (*Photobacterium phosphoreum*). This makes the bioassay applicable anytime, anywhere, without the need for continuous culturing of the test species.

The second breakthrough in cost-effective toxicity screening was made through the development of "cyst-based" toxicity tests [158,159]. The new approach is based on the use of cryptobiotic stages (generally called cysts) of selected aquatic invertebrate species; the cysts are used as the "dormant" biological material from which live test organisms can easily be hatched. Like seeds of plants, "resting eggs" can be stored for long periods of time without losing their viability, and can be hatched "on demand" within 24 hours. The continuous availability of live test organisms through hatching of cysts eliminates all the problems inherent or related to continuous recruitment or culturing of live stocks, and solves one of the major bottlenecks in routine ecotoxicological testing. Commercial products for toxicity measurement on liquid and solid samples are already available (Table 4).

2.8 APPLICATION OF THE BIOASSAYS FOR ASSESSMENT OF TOXICITY OF SOLID WASTE

The ecological risk assessment of toxicants in waste requires reproducible and relevant test systems using a wide range of species. It is generally acknowledged by ecotoxicologists and environmental legislators that single species toxicity tests provide an adequate first step toward the ecological risk assessment of toxicants in soil and water [116,161].

2.8.1 Application of Single Species Bioassays

Use of tests based on luminescence is proposed by Carlson-Ekvall and Morrison [162] for estimation of the copper in the presence of organic substances in sewage sludge. The authors applied the Microtox toxicity test and Microtox solid-phase method and revealed that copper toxicity in sewage sludge can increase dramatically in the presence of certain organic substances (linear alkylbenzene sulfonate, caffeine, myristic acid, palmitic acid, nonylphenol, ethyl xanthogenate, and oxine) in sewage sludge. They attributed this effect to synergism and potentially the formation of lipid-soluble complexes. Based on the results of the toxicity found in this study they concluded that all organic substances tested in some way affected copper toxicity, and measurements of total metal concentration in sewage sludge is insufficient for decision making concerning the suitability of sludge for soil amendment.

The Microtox test has been used for determination of toxicity of wastewater effluents, complex industrial wastes (oil refineries, pulp and paper), fossil fuel process water, sediments extracts, sanitary landfill, and hazard waste leachates [19].

The contribution of polycyclic aromatic hydrocarbons present in sewage sludge to toxicity measured with the ToxAlert® bioassay has been investigated by a Spanish group [163]. A ToxAlert® bioassay based on the inhibition of *V. fischeri* and chemical analysis using gas chromatography–mass spectrometry was applied to sludge extracts after purification by column chromatography. The toxicity data can be explained by the levels and composition of different

Table 4 Commercially Available Toxicity Tests

Testkit	Test organism and test process	References
Bacteria		
BioTox Kit	*Vibrio fischeri*, luminiscence	[38]
Microtox	*Vibrio fischeri*, luminiscence	[19,32–37,39]
Microtox Solid-Phase Test		
ECHA Biocide Monitor	*Bacillus* sp., inhibition of dehydrogenase activity	[29,41,42]
MetPAD	*E. coli*, mutant strain, inhibition of β-galactosidase activity	[19,29,41]
MetPLATE Kit	*E. coli*, inhibition of β-galactosidase activity	[19,40]
Toxi-Chromotest Kit	*E. coli* mutant strain, inhibition of β-galactosidase activity	[19,25,39,43,160]
MetSoil	*E. coli*, mutant strain, inhibition of β-galactosidase activity	[40]
Toxi-ChromoPad Kit	*E. coli*, inhibition of the *de novo* synthesis of β-galactosidase	[25,26,32,39]
Polytox	Blend of bacterial strains originally isolated from wastewater, reduction of respiratory activity	[19]
Muta-Chromoplate Kit	Modified version of Ames test	[55]
Mutatox	Dark mutant strain of *Photobacterium phosphoreum* (*V. fisheri*), genotoxicity	[54]
SOS-Chromotest Kit E	Mutant strain of *E. coli*, genotoxicity	[41,42,53]
Invertebtates		
Daphnotoxlit F magna	Cladoceran crustacean, *Daphnia magna*	[91]
Daphnotoxkit F pulex	Cladoceran crustacean, *Daphnia pulex*	[91]
IQ Toxicity Test Kit	*Daphnia magna*	[94]
Artoxkit F	Anostracan crustacean, *Artemia franciscana* (formerly *A. salina*)	[91]
Thamnotoxkit F	Crustacean *Thamnocephalus platyurus*	[161]
Rotoxkit F	Rotifer *Branchionus calyciflorus*	[130]
Protozoa		
Protoxkit F	Ciliate, *Tetrachymena thermophila*	[91]
Algae		
Algaltoxkit F	Algal growth test, *Selenastrum capricornutum*	[136]

polycyclic aromatic hydrocarbons in sewage sludge samples. It is the authors' opinion that the present approach can contribute to evaluating the toxicity of sewage sludge. Furthermore, these bioassays may help researchers in developing processes that produce ecologically sustainable soils [164].

Genotoxicity is one of the most important characteristics of toxic compounds in waste. For studying genotoxicity of waste, contaminated soil, sewage sludge, and sediments the conventional Ames test with *Salmonella* is usually used together with SOS-Chromotest^{TM} and Mutatox^{TM} [11,19,50–52,165–167].

2.8.2 Application of the Battery of Toxicity Tests

In many studies on solid waste in which ecotoxicological tests have been used, little attention has been given to such aspects as the selection of test species, sensitivity of the tests, and the simplicity and cost of the assays. Very few serious endeavors have been made to determine the minimum battery of the test required [10,168]. The potential toxicity of the product of composting pulp and paper sewage sludge has been determined using a battery of toxicity tests [11]. The tests were the bioluminescent bacteria test, the flash method, Mutatox™, MetPLATE™, MetPAD™, ToxiChromotest™, the reverse electron transfer (RET) test, and seed germination with red clover. Differences in sensitivity were found between the tested parameters. The high concentration of organic matter masked the toxicity effect due to the activation of bacterial metabolism and enzymatic reaction. Another disturbing factor was color, especially for the bioluminescence test. The flash method was found to be more sensitive than the traditional luminescent bacteria test and, in addition, the most sensitive test for solid samples.

A Russian group has suggested using a battery of biotests for toxicity estimation of ash from a power plant [169]. The ash of six power plants was intended for use in organo-mineral fertilizers. However, the presence of metals (Mn, Cu, Str, Ni, Mg, Cr, Zn, Co, Cd, Pb, Fe) required the performance of an investigation into their biological effects and safety. The battery included tests with the protozoan *Tetrachymena piriformis*, the water flea *Daphnia magna*, the algae *Scenedesmus quadricauda*, and barley seeds. It was established that the sensitivity of the tests varies. Results of the bioassays are presented in Table 5. The algae test and the water flea test were found to be more sensitive. It is the authors' opinion that a bioassay using such a battery of tests utilizing different kinds of organisms is needed for the estimation of biological effects of the ash and its suitability for agriculture.

A battery of toxicity tests has been used to study decontamination in the composting process of heterogenous oily waste [10]. This particular waste from an old dumping site was composted in three windrows with different proportions of waste, sewage sludge, and bark. Samples from the windrow having intermediate oil concentrations were tested with toxicity tests based on microbes (*Pseudomonas putida* growth inhibition test, ToxiChromotest, MetPLATE, and three different modifications of luminescent bacterial tests: BioTox, the bioluminescent direct contact test, and the bioluminescent direct contact flash test), Mutatox genotoxicity assay, enzyme inhibition (reverse electron transport), plants (duckweed growth inhibition and red clover seed germination), and soil animals (*Folsomia candida*, *Enchytraeus albidus*, and *Enchytraeus* sp.). The luminescent bacterial tests were used as prescreening tests. The bioassays were accompanied by chemical analysis. As a consequence of the investigation the authors concluded that the most sensitive tests, which also correlated with the oil hydrocarbon reduction, were the RET assay, the BioTox test, the bioluminescent direct contact test, the bioluminescent

Table 5 Bioassay of Water Extracts of the Ash Produced in Power Plants

	Value for the dilution factor of water extract, which exhibits 50% inhibition of the estimating function			
Power plant	Barley seeds	*Scenedesmus quadricauda*	*Daphnia magna*	*Tetrachymena piriformis*
Shaturskaya	–[a]	1:4	1:4	–
Azeiskaya	–	1:4	1:4	–
Kuzneckaya	–	1:2	1:2	1:0
CZKK	–	1:5	1:0	1:0
Irsha-Borodinskaya	1:0	1:5	1:3	1:0
Stupinskaya		1:5	1:5	1:0

[a]Indicates absence of toxic effect.

flash test, the red clover seed germination test, the test with soil arthropod *F. candida*, and the test with *Enchytraeus* sp. These tests represent different trophic levels and also assess the effects of solid samples and extracts. It is the authors' opinion that one test of each category should be used to assess the environmental impact of the composted product. The Mutatox assay can also be included in the battery to assess the disappearance of genotoxicity. Note that one biotest is sufficient if only process monitoring is concerned. The most suitable test for screening and monitoring during composting was the luminescent bacterial test, in particular flash modification.

An integrative approach using toxicological and chemical analyses to screen toxic substances that could be added to the septic sludge obtained at the wastewater treatment plant was proposed by Robidoux et al. [170,171] to assist in the management of septic sludge. The necessity of the development of this ecotoxicological procedure was provoked by the temptation for producers of toxic substances to mix their hazard waste with chemical-toilet sludge shipments. At the first stage, four toxicity tests (Microtox, bacterial respiration, root elongation, and seed germination tests) were used to estimate the toxicity range of a "normal" sludge and for determination of the threshold limits criteria. These detection criteria can be used with relative efficiency and confidence to determine whether a sludge sample is contaminated or not. Taken individually, the seed germination test was the least discriminating toxicological method (detecting only 10% of the spiked samples). The bacterial respiration test was relatively better (detecting 72% of the spiked samples). As a whole, the battery of toxicity tests detected at least 93% of the spiked samples. Using a limited battery of two toxicity tests (Microtox and respiration test), the identification of contaminated chemical-toilet sludge can be detected with good efficiency and possibly greater reliability (more than 80% of spiked samples). An integrated ecotoxicological approach to screen for illicit discharge of toxic substances in chemical-toilet sludge received at a wastewater treatment plant is proposed by the authors based on chemical and toxicological analyses (Fig. 2). After sampling the sludge received at the wastewater plant, a 1 L sample is sent to the laboratory for toxicological characterization and Microtox and bacterial respiration analyses performed. A result below one of the following criteria would indicate "abnormal" sludge. For the Microtox assay, the two lower criteria suggested by these authors are: an IC_{50}-5 minute value of 0.20% (w/w), and IC_{50}-30 minute value of 0.10% (w/w). Microtox IC_{50} values higher than 0.51% w/w (5 minute) or 0.22% w/w (30 minute) would indicate that the sludge could be considered normal. For the bacterial respiration test an oxygen consumption rate less than 14.4 mg/L hour would be considered "abnormal." The sludge would be considered normal if its respiration test rate is higher than 49.2 mg/L hour. Results lying between the two criteria for each test would be considered dubious. The sludge in this latter range is "probably abnormal" and would necessitate an investigation and closer monitoring by the manager to avoid subsequent illicit discharge of contaminants. In the absence of additional incriminating information, the suspicious sludge otherwise should be considered "normal."

In Russia the disposal cost of waste depends on the class of hazard. For sewage sludge the ecotoxicological procedure has been outlined for its attribution to different classes of hazard (nonhazard, low hazard, moderate hazard, and hazard) [172,173]. This approach combines chemical analysis with bioassay. The data of chemical analysis are used for the determination of the class of hazard by a method of calculation. However, all compounds could not be taken into account. Therefore the bioassay of sewage sludge was added. The battery of biotests employed the protozoan *Paramecium caudatum*, the bacterium *Pseudomonas putida*, the higher plant *Raphanus sativus*, and water flea *Daphnia magna.* These organisms are relevant to overall assessment endpoints, representative of functional roles played by resident organisms, and sensitive to the contaminants present. In addition, they are characterized by rapid life-cycles,

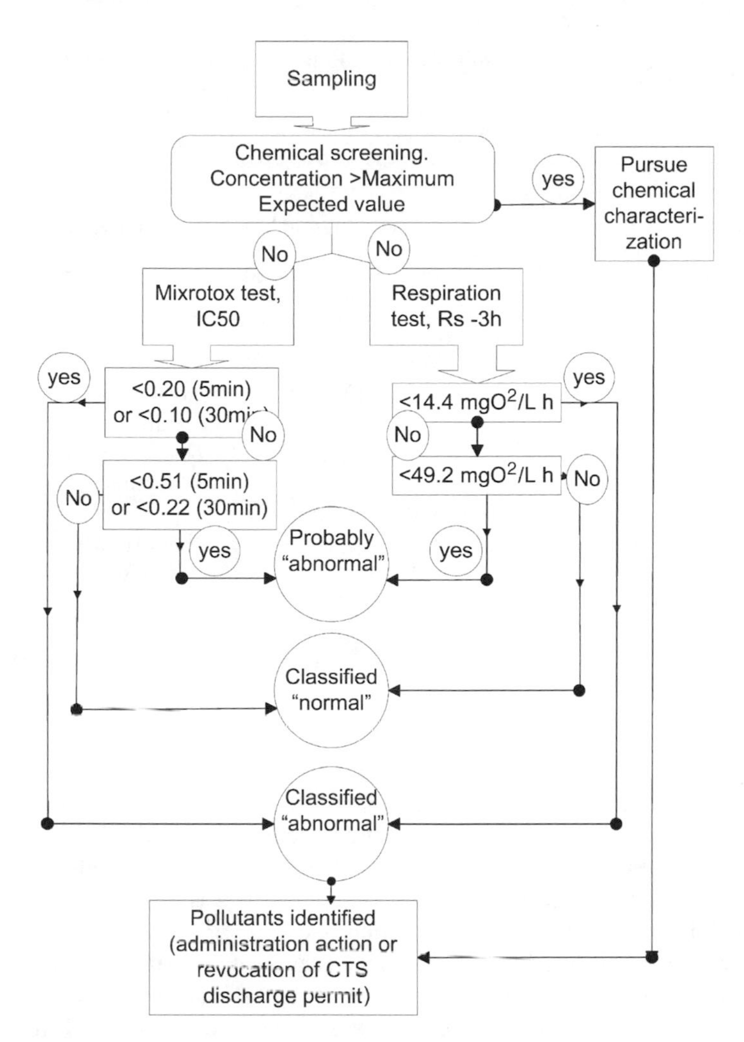

Figure 2 Proposed ecotoxicological procedure to screen for illicit discharge to toxic substances in chemical-toilet sludge.

uniform reproduction and growth, ease of culturing and maintenance in the laboratory, uniformity of population-wide phenotypic characteristics, and similar routes of exposure to those encountered in the field. The first stage consisted of the spiking of three samples of real sewage sludge with inorganic contaminants (metals) in such a manner as to create the samples on the bounds of the different classes of hazard. According to Russian legislation it is the metal content in the sewage sludge that defines the method of its disposal and its attribution to classes of hazards. Later the threshold limits criteria of these samples were established by determination of the lowest value for the dilution factor (LID_{10}) and the toxicity unit (TU) of the water extract, which exhibits less than 10% inhibition of the estimating function. Thus, the attribution of the sewage sludge samples to different classes of hazard includes the chemical analysis and the following calculation of the class of hazard and their simultaneous bioassay with the following attribution to the classes of hazard on the basis of the TU determined (Table 6). As a whole the sample of sewage sludge is attributed to the hazardous class by experimental and calculation methods (Fig. 3). In the following, for the attribution of the real waste to the classes of hazard the

Table 6 Attribution of the Sewage Sludge to the Classes of Hazard in Relation to the Results of the Bioassays, Expressed as TU $[(LID_{10})^{-1} \times 100]$

	Class of hazard			
Indexes	Hazard	Moderate hazard	Low hazard	Nonhazard
Index of hazard (K) calculated on the basis of the analytical data	$K > 1000$	$1000 \geq K > 100$	$100 \geq K > 10$	$10 \geq K$
Pseudomonas putida bioassay	$LID_{10} < 0.15$	$0.15 \leq LID_{10} < 4$	$LID_{10} \geq 4$	Nontoxic without dilution
Paramecium caudatum bioassay	$LID_{10} < 0.07$	$0.07 \leq LID_{10} < 0.6$	$ID_{10} \geq 0.6$	Nontoxic without dilution
Daphnia magna bioassay	$LID_{10} < 0.07$	$0.07 \leq LID_{10} < 0.43$	$ID_{10} \geq 0.43$	Nontoxic without dilution
Raphanus sativus bioassay	$LID_{10} < 1$	$1 \leq LID_{10} < 17$	$ID_{10} \geq 17$	Nontoxic without dilution

TU, toxic unit.

following procedure was carried out. After sampling of about 5 kg of the waste, the sample was divided into two parts. In one part, the pollutants were analyzed by chemical methods then the class of hazard calculated. The second part of the sample was analyzed by biological methods using bioassays with four test organisms. For this, the water extract (1 : 10) was produced, the series of dilutions obtained, and the toxicity measurement carried out.

The germination experiments (in quadruplicate) were carried out on filter paper in petri dishes. The corresponding water extracts (5 mL) (1/10) from the sewage sludge or soils were introduced into the dishes, with distilled water as the control in other dishes. Twenty-five radish seeds (*Raphanus sativus*) were then placed on the filter paper and the dishes placed in a germination chamber maintained at 20°C. The root lengths were measured after three days.

The tests with *Daphnia magna* were performed in 50 mL beakers. They were filled with 20 mL test solution and five animals (aged 6–24 hours) were added to each solution. For each dilution of the extract 2×5 daphnids were applied in parallel samples. The daphnids were incubated without feeding. After 96 hours the number of immobilized specimens was determined visually.

The toxicity tests with *Paramecium caudatum* were carried out in a special plate and examined under a Laboval microscope (Carl Zeiss, Jena). The test reaction was the death of the test organisms when exposed to 0.3 mL of test solution for 1 hour, using 10 individuals of *Paramecium*. Analysis was conducted five times simultaneously.

For toxicity testing with *Pseudomonas putida*, the inoculum, which has been adjusted to a specific turbidity, is added to the culture flask filled with the cultural medium and the test sample. Each dilution step should encompass three parallel batches. After an incubation period of 16 ± 1 hours at a constant temperature of 23°C in the dark, the measurement of turbidity, after homogenization by shaking, was carried out.

In all cases, the percent of inhibition (I%) was determined by comparing the response given by a control solution to the sample solution. After that, an inhibition curve was fitted to

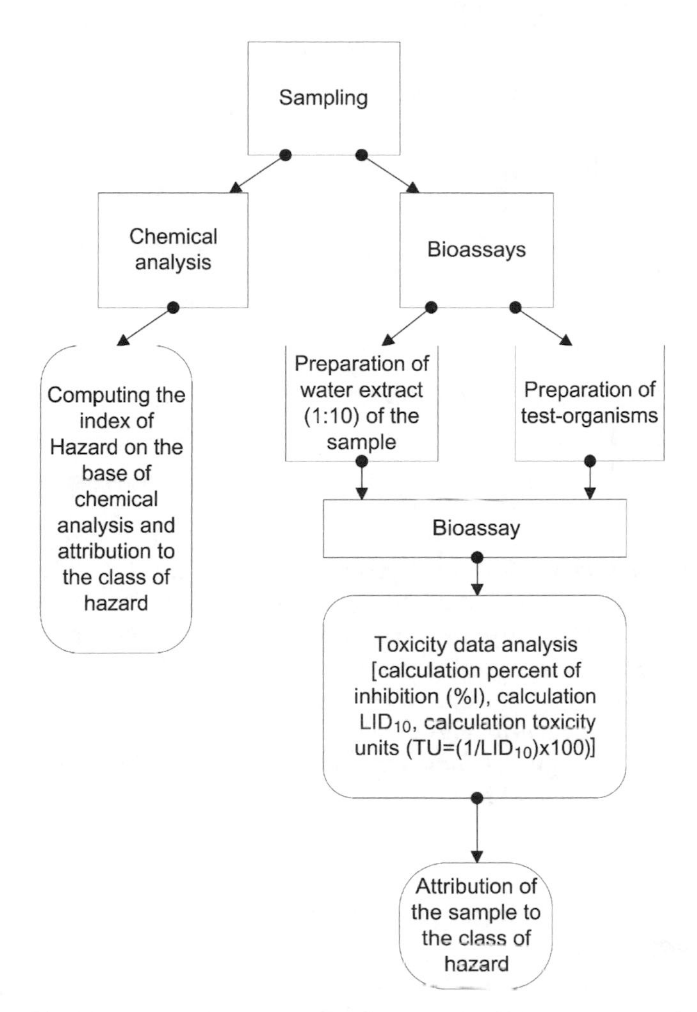

Figure 3 Proposed ecotoxicological procedure for assessment of solid waste toxicity and calculation of the classes of hazard.

calculate the 10% value for the dilution factor (LID_{10}) of the extract (Fig. 4). Acute toxicity was calculated in toxicity units (TU) according to the following formula:

$$TU = (LID_{10})^{-1} \times 100$$

This toxicity unit reflects the total toxicity of all toxic substances in the sample.

The examples of the attribution of the real sewage sludge formed on different treatment plants to the classes of hazard are presented in Table 7. The same approach was adopted in Russia for the attribution of the waste as a whole to the classes of hazard. The only difference is the use of the test organisms representing water life (water flea, algae, protozoa) [174].

A similar regulation concerning solid waste is applied in Hungary. Evaluation of hazard of the waste and the establishment of fines are based on the results of ecotoxicological tests. Classification of wastes is based on the results of toxicological tests (algal test, *Selenustrum capricornutum*; seeding test, *Sinapis alba*; crustacean, *Daphnia magna*; fish, *Zebradanio rerio*;

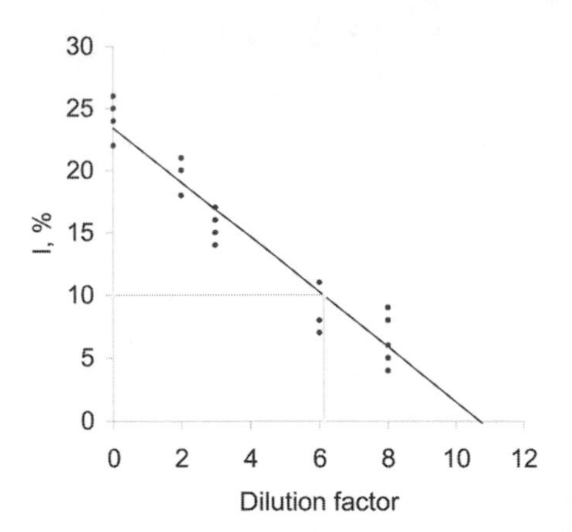

Figure 4 Example of the calculation of LID_{10}.

bacteria, *Azotobacter agile*, *Pseudomonas fluorescens*, Terravita mixed microflora) [175]. If at least one of above-mentioned tests is positive in 10-fold dilution, the waste is valued for hazard.

The use of a battery of environmental bioassays for the management of hazardous wastes is applied in the Czech Republic [176]. This battery of environmental bioassays has included representatives of producers, consumers, and destructors: *D. magna* (possible substitution by *D. pulex*), acute, reproduction, chronic test; *Scenedesmus quadricauda* (*S. capricornutum*), as bottle test or in microwell plates; *Poecillia reticulate* (*Danio rerio*), acute, chronic, embryolarval tests; *S. alba* (*Lactuca sativa*), germination test, 72 hours.

2.9 APPLICATION OF THE BIOASSAYS FOR REGULATORY REQUIREMENTS OF EFFLUENTS

Chapman [177] described in his paper a historical aspect of biotesting application, and wrote that we could date toxicity tests back at least to Aristotle, who collected "bloodworms" (most probably chironomids) from freshwater muds downstream of where Athenians discharged their sewage and observed the responses of these animals when placed into salt water. Similar experimentation has occurred on an investigator-specific basis through to the present century [178]. Effluent toxicity testing in support of organized efforts to assess and control water pollution began in the 1940s; the first attempt at standardizing effluent toxicity tests occurred in the 1950s [179]. In 1985, whole effluent toxicity (WET) testing was formalized by the U.S. Environmental Protection Agency (USEPA), with the intent: "To identify, characterize, and eliminate toxic effects of discharges on aquatic resources" [180]. Whole effluent toxicity testing is clearly a useful tool [181], but has a number of imperfections.

Among the objectives that strongly relate to the control function of wastewater biomonitoring are (1) the prevention/reduction of effects occurring in receiving water bodies; (2) to permit compliance testing as a part of the permit formulation; and (3) testing and steering the progress of technology based on improvement of effluent quality. Early warning of disasters and accident spills together with the prediction of effects occurring in receiving water bodies are mainly related to the alarm and the prediction function, respectively [177].

In order to use effluent toxicity data for pollution control purposes, it is necessary to test effluent samples that are representative of the characteristics of the effluent. Because an effluent

Table 7 Attribution of the Real Sewage Sludge Formed on Treatment Plants of Different Cities in Russia to the Classes of Hazard on the Basis of the Bioassay Data

	Toxicity expressed as TU of the water extract				
Type of sewage sludge	*Daphnia magna*	*Paramecium caudatum*	*Pseudomonas putida*	*Raphanus sativus*	Class of hazard
Raw mixture of primary and secondary sewage sludge (municipal treatment plant, Zelenodolsk)	0	0	0	0	Nonhazard
Anaerobically digested fresh sewage sludge (municipal treatment plant, Nabereznie Chelni)	0.7	0	4	33	Low hazard
Anaerobically digested sewage sludge, stored (municipal treatment plant, Nabereznie Chelni)	0	0	10	0	Low hazard
Raw secondary sewage sludge (municipal treatment plant, Kogalim)	0.4	0	0	0	Moderate hazard
Raw primary sewage sludge (municipal treatment plant, Kogalim)	0.75	2.6	4.2	25	Low hazard
Mixture of primary and secondary sewage sludge, treated with filter press (municipal treatment plant, Kazan)	1	10	3.4	8.3	Moderate hazard
Mixture of primary and secondary sewage sludge, stored in landfill (municipal treatment plant, Kazan)	5	0	20	50	Moderate hazard
Mixture of primary and secondary sewage sludge (municipal treatment plant, Usadi)	0	0	0	0	Nonhazard

may vary significantly in quantity and toxicity either randomly or with regular cycles, the design of an appropriate sampling regime is difficult, as illustrated in Figure 5. The variability of toxicity of samples from the Kazan municipal treatment plant is strongly dependent on the time and intervals of sampling.

Whole effluent toxicity test species are generally not the same as the resident species that the results of WET testing are aimed at protecting, particularly where nontemperate environments (e.g., tropical and Arctic environments) are concerned, or for estuaries [177]. Also, not all resident species have the same sensitivities to individual or combined contaminants in effluents. Further, differences exist between sensitivities and tolerances of WET species. Such differences are not unexpected; hence, it is desirable to use more than one toxicity test organism and endpoint to assess effluent toxicity.

Pontasch et al. [182] summarize the shortcomings of single species tests as follows:

1. They do not take into account interactions among species;
2. They utilize genetically homogeneous laboratory stock test populations;
3. They utilize species of unknown relative sensitivities;
4. They are mostly conducted under experimental conditions that lack similarity to natural habitats;
5. They utilize species that are not usually indigenous to the receiving ecosystem.

Indeed, toxicity assays are performed on a very limited set of species, and thus only represent a small fraction of the phylogenetic assemblages that characterize natural systems.

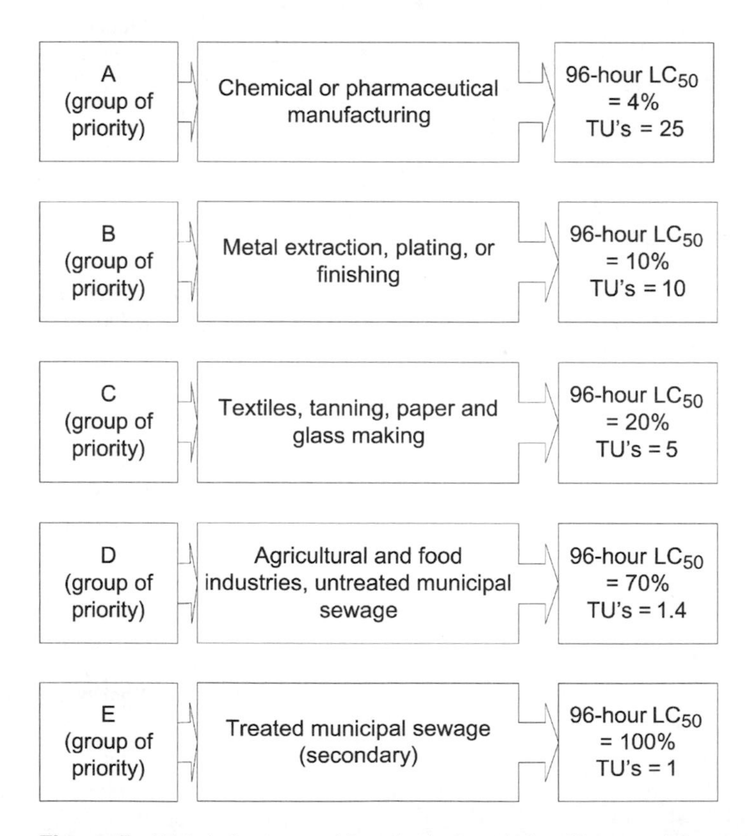

Figure 5 Irish industry specific criteria for whole effluent toxicity.

Test species currently used are those that are easily cultured and/or maintained in the laboratory. However, because of their much broader tolerances to natural environmental stressors, these biota may be poor predictors of the responses of organisms growing in a more delicately balanced and biologically inter-related environment. However, the search for the most sensitive taxon fortunately died a natural death when it was realized that different types of toxicants have different modes of action, and that no general toxicological relationship exists that is applicable to all categories of chemicals, and for all species. Yet, the question that has not yet been solved is how many species and what types of species need to be tested to adequately represent the whole range of indigenous biota of natural systems. For waste mixtures, the suite of biota must cover a much broader range of phylogenetic groups, unless it can be demonstrated that particular groups of biota are much less sensitive than others, and can be excluded from the battery [183]. In order to take the (often neglected) ecological realism in toxicity testing into consideration, the battery of bioassays are composed of test species belonging to the three trophic levels of aquatic food chains: producers, consumers, and decomposers. Bernard *et al.* [168] used *Scenedesmus subspicatus* (micro-algae) and *Lemna minor* (duckweed) for producers; *Brachionus calyciflorus* (rotifers) and *Daphnia magna* for consumers; and *Ceriodaphnia dubia* and *Thamnocephalus platyurus* (crustaceans), for the decomposers. *Vibrio fisheri* (bacteria) and *Spirostomum ambiguum* (ciliate protozoan) were used for testing such complex effluents as landfill leachates. Based on the results of their investigations the authors made recommendation for using a further such battery of tests: prokaryotes (*V. fisheri*), unicellular animal eukaryotes (*S. ambiguum*), unicellular plant eukaryotes (*S. subspicatus*), and one representative of either a multicellular plant or various groups of animal eukaryotes.

In Russian legislation there is a requirement for using two test organisms from different trophic levels in a battery of recommended test species: decomposer, bacteria *V. fisheri* and *E. coli* (Toxi-Chromotest), unicellular animal eucaryotes *P. caudatum*; producers, unicellular plant eukaryotes *Chlorella vulgaris*, *Scenedesmus quadricauda*; and consumers, multicellular animal eukaryotes *D. magna*, *Ceriodaphnia dubia (affinis)*. Sensitivity of test species depends on the wastewater composition, but sensitivity is often accorded in decreasing order as: *S. quadricauda* (*C. vulgaris*) → *P. caudatum* → *V. fisheri* → *D. magna* [*C. dubia* (*affinis*)]. In particular we would like to underline the ability of microalgae to increase biomass during wastewater testing. This effect is well known as stimulated, which necessitates eutrofication of the recipient water body. The use of ciliate *P. caudatum* has low sensitivity, but due to high expression (1 hour) is a very popular test, in particular for toxicity screening of wastewater in the sewage system before biological treatment. The test permits the most toxic wastewaters to be analyzed rapidly and cost-effectively.

2.9.1 Control of the Toxicity of Industrial Discharges

In most European countries, the control of toxicity of industrial discharges is carried out, to date, almost exclusively through quantitative chemical analysis of each compound for which a limit value has been set. Unfortunately, this practice is not very efficient from the point of view of protection of the aquatic ecosystem for the following two major reasons:

1. Chemical analyses are limited to a restricted number of compounds, which do not necessarily reflect the qualitative nor quantitative "overall" composition of the waste.
2. Wastes are very often complex mixtures of substances, each of which are present in a different concentration [156].

With regard to the first reason, it must be stressed that whereas each legislation prescribes explicitly that an industrial discharge should not affect the biota of the receiving waters, the practical implementation totally overlooks the (potential) toxic effects of compounds for which no limit values have been set, but which may make up a substantial part of the effluent. With regard to the second reason, it is virtually impossible to calculate the ultimate toxicity of a (complex) waste from the individual toxicities of each chemical present. A simple comparison to illustrate the latter statement is the impossibility "to predict" (at least with a certain degree of precision) the final color of a set of different dyes, to be mixed in different proportions. The only valid approach to determine the final color (i.e., in the case of hazard assessment: the ultimate toxicity) is the "experimental" way, namely by ecotoxicological testing [156].

Although ecotoxicological testing is the only valid approach to establish the real hazard of effluent discharges, it is seldom practiced in routine unless it is explicitly imposed by legislation, which is the case in only a few countries.

The data concerning the use of bioassays in the biomonitoring of liquid waste are presented in different reviews [12,19,184]. Hereafter we represent some information from these reviews.

2.9.2 Canada

Environment Canada recently developed an evaluation system based on effluent toxicity testing, capable of ranking the environmental hazards of industrial effluents [185]. This so-called Potential Ecotoxic Effects Probe (PEEP) incorporates the results of a variety of small-scale toxicity tests into one relative toxicity index to prioritize effluents for sanitation. In the index no allowance has been made for in-stream dilution, therefore the actual risk for environmental effects is not modeled. The tests performed on each effluent are the following: bacterial assay [*V. fisheri* (*P. phosphoreum*), *Microtox*], microalgal assay (*S. capricornutum*); crustacean assay (*C. dubia*); and bacterial genotoxicity test (*E. coli*, SOS-test).

All test results are expressed as threshold values (LOECs), and subsequently transformed to toxic units (TUs). The entire scheme results in a total number of 10 TUs per effluent. The results are put through the following calculation to produce the PEEP index.

$$PEEP = \log_{10}\left[1 + n\left(\frac{\sum_{i=1}^{N} TU_i}{N}\right)Q\right]$$

where N is the total number of bioassays performed, n is the number of bioassays indicating toxicity, and Q is the flow rate of the effluent in m^3/hour.

Based on the correlation matrix of all bioassays data obtained with 37 effluents, it can be concluded that none of the bioassays produces data that are redundant. In other words, all bioassay procedures add to the information content of the PEEP index.

In the 37-effluent study, the effluents of the pulp and paper industry proved to be consistently far more toxic than those of other types of industries (PEEP > 5). The same study revealed that approximately 90% of the total toxic discharge is caused by the added toxicity of only three effluents of the 37. The effluent pipes for these are clearly considered the most rewarding for counteractive measures [12].

2.9.3 USA

In 1984, the U.S. Environmental Protection Agency (EPA) [186] recommended the use of "biological techniques as a complement to chemical-specific analysis to assess effluent discharges and express permit limitations." Already in 1985 [187] a guidance document had been produced on the use of effluent toxicity test results in the process of granting permits for discharge. The Organization for Economic Cooperation and Development (OECD) [188,189] in 1987 and 1991 fully adhered to the guidelines provided by the EPA. Discharging industries are required to provide quality-assured data on toxicity according to a tiered approach, where the in-stream dilution is the first screening level, and increasing toxicity requires more complicated and definitive testing with increasing numbers of species from different trophic levels, at increasing frequencies. The permit requirements are set to the level where there is a minimal risk for ecosystem damage outside the in-stream mixing zone. Inside the mixing zone some nonlethal effects are allowed to occur, depending on the types of organisms and their duration of residence in the dilution plume. The 1985 scheme was rather complicated with respect to determining the balance between the projected in-stream toxicity and uncertainty/reliability. Since new policies and regulations have been promulgated and a vast amount of knowledge and experience has been gained in controlling toxic pollutants, the testing and evaluation scheme was greatly simplified, while retaining its integrity, in 1991 [190]. Genotoxicity is addressed in a chemical-specific way with respect to human health only, based on the average daily intake (ADI) with drinking water and the ADI with fish consumption. The aspect of bioaccumulative capacity is also dealt with in a chemical-specific way.

The biological approach (whole effluent) to toxics control for the protection of aquatic life involves the use of acute and chronic toxicity tests to measure the toxicity of wastewaters. Whole effluent tests (WET) employ the use of standardized, surrogate freshwater or marine (depending on the mixture of effluent and receiving water) plants (algae), invertebrates, and vertebrates.

The evaluation strategy applied to the combined data on in-stream dilution and multiple data on effluent toxicity involves a comparison of the calculated concentration of the effluent in the receiving water under worst-case conditions (RWC = receiving water concentration) with statistically derived "safe" concentrations of that specific effluent [the critical continuous concentration (CCC), based on chronic testing, and the critical maximum concentration (CMC), based on acute testing]. RWC, as well as CCC and CMC, is expressed as TUs. Action is taken when RWC > CCC or RWC > CMC. As a minimum input from toxicity testing it is required to perform acute toxicity tests on three different species quarterly for a period of at least one year. Additionally, some extrapolation to chronic toxicity has to be provided, or chronic toxicity has to be tested, depending on the rate of in-stream dilution. If the dilution is less than 1 : 100, chronic toxicity is required. If neither of the CCC or CMC are violated and the dilution is less than 1%, then it has to be demonstrated that combination effects will not occur in the receiving water (use up-stream dilution water in toxicity tests), and that the toxicity is nonpersistent (repeatedly test effluent/up-stream water samples after progressive storage under realistic conditions).

The EPA realized that setting water quality criteria with respect to toxic load, although playing an important role in assuring a healthy aquatic environment, has not been sufficient to ensure appropriate levels of environmental protection. The primary objective of the U.S. Clean Water Act (1987) is "... the restoration and maintenance of the chemical, physical, and biological integrity of the Nation's waters." To meet this objective, EPA rightfully states that water quality criteria should address biological integrity. Therefore, the Agency recommends that the water quality authorities begin to develop and implement biological criteria in their water quality standards. In order to verify the compliance of water bodies to their assigned standards, ecosystem monitoring is considered a necessity. In the guidance document on water quality based toxics control [190], it is explicitly stated that the chemical-specific and the whole effluent approaches for controlling water quality should eventually be integrated with ecological bioassessment approaches [12].

2.9.4 Argentina

As in many countries, the first attempts at understanding the effects of pollution on aquatic ecosystems in Argentina began within the academic and scientific community [191]. A systematic approach using toxicity tests with aquatic organisms is applied only in scientific laboratories.

2.9.5 Chile

The use of bioassays in environmental monitoring has not been developed in Chile [191]. In 1998 the Ministry of Agriculture started to set up a bioassay laboratory for evaluation of the presence of toxic substances in water for irrigation and animal consumption. This ministry is now in the process of implementation of EPA standardized crustacean and algal tests with *Daphnia* and *Selenastrum capricornutum*, respectively. There is no governmental wastewater bioassay monitoring.

In 1998, two bioassay methods were considered by the Chilean Regulation Institute (INN) as the first attempts for the introduction of microbioassays for routine testing in Chilean regulations: (1) the *Bacillus subtilis* growth inhibition test for toxicity evaluation of industrial effluents discharged into sewers, to detect interference with the BOD, is near endorsement; and (2) the assessment of acute toxicity in receiving waters using *D. pulex* is presently under discussion.

2.9.6 Columbia

The use of bioassays as an analytical tool for the assessment of environmental pollution is relatively new in Columbia. Even though the Ministry of Health established in Decree 1594 (1984) that environmental control agencies should propose acceptable LC_{50} values for 22 substances of ecotoxicological interest in order to protect fauna and flora, none of the entities has carried out this action up to mid-1998.

The control of toxic substances by means of bioassays at a governmental level has had little development. Even though there has been no great industrialization in this country, control of industrial contamination has centered on the implementation of treatment systems to remove organic material and bacteria. Consequently, although it is well-known that 85% of industrial effluents are discharged into continental waters and seas without any treatment, and that 74% of them are found around the Caribbean basin, currently proposed monitoring programs are centered on physico-chemical evaluation and the reduction of organic and bacteriological contamination [191].

2.9.7 Japan

In Japan, many chemicals are monitored at specific sites in rivers, lakes, and coastal areas, and data are published through the Japanese Environmental Agency. Environmental standards of water quality were revised in 1993 and over 50 chemicals were added to the list. Ecotoxicological monitoring is now considered to be very important for risk assessment of chemicals, and guidelines for ecotoxicological evaluation of chemicals are presently under examination at the level of the Japanese Government [192]. The methods that will be taken into consideration are in most cases in accordance with OECD Guidelines [79,133]. From the 10 toxicity tests described in the OECD Guidelines, the algal growth inhibition test, the *Daphnia* acute immobilization and reproduction test, and the fish toxicity test have been selected and the PNEC values from literature sources are compared with environmental concentrations. However, bioassays are not yet endorsed legally as a tool for environmental monitoring and hazard assessment in Japan. Toxic hazard is still only evaluated through chemical analysis.

2.9.8 France

In France, industrial effluents are regularly monitored for toxicity with daphnids. The toxicity data are used as a base for discharge taxation [193]. The Microtox test, chronic toxicity test, and a test on mutagenicity to the set of required bio-criteria are also used for wastewater monitoring [12,194].

2.9.9 Germany

German water authorities adopted a permit system for effluent emission where the requirements are based on fish toxicity [195]. Daphnia, algae, and luminescent bacteria are including for a screening additionally to the fish test. In this scheme the fish test (*Goldorfo*; *Leuciscus idus*) is still considered to be the only test producing definitive results.

The toxicity requirements are established per type of industry, in terms of the maximum number of times the effluents needs to be diluted to produce a no observed effect concentration (NOEC), defined as Gf for fish, Gd for daphnia, Ga for algae, and Gl for luminescent bacteria. Testing is limited to the exposure to only the appropriate Gx level, which should not produce any observed effect [the G-value corresponds with the dilution of the effluent, expressed as the lowest dilution factor $(1, 2, 4, \ldots)$ causing less than 10% mortality]. The level of maximum allowable toxicity per industrial branch is based on the level that is considered to be attainable with state-of-the-art process and/or treatment technology. Violating the toxicity requirements results in a levy, which makes state-of-the-art compliance a more economic option [12].

2.9.10 Ireland

In Ireland, compliance with toxicity limits for selected industries is ascertained by annual or biannual test on representative samples of effluent. The test species most commonly used is the rainbow trout (*Salmo gairdneri*). Control authorities normally require results from 96-hour tests. The toxicity values are expressed as the minimum acceptable proportion of effluent (as a percentage) in a test resulting in 50% fish mortality after 96 hours of exposure. The toxic units (TU) are defined as the maximum number of times an effluent may be diluted to produce the test criteria ($TU = 100/96\text{-hour } LC_{50}$, with LC_{50} expressed as the percentage of effluent in the test) (Fig. 5).

In order to encourage the optimum selection of sites for new industries, it is recommended that receiving waters at all times must provide a minimum of 20 dilutions in the immediate vicinity of the discharge for each toxic unit discharged. Flow measurements, mixing and dispersion studies are therefore a necessary addition to monitoring toxicity limits of effluents [12].

2.9.11 The Netherlands

For the control of water quality, the Netherlands government identified two pathways in a tiered procedure. The first path, the emission approach, requires dischargers to apply best available and/or best affordable technologies for the reduction of the environmental risk of their effluents with respect to good housekeeping, process control, choice of (raw) materials, and effluent pretreatment. Currently, this process is only iteratively guided by chemical-specific evaluation of effluent quality. In a combined effort, the Ministry of Housing, Spatial Planning and the Environment, together with the Ministry of Transport, Public Works and Water Management, are in the process of developing a whole effluent evaluation system that will complement the chemical-specific approach. The whole effluent evaluation method will only be applied to selected effluents (large quantities, high risk) to assist in formulating additional pollution reduction strategies. The method will be comprised of effluent tests on mutagenicity, persistence, chemical and biological oxygen demand (COD and BOD), acute and chronic toxicity, and bioaccumulation as intrinsic properties of the effluent (Fig. 6) [12,196,197].

Once effluent quality is considered to be acceptable, the water quality based approach will be followed, in which the remaining risks for effects in the receiving water are evaluated. In this framework, ambient water quality, inside and outside the mixing zone, will be verified against compound-specific water quality objectives, designated use requirements, the presence of actual toxicity (TRIAD), and biological integrity (biological water quality objectives). The results of the remaining risk evaluation may lead to the requirement of further risk reducing measures in the effluent. Additionally, the possibilities for setting permit limit requirements in the sense of whole effluent toxicity are also being evaluated.

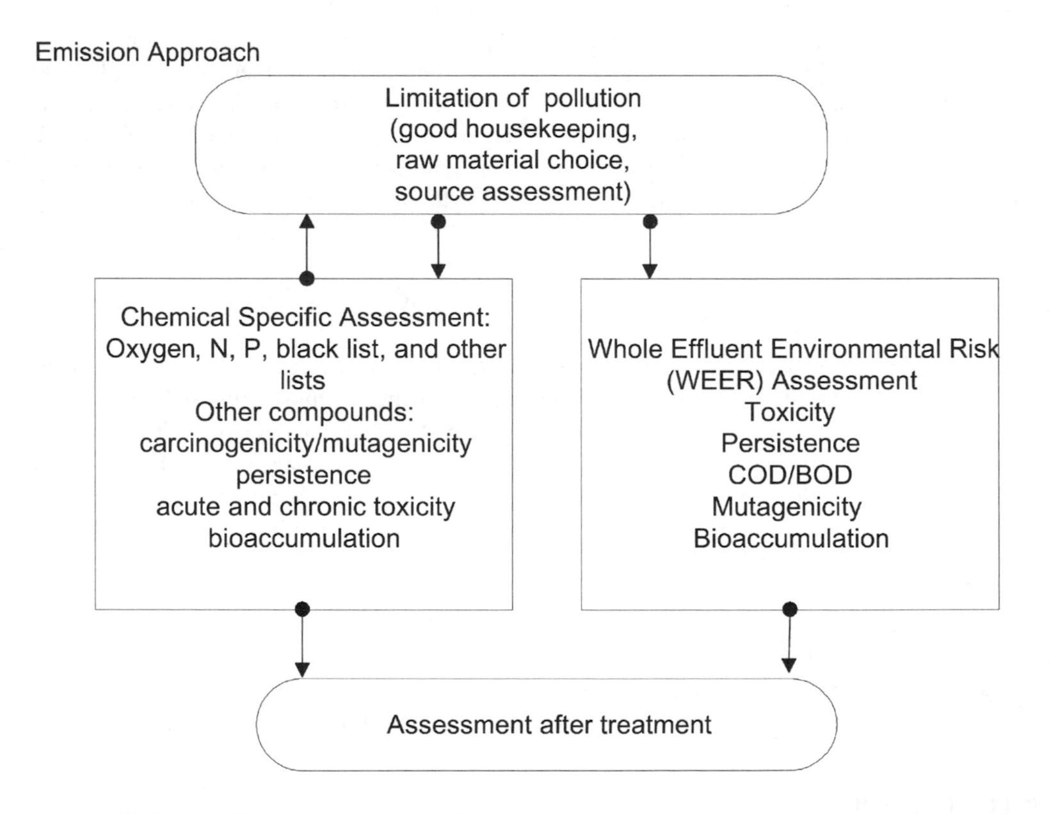

Figure 6 The Netherlands system for water quality control.

2.9.12 United Kingdom

The biological testing of wastewater initially only consists of acute toxicity screening with luminescent bacteria (Microtox) and a 24 hour *Daphnia* lethality test for freshwater or a 24 hour *Oyster* larvae test for estuarine or marine waters to reveal the need for further testing [12,198,199]. The results of these tests classify the permit requirements for the effluent in four categories. The most stringent class requires the effluent to be monitored with three or four acute toxicity tests (Freshwater: 72/96 hour algal growth inhibition test with *Selenastrum*, 48 hour *Daphnia* lethality test, and a 96 hour fish lethality test with *Salmo trutta*, *Oncorhynchus mykiss*, or *Cyprinus carpio*. Marine/estuarine water: 72/96 hour algal growth inhibition test with *Pheodactylum* or *Skeletonema*, 48 hour *Oyster* embryo/larvae development test, and a 96 hour fish lethality test with *Pleuronectes paltessa* or *Scopthalmus maximus*). The second stringent class prescribes effluent monitoring with one of the screening tests after verification with the above-mentioned three or four acute tests. A third lower level of toxicity leaves the obligation for toxicity monitoring to one of the screening tests only, and at the fourth level no toxicity monitoring will be required. Measurements of chronic toxicity are not considered, neither are evaluations of accumulation, persistency, degradability, and genotoxicity.

2.9.13 Sweden

In Sweden, industrial effluents are to be characterized by chemical composition, toxicity, bioaccumulative capacity, and degradability [200]. The evaluation is performed according to the following tiered procedure:

Step 1

- Degradability is measured as BOD_7/COD;
- Acute toxicity is evaluated for fish, crustacea, algae, and higher plants (model organisms);
- Bioaccumulation capacity is estimated by extraction with an organic solvent, followed by the separation of the lipophilic compounds with thin layer chromatography. The migration distances give information on possible bioconcentration factors. The compounds of interest can be isolated from the TLC-plate and analyzed by GC/MS;
- Chemical analysis, including group variables like absorbable organic halogenids or total organic chlorine.

Step 2

- Degradability – added test with possibly a characterization (toxicity or bioaccumulation) or identification of the nondegradable fraction;
- Biological effects measurements – chronic toxicity and mutagenicity tests;
- Bioaccumulation and chemical evaluation – involve more and more elaborate analysis.

Step 3

- Step 3 is only prescribed in general terms, but should be tailored for the specific effluent on the basis of the results from tier 1 and 2.

2.9.14 Norway

Norway has a standardized test program for permit derivation, comprising the Ames mutagenicity test, acute and chronic toxicity tests, and a biodegradation test. For monitoring purposes, it is advised to start screening the toxicity of an effluent with a comparatively large diversity of tests. The determination of precise concentration–effect relationships can then be restricted to the most sensitive types of organisms [12,201].

2.9.15 Poland

The first Polish toxicity standards were elaborated more than 25 years ago [202]:

- determination of acute toxicity to *Chlorella*;
- determination of acute toxicity to *Daphnia magna*;
- determination of acute toxicity to *Lebistes reticulatus.*

Those standards were modified and adapted to ISO standards 10 years ago. Standard bioassays used are mainly *D. magna*, luminescent bacteria (Microtox^{TM}, Lumistox), and Spirotox (protozoan *Spirostomum ambiguum*).

2.9.16 Estonia

In Estonia the monitoring of effluents is based on chemical analysis. The list of controlled water quality parameters depends on the type of industry. Bioassays are not used as a monitoring tool. However, according to HELCOM Recommendations No. 16/5, "Requirement for discharging of waste water from the chemical industry," and No. 16/10, "Reduction of discharges and emission from production of textiles," the toxicity effect of discharges into water bodies should be determined by (at least) two toxicity tests, which could be chosen out of the following four toxicity tests [203]:

- toxicity to fish;
- toxicity to algae;
- toxicity to invertebrates (*Daphnidae*);
- toxicity to bacteria.

2.9.17 Hungary

Chemical analyses are mainly used for detecting hazard of liquid and solid wastes [175]. Governmental orders and laws regulate the evaluation of hazard of effluent by toxicological tests. Waste control includes the determination of 30 chemical parameters, coliform count, and the result of ecotoxicological test (*D. magna* test). Category of toxicity:

- $>$100-fold dilution $\rightarrow$ strongly ecotoxic;
- 50 to 100-fold dilution $\rightarrow$ ecotoxic;
- 10 to 50-fold dilution $\rightarrow$ lightly ecotoxic;
- $<$10-fold dilution $\rightarrow$ nontoxic.

2.9.18 Czech Republic

The official use of bioassays for the environmental management of hazardous wastes and chemicals is requested in Law No. 157/98 (chemicals) and 132/97 (industrial and domestic wastes). As well as other laws (protected areas, reservoirs conservation, etc.) there is also a system of regulations, directives of the Ministry of the Environment (mostly wastes, monitoring system, etc.), agriculture (drinking water, agricultural soil, and water for animals) and guidelines of local governments. The philosophy is led by the idea that every battery of environmental bioassays includes representatives of producers, consumers, and destructors [176].

Here is an overview of ecotoxicological bioassays cited in the above-mentioned legislation:

- *Daphnia magna* (*Thamnocephalus platyurus*);
- *Scenedesmus quadricauda (Selenastrum capricornutum)* as bottle test or in microwell plates;
- *Poecillia reticulate (Danio rerio)* acute, chronic, embryolarval tests;
- Microtox test (or equivalent);
- Activated sludge respiration test.

2.9.19 Slovenia

Routine ecotoxicological tests are regulated for some wastewaters by a law that controls emission and taxation. For wastewaters that flow into receptacles, the inhibition of the mobility of *Daphnia magna* Straus (Cladocera, crustacea) acute toxicity test is obligatory [79]. For some types of wastewaters flowing through biological purification plants the evaluation of aerobic biodegradability of organic compounds in an aqueous medium/static test (Zahn–Wellens method) is likewise obligatory [204].

2.9.20 Lithuania

According to wastewater requirements, the water quality of effluents should not be toxic on the basis of results of two acute toxicity tests. The following tests can be applied: toxicity to fish, toxicity to daphnia, toxicity to luminescent bacteria, toxicity to green algae [205].

2.9.21 Russia

The battery of tests used involves the conventional crustacean test (*D. magna*, *C. dubia*), conventional algal test (*S. quadricauda*, *C. vulgaris*), protozoan tests (*P. caudatum*, death and chemotaxis) and ToxiChromotest, which are applied for taxation of discharge water [206,207]. Researchers should choose no less than two of the bioassays. The level of the tax depends on the level of the toxicity of the discharged water and is calculated by multiplication of the basal tax by a special coefficient, which is determined on the basis of the dilution factor, expressed as toxic units (Table 8). Analysis of data from 13 regions of Russia [207] has shown that only 47% of all samples of wastewater were nontoxic and 5% of them were extremely toxic (Fig. 7). Analysis of toxicity of industrial discharges of the Tatarstan region has shown that toxicological load on the recipient water body is proportional of wastewater toxicity. Thus, about 0.03% of the water capacity of the River Kazanka is necessary for dilution of toxic wastewaters every day (Table 9).

Table 8 Classification of Effluents Based on Toxicity Assessment

Level of toxicity	Toxic unit	Multiplying coefficient
Low toxicity	1.1–16	1.3
Moderate toxicity	16–50	1.5
High toxicity	50–90	1.8
Extremely high toxicity	>99	2.0

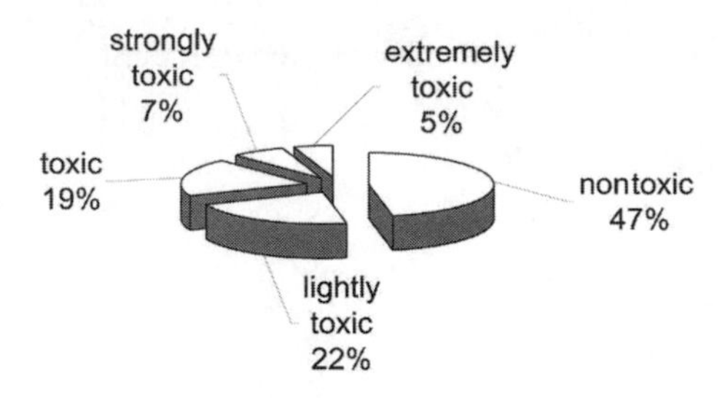

Figure 7 Wastewater classification on the basis of toxicological results (data analysis of 13 Russian regions).

Gelashvilly *et al.* [208] showed distribution of toxic input on the basis of types of industrial wastewaters for the Nizhni Novgorod region (Fig. 8). Toxicological contribution of different industrial wastewaters on recipient water bodies was calculated using the following equation:

$$T = TU \cdot Q \cdot t$$

Table 9 Toxicity Control of the Wastewaters Discharged into Recipient Water Bodies (Tatarstan Region, Russia)

Source of industrial discharge	TU_a	Capacity of discharge in m^3/day	Volume of natural water for decreasing acute toxicity of discharge (m^3/day)	Name of the recipient river
Kazan municipal treatment plant	1.3	572,000	743,600	Volga
Zainsk municipal treatment plant	3.75	8,000	30,000	Zai
Hennery "Yudinski"	12	240	2,880	Volga
Milk plant (Sabinsk)	7.2	280	2,016	Sabinka
Alcagol distilling plant (Usadi)	1	1,895	1,895	Kazanka
Optic-mechanical plant	6	253.44	1,774.1	Kazanka
Milk plant (Arsk)	6	200	1,200	Kazanka
Hennery (Laishevo)	1	800	800	Pond
Municipal treatment plant (Pestresi)	1	700	700	Miesha
Sanatorium "Krutushka"	4	150	600	Kazanka
Agricultural firm "Serp i Molot"	1	400	400	Kazanka
Municipal treatment plant (Laishevo)	1	300	300	Kama
Building plant (Kurkachi)	1	250	250	Kazanka
Kazan tuberculosis hospital	1.5	90	135	Kazanka
Sanatorium "Vasilievo"	1	90	90	Volga
Nutritive plant (Laishevo)	1	15	15	Kama

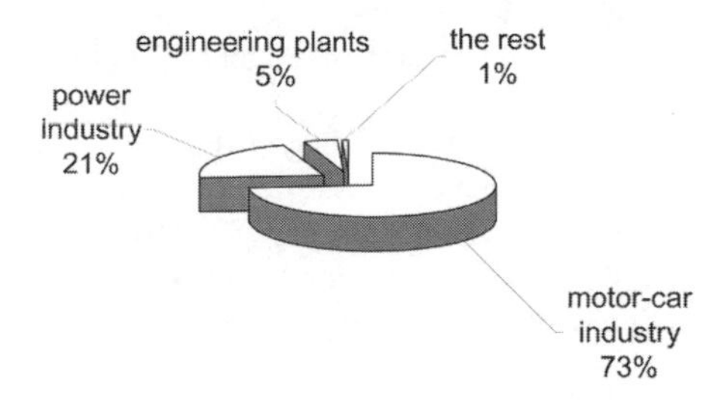

Figure 8 Toxic contribution of the different industrial wastewaters to the recipient water bodies of the Nizhni Novgorod region.

where T is toxicological load on the water body in $m^3/year$, TU is acute toxicity results of the most sensitive tests, Q is capacity of industrial discharge in m^3, and t is period of time (year).

The average toxicity of wastewaters decreased more than four-fold in the Nizhni Novgorod region as a result of the economic mechanism of tax collection.

2.10 CONCLUSION

Interest in bioassays as a tool in risk assessment of waste has definitely grown in recent years. There are many different toxicological bioassays, but most of them are developed for liquid waste or environments, with only a few standard tests being available for solid waste. What is more, very few have been widely adopted for routine toxicity evaluation. The main reason for this is difficulty in maintaining healthy laboratory cultures of the proposed organisms for long periods, resulting in a low degree of standardization of toxicity tests, which are thus not accepted by standardization and/or regulatory organizations. Another problem is the choice of suitable bioassays to produce a good toxicity test battery. The choice of toxicity method should be based on screening, regulatory requirements, or predictive hazard assessment. Each type of bioassay can have its own merits when properly used in the correct context.

Toxicity tests integrate interactions among complex mixtures of contaminants. They measure the total toxic effect, regardless of physical and chemical composition. As such, these tests are a useful tool. But they are not a perfect tool, particularly because they are commonly applied to conditions that do not reflect the test exposures. No single, perfect, universal tool exists; all tools have advantages, disadvantages, and assumptions [174]. Alone, bioassays cannot fulfill their stated purpose "to identify, characterize, and eliminate toxic effects of discharges on aquatic resources." However, together with other appropriate tools in a risk assessment framework (i.e., joint, not independent, applicability), toxicity testing is essential for ultimately achieving this purpose.

REFERENCES

1. Markwiese, J.T.; Ryti, R.T.; Hooten, M.M.; Michael, D.I.; Hlohowskyj, I. Toxicity bioassays for ecological risk assessment in arid and semiarid ecosystems. Rev. Environ. Contam. Toxicol. **2001**, *168*, 43–98.
2. Jouany, J.M. Ecologie et nuisances. Actual Pharmaceut. **1971**, *69*, 12–22.
3. Butler, G.C. *Principles of Ecotoxicology*; Wiley: Chichester, UK, 1978; SCOPE Vol. 12.
4. Ramade, F. *Ecotoxicologie*; Masson: Paris, France, 1979; Collection d'ecologie, No. 9.

5. Chapman, P.M. Sediment quality criteria from the sediment quality triad: an example. Environ. Toxicol. Chem. **1986**, *5*, 957–964.
6. Thomas, J.M.; Skalski, J.R.; Cline, J.F.; McShane, M.C.; Simpson, J.C.; Miller, W.E.; Peterson, S.A.; Callahan, C.A.; Green, J.C. Chemical characterization of chemical wastesite contamination and determination of its extent using bioassays. Environ. Toxicol. Chem. **1986**, *5*, 487–501.
7. Blaise, C.; Sergy, G.; Bermingham, N.; Van Coillie, R. Biological testing – development, application and trends in Canadian Environmental Protection Laboratories. Tox. Assess. Int. J. **1988**, *3*, 385–406.
8. Dutka, B. Priority setting of hazards in waters and sediments by proposed ranking scheme and battery of tests approach. German J. Appl. Zool. **1988**, *75*, 303–316.
9. Kreysa, G.; Wiesner, J., Eds. *Bioassays for Soils*; Schon & Wetzel GmbH: Frankfurt am M., Germany, 1995.
10. Juvonen, R.; Martikainen, E.; Schultz, E.; Joutti, A.; Ahtiainen, J.; Lehtokari, M. A battery of toxicity tests as indicators of decontamination in composting oily waste. Ecotox. Environ. Safe. **2000**, *47*, 156–166.
11. Kapanen, A.; Itavaara, M. Ecotoxicity tests for compost applications. Ecotox. Environ. Safe. **2001**, *49*, 1–16.
12. Zwart, D. *Monitoring Water Quality in the Future*; Ministry of Housing, Spatial Planning and the Environment, Department for International Relations: The Netherlands. Vol. 3: Biomonitoring, 1995; 83 pp.
13. USEPA. *Ecological Risk Assessment Guidance for Superfund: Process for Designing and Conducting Ecological Risk Assessments*; U.S. Environmental Protection Agency: Washington, DC, 1997.
14. Haanstra, L.; Doelman, P. An ecological dose response model approach of short and long-term effects of heavy metals on arylsulphatase activity in soil. Biol. Fert. Soils **1991**, *11*, 18–23.
15. Van Beelen, P.; Doelmann, P. Significance and application of microbial toxicity tests in assessing ecotoxicological risks of contaminants in soil and sediment. Chemosphere **1997**, *34*, 455–499.
16. Tubbing, D.; Santhagens, L.R.; Admiraal, W.; Van Beelen, P. Biological and chemical aspects of differences in sensitivity of natural populations of aquatic bacterial communities exposed to copper. Environ. Toxic. Water Quality **1993**, *S8*, 191–205.
17. Ronnpagel, K.; Janssen, E.; Ahlf, W. Asking for the indicator function of bioassays evaluating soil contamination: Are bioassay results reasonable surrogates of effects on soil microflora? Chemosphere **1998**, *36*, 1291–1304.
18. Selivanovskaya, S.Y.; Petrov, A.M.; Egorova, K.V.; Naumova, R.P. Protozoa and metazoa communities treating a simulated petrochemical industry wastewater in rotating disc biological reactor. World J. Microb. Biot. **1997**, *18*, 197–204.
19. Bitton, G. Toxicity testing in wastewater treatment plants using microorganisms. In *Wastewater Microbiology*; Wiley-Liss: New York, 1994, 478 pp.
20. Ongley, E.D.; Birkholz, D.A.; Carey, J.H.; Samoiloff, M.R. Is water a relevant sampling medium for toxic chemicals? An alternative environmental sensing strategy. J. Environ. Qual. **1988**, *17*, 391–401.
21. Ronnpagel, K.; Liss, W.; Ahlf, W. Microbial bioassays to assess the toxicity of solid-associated contaminants. Ecotox. Environ. Saf. **1995**, *31*, 99–103.
22. ISO 11269-1. *Soil quality – Determination of the Effects of Pollutants on Soil Flora – Part 1: Method for Measurement of Inhibition of Root Growth*; International Organization for Standardization: Geneve, 1993; 9 pp.
23. ISO 11269-2. *Soil Quality – Determination of the Effects of Pollutants on Soil Flora – Part 2: Effects of Chemicals on the Emergence and Growth of Higher Plants*; International Organization for Standardization, Geneve, 1995, 7 pp.
24. Brouwer, H.; Murphy, T.; McArdle, L. A sediment-contact bioassay with *Photobacterium phosphoreum*. Environ. Toxicol. Chem. **1990**, *9*, 1353–1358.
25. Kwan, K.K.; Dutka, B.J. A novel bioassay approach: direct application of Toxi-Chromotest and SOS Chromotest to sediments. Environ. Toxic. Water Quality **1992**, *7*, 49–60.

26. Kwan, K.K. Direct assessment of solid phase samples using the Toxi-Chromotest kit. Environ. Toxic. Water Quality **1993**, *8*, 223–230.
27. Prokop, Z.; Holoubek, I. The use of a microbial contact toxicity test for evaluating cadmium bioavailability in soil. J. Soil Sediment **2001**, *1*, 21–24.
28. Blaise, C. Microbiotests in aquatic ecotoxicology: characteristics, utility, and prospects. Environ. Toxic. Water Quality **1991**, *6*, 145–155.
29. Bitton, G.; Koopman, B. Bacterial and enzymatic bioassays for toxicity testing in the environment. Rev. Environ. Contam. Toxicol. **1992**, *125*, 1–22.
30. Dutka, B.J.; Nyholm, N.; Petersen, J. Comparison of several microbiological toxicity screening tests. Water. Res. **1983**, *17*, 1363–1368.
31. Ribo, J.M.; Kaiser, K.L.E. Photobacterium phosphoreum toxicity bioassay. I. Test procedures and application. Toxic. Assess. **1987**, *2*, 305–323.
32. Kwan, K.K.; Dutka, B.J. Evaluation of the Toxi-Chromotest direct sediment toxicity testing procedure and Microtox solid-phase testing procedure. B. Environ. Contam. Tox. **1992**, *49*, 656–662.
33. Day, C.; Dutka, B.J.; Kwan, K.K.; Batista, N.; Reynoldson, T.B.; Metcalfe-Smith, J.L. Correlations between solid-phase microbial screening assays, whole-sediment toxicity tests with macroinvertebrates and in situ benthic community structure. J. Great Lakes Res. **1995**, *21*, 192–206.
34. Bulich, A.A.; Green, M.M. *The use of luminescent bacteria for biological monitoring of water quality*. In Proceedings of the International Symposium on the Analysis and Application of Bioluminescence and Chemiluminescence; Schram, E., Philip, Eds.; Schram, State Printing and Publ. Inc., 1979; 193–211.
35. Bulich, A.A.; Isenberg, D.L. Use of the luminescent bacteria systems for rapid assessment in aquatic toxicology. Adv. Instrument **1980**, *35*, 35–40.
36. Bulich, A.A. A practical and reliable method for monitoring the toxicity of aquatic samples. Proc. Biochem. **1982**, *17*, 45–57.
37. Blaise, C. Canadian application of microbiotests to assess the toxic potential of complex liquid and solid media. In *New Microbiotests for Routine Toxicity Screening and Biomonitoring*; Persoone, G., Janssen, C., De Coen, W., Eds.; Kluwer Academic/Plenum Publishers, New York, 2000, 312.
38. Lappalainen, J.; Juovonen, R.; Vaajasaari, K.; Karp, M. A new flash method for measuring the toxicity of solid and colored samples. Chemosphere **1999**, *38*, 317–328.
39. Kwan, K.K. Direct sediment toxicity testing procedure using Sediment-Chromotest Kit. Environ. Toxil. Water Quality **1995**, *9*, 193–196.
40. Bitton, G.; Jung, K.; Koopman, B. Evaluation of a microplate assay specific for heavy metal toxicity. Arch. Environ. Con. Tox. **1994**, *27*, 25–28.
41. Ronco, A.E.; Sorbero, M.C.; Rossini, G.D.; Alzuet, P.R.; Dutka, B.J. Screening for sediment toxicity in the Rio Santiago basin: a baseline study. Environ. Toxic. Water Quality **1995**, *10*, 35–39.
42. Dutka, B.J.; McInnis, R.; Jurkovic, A.; Liu, D.; Castillo, G. Water and sediment ecotoxicity studies in Temulko and Rapel river basin, Chile. Environ. Toxic. Water Quality **1996**, *11*, 237–247.
43. Reinhartz, A.; Lampert, I.; Herzberg, M.; Fish F. A new, short-term sensitive, bacterial assay kit for the detection of toxicants. Toxic. Assess. **1987**, *2*, 193–206.
44. Dutton, R.J.; Bitton, G.; Koopman, B. Enzyme biosynthesis versus enzyme activity as a basis for microbial toxicity testing. Toxic. Assess. **1988**, *3*, 245–253.
45. Bringmann, G.; Kuhn, R. Limiting values for damaging action of water pollutions to bacteria *Pseudomonas putida* and green algae *Scenedesmus quadricauda* in cell multiplication inhibition test. Z. Wasser Abwasses-Forschung **1977**, *10*, 87–98.
46. ISO/DIS 10712. *Water Quality – Pseudomonas putida growth inhibition test (Pseudomonas Cell Multiplication Inhibition Test)*; International Organization for Standardization: Geneve, 1995; 14 pp.
47. Liu, D.; Chau, Z.K.; Dutka, B.J. Rapid toxicity assessment of water-soluble and water-insoluble chemicals using a modified agar plate method. Water Res. **1989**, *23*, 333–339.
48. Kilroy, A.; Gray, N.F. Treatability, toxicity and biodegradability test methods. Biol. Rev. **1995**, *70*, 243–275.

49. Maron, D.M.; Ames, B.N. Revised methods for the Salmonella mutagenicity test. Mutat. Res. **1983**, *113*, 174–210.
50. Donelly, K.C.; Brown, K.W.; Andersson, C.S.; Thomas, J.C.; Scott, B.R. Bacterial mutagenity and acute toxicity of solvent and aqueous extracts of soil samples from a chemical manufacturing site. Environ. Toxicol. Chem. **1991**, *10*, 1123–1131.
51. Ivanchenko, O.B.; Ilyinskaya, O.N.; Karamova, N.S.; Kostyukevich, I.I. Ecological and genetic description of the technogenic soils. Soil Sci. **1996**, *11*, 1394–1398.
52. Ivanchenko, O.B.; Karamova, N.S.; Schmidt, M.A.; Ilyinskaya, O.N. Toxicogenetic aspects in regulating the application of the sewage sludge. Toxicol. Rev. **2001**, *3*, 22–26 (in Russian).
53. Fish, F.; Lampert, I.; Halachmi, A.; Riesenfeld, G. The SOS Cromotest Kit: a rapid method for detection of genotoxicity. Toxic. Assess. **1987**, *2*, 135–147.
54. Kwan, K.K.; Dutka, B.J. Mutatox test: a new test for monitoring environmental mutagenic agents. Environ. Pollut. **1990**, *65*, 323–332.
55. Rao, S.S.; Lifshitz, R. The Muta-ChromoPlate Method for measuring mutagenicity of environmental samples and pure chemicals. Environ. Toxic. Water Quality **1995**, *10*, 307–313.
56. Garcia, C.; Hernandes, T. Effect of bromacil and sewage sludge addition on soil enzymatic activity. Soil Sci. Plant Nutr. **1996**, *42*, 191–195.
57. Pascual, J.A.; Auso, M.; Garcia, C.; Hernandez, T. Characterization of urban wastes according to fertility and phytotoxicity parameters. Waste Mgt. Res. **1997**, *15*, 103–112.
58. Pascual, J.A.; Garcia, C.; Hernandes, T.; Ayso, M. Changes in the microbial activity of the arid soil amended with urban organic wastes. Biol. Fert. Soil **1997**, *24*, 429–434.
59. Khan, M.; Scullion, J. Effect of soil on microbial responses to metal contamination. Environ. Pollut. **2000**, *110*, 115–125.
60. McGrath, S.P. Effects of heavy metals from sewage sludge on soil microbes in agricultural ecosystems. In *Toxic Metals in Soil-Plant Systems*; Ross, S.M., Ed.; John Wiley & Sons Ltd: New York, 1994; 247–274.
61. Johansson, M.; Stenberg, B.; Torstensson, L. Microbial and chemical changes in two arable soils after long-term sludge amendments. Biol. Fert. Soil **1999**, *30*, 160–167.
62. Wetzel, A.; Werner, D. Ecotoxicological evaluation of contaminated soil using the legume root nodule symbiosis as effect parameters. Environ. Toxic. Water Quality **1995**, *10*, 127–133.
63. Flie*β*bach, A.; Martens, R.; Reber, H.H. Soil microbial biomass and microbial activity in soils treated with heavy metal contaminated sewage sludge. Soil Biol. Biochem. **1994**, *26*, 1201–1205.
64. Hue, N.V. Sewage sludge. In *Soil Ammendments and Environmental Quality*; Rechcigl, J.E., Ed.; Lewis Publishers: Boca Raton, FL, 1995; 199–247.
65. Dar, G.H. Impact of lead and sewage sludge on soil microbial biomass and carbon and nitrogen mineralization. B. Environ. Contam. Tox. **1997**, *58*, 234–240.
66. Baath, E.; Diaz-Ravina, M.; Frostegard, A.; Campbell, C.D. Effect of metal-rich sludge amendments on the soil microbial communities. Appl. Environ. Microb. **1998**, *64*, 238–245.
67. Moreno, J.L.; Hernandez, T.; Garcia, C. Effects of cadmium-contaminated sewage sludge compost on dynamics of organic matter and microbial activity in an arid soil. Biol. Fert. Soil **1999**, *28*, 230–237.
68. Kelly, J.J.; Haggblom, M.; Robert, L.T. Effects of the land application of sewage sludge on soil heavy metal concentrations and soil microbial communities. Soil Biol. Biochem. **1999**, *31*, 1467–1470.
69. Chander, K.; Dyckmans, J.; Joergensen, R.; Meyer, B.; Raubuch, M. Different sources of heavy metals and their long-term effects on microbial properties. Biol. Fert. Soil. **2001**, *34*, 241–247.
70. Persoone, G.; Janssen, C.R. Freshwater invertebrate toxicity tests. In *Handbook of Ecotoxicology*; Calow, P., Ed.; Blackwell Scientific Publications, 1994; Vol. 1, Chap. 4, 51–65.
71. Horning, W.B.; Weber, C.I. *Short-Term Methods for Estimating the Chronic Toxicity of Effluents and Receiving Waters to Freshwater Organisms*, U.S. EPA, Report EPA/600/4-85/014, 1985; 161 pp.

72. Buikema, A.L.; Geiger, J.C.; Lee, D.R. *Daphnia* toxicity tests. In *Aquatic Invertebrate Bioassays*; Buibema, A.L., Cairns, J., Eds.; ASTM: Philadelphia, 1980; ASTM STP 715, 48–69.
73. Baudo, R. Ecotoxicological testing with *Daphnia*. In *Daphnia*; Peters, R.H., de Bernardi, R., Eds.; Mem. Ist. Ital. Idrobiol. **1987**, *45*, 461–482.
74. British Health and Safety Commission (BHSC). Testing for acute toxicity to *Daphnia*. In *Methods for the Determination of Ecotoxicity*; BHSC: London, 1982; 14–18.
75. ISO. *Water Quality – Determination of the Inhibition of Mobility of Daphnia magna Straus (Cladocera, Crustacea)*, 1st Ed.; Report 6341–1982; ISO: Geneva, 1982.
76. EPA, U.S. Environmental Protection Agency. *Daphnid Acute Toxicity Test*; U.S. EPA, Office of Toxic Substances, Document EG-1/ES-1: Washington, DC, 1982.
77. Association francaise de normalisation (AFNOR). Essais des eaux, determination de l'inhibition de la mobilite de *Daphnia magna*. Norme francaise homologuee, NF T90-301. Association francaise de normalisation: Paris, France, 1983.
78. American Society for Testing and Materials (ASTM). Standard practice for conducting static acute toxicity tests on waste water with *Daphnia*. In *Annual Book of ASTM Standards*; ASTM: Philadelphia, 1984; Vol. 11.01, D4229, 27–39.
79. Organization for Economic Cooperation and Development (OECD). *Daphnia spp., Acute Immobilization and Reproduction Test. OECD Guideline for Testing Chemicals*; Organization for Economic Cooperation and Development: Geneva, 1984; Vol. 202.
80. Peltier, W.H.; Weber, C.I. *Methods for Measuring the Acute Toxicity to Effluents of Freshwater and Marine Organisms*; U.S. EPA/600/4-85/013, Environmental Protection Agency: Cincinatti, 1985; 216 pp.
81. Green, J.C.; Bartels, C.L.; Warren-Hicks; Parkhurst, B.P.; Linder, G.L.; Peterson, S.A.; Miller, W.E. *Protocol of Short-Term Toxicity Screening of Hazardous Waste Sites*; U.S. EPA: Corvallis, OR, 1988.
82. Poirier, D.G.; Westlake, G.F.; Abernethy, S.G. *Daphnia magna Acute Lethalithy Toxicity Test Protocol*; Ontario Ministry of Environment, Aquatic Toxicity Unit, Water Res. Br.: Rexbale, Ontario, Canada, 1988.
83. Environmental Protection Series (EPS). *Biological Test Method: Acute Lethality Test Using Daphnia* spp.; Report EPS 1/RM/11, Environment Canada, 1990; 57 pp.
84. Rue, W.J.; Flava, J.A.; Grothe, D.R. A review of inter- and intralaboratory effluent toxicity test method variability. In *Aquatic Toxicology and Hazard Assessment*; Adams, W.J., Chapman, G.A., Landis, W.G., Eds.; ASTM: Philadelphia, 1988; Vol. 10, ASTM STP 971, 190–203.
85. Grothe, D.R.; Kimerle, R.A. Inter- and intra-laboratory variability in *Daphnia magna* effluent toxicity test results. Environ. Toxicol. Chem. **1988**, *4*, 189–192.
86. Canton, J.H.; Adema, D.M.M. Reproducibility of short-term and reproduction toxicity experiments with *Daphnia magna* and comparison of the sensitivity of *Daphnia magna* with *Daphnia pulex* and *Daphnia cucullata* in short-term experiments. Hydrobiologia **1978**, *59*, 135–140.
87. Parker, W.R. *Results of an Interlaboratory Study on the Toxicity of Potassium Dichromate to Daphnia*; EPS Report; Environmental Protection Service, Environment Canada: Canada, 1983.
88. Mount, D.I.; Norberg, T.J. A seven-day life-cycle cladoceran toxicity test. Environ. Toxicol. Chem. **1984**, *3*, 425–434.
89. Cowgill, U.M.; Keating, K.I.; Takahashi, I.T. Fecundity and longevity of *Ceriodaphnia dubia/affinis* in relation to diet at two different temperatures. J. Crustacean Biol. **1985**, *5*, 420–429.
90. American Society for Testing and Materials (ASTM). Standard guide for conducting three-brood, renewal toxicity tests with *Ceriodaphnia dubia*. In *Annual Book of ASTM Standards*; ASTM: Philadelphia, 1989; Vol. 11.04, 360–379.
91. Persoone, G. Development and validation of Toxkit microbiotests with invertebrates, in particular crustaceans. In *Microscale Testing in Aquatic Toxicology: Advantages, Techniques and Practice*; Wells, P.G., Lee, K., Blaise, C., Eds.; CRC Press LLC: Boca Raton, FL, USA, 1998, 437–449.
92. Daphtoxkit F^{TM} magna. *Crustacean Toxicity Screening Test for Freshwater. Standard Operational Procedure*; Creasel: Deinze, Belgium, 1996; 16 pp.

93. Daphtoxkit F^{TM} pulex. *Crustacean Toxicity Screening Test for Freshwater. Standard Operational Procedure*; Creasel: Deinze, Belgium, 1996; 17 pp.
94. Janssen, C.R.; Persoone, G. Rapid toxicity screening tests for aquatic biota: I. Methodology and experiments with *Daphnia magna.* Environ. Toxicol. Chem. **1993**, *12*, 711–717.
95. Bitton, G.; Rhodes, K.; Koopman, B.; Cornejo, M. Short-term toxicity assay based on daphnid feeding behavior. Water Environ. Res. **1995**, *67*, 290–293.
96. Bitton, G.; Rhodes, K.; Koopman, B. Ceriofast^{TM}: an acute toxicity test based on *Ceriodaphnia dubia* feeding behavior. Environ. Toxicol. Chem. **1996**, *15*, 123–125.
97. Dive, D.; Persoone, G. Protozoa as test organisms in marine ecotoxicology: luxury or necessity? In *Ecotoxicological Testing for the Marine Environment*; Persoone, G., Jaspers, E., Claus, C., Eds.; State Univ. Ghent and Inst. Mar. Sci. Res.: Belgium, 1984; Vol. 1, 281–306.
98. Dive, D. Nutrition et croissance de Colpodium campylum. Contribution experimentale. Possibilites d'application en ecotoxicologie; Universite Sci. Techn.: Lille, France, 1981; 295 pp. These Doctorale.
99. Dive, D.; Blaise, C.; Le Due, A. Standard protocol proposal for undertaking the *Colpodium campylum* ciliate protozoan growth inhibition test. Z. Angew. Zool. **1991**, *78*, 1.
100. Protoxkit^{TM}. *Freshwater Toxicity Test with a Ciliate Protozoan. Standard Operational Procedure*; Creasel: Deinze, Belgium, 1998; 18 pp.
101. Selivanovskaya, S.Yu.; Petrov, A.M.; Egorova, K.V.; Naumova, R.P. Forming of immobilized communities treated the waste water from the organic compounds. Chem. Technol. Water **1995**, *17*, 618–622 (in Russian).
102. Petrov, A.; Stepanova, N.; Gabaydullin, A.; Shagidullin, R. Use of *Paramecium caudatum* for toxicity screening of a local industrial flow into the sewage system before biological treatment. In *International Symposium on New Microbiotests for Routine Toxicity Screening and Biomonitoring*, Brno, Czech Republic, 1–3 June, 1998; 83 pp.
103. Selivanovskaya, S.Yu.; Maslov, A.P.; Naumova, R.P. Toxicological testing of waste water subjected to biological treatment by means of ciliate infusoria. Chem. Technol. Water **1993**, *15*, 286–290 (in Russian).
104. Fu, L.J.; Staples, R.E.; Stahl, R.G., Jr. Assessing acute toxicities of pre- and post-treatment industrial wastewaters with *Hydra attenuata*: a comparative study of acute toxicity with the fathead minnow *Pimephales promelas.* Environ. Toxicol. Chem. **1994**, *13*, 563–569.
105. Kusui, T.; Blaise, C. Ecotoxicological assessment of Japanese industrial effluents using a battery of small-scale toxicity tests. In *Impact Assessment of Hazardous Aquatic Contaminants: Concept and Approaches*; Salem, R., Ed.; Ann Arbor Press: Michigan, USA, 1998; 161–181.
106. Organization for Economic Cooperation and Development (OECD). *Draft Update OECD Guideline or Testing of Chemicals – Fish Early-Life Stage Toxicity Test*; Organization for Economic Cooperation and Development: Paris, France, 1991; OECD guidelines, TGP/145, 22 pp.
107. Organization for Economic Cooperation and Development (OECD). *Draft Updated OECD Guideline for Testing of Chemicals 203 – Fish, Acute Toxicity Test*; OECD: Paris, France, 1991; TGP/148, 10 pp.
108. ASTM *Standard Guide for Conducting Early Life-Stage Toxicity Tests With Fish*; Annual Book of ASTM Standards; American Society for Testing of Materials: Philadelphia, USA, 1991; Vol. 11.04, 857–882.
109. ISO/FDIS 11267. *Soil Quality – Inhibition of Reproduction of Collembola (Folsomia candida) by Soil Pollutants*; 1998.
110. ISO WD 16387. *Soil Quality – Effects on Pollutants on Enchytraedae (Enchytraeus* sp.*): Determination of Effects on Reproduction*; 1999.
111. Juvonen, R.; Martikainen, E.; Schultz, E.; Joutti, A.; Ahtiainen, J.; Lehtokari, M. A battery of toxicity tests as indicators of decontamination in composting oily waste. Ecotox. Environ. Safe. **2000**, *47*, 156–166.
112. Alekperov, I.Ch.; Kasimov, R.U. Infusorian for biotesting of the boring sludge. Hydrobiological J. **1986**, *22*, 96–98 (in Russian).

113. Cronin, M.T.D.; Schultz, T.W. Structure–toxicity relationships for phenols to *Tetrahymena pyrifirmis*. Chemosphere **1999**, *32*, 1453–1468.
114. Sauvant, M.P.; Pepin, D.; Bohatier, J.; Groliere, C.A.; Guillot, J. Toxicity assessment of 16 inorganic environmental pollutants by six bioassays. Ecotox. Environ. Safe. **1997**, *37*, 131–140.
115. Campbell, C.; Warren, A.; Cameron, C.; Hope, S. Direct toxicity assessment of two soils amended with sewage sludge contaminated with heavy metals using protozoan (*Colpoda steinii*) bioassay. Chemosphere **1997**, *34*, 501–514.
116. Bogaerts, P.; Bohatier, J.; Bonnemoy, F. Use of the ciliated protozoan *Tetrahymena pyriformis* for the assessment of the toxicity and quantitative structure–activity relationships of xenobiotics: comparison with the microtox test. Ecotox. Environ. Safe. **2001**, *49*, 293–301.
117. Nicolau, A.; Dias, N.; Mota, M.; Lima, N. Trends in the protozoa in the assessment of wastewater treatment. Res. Microbiol. **2001**, *152*, 621–630.
118. Kammenga, J.E.; Koert, P.; Riksen, J.; Korthals, G.W.; Bakker, J. A toxicity test in artificial soil based on the life-history strategy of the nematode *Plectus acuminatys*. Environ. Toxicol. Chem. **1996**, *15*, 722–727.
119. Traunspurger, W.; Haitzer, M.; Hoss, S.; Beier, S.; Ahlf, W.; Steinberg, C. Ecotoxicological assessment of aquatic sediments with *Caenorhabditis elegans* (nematoda) – a method for testing liquid medium and a whole-sediment samples. Environ. Toxicol. Chem. **1997**, *16*, 245–250.
120. Gratzer, H.; Ahlf, W. Adjustment of a formulated sediment for sediment testing with *Caenorhabditis elegans* (nematoda). Acta Hydrochim. Hydrobiol. **2001**, *29*, 41–46.
121. Samoiloff, R.; Schulz, S.; Jordan, Y.; Arnott, E. A rapid simple long-term toxicity assay for aquatic contaminants using the nematode *Panagrellus redivivus*. Can. J. Fish. Aquat. Sci. **1980**, *37*, 1167–1174.
122. Callahan, C.A. Earthworms as ecotoxicological assessment tools. In *Earthworms in Waste and Environmental Management*; Edwards, C.A., Neuhauhauser, E.F., Eds.; SPB Academic: The Hague, 1998; 295–301.
123. Staint-Dernis, M.; Narbonne, J.F.; Arnaud, C.; Ribera, D. Biochemical responses of the earthworm *Eisenia fetida andrei* exposed to contaminated artificial soil: effects of lead acetate. Soil Biol. Biochem. **2001**, *33*, 395–404.
124. Pallant, E.; Hilster, L.M. Earthworm response to 10 weeks of incubation in a pot with acid mine spoil, sewage sludge and lime. Biol. Fert. Soil. **1996**, *22*, 355–358.
125. Berry, E.C.; Jordan, D. Temperature and soil moisture content effects on the growth of *Lumbricus terrestris (Oligochaeta: Lunbricidae)* under laboratory conditions. Soil Biol. Biochem. **2001**, *33*, 133–136.
126. Snell, T.W.; Janssen, C.R. Rotifers in ecotoxicology: a review. Hydrobiologia **1995**, 313–314, 231–247.
127. Persoone, G.C.; Janssen, C.R. Freshwater invertebrate toxicity tests. In *Handbook of Ecotoxicology*; Calow, P., Ed.; Blackwell Publishers: UK, 1993; 51–66.
128. Rotoxkit F^{TM} *Rotifer Toxicity Screening Test for Freshwater. Standard Operation Procedure*; Creasel: Deinze, Belgium, 1992; 22 pp.
129. Snell, T.W.; Janssen, C.R. Microscale toxicity testing with rotifers. In *Microscale Testing in Aquatic Toxicology: Advantages, Techniques and Practice*; Wells, P.G., Lee, K., Blaise, C., Eds.; CRC Press LLC: Boca Raton, FL, 1998; 409–422.
130. APHA. *Toxicity Testing with Phytoplancton. Standard Methods for the Examination of Water and Wastewater*, 17th Ed.; American Public Health Association: Washington, DC, 1989.
131. ASTM. *Standard Guide for Conducting Static 96h Toxicity Tests with Microalgae*, E1218-90; American Society for Testing and Materials: Philadelphia, PA, 1990.
132. ISO. *Algal Growth Inhibition Test*; Draft ISO Standard ISO/DIS 10253.2; International Organization for Standardization: Paris, France, 1987.
133. Organization for Economic Cooperation and Development (OECD). *Algal Growth Inhibition Test, OECD Guideline for Testing Chemicals*, No. 201; Organization for Economic Cooperation and Development: Geneva, Switzerland, 1984; Vol. 201.

134. Radetski, C.M.; Ferard, J.M.; Blaise, C. A semistatic microplate based phytotoxicity test. Environ. Toxicol. Chem. **1995**, *14*, 299–302.
135. Persoone, G. Development and first validation of a "Stock culture free" algal microbiotest: the Algaltoxkit. In *Microscale Testing in Aquatic Toxicology; Advantages, Techniques and Practice*; Wells, P.G., Lee, K., Blaise, C., Eds.; CRC Press Boca Raton, FL, 1998a; 311–320.
136. Blaise, C.; Legault, R.; Bermingham, N.; Van Coille, R.; Vasseur, P. A simple microplate algal assay for aquatic toxicity assessment. Tox. Assess. Int. J. **1986**, *1*, 261–281.
137. Bozeman, J.; Koopman, K.; Bitton, G. Toxicity testing using immobilized algae. Aquat. Toxicol. **1989**, *9*, 345–352.
138. Gala, W.R.; Giesy, J.P. Flow cytometric techniques to assess toxicity to algae. In *Aquatic Toxicology and Risk Assessment*; Landis, W., Vander Schalie, W.H., Eds.; ASTM: Philadelphia, PA, 1993; Vol. 13, 237–246.
139. Wren, M.J.; McCaroll, D. A simple and sensitive bioassay for the detection of toxic materials using a unicellular green alga. Environ. Pollut. **1990**, *64*, 87–91.
140. Amparado, R.F. *Development and Application of a Cost-Effective Algal Growth Inhibition Test with the Green Alga Selenastrum capricornutum (Printz)*; University of Gent: Belgium, 1995; 217 pp, Ph.D. Thesis.
141. Caux, P.Y.; Blaise, C.; Le Blanc, P.; Tache, M. A phytoassay procedure using fluorescence induction. Environ. Toxicol. Chem. **1992**, *11*, 549–557.
142. Willemsen, A.; Vaal, M.A.; de Zwart, D. *Microbiotests as Tools for Environmental Monitoring*; Report No 9, 607042005; National Institute of Public Health and Environmental Planning (RIVM): The Netherlands, 1995; 39 pp.
143. Blaise, C.; Sergy, G.; Bermingham, N.; Van Coillie, R. Biological testing-development, application and trends in Canadian Environmental Protection Laboratories. Tox. Assess. Int. J. **1988**, *3*, 385–406.
144. Blaise, C. Microbiotests in aquatic ecotoxicology: characteristics, utility and prospects. Environ. Toxic. Water Quality **1991**, *6*, 145–156.
145. Algaltoxkit F^{TM}. *Freshwater Test with Microalgae. Standard Operational Procedure*; Creasel: Deinze, Belgium, 1996; 28 pp.
146. Ostroumov, S.A. Some aspects of the estimation of biological activity of xenobiotics. Bulletin of Moscow University **1990**, 2, 27–34.
147. Iannotti, D.A.; Grebus, M.E.; Toth, B.L.; Madden, L.V.; Hoitink, H.A.G. Oxygen respirometry to assess stability and maturity of composted municipal solid waste. J. Environ. Qual. **1994**, *23*, 1177–1183.
148. Kapustka, L.A.; Lipton, J.; Galbraith, A.; Cacela, O.; Leyeune, K. Metal and arsenic impacts to soils, vegetation communities and wildlife habitat in southwest Montana uplands contaminated by smelter emissions. 2. Laboratory phytotoxicity studies. Environ. Toxicol. Chem. **1995**, *14*, 1905–1912.
149. Boelens, J.; De Wilde, B.; De Baere, L. Comparative study on biowaste definition: effects on biowaste collection, compost process and compost quality. Compost. Sci. Util. **1996**, *4*, 60–72.
150. Campbell, A.; Zhang, X.; Tripepi, R.R. Composting and evaluation a pulp and paper sludge for use as a soil amendment/mulch. Compost. Sci. Util. **1995**, *84*, 84–95.
151. Lau, S.; Fang, M.; Wong, J. Effects of composting process and flu ash amendment on phytotoxicity of sewage sludge. Arch. Environ. Con. Tox. **2001**, *40*, 184–191
152. Markwiese, J.T.; Ryti, R.T.; Hooten, M.M.; Michael, D.I.; Hlohowskyj, I. Toxicity bioassays for ecological risk assessment in arid and semiarid ecosystems. Rev. Environ. Contam. Tox. **2001**, *168*, 43–98.
153. Keeling, A.A.; Griffiths, B.S.; Ritz, K.; Myers, M. Effects of compost stability on plant growth, microbiological parameters and nitrogen availability in media containing mixed garden waste compost. Bioresource Technol. **1995**, *54*, 279–284.
154. Itavara, M.; Vilman, M. Venelampi, O. Windrow composting of biodegradable packaging materials. Compost. Sci. Util. **1997**, *5*, 84–92.

155. Maynard, A.A. Utilization of MSW compost in nursery stock production. Compost. Sci. Util. **1998**, *6*, 38–44.
156. Persoone, G. Ecotoxicology and water quality standards. In *River Water Quality – Ecological Assessment and Control*; Newman, P., Piavaux, A., Sweeting, R., Eds.; 1992; 751 pp.
157. Persoone, G.; Van de Vel, A. *Cost-Analysis of Five Current Aquatic Toxicity Tests*; Report EUR 11342 EN, Commission of the European Communities, 1988; 119 pp.
158. Persoone, G. Cyst-based toxicity tests. I. A promising new tool for rapid and cost-effective toxicity screening of chemicals and effluents. Zeitschr. Für Angewandte Zoologie **1991**, *78*, 235–241.
159. Janssen, C.R.; Vangheluwe, M.; Van Sprang, P. A brief review and critical evaluation of the status of microbiotests. In *New Microbiotests for Routine Screening and Biomonitoring*; Persoone, G., Janssen, C., Coen, W., Eds.; Kluwer Academic/Plenum Publishers: New York, 2000; 27–37.
160. Thamnotoxkit F^{TM} *Crustacean Toxicity Screening Test for Freshwater. Standard Operational Procedure*; Creasel: Deinze, Belgium, 1995; 23 pp.
161. Lokke, H. Ecotoxicological extrapolation: tool or toy? In *Ecotoxicology of Soil Organisms*; Donker, M.H., Eijsakers, H., Heimbach, F., Eds.; SETAC Special Publication. Lewis: Boca Raton, FL, 1994; 411–425.
162. Carlson-Ekvall, C.E.A.; Morisson, G.M. Toxicity of copper in the presence of organic substances in sewage sludge. Environ. Technol. **1995**, *16*, 243–251.
163. Perez, S.; Farre, M.; Garcia, M.J.; Barcelo, D. Occurrence of polycyclic aromatic hydrocarbons in sewage sludge and their contribution to its toxicity in the ToxAlert® 100 bioassay. Chemosphere **2001**, *45*, 705–712.
164. Sayles, D.; Achenson, C.M.; Kupferle, M.J.; Brenner, R.C. Land treatment of PAH contaminated soil: performance measured by chemical and toxicity assays. Environ. Sci. Technol. **1999**, *33*, 4310–4317.
165. Canna-Michaelidou, S.; Nicolaou, A.S.; Neopfytou, E.; Christodoulidou, M. The use of a battery of microbiotests as a tool for integrated pollution control: evaluation and perspectives in Cyprus. In *New Microbiotests for Routine Toxicity Screening and Biomonitoring*; Persoone, G., Janssen, C., De Coen, W., Eds.; Kluwer Academic/Plenum Publishers: New York, 2000; 39–48.
166. Ehrlichmann, H.; Manh, B.; Dott, W.; Eisentraeger, A. Development of a miniaturized *Salmonella typhimurium* reversion test with kinetic data acquisition. In *New Microbiotests for Routine Toxicity Screening and Biomonitoring*; Persoone, G., Janssen, C., De Coen, W., Eds.; Kluwer Academic/Plenum Publishers: New York, 2000; 503–510.
167. Ivanchenko, O.; Ilinskaya, O.; Kruglova, Z.; Petrov, A. Genotoxicity monitoring of environmental samples in Tatarstan, Russia. In *New Microbiotests for Routine Toxicity Screening and Biomonitoring*; Persoone, G., Janssen, C., De Coen, W., Eds.; Kluwer Academic/Plenum Publishers: New York, 2000; 511–516.
168. Bernard, C.; Persoone, G.; Colin, J.; Le Du-Delepierre, A. Estimation of the hazard of landfills through toxicity testing of leachates: determination of leachate toxicity with a battery of acute tests. Chemosphere **1996**, *33*, 2203–2230.
169. Pryadko, A.L.; Alekseeva, T.V. The using of biotesting for estimation of toxicity the ash of power plant. Hygiene and Sanitation **1992**, *3*, 69–71 (in Russian).
170. Robidoux, P.Y.; Lopes-Gastey, J.; Choucri, A.; Sunahara, G.I. Procedure to screen illicit discharge of toxic substances in septic sludge received at a wastewater treatment plant. Ecotox. Environ. Safe. **1998**, *39*, 31–40.
171. Robidoux, P.Y.; Lopes-Gastey, J.; Choucri, A.; Sunahara, G.I. Screening of illicit toxic substances discharged in chemical toilet sludge. Qual. Assur. **1999**, *6*, 23–44.
172. Selivanovskaya, S.Yu.; Latypova, V.Z. Substantiation of the system for experimental estimation of classes of hazard of sewage sludge and the selection of the way of its disposal. Ecol. Chem. **2001**, *10*, 124–134 (in Russian).

173. Semanov, D.A.; Ravzieva, G.M.; Chabibullin, D.I.; Latypova, V.Z.; Selivanovskaza, S.Yu. Comparative analysis of the approaches to the definition of toxicity classes of sewage sludge. Toxicol. Rev. **2001**, *3*, 2–6 (in Russian).
174. Ministry of Natural Resources in Russia. Criteria for the attribution of the hazardous solid waste to classes of hazard. Ministry of Natural Resources of Russia, 15.06.2001. Ecol. Consulting **2001**, *2*, 30–34 (in Russian).
175. Törökne, A. *State of Environmental Pollution and Toxicity Testing/Monitoring in Hungary*, International Workshop FITA 4 Programme, Tallin, Estonia. September 10–11, 1999.
176. Marsalek, B. *Ecotoxicological Bioassays in the Czech Republic*, International Workshop FITA 4 Programme, Tallin, Estonia. September 10–11, 1999.
177. Chapman, P.M. Whole effluent toxicity testing – usefulness, level of protection, and risk assessment. Envir. Toxicol. Chem. **2000**, *19*, 3–13.
178. Anderson, B.G. Aquatic invertebrates in tolerance investigations from Aristotle to Naumann. In *Aquatic Invertebrate Bioassays*; Buikema, A.L. Jr., Cairns, J. Jr., Eds.; American Society for Testing and Materials: Philadelphia, PA, 1980; Vol. 3, 3–35.
179. American Public Health Association. *Standard Methods for the Examination of Water, Sewage and Industrial Wastes*; American Public Health Association, American Water Works Association, Water Pollution Control Federation: Washington, DC, 1955.
180. EPA (U.S. Environmental Protection Agency) *Regions 9 and 10 Guidance for Implementing Whole Effluent Toxicity Testing Programs*, Technical Report; Seattle, WA, 1996.
181. Grothe, D.R.; Johnson, D.E. Bacterial interferences in whole effluent toxicity tests. Environ. Toxicol. Chem. **1996**, *15*, 761–764.
182. Pontasch, K.W.; Niederlehner, B.R.; Cairns, J. Jr. Comparisons of single species, microcosms and field responses to a complex effluent. Environ. Toxicol. Chem. **1989**, *8*, 521–532.
183. Persoone, G.; Janssen, C.R. Field validation of predictions based on laboratory toxicity tests. In *Freshwater Field Tests for Hazard Assessment of Chemicals*; Hill, I.A., Helmbach, F., Leeuwangh, P., Matthiessen, P., Eds.; CRC Press, Inc.: Boca Raton, FL, 1994; 379–397.
184. Metcalf & Eddy, Inc. *Wastewater Treatment Engineering: Treatment, Disposal and Reuse*; Singapore, 1991; 102–108.
185. Costan, G.; Bermingham, N.; Blaise, C.; Ferard, J.F. Potential Ecotoxic Effects Probe (PEEP): a novel index to assess and compare the toxic potential of industrial effluents. Environ. Toxic. Water Quality **1993**, *8* (1).
186. EPA (U.S. Environmental Protection Agency). *Policy for the Development of Water Quality-Based Limitations for Toxic Pollutants*; U.S. Environmental Protection Agency: Washington DC, 1984; EPA-49-FR-9016.
187. EPA (U.S. Environmental Protection Agency). *Technical Support Document for Water Quality-Based Toxics Control*; U.S. Environmental Protection Agency, Office of Water: Washington, DC, 1985; EPA-440/4-85-032.
188. OECD. *The Use of Biological Tests for Water Pollution Assessment and Control*; Organisation for Economic Cooperation and Development: Paris, France, Environment Monographs, No. 11, 1987.
189. Hanmer, R.W. *Biological Testing of Complex Effluents in Wastewater Regulation: OECD Work and Implementation in the United States*, International Conference on River Water Quality – Ecological Assessment and Control, Palias des Congres: Brussels, December 16–18, 1991.
190. EPA (U.S. Environmental Protection Agency). *Technical Support Document for Water Quality-Based Toxics Control*; U.S. Environmental Protection Agency, Office of Water: Washington, DC, USA, 1991; EPA/505/2-90-001.
191. Ronco, A.E.; Castillo, G.; Diaz-Baez, M.C. Development and application of microbioasays for routine testing and biomonitoring in Argentina, Chile and Colombia. In *New Microbiotests for Routine Screening and Biomonitoring*; Persoone, G., Janssen, C., Coen, W., Eds.; Kluwer Academic/Plenum Publishers: New York, 2000; 49–61.

192. Aoyama, I.; Okamura, H.; Rong, L. Toxicity testing in Japan and the use of Toxkit microbiotests. In *New Microbiotests for Routine Screening and Biomonitoring*; Persoone, G., Janssen, C., Coen, W., Eds.; Kluwer Academic/Plenum Publishers: New York, 2000; 123–133.
193. Garric, J.; Vindimian, E.; Ferard, J.F. *Ecotoxicology and Wastewater: Some Practical Applications*; Secotox: Amsterdam 1992. The Science of the Total Environment, Supplement, 1993; 1085–1103.
194. Vasseur, P.; Ferard, J.F.; Babut, M. The biological aspects of the regulatory control of industrial effluents in France. Chemosphere **1991**, *22* (5), 625–633.
195. Steinhäuser, K.G.; Hansen, P.D. *Biologische Testverfahren*; Gustav-Fisher Verlag: Stuttgart, 1992; 884 pp.
196. Tonkes, M.; Botterweg, J. *Totaal Effluent Milieubezwaarlijkheid.* RIZA-nota, AquaSense-rapport 93.0435, Rijksinstituut voor Integraal Zoetwaterbeheer en Afvalwaterbehandeling: Lelystad, The Netherlands, 1994; 157 pp.
197. Heinis, F.; Brils, J.M.; Klapwijl, S.P.; Poorter, L.R.M. In *New Microbiotests for Routine Screening and Biomonitoring*; Persoone, G., Janssen, C., Coen, W., Eds.; Kluwer Academic/Plenum Publishers: New York, 2000; 65–72.
198. Hunt, D.T.E.; Johnson, I.; Milne, R. *The Control and Monitoring of Discharges by Biological Techniques*; IWEM 91 Conf. paper. J. IWEM **1992**; *6*, 269–277.
199. Crawshaw, T. *Pre-Congress Workshop: SETAC Effluent Toxicity Program, Implementation, Compliance and Enforcement*; National Rivers Authority (NRA): Worthing, West Sussex, UK, March 28, 1993. SETAC, Lissabon 1993.
200. SNV. *Biological-chemical Characterization of Industrial Waste Water. Application When Granting Permits and Exercising Supervisory Authority for Activities Harmful to the Environment*; Swedish Environmental Protection Agency: Solna, Sweden, 1990.
201. Tapp, J.F.; Williams, B.R.H. *An Assessment of the Application of Acute Toxicity Testing for the Monitoring and Control of Oil Refinery Effluents*; Conservation Clean Air and Water Europe: Brussels, Belgium, CONCAWE Report No. BL/A/2894, 1986; 96 pp.
202. Nalecz-Jawecki, G. *Environmental (Water) Pollution in Poland. Ecotoxicological Bioassays in Poland*, International Workshop FITA 4 Programe, Tallin, Estonia, September 10–11, 1999.
203. Kahru, A.; Blinova, I. *Monitoring of Surface Water in Estonia*, International Workshop FITA 4 Programe, Tallin, Estonia. September 10–11, 1999.
204. Kolar, B. *The State of Art of Environmental Pollution, Toxicity Testing and Hazard Monitoring in Slovenia*, International Workshop FITA 4 Programe, Tallin, Estonia, September 10–11, 1999.
205. Manusadzianas, L. *General Requirements for Treated Wastewaters in Lithuania*, International Workshop FITA 4 Programe, Tallin, Estonia, September 10–11 1999.
206. Zmur, N.S. Monitoring problems of sources of natural water contaminations: conditions of decision making and some perspectives. Ecol. Chem. **1998**, *7*, 191–199 (in Russian).
207. Stepanova, N.; Latypova, V. Chemical structure and waste water toxicity: several results of economic experiment in Republic of Tatarstan. Ecol. Consulting **2001**, *3*, 17–20 (in Russian).
208. Gelashvilly, D.B.; Bezrukova, N.V.; Bezrukov, M.E. Ecotoxicological analysis of toxic load of industrial enterprises of Nizhni Novgorod to water bodies of river part of Cheboksarski reservoir. News of Samara Scientific Center of Russian Academy of Science **2000**, *2*, 244–251 (in Russian).

3

In-Plant Management and Disposal of Industrial Hazardous Substances

Lawrence K. Wang

Lenox Institute of Water Technology and Krofta Engineering Corporation, Lenox, Massachusetts and Zorex Corporation, Newtonville, New York, U.S.A.

3.1 INTRODUCTION

If the hazardous substances at industrial, commercial, and agricultural sites can be properly handled, stored, transported, and/or disposed of, there will be no environmental pollution, and no need to embark on any site remediation. With this concept in mind, the goal of in-plant hazardous waste management is to achieve pollution prevention and human-health protection at the sources where there are hazardous substances. This chapter begins with hazardous waste terminologies and characteristics. Special emphasis is placed on the manifest system, hazardous substances storage requirements, underground storage tanks, above-ground storage tanks, hazardous substances transportation, hazardous waste handling, and disposal.

3.1.1 General Introduction and Objectives

Most hazardous wastes are produced in the manufacturing of products for domestic consumption, or various industrial applications. Rapid development and improvement of industrial technologies, products, and practices frequently increase the generation rate of hazardous substances (including both useful materials and waste materials). These hazardous substances, which can be in the form of gas, liquid, or solid, must be properly handled in order to protect the plant personnel, the general public, and the environment.

The term "hazardous substance" refers to any raw materials, intermediate products, final products, spent wastes, accidental spills, leakages, and so on, that are hazardous to human health and the environment. Technically speaking, all ignitable, corrosive, reactive (explosive), toxic, infectious, carcinogenic, and radioactive substances are hazardous [1–3].

Legally radioactive substances (including radioactive wastes) are regulated by the Nuclear Regulatory Commission (NRC), while all other hazardous substances (excluding radioactive substances) are mainly regulated by the U.S. Environmental Protection Agency (USEPA), the Occupational Safety and Health Administration (OSHA), and the state environmental protection agencies [4–22]. Guidelines and recommendations by the National Institute for Occupational Safety and Health (NIOSH), the American Conference of Governmental Industrial Hygienists (ACGIH), American Water Works Association (AWWA), American Public Health Association (APHA), Water Environmental Federation (WEF), American Institute of Chemical

Engineers (AIChE), and the American Society of Civil Engineers (ASCE) are seriously considered by practicing environmental engineers and scientists (including chemical/civil/mechanical engineers, biologists, geologists, industrial hygienists, chemists, etc.) in their decision-making process when managing, handling, and/or treating hazardous substances.

In the past 25 years, industry, government, and the general public in the industrially developed as well as developing countries have become increasingly aware of the need to respond to the industrial hazardous substance problems.

Some hazardous wastes, or mixture of hazardous wastes (such as cyanides, hydrogen sulfide, and parathion) are extremely or acutely hazardous because of their high acute toxicity. These extremely hazardous wastes, if human exposure should occur, may result in disabling personal injury, illness, or even death.

Dioxin-contaminated sites, which pose a human health threat, have been the subject of recent analyses by the Centers for Disease Control (CDC) in Atlanta, GA. It has been determined by CDC that 1 ppb of dioxin is detrimental to public health and that people should be dissociated from the hazard. A level of 1 ppb of dioxin (2,3,7,8-TCDD) in soil is recommended as an action level. In cases where soil concentrations exceed 1 ppb, it is recommended by CDC that potential human exposure to the contamination be examined further. If there is human exposure to 1 ppb or higher on a regular basis, cleanup is indicated. A substance that may be more toxic and hazardous than dioxin is expected to be discovered in the near future.

Although the properties of hazardous substances may sound alarming, the managerial skills and technologies used to handle, store, or treat hazardous substances are available. Modern technology exists to build and maintain environmentally sound industrial facilities that effectively produce useful products and, at the same time, render hazardous waste inert. Environmental laws, rules, regulations, and guidelines also exist to ensure that the modern technology will be adopted by owners or plant managers of industrial facilities for environmental protection.

This chapter is intended for the plant owner, the plant engineer/manager, their contractors, their consulting engineers, and the general public. This chapter may be used:

1. As a management and planning tool by industrial and technical personnel; and
2. As a reference document and an educational tool by any individuals who want to review important aspects of in-plant air quality, water quality, safety, and health protection at industrial sites having hazardous substances.

This chapter is not a comprehensive information source on occupational safety and health. It provides a general guideline for industrial and technical personnel at industrial sites to understand or familiarize themselves with:

- hazardous substance classification;
- environmental hazards and their management;
- hazardous air quality management;
- hazardous water quality management;
- hazardous solid waste (including asbestos) management;
- monitoring and analysis of hazardous samples;
- measuring instruments for environmental protection;
- hazardous waste generator status, and the regulatory requirements;
- hazardous waste and waste oil documentation requirements;
- hazardous waste and waste oil storage and shipping requirements;
- emergency preparation and response procedures;
- responsibilities and management strategies of very small quantity generator (VSQG), small quantity generator (SQG), and large quantity generator (LQG) of hazardous wastes;
- an example for managing hazardous wastes generated at medical offices;

- an example for managing hazardous wastes generated at graphic artists, printers, and photographers; and
- two case histories for disposing of photographic wastes by a very small quantity generator (VSQG) and a large quantity generator (LQG).

3.1.2 Hazardous Waste Classification

The first step of site management is to determine whether or not the waste generated or an accidental release (i.e., spill of leaks of chemical/biological substances) occurring on an industrial site is hazardous.

Common hazardous wastes include: (a) waste oil, (b) solvents and thinners, (c) acids and bases/alkalines, (d) toxic or flammable paint wastes, (e) nitrates, perchlorates, and peroxides, (f) abandoned or used pesticides, and (g) some wastewater treatment sludges. Special hazardous wastes include: (a) industrial wastes containing the USEPA priority pollutants, (b) infectious medical wastes, (c) explosive military wastes, and (d) radioactive wastes or releases.

In general, there are two ways a waste or a substance may be identified as hazardous – it may be listed in the Federal and/or the State regulations or it may be defined by its hazardous characteristics.

Hazardous waste may be a listed discarded chemical, an off-specification product, an accidental release, or a liquid or solid residue from an operation process, which has one or more of the characteristics below:

- ignitable (easily catches fire, flash point below 140°F);
- corrosive (easily corrodes materials or human tissue, very acidic or alkaline, pH of <2 or >12.5);
- reactive (explosive, produces toxic gases when mixed with water or acid);
- toxic (can leach toxic chemicals as determined by a special laboratory test); and
- radioactive.

The hazardous waste identification regulations that define the characteristics of toxicity, ignitability, corrosivity, reactivity, and the tests for these characteristics, differ from state to state. In addition, concentration limits may be set out by a state for selected persistent and bioaccumulative toxic substances that commonly occur in hazardous substances. For example, the California Hazardous Waste Control Act requires the California State Department of Health Services (CDHS) to develop and adopt by regulation criteria and guidelines for the identification of hazardous wastes and extremely hazardous wastes.

In the State of California, a waste or a material is defined as hazardous because of its toxicity if it meets any of the following conditions: (a) acute oral LD_{50} of less than 5000 mg/kg; (lethal oral dose for 50% of an exposed population); (b) acute dermal LD_{50} of less than 4300 mg/kg; (c) acute 8 hour inhalation LC_{50} of less than 10,000 ppm; (d) acute aquatic 96 hour LC_{50} of less than 500 mg/L measured in waste with specified conditions and species; (e) contains 0.001% by weight, or 10 ppm, of any of 16 specified carcinogenic organic chemicals; (f) poses a hazard to human health or the environment because of its carcinogenicity, acute toxicity, chronic toxicity, bioaccumulative properties, or persistence in the environment; (g) contains a soluble or extractable persistent or bioaccumulative toxic substance at a concentration exceeding the established Soluble Threshold Limit Concentration (STLC); (h) contains a persistent or bioaccumulative toxic substance at a total concentration exceeding its Total Threshold Limit Concentration (TTLC); (i) is a listed hazardous waste (California list consistent with the Federal RCRA list) designated as toxic; and (j) contains one or more materials with an 8 hour LC_{50} or LCLo of less than 10,000 ppm and the LC_{50} or LCLo is exceeded in the head space vapor (lethal inhalation concentration for 50% of an exposed population).

A waste or a material is designated as "extremely hazardous" in the State of California if it meets any of the following criteria: (a) acute oral LD_{50} of less than or equal to 50 mg/kg; (b) acute dermal LD_{50} of less than or equal to 50 mg/kg; (c) acute inhalation LC_{50} of less than or equal to 100 ppm; (d) contains 0.1% by weight of any of 16 specified carcinogenic organic chemicals; (e) has been shown through experience or testing to pose an extreme hazard to the public health because of its carcinogenicity, bioaccumulative properties, or persistence in the environment; (f) contains a persistent or bioaccumulative toxic substance at a total concentration exceeding its TTLC as specified for extremely hazardous waste; and (g) is water-reactive (i.e., has the capability to react violently in the presence of water and to disperse toxic, corrosive, or ignitable material into the surroundings).

The carcinogenic substances specified in the California criteria for hazardous and extremely hazardous materials have been designated potential carcinogens by OSHA. Under the California criteria, these substances cause a material to be designated as hazardous if they are present at a concentration of 0.001% by weight (10 ppm). A material containing 0.1% of these substances is designated extremely hazardous. The carcinogenic chemicals are the following: 2-acetylaminofluorence, acrylonitrile, 4-aminodiphenyl, benzidine and its salts, bis(chloromethyl) ether (CMME), 1,2-dibromo-3-chloropropane (DBCP), 3,3-dichlorobenzidine and its salts (DCB), 4-dimethylaminoazobenzene (DAB), ethyleneimine (EL), alpha-naphthylamine (1-NA), beta-naphthylamine (2-NA), 4-nitrobiphenyl (4-NBP), n-nitrosodimethylamine (DMN), beta-propiolactone (BPL), and vinyl chloride (VCM).

California criteria for defining hazardous wastes that are ignitable and reactive are identical to Federal criteria for hazardous wastes under RCRA defined at 40 CFR, Part 261. The California corrosivity criteria differ from the Federal criteria only in the addition of a pH test for nonaqueous wastes.

Because each state has its own criteria for defining hazardous wastes, the plant manager of an industrial site having hazardous substances should contact the local state environmental protection agency for the details.

In the State of Massachusetts, the waste generated on the site is considered "acutely hazardous" (equivalent to "extremely hazardous" as defined by the State of California) if it is on the list of "acutely hazardous wastes" published by the State of Massachusetts and/or Federal governments. These acutely hazardous wastes are extremely toxic or reactive and are regulated more strictly than other hazardous wastes. In order to find out if the waste on the site is hazardous, or even acutely hazardous, a plant manager may also check with: (a) the supplier of the product (request a hazardous material safety data sheet); (b) laboratories; (c) trade associations; and/or (d) environmental consulting engineers and scientists. In addition, self-reviewing the State and/ or Federal hazardous waste regulations for the purpose of verification is always required.

Radioactive wastes are, indeed, hazardous, but are only briefly covered in this chapter. The readers are referred elsewhere [23–25] for detailed technical information on management of radioactive wastes.

Noise hazard at an industrial site should also be properly controlled. The readers are referred to another source [26] for detailed noise control technologies.

3.2 MANAGEMENT OF ENVIRONMENTAL HAZARDS AT INDUSTRIAL SITES

Environmental hazards are a function of the nature of the industrial site as well as a consequence of the work being performed there. They include (a) chemical exposure hazards, (b) fire and explosion hazards, (c) oxygen deficiency hazards, (d) ionizing radiation hazards, (e) biological hazards, (f) safety hazards, (g) electrical hazards, (h) heat stress hazards, (i) cold exposure hazards, and (j) noise hazards. Both the hazards and the solutions are briefly described in this section [21].

3.2.1 Chemical Exposure Hazards

Preventing exposure to hazardous industrial chemicals is a primary concern at industrial sites. Most sites contain a variety of chemical substances in gaseous, liquid, or solid form. These substances can enter the unprotected body by inhalation, skin absorption, ingestion, or through a puncture wound (injection). A contaminant can cause damage at the point of contact or can act systemically, causing a toxic effect at a part of the body distant from the point of initial contact.

Chemical exposure hazards are generally divided into two categories: acute and chronic. Symptoms resulting from acute exposures usually occur during or shortly after exposure to a sufficiently high concentration of a hazardous contaminant. The concentration required to produce such effects varies widely from chemical to chemical. The term "chronic exposure" generally refers to exposures to "low" concentrations of a contaminant over a long period of time. The "low" concentrations required to produce symptoms of chronic exposure depend upon the chemical, the duration of each exposure, and the number of exposures. For either chronic or acute exposure, the toxic effect may be temporary and reversible, or may be permanent (disability or death). Some hazardous chemicals may cause obvious symptoms such as burning, coughing, nausea, tearing eyes, or rashes. Other hazardous chemicals may cause health damage without any such warning signs (this is a particular concern for chronic exposures to low concentrations). Health effects such as cancer or respiratory disease may not become manifest for several years or decades after exposure. In addition, some hazardous chemicals may be colorless and/or odorless, may dull the sense of smell, or may not produce any immediate or obvious physiological sensations. Thus, a worker's senses or feelings cannot be relied upon in all cases to warn of potential toxic exposure to hazardous chemicals.

Many guidelines for safe use of chemicals are available in the literature [27,28].

3.2.2 Explosion and Fire Hazards

There are many potential causes of explosions and fires at industrial sites handling hazardous substances: (a) chemical reactions that produce explosion, fire, or heat; (b) ignition of explosive or flammable chemicals; (c) ignition of materials due to oxygen enrichment; (d) agitation of shock- or friction-sensitive compounds; and (e) sudden release of materials under pressure [21,29].

Explosions and fires may arise spontaneously. However, more commonly, they result from site activities, such as moving drums, accidentally mixing incompatible chemicals, or introducing an ignition source (such as a spark from equipment) into an explosive or flammable environment. At industrial sites, explosions and fires not only pose the obvious hazards of intense heat, open flame, smoke inhalation, and flying objects, but may also cause the release of hazardous chemicals into the environment. Such releases can threaten both plant personnel on site and members of the general public living or working nearby.

To protect against the explosion and fire hazard, a plant manager should (a) have qualified plant personnel field monitor for explosive atmospheres and flammable vapors, (b) keep all potential ignition sources away from an explosive or flammable environment, (c) use non-sparking, explosion-proof equipment, and (d) follow safe practices when performing any task that might result in the agitation or release of chemicals.

3.2.3 Oxygen Deficiency Hazards

The oxygen content of normal air at sea level is approximately 21%. Physiological effects of oxygen deficiency in humans are readily apparent when the oxygen concentration in the air decreases to 16%. These effects include impaired attention, judgment, and coordination, and increased breathing and heart rate. Oxygen concentrations lower than 16% can result in nausea and vomiting, brain damage, heat damage, unconsciousness, and death. To take into account individual physiological responses and errors in measurement, concentrations of 19.5% oxygen or lower are considered to be indicative of oxygen deficiency.

Oxygen deficiency may result from the displacement of oxygen by another gas, or the consumption of oxygen by a chemical reaction. Confined spaces or low-lying areas are particularly vulnerable to oxygen deficiency and should always be monitored prior to entry. Qualified plant personnel should always monitor oxygen levels and should use atmosphere-supplying respiratory equipment [21].

3.2.4 Ionizing Radiation Hazards

Radioactive materials emit one or more of three types of harmful radiation: alpha, beta, and gamma. Alpha radiation has limited penetration ability and is usually stopped by clothing and the outer layers of the skin. Alpha radiation poses little threat outside the body, but can be hazardous if materials that emit alpha radiation are inhaled or ingested. Beta radiation can cause harmful "beta burns" to the skin and damage the subsurface blood system. Beta radiation is also hazardous if materials that emit beta radiation are inhaled or ingested. Use of protective clothing, coupled with scrupulous personal hygiene and decontamination, affords good protection against alpha and beta radiation.

Gamma radiation, however, easily passes through clothing and human tissue and can also cause serious permanent damage to the body. Chemical-protective clothing affords no protection against gamma radiation itself; however, use of respiratory and other protective equipment can help keep radiation-emitting materials from entering the body by inhalation, ingestion, infection, or skin absorption.

If levels of radiation above natural background are discovered, a plant manager should consult a health physicist. At levels greater than 2 mrem/hour, all industrial site activities should cease until the site has been assessed by an industrial health scientist or licenced environmental engineers.

3.2.5 Biological Hazards

Wastes from industrial facilities, such as a biotechnology firms, hospitals, and laboratories, may contain disease-causing organisms that could infect site personnel. Like chemical hazards, etiologic agents may be dispersed into the environment via water and wind. Other biological hazards that may be present at an industrial site handling hazardous substances include poisonous plants, insects, animals, and indigenous pathogens. Protective clothing and respiratory equipment can help reduce the chances of exposure. Thorough washing of any exposed body parts and equipment will help protect against infection [30,31].

3.2.6 Safety Hazards

Industrial sites handling hazardous substances may contain numerous safety hazards, such as (a) holes or ditches, (b) precariously positioned objects, such as drums or boards that may fall,

(c) sharp objects, such as nails, metal shards, and broken glass, (d) slippery surfaces, (e) steep grades, (f) uneven terrain, and (g) unstable surfaces, such as walls that may cave in or flooring that may give way.

Some safety hazards are a function of the work itself. For example, heavy equipment creates an additional hazard for workers in the vicinity of the operating equipment. Protective equipment can impair a worker's ability, hearing, and vision, which can result in an increased risk of an accident.

Accidents involving physical hazards can directly injure workers and can create additional hazards, for example, increased chemical exposure due to damaged protective equipment, or danger of explosion caused by the mixing of chemicals. Site personnel should constantly look out for potential safety hazards, and should immediately inform their supervisors of any new hazards so that proper action can be taken [1,21,31].

3.2.7 Electrical Hazards

Overhead power lines, downed electrical wires, and buried cables all pose a danger of shock or electrocution if workers contact or sever then during site operations. Electrical equipment used on site may also pose a hazard to workers. To help minimize this hazard, low-voltage equipment with ground-fault interrupters, and water-tight, corrosion-resistant connecting cables should be used on site. In addition, lightning is a hazard during outdoor operations, particularly for workers handling metal containers or equipment. To eliminate this hazard, weather conditions should be monitored and work should be suspended during electrical storms. An additional electrical hazard involves capacitors that may retain a charge. All such items should be properly grounded before handling. OSHA's standard 29 CFR, Part 1910.137, describes clothing and equipment for protection against electrical hazards.

3.2.8 Heat Stress Hazards

Heat stress is a major hazard, especially for workers wearing protective clothing. The same protective materials that shield the body from chemical exposure also limit the dissipation of body heat and moisture. Personal protective clothing can therefore create a hazardous condition. Depending on the ambient conditions and the work being performed, heat stress can occur within as little as 15 minutes. It can pose as great a danger to worker health as chemical exposure. In its early stages, heat stress can cause rashes, cramps, discomfort, and drowsiness, resulting in impaired functional ability that threatens the safety of both the individual and coworkers.

Continued heat stress can lead to stroke and death. Careful training and frequent monitoring of personnel who wear protective clothing, judicious scheduling of work and rest periods, and frequent replacement of fluids can protect against this hazard [21].

3.2.9 Cold Exposure Hazards

Cold injury (frostbite and hypothermia) and impaired ability to work are dangers at low temperatures and when the wind-chill factor is low. To guard against them, the personnel at an industrial site should (a) wear appropriate clothing, (b) have warm shelter readily available, and (c) carefully schedule work and rest periods, and monitor workers' physical conditions.

3.2.10 Noise Hazards

Work around large equipment often creates excessive noise. The effects of noise can include (a) workers being startled, annoyed, or distracted, (b) physical damage to the ear, pain, and temporary and/or permanent hearing loss, and (c) communication interference that may increase potential hazards due to the inability to warn of danger and the proper safety precautions to be taken.

If plant workers are subjected to noise exceeding an 8 hour, time-weighted average sound level of 90 dBA (decibels on the A-weighted scale), feasible administrative or engineering controls must be utilized. In addition, whenever employee noise exposure equals or exceeds an 8 hour, time-weighted average sound level of 85 dBA, workers must administer a continuing, effective hearing conservation program as described in OSHA regulation 29 CFR, Part 1910.95, [1,21,26].

3.3 MANAGEMENT OF AIR QUALITY AT INDUSTRIAL SITES

3.3.1 Airborne Contaminants

The U.S. Environmental Protection Agency (USEPA) has estimated that about 30% of commercial and industrial buildings cause "sick building syndrome." Alternatively the health problems associated with such buildings can also be called "building syndrome," "building-related illness," or "tight building syndrome." As a rule of thumb, to be considered as causing "sick building syndrome" a commercial/industrial building must have at least 20% of its occupants' complaints last for more than two weeks, with symptom relief when the occupants leave the sick building.

At an industrial site, occupants complain when they experience respiratory problems, headache, fatigue, or mucous membrane irritation of their eyes, noses, mouths, and throats.

The following contaminants in air are caused by the building materials [1,32,33,61]:

- Formaldehyde: from particle board, pressed wood, urea-formaldehyde foam insulation, plywood resins, hardwood paneling, carpeting, upholstery;
- Asbestos: from draperies, filters, stove mats, floor tiles, spackling compounds, older furnaces, roofing, gaskets, insulation, acoustical material, pipes, etc.;
- Organic vapors: from carpet adhesives, wool finishes, etc.;
- Radon: from brick, stone, soil, concrete, etc.;
- Synthetic mineral fibers: from fiberglass insulation, mineral wood insulation, etc.; and
- Lead: from older paints.

The following contaminants in air are caused by the use of various building equipments [33–36,66,70–75,79–81]:

- Ammonia: from reproduction, microfilm, and engineering drawing machines;
- Ozone: from electrical equipment and electrostatic air cleaners;
- Carbon monoxide, carbon dioxide, sulfur dioxide, hydrogen cyanide, particulates, nitrogen dioxide, benzoapryene, etc.: from combustion sources including gas ranges, dryers, water heaters, kerosene heaters, fireplaces, wood stoves, garage, etc.;
- Aminos: from humidification equipment;
- Carbon, powder, methyl alcohol, trinitrofluorene, trinitrofluorenone: from photocopying machines;
- Methacrylates: from signature machines;
- Methyl alcohol: from spirit duplication machines;
- Dusts: from various industrial equipments; and

- Microorganisms including bacteria, protozoa, virus, nematodes, and fungi: from stagnant water in central air humidifier, microbial slime in heating, ventilation, and air conditioning (HVAC) systems, fecal material of pigeons in HVAC units, etc.

Certain common contaminants in air are caused by the building inhabitants and hazardous substance releases:

- Formaldehyde: from smoking, waxed paper, shampoo, cosmetics, and medicine products, etc.;
- Acetone, butyric acid, ethyl alcohol, methyl alcohol, ammonia, odors: from biological effluents;
- Asbestos: from talcum powder, hot mittens;
- Nicotine, acrolein, carbon monoxide: from smoking;
- Vapors and dusts: from personal care products, cleaning products, fire retardants, insecticides, fertilizers, adhesives, carbonless paper products, industrial hazardous substance releases, etc.;
- Vinyl chloride: from aerosol spray; and
- Lead: from lead-containing gasoline.

Any real property, the expansion, redevelopment, or reuse of which may be complicated by the presence of one or more of the above hazardous substances is termed "brownfield" [37,38,70,84].

3.3.2 Health Effects

Various airborne contaminant sources and the health effects of each specific pollutant are described below in detail.

Carbon Monoxide

Carbon monoxide (CO) is a common colorless and odorless pollutant resulting from incomplete combustion. One of the major sources of CO emission in the atmosphere is the gasoline-powered internal combustion engine. The chemical can be a fatal poison. It can be traced to many sources, including incomplete incineration, unvented gas appliances and heaters, malfunctioning heating systems, kerosene heaters, and underground or connected garages. Environmental tobacco smokes is another major source of CO. The gas ties up hemoglobin from binding oxygen and may cause asphyxiation. Fatigue, headache, and chest pain are the result of repeated exposure to low concentrations. Impaired vision and coordination, dizziness, confusion, and death may develop at the high concentration exposure levels [32,33].

Carbon Dioxide

Carbon dioxide (CO_2) is a colorless and odorless gas. It is an asphyxiant-causing agent. A concentration of 10% can cause unconsciousness and death from oxygen deficiency. The gas can be released from industrial studies [39], automobile exhaust, environmental tobacco smoke (ETS), and inadequately vented fuel heating systems. It is heavy and accumulates at low levels in depressions and along the floor.

Nitrogen Oxides

Nitrogen oxides, which are mainly released from industrial stacks, include nitrous oxide (N_2O), nitric oxide (NO), nitrogen dioxide (NO_2), nitrogen trioxide (N_2O_3), nitrogen tetraoxide (N_2O_4), nitrogen pentoxide (N_2O_5), nitric acid (HNO_5), and nitrous acid (HNO_2). Nitrogen dioxide is the most significant pollutant. The nature of the combustive process varies with the concentration of nitrogen oxides. Inhalation of nitrogen oxides may cause irritation of the eyes and mucous membranes. Prolonged low-level exposure may stain skin and teeth yellowish and brownish. Chronic exposure may cause respiratory dysfunction. Nitrogen oxides partially cause acid rains.

Sulfur Dioxide

Sulfur dioxide (SO_2) is a colorless gas with a strong odor and is the major substance causing acid rains. The major emission source of the gas is fuel or rubber tire combustion from industry [40]. Excess exposure may occur in industrial processes such as ore smelting, coal and fuel oil combustion, paper manufacturing, and petroleum refining. The chemical has not been identified as a carcinogen or co-carcinogen by the data, but short-term acute exposures to a high concentration of sulfur dioxide suggest adverse effects on pulmonary function [33].

Ozone

Ozone (O_3) is a powerful oxidizing agent. It is found naturally in the atmosphere by the action of electrical storms. The major indoor source of ozone is from electrical equipment and electrostatic air cleaners. The indoor ozone concentration is determined by ventilation. It depends on the room volume, the number of air changes in the room, room temperature, materials, and the nature of surfaces in the room. Ozone is irritating to the eyes and all mucous membranes. Pulmonary edema may occur after exposure has ceased [32,33].

Radon

Radon is a naturally occurring radioactive decay product of uranium. A great deal of attention centers around radon^{222}, which is the first decay product of radium^{228}. Radon and radon daughters have been found to contribute to lung cancer; USEPA estimates that radon may cause 5000 to 20,000 lung cancer deaths per year in the United States. The released energy from radon decay may damage lung tissue and lead to lung cancer. Smokers also may have a higher risk of developing lung cancer induced by radon.

Radon is present in the air and soil. It can leak into the indoor environment through dirt floors, cracks in walls and floors, drains, joints, and water seeping through walls. Radon can be measured by using charcoal containers, alpha-track detectors, and electronic monitors. Results of the measurement of radon decay products and the concentration of radon gas are reported as "working levels (WL)" and "picocuries per liter" (pCi/L), respectively. The continuous exposure level of 4 pCi/L or 0.02 WL has been used by USEPA and CDC as a guidance level for further testing and remedial action [33].

Once identified, the risk of radon can be minimized through engineering controls and practical living methods. The treatment techniques include sealing cracks and other openings in basement floors, and installation of sub-slab ventilation. Crawl spaces should also be well ventilated. Radon-contaminated groundwater can be treated by aerating [41–43] or filtering through granulated activated carbon [43,44].

Asbestos

Asbestos is a naturally occurring mineral and was widely used as an insulation material in building construction [35]. Asbestos possesses a number of good physical characteristics that make it useful as thermal insulation and fire-retardant material. It is electrically nonconductive, durable, chemical resistant, and sound absorbent. However, lung cancer and mesothelioma have been found to be associated with environmental asbestos exposure. USEPA has listed asbestos as a hazardous air pollutant since 1971. The major route of exposure is the respiratory system. Adverse health effects include asbestosis, lung cancer, mesothelioma, and other diseases. The latency period for asbestos diseases varies from 10 to 30 years [33].

Formaldehyde

Formaldehyde (HCHO) is a colorless gas with a pungent odor. Formaldehyde has found wide industrial usage as a fungicide and germicide, and in disinfectants and embalming fluids. The serious sources of indoor airborne formaldehyde are furniture, floor underlayment insulation, and environmental tobacco smoke. Urea formaldehyde (UF) is mixed with adhesives to bond veneers, particles, and fibers. It has been identified as a potential hazardous source.

Formaldehyde gas may cause severe irritation to the mucous membranes of the respiratory tract and eyes. Repeated exposure to formaldehyde may cause dermatitis either from irritation or allergy. The gas can be removed from the air by an absorptive filter of potassium permanganate-impregnated alumina pellets or fumigation using ammonia. Exposure to formaldehyde may be reduced by using exterior grade pressed wood products that contain phenol resins. Maintaining moderate temperature and low humidity can reduce emissions from formaldehyde-containing material. The chemical is intensely irritating to mucous membranes of the upper respiratory tract, the eyes, and skin. Repeated exposure may cause dermatitis and skin sensitization. This substance has been listed as a carcinogen.

Pesticides

Pesticides are used to kill household insets, rats, cockroaches, and other pests. Pesticides can be classified based on their chemical nature or use as organophosphates, carbonates, chlorinated hydrocarbons, bipyridyls, coumarins and indandiones, rodenticides, fungicides, herbicides, fumigants, and miscellaneous insecticides. The common adverse effects are irritation of the skin, eyes, and upper respiratory tract. Prolonged exposure to some chemicals may cause damage to the central nervous system and kidneys [32,33].

Volatile Organic Compounds

The sources of volatile organic compounds (VOCs) include building materials, maintenance materials, building inhabitants, and gasoline spills/leaks. Building materials include carpet adhesives and wool finishes. Maintenance materials include varnishes, paints, polishes, and cleaners. Volatile organic compounds may pose problems for mucous surfaces in the nose, eyes, and throat. Chemicals that have been recognized as a cancer-causing agent include, at least, perchloroethylene used in dry cleaning, chloroform from laboratories, gasoline from gas stations, etc. [33,42].

Lead

Lead has been widely used in the storage battery industry, the petroleum industry, pigment manufacturing, insecticide production, the ceramics industry, and the metal products industry. Most of the airborne lead that has been identified comes from combustion of gasoline [33,79] and removal of lead paint [34].

Respirable Particles

Respirable particles are 10 or less micrometers in aerodynamic diameter. The sources of respirable particles include kerosene heaters, paint pigments, insecticide dusts, radon, and asbestos. The particles may irritate the eyes, nose, and throat and may contribute to respiratory infections, bronchitis, and lung cancer.

Tobacco Smoke

Environmental tobacco smoke (ETS) is a major indoor pollutant. Both the National Research Council (NRC) and USEPA have indicated that passive smoking significantly increases the risk of lung cancer in adults and respiratory illness in children. It is composed of irritating gases and carcinogenic tar particles. Nonsmokers breathing ETS are called "involuntary smokers," "passive smokers," or "second-hand smokers." There are more than 4700 chemical compounds in cigarette combustion products, such as carbon monoxide, carcinogenic/tars, hydrogen cyanide, formaldehyde, and arsenic. Of the chemicals, 43 have been recognized as carcinogens.

Environmental tobacco smoke (ETS) is a suspected source of many pollutants causing impaired health. A plant manager should either ban indoor smoking, or assign smoking areas at an industrial site. The most common impact in children from ETS is the development of wheezing, coughing, and sputum. According to 1986 reports by NRC, the risk of lung cancer is about 30% higher for nonsmoking spouses of smokers than for nonsmoking spouses of nonsmokers. Some studies also showed that ETS has been associated with an increased risk of heart disease [33].

PCB (Polychlorinated Biphenyl)

Polychlorinated biphenyls (PCBs) are a family of compounds that were used extensively in electrical equipment, such as transformers, because of their insulating and heat transferring qualities. They are suspected human carcinogens and have been linked to liver, kidney, and other health problems. It is known that PCBs can be transported by air, and this is thought to be one of the major ways in which they circulate around the world, explaining why they are found in the Arctic and Antarctic. Indian women dwelling on Cornwall Island located in the Canadian portion of the reservation have elevated levels of toxic PCBs in their breast milk. The PCB contamination does not appear to come from fish, but from air the women breathe every day [45].

Chlorofluorocarbon (CFC) and Freon

Freon is a commercial trademark for a series of fluorocarbon products used in refrigeration and air-conditioning equipment, as aerosol propellants, blowing agents, fire extinguishing agents, and cleaning fluids and solvents. Many types contain chlorine as well as fluorine, and should be called chlorofluorocarbons (CFCs) [85,86].

According to USEPA, roughly 28% of the ozone depletion attributed to chlorofluorocarbon (CFC) is caused by coolants in refrigerators and mobile air-conditioners. This being the case, it is necessary to analyze such issues as the refrigerants themselves used in air-conditioners, the types of air-conditioning resulting in CFC emissions, and the environmental fate, human toxicity, and legislation applying to these refrigerants.

The two most common CFC refrigerants in use today for air-conditioning purposes are Refrigerant 12 (CCl_2F_2) and Refrigerant 22 ($CHClF_2$). Refrigerant 12 was the first fluorocarbon-type refrigerant developed and used commercially. Its high desirability in air-conditioning applications arises from its extremely low human toxicity, good solubility, lack of effect on elastomers and other plastics, and reasonable compression ratio. Refrigerant 22, another commonly used air-conditioning coolant, although much safer to stratospheric ozone (because of the hydrogen molecule contained), tends to enlarge elastomers and weaken them, thus causing leakage wherever there is a rubber seal [46]. Of the CFC-12 used for refrigeration in the United States, 41% is used by vehicle air-conditioners. However, because vehicle air-conditioners are particularly prone to leaks and need frequent replacements of refrigerant, they use 75% of the country's replacement CFC-12.

The acute health effects of Refrigerant 12 are (a) irritation of mouth, nose, and throat; (b) irregular heart beat; and (c) dizziness and light headiness. Chronic health effects are not known at this time. The acute health effects of Refrigerant 22 are (a) heart palpitations; (b) tightness in the chest; and (c) difficulty in breathing. Chronic health effects include irregular heat rhythms and skipped beats, and possible damage to the liver, kidneys, and blood.

Dioxins

Dioxins form a family of aromatic compounds known chemically as di-benzo-p-dioxins. Each of these compounds has a nucleus triple ring structure consisting of two benzene rings interconnected to each other through a pair of oxygen atoms. Dioxin compound generally exists as colorless crystalline solid at room temperatures, and is only slightly soluble in water and most organic liquids. They are usually formed through combustion processes involving precursor compounds. Once formed, the dioxin molecule is quite stable.

Dioxins are not decomposed by heat or oxidation in a 700°C incinerator, but pure compounds are largely decomposed at 800°C. Chlorinated dioxins lose chlorine atoms on exposure to sunlight and to some types of gamma radiation, but the basic dioxin structure is largely unaffected. The biological degradation rate of chlorinated dioxins is slow, although measured rates differ widely.

Incineration has been well organized as one of the best demonstrated and available technologies for waste destruction by direct heat, thus the volume and toxicity of the remaining residuals can be reduced.

Most interest has been directed toward the isomer 2,3,7,8-TCDD, which is among the most toxic compounds known. Experimental animals are exceedingly sensitive to TCDD. The LD_{50}, the dose that kills half of a test group, for 2,3,7,8-TCDD is 0.6 μ/kg of body weight for male guinea pigs. Humans exhibit symptoms effecting on enzyme and nervous systems, and muscle and joint pains [46].

Dioxin can enter a person through (a) dermal contact, absorption through skin; (b) inhalation, breathing of contaminated air; and (c) ingestion, eating contaminated materials such as soil, food, or drinking water contaminated by dioxin. In assessing these three routes, control of the physical and chemical properties of TCDD in the environment are containment, capping, and monitoring.

Under existing USEPA regulations, dioxin-bearing wastes may be stored in tanks, placed in surface impoundments and waste piles, and placed in landfills. However, in addition to meeting the Resource Conservation and Recovery Act (RCRA) requirements for these storage and disposal processes, the operators of these processes must operate in accordance with a management plan for those wastes that is approved by USEPA. Factors to be considered include: (a) volume, physical, and chemical characteristics of wastes, including their potential to migrate through soil or to volatilize or escape into the atmosphere; (b) the alternative properties of underlying and surrounding soils or other materials; (c) the mobilizing properties of other materials codisposed with these wastes; and (d) the effectiveness of additional treatment, design, or monitoring techniques.

Additional design, operating, and monitoring requirements may be necessary for facilities managing dioxin wastes in order to reduce the possibility of migration of these wastes to groundwater, surface water, or air so as to protect human health and the environment.

3.3.3 Air Emission Control

Air emission control technologies reduce levels of particulate emission and/or gaseous emission. Some air emission control equipment, such as dry injection units, fabric filters, cyclones, and electrostatic precipitators, are mainly designed to control particulate emissions. Others, such as dry scrubbers, thermal oxidizers, granular activated carbon, adsorption filters, and coalescing filters, control mainly gaseous pollutants including oily vapor. Air emission control equipment such as wet scrubbers and cartridge filters can control both particulate and gaseous emissions. Any gaseous effluent discharge at an industrial site that handles hazardous substances will normally require a discharge permit from one or more regular agencies.

For indoor air quality control, in addition to the air emission control technologies identified above, ventilation and air conditioning are frequently adopted by plant managers [36,85,86].

3.4 MANAGEMENT OF WATER QUALITY AT INDUSTRIAL SITES

3.4.1 Waterborne Contaminants and Health Effects

All point source and nonpoint source wastewaters at an industrial site must be properly managed for source separation, waste minimization, volume reduction, collection, pretreatment, and/or complete end-of-pipe treatment [39,47]. When industrial waste is not disposed of properly, hazardous substances may contaminate a nearby surface water (river, lake, sea, or ocean) and/or groundwater. Any hazardous substance release, either intentionally or unintentionally, increases the risk of water supply contamination and human disease. Major waterborne contaminants and their health effects are listed below.

Arsenic (As)

Arsenic occurs naturally and is also used in insecticides. It is found in tobacco, shellfish, drinking water, and in the air in some locations. The standard allows for 0.05 mg of arsenic per liter of water. If persons drink water that continuously exceeds the standard by a substantial amount over a lifetime, they may suffer from fatigue and loss of energy. Extremely high levels can cause poisoning.

Barium

Although not as widespread as arsenic, barium also occurs naturally in the environment in some areas. It can also enter water supplies through hazardous industrial waste discharges or releases. Small doses of barium are not harmful. However, it is quite dangerous when consumed in large quantities. The maximum amount of barium allowed in drinking water by the standard is 1.0 mg/L of water.

Cadmium

Only minute amounts of cadmium are found in natural waters in the United States. Hazardous waste discharges from the electroplating, photography, insecticide, and metallurgy industries can increase cadmium levels. Another common source of cadmium in drinking water is from galvanized pipes and fixtures if the pH of a water supply is not properly controlled. The sources of cadmium exposure are the foods we eat and cigarette smoking. The maximum amount of cadmium allowed in drinking water by the standard is 0.01 mg/L of water.

Chromium

Chromium is commonly released to the environment from the electroplating industry and is extremely hazardous. Some studies suggest that in minute amounts, chromium may be essential to human beings, but this has not been proven. The standard for chromium is 0.05 mg/L of water [76].

Lead

Lead sources include lead and galvanized pipes, auto exhausts, and hazardous waste releases. The maximum amount of lead permitted in drinking water by the standards is 0.05 mg/L of water. Excessive amounts well above this standard may result in nervous system disorders or brain or kidney damage [69].

Mercury

Large increases in mercury levels in water can be caused by industrial and agricultural use and waste releases. The health risk from mercury is greater from mercury in fish than simply from water-borne mercury. Mercury poisoning may be acute, in large doses, or chronic, from lower doses taken over an extended time period. The maximum amount of mercury allowed in drinking water by the standard is 0.002 mg/L of water. That level is 13% of the total allowable daily dietary intake of mercury.

Selenium

Selenium is found in meat and other foods due to water pollution. Although it is believed to be essential in the diet, there are indications that excessive amounts of selenium may be toxic. Studies are under way to determine the amount required for good nutrition and the amount that may be harmful. The standard for selenium is 0.01 mg/L of water. If selenium came only from drinking water, it would take an amount many times greater than the standard to produce any ill effects.

Silver

Silver is some times released to the environment by the photographic industry, and is considered to be toxic at high concentration. Because of the evidence that silver, once absorbed, is held indefinitely in tissues, particularly the skin, without evident loss through usual channels of elimination or reduction by transmigration to other body sites, and because of other factors, the maximum amount of silver allowed in drinking water by the standard is 0.05 mg/L of water.

Fluoride

High levels of fluoride in drinking water can cause brown spots on the teeth, or mottling, in children up to 12 years of age. Adults can tolerate ten times more than children. In the proper amounts, however, fluoride in drinking water prevents cavities during formative years. This is why many communities add fluoride in controlled amounts to their water supply. The maximum amount of fluoride allowed in drinking water by the standard ranges from 0.4 to 2.4 mg/L depending on average maximum daily air temperature. The hotter the climate, the lower the amount allowed, for people tend to drink more in hot climates. In this hot area, the maximum contaminant level for fluoride is 2.0 mg/L of water.

Nitrate

Nitrate in drinking water above the standard poses an immediate threat to children under three months of age. In some infants, excessive levels of nitrate have been known to react with the hemoglobin in the blood to produce an anemic condition commonly known as "blue baby." If the drinking water contains an excessive amount of nitrate, it should not be given to infants under three months of age and should not to be used to prepare formula. The standard allows for 10.0 mg of nitrate (as N) per liter of water. Nitrate can be removed from water by ion exchange, RO, or distillation [48].

Pesticides

Millions of pounds (1 lb = 0.454 k) of pesticides are used on croplands, forests, lawns, and gardens in the United States each year. A large quantity of hazardous pesticides is also released by the pesticide industry to the environment. These hazardous pesticides drain off into surface waters or seep into underground water supplies. Many pesticides pose health problems if they get into drinking water and the water is not properly treated. The maximum limits for pesticides in drinking water are: (a) endrin, 0.0002 mg/L; (b) lindane, 0.004 mg/L; (c) methoxychlor, 0.1 mg/L; (d) toxaphene, 0.005 mg/L; (e) 2,4-D, 0.1 mg/L; and (f) 2,4,5-TP silvex, 0.01 mg/L.

Priority Pollutants

Many toxic organic substances, known as the USEPA priority pollutants, are cancer-causing substances and, in turn, are hazardous substances. Both the U.S. Drinking Water Standards and the Massachusetts Drinking Water Standards give maximum contaminant levels (MCL) for benzene, carbon tetrachloride, p-dichlorobenzene, 1,2-dichloroethane, 1,2-dichloroethylene, 1,1,1-trichloroethane, trichloroethylene (TEC), vinyl chloride, and total trihalomethanes (TTHM) in drinking water. In Massachusetts, monitoring for 51 unregulated VOCs is also required. In addition, the State of Massachusetts has announced the Massachusetts Drinking Water Guidelines, giving the lowest practical quantization limit (PQL) for 40 contaminants that have no regulated MCLs, but are evaluated on a case-by-case, on-going basis. More toxic priority pollutants may be incorporated into this list for enforcement by the State. Plant managers and consulting engineers should contact the home state for specific state regulations.

Microorganisms

Pathogenic microorganisms from the biotechnology industry, agricultural industry, hospitals, and so on may cause waterborne diseases, such as typhoid, cholera, infectious hepatitis, dysentery, etc. Coliform bacteria regulated by both the Federal and the State governments are only an indicator showing whether or not the water has been properly disinfected. For a disinfected water, a zero count on coliform bacteria indicates that the water is properly disinfected, and other microorganisms are assumed to be sterilized.

Radionuclides

Gross alpha particle activity, gross beta particle activity, and total radium 226 and 228 are found from radioactive wastes, uranium deposits, and certain geological formations, and are a cancer-causing energy. The MCLs for gross alpha particle activity, gross beta particle activity, and total radium 226 and 228 are set by the USEPA at 15 pCi/L, 4 mrem/year, and 5 pCi/L, respectively. Again the Massachusetts Drinking Water Guidelines are more stringent, and include additional photon activity, tritium, strontium-90, radon-222, and uranium for State enforcement. Radon in groundwater can be effectively removed by granular activated carbon [44]. In a recent decision having potentially broad implications, a U.S. Federal Court of Appeals has upheld USEPA regulations establishing standards for radionuclides in public water systems [49].

PCBs, CFCs, and Dioxin

Polychlorinated biphenyls (PCBs), CFCs, petroleum products, and dioxin are major toxic contaminants in air (Section 3.3.2), soil (Section 3.5.3), and also in water. The readers are referred to Sections 3.3.2 and 3.5.3 for details about PCB characteristics, health effects, treatment technologies, and so on. For water quality management, they have been included in the list of the USEPA priority pollutants [86].

Asbestos

Asbestos is an airborne contaminant (Section 3.3.2), a hazardous solid waste (Section 3.5), and also a waterborne contaminant, regulated by many states. The health effect of asbestos in water, however, is not totally known.

3.4.2 Water Pollution Prevention and Control

Depending on the state where the industrial plant is located, an aqueous effluent from a pretreatment facility or a complete end-of-pipe treatment facility can be discharged into a river, a lake, or an ocean, only if it meets the pretreatment standards and the effluent discharge standards established by the regulatory agencies, in accordance with the National Pollutant Discharge Elimination System (NPDES) or the State Pollutant Discharge Elimination System (SPDES). The standards can be industry-specific, chemical-specific, or site-specific, or all three. The readers are referred to other chapters of this handbook series for the details.

The plant manager of an industrial site having hazardous substances must establish an in-plant hazardous substance management program to ensure that the plant's hazardous substances will not be released by accident, or by neglect, to the plant's soil and groundwater.

Once a groundwater or a surface water is contaminated, the cleanup cost is very high. In general, a contaminated groundwater or surface water must be decontaminated to meet the Federal and the State drinking water standards and the State Guidelines if the groundwater or surface water source is also a potable water supply source. Even if a receiving water (either a surface water or a groundwater) is not intended to be used as a water supply source, the cleanup cost and the loss of revenue can be as high as hundreds of millions of dollars. Pollution prevention before contamination occurs is always better and more economical than pollution control after contamination occurs.

3.4.3 A Case History of Water Pollution by PCB Release

Polychlorinated biphenyls (PCBs) are colorless toxic organic substances that cause cancer and birth defects. There are more than 200 different types of PCBs, ranging in consistency from heavy, oily liquids to waxy solids, and each type further varying in the number and location of chlorine atoms attached to its molecular carbon rings. They are fire resistant and do not conduct heat or electricity well. Accordingly they have numerous commercial applications as insulation in electrical systems, for example, for transformers.

Owing to a lack of environmental knowledge and governmental guidance, General Electric Company released about 500,000 lb of hazardous PCBs into Hudson River in New York State between 1947 and 1976 from its plants in Fort Edward and Hudson Falls. Hudson River is one of North America's great mountain streams, cruising through gorges, crashing over boulders, churning into a white-water delight, and eventually reaching the great Atlantic Ocean. For centuries, the great Hudson has been a reliable water resource for navigation, fishing, boating, swimming, winter sports, water supply, and natural purification. Around Glens Falls, the Hudson runs into civilization, into industry, and, in turn, into an industrial disaster: the pollution of more than 185 miles of the river with over half a million pounds of hazardous and poisonous PCBs.

In 1977, PCB production was banned in the United States, and its release to the Hudson was stopped. Since 1976, the State of New York has banned all fishing on the river between Bakers Fall in the Village of Hudson Fall and the Federal Dam at Troy. Most affected has been the commercial striped bass fishery, which once earned New Yorkers $40 million a year. Now the river is no longer suitable for swimming or any water contact sports, and of course, definitely not suitable for domestic water supply. The loss of its recreation and water supply revenues is simply too high to be priced. In 1983, the USEPA declared the Hudson River, from Hudson Falls to New York City, one of the Nation's largest and most complicated Superfund toxic-waste sites.

Now the New York State Department of Environmental Conservation and some environmental groups have advocated dredging the PCB-contaminated river bottom and transferring the PCB-containing sediment to a landfill site. Even though the cleanup costs, now estimated to run as high as $300 million U.S. dollars, are acceptable to U.S. tax payers, a landfill site to receive the PCB-contaminated sediment still cannot be found because of public resistance [50].

This is a typical environmental disaster that the industry must not forget and must not repeat. For more information on PCB pollution and management, the readers are referred to the literature [46,51].

3.5 MANAGEMENT OF HAZARDOUS SOLID WASTES AT INDUSTRIAL SITES

3.5.1 Disposal of a Large Quantity of Hazardous Solid Wastes

When disposed of improperly, hazardous solid wastes may contaminate air, soil, and/or groundwater, and increase the risk of human disease and environmental contamination.

Inevitably, some hazardous solid wastes generated at an industrial site must be discarded. Rusted, old containers or equipment might be targets for plantwide cleaning. Some industrial materials or products, such as half-used cans of paint or chemical, might be discarded. Or the owner or plant manager might want to dispose of some products that are too old to be sold, or some building material (such as asbestos) that is too hazardous for everyday use.

A large quantity of any hazardous solid wastes can only be properly transported or disposed of by licenced or certified environmental professionals. Small quantities of hazardous wastes, however, can be handled by a plant manager.

3.5.2 Disposal of a Small Quantity of Hazardous Solid Wastes

Right now there is no easy way to dispose of very small quantities of hazardous household products, such as pesticides, batteries, outdated medicines, paint, paint removals, used motor oil, wool preservatives, acids, caustics, and so on. There are no places that accept such small quantities of wastes as generated by a small industrial/commercial site. For now, the best disposal techniques are listed in Table 1, which is recommended by the Massachusetts Department of Environmental Management, Bureau of Solid Waste Disposal.

Table 1 Methods for Disposal of Small Quantities of Common Hazardous Wastes

Product	Take to a hazardous waste collection site (or store until available)	Wrap in plastic bag, put in trash, and alert the collector	Wash down drain with lots of water	Take to a special recycling center (not paper recycling)	Give to a friend to use, with careful instructions	Return to the manufacturer or to the retailer
Acids (strong)	Best	Never	Never	Unavailable	Impractical	Impractical
Acids (weak)	Best	4th best	3rd best	Unavailable	2nd best	Impractical
Banned pesticides	2nd best	Never	Never	Never	Never	Best
Batteries	3rd best	Never	Impractical	Best	Never	2nd best
Caustics	Best	3rd best	4th best	Unavailable	2nd best	Impractical
Pesticide containers	Best	2nd best	Impractical	Unavailable	Impractical	Impractical
Flammables	Best	3rd best	Never	Unavailable	2nd best	Impractical
Outdated medicines	Best	3rd best	2nd best	Never	Never	Impractical
Paint	2nd best	3rd best	Never	Unavailable	Best	Impractical
Paint remover	Best	Never	Never	Unavailable	2nd best	Impractical
Pesticides	Best	3rd best	Never	Unavailable	2nd best	Impractical
Used motor oil	3rd best	Never	Never	Best	Never	2nd best
Wood preservatives	Best	2nd best	Never	Unavailable	3rd best	Impractical

Note: Strong acids include battery acid, murintic acid, and hydrochloric acid. Weak acids include acetic acid, toilet bowl cleaner, and lactic acid. Banned pesticides include Silvex, Mirex, Aldrin, Chlordane, DDT, and Heptachlor. Caustics include oven cleaner and drain cleaner. Flammables include alcohol, acetone, turpentine, lacquer, and paint thinner. Pesticides include rodent poisons, insecticides, weed killer, and other herbicides and fungicides. Pesticide containers should be triple-rinsed, and the contents sprayed on crops or yard, before discarding.

Small quantities of hazardous solid wastes (such as potassium dichromate, lead nitrate, silver nitrate, asbestos, etc.), liquid chemicals (such as chloroform, PCB, methylene chloride, etc.), petrochemicals (such as gasoline, No. 2 fuel oil, etc.), or pure metals (such as mercury, sodium, etc.), which are stored in bottles or cans, however, are not considered to be hazardous "household products." Accordingly these nonhousehold hazardous solid wastes, even in small quantities, can only be properly disposed of by licenced or certified environmental professionals.

3.5.3 Hazardous and Infectious Solid Wastes

A few selected hazardous solid wastes, and hazardous liquid wastes stored in drums/tanks, are described below for reference.

Infectious and Hazardous Medical Wastes

In a 1987 *Federal Register* notice, USEPA first defined the three waste categories (pathological waste, laboratory waste, isolation waste) below, which should be treated as infectious:

1. Pathological waste: Surgical or operating room specimens (like body parts) and other potentially contaminated waste from outpatient areas and emergency rooms.
2. Laboratory waste: Pathological specimens (all tissues, blood specimens, excreta, and secretions obtained from patients or laboratory animals) and other potentially contaminated wastes.
3. Isolation waste: Disposable equipment and utensils (like syringes and swabbing) from rooms of patients suspected to have a communicable disease.
4. General hospital waste: Cafeteria garbage, disposal gowns, drapes, packaging, etc., representing about 85% of total hospital waste.
5. Hazardous waste: Dental clinics, chemotherapy wastes (some) listed as hazardous by USEPA, and low-level radioactive waste.

Incineration has been common practice in hospitals for decades. It is quick, easy, and especially handy for rendering the more repulsive wastes unrecognizable. It also reduces waste volume by up to 90%, leaving mostly ashes behind, for landfilling. Because of their comparatively small size, hospital incinerators have until recently been exempted from federal rules that control air emissions of larger incinerators, like mass-burn facilities. According to the November 1987 USEPA report, there were 6200 hospital incinerators around the United States. Only 1200 are "controlled-air" incinerators, a relatively new design that limits the air in the burn chamber, ensuring more complete incineration. However, even the 1200 controlled-air models do not necessarily have stacks equipped with scrubbers to prevent acid gas and dioxin emissions [46,52].

In many states, regulations only require that hospital incinerators not create a public nuisance usually recognized as odors and smoke opacity. Disposal costs for these medical wastes are becoming stiffer, just as surely as they are for infectious and other hazardous/toxic wastes. This adds another incentive to incinerate. It may be possible that a good deal of hospital waste could be separated, reduced, and recycled. While infectious waste is obviously not recyclable, the amount of waste designated infectious can be greatly reduced by separating materials to avoid excess contamination [74].

Health officials are increasingly concerned about disposal of infectious, radioactive, and toxic medical wastes that have become major components in the treatment and diagnosis of many diseases. Legal complications in handling medical wastes are another issue. There are, for example, no federal regulations for disposal of medical waste. State and local regulations are widely divergent.

Petroleum Contaminated Soil

Petroleum (crude oil) is a highly complex mixture of paraffinic, cycloparaffinic (naphthenic), and aromatic hydrocarbons, containing low percentages of sulfur and trace amounts of nitrogen and oxygen compounds. The most important petroleum fractions, obtained by cracking or distillation, are various hydrocarbon gases (butane, ethane, propane), naphtha of several grades, gasoline, kerosene, fuel oils, gas oil, lubricating oils, paraffin wax, and asphalt. From the hydrocarbon gases, ethylene, butylene, and propylene are obtained. About 5% of the petroleum (crude oil) consumed in the United States is used as feedstocks by the chemical industries. The rest is consumed for production of various products, such as gasoline, fuel oils, and so on, introduced above. The crude oil, when spilled or leaked, will contaminate the soil because it is flammable, and moderately toxic by ingestion. One of the major components of petroleum product is benzene, which is a known human carcinogen.

Gasoline, fuel oils, and lubricating oils are three major pollutants among the petroleum family members, and are therefore introduced in more detail.

Gasoline is a mixture of volatile hydrocarbons suitable for use in a spark-ignited internal combustion engine and having an octane number of at least 60. The major components are branched-chain paraffins, cycloparaffins, and aromatics. The present source of gasoline is petroleum, but it may also be produced from shale oil and Athabasca tar sands, as well as by hydrogenation or gasification of coal. There are many different kinds of gasolines:

- Antiknock gasoline: a gasoline to which a low percentage of tetra-ethyl-lead, or similar compound, has been added to increase octane number and eliminate knocking. Such gasolines have an octane number of 100 or more and are now used chiefly as aviation fuel.
- Casinghead gasoline: see natural gasoline (below).
- Cracked gasoline: gasolines produced by the catalytic decomposition of high-boiling components of petroleum, and having higher octane ratings (80–100) than gasoline produced by fractional distillation. The difference is due to the prevalence of unsaturated, aromatic, and branched-chain hydrocarbons in the cracked gasoline.
- High-octane gasoline: a gasoline with an octane number of about 100.
- Lead-free gasoline: an automotive fuel containing no more than 0.05 g of lead per gallon, designed for use in engines equipped with catalytic converters.
- Natural gasoline: a gasoline obtained by recovering the butane, pentane, and hexane hydrocarbons present in small proportions in certain natural gases. Used in blending to produce a finished gasoline with adjusted volatility, but low octane number. Do not confuse with natural gas (q.v.).
- White gasoline: an unleaded gasoline especially designed for use in motorboats; it is uncracked and strongly inhibited against oxidation to avoid gum formation, and is usually not colored to distinguish it from other grades. It also serves as a fuel for camp lanterns and portable stoves.
- Polymer gasoline: a gasoline produced by polymerization of low-molecular-weight hydrocarbons such as ethylene, propane, and butanes. It is used in small amounts for blending with other gasoline to improve its octane number.
- Pyrolysis gasoline: gasoline produced by thermal cracking as a byproduct of ethylene manufacture. It is used as a source of benzene by the hydrodealkylation process.
- Reformed gasoline: a high-octane gasoline obtained from low-octane gasoline by heating the vapors to a high temperature or by passing the vapors through a suitable catalyst.
- Straight-run gasoline: gasoline produced from petroleum by distillation, without use of cracking or other chemical conversion processes. Its octane number is low.

Fuel oil is any liquid petroleum product that is burned in a furnace for the generation of heat, or used in an engine for the generation of power, except oils having a flash point below 100°F and oil burned in cotton or wool burners. The oil may be a distillated fraction of petroleum, a residuum from refinery operations, a crude petroleum, or a blend of two or more of these.

ASTM has developed specifications for six grades of fuel oil. No. 1 is a straight-run distillate, a little heavier than kerosene, used almost exclusively for domestic heating. No. 2 (diesel oil) is a straight-run or cracked distillate used as a general purpose domestic or commercial fuel in atomizing-type burners. No. 4 is made up of heavier straight-run or cracked distillates and is used in commercial or industrial burner installations not equipped with preheating facilities. The viscous residuum fuel oils, Nos. 5 and 6, sometimes referred to as bunker fuels, usually must be preheated before being burned. ASTM specifications list two grades of No. 5 oil, one of which is lighter and under some climatic conditions may be handled and burned without preheating. These fuels are used in furnaces and boilers of utility power plants, ships, locomotives, metallurgical operations, and industrial power plants.

Lubrication oil is a selected fraction of refined mineral oil used for lubrication of moving surfaces, usually metallic, and ranging from small precision machinery (watches) to the heaviest equipment. Lubricating oils usually have small amounts of additives to impart special properties such as viscosity index and detergency. They range in consistency from thin liquids to greaselike substances. In contract to lubricating greases, lube oils do not contain solid or fibrous minerals.

The major petroleum release sources are bulk gasoline terminals, bulk gasoline plants, service stations, and delivery tank trucks. USEPA estimates there are approximately 1500 bulk terminals, 15,000 bulk plants, and 390,000 gasoline service stations in the United States, of which some 180,000 are retail outlets [46]. Fuel oil release is mainly caused by underground storage tank leakage. Lubricating oil release, however, is mainly caused by neglect or intentional dump.

Release of gasoline, lubricating oil, and fuel oils to the soil occurs from spills, leaks, loading and unloading operations. Disposal of petroleum-contaminated soil is now one of the major environmental tasks.

Dioxin

Dioxin (2,3,7,8-tetrachlorodibenzo-p-dioxin; TCDD) is among the most toxic compounds known today. It is an airborne contaminant from an incineration process, which has been described in Section 3.3.2. Dioxin also frequently occurs as an impurity in the herbicide 2,4,5-T. Accordingly, when the herbicide 2,4,5-T is applied to crops, dioxin is also released to the soil. Any spills of dioxin also cause soil contamination. It may be removed by extraction with coconut-activated carbon. Its half-life in soil is about one year.

PCBs

Polychlorinated diphenyl (PCB) is an airborne contaminant (Section 3.3.2), a waterborne contaminant (Section 3.4.1), and also a contaminant in soil due to PCB releases, such as spills, leakages, and landfills. Before the United States banned manufacture of PCBs in 1979, Monsanto had produced more than 1 billion pounds. Practices one thought acceptable and hazard-free in the past have led to PCB releases into the environment. Such practices were conducted by industries using PCBs in processes and products and discharging the PCB-containing waste into rivers and streams. Other PCB-containing waste was disposed of in landfills. When used in transformers and electrical capacitors, PCB compartments are sealed and in place for the life of the equipment. Occasionally seals will leak or external structures are damaged, resulting in leakage. The following are applications in which PCBs have been found and hence are potential sources: (a) cooling and insulating fluids for transformers; (b) dielectric impregnating for capacitors; (c) flame retardants for resins and plastics in the electrical industry; (d) formulations in paints and printing inks; (e) water-repellent additives; (f) dye carrier for pressure-sensitive copy paper; (g) incombustible hydraulic fluids; and (h) dust control agents for road construction.

Other Organic and Inorganic Contaminants

In addition to gasoline, CFC, and so on, various other organic and inorganic compounds such as heavy metals, sulfides, and cyanides on the USEPA Priority Pollutants List, and subject to various water quality criteria, guidelines, etc., when released can also contaminate the soil. The contaminated soil then becomes a hazardous solid waste which must be properly disposed of [63–86].

3.5.4 Disposal of Hazardous and Infectious Wastes

Incineration has been used extensively in hospitals for disposal of hospital wastes containing infectious and/or hazardous substances. Most hospital incinerators (over 80%), however, are outdated or poorly designed. Modern incineration technology, however, is available for complete destruction of organic hazardous and infectious wastes. In addition, adequate air pollution control facilities, such as scrubbers, secondary combustion chambers, stacks, and so on, are needed to prevent acid gas, dioxin, and metals from being discharged from the incinerators.

The same modern incinerators equipped with scrubbers, bag-filters, electro-precipitators, secondary combustion chambers, stacks, etc., are equally efficient for disposal of hazardous PCBs, dioxin, USEPA priority pollutants, and so on, if they are properly designed, installed, and managed. Incineration technology is definitely feasible, and should not be overlooked. The only residues left in the incinerators are small amount of ashes containing metals. The metal-containing ashes may be solidified and then disposed of on a landfill site.

Environmentalists and ecologists, however, oppose construction of any new incinerators and landfill facilities. They would like to close all existing incineration and landfill facilities, if possible. They are wrong. Unless human civilization is to go backward, there will always be hazardous and infectious wastes produced by industry. These wastes must go somewhere. A solution must be found.

It is suggested that waste minimization, spill prevention, leakage prevention, volume reduction, waste recycle, energy conversion, and conservation be practiced by the industry as well as the community. Innovative technology must be developed, and good managerial methods must be established for this practice. With all these improvements, modern incinerators and landfill facilities may still be needed, but their numbers and sizes will be significantly reduced.

Section 3.15 introduces a case history showing how an organic hazardous waste can be reused as a waste fuel in the cement industry. A cement plant is a manufacturing plant needed by our civilization. With special managerial arrangements and process modification, a cement kiln can be operated for production of cement as well as for incineration of hazardous waste. Because hazardous waste can replace up to 15% of fuel for this operation, the industry not only saves 15% of energy cost, but also solves a hazardous waste disposal problem. It should be noted that modern incineration and air purification technologies are still required. In this case the cement kiln acts like an incinerator. It is not necessary for the community or the waste-producing industry to build an incinerator solely for waste disposal.

Section 3.14 presents two case histories: (a) disposal of photographic wastes by a large quantity generator; and (b) disposal of photographic wastes by a small quantity generator. In general, it is economically feasible for a large quantity generator to pretreat its wastes, aiming at regulatory compliance. A small quantity generator with in-house engineering support may also pretreat its wastes, and discharge the pretreated effluent to a receiving water or a POTW. Without in-house engineering support, it would be more cost-effective for the small quantity generator to hire an outside engineering consultant and/or an outside general contractor for proper onsite storage of its hazardous/infectious wastes, subsequent transportation of its wastes by a licenced transporter, and final offsite disposal of its wastes by a licenced facility.

Section 3.13 presents an example showing how a medical office manages its hazardous wastes and what the regulatory requirements are.

Friable asbestos is hazardous, and should be properly disposed of following governmental requirements and guidelines presented in Section 3.6.

3.6 DISPOSAL OF HAZARDOUS ASBESTOS

3.6.1 Asbestos, Its Existence and Releases

The term "asbestos" describes six naturally occurring fibrous minerals found in certain types of rock formations. Of that general group, the minerals chrysolite, amosite, and crocidolite have been most commonly used in building products. Under the Clean Air Act of 1970, the USEPA has been regulating many asbestos-containing materials (ACM), which, by USEPA definition, are materials with more than 1% asbestos. "Friable asbestos" includes any materials that contain greater than 1% asbestos, and that can be crumbled, pulverized, or reduced to powder by hand pressure. This asbestos may also include previously nonfriable material that becomes broken or damaged by mechanical force. The Occupational Safety and Health Administration's (OSHA) asbestos construction standard in Section K, "Communication of Hazards to Employees," specifies labeling many materials containing 0.1% or more asbestos [20,22,53].

Asbestos became a popular commercial product because it is strong, will not burn, resists corrosion, and insulates well. When mined and processed, asbestos is typically separated into very thin fibers. When these fibers are present in the air, they are normally invisible to the naked eye. Asbestos fibers are commonly mixed during processing with material that binds them together so that they can be used in many different products. Because these fibers are so small and light, they remain in the air for many hours if they are released from ACM in a building. When fibers are released into the air they may be inhaled by people in the building.

In July 1989, USEPA promulgated the Asbestos Ban and Phase-down Rule. The rule applies to new product manufacture, importation, and processing, and essentially bans almost all asbestos-containing products in the United States by 1997. This rule does not require removal of ACM currently in place in buildings. In fact, undisturbed materials generally do not pose a health risk; they may become hazardous when damaged, disturbed, or deteriorate over time and release fibers into building air. Controlling fiber release from ACM in a building or removing it entirely is termed "asbestos abatement," aiming at mainly friable asbestos.

Asbestos has been mainly used as building construction materials for many years. Their applications and releases include the following situations.

Vinyl Floor Tiles and Vinyl Sheet Flooring

Asbestos has been added to some vinyl floor tiles to strengthen the product materials, and also to decorate the exposed surfaces. Asbestos is also present in the backing in some vinyl sheet flooring. The asbestos is often bound in the tiles and backing with vinyl or some type of binder. Asbestos fibers can be released if the tiles are sanded or seriously damaged, or if the backing on the sheet flooring is dry-scraped or sanded, or if the tiles are severely worn or cut to fit into place.

Pipe Insulation

Hot water and steam pipes in some older homes may be covered with an asbestos-containing material, primarily as thermal insulation to reduce heat loss, and to protect nearby surfaces from the hot pipes. Pipes may also be wrapped in an asbestos "blanket" or asbestos paper tape.

Asbestos-containing insulation has also been used on furnace ducts. Most asbestos pipe insulation in homes is preformed to fit around various diameter pipes. This type of asbestos-containing insulation was manufactured from 1920 to 1972. Renovation and home improvements may expose and disturb the asbestos-containing materials.

Wall and Ceiling Insulation

Buildings constructed between 1930 and 1950 may contain insulation made with asbestos. Wall and ceiling insulation that contains asbestos is generally found inside the wall or ceiling ("sandwiched" behind plaster walls). The asbestos is used as material for thermal insulation, acoustical insulation, and fire protection. Renovation and home improvements may expose and disturb the materials.

Appliances

Some appliances, such as toasters, popcorn poppers, broilers, dishwashers, refrigerators, ovens, ranges, clothes dryers, and electric blankets are, or have been, manufactured with asbestos-containing parts or components for thermal insulation. As a typical example, hair dryers with asbestos-containing heat shields were only recalled in 1979. Laboratory tests of most hair dryers showed that asbestos fibers were released during use.

Roofing, Shingles, and Siding

Some roofing shingles, siding shingles, and sheets have been manufactured with asbestos using Portland cement as a binding agent. The purposes for the addition of asbestos are strength enhancement, thermal insulation, acoustical insulation, and fire protection. Because these products are already in place and outdoors, there is likely to be little risk to human health. However, if the siding is worn or damaged, asbestos may be released.

Ceilings and Walls with Patching Compounds and Textured Paints

Some large buildings built or remodeled between 1978 and 1987 may contain a crumbly, asbestos-containing material that has been sprayed onto the ceiling or walls. Some wall and ceiling joints may be patched with asbestos-containing material manufactured before 1977. Some textured paint sold before 1978 contained asbestos. Sanding or cutting a surface with the building materials that may contain asbestos will release asbestos to the air, and thus should be avoided.

Stoves, Furnaces, and Door Gaskets

Asbestos-containing cement sheets, millboard, and paper have been used frequently in buildings when wood-burning stoves have been installed. These asbestos-containing materials were used as thermal insulation to protect the floor and walls around the stoves. On cement sheets, the label may tell the plant manager if they contains asbestos. The cement sheet material will probably not release asbestos fibers unless scraped. This sheet material may be coated with a high temperature paint, which will help seal any asbestos into the material. Asbestos paper or millboard was also used for this type of thermal insulation. If these materials were placed where they are subjected to wear, there is an increased possibility that asbestos fibers may be released. Damage or misuse of the insulating material by sanding, drilling, or sawing will also release asbestos fibers.

Oil, coal, or wood furnaces with asbestos-containing insulation and cement are generally found in some older buildings. Updating the system to oil or gas can result in removal or damage to the old insulation. If the insulation on or around the furnaces is in good condition, it is best to leave it alone. If the insulation is in poor condition, or pieces are breaking off, there will be an asbestos release.

Some door gaskets in furnaces, ovens, and wood and coal stoves may contain asbestos. The asbestos-containing door gaskets on wood and coal-burning stoves are subject to wear and can release asbestos fibers under normal use conditions. Handle the asbestos-containing material as little as possible.

3.6.2 Health Risk of Asbestos

Asbestos has been shown to cause cancer of the lung and stomach according to studies of workers and others exposed to asbestos. There is no level of exposure to asbestos fibers that experts can assume is completely safe.

Some asbestos materials can break into small fibers that can float in the air, and these fibers can be inhaled. These tiny fibers are small, cannot be seen, and can pass through the filters of normal vacuum cleaners and get back into the air. Once inhaled, asbestos fibers can become lodged in tissue for a long time. After many years, cancer or other sickness can develop. In order to be a health risk, asbestos fibers must be released from the material and be present in the air for people to breathe. A health risk exists only when asbestos fibers are released from the material or product. Soft, easily crumbled asbestos-containing material, previously defined as "friable asbestos," has the greatest potential for asbestos release and therefore has the greatest potential to create health risks.

Asbestos fibers, in particular in friable asbestos, can cause serious health problems. If inhaled, they can cause diseases that disrupt the normal functioning of the lungs. Three specific diseases – asbestoses (a fibrous scarring of lungs), lung cancer, and mesothelioma (a cancer of the lining of the chest or abdominal cavity) – have been linked to asbestos exposure. These diseases do not develop immediately after inhalation of asbestos fibers; it may be 20 years or more before symptoms appear. In general, as with cigarette smoking and the inhalation of tobacco smoke, the more asbestos fibers a person inhales, the greater the risk of developing an asbestos-related disease.

3.6.3 Identification of Asbestos

Plumbers, building contractors, or heating contractors are often able to make a reasonable judgment about whether or not a product contains asbestos, based on a visual inspection. In some cases, the plant manager may want to have the material analyzed. Such analysis may be desirable if the industrial plant has a large area of damaged material or if the plant manager is preparing a major renovation that will expose material contained behind a wall or other barrier.

A list of 221 laboratories receiving initial accreditation to perform bulk asbestos analysis during the second quarter of 1989 has been released by the National Institute of Standards and

Technology, Gaithersburg, MD. There are two types of air sampling techniques:

1. Personal air sampling (required by OSHA) is designed to measure an individual worker's exposure to fibers while the worker is conducting tasks that may disturb ACM. The sampling device is worn by the worker and positioned so that it samples air in the worker's breathing zone.
2. Area (or ambient) air sampling is conducted to get an estimate of the numbers of airborne asbestos fibers present in a building. It is used as an assessment tool in evaluating the potential hazard posed by asbestos to all building occupants.

3.6.4 Operation and Maintenance (O&M) Program

The principal objective of an O&M program is to minimize exposure of all building occupants to asbestos fibers. To accomplish this objective, an O&M program includes work practices to (a) maintain ACM in good condition, (b) ensure proper cleanup of asbestos fibers previously released, (c) prevent further release of asbestos fibers, and (d) monitor the condition of ACM.

The methods for monitoring/correcting the condition of ACM include: (a) "surfacing ACM" (asbestos-containing material that is sprayed on or otherwise applied to surfaces, such as acoustical plaster on ceilings and fireproofing materials on structural members, or other materials on surfaces for acoustical, fireproofing, or other purposes); (b) "thermal system insulation" (TSI) (asbestos-containing material applied to pipes, fittings, boiler, breaching, tanks, ducts, or other interior structural components to prevent heat loss or gain or water condensation); and (c) "miscellaneous ACM" (interior asbestos-containing building material on structural components, structural members or fixtures, such as floor and ceiling tiles; does not include surfacing material or thermal system insulation).

The O&M program can be divided into three types of projects: (a) those that are unlikely to involve any direct contact with ACM; (b) those that may cause accidental disturbance of ACM; and (c) those that involve relatively small disturbances of ACM.

First, a person who may be the plant manager, a principal member of staff, or an outside asbestos consultant should be installed as the Asbestos Program Manager in order to establish and implement an O&M program. The appointed Asbestos Program Manager shall have overall responsibility for the asbestos control program. He/she may develop and implement the O&M program, establish training and experience requirements for contractors' workers, supervise and enforce work practices with assistance of work crew supervisors, and conduct periodic reinspections and be responsible for record keeping. This Asbestos Program Manager should be properly trained in O&M program development and implementation. An asbestos contractor may be hired to provide services for ACM abatement and for building decontamination following a fiber release episode. In addition to the above-mentioned Asbestos Program Manager, the plant manager, asbestos consultant, asbestos contractor, a communications person, a record-keeping person, a lawyer, and the federal, state, and local government advisors may also get involved in the O&M program. Secondly, a physical and visual inspection of the building is to be conducted and bulk samples of such materials are to be taken to determine if ACM is present. Then an ACM inventory can be established, and the ACM's condition and potential for disturbance can be assessed.

An official O&M program is to be developed based on the inspection and assessment data, as soon as possible if ACM is located. Either the Asbestos Program Manager or a qualified consultant should develop the O&M program. The written O&M program should state clearly the O&M policies and procedures for that building, identify and describe the administrative line of authority for that building, and should clearly define the responsibilities of key participants, such as the Asbestos Program Manager and custodial and maintenance supervisors and staff. The written O&M program should be available and understood by all participants involved in the management and operations of the building.

In general the O&M program developed for a particular building should include the following O&M program elements:

- Notification: a program to tell workers, tenants, and building occupants where ACM is located, and how and why to avoid disturbing the ACM. All persons affected should be properly informed.
- Surveillance: regular ACM surveillance to note, assess, and document any changes in the ACM's condition by trained workers or properly trained inspectors. Air monitoring to detect airborne asbestos fibers in the building may provide useful supplemental information when conducted along with a comprehensive visual and physical ACM inspection/reinspection program. Air samples are most accurately analyzed using transmission electron microscopy (TEM).
- Controls: work control/permit system to control activities that might disturb ACM. This system requires the person requesting work to submit a job request form to the Asbestos Program Manager before any work is begun.
- Work practices: O&M work practices to avoid or minimize fiber release during activities affecting ACM.
- Record keeping: to document O&M activities. OSHA and USEPA have specific requirement for workers exposed to asbestos.
- Worker protection: medical and respiratory protection programs, as applicable.
- Training: the Asbestos Program Manager, and custodial and maintenance staff training. The building owner should make sure that the O&M program developed is site-specific and tailored for the building. The O&M program should take into account use, function, and design characteristics of a particular building.

The O&M program once established shall be implemented and managed conscientiously and reviewed periodically. Alternatives on control options that may be implemented under an O&M program include: (a) repair, (b) encapsulation, (c) enclosure, (d) encasement, and (e) minor removal. The abatement actions other than O&M can also be selected when necessary. For instance, removal of ACM before renovations may be necessary in some instances.

3.6.5 O&M Training Program

Properly trained custodial and maintenance workers are critical to a successful A&M program. The following items are highlighted training requirements:

1. OSHA and USEPA require a worker training program for all employees exposed to fiber levels at or above the action level (0.1 f/cc, 30 min time-weighted average or TWA).
2. Some states and municipalities may have specific work training requirements.
3. At least three levels of maintenance worker training can be identified: (a) Level 1 Awareness training for workers involved in activities where ACM may be accidentally disturbed (may range from 2 to 8 hours); (b) Level 2 Special O&M training for maintenance workers involved in general maintenance and incidental ACM repair tasks (at least 16 hours); (c) Level 3 Abatement worker training for workers who may conduct asbestos abatement. This work involves direct, intentional contact with ACM. "Abatement worker" training courses that involve 24 to 32 hours of training fulfill this level of training.

3.6.6 General Guidelines for Handling Asbestos-Containing Materials

If the plant manager thinks that a material contains asbestos, and the material must be banned, rubbed, handled, or taken apart, he/she should hire a trained, asbestos-removal contractor before taking any risky action. In order to determine the experience and skill of a prospective asbestos-removal contractor, the contractor should be asked these questions:

1. Is the contractor certified? (Ask to see the certificate).
2. Have the contractor and the contractor's workers been trained?
3. Does the contractor have experience of removing asbestos from buildings?
4. Will the contractor provide a list of references from people for whom he/she has worked with asbestos?
5. Will the contractor provide a list of places where he/she has worked with asbestos?
6. Will the contractor use the "wet method" (water and detergent)?
7. Will the contractor use polyethylene plastic barriers to contain dust?
8. Will the contractor use a HEPA (high efficiency particulate air) filter vacuum cleaner?
9. Will the contractor's workers wear approved respirators?
10. Will the contractor properly dispose of the asbestos and leave the site free of asbestos dust and debris?
11. Will the contractor provide a written contract specifying these procedures?

The plant manager or the owner of an industrial site must make sure to hire a certified, trained, and experienced asbestos contractor who follows the following General Guidelines for Handling Products Containing Asbestos established by the U.S. Consumer Product Safety Commission and the U.S. Environmental Production Agency [22]:

1. The contractor should seal off the work area from the rest of the residence and close off the heating/air conditioning system. Plastic sheeting and duct tape may be used, which can be carefully sealed with tape when work is complete. The contractor should take great care not to track asbestos dust into other areas of the residence.
2. The work site should be clearly marked as a hazard area. Only workers wearing disposable protective clothing should have access. Household members and their pets should not enter the area until work is completed and inspected.
3. During the removal of asbestos-containing material, workers should wear approved respirators appropriate for the specific asbestos activity. Workers should also wear gloves, hats, and other protective clothing. The contractor should properly dispose of all of this equipment (along with the asbestos material) immediately after using it.
4. The contractor should wet the asbestos-containing material with a hand sprayer. The sprayer should provide a fine mist, and the material should be thoroughly dampened, but not dripping wet. Wet fibers do not float in the air as readily as dry fibers and will be easier to clean up. The contractor should add a small amount of a low sudsing dish or laundry detergent to improve the penetration of the water into the material and reduce the amount of water needed.
5. The contractor should assure that if asbestos-containing material must be drilled or cut, it is done outside or in a special containment room, with the material wetted first.
6. The contractor should assure that, if the material must be removed, it is not broken into small pieces, as asbestos fibers are more likely to be released. Pipe insulation is usually installed in preformed blocks and should be removed in complete pieces.

7. The contractor should place any material that is removed and any debris from the work in sealed, leak-proof, properly labeled, plastic bags (6 mm thick) and should dispose of them in a proper land-fill. The contractor should comply with Health Department instructions about how to dispose of asbestos-containing material.
8. The contractor should assure that after removal of the asbestos-containing material, the area is thoroughly cleaned with wet mops, wet rags, or sponges. The cleaning procedure should be repeated a second time. Wetting will help reduce the chance that the fibers are spread around. No asbestos material should be tracked into other areas. The contractor should dispose of the mop heads, rags, and sponges in the sealed plastic bags with the removed materials.
9. Plant personnel, if trained but not certified, can perform minor repairs (approximately the size of a hand), taking special precautions regarding dust, sweep, or vacuum particles suspected of containing asbestos. The fibers are so small that they cannot be seen and can pass through normal vacuum cleaner filters and get back into the air. The dust should be removed by a wet-mopping procedure or by specially designed "HEPA" vacuum cleaners used by trained asbestos contractors.

3.6.7 Environmental Regulations on ACM Mandatory Requirements

Regulations

There are several important OSHA and USEPA regulations that are designed to protect workers. They are summarized here, as guidance. OSHA has specific requirements concerning worker protection and procedures used to control ACM. These include the OSHA construction industry standard for asbestos (29 CFR1926.58), which applies to O&M work, and the general industry asbestos standard (29 CFR1910.1001). State-delegated OSHA plans, as well as local jurisdictions, may impose additional requirements.

The OSHA standards generally cover private sector workers and public sector employees in states that have an OSHA state plan. Public sector employees, or certain school employees, who are not already subject to a state OSHA plan are covered by the USEPA "Worker Protection Rule" (Federal Register: February 25, 1987; 40 CFR 763, Subpart G, Abatement Projects; Worker Protection, Final Rule).

The OSHA standards and the USEPA Worker Protection Rule require employers to address a number of items, which are triggered by exposure of employees to asbestos fibers. Exposure is discussed in terms of fibers per cubic centimeter (cc) of air. A cc is a volume approximately equivalent to that of a sugar cube.

Two main provisions of the regulations fall into the federal category of "Permissible Exposure Limits" (PELs) to airborne asbestos fibers. They are:

1. An 8 hour time-weighted average limit (TWA) of 0.2 fiber per cubic centimeter (f/cc) of air based on an 8 hour time-weighted average (TWA) sampling period. This is the maximum level of airborne asbestos, on average, that any employee may be exposed to over an 8 hour period (normal work shift).
2. Excursion limit (El): 1.0 f/cc as averaged over a sampling period of 30 minutes.

These levels trigger mandatory requirements, which include the use of respirators and protective clothing, the establishment of "regulated areas," the posting of danger signs, as well as the use of engineering controls and specific work practices [20,53].

OSHA regulations also establish an "action level": 0.1 f/cc for an 8 hour TWA. Employee training is required once an action level of 0.1 f/cc and/or the "excursion limit" is reached.

This training must include topics specified by the OSHA rules. If an employee is exposed at or above the action level for a period of 30 days or more in a calendar year, medical surveillance is required according to the OSHA construction industry asbestos standard.

Medical Examination and Medical Surveillance

OSHA also requires medical examinations under its "General Industry Standard" for any employee exposed to fiber levels in the air at or above the OSHA "action level" (0.1 f/cc) and/or the "excursion limit" (1.0 f/cc). In both cases – the action level and excursion limit – the OSHA medical examination requirement applies if the exposure occurs for at least one day per year.

Medical surveillance is defined as "a periodic comprehensive review of a worker's health status." The required elements of an acceptable medical surveillance program are listed in the OSHA standards for asbestos. According to those regulations, participation in a medical surveillance program is required for any employee who is required to wear a negative pressure, air-purifying respirator. Replacement, annual, and termination physical exams are also required for these employees. However, a termination exam is only necessary under the construction industry standard (which applies to custodial and maintenance employees) if a physician recommends it. While not mandatory, USEPA and NIOSH recommend physical examinations, including cardiac and pulmonary tests, for any employee required to wear a respirator by the building owner. These tests determine whether workers will be unduly stressed or uncomfortable when using a respirator [20].

3.6.8 Notification Requirements

USEPA or the State [if the State has been delegated authority under National Emission Standards for Hazardous Air Pollutants (NESHAP)] must be notified before a building is demolished or renovated. The following information is required on the NESHAP notice: (a) name and address of the building owner or manager; (b) description and location of the building; (c) estimate of the approximate amount of friable ACM present in the facility; (d) scheduled starting and completion dates of ACM removal; (e) nature of planned demolition or renovation and method(s) to be used; (f) procedures to be used to comply with the requirements of the regulation; and (g) name, address, and location of the disposal site where the friable asbestos waste material will be deposited.

The notification requirements do not apply if a building owner plans renovation projects that will disturb less than the NESHAP limits of 160 square feet of friable ACM on facility components or 260 linear feet of friable ACM on pipes (quantities involved over a one-year period). For renovation operations in which the amount of ACM equals or exceeds the NESHAP limits, notification is required as soon as possible.

3.6.9 Emissions Control, Waste Transportation, and Waste Disposal

The NESHAP asbestos rule prohibits visible emissions to the outside air by requiring emission control procedures and appropriate work practices during collection, packaging, transportation, or disposal of friable ACM waste. All ACM must be kept wet until sealed in a leak-tight container that includes the appropriate label. The following table provides a simplified reference for building owners regarding the key existing NESHAP requirements.

Under the expanded authority of RCRA, a few states have classified asbestos-containing waste as a hazardous waste, and require stringent handling, manifesting, and disposal procedures. In those cases, the state hazardous waste agency should be contacted before disposing of asbestos for approved disposal methods and record-keeping requirements, and for a list of approved disposal sites.

Friable asbestos is also included as a hazardous substance under USEPA's CERCLA regulations. The owner or manager of a facility (e.g., building, installation, vessel, landfill) may have some reporting requirements, for example, the U.S. Department of Transportation (USDOT) requirements for asbestos transport activities under the Hazardous Materials Transportation Act of 1975 (HMTA). The HMTA regulatory program applies to anyone who transports hazardous materials, or arranges for their transportation or shipment, and to anyone who manufactures, reconditions, repairs, tests, or marks packages or containers for use in the transportation of hazardous materials [49 USC Sec. 1804(a)].

USDOT has designated asbestos as a hazardous material for the purposes of transportation, and has issued requirements for shipping papers, packaging, marking, labeling, and transport vehicles applicable to shipment and transportation of asbestos materials (49 CFR 173.101). Commercial asbestos must be transported in rigid, leak-tight packages: in bags or other nonrigid packaging in close freight containers, motor vehicles or rail cars loaded by the consignor and unloaded by the consignee exclusively, or bags or other nonrigid packages that are dust- and sift-proof in strong fiberboard or wooden boxes (49 CFR 173.1090).

Specific regulations exist for the transport of asbestos materials by highway [53]. Asbestos must be loaded, handled, and unloaded using procedures that minimize occupational exposure to airborne asbestos particles released in association with transportation. Any asbestos contamination of transport vehicles also must be removed using such procedures (49 CFR 177.844). Additional motor carrier's safely regulations apply to common, contract, and private carriers of property by motor vehicle, as defined under these regulations (49 CFR Parts 390–397).

3.7 MONITORING AND ANALYSIS OF AIR, WATER, AND CONTAMINATED MATERIALS

3.7.1 General Approach

Because airborne and volatile contaminants can present a significant threat to industrial workers' health and safety, identification and quantification of these airborne and volatile contaminants through air/soil monitoring is an essential component of a health and safety program at an industrial site having hazardous substances. The purpose of air and soil monitoring is to identify and quantify airborne and volatile hazardous contaminants in order to determine the level of plant worker's protection needed.

In general, there are two principal approaches available for identifying and/or quantifying airborne contaminants as well as volatile contaminants in soil:

1. The first approach: onsite use of direct-reading instruments as initial qualitative identification or screening (note: the airborne/volatile contaminant, or the class to which it belongs, is demonstrated to be present but quantitative determination of its exact concentration must await subsequent testing); and
2. The second approach: laboratory analysis of air and/or soil samples (note: the air sample can be obtained by gas sampling bag, filter, sorbent, and wet-contaminant collection methods).

Care must be taken in sampling of contaminated air, soil, water, or materials in order to obtain representative samples, and, in turn, to gain meaningful results. In general, the onsite use of direct-reading instruments for qualitative analysis and the onsite sampling of contaminated air, soil, water, or materials are performed by a licenced engineer, a licenced geologist, or a certified technician. The subsequent quantitative laboratory analysis, if required, can be performed by either a certified laboratory or a licenced engineering firm, depending on the environmental quality parameters.

For instance, air samples and the building material samples contaminated by formaldehyde and lead are routinely sampled by an engineering technician under the supervision of a licenced engineer. The samples are shipped to a certified laboratory for quantitative analysis by the licenced engineer.

In another common case, soil that may be contaminated by volatile gasoline is routinely qualitatively tested with a direct-reading instrument and sampled by an engineer/scientist under the supervision of either a licenced engineer or geologist. The contaminated soil is qualitatively identified and/or documented and shipped by the engineer/scientist, quantitatively analyzed by a certified laboratory, and its quantitative data interpreted by the licenced engineer/geologist.

In New York and Massachusetts where PCB contamination is always a possibility, the laboratory tests required by the state environmental protection agencies for analysis of a petroleum-contaminated soil are as follows: (a) flash point; (b) total petroleum hydrocarbon (TPH); (c) PCB screening; (d) total organic halides (TOH); (e) reactivity of cyanide and sulfide; (f) BTEX or equivalent; (g) eight metals under TCLP (Toxicity Characteristics Leaching Procedure) for USTs; and (h) full range of tests under TCLP for ASTs and spills.

In still another case, airborne asbestos is frequently qualitatively identified and/or sampled by either a licenced engineer or a certified asbestos contractors, and quantitatively analyzed by a certified laboratory. The building material, such as the insulation for the plumbing system, however, can only be removed by a State-certified asbestos contractor. The readers are referred to Section 3.6.3 for air sampling and identification of asbestos-containing materials.

A continuous contaminant source monitor can provide both industrial plants and regulatory agencies with numerous benefits. A properly installed and operated continuous monitoring system can yield a large amount of data on source air emissions or source effluent discharges. This information is beneficial, because it establishes a reliable foundation upon which important decisions can be made.

3.7.2 Measuring Instruments

Reliable measurements of airborne volatile or hazardous substances in the field using onsite instruments are useful for: (a) selecting personal protective equipment at an industrial site; (b) delineating areas where protection is needed; (c) assessing the potential health effects of hazardous exposure; (d) determining the need for specific medical monitoring; and (e) providing an early warning for personnel evacuation due to contamination, when necessary.

The National Pollutant Discharge Elimination System (NPDES) reporting requirements for effluent testing allow alternate methods of analysis to be substituted for the prescribed methods if prior approval has been obtained from the U.S. Environmental Protection Agency (USEPA) regional administrator having jurisdiction where the discharge occurs.

Steps an individual permit holder must take to use an alternate test procedure for regulatory reporting of specific discharges follow. An alternate test procedure differs from those published in the Federal Register for NPDES-certification purposes (Source: Federal Register, Title 40, Chapter 1, Subchapter D, Part 136: Vol. 38, No. 199, Oct. 16, 1973; Vol. 41, No. 232, Dec. 1, 1976). Many Hach methods (Hach Company, Loveland, CO, USA) are identical to these published methods and thus are approved by USEPA and highly recommended by the authors for rapid field testing of effluent samples.

Direct-reading instruments have been developed as early warning devices for use at various industrial sites, where a leak or an accident could release a high concentration or high dose of a known chemical or known radiation into the environment. They provide information on flammable or explosive atmospheres, oxygen deficiency, certain gases and vapors, or ionizing radiation, at the time of measuring, enabling rapid decision making by the plant managers. Direct-reading instruments, which can be either batch monitoring systems or continuous monitoring systems, are the primary tools of initial site characterization. The readers are referred to Chapter 1 entitled "Onsite Monitoring and Analyses of Industrial Pollutants" for more information on several common direct-reading field instruments and their conditions and/or hazardous substances they measure.

As a minimum, the flame ionization detector (FID) or the photo-ionization detector (PID) must be available at industrial sites handling hazardous substances.

3.8 HAZARDOUS WASTE GENERATOR STATUS AND REGULATORY REQUIREMENTS

3.8.1 Hazardous Waste Generators

Regulations

In general, two activities determine the generator category of an industrial plant: the rate at which the plant generates and how much the plant stores (accumulates). Under new, more flexible regulations, the amount and length of time an industrial plant can accumulate wastes may vary according to the type of waste. In the State of Massachusetts, there are three generator statuses, which are introduced below as a typical example.

1. Large Quantity Generator (LQG): generates more than 1000 kg (2200 lb) of hazardous waste in a month; once the first 1000 kg has been accumulated, the waste must be shipped within 90 days; there is no limit to the amount that can be accumulated.
2. Small Quantity Generator (SQG): generates less than 1000 kg of hazardous waste in a month, and/or less than 1 kg of acutely hazardous waste (acutely hazardous waste is listed in the State regulations).
3. Very Small Quantity Generator (VSQG): generates less than 100 kg of hazardous waste in a month, and generates no acutely hazardous waste.

Other State governments in the United States have similar regulatory requirements. The maximum monthly volume of waste oil and maximum monthly volume of all other hazardous waste generated at an industrial plant site can be estimated and regulated according to the State of Massachusetts "Guide to Determining Status and Regulatory Requirements" (Table 2).

An Example in Massachusetts

An industrial plant in Massachusetts generates 60 gallons of spent solvent and 550 gallons (2081.75 L) of waste oil in a month. According to the Guide (Table 2), the plant is a Small Quantity Generator (SQG) of hazardous waste because it produces more than 100 kg but less than 1000 kg, and the plant is also a Large Quantity Generator (LQG) of waste oil because the plant produces more than 1000 kg. The plant's regulatory status is found in Table 2, under line 5 (SQG for HW; LQG for WO).

Reading across the columns, on line 5, the plant may accumulate its solvent for as long as 180 days, or until the plant has reached a volume of 2000 kg (500 gallons; 1892.5 L) in

Table 2 Guide to Determining Status and Regulatory Requirements for Hazardous Waste Management

Regulatory status of co.

Hazardous waste (HW)[c]	Waste oil (WO)[c]	Accumulation time, HW (days)	Accumulation HW volume in tanks (kg)	Accumulation HW volume in containers (kg)	Manifest usage requirement	Permission for self-transport HW
LQG	LQG	90	No limit	No limit	Yes	No
LQG	SQG	90	No limit	No limit	Yes	No
LQG	VSQG	90	No limit	No limit	Yes[a]	No
LQG	None	90	No limit	No limit	Yes	No
SQG	LQG	180	6000[b]	2000	Yes	No
SQG	SQG	180	6000[b]	2000	Yes	No
SQG	VSQG	180	6000[b]	2000	Yes[a]	No
SQG	None	180	6000[b]	2000	Yes	No
VSQG	LQG	No limit	600	600	Yes[a]	Yes
None	LQG	N/A	N/A	N/A	Yes	No
VSQG	SQG	No limit	600	600	Yes[a]	Yes
VSQG	VSQG	No limit	600	600	Yes[a]	Yes
VSQG	None	No limit	600	600	Yes[a]	Yes
None	SQG	N/A	N/A	N/A	Yes	No
None	VSQG	N/A	N/A	N/A	Yes[a]	Yes

Regulatory status of co.

Hazardous waste (HW)[c]	Waste oil (WO)[c]	Accumulation time, WO (days)	Accumulation WO volume in tanks (kg)	Accumulation WO volume in containers (kg)	Manifest usage requirement	Permission for self-transport WO
LQG	LQG	90	No limit	No limit	Yes	No
LQG	SQG	180	6000[b]	2000	Yes	No
LQG	VSQG	No limit	600	600	Yes[a]	Yes
LQG	None	N/A	N/A	N/A	Yes	No
SQG	LQG	90	No limit	No limit	Yes	No
SQG	SQG	180	6000[b]	2000	Yes	No
SQG	VSQG	No limit	600	600	Yes[a]	Yes
SQG	None	N/A	N/A	N/A	Yes	No
VSQG	LQG	90	No limit	No limit	Yes[a]	No
None	LQG	90	No limit	No limit	Yes	No
VSQG	SQG	180	6000[b]	2000	Yes[a]	No
VSQG	VSQG	No limit	600	600	Yes[a]	Yes
VSQG	None	N/A	N/A	N/A	Yes[a]	Yes
None	SQG	180	6000[b]	2000	Yes	No
None	VSQG	No limit	600	600	Yes[a]	Yes

Note: This matrix guide does not reflect **acutely** hazardous wastes.
[a]A manifest must be used for the VSQG category unless self-transported.
[b]When accumulating in both tanks and containers, the total accumulation cannot exceed 6000 kg and the container accumulation cannot exceed 2000 kg.
[c]LQG = 1000 or more kg per month of waste generation; SQG = 100–999 kg per month of waste generation; VSQG = less than 100 kg per month of waste generation.

containers (Table 2), whichever happens first (column 3). The plant must ship its waste oil every 90 days regardless of the volume. The plant manager must obtain an USEPA Identification Number and use a manifest for both wastes. The plant manager must manage his/her waste according to the accumulation area standards and must fulfill the emergency preparation and response requirements listed in subsequent sections. The plant manager, however, is not required to file an annual report or a contingency plan or provide full personnel training, which is necessary for larger generators.

3.8.2 Hazardous Waste and Waste Oil Documentation Using a Manifest

As a generator, an industrial plant always retains responsibility for hazardous waste. If the plant's waste is dumped or disposed of improperly, the plant manager and the owner will be held responsible. It is therefore important that the plant manager or the owner knows where the plant's waste is going and whether or not it is handled properly and safely [73].

U.S. Federal law (the Recourse Conservation and Recovery Act of 1976, known as RCRA) requires a national "cradle to grave" tracking system for hazardous waste. In the State of Massachusetts, for instance, every shipment of hazardous waste by a large or small generator must be transported by a licenced hauler and sent to a licenced treatment, storage, or disposal facility (TSD) or a permitted recycling facility, and it must be accompanied by a multipart shipping document, called the Uniform Hazardous Waste Manifest.

In the State of Massachusetts, the plant manager or a designated consulting engineer must use the Massachusetts Manifest form unless the plant is sending its waste to a facility out of state, in which case the plant manager should contact the other state to find out which form to use. The plant manager or the plant's consulting engineer will be responsible for completing the generator portion of the manifest. Directions for the distribution of the copies are printed on the manifest. A copy will be returned to the industrial plant when the disposal facility or the recycling facility has accepted its shipment.

If the industrial plant's manager or consulting engineer does not receive a copy of the manifest from the receiving facility (i.e., the disposal facility and/or the recycling facility) within 35 days of the date when the plant's waste was shipped, the transporter or the operator of the facility must be contacted to determine the status of the waste. If the plant has still not received the manifest within 45 days, an Exception Report, explaining the efforts the plant has taken, must be filed with the State's Division of Hazardous Waste and with the State where the designated facility is located.

For all generators, copies of all manifests and any records of tests and analyses carried out on the hazardous waste must be kept for at least three years, and for the duration of any enforcement action.

The most common problems in completing the manifest are clerical. For clarity, because this is a multiple carbonless copy form of about eight pages, typing is strongly recommended. The generator should check for legibility of all copies before transferring the manifest to the transporter at the time of shipment. The generator must ensure that all information is complete and accurate by reviewing the following summary when completing the manifest.

1. The plant's federal Identification (ID) Number must be correctly stated. The plant's specific location must have an ID number to use the manifest.
2. The identification number of the transporter and the receiving facility and their valid hazardous waste licenses must be double checked with the State's regulatory agency.
3. If there is a second transporter, the generator has the responsibility to select this second transporter and both the generator and the second transporter must complete certain portions of the manifest.
4. The generator shall have a program to reduce the volume and toxicity of waste generated, which is a national requirement of all generators and is intended to encourage good management practise. Large quantity generators are required to report how they are reducing waste in their annual report.

5. The contents of the shipment must be fully and accurately described, packed, marked, and labeled.
6. Any special handling instructions must be clearly given. The generator can list an alternative receiving facility and must list, in the case of an international shipment, the city and state at which the shipment leaves the United States.
7. If more than four wastes are included in a single shipment, a second prenumbered manifest must be used. When more than two transporters are used for one shipment, the State requirements must be reviewed. In the State of Massachusetts an eight-part Massachusetts Continuation Sheet, numbered to match the first manifest, should be used.
8. Instructions regarding the use and distribution of the manifest copies that are stated on the manifest must be reviewed. The generator retains certain copies at the time of shipment. One copy should be mailed to the manifest office of the State in which the destination facility is located. One copy is returned to the generator by the receiving facility when the shipment arrives. The generator copies must be kept in the file for at least three years. If a signed manifest copy from the destination facility is not received within 35 days, the generator must investigate and file an Exception Report with the State Enforcement Section within 45 days of shipment if the signed copy has still not been received.

When a small or a very small quantity generator is to ship only waste oil or a very small quantity generator is to spill other waste, a transporter's log instead of a manifest may be used for that shipment. However, the generator must register on a prescribed form with the State of Massachusetts.

3.8.3 The USEPA Identification Number (USEPA-ID)

In order to have an industrial hazardous waste accepted by a licenced hauler or treatment/storage facility, the industrial plant (i.e., the generator) must be assigned a number, with a special prefix for the plant location. This number will be entered on each manifest.

In order to get a USEPA-ID, the plant manager shall call or contact the State government for an application for a USEPA Identification Number. The completed application should be mailed to the state office listed in the instruction. While a plant is waiting for a permanent USEPA-ID number, the plant can obtain a temporary USEPA-ID number over the telephone.

The USEPA-ID number is site-specific. The State Division of Hazardous Waste must be notified in writing, or on a specified form, of any change in the generator's address, contact person, or generator status.

3.8.4 Shipping Hazardous Waste

All hazardous waste must be transported in containers [24,54] that are labeled with the words HAZARDOUS WASTE, the name of the waste, type of hazard (e.g., toxic, flammable), and generator's name, address, and USEPA-ID number.

A list of licenced transporters and facilities for treatment, storage, or disposal is always available from the State government. Many transporters are authorized to assist the plant manager in preparing the plant's hazardous waste for shipment.

A summary of recommended procedures for shipping hazardous wastes from an industrial plant to another location is now given below:

1. Select a licenced transporter and a hazardous waste facility that will receive the plant's waste;

2. Identify the waste based on a licenced engineer's testing or a certified laboratory testing prior to shipping the waste;
3. Obtain a federal identification (USEPA-ID) number by requesting a required form (such as Notification of Hazardous Waste Activity Form) from a State regulatory agency (note: the identification number is specific to the location, not the hazardous waste);
4. Obtain a manifest for a shipment of waste destined for disposal in a State (note: this specific State's manifest form with a preprinted State document number is required); and
5. Ship the plant's waste in accordance with federal transportation regulations (CFR Title 49, Part 100–177).

3.8.5 Hazardous Waste Storage Standards for an Accumulation Area

The accumulation or storage area of an industrial plant (i.e., a generator) must meet the following conditions for both containers and tanks in accordance with the home State regulations. The Massachusetts hazardous waste regulations (310 CMR 30.000) are listed below as a reference:

1. Above-ground tanks and containers must be on a surface that does not have any cracks or gaps and is impervious to the hazardous wastes being stored;
2. The area must be secured against unauthorized entry;
3. The area must be clearly marked (e.g., by a visible line or tape, or by a fence) and be separate from any points of generation;
4. The area must be posted with a sign "HAZARDOUS WASTE" in capital letters at least one inch high (1 in. = 2.54 cm);
5. An outdoor area must have secondary containment, such as a dike, which will hold any spill or leaks at (a) 10% of the total volume of the containers, or (b) 110% of the volume of the largest container, whichever is larger; and
6. Any spillage must be promptly removed: in general, if the hazardous waste being stored has no free liquids, no pad is required, provided that the accumulation area is sloped, or the containers are elevated.

3.8.6 Standards for Waste Containers and Tanks

General Massachusetts standards (310 CMR 30.680–30.690) for waste containers and tanks in accordance with the same Massachussetts hazardous waste regulations (310 CMR 30.000) are given below as a reference:

1. Each container and tank must be clearly and visibly labeled throughout the period of accumulation with the following:
   (a) the words "HAZARDOUS WASTE,"
   (b) the name of the waste (e.g., waste oil, acetone),
   (c) the type of hazard(s) (e.g., ignitable, toxic, dangerous when wet, corrosive), and
   (d) the date on which the accumulation begins;
2. Each container must be in good condition;
3. Wastes of different types must be segregated; for example, this includes not mixing waste oil or used fuel oil with other wastes; be careful not to put incompatible wastes in the same container or put wastes in unwashed containers that previously stored incompatible wastes;
4. Separate containers of incompatible wastes by a dike or similar structure;

5. Each container holding hazardous wastes must be tightly closed throughout the period of accumulation, except when the waste is being added or removed;
6. Containers holding ignitable or reactive wastes must be at least 15 m (50 ft) away from the property line; if this is not possible or practical, the plant manager representing the generator must store such containers in compliance with all applicable local ordinances and bylaws; and
7. Inspect the accumulation area at least once a week for any leaking or deterioration of all containers; there must be enough aisle space between the containers to allow for inspections.

3.8.7 Criteria for Accumulation Time Limits

If an industrial plant is classified as a small quantity generator (SQG), the plant manager may accumulate up to 2000 kg or 4400 lb in containers, or up to 6000 kg (approximately 1650 gal or 6245 L) in tanks for as long as 180 days according to Massachusetts regulations 310 CMR 30.351. If both tanks and containers are used to store hazardous waste and/or waste oil, the total waste that can be accumulated at any one time may not be determined by adding the two limits. The 180 day clock may be started when a total of 100 kg, (approximately 25 gal or 94.63 L) is accumulated, if the containers are redated at that time.

3.8.8 Criteria for Satellite Accumulation

Additional flexibility is offered by allowing an industrial plant to accumulate up to 55 gal (or 208.18 L) of hazardous waste, or one quart (or 1 L) of acutely hazardous waste, at each point where the plant generates its waste if the plant meets the following conditions:

1. The waste must be generated from a process at the location of the satellite accumulation;
2. Each satellite accumulation area can have only one container for each waste stream in use at a time;
3. Each satellite accumulation area must be managed by a person who is directly responsible for the process producing the waste; and
4. The waste must be moved to the main designated accumulation area within three days after the container is full.

3.8.9 Criteria for Accumulation of Waste Oil in Underground Storage Tanks

The Massachusetts criteria (310 CMR 30.690) for accumulating waste oil in underground storage tanks (USTs), including those resting directly on the ground, are generalized below:

1. For leak detection in old tanks containing waste oil that were installed before October 15, 1983 under a grandfather clause, a dipstick test must be conducted every 30 days; a more than $\frac{1}{2}$ in. (1.27 cm) difference in level within a 24 hour period must be reported to the State government; underground tanks containing other hazardous wastes must undergo a tightness test, and must be monitored on a daily basis;
2. Tanks installed after the effective date (October 15, 1983) of a new Massachusetts law regarding underground storage tanks must have secondary containment and a monitoring system or be constructed of a corrosion-resistant material; and
3. A log must be kept of all test results for at least three years.

3.9 STORAGE TANK INSPECTION AND LEAK DETECTION

3.9.1 Requirements for Underground Storage Tanks

The State of New York [11–14,55] has promulgated rules and regulations for the early detection of leaks or potential leaks of petroleum bulk storage by plant owners and operators. In the State of New York [14], underground tanks shall be checked for leakage using one or more of the following:

1. Inventory monitoring may be used if it detects a leak of one percent (1%) of flowthrough plus 130 gal on a monthly basis and is coupled with an annual tightness test. Inventory monitoring must be done.
2. Weekly monitoring of the interstitial space of a double-walled tank may be practiced using pressure monitoring, vacuum monitoring, electronic monitoring, or manual sampling.
3. Vapor wells for monitoring soils in the excavation zone may be used. Vapor monitoring systems must be designed and installed by a qualified engineer or technician in accordance with generally accepted practices. Wells must be protected from traffic, permanently labeled as a "monitoring well" or "test well – no fill" and equipped with a locking cap, which must be locked when not in use so as to prevent unauthorized access and tampering. Vapor monitoring may be used only under the following conditions: (a) soils in the excavation zone must be sufficiently porous to allow for the movement of the vapors from the tank to the vapor sensor; gravel, coarse and crushed rocks are examples of porous soils; (b) the stored substance or a tracer compound placed in the tank must be sufficiently volatile so as to be detectable by the vapor sensor; (c) vapor monitoring must not be hindered by groundwater, rainfall, or soil moisture such that a release could go undetected for more than 30 days; (d) background contamination must not mask or interfere with the detection of a release; (e) the system must be designed and operated to detect increases in vapors above background levels; monitoring must be carried out at least once per week; and (f) the number and positioning of vapor monitoring wells must be sufficient to ensure detection of releases from any portion of the tank and must be based on a scientific study; wells must be at least four inches in diameter.
4. Groundwater monitoring wells designed and installed by a qualified engineer or technician may be used. Wells must be protected from traffic, permanently labeled as a "monitoring well" or "test well – no fill," and equipped with a locking cap that must be locked when not in use to prevent unauthorized access and tampering. Groundwater monitoring may be used only under the following conditions: (a) the substance stored must be immiscible in water and have a specific gravity of less than one; (b) the groundwater table must be less than 20 ft from the ground surface; the hydraulic conductivity of the soil between the tank and well must not be less than one hundredth (0.01) cm/s; gravel and coarse to medium sand are examples of such soil; (c) the slotted portion of the well casing must be designed to prevent migration of soil into the well and must allow entry of the hazardous substances into the well under both high and low groundwater conditions; (d) wells must be at least four inches in diameter and be sealed from the ground surface to the top of the filter pack to prevent surface water from entering the well; (e) wells must be located within the excavation zone or as close to it as technically feasible; (f) the method of monitoring must be able to detect at least one-eighth ($\frac{1}{8}$) of an inch of free product on top of the groundwater; monitoring must be carried out once per week; and (g) the number and positioning of

the groundwater monitoring well(s) must be sufficient to ensure detection of releases from any portion of the tank and must be based on a scientific study.

5. Automatic tank gauging equipment may be used if it can detect a leak of two-tenths (0.2) of a gallon per hour or larger with a probability of detection of 95% and probability of false alarm of 5% or less. Monitoring must be carried out once per week; or
6. Other equivalent methods as approved by the Department if the method can detect a leak of two-tenths (0.2) of a gallon per hour with a 95% probability of detection and probability of false alarm of 5%.

In the State of New York, underground and on-ground piping shall also be checked for leakage by the owner or the plant manager according to the general guidelines established by the Department of Environmental Conservation [14].

3.9.2 Requirements for Aboveground Storage Tanks

While leak detection is not emphasized for aboveground storage tanks (ASTs), daily inspections, monthly inspections, annual inspections, and five-year inspections are legally required by the State of New York for AST owners or operators [14].

Daily Inspection

The owner or operator must visually inspect the aboveground storage equipment for spills and leaks each operating day. In addition, the owner or operator must check to ensure that drain valves are closed if not in use and there are no unpermitted discharges of contaminated water or hazardous substances.

Monthly Inspections

The owner or operator must conduct comprehensive monthly inspections of aboveground storage equipment. This inspection includes: (a) identifying cracks, area of wear, corrosion, poor maintenance and operating practises, excessive settlement of structures, separation or swelling of tank insulation, malfunctioning equipment, safety interlocks, safety trips, automatic shutoffs, leak detection, and monitoring, warning, or gauging equipment that may not be operating properly; (b) visually inspecting dikes and other secondary containment systems for erosion, cracks, evidence of releases, excessive settlement, and structural weakness; (c) checking on the adequacy of exterior coatings, corrosion protection systems, exterior welds and rivets, foundations, spill control equipment, emergency response equipment, and fire extinguishing equipment; (d) visual checking of equipment, structure, and foundations for excessive wear or damage; (e) reviewing the State compliance; and (f) performing monthly release detection, which meets the performance standards established by the State.

Annual Inspections

The structure-to-electrolyte potential of corrosion protection systems used to protect aboveground tank bottoms and connecting underground pipes must be inspected annually.

Five-Year Inspections

The owner or operator must inspect aboveground piping systems and all aboveground tanks; the inspection must be consistent with a consensus code, standard, or practice and be developed by a nationally recognized association or independent testing laboratory and meet the specifications of this subdivision; based on the inspection, an assessment and evaluation must be made of system tightness, structural soundness, corrosion, wear and operability; reinspection is required no later than every five years from the date of the initial inspection or regulatory deadline, whichever occurs first, except as follows. If thinning of 1 mL per year or greater occurs on the pipe or tank walls, or the expected remaining useful life as determined by the above inspections is less than ten years, then reinspection must be performed on the tank or pipe at one-half of the remaining useful life.

3.9.3 Tank and Pipeline Leak Tests

Tracer tank and pipeline leak tests developed by Tracer Research Corporation do not require that tanks or pipelines be taken out of service during any testing procedures.

The leak tests have demonstrated the capability for unambiguously detecting, quantifying, and locating leaks as small as 0.05 gal/hour in underground and aboveground storage tanks and pipelines. Storage tanks containing fuels, lubricants, heating oils, solvents, wastewater, volatile or nonvolatile chemicals, and hazardous wastes are easily tested regardless of size or type.

Leak Testing for Underground Storage Tanks

This section introduces a five-step procedure developed by Tracer Research Corporation for conducting the leak testing for an underground storage tank.

1. Step 1. Leak testing is performed by adding a small amount of a special volatile chemical tracer to the contents of a tank or pipeline; these chemicals are selected for their compatibility with tank and pipeline systems, as well as the lack of their presence in the environment around the tank; the tracer is added at a concentration of only a few PPM, and thus has no impact on the physical properties of tank and pipeline contents.
2. Step 2. Tracer mixes evenly in tanks, pipelines, and the vapor space inside a tank, by diffusion and product use.
3. Step 3. If a tank or pipeline leaks, the tracer is released into the surrounding soil where it rapidly volatizes; after the tracer has had time to disperse and migrate through the soil away from the leak (usually about two weeks), soil gas samples are collected from the probes surrounding the tanks and pipelines.
4. Step 4. Samples are analyzed for tracer and hydrocarbon vapors by means of a very sensitive gas chromatograph; the presence of tracer vapors, which can be detected in the low parts-per-trillion, provides unambiguous information about the occurrence of leakage and its location.
5. Step 5. Because information about site contamination is important, the plant manager is provided with a hydrocarbon site survey at the same time; hydrocarbon vapor maps serve to show the magnitude of leakage and the extent of the contamination if leaks are detected; if no leaks are detected, the absence of hydrocarbons confirms this finding.

Leak Testing for Pipelines

The tracer pipeline leak testing, which is similar to the tracer tank leak testing, is effective for locating leaks in all types of pipeline installations, including pipe buried under pavement, airline runways, buildings, or underwater. Where leaks are known to exist, the tracer leak test is effective in determining their location without expensive excavation.

The testing method can be retrofitted to existing underground piping. Where the pipeline runs under soil cover, a special leak detection hose that is permeable to the tracer is buried approximately 0.61 m (2 ft) deep in a ditch running above the pipeline. One sample from the hose can provide monitoring coverage of up to 152.4 m (500 ft) of pipeline. At new installations, the leak detection hose is installed adjacent to the pipe at the time of burial. This installation is very low cost and provides unique sensitivity. When a pipeline runs under concrete or pavement, it is monitored by a series of probes placed 7.62 m (25 ft) apart, installed through the pavement.

Leak Testing for Aboveground Storage Tanks

Aboveground tank testing is performed by inserting vapor sampling probes under the tank bottom. To ensure detection of leakage from any point on the tank floor, evacuation probes are placed under the perimeter of the tank and one or more air injection probes are placed beneath the center of the tank. A program of air injection and/or evacuation is initiated to collect samples from under the tank. These samples are analyzed for the presence of tracer.

In the case of a facility that has multiple tanks in close proximity to each other, different tracer compounds can be used so that sample analysis will rapidly identify a specific leaking tank.

Leak Testing for Tank Farms

The tracer leak tests are also economical means for testing aboveground storage tanks at large tank installations, such as jet fuel systems at military bases, large airport hydrant fuel systems, terminals, and refineries. Important benefits result from the fact that the testing is implemented by placing tracer in the receiving tanks where incoming product is stored. The product is released to other parts of the system and the same tracer is used to test all the portions of the system that contain or transport the product.

3.10 EMERGENCY PREPARATION AND RESPONSE

3.10.1 Emergency Equipment

To minimize the risk of fire, explosion, or release of hazardous wastes that may contaminate the environment, an industrial plant classified as a generator is required to have the following on site, and immediately accessible to its hazardous waste handling area:

- an alarm or communication system that can provide emergency instruction to employees;
- a telephone, two-way radio, or other device that can summon police, fire, or emergency response teams;
- portable fire extinguishers and/or fire control equipment (e.g., foam, inert gas), spill control equipment, and decontamination equipment; and
- adequate supply and pressure of water, automatic sprinklers or water sprays, or foam-producing equipment.

All equipments identified above are required unless the hazards posed by the plant's wastes do not require one of them. In such a case, an approval from the regulatory agency is required. The equipment, when provided, must be periodically tested and properly maintained so it will work during an emergency.

3.10.2 Emergency Preparation

An industrial plant classified as a generator must thoroughly familiarize each of its employees with all the waste handling and emergency procedures that may be needed for each of their jobs. An employee must have immediate access to alarm or communication devices, either directly or through another employee, whenever hazardous waste is being handled. If the plant's operation is at any time being handled by a single employee, that person must have immediate access to a telephone or two-way radio.

For easy movement of employees and emergency equipment, the plant manager must mark all exits clearly and maintain adequate aisle space in the area of hazardous waste handling.

3.10.3 Liaison with Local Authorities

A generator and its designated consulting engineer must make every reasonable attempt to carry out the following arrangements, with regard to the waste produced by the generator:

- Familiarizing the plant's local police department, fire department, local boards of health, and any emergency response teams with the hazardous nature of the plant's waste; the layout of the plant site, including entrances and evacuation routes, and the location where the plant's employees usually work;
- Familiarizing local hospitals with the hazards of the plant's waste and the types of injuries that could result from any accidents;
- Obtaining agreements with emergency response teams and contractors, and local boards of health; and
- Making an agreement with the regulatory agency and service agency that will have primary emergency authority, and specifying others as support, if more than one police and/or fire department might respond to an emergency.

If such arrangements cannot be made, a copy of a signed and dated letter from the plant, the generator, to the State or local entity, which demonstrates an effort to make these arrangements, must be considered sufficient, if an approval from the State or local entity can be obtained.

3.10.4 Emergency Coordinator

The industrial plant, the generator, must designate at least one employee to be on call (or on the premises) at all times. This person is the emergency coordinator and is responsible for coordinating all emergency response measures. Alternatively, a licenced consulting engineer can also be retained by the generator to be its emergency coordinator.

3.10.5 Emergency Response

It is generally required by the State regulations that the generator have posted next to each telephone near the plant's waste generation area the following:

- Name(s) and telephone number(s) of the plant's emergency coordinator(s);
- Location(s) of the fire control equipment and any fire alarms;
- Telephone numbers of the National Response Center, the fire department, the police department, and the ambulance department, or if there is a direct alarm system, instructions on how to use it; and
- Evacuation routes, where applicable.

If any of the following emergencies occur, the plant manager or the assigned emergency coordinator should immediately perform the following:

- Fire. Attempt to extinguish the fire and/or calling the fire department;
- Hazardous chemical/oil spill or leak. Contain the flow as quickly as possible and as soon as possible clean up the waste and any soil or other materials that may have become contaminated with waste;
- A hazardous chemical/oil release (spill or leak) or threat of release, fire or explosion of hazardous waste that may threaten human health or the environment. (a) Call the appropriate State environmental protection agency's regional office, or (b) Call the State police if the incident occurs after 5 p.m., or on a day that the State environmental protection agency is closed, and (c) Call the National Response Center, which usually has a 24-hour toll-free number.

3.11 MANAGEMENT OF AN INDUSTRIAL SITE CLASSIFIED AS A VERY SMALL QUANTITY GENERATOR

3.11.1 Registration

If an industrial plant in Massachusetts generates less than 100 kg a month of hazardous waste, and no acutely hazardous waster, the plant is eligible to register as a very small quantity generator. To qualify as a very small quantity generator (VSQG), the plant manager must register a waste management plan with the appropriate State environmental protection agency. If the plant does not register as a VSQG, it will be subject to the more stringent SQG regulations.

3.11.2 Treatment/Disposal Options

As a registered VSQG, an industrial plant has the following options for handling the waste:

1. The plant may recycle or treat its waste, provided the process described in the plant's registration is acceptable to the appropriate State environmental protection agency;
2. The plant may transport its waste to another generator who is in compliance with the regulations and who will count the plant's waste as part of their generation; or
3. The plant may transport its waste in the plant's own vehicle to a licenced treatment, storage, or disposal facility, or permitted recycling facility, or use a licenced transporter and a manifest form, which requires a USEPA-ID number; or
4. The plant may use a licenced transporter and a manifest form, which requires a USEPA-ID number.

3.11.3 Self-Transport Option

As a registered VSQG, an industrial plant may transport its own hazardous waste under certain conditions in accordance with the appropriate State regulations. The following are the Massachusetts regulations (310 CMR 30.353), which are presented as a typical example:

1. The plant transports only the waste that the plant generated on its premises.
2. The plant does not transport more than 200 kg at one time.
3. The plant's waste must be in containers that are:
   (a) no larger than 55 gal or 208.18 L in volume;
   (b) compatible with the waste;

(c) tightly sealed;
(d) labeled as "HAZARDOUS WASTE";
(e) labeled with the name of the waste and the type of hazard (i.e., ignitable); and
(f) tightly secured to the vehicle.

4. The plant does not transport incompatible wastes in the same shipment.
5. In the event of a spill or leak of hazardous waste that may threaten human health or the environment, the plant or its designated consulting engineer should notify the appropriate State environmental protection agency, the State police, the local fire department, and the National Response Center, as described previously.
6. The plant must have a copy of its registration with the State in the vehicle.
7. The plant must be in compliance with the Federal Department of Transportation and State Department of Public Safety requirements, if any.

3.11.4 Record-Keeping

If an industrial plant in Massachusetts, for instance, is not using a licenced transporter but is transporting its own wastes, this plant does not need a USEPA-ID number or manifest form. The plant must, however, keep a record of the type and quantity, as well as the date, method of transport, and treatment/disposal of its waste(s). The plant manager needs proof of the receipt of the waste by the facility and/or generator.

All generators must keep receipts or manifests of waste shipped, and records of waste analysis for at least three years, or for the duration of any enforcement action by the appropriate State environmental protection agency.

3.11.5 Accumulation Limits

The plant as a very small quantity generator (VSQG) in Massachusetts may accumulate up to 600 kg (approximately 165 gal or three 55 gal drums) of hazardous waste in containers that meet the standards introduced previously, with no time limit.

3.12 MANAGEMENT OF AN INDUSTRIAL SITE CLASSIFIED AS A LARGE QUANTITY GENERATOR OR A SMALL QUANTITY GENERATOR

3.12.1 Registration

The amount and length of time a large industrial plant accumulates its wastes may vary according to the type of waste. The Massachusetts Guide to Determining Status and Regulatory Requirements (Table 2) or equivalent should be used as a guide to determine the plant's generator category (Regulatory Status) for hazardous waste and waste oil [6].

For example, a plant in Massachusetts must be registered as a Large Quantity Generator (LQG) if it produces more than 2200 lb (1000 kg) of hazardous waste, not including waste oil, or one quart (1 kg) or more of acutely hazardous waste, as defined in the December 1992 Massachusetts regulations [4–10], in a month's time. There is no limit to the amount that can be accumulated by the plant, but the waste must be shipped within 90 days. A generator not in Massachusetts must contact the local State agency in order to obtain the most recent regulations for its home state.

If a Massachusetts plant produces less than this amount each month, the plant is classified as a small (SQG) or very small (VSQG) quantity generator and is subject to less stringent requirements, as discussed previously.

If a Massachusetts plant produces more than 1000 kg (approximately 265 gal) of waste oil in a month, the plant's waste oil must be shipped within 90 days but the plant is not subject to certain written plans and reports under Massachusetts Management Requirements. The plant may, however, be classed as a small quantity generator (SQG) or very small quantity generator (VSQG) of other hazardous wastes.

As a large (LQG) or small (SQG) quantity generator of hazardous waste in the State of Massachusetts, the plant is required to:

1. Notify the U.S. Environmental Protection Agency (USEPA), and obtain a USEPA Identification Number for the industrial site;
2. Identify and segregate the plant's hazardous wastes;
3. Label the plant's waste as hazardous waste, describing the waste and the hazards associated with it and the date when accumulation began in each container;
4. Store the plant's waste by type in separate containers that are tightly sealed, and provide appropriate aisle space to meet fire codes;
5. Use a licenced hazardous waste transporter and/or a licenced treatment, storage, or disposal facility, under the condition that the plant, as a generator, has ultimate legal responsibility for the plant's hazardous wastes;
6. Use a uniform hazardous waste manifest as a shipping document for all plant wastes, including waste oil; and
7. Keep records of waste analyses, reports, and manifests for at least three years.

In Massachusetts, there are additional requirements for Large Quantity Generators (LQG) of hazardous waste (but not waste oil):

1. Each manifest must contain a certification that the plant has a program in place to reduce the volume and toxicity of waste generated, as much as is economically practicable;
2. A Biennial Report summarizing the plant's manifest shipments for the previous years must be submitted to the State environmental protection agency;
3. A training program is required for all personnel involved in managing hazardous waste. A written plan is required to specify how the plant's personnel will be familiarized with procedures for using and repairing emergency and monitoring equipment, how the plant's personnel will respond to fire or explosions, potential groundwater or surface water contamination, how to shut down operations, what the job title and description of each position will be related to hazardous waste management with the requisite qualifications and duties, what training will be provided, and what the qualifications of the relevant training personnel will be; and
4. A written contingency plan is prepared based on a Spill Prevention Control and Countermeasures (SPCC) Plan, or similar emergency plan, which describes the layout of the plant, emergency equipment and handling procedures, places where the plant personnel would normally be working, entrances and exits, and evacuation routes; a list including the names, addresses, and telephone numbers of the emergency coordinator(s) must be distributed to local fire and police departments, the mayor, board of health, and emergency response teams.

3.13 MANAGEMENT OF HAZARDOUS INDUSTRIAL WASTE FROM MEDICAL OFFICES: AN EXAMPLE

3.13.1 Hazardous and Infectious Wastes from Medical Offices

Federal and State laws define waste as "hazardous" if it is ignitable, corrosive, reactive, or toxic. Other wastes are listed by name. These may differ from lists of hazardous materials, which are regulated by OSHA and Right-to-Know. The Standard Industrial Classifications (SICs) of a physician's medical office and a dentist's office are 8011 and 8021, respectively.

If a medical office has photoprocessing waste, typically from x-ray processes, which leaches silver in a concentration of 5 mg/L or more, or has a dental waste which leaches mercury in a concentration of 0.2 mg/L or more, this medical office is a "generator" of hazardous waste, of which concentrations are determined by an extraction procedure toxicity test.

Syringes, sharps, blood products, and the like from hospitals are considered infectious waste and are regulated by the U.S. Department of Public Health. It is recommended that any infectious waste from a medical office be placed in rigid containers and steam-sterilized or autoclaved. A method and a facility for disinfecting and compacting infectious wastes, such as disposal diapers, animal beddings, and so on, have recently been developed by Wang and Wang [59].

3.13.2 Waste Disposal

If the amount of hazardous waste a medical office produces in a month is less than 25 gal (95 L), this medical office qualifies as a very small quantity generator (VSQG) in Massachusetts. As a VSQG, the medical office is required to register with the State regulatory agency, label its wastes as hazardous, and ship it with a licenced hazardous waste hauler or precious metal transporter to a licenced treatment or disposal facility.

The disposal options of this medical office as a VSQG are listed below [5]:

1. The generator may want to reclaim silver from the x-ray waste itself. If its silver recovery equipment is hard-piped and connected to the photoprocessor, this generator is currently exempt from recycling permits.
2. The generator may ship its silver waste to a reclaimer. Be aware that it is the generator's responsibility as the generator of the waste to know where its waste is going and how it is handled.
3. If the generator is a registered VSQG, it may transport its waste to another generator or a receiving facility as long as it carries its VSQG registration in its vehicle, does not transport more than 55 gal (208.18 L) at a time, obtains a receipt for its waste, and retain the receipts for at least three years.
4. Some liquid residues can be discharged to the sewer if they are not classified as hazardous waste. The generator should call its local sewer authority for information. If the generator discharges the waste to a septic tank or dry well, it needs a groundwater discharge permit from the State regulatory agency.

3.14 MANAGEMENT OF HAZARDOUS INDUSTRIAL WASTES FROM GRAPHIC ARTS, PRINTERS, AND PHOTOGRAPHERS: AN EXAMPLE

3.14.1 Requirements

Each State has its own requirements and regulations for management of hazardous wastes at industrial sites. This section presents the Massachusetts requirements for graphic artists, printers, and photographers as a typical example.

Massachusetts Law requires industrial plants that produce hazardous waste to: (a) identify their wastes; (b) count their wastes to determine monthly quantities; (c) manage their wastes properly, based on the State requirements on monthly quantities of hazardous wastes that can be stored; (d) apply for a federal USEPA-ID if the industrial plant is a small quantity generator (SQG) or very small quantity generator (VSQG); or (e) register with the State Division of Hazardous Waste if the plant qualifies as a very small quantity generator (VSQG).

3.14.2 Hazardous Waste Identification

The hazardous wastes generated from graphic artists, printers, and photographers can be identified by their specific wastes number, hazard condition, and SIC as shown in Tables 3 and 4.

To identify other hazardous wastes in shop, the three types of the material safety data sheet (MSDS) provided by the supplier of the product should be reviewed. A plant manager can also find out the hazardous ingredients in the processing chemical and refer to the State Hazardous Waste Regulations or call the State Division of Hazardous Waste.

It should be noted that ink and paint wastes may contain certain metals that make the waste "FP Toxic." For more information, the readers are referred to the MSDS, to talk to the manufacturer or an environmental consultant, or have a certified laboratory conduct an Extraction Procedure Toxicity Test on the waste in question.

The following summarizes the current Massachusetts regulations for disposal of photographic wastes containing silver (waste number D011 shown in Table 3).

If less than 25 gal (95 L) of spent fixer is generated each month (assuming there are no other wastes, or if there are other wastes, the total quantity, excluding waste oil, does not exceed 25 gal or 95 L), a generator will face the following situations:

1. The generator may register as a very small quantity generator, and/or obtain a USEPA-ID;
2. No recycling permit is required for the generator; registration is sufficient;
3. The generator may treat spent fixer for reclaiming the silver at the site of generation, or the generator may ship the spent fixer waste offsite with a licenced hazardous waste transporter or a State-approved precious metal transporter and recycling facility, or self-transport up to 55 gal at a time to another generator or a receiving facility;

Table 3 Identification of Hazardous Wastes

Typical waste	Waste number	Hazard
A. Spent solvents		
Ethyl alcohol, isopropanol	D001	Ignitable
Methylene chloride, Trichloroethylene	F001	Toxic
Ethyl benzene	F003	Toxic
B. Ink/paint wastes	D001	Ignitable
C. Ink/paint wastes containing metals such as:		
Chromium	D007	EP Toxic
Lead	D008	EP Toxic
D. Etch and acid baths	D002	Corrosive
E. Spent photographic wastes containing silver	D011	Toxic

Table 4 Standard Industrial Classification (SIC)

Type of business	SIC
Graphic arts, photographic labs	7333
Advertising/art	7311
Commercial printing	2751
Miscellaneous publishing	2741

4. If the generator transports to another generator or an authorized facility, the generator must obtain a receipt for the generator's waste and retain records for a minimum of three years.

If more than 25 gal (95 L) of hazardous waste, including spent fixer, is generated in a month, the generator will face different situations:

1. The generator needs a USEPA-ID number and must use a manifest if it ships its waste offsite and may use a licenced hazardous waste transporter or a precious metal transporter and recycling facility;
2. The generator can use a recovery device directly connected by pipe to the film processor at the site of generation (no recycling permit is required: operation can begin within 10 days of the receipt of the application if applicant does not hear from the State); and
3. The generator must meet the concentration limits of the local sewer authority if discharging the waste to a sewer system is intended, or obtain a groundwater discharge permit from the State Division of Water Pollution Control if the generator discharges to a septic system or other groundwater disposal. If the waste or its pretreated effluent meets silver concentration limits of less than 5 mg/L of silver, the waste or the effluent is not classified as a hazardous waste.

3.14.3 A Case History for Disposal of Photographic Wastes by a Large Quantity Generator (LQG)

A graphic arts company in Farmingdale, New York, produced four wastewater streams at Outfall Nos. 001, 002, 003, and 004, as shown in Figure 1.

Outfall	Wastewater flow (gal/day)
001	2000 (average)
002	30,000 (average)
003	5000 (average)
004	5000 (average)

Note: 1 gal/day = 3.785 L/day.

The four wastewater streams discharged at Outfall Nos. 001, 002, 003, and 004 were photographic process wastewater, cooling water (noncontact), sanitary waste A, and sanitary waste B, respectively. A State Pollutant Discharge Elimination System (SPDES) discharge permit was issued to the company in compliance with the Environmental Conservation Law of

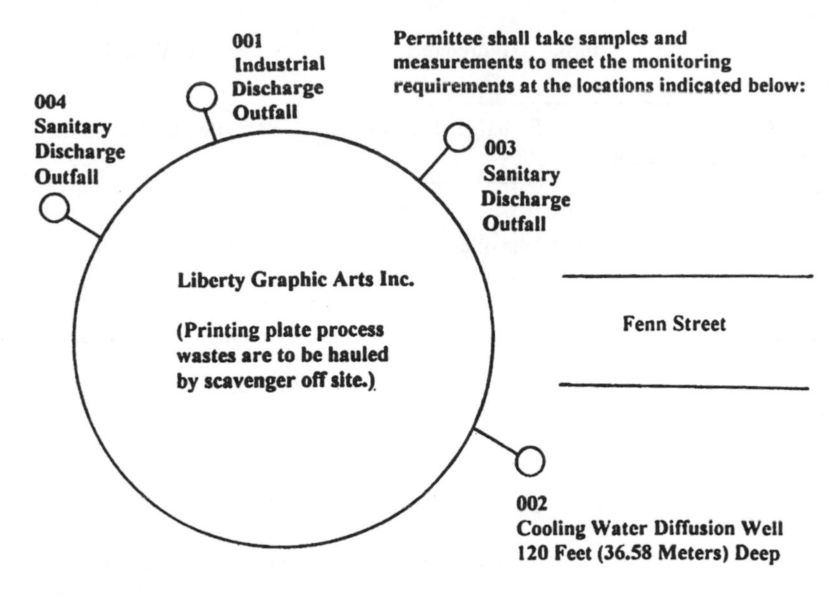

Figure 1 Monitoring locations at Liberty Graphic Arts, Inc.

New York State [13] and the Federal Clean Water Act, as amended. Specifically, the company was authorized to discharge its treated effluents from the company's facility to a nearby receiving water. Table 5 indicates the final effluent limitations and monitoring requirements specified by the SPDES discharge permit.

The company was required to take samples and measurements to meet the monitoring requirements at the outfall location Nos. 001, 002, 003 and 004, as indicated in Figure 1.

The photographic wastewater from the company was collected for treatment in accordance with the technologies described by Bober *et al.* [57] before being discharged to a State-approved receiving water.

A small quantity of wastewater containing extremely toxic pollutants, however, was held and hauled by an approved scavenger.

3.14.4 A Case History for Disposal of Photographic Wastes by a Very Small Quantity Generator (VSQG)

Environmental Situations

A small printing company in Lenox, Massachusetts, United States, produced 24 gal/month (91 L/month) of industrial wastewater mainly consisting of the following two spent chemicals:

- Spent Kodak ultratek fixer and replenisher were accumulated (3 parts water per 1 part fixer dilution; about 100 gal or 378.5 L accumulated); and
- Spent Kodak ultratec developer and replenisher were accumulated (1 part water per 5 parts developer; about 100 gal or 378.5 L accumulated).

The MSDSs of both diluted spent chemicals were obtained from the chemical supplier for review by the company's consulting engineer. The following are the chemical descriptions and disposal methods from the MSDSs:

- Kodak ultratek fixer and replenisher. This chemical formulation has a high biological oxygen demand, and it is expected to cause significant oxygen depletion in aquatic systems. It is expected to have a low potential to affect aquatic organics. It is expected

Table 5 Final Effluent Limitations and Monitoring Requirements Specified by the New York State SPDES Discharge Permit

During the period April 1, 1981, to April 1, 1986, the discharges from the permitted facility shall be limited and monitored by the permittee as specified below:

Outfall number	Effluent parameter	Discharge limitations other units (specify)		Monitoring requirements measurement sample	
		Daily avg.	Daily max.	Frequency	Type
001	Process flow, photographic waste		–	Continuous	Recorded
	Nitrogen, total – monitor only		10 mg/L	Monthly	Composite
	Cadmium, total – monitor only		–	Monthly	Composite
	Dissolved solids, total		1000 mg/L	Monthly	Composite
	Chemical oxygen demand		150 mg/L	Monthly	Composite
	Color, units – monitor only		–	Monthly	Composite
	Iron, total		0.6 mg/L	Monthly	Composite
	Phenols		0.002 mg/L	Monthly	Composite
	Zinc, total		5.0 mg/L	Monthly	Composite
	pH Units – Range	6.5–8.5		Daily	Grab
002	Cooling water, noncontact – no chemical treatment allowed				
003	Sanitary wastes only				
004	Sanitary wastes only				

Notes:
1. All wastewater discharges from the printing plate process are held and hauled by an approved scavenger.
2. Approximate flows are as follows:

Outfall	Flow (gal/day)
001	2000
002	30,000
003	5000
004	5000

to have a moderate potential to affect secondary waste treatment microorganisms. It is expected to have a moderate to high potential to affect the germination and growth of some plants. The components of this chemical formulation are biodegradable and are not likely to bioconcentrate. If diluted with a large amount of water, this chemical formulation released directly or indirectly into the environment is not expected to have a significant impact.

- Kodak ultratec developer and replenisher. This formulation is a strongly alkaline aqueous solution, and this property may cause adverse environmental effects. It has a low biological oxygen demand and is expected to cause little oxygen depletion in aquatic systems. It is expected to have a high potential to affect aquatic organisms and a moderate potential to affect secondary waste treatment microorganisms and the germination and growth of some plants. The organic components of this chemical formulation are readily biodegradable and are not likely to bioconcentrate. The direct instantaneous discharge to a receiving body of water of an amount of this chemical formulation that will rapidly produce, by dilution, a final concentration of 0.1 mg/L or less is not expected to cause an adverse environmental effect. After dilution with a large amount of water, followed by a secondary waste treatment, the chemicals in this formulation are not expected to have any adverse environmental impact.

Both spent chemicals were analyzed by a certified laboratory. The analytical data of the two spent chemicals were:

- Spent Kodak ultratek fixer and replenisher (i.e., spent fixer): COD = 161,000 mg/L, silver = 1384 mg/L, total solid = 6%; and
- Spent Kodak ultratec developer and replenisher (i.e., spent developer): COD = 103,000 mg/L, silver = 0 mg/L.

The company produced less than 100 kg (220 lb or approximately 25 gal) of hazardous waste a month. Accordingly, it was eligible to be registered as a very small quantity generator (VSQG).

To qualify as a VSQG, the company owner notified the Massachusetts Department of Environmental Protection (DEP) on a two-part registration form, which listed the types of hazardous waste generated, the amount of each in gallons per month, and the proposed disposal, treatment, storage, and/or recycling destination of the waste.

The owner's registration was effective as soon as it was received by the State. Renewal would occur after one year.

As a VSQG, the company had never accumulated more than 165 gal (624.5 L) at any one time. As a registered VSQG, the company's owner also tried four different treatment and disposal methods for his hazardous wastes.

Options for Recycling, Treatment, and POTW Discharge

Initially the owner tried to recycle or treat his wastes, because the process described in his registration was acceptable to the Massachusetts DEP. It appeared that the silver in the spent fixer had to be removed, and the remaining pollutants were mainly biodegradable organics. Four silver removal methods were considered by the owner [58].

Chemical Recovery Cartridge Metallic Replacement Method

In this method, a metal (usually iron) in a chemical recovery cartridge (CRC) reacts with the silver thiosulfate in the spent fixer and goes into solution. The less active metal (silver) settles out as a solid. To bring the silver into contact with the iron, the spent fixer is passed through the CRC container, which is filled with steel wool. The steel wool provides the source of iron to replace the silver. The main advantages of this CRC method are the very low initial cost (cartridges cost about US$60) and the simplicity of installation; only a few simple plumbing connections (shown in Fig. 2) are required.

The main disadvantages, compared to the electrolytic method, are that the silver is recovered as a sludge, making it more difficult to determine the exact amount recovered. The recovered sludge containing silver requires more refining processes than the plate silver obtained

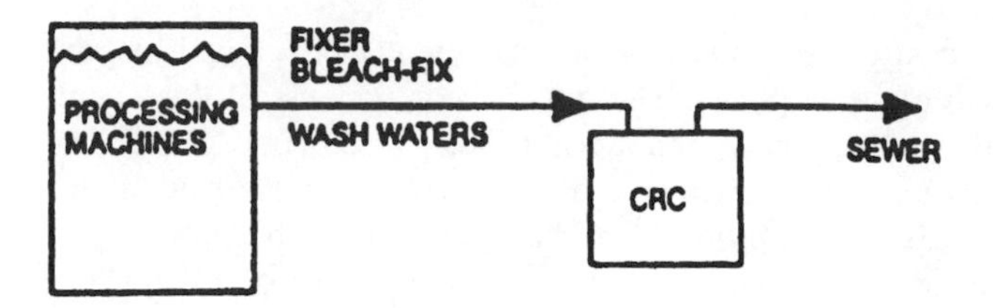

Figure 2 Chemical recovery cartridges (CRCs). (Courtesy of Eastman Kodak Co., NY.)

from electrolytic methods, if silver recovery and silver refining are both intended. Also, CRCs cannot be reused. They must be replaced when they are exhausted.

In summation, the silver chemical recovery cartridge method (Fig. 2) can achieve silver recovery efficiencies of greater than 90%. However, it is difficult to achieve this level of recovery consistently, making it an unreliable choice if the operator needs to meet low silver discharge limits. Another problem with the chemical recovery cartridge method is that as silver is recovered, the steel wool becomes soluble, producing iron levels in the effluent as high as 3000 mg/L. Iron is regulated to levels well below those concentrations by many sewer codes.

Electrolytic Silver Recovery Method

In this method, the silver-bearing solution is passed between two electrodes through which a controlled direct electric current flows as shown in Figure 3. Silver plates out on the cathodes as almost pure metal.

The advantages of the electrolytic method is that silver is recovered in an almost pure form, making it easier to handle and less costly to refine. With careful monitoring, it also permits fixer reuse for some processes. It also avoids the need to store and replace cartridges, as with the

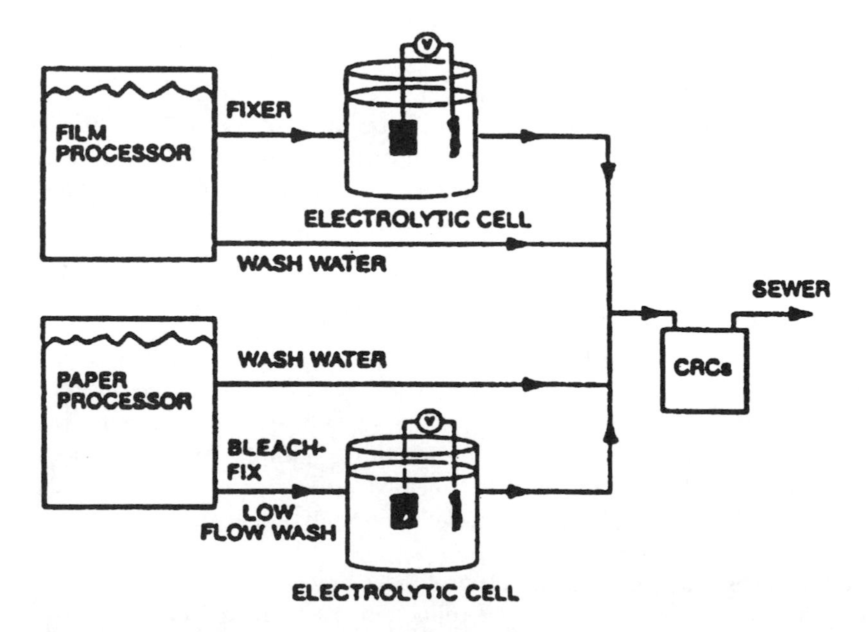

Figure 3 Electrolytic cells plus CRCs. (Courtesy of Eastman Kodak Co., NY.)

metallic replacement method. Recovery efficiency is typically 93–97%, and by maintaining the correct mix of processing effluent, can be as high as 99%.

The disadvantages of electrolytic methods are the difficulty in reducing silver in the effluent to very low levels, and the careful monitoring required to avoid silver sulfide formation. Initial capital investment is high. None of these disadvantages is a serious deterrent. The concentration that can be achieved depends on how low the current density can be set with the unit. As the silver concentration gets lower, the current density can be set lower to prevent silver sulfide from forming. With low current densities, a large cathode area is needed to achieve the necessary silver recovery rate. In order to reduce the residual silver concentration in the electrolytic cell effluent further, at least one CRC is used for finally polishing the electrolytic cell effluent (Fig. 3).

Conventional Ion-Exchange Method

There are two ion-exchange methods that have been used in photoprocessing laboratories to recover silver from dilute solutions: conventional ion exchange (Fig. 4) and in situ ion exchange (Fig. 5). With both of these ion-exchange methods, the silver is removed by pumping it through a column of anion-exchange resin. The difference between the two ion-exchange methods is the regeneration step.

In the conventional ion-exchange method (Fig. 4), the silver is removed from the resin by regenerating it with thiosulfate solution. The silver is then removed from the regenerant by running it through an electrolytic cell. The greatest advantage of using the conventional ion-exchange method for silver recovery is that the operator can reduce the silver in the processing effluent to very low levels (0.1–2 mg/L). In areas that strictly regulate the discharge of silver, it may be the only recovery method that is satisfactory.

The conventional ion-exchange method also has some major disadvantages, such as the high capital investment (both an ion-exchange unit and an electrolytic unit are needed), and the increased complexity of operation (only a few high-volume laboratories have used this method successfully). However, it remains an option for those laboratories that must meet strict limits on

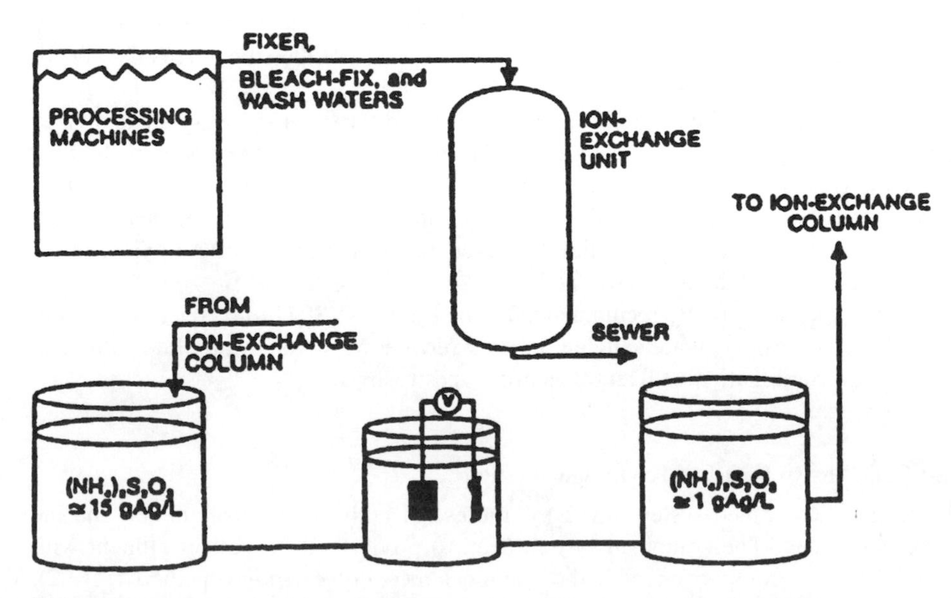

Figure 4 Conventional ion exchange. (Courtesy of Eastman Kodak Co., NY.)

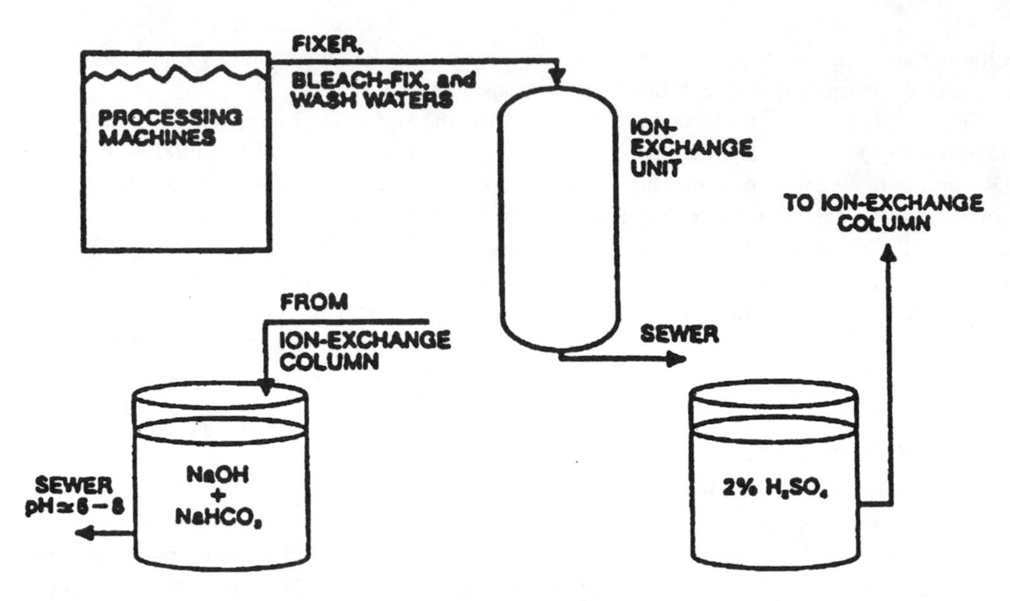

Figure 5 In situ precipitation. (Courtesy of Eastman Kodak Co., NY.)

the amount of silver discharged. It is also critical that the operator dilutes the concentrate with the proper amount of wash water prior to ion-exchange treatment; too high a thiosulfate concentration in the solution being treated will cause silver to leak through the column.

In Situ Ion-Exchange Method

With the in situ ion-exchange method (Fig. 5), dilute sulfuric acid is used to precipitate the silver in the resin beads as silver sulfide instead of removing it with regenerant. The resin that is inside the ion exchange unit is used for many cycles without a loss in capacity. When the resin eventually loses its capacity to recover silver, or when there is sufficient silver to make recovery worthwhile, it is sent to a silver refiner who incinerates it to remove the silver. This may occur after between six months and a year.

The advantages of using the in situ ion exchange method for silver recovery are similar to that of using the conventional ion exchange method. The disadvantages of the in situ method are that it requires a greater capital investment, and more chemical handling than with either chemical recovery cartridges or electrolytic cells. Also, the pH of the spent regenerant must be adjusted as it is discharged from the columns to prevent the formation of sulfur dioxide, and to be sure the discharge meets the local sewer codes.

Ion-exchange methods are not recommended to be used by VSQG to recover silver from spent fixers or bleach-fixes. They are suitable only for recovering silver from dilute solutions, like washwater, or a combination of fixer, bleach-fix, and washwater.

Electrolytic Cell Plus In Situ Ion-Exchange Method

Figure 6 illustrates a combined system involving the use of both the electrolytic cell and the in situ ion-exchange unit. The combined system (Fig. 6) produces an excellent effluent with lower residual silver in comparison with the chemical recovery cartridge method (Fig. 2), electrolytic silver recovery method (Fig. 3), the conventional ion-exchange method (Fig. 4), and

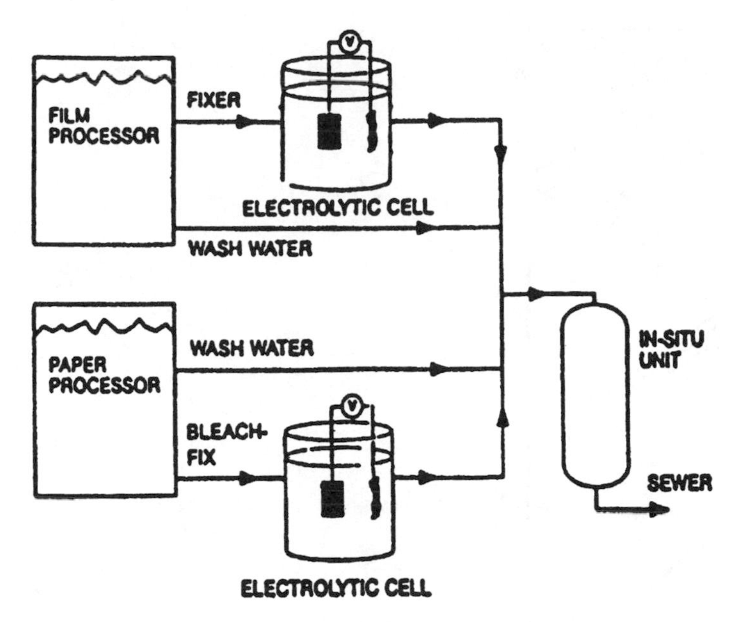

Figure 6 Electrolytic plus in situ precipitation. (Courtesy of Eastman Kodak Co., NY.)

the in situ ion exchange (Fig. 5). The disadvantage of the combined system is its high capital and operating costs.

Hydroxide Precipitation Method

In this method, sodium hydroxide, potassium hydroxide, calcium hydroxide, magnesium hydroxide, or sodium aluminate can be fed to the spent fixer for precipitation of silver ions as insoluble silver hydroxide precipitates. Figure 7 indicates that the residual silver concentration in the hydroxide precipitation treated effluent can be about 1 mg/L at pH 12 [19].

The advantage of this method is its low cost. Its disadvantages are: (a) the residual silver concentration in the treated effluent (about 1 mg/L at pH 12) may exceed the local regulatory agency's effluent limit on silver; (b) the hydroxide sludges produced in the hydroxide precipitation method require further thickening and dewatering treatment and final disposal; and (c) refining silver from the precipitated silver hydroxide sludges is difficult [87].

Sulfide Precipitation Method

In this method, sodium sulfide, potassium sulfide, and/or ferrous sulfide can be dosed to the spent fixer during mixing at an alkaline pH range, for precipitation of silver ions as insoluble silver sulfide precipitates [87]. Figure 7 [19,59] indicates that the residual silver concentration in the sulfide precipitation treated effluent can be below 10^{-9} mg/L in the entire alkaline range, and can be as low as 10^{-12} mg/L at pH 10.5.

There are two advantages for this method: (a) the capital and operating costs are low; and (b) silver removal efficiency is extremely high. There are a few disadvantages for the method: (a) the sulfide sludges produced in the sulfide precipitation method require further thickening and dewatering treatment, and final disposal; (b) refining of silver from the precipitated silver sulfide sludges is not easy; and (c) hydrogen sulfide toxic gas may be produced from the sulfide precipitation process system if the pH of the spent fixer is controlled in an acid range by accident.

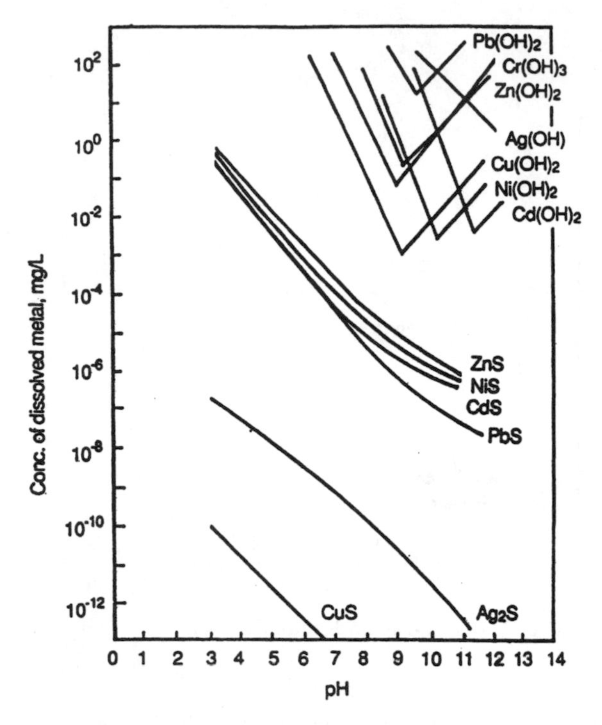

Figure 7 Solubilities of metal hydroxides and metal sulfides. (Courtesy of USEPA.)

Process Comparison and Selection

Selection of a suitable method for silver removal depends on many factors: what processes the company uses, what volume of wastes the company produces, what kind of training and technical knowledge the company's personnel has, whether the company wants to reuse the company's fixer or bleach-fix, how much the company wants to spend for recovery equipment, and what the environmental concerns are, such as how strict the effluent discharge limits are. Just considering these factors makes choosing a silver recovery method very much an individual decision for each company.

Table 6 summarizes the silver removal efficiencies of the various process methods identified above. The silver concentration that can be discharged to a treatment plant or to a receiving body of water is often regulated even though silver in photographic effluent is in a form

Table 6 Typical Silver Concentrations in Effluent after Recovery

Recovery method	Silver concentration (mg/L)
Chemical recovery cartridges (CRCs)	10–20
Electrolytic (with tailing CRCs)	1–5
In situ ion exchange	0.1–2
Conventional ion exchange	0.5–2
Electrolytic (with tailing in situ ion exchange)	<0.1–2
Hydroxide precipitation	1–5
Sulfide precipitation	<0.1

that is not harmful. Therefore, cost is only one of the primary considerations in choosing the company's silver-recovery method.

The company had considered all silver recovery options. The main factor was the company's processing volume. Other economic considerations were the price of silver and operating and refining costs. If the company's processing volume were to be high, the company would probably want to make frequent cartridge replacements or set aside a large amount of storage area for spent and replacement cartridges. Although initial capital investment with electrolytic recovery cells was higher than with chemical recovery cartridges, there would not be the recurring cost of equipment replacement. If the company were to use an electrolytic cell, the company's refining costs for the recovered silver would be much lower than with other methods because the silver-plated out would usually be more than 95% pure.

If the company were a large-volume operation, the in situ ion exchange would also be an option. The company could use this method for primary treatment. The company could also use it to tail an electrolytic unit if the company were first to dilute the discharge from the electrolytic cell with washwater. Using this method would enable the company to recover the maximum amount of silver and to minimize the amount of silver discharged. It would require a greater capital investment and more chemical handling than with electrolytic cells or chemical recovery cartridges in accordance with the information from Kodak Company [58].

However, the company's wastewater volume was actually very low, and chemical recovery cartridges, hydroxide precipitation tanks, and sulfide precipitation tanks became reasonable choices for silver recovery. Chemical recovery cartridges and the two types of precipitation tanks were all very simple to install. The costs for purchasing, installing, operating, and monitoring this equipment are very low compared with other methods.

In comparison with the silver recovery/removal efficiencies of the chemical recovery cartridge (CRC) method, the hydroxide precipitation method, and the sulfide precipitation method shown in Table 6, the two precipitation methods appeared to be a better choice than the CRC method.

Considering the silver removal efficiencies of the various process methods in Table 6 (the lower the residual silver concentration in effluent after treatment, the higher the silver removal efficiency), the company's local code limits for silver, the ease of process operation, the safety, the costs, the volume of waste production, and the silver content in the spent fixer, the company finally selected the hydroxide precipitation method as the first-stage treatment, and the sulfide precipitation method as the second-stage treatment for silver removal from the spent fixer. It should be noted, however, that sulfide precipitation alone would have been sufficient. After silver was significantly removed from the spent fixer, both the treated spent fixer and the untreated spent developer were mixed together, forming a pretreated combined wastewater for possible discharge to a POTW. The analytical data of the pretreated combined wastewater are: COD = 132,000 mg/L; silver = <0.1 mg/L; and pH = 9.5. At this stage, the pretreated combined wastewater was no longer considered to be hazardous because it contained only a high concentration of biodegradable organics in terms of 132,000 mg/L of COD.

It is important to note that if the precipitation tanks were hard-piped and connected to the company's processing units, the pretreated combined wastewater would not be considered to be a hazardous waste legally, and would be allowed to be discharged into the POTW without any legal problems. The precipitation tanks of the company, however, were not hard-piped to the company's processing units. An application for a Permit for Sewer System Extension or Connection was then officially filed at the local town of Lenox, Massachusetts, by the consulting engineer, on behalf of the company. It was proposed that a permit be issued by the town and the State for the company to discharge a design flow ranging from 5 gpd (average) to 20 gpd (maximum) of the aforementioned pretreated combined wastewater into an existing Lenox

POTW system (1 gpd = 3.785 L/day). It should be noted that actual company's wastewater flow was only 24 gal/month (90.84 L/month). The average sewage flow of the Lenox POTW was 0.4 MGD (1514 m^3/day), and the BOD/COD ratio of the waste was determined to be 0.65. If a permit were issued to the company for discharging the pretreated wastewater to the Lenox POTW, an increase in silver concentration would be negligible, and an increase in BOD in the Lenox POTW would only be by about 1 mg/L during discharging of the pretreated wastewater at an instantaneous flow as high as 5 gpm (18.93 L/min). Besides, the organics in the pretreated wastewater were biodegradable in accordance with the MSDS. Under normal situation, a sewer discharge permit could have been granted because the company's pretreated wastewater would not adversely affect the normal operation of biological wastewater treatment at the Lenox POTW. The town of Lenox was too small to have a licenced engineer to handle the legal case. The company was advised by the town to haul their small quantity of pretreated wastewater to the nearby city of Pittsfield's POTW for disposal because there was an agreement between the town and the city.

The transportation of the pretreated wastewater from the company to the city, which was only 6 miles away, had to comply with all government rules and regulations because the pretreated wastewater was legally considered to be a hazardous waste, although technically it was not. The company faced a transportation problem because of its high cost.

Option of Transporting Wastes to Another Generator

The company could transport its untreated wastewaters or pretreated wastewater(s) to another generator who is in compliance with the regulations and who will count the company's waste as part of another generator's waste. Another generator, J.F. Co., Inc., was found in Springfield, Massachusetts, which was about 80 miles away from the company. J.F. Co., Inc., agreed to accept the spent Kodak Ultratec Fixer and Replenisher (i.e., the spent fixer) containing 1400 mg/L silver for silver recovery at a cost to the Lenox Company of US$2.00 per gallon, delivered to J.F. Co., Inc., in Springfield. While the cost quoted for disposal of the spent fixer was reasonable, the company in Lenox faced two other problems: (a) the spent developer containing no silver but a high concentration of dissolved organic carbon (DOC) still needed to be disposed of; and (b) the Lenox company, which was the original generator, would have to take full responsibility for whatever the actions were to be taken by the other generator, which was not well-known in the field.

The company was discouraged by the option.

Option of Transporting Wastes to a Licenced Facility

The company in Lenox also had an option to transport its wastes to a licenced treatment, storage or disposal facility, or permitted recycling facility, with the facility's permission. There were many licenced facilities in Massachusetts that were willing to accept the company's spent fixer and spent developer for final disposal.

As a registered VSQG, the company might transport its own hazardous waste under the following conditions: (a) the company transports only the waste that is generated on its premises (no problem); (b) the company does not transport more than 55 gal or 208.2 L at one time (no big problem but time consuming); (c) the company does not transport incompatible wastes in the same shipment (no problem); (d) the company's waste is in containers that are tightly sealed, labeled as "HAZARDOUS WASTE," with the name of the waste and the type of hazard, and are tightly secured to the vehicle (no problem); (e) the company keeps a copy of its registration as a VSQG in the vehicle while transporting its waste (no problem); (f) the company is in compliance with all USDOT and Massachusetts Department of Public Safety requirements (a problem to

VSQG); and (g) in the event of a spill or leak of hazardous waste that may threaten humans or the environment, the company shall notify the Massachusetts DEP or the State Police (no problem).

The company's official Massachusetts DEP assigned SIC number was 2751. The two wastewater classifications and waste numbers were as follows: (a) spent fixer: waste number is D011, classified as "toxic"; and (b) spent developer: no waste number, classified as "corrosive," neutralization is recommended but not required before transportation.

If the company decided not to use a licenced transporter but would be transporting its own wastes, the company did not need a USEPA-ID number or manifest form. The company must, however, keep a record of the type and quantity, as well as the date, method of transport, and treatment/disposal of its waste(s). The company would need proof of the receipt of the waste by the treatment/disposal facility. The company or its consulting engineer must keep receipts or manifests of waste shipped and records of waste analysis for at least three years, or for the duration of any enforcement action by the Massachusetts DEP.

Apparently this option was technically and economically feasible for the small company in Lenox. Unfortunately small (SQG) or very small (VSQG) generators similar to this Lenox company simply cannot find out the latest USDOT requirements, the Massachusetts Department of Public Safety requirements, and the Massachusetts Department of Environmental Protection requirements by themselves without hiring a consulting engineer. The State regulatory agency(s) should provide more technical assistance to VSQG and SQG, whenever possible.

Option of Using a Licenced Transporter and Facility

The company's owner finally decided to use a licenced transporter and a licenced facility for transportation, treatment, and disposal of its untreated wastes, even though the option of recycling, treatment, POTW discharge, and the option of self-transportation to a licenced facility were equally feasible. The licenced transporter and the licenced facility can be owned by two different firms or by one firm. In this particular case, the company in Lenox, which was a VSQG, selected an environmental service company in Albany, New York, which was a licenced transporter as well as a licenced facility. The costs for picking up, transporting, and disposing of six 55 gal drums of photographic developer and fixer solution in 1991 are documented in Table 7.

The prices given in Table 7 were based upon the following conditions:

- Free and easy access for the transporter/facility personnel to work site;
- Applicable taxes and state regulatory fees are not included in quoted process;
- A fuel usage surcharge of half of 1% will be added to the invoice total to cover rising fuel cost;

Table 7 Costs for Picking Up, Transporting, and Disposing of Chemicals

Method	Waste material	Container	Charge
A. Picking up for disposal	Spent Kodak ultratec developer and replenisher	55 gal drum	$145/drum
B. Picking up for disposal	Spent Kodak ultratek fixer and replenisher	55 gal drum	$130/drum
C. Transportation of drum to the Massachusetts disposal facility (maximum 6 drums per load)			US$300.00/load

- Waste material conforms to waste profile sheets;
- All drums are centrally located and DOT approved;
- Drums are in shippable condition; and
- Transportation rate allows one hour for loading time; additional time required will be billed at US$75.00 per hour.

3.15 RECYCLING OF HAZARDOUS INDUSTRIAL WASTES AS WASTE-DERIVED FUELS

3.15.1 Introduction and Objective

Southdown, Inc., Houston, TX, engages in the cement, ready-mixed concrete, concrete products, construction aggregates, and hazardous waste management industries throughout the United States. According to Southdown, they are making a significant contribution to both the environment and energy conservation through the utilization of waste-derived fuels as a supplemental fuel source. Cement kiln energy recovery is an ideal process for managing certain organic hazardous wastes. The burning of wastes or hazardous wastes as supplemental fuel in the cement and other industries is their engineering approach.

By substituting only 15% of its fossil fuel needs with solid hazardous waste fuel, a modern dry-process cement plant with an annual production capacity of 650,000 tons of clinker can save the energy equivalent of 50,000 barrels of oil (or 12,500 tons of coal) a year. Southdown typically replaces 10–20% of the fossil fuels it needs to make cement with hazardous waste fuels.

By using hazardous waste fuels, the nation's hazardous waste (including infectious waste) problem can be at least partially solved economically.

3.15.2 Cement Kiln Energy Recovery System

The cement kiln is a long, inclined cylinder that can be hundreds of feet in length and up to 15 ft in diameter. Raw materials, such as limestone, clay, sand, and a small amount of iron-containing substances enter at one end and cement clinker, the product, exits at the other. Material temperatures required to make cement clinker must be maintained at a minimum of 2450°F while gas temperatures inside the kiln can reach 3500°F. During operation, the kiln slowly rotates to ensure a thorough blending and "cooking" of the raw materials. These raw materials are heated using fossil fuels (about 85%) along with hazardous waste fuels (about 15%) in the huge kiln at such high temperature until they chemically combine to become marble-sized nodules called "clinker." The clinker is then mixed with gypsum and ground to a fine powder to make cement. Cement, in turn, is a key ingredient in concrete, which is a vital component of the world's roads, buildings, houses, and offices.

Cement kilns manage destruction of organics in hazardous waste through a high-temperature combustion. This involves heating the waste to a sufficient temperature, keeping it in the kiln for enough time, and providing the waste with sufficient oxygen. Because this method destroys organic chemical wastes, such as paint thinners, printing inks, and industrial cleaning solvents, combustion has become the preferred method of managing them and utilizing their BTU value. The conditions in the kiln ensure, and U.S. Environmental Protection Agency (USEPA) regulations require, that 99.99% or more of organic hazardous wastes are destroyed, that is, converted to carbon dioxide and water vapor.

Exhaust gases leaving the kiln pass through highly efficient air pollution control devices such as baghouse filters or electrostatic precipitators. The high temperatures required to make cement destroy 99.99% or more of the organic hazardous wastes. The content of hydrocarbons and carbon monoxide in stack emissions is monitored to ensure that the combustion process is optimized. When combustion is efficient, emission of carbon monoxide is minimized and hydrocarbons disappear. In this way, operators are assured that a destruction efficiency of 99.99% or more is always maintained and that they stay within the stringent limits on emissions set by USEPA, which has identified ten metals that must be controlled. The list includes antimony, arsenic, barium, beryllium, cadmium, chromium, lead, mercury, silver, and thallium. All cement kilns that want to recycle hazardous waste as fuels will have to meet stringent limits on emissions of these metals.

3.15.3 Cement Kiln Monitoring and Control

Under the newly adopted federal regulations for facilities using hazardous waste fuels, cement kilns must comply with stringent testing and permitting requirements before they can recycle the wastes. These procedures ensure that cement companies wanting to recycle hazardous wastes as fuel will do so safely. Those facilities unable to meet the rigorous RCRA standards will not be allowed to burn hazardous waste fuel. Under the USEPA BIF rule ("Burning Hazardous Waste in Boilers and Industrial Furnaces"), cement kilns recycling hazardous waste as fuel are now perhaps the most regulated form of thermal treatment. Major components of the regulatory approach include monitoring and control – allowing operators to detect problems in the process and control the system on a continuing basis. This ensures the process always stays within a safe window of operating conditions – and that emissions always remain within the strict limits prescribed by USEPA.

Under USEPA's BIF rule, manufacturers are required to closely monitor numerous conditions in the kiln and to observe limits on the following aspects of the process: (a) the maximum feed rate of hazardous waste fuel; (b) the maximum feed rate of metals from both raw materials and fuels; (c) the maximum feed rate of chlorine from raw materials and fuels; (d) the maximum feed rate of raw materials; (e) the maximum temperature at the inlet to the air pollution control devices; (f) the maximum concentration of carbon monoxide and total hydrocarbons in the flue gas; (g) the maximum temperature in the combustion zone or minimum temperature at the kiln inlet; and (h) any decrease of pressure at the baghouses or any decline in the strength of the electric field of electrostatic precipitators (both are types of air pollution control devices).

Cement manufacturers use a number of quality control measures. Key among these is careful selection of fuels for recycling. Fuels that contain metals above specified levels, for instance, will be rejected. For that reason, each shipment of fuel is carefully analyzed to determine its ingredients. If the fuel fails to meet predetermined specifications, it will not be used.

There are two primary sources of controls on recycling hazardous waste fuels in cement kilns. First, cement kiln operations are tightly regulated on both the Federal and State level. These regulations cover everything from the transportation of the fuel to the conditions that must be maintained in the kiln. Using USEPA's new, highly sensitive Toxicity Characteristics Leaching Procedure (TCLP), scientists have confirmed that the chemical reactions that must take place in order to make cement prevent unacceptable concentrations of metals from being released from cement or the concrete. Second, because the chemistry of cement-making is both sensitive and precise, manufacturers cannot afford to put anything into their kilns that could produce variations in the clinker. If they did, the cement would not meet the rigorous, industry-wide product quality standards set by the American Society of Testing and Materials (ASTM). ASTM specifies tests and test methods [60] to ensure uniform controls on cement producers nationwide. Before the product can be called Portland cement, tests must show it has the required chemical composition. It also must pass tests measuring physical qualities, such as strength and particle fineness. In this way, product quality is assured regardless of what raw materials or fuels are used.

3.15.4 Permit System for Process Operation, Waste Transportation, and In-Plant Waste Handling

All cement kilns burning hazardous waste fuels will have to obtain a permit from USEPA and local regulatory agencies. Because the permitting process can be lengthy, cement kilns already burning hazardous wastes will be subject to regulation almost immediately under what is known as "interim status." Interim status is a standard regulatory approach used when new regulations are approved under RCRA to bring existing facilities under the new regulations without delay.

Trucks transporting hazardous wastes to cement kilns are regulated by Federal and State transportation agencies. This means they are controlled every step of the way. All trucks must meet U.S. Department of Transportation standards, which require all hazardous wastes to be transported under strict conditions in specially designed containers. State transportation agencies test and licence truck drivers to ensure they understand the precautions required with these fuels.

Both Federal and State regulations under the RCRA specify storage and handling requirements designed to ensure safe operations. For example, the facilities for unloading, storing, and transporting hazardous wastes to the kiln are built with government-approved systems designed to prevent sparks and accidental fires. Such areas are also designed to meet or exceed Federal and State standards for environmental safety, including secondary containment in the unlikely event of a spill.

3.15.5 Health Effects and Risks

Because cement kilns effectively destroy more than 99.99% of organic chemical wastes and emissions are tightly controlled by the BIF rule and other regulations, only minute amounts of organic compounds are emitted and testing has indicated that these emissions are independent of fuel type. In fact, organic emissions are sometimes reduced through the use of waste fuels. The quantity is so small that it does not present a perceptible increase in risks to public health or the environment. Cement kiln exhaust gases typically contain less than one-tenth of the hydrocarbons present in automobile exhaust gases.

Because cement kilns are so good at destroying organic chemical wastes, emissions of dioxins – or any other type of products of incomplete combustion (PIC) – are so low they pose no danger to the environment. In the case where some of the hazardous waste fuels used contain toxic dioxin, the cement kiln temperatures of 1650°F will destroy dioxins in less than one second. Because cement kilns operate at much higher temperatures (at least 2450°F), and because the burning wastes have an average residence time in the kiln of at least two seconds, any dioxins are destroyed. However, dioxin waste is never accepted by Southdown for use in its cement kilns.

Cement made with hazardous waste fuels contains essentially the same amount of metals as cement made using traditional fossil fuels, such as coal, coke, or oil. Also, tests show cement made with hazardous waste fuels has essentially identical leaching characteristics as those of cement produced solely with traditional fuels. This means the metals are no more likely to leach out of the cement made using hazardous waste fuels than if it were made using coal, coke, or oil.

The TCLP tests are performed by subjecting samples to a much harsher environment than would be encountered in natural surroundings. The samples of concrete are pulverized to maximize exposure to the acid used. Next, a particularly harsh solution of acetic acid is employed. Acid can leach out much higher concentrations of metals than liquids to which concrete is normally exposed, such as rain or groundwater. Finally, the amount of acid solution used is very large in comparison to the amount of concrete.

Health risks to residents near cement kilns may actually decrease when hazardous waste fuels are used. This is because the permit needed to recycle hazardous waste fuels requires more stringent emissions controls than those for cement kilns using only fossil fuels. Also, fossil fuels contain natural impurities that are reduced or no longer emitted when some types of hazardous waste fuels are used.

3.15.6 Southdown Experience in Waste Fuel Selection

Production of high-quality cement and compliance with environmental regulations are Southdown's top priorities. Therefore, great care is taken to ensure that only those wastes that can be safely recycled and that are compatible with the cement manufacturing process are used. Cement production requires fuels with a high energy value. Waste materials that provide enough heat include such familiar items as paint thinners, printing inks, paint residues, and industrial cleaning solvents. Cement kilns can also help alleviate one of the most difficult solid waste problems – scrap tires, which take up valuable landfill space. Tires (also known as "tire-derived fuel") can be used as an efficient fuel in cement kilns.

Before Southdown accepts any waste materials for recycling as fuel, a chemical analysis must be performed to identify their chemical composition. Wastes that cannot be blended to meet standards for content, heat value, and compatibility with cement production are not accepted. For instance, cement cannot be made with fuels that have a high chlorine content.

Both fossil fuels and hazardous waste fuels used in Southdown cement kilns contain metals. The raw materials (limestone, clay, sand) used to make cement clinker also contain metals. In fact, certain metals, such as iron and aluminum, are essential components of the final product. While metals cannot be destroyed, the Southdown cement kiln process effectively manages them in the following ways: (a) cement kiln operators limit emissions by carefully restricting the metals content in wastes accepted for recycling; (b) dust particles containing metals are returned to the kiln through closed-loop mechanisms, where metals are chemically bonded into the cement clinker; (c) particles not returned to the kiln are captured in state-of-the-art pollution control devices; and (d) small amounts are emitted from the stack in quantities strictly limited by USEPA's BIF rule.

Electrostatic precipitators and baghouses are used to catch dust particles containing metals. Electrostatic precipitators use an electrical field to remove the particles. Baghouses use fiberglass filters, similar to vacuum cleaner bags, to catch them. The majority of theses particles, called cement kiln dust (CKD), are trapped by this equipment and returned to the kiln for incorporation into the cement clinker. Under USEPA's BIF rule, Southdown tests its cement kiln dust to judge whether it is hazardous. If the CKD does not meet the standards set under the BIF rule, it must be disposed of in accordance with USEPA's strict hazardous waste regulations. For that reason, Southdown does not accept fuels that would cause the waste CKD to fail this test.

3.15.7 Southdown Experience in Product Quality Monitoring

A concrete made from Southdown cement is called a "Southdown concrete." Even under the TCLP testing extreme conditions, the amount of metals that leached out of the Southdown concrete were many orders of magnitude below the standards set by USEPA. In many cases the levels were, in fact, below the limits of detection for the test. One historical use of Southdown concrete has been for pipes used to transport drinking water. Drinking water is routinely tested to show that it meets Federal standards for a wide variety of contaminants, including metals. If metals leaching from the concrete pipes were a concern after many years of use, either USEPA or another recognized scientific organization would have sounded a warning.

The water distribution system in the city of Dayton, OH, uses Southdown concrete water mains to deliver water to its citizens. Routine sampling and testing of Dayton's water supply by the city's Department of Water consistently shows that the levels of metals are well below the Ohio EOA Community Drinking Water Standards, and that these levels have remained constant throughout a nine-year testing period from 1982 to 1990. Because metal leaching has not occurred, there is no reason for concern over the safety of Southdown concrete pipes to transport drinking water.

REFERENCES

1. Wang, L.K.; Wang, M.H.S.; Wang, P. *Management of Industrial Hazardous Substances at Industrial Sites*; United Nations Industrial Development Organization (UNIDO): Vienna, Austria, 1995; Training Manual no. 4-4-95, 104 p.
2. Wang, L.K. *Case Studies of Cleaner Production and Site Remediation.* United Nations Industrial Development Organization (UNIDO): Vienna, Austria, April 1995; Training Manual no. 5-4-95, 136 p.
3. WPCF. *Hazardous Waste Treatment Processes*; Water Pollution Control Federation: Alexandria, VA, 1990.
4. Massachusetts DEP. *Massachusetts Hazardous Waste Regulations, 310CMR30.000*; Massachusetts Department of Environmental Protection: Boston, MA, 2000.
5. Massachusetts DEP. *Massachusetts Hazardous Waste Information for Medical Offices*; Massachusetts Department of Environmental Protection: Boston, MA, 1988.
6. Massachusetts DEP. *Large Quantity Generator Fact Sheet*; Massachusetts Department of Environmental Protection: Boston, MA, 2000.
7. Massachusetts DEP. *Small Quantity Generator Fact Sheet*; Massachusetts Department of Environmental Protection: Boston, MA, 2000.
8. Massachusetts DEP. *Application for an USEPA Identification Number*; Massachusetts Department of Environmental Protection: Boston, MA, 1992.
9. Massachusetts DEP. *Guide for Determining Status and Regulatory Requirements*; Massachusetts Department of Environmental Protection: Boston, MA, 1992.
10. Massachusetts DEP. *How Many of Your Common Household Products Are Hazardous?* Massachusetts Department of Environmental Protection: Boston, MA, January 1983.
11. New York DEC. *Petroleum Bulk Storage*, Parts 612, 613, and 614; NYS Department of Environmental Conservation: Albany, NY, December 27, 1985.
12. New York DEC. *Supporting Documents for Chemical Bulk Storage Regulations*, Parts 595, 596, and 597; NYS Department of Environmental Conservation: Albany, NY, October 1987.
13. New York DEC. *Water Pollution Control and Enforcement Laws, and Environmental Conservation Law of the State of New York*; NYS Department of Environmental Conservation: Albany, NY, 1992.
14. New York DEC. *Chemical Bulk Storage*, Parts 595, 596, 597, 598, and 599; NYS Department of Environmental Conservation: Albany, NY, May 1993.
15. USEPA. *Scientific and Technical Assessment Report on Cadmium*, USEPA-600/6-75-003; U.S. Environmental Protection Agency: Washington, DC, 1975.
16. USEPA. *Federal Register*; U.S. Environmental Protection Agency: Washington, DC, November 28, 1980.
17. USEPA. *Field Standard Operating Procedures for the Decontamination of Response Personnel*, Publication No. FSOP-7; U.S. Environmental Protection Agency: Washington, DC, January 1985.
18. USEPA. *Reclamation and Redevelopment of Contaminated Land: U.S. Case Studies*, Report No. USEPA/600/2-86/066; U.S. Environmental Protection Agency: Washington, DC, August 1986.

19. USEPA. *Solid Waste and Emergency Response*, Report No. USEPA/625/6-87-015; U.S. Environmental Protection Agency: Washington, DC, 1987.
20. USEPA. *Managing Asbestos in Place – A Building Owner's Guide to Operations and Maintenance Programs for Asbestos-Containing Materials*, Report No. TS-799; U.S. Environmental Protection Agency: Washington, DC, July 1990.
21. USGPO. *Occupational Safety and Health Guidance Manual for Hazardous Waste Site Activities*, Publication No. DHHS-NIOSH-85-115; U.S. Government Printing Office: Washington, DC, October 1985.
22. USGPO. *Asbestos in the Home*; U.S. Government Printing Office: Washington, DC, August 1989.
23. Aulenbach, D.B.; Ryan, R.M. Management of radioactive wastes. In *Handbook of Environmental Engineering, Volume 4, Water Resources and Natural Control Processes*; Wang, L.K., Pereira, N.C. Eds.; The Humana Press: Clifton, NJ, 1986; 283–372.
24. Centofanti, L.F. Halting the legacy. Environ. Protection **2002**, *13* (*5*), 30.
25. Wang, L.K. *Biological Process for Groundwater and Wastewater Treatment*. U.S. Patent No. 5,451,320, September 19, 1995.
26. Jensen, P. Noise control. In *Handbook of Environmental Engineering, Volume 1, Air and Noise Pollution Control*; Wang, L.K., Pereira, N.C., Eds.; Clifton, NJ: 1979, 411–474.
27. Gallery, A.G. Disinfect with sodium hypochlorite – safety guidelines. Chem. Engng Prog. **2003**, *99* (*3*), 42–47.
28. USEPA. *Chemical Aids Manual for Wastewater Treatment Facilities*; U.S. Environmental Protection Agency: Washington, DC, 1979; 430/9-79-018.
29. Swichtenberg, B. Firefighting. Water Engng. and Mgnt. **2003**, *150* (*3*), 8–9.
30. NRC. *Arsenic*; Committee on Medical and Biological Effects of Environmental Pollutants, National Research Council, National Academy of Sciences: Washington, DC, 1977; ISBN 0-709-02604-0.
31. Wirth, N. Hazardous chemical safety. Operations Forum, Water Environment Federation, Alexandria, VA, **1998**, *10* (*8*).
32. Cook, C.; McDaniel, D. IAQ problem solving. Environ. Protection **2002**, *13* (*5*), 24.
33. Ferrante, L.M. Indoor/in-plant air quality. Natl. Environ. J. **1993**, *March–April*, 36–40.
34. Wang, L.K.; Wu, B.C.; Zepka, J. *An Investigation of Lead Content in Paints and PCB content in Water Supply for Eagleton School*; U.S. Department of Commerce, National Technical Information Service: Springfield, VA, 1984; Report No. PB86-169315, 23 p.
35. Wang, L.K.; Zepka, J. *An Investigaiton of Asbestos Content in Air for Eagleton School*; U.S. Department Commerce, National Technical Information Service: Springfield, VA, 1984; Report No. PB86-194172/AS, 17 p.
36. Wang, M.H.S.; Wang, L.K. Ventilation and air conditioning. In *Handbook of Environmental Engineering, Volume 1, Air and Noise Pollution Control*; Wang, L.K., Pereira, N.C., Eds.; The Humana Press: Clifton, NJ, 1979; 271–353.
37. Nevius, J.G. Brownfields and green insurance. Environ. Protection **2003**, *14* (*2*), 27–27.
38. Addlestone, S.I. Waste makes haste: top issues in waste management in 2003. Environ. Protection **2003**, *14* (*1*), 34–37.
39. Wang, L.K.; Wang, M.H.S. *Control of Hazardous Wastes in Petroleum Refining Industry, Symposium on Environmental Technology and Managements*; U.S. Department of Commerce, National Technical Information Service: Springfield, VA, 1982; Report No. PB82-185273, 60–77.
40. Wang, L.K.; McGinnis, W.C.; Wang, M.H.S. *Analysis and Formulation of Combustible Components in Wasted Rubber Tires*; U.S. Department of Commerce, National Technical Information Service: Springfield, VA, 1985; Report No. PB86-169281/AS, 1985; 39 p.
41. Hrycyk, O.; Kurylko, L.; Wang, L.K. *Removal of Volatile Compounds and Surfactants from Liquid*. U.S. Patent No. 5,122,166, June 1992.
42. Wang, L.K.; Hrycyk, O.; Kurylko, L. *Removal of Volatile Compounds and Surfactants from Liquid*. U.S. Patent No. 5,122,165, June 1992.
43. Wang, L.K.; Kurylko, L.; Wang, M.H.S. *Contamination Removal System Employing Filtration and Plural Ultraviolet and Chemical Treatment Steps and Treatment Mode Controller*. U.S. Patent No. 5,190,659, March 1993.

44. Wang, L.K.; Kurylko, L.; Wang, *M.H.S. Method and Apparatus for Filtration with Plural Ultraviolet Treatment Stages*. U.S. Patent No. 5,236,595, August 17, 1993.
45. Clabby, C. PCBs threaten a way of life. Times Union **1993**, July 15, A-1, A-12.
46. Cheremisinoff, P.N. Focus on high hazard pollutants. Pollut. Engng **1990**, *22* (*2*), 67–79.
47. Krofta, M.;Wang, L.K. *Hazardous Waste Management in Institutions and Colleges*; U.S. Department of Commerce, National Technical Information Service: Springfield, VA, 1985; Report No. PB86-194180/AS, June 1985.
48. Desilva, F.J. Nitrate removal by ion exchange. Water Qual. Prod. **2003**, *8* (*4*), 9–30.
49. Kucera, D. Court upholds U.S. EPA's radionuclide rules. Water Engng. Mgnt. **2003**, *150* (*4*), 7.
50. Environmental Control Library. PCB. *Current Awareness* **1993**, *93* (*13*), 5–7.
51. Martin, W.H. Risk management of PCB transformers. Pollut. Engng. **1990**, *March*, 74–77.
52. Cheremisinoff, P.N. Spill and leak containment and emergency response. Pollut. Engng. **1989**, *21* (*13*), 42–51.
53. Steinway, D.M. Scope and numbers of regulations for asbestos-containing materials, abatement continue to grow. Hazmat World **1990**, *April*, 32–58.
54. Hannak, W.R. Hazardous waste shipping containers. Environ. Protection **2002**, *13* (*1*), 34.
55. Newton, J.J. Are you an SQG? Pollut. Engng. **1989**, *21* (*13*), 64–66.
56. Wang, L.K.; Cheryan, M. Application of membrane technology in food industry for cleaner production. Water Treatment **1995**, *10* (*4*), 283–298.
57. Bober, T.W.; Dagon, T.J.; Fowler, H.E. Treatment of photographic processing wastes. In *Handbook of Industrial Waste Treatment*; Wang, L.K., Wang, M.H.S., Eds.; Marcel Dekker, Inc.: New York, NY, 1992; 173–227.
58. Kodak Company. *Choices: Choosing the Right Silver Recovery Method for Your Need–Environment*; Kodak Company: Rochester, NY,1987.
59. Wang, L.K.; Wang, M.H.S. *Method and Apparatus for Purifying and Compacting Solid Wastes*. U.S. Patent No. 5,232,584, August 7, 1983.
60. ASTM. *ASTM Standards on Environmental Site Characterization*; ASTM, 2002; 1827 p.
61. DeChacon, J.R.;Van Houten, N.J. Indoor air quality. Natl. Environ. J. **1991**, *16–18 Nov*.
62. Drill, S.; Konz, J.H. Mahar; Morse, M. *The Environmental Lead Problem: An Assessment of Lead in Drinking Water from a Multi-Media Perspective*; U.S. Environmental Protection Agency, Criteria and Standards Division, PB-296 566, May 1979.
63. Hall, S.K. Oil spills at sea. Pollut. Engng. **1989**, *21* (*13*), 59–63.
64. Heinold, D.; Smith, D. Finding the weak links. Environ. Protection **2003**, *14* (*2*), 56–65.
65. Munshower, F.F. Microelements and their role in surface mine planning. In *Coal Development: Collected Papers, Volume II*; Coal Development Workshops in Grand Junction, Colorado and Casper, Wyoming, Bureau of Land Management, July 1983.
66. Nagl, G.J. Air: controlling hydrogen sulfide emissions. Environ. Technol. **1999**, *9* (*7*), 18–23.
67. NRC. *Nickel*; Committee on Medical and Biological Effects of Environmental Pollutants, Division of Medical Sciences, National Research Council, National Academy of Sciences: Washington, DC, 1975; ISBN 0-309-02314-9.
68. NRC. *Zinc*; Committee on Medical and Biological Effects of Environmental Pollutants, National Research Council, National Academy of Sciences, prepared for the U.S. Environmental Protection Agency: Washington, DC, 1978; USEPA-600/1-78-034.
69. NRC. *Lead in the Human Environment*; Committee on Lead in the Human Environment, National Research Council, National Academy of Sciences: Washington, DC, 1980.
70. Qudir, R.M. The brownfield challenge. Environ. Protection **2002**, *13* (*3*), 34–39.
71. Ryan, J.A. Factors affecting plant uptake of heavy metals from land application of residuals. In *Proceedings of the National Conference on Disposal of Residues on Land*, St. Louis, Missouri, September 13–15, 1976.
72. Spicer, S. Accident releases lethal gas. Operations Forum, Water Environment Federation, Alexandria, VA, **2001**, *13* (*8*).
73. Turner, P.L. Preparing hazardous waste transport manifests. Environ. Protection **1987**, *3* (*10*), 12–16.

74. VanDenBos, A.; Izadpanah, A. A new partnership for handling medical waste. Environ. Protection **2002**, *13* (*5*), 26.
75. Wang, L.K. *Prevention of Airborn Legionnaires' Disease by Formation of a New Cooling Water for Use in Central Air Conditioning Systems*; U.S. Department of Commerce, National Technical Information Service, Report No. PB85-215317/AS, 1984; 92 p.
76. Wang, L.K. *Design of Innovative Flotation – Filtration Wastewater Treatment Systems for a Nickel–Chromium Plating Plant*; U.S. Department of Commerce, National Technical Information Service, Report No. PB88-200522/AS, 1984; 50 p.
77. Wang, L.K.; Pressman, M.; Shuster, W.W.; Shade, R.W.; Bilgen, F.; Lynch, T. Separation of nitrocellulose fine particles from industrial effluent with organic polymers. Can. J. Chem. Eng. **1982**, *60*, 116–122.
78. Wang, L.K.; Wu, B.C. Treatment of groundwater by dissolved air flotation systems using sodium aluminate and lime as flotation aids. OCEESA J. **1984**, *1* (*3*), 5–18; NTIS Report No. PB85-167229/AS.
79. Wang, M.H.S.; Wang, L.K.; Simmons, T.; Bergenthal, J. Computer-aided air quality management. J. Environ. Mgnt. **1979**, *9*, 61–87.
80. Wang, L.K. *Identification, Transfer, Acquisition and Implementation of Environmental Technologies Suitable for Small and Medium Size Enterprises*; United Nations Industrial Development Organization (UNIDO), Vienna, Austria, 1995; Technical paper no. 9-9-95, 5 p.
81. Wang, L.K. *Liquid Treatment System with Air Emission Control.* U.S. Patent No. 5,399,267, March 21, 1995.
82. Wastewater Engineers, Inc. Case histories in wastewater recycling. Environ. Technol. **1999**, *9* (*7*), 24–26.
83. Wang, L.K.; Wang, P.; Celesceri, N.L. Groundwater decontamination using sequencing batch processes. Water Treatment **1995**, *10* (*2*), 121–134.
84. Wang, L.K. *Site Remediation Technology*. U.S. Patent No. 5,552,051, September 3, 1996.
85. Wang, L.K.; Pereira, N.C.; Hung, Y. *Air Pollution Control Engineering*. Humana Press, Totowa, NJ. 2004.
86. Wang, L.K.; Pereira, N.C.; Hung, Y. *Advanced Air and Noise Pollution Control.* Humana Press, Totowa, N.J. 2004.
87. Wang, L.K.; Hung, Y.; Shammas, N.K. *Physicochemical Treatment Processes*. Humana Press, Totowa, NJ. 2004.

4

Application of Biotechnology for Industrial Waste Treatment

Joo-Hwa Tay, Stephen Tiong-Lee Tay, and Volodymyr Ivanov
Nanyang Technological University, Singapore

Yung-Tse Hung
Cleveland State University, Cleveland, Ohio, U.S.A.

4.1 BIOTREATABILITY OF INDUSTRIAL HAZARDOUS WASTES

Environmental biotechnology concerns the science and practical knowledge relating to the use of microorganisms and their products. Biotechnology combines fundamental knowledge in microbiology, biochemistry, genetics, and molecular biology, and engineering knowledge of the specific processes and equipment. The main applications of biotechnology in industrial hazardous waste treatment are: prevention of environmental pollution through waste treatment, remediation of polluted environments, and biomonitoring of environment and treatment processes. The common biotechnological process in the treatment of hazardous waste is the biotransformation or biodegradation of hazardous substances by microbial communities.

Bioagents for hazardous waste treatment are biotechnological agents that can be applied to hazardous waste treatment including bacteria, fungi, algae, and protozoa. Bacteria are microorganisms with prokaryotic cells and typically range from 1 to 5 μm in size. Bacteria are most active in the biodegradation of organic matter and are used in the wastewater treatment and solid waste or soil bioremediation. Fungi are eukaryotic microorganisms that assimilate organic substances and typically range from 5 to 20 μm in size. Fungi are important degraders of biopolymers and are used in solid waste treatment, especially in composting, or in soil bioremediation for the biodegradation of hazardous organic substances. Fungal biomass is also used as an adsorbent of heavy metals or radionuclides. Algae are saprophytic eukaryotic microorganisms that assimilate light energy. Algal cells typically range from 5 to 20 μm in size. Algae are used in environmental biotechnology for the removal of organic matter in waste lagoons. Protozoa are unicellular animals that absorb organic food and digest it intracellularly. Typical cell size is from 10 to 50 μm. Protozoa play an important role in the treatment of industrial hazardous solid, liquid, and gas wastes by grazing on bacterial cells, thus maintaining adequate bacterial biomass levels in the treatment systems and helping to reduce cell concentrations in the waste effluents.

Microbial aggregates used in hazardous waste treatment. Microorganisms are key biotechnology agents because of their diverse biodegradation and biotransformation abilities and their small size. They have high ratios of biomass surface to biomass volume, which ensure high rates of metabolism. Microorganisms used in biotechnology typically range from 1 to 100 μm in size. However, in addition to individual cells, cell aggregates in the form of flocs, biofilms, granules, and mats with dimensions that typically range from 0.1 to 100 mm may also be used in biotechnology. These aggregates may be suspended in liquid or attached to solid surfaces. Microbial aggregates that can accumulate in the water–gas interface are also useful in biotechnology applications in hazardous waste treatment.

Microbial communities for hazardous waste treatment. It is extremely unusual for biological treatment to rely solely on a single microbial strain. More commonly, communities of naturally selected strains or artificially combined strains of microorganisms are employed. Positive or negative interactions may exist among the species within each community. Positive interactions, such as commensalism, mutualism, and symbiosis, are more common in microbial aggregates. Negative interactions, such as amensalism, antibiosis, parasitism, and predation, are more common in natural or engineering systems with low densities of microbial biomass, for example, in aquatic or soil ecosystems.

4.1.1 Industrial Hazardous Solid, Liquid, and Gas Wastes

Hazardous Waste

Industrial wastes are identified as hazardous wastes by the waste generator or by the national environmental agency either because the waste component is listed in the List of Hazardous Inorganic and Organic Constituents approved by the national agency or because the waste exhibits general features of hazardous waste, such as harming human health or vital activity of plants and animals (acute and chronic toxicity, carcinogenicity, teratogenicity, pathogenicity, etc.), reducing biodiversity of ecosystems, flammability, corrosive activity, ability to explode, and so on. The United States annually produces over 50 million metric tonnes of federally regulated hazardous wastes [1].

Hazardous Substances

It is estimated that approximately 100,000 chemical compounds have been produced industrially [2,3] and many of them are harmful to human health and to the environment. However, only 7% of the largest-volume chemicals require toxicity screening [2]. In the United States, the Agency for Toxic Substances and Disease Registry (ATSDR) and the Environmental Protection Agency (EPA) maintain a list, in order of priority, of substances that are determined to pose the most significant potential threat to human health due to their known or suspected toxicity. This Comprehensive Environmental Response, Compensation, and Liability Act (CERCLA) Priority List of Hazardous Substances was first issued in 1999 and includes 275 substances (www.atsdr.cdc.gov/clist.html).

Application of Biotechnology in the Treatment of Hazardous Substances from the CERCLA Priority List

The CERCLA Priority List of Hazardous Substances has been annotated with information on the types of wastes and the possible biotechnological treatment methods, as shown in Table 1. The remarks on biotreatability of these hazardous substances are based on data from numerous papers, reviews, and books on this topic [4–8]. Databases are available on the biodegradation of hazardous substances. For example, the Biodegradative Strain Database [9] (bsd.cme.msu.edu) can be used to select suitable microbial strains for biodegradation applications, while the

Table 1 Major Hazardous Environmental Pollutants and Applicability of Biotechnology For Their Treatment

1999 Rank	Substance name	Type of waste (S = solid, L = liquid, G = gas)	Biotechnological treatment with formation of nonhazardous or less hazardous products
1	Arsenic	S, L	Bioreduction/biooxidation following immobilization or dissolution
2	Lead	S,L	Bioimmobilization, biosorption, bioaccumulation
3	Mercury	S,L,G	Bioimmobilization, biovolatilization, biosorption
4	Vinyl chloride	L,G	Biooxidation by cometabolization with methane or ammonium
5	Benzene	L,G	Biooxidation
6	Polychlorinated biphenyls	S,L	Biooxidation after reductive or oxidative biodechlorination
7	Cadmium	S,L	Biosorption, bioaccumulation
8	Benzo(A)pyrene	S,L	Biooxidation and cleavage of the rings
9	Polycyclic aromatic hydrocarbons	S,L,G	Biooxidation and cleavage of the rings
10	Benzo(B)fluoranthene	S,L	Biooxidation and cleavage of the rings
11	Chloroform	L,G	Biooxidation by cometabolization with methane or ammonium
12	DDT, P,P′-	S,L	Biooxidation after reductive or oxidative biodechlorination
13	Aroclor 1260	S,L	Biooxidation after reductive or oxidative biodechlorination
14	Aroclor 1254	S,L	Biooxidation after reductive or oxidative biodechlorination
15	Trichloroethylene	L,G	Biooxidation by cometabolization with methane or ammonium
16	Chromium, hexavalent	S,L	Bioreduction/bioimmobilization, biosorption
17	Dibenzo(A,H)anthracene	S,L	Biooxidation and cleavage of the rings
18	Dieldrin	S,L	Biooxidation after reductive or oxidative biodechlorination
19	Hexachlorobutadiene	L,G	Biooxidation after reductive or oxidative biodechlorination
20	DDDE, P,P′-	S,L	Biooxidation after reductive or oxidative biodechlorination
21	Creosote	S,L	Biooxidation and cleavage of the rings
22	Chlordane	S,L	Biooxidation after reductive or oxidative biodechlorination
23	Benzidine	L,G	Biooxidation and cleavage of the rings
24	Aldrin	S,L	Biooxidation
25	Aroclor 1248	S,L	Biooxidation after reductive or oxidative biodechlorination
26	Cyanide	S,L,G	Removal by ferrous ions produced by bacterial reduction of Fe(III)

(continues)

Table 1 Continued

1999 Rank	Substance name	Type of waste (S = solid, L = liquid, G = gas)	Biotechnological treatment with formation of nonhazardous or less hazardous products
27	DDD, P,P′-	S,L	Biooxidation after reductive or oxidative biodechlorination
28	Aroclor 1242	S,L	Biooxidation after reductive or oxidative biodechlorination
29	Phosphorus, white	S,L,G	
30	Heptachlor	L,G	
31	Tetrachloroethylene	L,G	Biooxidation by cometabolization with methane or ammonium
32	Toxaphene	S,L	Reductive (anaerobic) dechlorination
33	Hexachlorocyclohexane, gamma-	S,L,G	Biooxidation by white-rot fungi
34	Hexachlorocyclohexane, beta-	S,L,G	Biooxidation by white-rot fungi
35	Benzo(A)Anthracene	S,L	Biooxidation and cleavage of the rings
36	1,2-Dibromoethane	L,G	Biooxidation by cometabolization with methane or ammonium
37	Disulfoton	S,L	Biooxidation
38	Endrin	S,L	Biooxidation
39	Beryllium	S,L	Biosorption
40	Hexachlorocyclohexane, delta-	S,L,G	Biooxidation by white-rot fungi; biooxidation after reductive or oxidative biodechlorination
41	Aroclor 1221	S,L	Biooxidation after reductive or oxidative biodechlorination
42	Di-*N*-Butyl phthalate	L,G	Biooxidation
43	1,2-Dibromo-3-chloropropane	L,G	Biooxidation after reductive or oxidative biodechlorination
44	Pentachlorophenol	L,G	Biooxidation after reductive or oxidative biodechlorination
45	Aroclor 1016	S,L	Biooxidation after reductive or oxidative biodechlorination
46	Carbon tetrachloride	L,G	Biodechlorination and biodegradation
47	Heptachlor epoxide	L,G	
48	Xylenes, total	S,L,G	Biooxidation
49	Cobalt	S,L	Biosorption
50	Endosulfan sulfate	S,L	Biosorption
51	DDT, O,P′-	S,L	Biooxidation by white-rot fungi
52	Nickel	S,L	Biosorption
53	3,3′-Dichlorobenzidine	L,G	Biooxidation after reductive or oxidative biodechlorination

54	Dibromochloropropane	L,G	Biooxidation after reductive or oxidative biodechlorination
55	Endosulfan, alpha	S,L	Biooxidation by fungi or bacteria
56	Endosulfan	S,L	Biooxidation by fungi or bacteria
57	Benzo(K)fluoranthene	S,L	Biooxidation and cleavage of the rings
58	Aroclor	S,L	Biooxidation after reductive or oxidative biodechlorination
59	Endrin ketone	S,L	
60	Cis-Chlordane	S,L	Biooxidation after reductive or oxidative biodechlorination
61	2-Hexanone	L,G	
62	Toluene	L,G	Biooxidation and cleavage of the ring
63	Aroclor 1232	S,L	Biooxidation after reductive or oxidative biodechlorination
64	Endosulfan, beta	S,L	Biooxidation by fungi and bacteria
65	Methane	G	Biooxidation by methanotrophic bacteria
66	Trans-Chlordane	S,L,G	
67	2,3,7,8-Tetrachlorodibenzo-p-dioxin	S,L	Biooxidation after reductive or oxidative biodechlorination
68	Benzofluoranthene	S,L	Biooxidation and cleavage of the rings
69	Endrin aldehyde	S,L	
70	Zinc	S,L	Microbial immobilization/solubilization
71	Dimethylarsinic acid	S,L	
72	Di(2-ethylhexyl)phthalate	S,L	Biooxidation and cleavage of the rings
73	Chromium	S,L	Microbial reduction/oxidation followed immobilization or solubilization
74	Methylene chloride	L,G	Biooxidation by cometabolization with methane or ammonium
75	Naphthalene	S,L,G	Biooxidation and cleavage of the rings
76	Methoxychlor	S,L	Biooxidation after reductive or oxidative biodechlorination
77	1,1-Dichloroethene	L,G	Biooxidation by cometabolization with methane or ammonium
78	Aroclor 1240	S,L	Biooxidation after reductive or oxidative biodechlorination
79	*Bis*(2-chloroethyl) ether	L,G	
80	1,2-Dichloroethane	L,G	Biooxidation by cometabolization with methane or ammonium
81	2,4-Dinitrophenol	S,L,G	Biooxidation
82	2,4,6-Trinitrotoluene	S, L,G	Biooxidation
83	2,4,6-Trichlorophenol	S,L,G	Biooxidation
84	Chlorine	L,G	Removal by ferrous or manganese ions produced by bacterial reduction of Fe(III) and Mn(IV)
85	Cyclotrimethylenetrinitramine (Rdx)	S,L	

(continues)

Table 1 Continued

1999 Rank	Substance name	Type of waste (S = solid, L = liquid, G = gas)	Biotechnological treatment with formation of nonhazardous or less hazardous products
86	1,1,1-Trichloroethane	L,G	Biooxidation by cometabolization with methane or ammonium
87	Ethylbenzene	L,G	Biooxidation and cleavage of the rings
88	1,1,2,2-Tetrachloroethane	L,G	Biooxidation by cometabolization with methane or ammonium
89	Thiocyanate	S,L	Removal by ferrous or manganese ions produced by bacterial reduction of Fe(III) and Mn(IV)
90	Asbestos	S,G	
91	4,6-Dinitro-o-cresol	S,L	Biooxidation
92	Uranium	S,L	Bioleaching of uranium from minerals
93	Radium	S,L	
94	Radium-226	S,L	
95	Hexachlorobenzene	L,G	
96	Ethion	S,L	
97	Thorium	S,L	
98	Chlorobenzene	S,L,G	Biooxidation after reductive or oxidative biodechlorination
99	Barium	S,L	Biosorption
100	2,4-Dinitrotoluene	S,L	Biooxidation
101	Fluoranthene	S,L	Biooxidation and cleavage of the rings
102	Radon	G	
103	Radium-228	S,L	
104	Thorium-230	S,L	
105	Diazinon	S,L	
106	Bromine	G	Binding with Fe or Mn reduced by bacteria
107	1,3,5-Trinitrobenzene	S,L,G	Biodegradation
108	Uranium-235	S,L	Biosorption/bioleaching and oxidation/reduction mediated by other elements oxidized or reduced by microorganisms
109	Tritium	S,L	
110	Uranium-234	S,L	Biosorption/bioleaching and oxidation/reduction mediated by other elements oxidized or reduced by microorganisms

111	Thorium-228	S,L	
112	*N*-Nitrosodi-*N*-propylamine	S,L,G	
113	Cesium-137	S,L	Bioimmobilization/biosorption
114	Hexachlorocyclohexane, alpha-	S,L	Biooxidation after reductive or oxidative biodechlorination
115	Chrysene	S,L	Biooxidation and cleavage of the rings
116	Radon-222	G	
117	Polonium-210	S,L	
118	Chrysotile asbestos	S,G	
119	Thorium-227	S,L	
120	Potassium-40	S,L	Bioaccumulation
121	Coal tars	S,L	Biooxidation
122	Plutonium-238	S,L	Biosorption
123	Thoron (Radon-220)	G	
124	Copper	S,L	Biosorption
125	Strontium-90	S,L	Bioimmobilization/solubilization
126	Cobalt-60	S,L	Biosorption
127	Methylmercury	L,G	Biodegradation
128	Chlorpyrifos	S,L	
129	Lead-210	S,L	Biosorption
130	Plutonium-239	S,L	Biosorption
131	Plutonium	S,L	Biosorption
132	Americium-241	S,L	
133	Iodine-131	S,L	
134	Amosite asbestos	S,G	
134	Guthion	S,L	
136	Bismuth-214	S,L	
136	Lead-214	S,L	Biosorption
138	Chlordecone	S,L	
138	Plutonium-240	S,L	Biosorption
138	Tributyltin	S,L	Biodetoxication
141	Manganese	S,L	Microbial reduction/oxidation
142	S,S,S-Tributyl phosphorotrithioate	S,L,G	
143	Selenium	S,L	Microbial reduction/oxidation

(continues)

Table 1 Continued

1999 Rank	Substance name	Type of waste (S = solid, L = liquid, G = gas)	Biotechnological treatment with formation of nonhazardous or less hazardous products
144	Polybrominated biphenyls	S,L	Biooxidation after reductive or oxidative biodechlorination
145	Dicofol	S,L	
146	Parathion	S,L	Biodegradation by enzymes of genetically engineered strains
147	Hexachlorocyclohexane, technical	S,L	Biooxidation after reductive or oxidative biodechlorination
148	Pentachlorobenzene	L,G	Biooxidation after reductive or oxidative biodechlorination
149	Trichlorofluoroethane	L,G	Biooxidation by cometabolization with methane or ammonium
150	Treflan (Trifluralin)	S,L	
151	4,4′-Methylenebis(2-chloroaniline)	S,L	
152	1,1-Dichloroethane	L,G	Biooxidation by cometabolization with methane or ammonium
153	DDD, O,P′-	S,L	Biooxidation after reductive or oxidative biodechlorination
154	Hexachlorodibenzo-p-dioxin	S,L	Biooxidation after reductive or oxidative biodechlorination
155	Heptachlorodibenzo-p-dioxin	S,L	Biooxidation after reductive or oxidative biodechlorination
156	2-Methylnaphthalene	S,L	Biooxidation and cleavage of the rings
157	1,1,2-Trichloroethane	L,G	Biooxidation by cometabolization with methane or ammonium
158	Ammonia	L,G	Biooxidation (nitrification) followed denitrification; bioremoval by combined IRB/IOB biotechnology
159	Acenaphthene	S,L	
160	1,2,3,4,6,7,8,9-Octachlorodibenzofuran	S,L	Biooxidation after reductive or oxidative biodechlorination
161	Phenol	L,G	Biooxidation and cleavage of the rings; anaerobic biodegradation
162	Trichloroethane	L,G	Biooxidation by cometabolization with methane or ammonium
163	Chromium(Vi) trioxide	S,L	
164	1,2-Dichloroethene, trans-	L,G	Biooxidation by cometabolization with methane or ammonium
165	Heptachlorodibenzofuran	S,L	Biooxidation after reductive or oxidative biodechlorination
166	Hexachlorocyclopentadiene	L,G	Biooxidation after reductive or oxidative biodechlorination
167	1,4-Dichlorobenzene	L,G	Biooxidation after reductive or oxidative biodechlorination
168	1,2-Diphenylhydrazine	L,G	
169	Cresol, para-	S,L,G	
170	1,2-Dichlorobenzene	L,G	Biooxidation after reductive or oxidative biodechlorination

171	Lead-212	S,L	
172	Oxychlordane	S,L	Biooxidation after reductive or oxidative biodechlorination
173	2,3,4,7,8-Pentachlorodibenzofuran	S,L	Biooxidation after reductive or oxidative biodechlorination
174	Radium-224	G	
175	Acetone	L,G	
176	Hexachlorodibenzofuran	S,L	Biooxidation after reductive or oxidative biodechlorination
177	Benzopyrene	S,L	Biooxidation and cleavage of the rings
177	Bismuth-212	S,L	
179	Americium	S,L	
179	Cesium-134	S,L	Biosorption
179	Chromium-51	S,L	Bioreduction/biooxidation
182	Tetrachlorophenol	L,G	Biooxidation after reductive or oxidative biodechlorination
183	Carbon disulfide	L,G	
184	Chloroethane	L,G	Biooxidation by cometabolization with methane or ammonium
185	Indeno(1,2,3-Cd)pyrene	S,L	Biooxidation and cleavage of the rings
186	Dibenzofuran	S,L	Biooxidation and cleavage of the rings
187	p-Xylene	L,G	Biooxidation and cleavage of the rings
188	2,4-Dimethylphenol	L,G	Biooxidation and cleavage of the rings
189	Aroclor 1268	S,L	Biooxidation after reductive or oxidative biodechlorination
190	1,2,3-Trichlorobenzene	L,G	Biooxidation after reductive or oxidative biodechlorination
191	Pentachlorodibenzofuran	S,L	Biooxidation after reductive or oxidative biodechlorination
192	Hydrogen sulfide	L,G	Biooxidation by aerobic or microaerophilic bacteria; binding with ferrous ions produced by iron-reducing bacteria; biooxidation by phototrophic bacteria
193	Aluminum	S,L	
194	Tetrachloroethane	L,G	Biooxidation by cometabolization with methane or ammonium
195	Cresol, Ortho-	L,G	Biooxidation and cleavage of the rings
196	1,2,4-Trichlorobenzene	L,G	Biooxidation after reductive or oxidative biodechlorination
197	Hexachloroethane	L,G	Biooxidation after reductive or oxidative biodechlorination
198	Butyl benzyl phthalate	S,L	Biooxidation and cleavage of the rings
199	Chloromethane	L,G	Biooxidation by cometabolization with methane or ammonium
200	Vanadium	S,L	Biosorption
201	1,3-Dichlorobenzene	L,G	Biooxidation after reductive or oxidative biodechlorination
202	Tetrachlorodibenzo-p-dioxin	S,L	Biooxidation after reductive or oxidative biodechlorination

(continues)

Table 1 Continued

1999 Rank	Substance name	Type of waste (S = solid, L = liquid, G = gas)	Biotechnological treatment with formation of nonhazardous or less hazardous products
203	2-Butanone	G	Biooxidation
204	*N*-Nitrosodiphenylamine	S,L	
205	Pentachlorodibenzo-p-dioxin	S,L	Biooxidation after reductive or oxidative biodechlorination
206	2,3,7,8-Tetrachlorodibenzofuran	S,L	Biooxidation after reductive or oxidative biodechlorination
207	Silver	S,L	Biosorption
208	2,4-Dichlorophenol	L,G	Biooxidation after reductive or oxidative biodechlorination
209	1,2-Dichloroethylene	L,G	Biooxidation after reductive or oxidative biodechlorination
210	Bromoform	L,G	Biooxidation by cometabolization with methane or ammonium
211	Acrolein	L,G	
212	Chromic acid	S,L	
213	2,4,5-Trichlorophenol	L,G	Biooxidation after reductive or oxidative biodechlorination
214	Nonachlor, trans-	S,L	
215	Coal tar pitch	S,L	Biooxidation and cleavage of the rings
216	Phenanthrene	S,L	Biooxidation and cleavage of the rings
217	Nitrate	S,L	Microbial denitrification
218	Arsenic trioxide	S,L	
219	Nonachlor, *cis*-	S,L	
220	Hydrazine	L,G	
221	Technetium-99	S,L	Biosorption
222	Nitrite	S,L	Microbial denitrification
223	Arsenic acid	S,L	Bioreduction
224	Phorate	S,L	
225	Bromodichloroethane	L,G	Biooxidation by cometabolization with methane or ammonium
225	Dimethoate	S,L	
227	Strobane	S,L	
228	Naled	S,L	
229	Arsine	S,L	Biooxidation
230	4-Aminobiphenyl	S,L	

230	Pyrethrum	S,L	
230	Tetrachlorobiphenyl	S,L	Biooxidation after reductive or oxidative biodechlorination
233	Dibenzofurans, chlorinated	S,L	Biooxidation after reductive or oxidative biodechlorination
233	Ethoprop	S,L	
233	Nitrogen dioxide	G	Bioreduction
236	Carbophenothion	S,L	
236	Thorium-234	S,L	
238	Dichlorvos	S,L	
238	Ozone	G	
238	Palladium	S,L	
241	Calcium arsenate	S,L	Bioreduction; bioaccumulation
241	Carbon-14	S,L,G	
241	Europium-154	S,L	
241	Krypton-85	G	
241	Mercuric chloride	S,L	Bioimmobilization; biomethylation
241	Sodium-22	S,L	Bioaccumulation
241	Strontium-89	S,L	Biosorption
241	Sulfur-35	S,L,G	Biooxidation/bioreduction
241	Uranium-233	S,L	Bioaccumulation/biosorption or bioleaching
250	2,4-D Acid	S,L	
251	Antimony	S,L	Biooxidation/bioreduction
252	Cresols	L,G	Biooxidation and cleavage of the rings
253	Pyrene	S,L	Biooxidation and cleavage of the rings
254	2-Chlorophenol	L,G	Biooxidation after reductive or oxidative biodechlorination
255	Dichlorobenzene	S,L,G	Biooxidation after reductive or oxidative biodechlorination
256	Formaldehyde	L,G	
257	*N*-Nitrosodimethylamine	S,L	
258	Chlorodibromomethane	L,G	Biooxidation by cometabolization with methane or ammonium
259	Sutan	S,L	
260	Dichloroethane	S,L,G	Biooxidation by cometabolization with methane or ammonium
261	1,3-Dinitrobenzene	S,L,G	Biodegradation
262	Dimethyl formamide	S,L	Biodegradation
263	1,3-Dichloropropene, *cis-*	S,L,G	

(continues)

Table 1 Continued

1999 Rank	Substance name	Type of waste (S = solid, L = liquid, G = gas)	Biotechnological treatment with formation of nonhazardous or less hazardous products
264	Ethyl ether	L,G	
265	4-Nitrophenol	L,G	Biodegradation
266	1,3-Dichloropropene, *trans-*	L,G	
267	Trichlorobenzene	L,G	Biooxidation after reductive or oxidative biodechlorination
268	Fluoride	S,L	
269	1,2-Dichloropropane	L,G	Biooxidation after reductive or oxidative biodechlorination
270	2,6-Dinitrotoluene	L,G	Biodegradation
271	Methyl parathion	S,L	
272	Methyl isobutyl ketone	L,G	
273	Octachlorodibenzo-p-dioxin	S,L	Biooxidation after reductive or oxidative biodechlorination
274	Styrene	S,L	Biooxidation and cleavage of the rings
275	Fluorene	S,L	Biooxidation and cleavage of the rings

University of Minnesota Biocatalysis/Biodegradation Database (umbbd.ahc.umn.edu) can be used to predict biodegradation pathways and biodegradation metabolites. Approximately two-thirds of the hazardous substances mentioned in the CERCLA Priority List of Hazardous Substances can be treated by different biotechnological methods.

Production of Hazardous Wastes

The toxic substances appear mostly in: (a) the waste streams of manufacturing processes of commercial products; (b) the wastes produced during the use of these products, or (c) the post-manufacturing wastes related to the storage of these commercial products. Some toxic substances appear as constituents of commercial products that are disposed of once their useful lives are over [2]. If these products are disposed of in a landfill, product deterioration will eventually lead to release of toxic chemicals into the environment. The annual world production of hazardous wastes is estimated to range from 20×10^6 to 50×10^6 metric tonnes. These hazardous wastes include oil-polluted soil and sludges, hydroxide sludges, acidic and alkaline solutions, sulfur-containing wastes, paint sludges, halogenated organic solvents, nonhalogenated organic solvents, galvanic wastes, salt sludges, pesticide-containing wastes, explosives, and wastewaters and gas emissions containing hazardous substances [3].

Secondary Hazardous Wastes

Secondary wastes are generated from the collection, treatment, incineration, or disposal of hazardous wastes, such as sludges, sediments, effluents, leachates, and air emissions. These secondary wastes may also contain hazardous substances and must be treated or disposed of properly to prevent secondary pollution of underground water, surface water, soil, or air.

Oil and Petrochemical Industries as Sources of Hazardous Organic Wastes

The petrochemical industry is a major source of hazardous organic wastes, produced during the manufacture or use of hazardous substances. The recovery, transportation, and storage of raw oil or petrochemicals are major sources of hazardous wastes, often produced as the consequence of technological accidents. Seawater and freshwater pollution due to oil and oil-product spills, underground or soil pollution due to land spills or leakage from pipelines or tanks, and air pollution due to incineration of oil or oil sludge are major cases of environmental pollution. Gasoline is the main product in the petrochemical industry and consists of approximately 70% aliphatic linear and branched hydrocarbons, and 30% aromatic hydrocarbons, including xylenes, toluene, di- and tri-methylbenzenes, ethylbenzenes, benzene, and others. Other pure bulk chemicals used for chemical synthesis include formaldehyde, methanol, acetic acid, ethylene and polyethylenes, ethylene glycol and polyethylene glycols, propylene, propylene glycol and polypropylene glycols, and such aromatic hydrocarbons as benzene, toluene, xylenes, styrene, aniline, phthalates, naphthalene, and others.

Hazardous Wastes of Other Chemical Industries

The hazardous substances contained in solid, liquid, or gaseous wastes may include products from the pesticide and pharmaceutical industries. The paint and textile industries produce hazardous solid, liquid, and gaseous wastes that contain diverse organic solvents, paint and fiber preservatives, organic and mineral pigments, and reagents for textile finishing [3]. The pulp industry generates wastewater that contains chlorinated phenolic compounds produced in the chlorine bleaching of pulp. Widely used wood preservatives are usually chlorinated or unchlorinated monocyclic and polycyclic aromatic hydrocarbons. The explosives industry generates wastes containing recalcitrant chemicals with nitrogroups [3].

Xenobiotics and Their Biodegradability

Organic substances, synthesized in the chemical industry, are often hardly biodegradable. The substances that are not produced in nature and are slowly/partially biodegradable are called xenobiotics. Vinylchloride (a monomer for the plastic industry), chloromethanes and chloroethylenes (solvents), polychlorinated aromatic hydrocarbons (pesticides, fungicides, dielectrics, wood preservatives), and organophosphate- and nitro-compounds are examples of xenobiotics. The biodegradability of xenobiotics can be characterized by biodegradability tests such as: rate of CO_2 formation (mineralization rate), rate of oxygen consumption (respirometry test), ratio of BOD to COD (oxygen used for biological or chemical oxidation), and by the spectrum of intermediate products of biodegradation.

Hazardous Wastes of Nonchemical Industries

The coal industry, mining industry, hydrometallurgy, and metal industry are sources of solid and liquid wastes that may contain heavy metals, sulfides, sulfuric and other acids, and some toxic reagents used in industrial processes. The electronics and mechanical production industries are sources of hazardous wastes containing organic solvents, surfactants, and heavy metals. Nuclear facilities produce solid and liquid wastes containing radionuclides. Large-scale accidents on nuclear facilities serve as potential sources of radioactive pollution of air and soil, and the polluted areas can be as large as the combined areas of several states.

4.1.2 Suitability of Biotechnological Treatment for Hazardous Wastes

Comparison of Different Treatments of Hazardous Wastes

Usually, the hazardous substance can be removed or treated by physical, chemical, physicochemical, or biological methods. Advantages and disadvantages of these methods are shown in Table 2. The advantages of biotechnological treatment of hazardous wastes are biodegradation or detoxication of a wide spectrum of hazardous substances by natural microorganisms and availability of a wide range of biotechnological methods for complete destruction of hazardous wastes without production of secondary hazardous wastes. However, to intensify the biotreatment, nutrients and electron acceptors must be added, and optimal conditions must be maintained. On the other hand, there may be unexpected or negative effects mediated by microorganisms, such as emission of odors or toxic gases during the biotreatment, and it may be difficult to manage the biotreatment system because of the complexity and high sensitivity of the biological processes.

Cases When Biotechnology is Most Applicable for the Treatment of Hazardous Wastes

The main considerations for application of biotechnology in hazardous waste treatment are as follows:

1. Reasonable rate of biodegradability or detoxication of hazardous substance during biotechnological treatment; such rates are derived from a knowledge of the optimal treatment duration;
2. Necessity to have low volume or absence of secondary hazardous substances produced during biotechnological treatment;

Table 2 Advantages and Disadvantages of Different Treatments of Hazardous Wastes

Method of treatment	Advantages	Disadvantages
Physical treatment (sedimentation, volatilization, fixation, evaporation, heat treatment, radiation, etc.)	• Required time is from some minutes to some hours.	• High expenses for energy and equipment.
Chemical treatment (oxidation, incineration, reduction, immobilization, chelating, transformation)	• Required time is from some minutes to some hours.	• High expenses for reagents, energy, and equipment; air pollution due to incineration.
Physico-chemical treatment (adsorption, absorption)	• Required time is from some minutes to some hours.	• High expenses for adsorbents; formation of secondary hazardous waste.
Biotechnological aerobic treatment (oxidation, transformation, degradation)	• Low volume or absence of secondary hazardous wastes. • Process can be initiated by natural microorganisms or small quantity of added microbial biomass. • Wide spectrum of degradable substances and diverse methods of biodegradability.	• Some expenses for aeration, nutrients, and maintenance of optimal conditions. • Required time is from some hours to days. • Unexpected or negative effects of microorganisms-destructors. • Low predictability of the system because of complexity and high sensitivity of biological systems.
Biotechnological anaerobic treatment (reduction, degradation)	• Low volume or absence of secondary hazardous wastes. • Process can be initiated by natural microorganisms or small quantity of added microbial biomass. • Wide spectrum of degradable substances and diverse methods of biodegradability.	• Required time is from some days to months. • Emission of bad smelling or toxic gases. • Unexpected or negative effects of microorganisms-destructors. • Low predictability of the system because of complexity and high sensitivity of biological systems.
Landfilling (as a combination of physical and biological treatment)	• Low expenses for landfilling.	• Harmful air emissions; leaching; expensive land use. • Required time is some years.

3. Biotechnological treatment is more cost-effective than other methods; the low cost of biotechnological treatment is largely attributed to the small quantities or total absence of added reagents and microbial biomass to start up the biotreatment process;
4. Public acceptance of biotechnological treatment is better than for chemical or physical treatment.

However, the efficiency of actual biotechnological application depends on its design, process optimization, and cost minimization. Many failures have been reported on the way from bench laboratory-scale to field full-scale biotechnological treatment because of variability, instability, diversity, and heterogeneity of both microbial properties and conditions in the treatment system [10].

Treatment Combinations

In many cases, a combination of physical, chemical, physico-chemical, and biotechnological treatments may be more efficient than one type of treatment (Table 3). Efficient pretreatment schemes, used prior to biotechnological treatment, include homogenization of solid wastes in water, chemical oxidation of hydrocarbons by H_2O_2, ozone, or Fenton's reagent, photochemical oxidation, and preliminary washing of wastes by surfactants.

Roles of Biotechnology in Hazardous Waste Management

Biotechnology can be applied in different fields of hazardous waste management (Table 4): hazardous waste identification by biotechnological tests of toxicity and pathogenicity; prevention of hazardous waste production using biotechnological analogs of products; hazardous

Table 3 Examples of Combinations of Different Treatments

Combination of treatments	Example of combination
Physical and biotechnological treatment	• Thermal pretreatment of waste can enhance the biodegradability of hazardous substance. • Homogenization/suspension of solid wastes or nondissolved sludges in water will increase the surface of the waste particles and, as result of this, the rate of biodegradation will also be increased.
Chemical and biotechnological aerobic treatment	• Preliminary chemical oxidation of aromatic hydrocarbons by H_2O_2 or ozone will improve the biodegradability of these hazardous substances because of the cleavage of aromatic rings.
Biotechnological and chemical treatment	• Reduction of Fe(III) from nondissolved iron hydroxides will produce dissolved Fe(II) ions, which can be used for the precipitation of organic acids or cyanides.
Physico-chemical and biotechnological treatment	• Preliminary washing of wastes polluted by hydrophobic substances by water or solution of surfactants will remove these molecules from the waste; thus, the hydrophobic substances of suspension will be degraded faster than if attached to the particles of hazardous waste.
Biotechnological anaerobic and aerobic treatment	• Anaerobic treatment will perform anaerobic dechlorination of hazardous substances; it will enhance following aerobic treatment.

Table 4 Roles of Biotechnological Applications in Hazardous Waste Management

Type of waste management	Examples of biotechnological application
Hazardous waste identification	• Detection of toxicity, mutagenicity, or pathogenicity by conventional methods or by fast biotechnological tests.
Prevention of hazardous waste production	• Production, trade, or use of specific products containing nonbiodegradable hazardous substances may be banned based on biotechnological tests of biodegradability and toxicity. • Selection of environmentally preferred products based on biotechnological tests of biodegradability and toxicity. • Replacement of chemical pesticides, herbicides, rodenticides, termiticides, fungicides, and fertilizers by biodegradable and nonpersistent in the environment biotechnological analogs.
Hazardous waste collection	• Production and use of biodegradable containers. • Biotechnological formation of chemical substances (H_2S, Fe^{2+}) used for the collection of hazardous substances.
Hazardous waste reduction	• Biotreatment and biodegradation of hazardous waste. • Immobilization of hazardous substances from the streams. • Solubilization of hazardous substances from waste.
Hazardous waste toxicity reduction	• Biodegradation of hazardous substances. • Immobilization/solubilization of hazardous substances. • Biotransformation and detoxication of hazardous substances.
Hazardous waste recycling	• Solubilization/precipitation and recycling of heavy metals from waste. • Bioassimilation, precipitation, and recycling of ammonia, nitrate, and nitrite.
Hazardous waste incineration	• Sorption of hazardous products of combustion and their biodegradation.
Hazardous waste landfilling	• Inoculation of landfill for faster biodegradation. • Biotreatment of landfill leachate.

waste collection in biodegradable containers; hazardous waste toxicity reduction by biotreatment/biodegradation/bioimmobilization of hazardous substances; and hazardous waste recycling by recycling of nutrients during hazardous waste treatment.

4.1.3 Biosensors of Hazardous Substances

An important application of biotechnology in hazardous waste management is the biomonitoring of hazardous substances. This includes monitoring of biodegradability, toxicity, mutagenicity, concentration of hazardous substances, and monitoring of concentration and pathogenicity of microorganisms in untreated wastes, treated wastes, and in the environment [11,12].

Whole-Cell Biosensors

Simple or automated offline or online biodegradability tests can be performed by measuring CO_2 or CH_4 gas production or O_2 consumption [13]. Biosensors may utilize either whole bacterial cells or enzymes to detect specific molecules of hazardous substances. Toxicity can be monitored specifically by whole-cell sensors whose bioluminescence may be inhibited by the presence of hazardous substance. The most popular approach uses cells with an introduced luminescent reporter gene to determine changes in the metabolic status of the cells following intoxication [14]. Nitrifying bacteria have multiple-folded cell membranes that are sensitive to all membrane-disintegrating substances. Therefore, respirometric sensors that measure the respiration rates of these bacteria can be used for toxicity monitoring in wastewater treatment [15]. Another approach involves amperometric measurements of oxidized or reduced chemical mediators as an indicator of the metabolic status of bacterial or eukaryotic cells [14]. Biosensors measuring concentrations of hazardous substances are often based on the measurement of bioluminescence [16]. This toxicity sensor is a bioluminescent toxicity bioreporter for hazardous wastewater treatment. It is constructed by incorporating bioluminescence genes into a microorganism. These whole-cell toxicity sensors are very sensitive and may be used online to monitor and optimize the biodegradation of hazardous soluble substances. Similar sensors can be used for the measurement of the concentration of specific pollutants. A gene for bioluminescence has been fused to the bacterial genes coding for enzymes that metabolize the pollutant. When this pollutant is degraded, the bacterial cells will produce light. The intensity of biodegradation and bioluminescence depend on the concentration of pollutant and can be quantified using fiber-optics online. Combinations of biosensors in array can be used to measure concentration or toxicity of a set of hazardous substances.

Microbial Test of Mutagenicity

The mutagenic activity of chemicals is usually correlated with their carcinogenic properties. Mutant bacterial strains have been used to determine the potential mutagenicity of manufactured or natural chemicals. The most common test, proposed by Ames in 1971, utilizes back-mutation in auxotrophic bacterial strains that are incapable of synthesizing certain nutrients. When auxotrophic cells are spread on a medium that lacks the essential nutrients (minimal medium), no growth will occur. However, cells that are treated with a tested chemical that causes a reversion mutation can grow in minimal medium. The frequency of mutation detected in the test is proportional to the potential mutagenicity and carcinogenicity of the tested chemical. Microbial mutagenicity tests are used widely in modern research [17–19].

Molecular Sensors

Cell components or metabolites capable of recognizing individual and specific molecules can be used as the sensory elements in molecular sensors [11]. The sensors may be enzymes, sequences of nucleic acids (RNA or DNA), antibodies, polysaccharides, or other "reporter" molecules. Antibodies, specific for a microorganism used in the biotreatment, can be coupled to fluorochromes to increase sensitivity of detection. Such antibodies are useful in monitoring the fate of bacteria released into the environment for the treatment of a polluted site. Fluorescent or enzyme-linked immunoassays have been derived and can be used for a variety of contaminants, including pesticides and chlorinated polycyclic hydrocarbons. Enzymes specific for pollutants and attached to matrices detecting interactions between enzyme and pollutant are used in online biosensors of water and gas biotreatment [20,21].

Detection of Bacterial DNA Sequences by Oligonucleotide Probe or Array

A useful approach to monitor microbial populations in the biotreatment of hazardous wastes involves the detection of specific sequences of nucleic acids by hybridization with complementary oligonucleotide probes. Radioactive labels, fluorescent labels, and other kinds of labels are attached to the probes to increase sensitivity and simplicity of the hybridization detection. Nucleic acids that are detectable by the probes include chromosomal DNA, extra-chromosomal DNA such as plasmids, synthetic recombinant DNA such as cloning vectors, phage or virus DNA, rRNA, tRNA, and mRNA transcribed from chromosomal or extra-chromosomal DNA. These molecular approaches may involve hybridization of whole intact cells, or extraction and treatment of targeted nucleic acids prior to probe hybridization [22–24]. Microarrays for simultaneous semiquantitative detection of different microorganisms or specific genes in the environmental sample have also been developed [25–27].

4.2 AEROBIC, ANAEROBIC, AND COMBINED ANAEROBIC/AEROBIC BIOTECHNOLOGICAL TREATMENT

Relation of Microorganisms to Oxygen. The evolution from an anaerobic atmosphere to an aerobic one resulted in the creation of anaerobic (living without oxygen), facultative anaerobic (living under anaerobic or aerobic conditions), microaerophilic (preferring to live under low concentrations of dissolved oxygen), and obligate aerobic (living only in the presence of oxygen) microorganisms. Some anaerobic microorganisms, called tolerant anaerobes, have mechanisms protecting them from exposure to oxygen. Others, called obligate anarobes, have no such mechanisms and may be killed after some seconds of exposure to aerobic conditions. Obligate anaerobes produce energy from: (a) fermentation (destruction of organic substances without external acceptor of electrons); (b) anaerobic respiration using electron acceptors such as CO_2, NO_3^-, NO_2^-, Fe^{3+}, SO_4^{2-}; and (c) anoxygenic ($H_2S \rightarrow S$) or oxygenic ($H_2O \rightarrow O_2$) photosynthesis. Facultative anaerobes can produce energy from these reactions or from the aerobic oxidation of organic matter. The following sequence arranges respiratory processes according to increasing energetic efficiency of biodegradation (per mole of transferred electrons): fermentation $\rightarrow$ CO_2 respiration ("methanogenic fermentation") $\rightarrow$ dissimilative sulfate reduction $\rightarrow$ dissimilative iron reduction ("iron respiration") $\rightarrow$ nitrate respiration ("denitrification") $\rightarrow$ aerobic respiration.

4.2.1 Aerobic Microorganisms and Aerobic Treatment of Solid Wastes

Such xenobiotics as aliphatic hydrocarbons and derivatives, chlorinated aliphatic compounds (methyl, ethyl, methylene, and ethylene chlorides), aromatic hydrocarbons and derivatives (benzene, toluene, phthalate, ethylbenzene, xylenes, and phenol), polycyclic aromatic hydrocarbons, halogenated aromatic compounds (chlorophenols, polychlorinated biphenyls, dioxins and relatives, DDT and relatives), AZO dyes, compounds with nitrogroups (explosive-contaminated waste and herbicides), and organophosphate wastes can be treated effectively by aerobic microorganisms.

Conventional Composting of Organic Wastes

Technologically, composting is the simplest way to treat solid waste containing hazardous substances. Composting converts biologically unstable organic matter into a more stable humuslike product that can be used as a soil conditioner or organic fertilizer. Additional benefits of composting of organic wastes include prevention of odors from rotting wastes, destruction of pathogens and parasites (especially in thermophilic composting), and retention of nutrients in the endproducts. There are three main types of composting technology: the windrow system, the static pile system, and the in-vessel system.

Windrow System

Composting in windrow systems involves mixing an organic waste with inexpensive bulking agents (wood chips, leaves, corncobs, bark, peanut and rice husks) to create a structurally rigid matrix, to diminish heat transfer from the matrix to the ambient environment, to increase the treatment temperature, and to increase the oxygen transfer rate. The mixed matter is stacked in 1–2 m high rows called windrows. The mixtures are turned over periodically (2 to 3 times per week) by mechanical means to expose the organic matter to ambient oxygen. Aerobic and partially anaerobic microorganisms, which are present in the waste or were added from previously produced compost, will grow in the organic waste. Owing to the biooxidation and release of energy, the temperature in the pile will rise. This is accompanied by successional changes in the dominant microbial communities, from less thermoresistant to more thermophilic ones. This composting process ranges from 30 to 60 days in duration.

Static Pile System

The static pile system is an intensive biotreatment because the pile of organic waste and bulking agent is intensively aerated using blowers and air diffusers. The pile is usually covered with compost to remove odor and to maintain high internal temperatures. The aerated static pile process typically takes 21 days, after which the compost is cured for another 30 days, dried, and screened to recycle the bulking agent.

In-Vessel Composting

In-vessel composting results in the most intensive biotransformation of organic wastes. In-vessel composting is performed in partially or completely enclosed containers in which moisture content, temperature, and oxygen content in the gas can be controlled. This process requires little space and takes some days for treatment, but its cost is higher than that of open systems.

Composting of Hazardous Organic Wastes

Hazardous wastes can be treated in all the systems mentioned above, but long durations are usually needed to reach permitted levels of pollution. The choice of the system depends on the required time and possible cost of the treatment. Time of the treatment decreases, but the costs increase in the following sequence: windrow system → static pile system → in-vessel system. To intensify the composting of hazardous solid waste, the following pretreatments can be used: mechanical disintegration and separation or screening to improve bioavailability of hazardous substances, thermal treatment, washing out of hazardous substances from waste by water or surfactants to diminish their content in waste, or application of H_2O_2, ozone, or Fenton's reagent as a chemical pretreatment to oxidize and cleave aromatic rings of hydrocarbons. There are many reports of successful applications of all types of composting for the treatment of crude-oil-impacted soil, petrochemicals-polluted soil, and explosives-polluted soil.

Application of Biotechnology in/on the Sites of Postaccidental Wastes

This direction of environmental biotechnology is known as soil bioremediation. There are many options in the process design described in the literature [7,28,29].

The main options tested in the field are as follows:

- Engineered in situ bioremediation (in-place treatment of a contaminated site);
- Engineered onsite bioremediation (the treatment of a percolating liquid or eliminated gas in reactors placed on the surface of the contaminated site). The reactors used for

this treatment are suspended biomass stirred-tank bioreactors, plug-flow bioreactors, rotating-disc contactors, packed-bed fixed-biofilm reactors (biofilter), fluidized bed reactors, diffused aeration tanks, airlift bioreactors, jet bioreactors, membrane bioreactors, and upflow bed reactors [30].

- Engineered ex situ bioremediation (the treatment of contaminated soil or water that is removed from a contaminated site).

The first option is used when the pollution is not strong, time required for the treatment is not a limiting factor, and there is no pollution of groundwater. The second option is usually used when the level of pollution is high and there is secondary pollution of groundwater. The third option is usually used when the level of pollution is so high that it diminishes the biodegradation rate due to toxicity of substances or low mass transfer rate. Another reason for using this option might be that the conditions insite or onsite (pH, salinity, dense texture or high permeability of soil, high toxicity of substance, and safe distance from public place) are not favorable for biodegradation.

Artificial Formation of Geochemical Barrier

One aim of using biotechnology is to prevent the dispersion of hazardous substances from the accident site into the environment. This can be achieved by creating physical barriers on the migration pathway with microorganisms capable of biotransforming the intercepted hazardous substances, for example, in polysaccharide (slime) viscous barriers in the contaminated subsurface. Another approach, which can be used to immobilize heavy metals in soil after pollution accidents, is the creation of biogeochemical barriers. These geochemical barriers could comprise gradients of H_2S, H_2, or Fe^{2+} concentrations, created by anaerobic sulfate-reducing bacteria (in the absence of oxygen and the presence of sulfate and organic matter), fermenting bacteria (after addition of organic matter and in the absence of oxygen), or iron-reducing bacteria [in the presence of Fe(III) and organic matter], respectively. Other bacteria can form a geochemical barrier for the migration of heavy metals at the boundary between aerobic and anaerobic zones. For example, iron-oxidizing bacteria will oxidize Fe^{2+} in this barrier and produce iron hydroxides that can diminish the penetration of ammonia, phosphate, organic acids, cyanides, phenols, heavy metals, and radionuclides through the barrier.

4.2.2 Aerobic Biotechnological Treatment of Wastewater

Treatment in Aerobic Reactors

Industrial hazardous wastewater can be treated aerobically in suspended biomass stirred-tank bioreactors, plug-flow bioreactors, rotating-disc contactors, packed-bed fixed-biofilm reactors (or biofilters), fluidized bed reactors, diffused aeration tanks, airlift bioreactors, jet bioreactors, membrane bioreactors, and upflow bed reactors [28,30].

One difference between these systems and the biological treatment of nonhazardous wastewater is that the exhaust air may contain volatile hazardous substances or intermediate biodegradation products. Therefore, the air must be treated as secondary hazardous wastes by physical, chemical, physico-chemical, or biological methods. Other secondary hazardous wastes may include the biomass of microorganisms that may accumulate volatile hazardous substances or intermediate products of their biodegradation. This hazardous liquid or semisolid waste must be properly treated, incinerated, or disposed.

Treatment of Wastewater with Low Concentration of Hazardous Substance

Wastewater with low concentrations of hazardous substances may reasonably be treated using biotechnologies such as granular activated carbon (GAC) fluidized-bed reactors or co-metabolism. Granulated activated carbon or other adsorbents ensure sorption of hydrophobic hazardous substances on the surface of GAC or other adsorbent particles. Microbial biofilms can also be concentrated on the surface of these particles and can biodegrade hazardous substances with higher rates compared to situations when both substrate and microbial biomass are suspended in the wastewater. Cometabolism refers to the simultaneous biodegradation of hazardous organic substances (which are not used as a source of energy) and stereochemically similar substrates, which serve as a source of carbon and energy for microbial cells. Biooxidation of the hazardous substance is performed by the microbial enzymes due to stereochemical similarity between the hazardous substance and the substrate. The best-known applications of cometabolism are the biodegradation/detoxication of chloromethanes, chloroethanes, chloromethylene, and chloroethylenes by enzyme systems of bacteria for oxidization of methane or ammonia as the main source of energy. In practice, the bioremediation is achieved by adding methane or ammonia, oxygen (air), and biomass of methanotrophic or nitrifying bacteria to soil and groundwater polluted by toxic chlorinated substances.

Combinations of Aerobic Treatment with Other Treatments

To intensify the biotreatment of hazardous liquid waste, the following pretreatments can be used: mechanical disintegration/suspension of hazardous hydrophobic substances to improve the reacting surface in the suspension and increase the rate of biodegradation; removal from wastewater or concentration of hazardous substances by sedimentation, centrifugation, filtration, flotation, adsorption, extraction, ion exchange, evaporation, distillation, freezing separation; preliminary oxidation by H_2O_2, ozone, or Fenton's reagent to produce active oxygen radicals; preliminary photo-oxidation by UV and electrochemical oxidation of hazardous substances.

Application of Microaerophilic Microorganisms in Biotechnological Treatment

Some aerobic microorganisms prefer low concentrations of dissolved oxygen in the medium for growth, for example, concentrations below 1 mg/L. These microorganisms include filamentous hydrogen sulfide-oxidizing bacteria (e.g., *Beggiatoa* spp.), pathogenic bacteria (e.g., *Campylobacter* spp., *Streptococcus* spp., and *Vibrio* spp.), microaerophilic spirilla (e.g., *Magnetospirillum* spp.), and neutrophilic iron-oxidizing bacteria. Iron-oxidizing bacteria can produce sheaths or stalks that act as organic matrices upon which the deposition of ferric hydrooxides can occur [31,32]. Some microaerophiles are active biodegraders of organic pollutants in postaccident sites [33], while other microaerophiles form H_2O_2 to oxidize xenobiotics. Sheaths of neutrophilic iron-oxidizing bacteria can adsorb heavy metals and radionuclides from hazardous streams.

4.2.3 Aerobic Biotechnological Treatment of Hazardous Waste Gas

Biodegradable Hazardous Gases

The CERCLA Priority List of Hazardous Substances contains many substances released in industry as gaseous hazards and which can be treated biotechnologically (Table 1), including the following: chloroform, trichloroethylene, 1,2-dibromoethane, 1,2-dibromo-3-chloropropane, carbontetrachloride, xylenes, dibromochloropropane, toluene, methane, methylene chloride, 1,1-dichloroethene, *bis*(2-chloroethyl) ether, 1,2-dichloroethane, chlorine, 1,1,2-trichloroethane, ethylbenzene, 1,1,2,2-tetrachloroethane, bromine, methylmercury, trichlorofluoroethane, 1,1-dichloroethane, 1,1,2-trichloroethane, ammonia, trichloroethane, 1,2-dichloroethene, carbon disulfide, chloroethane, p-xylene, hydrogen sulfide, chloromethane, 2-butanone, bromoform, acrolein, bromodichloroethane, nitrogen dioxide, ozone, formaldehyde, chlorodibromomethane, ethyl ether, and 1,2-dichloropropane.

Reactors

The common way to remove vaporous or gaseous pollutants from gas or air streams is to pass contaminated gases through bioscrubbers containing suspensions of biodegrading microorganisms or through a biofilter packed with porous carriers covered by biofilms of degrading microorganisms. Depending on the nature and volume of polluted gas, the biofilm carriers may be cheap porous substrates, such as peat, wood chips, and compost, or regular artificial carriers such as plastic or metal rings, porous cylinders and spheres, fibers, and fiber nets. The bioscrubber contents must be stirred to ensure high mass transfer between the gas and microbial suspension. The liquid that has interacted with the polluted gas is collected at the bottom of the biofilter and recycled to the top part of the biofilter to ensure adequate contact of polluted gas and liquid and optimal humidity of biofilter. Addition of nutrients and fresh water to the bioscrubber or biofilter must be made regularly or continuously. Fresh water can be used to replace water that has evaporated in the bioreactor. If the mass transfer rate is higher than the biodegradation rate, the absorbed pollutants must be biodegraded in an additional suspended bioreactor or biofilter connected in series to the bioscrubber or absorbing biofilter.

Applications

The main application of biotechnology for the treatment of hazardous waste gases is the bioremoval of biodegradable organic solvents. Other important applications include the biodegradation of odors and toxic gases such as hydrogen sulfide and other sulfur-containing gases from the exhaust ventilation air in industry and farming. Industrial ventilation air containing formaldehyde, ammonia, and other low-molecular-weight substances can also be effectively treated in the bioscrubber or biofilter.

4.2.4 Anaerobic Microorganisms and Anaerobic Biotechnological Treatment of Hazardous Wastes

Fermentation and Anaerobic Respiration

The main energy-producing pathways in anaerobic treatment are fermentation (intramolecular oxidation/reduction without external electron acceptor) or anaerobic respiration (oxidation by electron acceptor other than oxygen). The advantage of anaerobic treatment is that there is no need to supply oxygen in the treatment system. This is useful in cases such as bioremediation of clay soil or high-strength organic waste. However, anaerobic treatment may be slower than aerobic treatment, and there may be significant outputs of dissolved organic products of fermentation or anaerobic respiration.

Biotreatment by Facultative Anaerobic Microorganisms

Facultative anaerobic microorganisms may be useful when integrated together with aerobic and anaerobic microorganisms in microbial aggregates. However, this function is still not well studied. One interesting and useful feature in this physiological group is the ability in some representatives (e.g., *Escherichia coli*) to produce an active oxidant, hydrogen peroxide, during normal aerobic metabolism [34].

Biotreatment by Anaerobically Respiring Bacteria

Aerobic respiration is more effective in terms of output of energy per mole of transferred electrons than fermentation. Anaerobic respiration can be performed by different groups of prokaryotes with such electron acceptors as NO_3^-, NO_2^-, Fe^{3+}, SO_4^{2-}, and CO_2. Therefore, if the concentration of one such acceptor in the hazardous waste is sufficient for the anaerobic respiration and oxidation of the pollutants, the activity of the related bacterial group can be used for the treatment. CO_2-respiring prokaryotes (methanogens) are used for methanogenic biodegradation of organic hazardous wastes in anaerobic reactors or in landfills. Sulfate-reducing bacteria can be used for anaerobic biodegradation of organic matter or for the precipitation/immobilization of heavy metals of sulfate-containing hazardous wastes. Iron-reducing bacteria can produce dissolved Fe^{2+} ions from insoluble Fe(III) minerals. Anaerobic biodegradation of organic matter and detoxication of hazardous wastes can be significantly enhanced as a result of precipitation of toxic organics, acids, phenols, or cyanide by Fe(II). Nitrate-respiring bacteria can be used in denitrification, that is, reduction of nitrate to gaseous N_2. Nitrate can be added to the hazardous waste to initiate the biodegradation of different types of organic substances, for example, polycyclic aromatic hydrocarbons [35]. Nitrogroups of hazardous substances can be reduced by similar pathway to related amines.

Biotreatment of Hazardous Waste by Anaerobic Fermenting Bacteria

Anaerobic fermenting bacteria (e.g., from genus *Clostridium*) perform two important functions in the biodegradation of hazardous organics: they hydrolyze different natural polymers and ferment monomers with production of alcohols, organic acids, and CO_2. Many hazardous substances, for example, chlorinated solvents, phthalates, phenols, ethyleneglycol, and polyethylene glycols, can be degraded by anaerobic microorganisms [28,36–38]. Fermenting bacteria perform anaerobic dechlorination, thus enhancing further biodegradation of chlorinated organics. There are different biotechnological systems to perform anaerobic biotreatment of wastewater: biotreatment by suspended microorganisms, anaerobic biofiltration, and biotreatment in upflow anaerobic sludge blanket (UASB) reactors [7,28].

Application of Biotechnology in Landfilling of Hazardous Solid Wastes

Landfilled organic and inorganic wastes are slowly transformed by indigenous microorganisms in the wastes [39]. Organic matter is hydrolyzed by bacteria and fungi. Amino acids are degraded via ammonification with formation of toxic organic amines and ammonia. Amino acids, nucleotides, and carbohydrates are fermented or anaerobically oxidized with formation of organic acids, CO_2, and CH_4. Xenobiotics and heavy metals may be reduced, and subsequently dissolved or immobilized. These bioprocesses may result in the formation of toxic landfill leachate, which can be detoxicated by aerobic biotechnological treatment to oxidize organic hazards and to immobilize dissolved heavy metals.

4.2.5 Combined Anaerobic/Aerobic Biotreatment of Wastes

A combined anaerobic/aerobic biotreatment can be more effective than aerobic or anaerobic treatment alone. The simplest approach for this type of treatment is the use of aerated stabilization ponds, aerated and nonaerated lagoons, and natural and artificial wetland systems, whereby aerobic treatment occurs in the upper part of these systems and anaerobic treatment occurs at the bottom end. A typical organic loading is 0.01 kg BOD/m^3 day and the retention time varies from a few days to 100 days [30]. A more intensive form of biodegradation can be achieved by combining aerobic and anaerobic reactors with controlled conditions, or by integrating anaerobic and aerobic zones within a single bioreactor. Combinations or even alterations of anaerobic and aerobic treatments are useful in the following situations: (a) biodegradation of chlorinated aromatic hydrocarbons including anaerobic dechlorination and aerobic ring cleavage; (b) sequential nitrogen removal including aerobic nitrification and anaerobic denitrification; (c) anaerobic reduction of Fe(III) and microacrophilic oxidation of Fe(II) with production of fine particles of iron hydroxide for adsorption of organic acids, phenols, ammonium, cyanide, radionuclides, and heavy metals.

4.2.6 Biotechnological Treatment of Heavy Metals-Containing Waste and Radionuclides-Containing Waste

Liquid and solid wastes containing heavy metals may be successfully treated by biotechnological methods. The effects of microorganisms on metals are described below.

Direct Reduction/Oxidation, or Reduction/Oxidation Mediated by Other Metals or Microbial Metabolites

Some metals such as iron are reduced or oxidized by specific enzymes of microorganisms. Microbial metabolism generates products such as hydrogen, oxygen, H_2O_2, and reduced or oxidized iron that can be used for oxidation/reduction of metals. Reduction or oxidation of metals is usually accompanied by metal solubilization or precipitation.

Effect of Microbial Metabolites

Solubilization or precipitation of metals may be mediated by microbial metabolites. Microbial production of organic acids in fermentation or inorganic acids (nitric and sulfuric acids) in aerobic oxidation will promote formation of dissolved chelates of metals. Microbial production of phosphate, H_2S, and CO_2 will stimulate precipitation of nondissolved phosphates, carbonates, and sulfides of heavy metals, for example, arsenic, cadmium, chromium, copper, lead, mercury, and nickel; production of H_2S by sulfate-reducing bacteria is especially useful to remove heavy metals and radionuclides from sulfate-containing mining drainage waters, liquid waste of nuclear facilities, and drainage from tailing ponds of hydrometallurgical plants; wood straw or saw dust. Organic acids, produced during the anaerobic fermentation of cellulose, may be preferred as a source of reduced carbon for sulfate reduction and further precipitation of metals.

Biosorption

The surface of microbial cells is covered by negatively charged carboxylic and phosphate groups, and positively charged amino groups. Therefore, depending on pH, there may be significant adsorption of heavy metals onto the microbial surface [7]. Biosorption, for example by fungal fermentation residues, is used to accumulate uranium and other radionuclides from waste streams.

Degradation of Minerals

Metal-containing minerals, for example sulfides, can be oxidized and metals can be solubilized. This approach is used for the bioleaching of heavy metals from sewage sludge [40,41] before landfilling or biotransformation.

Volatilization

Some metals, arsenic and mercury for example, may be volatilized by methylation due to activity of anaerobic microorganisms. Arsenic can be methylated by methanogenic *Archaea* and fungi to volatile toxic dimethylarsine and trimethylarsine or can be converted to less toxic nonvolatile methanearsonic and dimethylarsinic acids by algae [42].

Combination of Methods

In some cases the methods may be combined. Examples would include the biotechnological precipitation of chromium from Cr(VI)-containing wastes from electroplating factories by sulfate reduction to precipitate chromium sulfide. Sulfate reduction can use fatty acids as organic substrates with no accumulation of sulfide. In the absence of fatty acids but with straw as organic substrate, the direct reduction of chromium has been observed without sulfate reduction [43].

Biodegradation of Organometals

Hydrophobic organotins are toxic to organisms because of their solubility in cell membranes. However, many microorganisms are resistant to organotins and can detoxicate them by degrading the organic part of them [5].

4.3 ENHANCED BIOTECHNOLOGICAL TREATMENT OF INDUSTRIAL HAZARDOUS WASTES

Several key factors are critical for the successful application of biotechnology for the treatment of hazardous wastes: (a) environmental factors, such as pH, temperature, and dissolved oxygen concentration, must be optimized; (b) contaminants and nutrients must be available for action or assimilation by microorganisms; (c) content and activity of essential microorganisms in the treated waste must be sufficient for the treatment.

4.3.1 Enhancement of Biotreatment by Abiotic Factors

Optimal Temperature

Psychrophilic microorganisms have optimal growth temperatures below 15°C. These organisms may be killed by exposure to temperatures above 30°C. Mesophilic microorganisms have optimal growth temperatures in the range between 20 and 40°C. Thermophiles grow best above 50°C. Some bacteria can grow up to temperatures where water boils; those with optimal growth temperatures above 75°C are categorized as extreme thermophiles. Therefore, the biotreatment temperature must be maintained at optimal growth temperatures for effective biotreatment by certain physiological groups of microorganisms. The heating of the treated waste can come from microbial oxidation or fermentation activities provided there is sufficient heat generation and good thermoisolation of treated waste from the cooler surroundings. The bulking agent added to solid wastes may also be used as a thermoisolator.

Optimum pH

The pH of natural microbial biotopes varies from 1 to 11: volcanic soil and mine drainage have pH values between 1 and 3; plant juices and acid soils have pH values between 3 and 5; fresh and sea water have pH values between 7 and 8; alkaline soils and lakes, solutions of ammonia, and rotten organics have pH values between 9 and 11. Most microbes grow most efficiently within the pH range 5–9. They are called neutrophiles. Species that have adapted to grow at pH values lower than 4 are called acidophiles. Species that have adapted to grow at pH values higher than 9 are called alkalophiles. Therefore, pH of the treatment medium must be maintained at optimal values for effective biotreatment by certain physiological groups of microorganisms. The optimum pH may be maintained physiologically, by addition of pH buffer or pH regulator as follows: (a) control of organic acid formation in fermentation; (b) prevention of formation of inorganic acids in aerobic oxidation of ammonium, elemental sulfur, hydrogen sulfide, or metal sulfides; (c) assimilation of ammonium, nitrate, or ammonium nitrate, leading to decreased pH, increased pH, or neutral pH, respectively; (d) pH buffers such as $CaCO_3$ or $Fe(OH)_3$ can be used in large-scale waste treatment; (e) solutions of KOH, NaOH, NH_4OH, $Ca(OH)_2$, HCl, or H_2SO_4 can be added automatically to maintain the pH of liquid in a stirred reactor. Maintenance of optimum pH in treated solid waste or bioremediated soil may be especially important if there is a high content of sulfides in the waste or acidification/alkalization of soil in the bioremediation process.

Enhancement of Biodegradation by Nutrients and Growth Factors

The major elements that are found in microbial cells are C, H, O, N, S, and P. An approximate elemental composition corresponds to the formula $CH_{1.8}O_{0.5}N_{0.2}$. Therefore, nutrient amendment may be required if the waste does not contain sufficient amounts of these macroelements. The waste can be enriched with carbon (depending on the nature of the pollutant that is treated), nitrogen (ammonium is the best source), phosphorus (phosphate is the best source), and/or sulfur (sulfate is the best source). Other macronutrients (K, Mg, Na, Ca, and Fe) and micronutrients (Cr, Co, Cu, Mn, Mo, Ni, Se, W, V, and Zn) are also essential for microbial growth and enzymatic activities and must be added into the treatment systems if present in low concentrations in the waste. The best sources of essential metals are their dissolved salts or chelates with organic acids. The source of metals for the bioremediation of oil spills may be lipophilic compounds of iron and other essential nutrients that can accumulate at the water–air interface where hydrocarbons and hydrocarbon-degrading microorganisms can also occur [44]. In some biotreatment cases, growth factors must also be added into the treated waste. Growth factors are organic compounds such as vitamins, amino acids, and nucleosides that are required in very small amounts and only by some strains of microorganisms called auxotrophic strains. Usually those microorganisms that are commensals or parasites of plant and animals require growth factors. However, sometimes these microorganisms may have the unique ability to degrade some xenobiotics.

Increase of Bioavailability of Contaminants

Hazardous substances may be protected from microbial attack by physical or chemical envelopes. These protective barriers must be destroyed mechanically or chemically to produce fine particles or waste suspensions to increase the surface area for microbial attachment and subsequent biodegradation. Another way to increase the bioavailability of hydrophobic substances is washing of waste or soil by water or a solution of surface-active substances (surfactants). The disadvantage of this technology is the production of secondary hazardous waste because chemically produced surfactants are usually resistant to biodegradation. Therefore, only easily biodegradable or biotechnologically produced surfactants can be used for pretreatment of hydrophobic hazardous substances.

Enhancement of Biodegradation by Enzymes

Extracellular enzymes produced by microorganisms are usually expensive for large-scale biotreatment of organic wastes. However, enzyme applications may be cost-effective in certain situations. Toxic organophosphate waste can be treated using the enzyme parathion hydrolase produced and excreted by a recombinant strain of *Streptomyces lividans*. The cell-free culture fluid contains enzymes that can hydrolyze organophosphate compounds [4]. Future applications may be related with cytochrome-P450-dependent oxygenase enzymes that are capable of oxidizing different xenobiotics [45].

Enhancement of Biodegradation by Aeration and Oxygen Supply

Concentrations of dissolved oxygen may be very low (7–8 mg/L) and can be rapidly depleted during waste biotreatment with oxygen consumption rates ranging from 10 to 2000 g O_2/L hour. Therefore, oxygen must be supplied continuously in the system. Supply of air in liquid waste treatment systems is achieved by aeration and mechanical agitation. Different techniques are employed to supply sufficient quantities of oxygen in fixed biofilm reactors, in viscous solid wastes, in underground layers of soil, or in aquifers polluted by hazardous substances. Very often the supply of oxygen is the critical factor in the successful scaling up of bioremediation technologies from laboratory experiments to full-scale applications [10]. Air sparging in situ is a commonly used bioremediation technology, which volatilizes and enhances aerobic biodegradation of contamination in groundwater and saturated soils. Successful case studies include a 6–12 month bioremediation project that targeted both sandy and silty soils polluted by petroleum products and chlorinated hydrocarbons [46]. Application of pure oxygen can increase the oxygen transfer rate by up to five times, and this can be used in situations with strong acute toxicity of hazardous wastes and low oxygen transfer rates, to ensure sufficient oxygen transfer rate in polluted waste.

Enhancement of Biodegradation by Oxygen Radicals

In some cases, hydrogen peroxide has been used as an oxygen source because of the limited concentrations of oxygen that can be transferred into the groundwater using aboveground aeration followed by reinjection of the oxygenated groundwater into the aquifer or subsurface air sparging of the aquifer. However, because of several potential interactions of H_2O_2 with various aquifer material constituents, its decomposition may be too rapid, making effective introduction of H_2O_2 into targeted treatment zones extremely difficult and costly [47]. Pretreatment of wastewater by ozone, H_2O, by TiO_2-catalyzed UV-photooxidation, and electrochemical oxidation can significantly enhance biodegradation of halogenated organics, textile dyes, pulp mill effluents, tannery wastewater, olive-oil mills, surfactant-polluted wastewater, and pharmaceutical wastes, and diminish the toxicity of municipal landfill leachates. In some cases, oxygen radicals generated by Fenton's reagent ($Fe^{2+} + H_2O_2$ at low pH), and iron peroxides [Fe(VI) and Fe(V)], can be used as the oxidants in the treatment of hazardous wastes.

Enhancement of Biodegradation by Biologically Produced Oxygen Radicals

Many microorganisms can produce and release to the environment such toxic metabolites of oxygen as hydrogen peroxide (H_2O_2), superoxide radical ($O_2^{\bullet}$), and hydroxyl radical ($OH^{\bullet}$).

Lignin-oxidizing "white-rot" fungi can degrade lignin and all other chemical substances due to intensive generation of oxygen radicals, which oxidize the organic matter by random incorporation of oxygen into the molecule. Not much is known about the biodegradation ability of H_2O_2-generating microaerophilic bacteria.

Enhancement of Biodegradation by Electron Acceptors Other Than Oxygen

Dissolved acceptors of electrons such as NO_3^-, NO_2^-, Fe^{3+}, SO_4^{2-}, and HCO_3^- can be used in the treatment system when oxygen transfer rates are low. The choice of the acceptor is determined by economical and environmental factors. Nitrate is often proposed for bioremediation [35] because it can be used by many microorganisms as an electron acceptor. However, it is relatively expensive and its supply to the treatment system must be thoroughly controlled because it can also pollute the environment. Fe^{3+} is an environmentally friendly electron acceptor. It is naturally abundant in clay minerals, magnetite, limonite, goethite, and iron ores, but its compounds are usually insoluble and it diminishes the rate of oxidation in comparison with dissolved electron acceptors. Sulfate and carbonate can be applied as electron acceptors in strictly anaerobic environments only. Another disadvantage of these acceptors is that these anoxic oxidations generate toxic and foul smelling H_2S or "greenhouse" gas CH_4.

4.3.2 Enhancement of Biotreatment by Biotic Factors

Reasons for Bioenhancement of the Treatment

Addition of microorganisms (inoculum) to start up or to accelerate the biotreatment process is a reasonable strategy under the following conditions:

1. If microorganisms that are necessary for hazardous waste treatment are absent or their concentration is low in the waste;
2. If the rate of bioremediation performed by indigenous microorganisms is not sufficient to achieve the treatment goal within the prescribed duration;
3. If the acclimation period is too long;
4. To direct the biodegradation/biotreatment to the best pathway from many possible pathways;
5. To prevent growth and dispersion in the waste treatment system of unwanted or nondetermined microbial strains, which may be pathogenic or opportunistic.

Application of Acclimated Microorganisms

A simple way to produce a suitable microbial inoculum is the production of an enrichment culture, which is a microbial community containing one or more dominant strains naturally formed during cultivation in a growth medium modeling the hazardous waste under defined conditions. If the cultivation conditions change, the dominant strains in the enrichment culture may also change. Another approach involves the use of part of the treated waste containing active microorganisms as inoculum to start up the process. Application of acclimated microorganisms in an enrichment culture or in biologically treated waste may significantly decrease the start-up period for biotechnological treatment. In cases involving treatment of toxic substances and high death rates of microorganisms during treatment, regular additions of active microbial cultures may be useful to maintain constant rates of biodegradation.

Selection and Use of Pure Culture

Notwithstanding the common environmental engineering practice of using part of the treated waste as inoculum, applications of defined pure starter cultures have the following theoretical advantages: (a) greater control over desirable processes; (b) lower risk of release of pathogenic or opportunistic microorganisms during biotechnological treatment; (c) lower risk of accumulation of harmful microorganisms in the final biotreatment product. Pure cultures that are most active in biodegrading specific hazardous substances can be isolated by conventional microbiological methods, quickly identified by molecular–biological methods, and tested for pathogenicity and biodegradation properties. The biomass of pure culture can be produced in a large scale in commercial fermenters, then concentrated and dried for storage before field application. Therefore, it is not only the biodegradation abilities of pure cultures, but also the suitability for industrial production of dry biomass that must be taken into account in the selection of pure culture for the biotechnological treatment of industrial hazardous waste. Generally, Gram-positive bacteria are more viable after drying and storage than Gram-negative bacteria. Spores of Gram-positive bacteria can form superstable inocula.

Construction of Microbial Community

A pure culture is usually active in the biodegradation of one type of hazardous substance. Wastes containing a variety of hazardous substances must be treated by a microbial consortium comprising a collection of pure cultures most active in the degradation of the different types of substances. However, even in cases involving a single hazardous substance, degradation rates may be higher for a collection of pure cultures acting mutually (symbiotically) than for single pure cultures. Mutualistic relationships between pure cultures in an artificially constructed or a naturally selected microbial community may be based on the sequential degradation of xenobiotic, mutual exchange of growth factors or nutrients between these cultures, mutual creation of optimal conditions (pH, redox potential), and gradients of concentrations. Mutualistic relationships between the microbial strains are more clearly demonstrated in dense microbial aggregates such as biofilms, flocs, and granules used for biotechnological treatment of hazardous wastes.

Construction of Genetically Engineered Microorganisms

Microorganisms suitable for the biotreatment of hazardous substances can be isolated from the natural environment. However, their ability for biodegradation can be modified and amplified by artificial alterations of the genetic (inherited) properties of these microorganisms. The description of the methods is given in many books on environmental microbiology and biotechnology [7,28]. Natural genetic recombination of the genes (units of genetic information) occurs during DNA replication and cell reproduction, and includes the breakage and rejoining of chromosomal DNA molecules (separately replicated sets of genes) and plasmids (self-replicating minichromosomes containing several genes). Recombinant DNA techniques or genetic engineering can create new, artificial combinations of genes, and increase the number of desired genes in the cell. Genetic engineering of recombinant microbial strains suitable for the biotreatment may involve the following steps: (a) DNA is extracted from the cell and cut into small sequences by specific enzymes; (b) the small sequences of DNA can be introduced into DNA vectors; (c) the vector (virus or plasmid) is transferred into the cell and self-replicated to produce multiple copies of the introduced genes; (d) the cells with newly acquired genes are selected based on activity (e.g., production of defined enzymes, biodegradation capability) and stability of acquired genes. Genetic engineering of microbial strains can create (transfer) the ability to biodegrade xenobiotics or amplify this ability through the amplification of related genes. Another approach is the construction of hybrid metabolic pathways to increase the range of biodegraded xenobiotics and the rate of biodegradation [48]. The desired genes for biodegradation of different xenobiotics can be isolated and then cloned into plasmids. Some plasmids have been constructed containing multiple genes for the biodegradation of several xenobiotics simultaneously. The strains containing such plasmids can be used for the bioremediation of sites heavily polluted by a variety of xenobiotics. The main problem in these applications is maintaining the stability of the plasmids in these strains. Other technological and public concerns include the risk of application and release of genetically modified microorganisms in the environment.

Application of Microbial Aggregates and Immobilized Microorganisms

Self-aggregated microbial cells of biofilms, flocs, and granules, and artificially aggregated cells immobilized on solid particles are often used in the biotreatment of hazardous wastes. Advantages of microbial aggregates in hazardous waste treatment are as follows: (a) upper layers and matrix of aggregates protect cells from toxic pollutants due to adsorption or detoxication; therefore microbial aggregates or immobilized cells are more resistant to toxic xenobiotics than suspended microbial cells; (b) different or alternative physiological groups of microorganisms (aerobes/anaerobes, heterotrophs/nitrifiers, sulfate-reducers/sulfur-oxidizers) may coexist in aggregates and increase the diversity of types of biotreatments, leading to higher treatment efficiencies in one reactor; (c) microbial aggregates may be easily and quickly separated from treated water. Microbial cells immobilized on carrier surfaces such as GAC that can adsorb xenobiotics will degrade xenobiotics more effectively than suspended cells [49]. However, dense microbial aggregates may encounter problems associated with diffusion limitation, such as slow diffusion both of the nutrients into and the metabolites out of the aggregate. For example, dissolved oxygen levels can drop to zero at some depth below the surface of microbial aggregates. This distance clearly depends on factors such as the specific rate of oxygen consumption and the density of biomass in the microbial aggregate. When the environmental conditions within the aggregate become unfavorable, cell death may occur in zones that do not receive sufficient nutrition or that contain inhibitory metabolites. Channels and pores in aggregate can facilitate transport of oxygen, nutrients, and metabolites. Channels in microbial spherical granules have been shown to penetrate to depths of 900 μm [50] and a layer of obligate anaerobic bacteria was detected below the channelled layer [51]. This demonstrates that there is some optimal size or thickness of microbial aggregates appropriate for application in the treatment of hazardous wastes.

REFERENCES

1. Levin, M.A.; Gealt, M.A. Overview of biotreatment practices and promises. In *Biotreatment of Industrial and Hazardous Wastes*; Levin, M.A., Gealt, M.A., Eds.; McGraw-Hill, Inc.: NY, 1993; 1–18.
2. Geizer, K. Source reduction: quantity and toxicity. Part 6B. Toxicity reduction. In *Handbook of Solid Waste Management*, 2nd ed.; Kreith, F., Tchobanoglous, G., Eds.; McGraw-Hill: NY, 2002; 6.27–6.41.
3. Swoboda-Goldberg, N.G. Chemical contamination of the environment: sources, types, and fate of synthetic organic chemicals. In *Microbial Transformation and Degradation of Toxic Organic Chemicals*; Young, L.Y., Cerniglia, C., Eds.; Wiley-Liss: NY, 1995; 27–74.

4. Coppella, S.J.; DelaCruz, N.; Payne, G.F.; Pogell, B.M.; Speedie, M.K.; Karns, J.S.; Sybert, E.M.; Connor, M.A. Genetic engineering approach to toxic waste management: case study for organophosphate waste treatment. Biotechnol. Prog. **1990**, *6*, 76–81.
5. Gadd, G.M. Microbial interactions with tributyltin compounds: detoxification, accumulation, and environmental fate. Sci. Total Environ. **2000**, *258*, 119–227.
6. Gealt, M.A.; Levin, M.A.; Shields, M. Use of altered microorganisms for field biodegradation of hazardous materials. In *Biotreatment of Industrial and Hazardous Wastes*; Levin, M.A., Gealt, M.A., Eds.; McGraw-Hill, Inc: NY, 1993; 197–206.
7. Moo-Young, M.; Anderson, W.A.; Chakrabarty, A.M. Eds. *Environmental Biotechnology: Principles and Applications*; Kluwer Academic Publishers: Dordrecht, 1996.
8. Sayles, G.D.; Suidan, M.T. Biological treatment of industrial and hazardous wastewater. In *Biotreatment of Industrial and Hazardous Wastes*; Levin, M.A., Gealt, M.A., Eds.; McGraw-Hill, Inc.: NY, 1993; 245–267.
9. Urbance, J.W.; Cole, J.; Saxman, P.; Tiedje, J.M. BSD: the biodegradative strain database. Nucl. Acids Res. **2003**, *31*, 152–155.
10. Talley, J.W.; Sleeper, P.M. Roadblocks to the implementation of biotreatment strategies. Ann. N. Y. Acad. Sci. **1997**, *829*, 16–29.
11. Burlage, R.S. Emerging technologies: bioreporters, biosensors, and microprobes. In *Manual of Environmental Microbiology*; Hurst, C.J., Crawford, R.L., McInerney, M.J., Eds.; ASM Press: Washington, DC, 1997; 115–123.
12. Wood, K.V.; Gruber, M.G. Transduction in microbial biosensors using multiplexed bioluminescence. Biosens. Bioelectron. **1996**, *11*, 207–214.
13. Reuschenbach, P.; Pagga, U.; Strotmann, U. A critical comparison of respirometric biodegradation tests based on OECD 301 and related test methods. Water Res. **2003**, *37*, 1571–1582.
14. Bentley, A.; Atkinson, A.; Jezek, J.; Rawson, D.M. Whole cell biosensors – electrochemical and optical approaches to ecotoxicity testing. Toxicol. in Vitro **2001**, *15*, 469–475.
15. Inui, T.; Tanaka, Y.; Okayas, Y.; Tanaka, H. Application of toxicity monitor using nitrifying bacteria biosensor to sewerage systems. Water Sci. Technol. **2002**, *45*, 271–278.
16. Lajoie, C.A.; Lin, S.C.; Nguyen, H.; Kelly, C.J. A toxicity testing protocol using a bioluminescent reporter bacterium from activated sludge. J. Microbiol. Methods **2002**, *50*, 273–282.
17. Czyz, A.; Jasiecki, J.; Bogdan, A.; Szpilewska, H.; Wegrzyn, G. Genetically modified *Vibrio harveyi* strains as potential bioindicators of mutagenic pollution of marine environments. Appl. Environ. Microbiol. **2000**, *66*, 599–605.
18. Hwang, H.M.; Shi, X.; Ero, I.; Jayasinghe, A.; Dong, S.; Yu, H. Microbial ecotoxicity and mutagenicity of 1-hydroxypyrene and its photoproducts. Chemosphere **2001**, *45*, 445–451.
19. Yamamoto, A.; Kohyama, Y.; Hanawa, T. Mutagenicity evaluation of forty-one metal salts by the umu test. Biomed. Mater. Res. **2002**, *59*, 176–183.
20. Dewettinck, T.; Van Hege, K.; Verstraete, W. The electronic nose as a rapid sensor for volatile compounds in treated domestic wastewater. Water Res. **2001**, *35*, 2475–2483.
21. Nielsen, M.; Revsbech, N.P.; Larsen, L.H.; Lynggaard-Jensen, A. On-line determination of nitrite in wastewater treatment by use of a biosensor. Water Sci. Technol. **2002**, *45*, 69–76.
22. Hatsu, M.; Ohta, J.; Takamizawa, K. Monitoring of *Bacillus thermodenitrificans* OHT-1 in compost by whole cell hybridization. Can. J. Microbiol. **2002**, *48*, 848–852.
23. Nogueira, R.; Melo, L.F.; Purkhold, U.; Wuertz, S.; Wagner, M. Nitrifying and heterotrophic population dynamics in biofilm reactors: effects of hydraulic retention time and the presence of organic carbon. Water Res. **2002**, *36*, 469–481.
24. Sekiguchi, Y.; Kamagata, Y.; Ohashi, A.; Harada, H. Molecular and conventional analyses of microbial diversity in mesophilic and thermophilic upflow anaerobic sludge blanket granular sludges. Water Sci. Technol. **2002**, *45*, 19–25.
25. Fredrickson, H.L.; Perkins, E.J.; Bridges, T.S.; Tonucci, R.J.; Fleming, J.K.; Nagel, A.; Diedrich, K.; Mendez-Tenorio, A.; Doktycz, M.J.; Beattie, K.L. Towards environmental toxicogenomics – development of a flow-through, high-density DNA hybridization array and its application to ecotoxicity assessment. Sci. Total. Environ. **2001**, *274*, 137–149.

26. Koizumi, Y.; Kelly, J.J.; Nakagawa, T.; Urakawa, H.; El-Fantroussi, S.; Al-Muzaini, S.; Fukui, M.; Urushigawa, Y.; Stahl, D.A. Parallel characterization of anaerobic toluene- and ethylbenzene-degrading microbial consortia by PCR-denaturing gradient gel electrophoresis, RNA–DNA membrane hybridization, and DNA microarray technology. Appl. Environ. Microbiol. **2002**, *68*, 3215–3225.
27. Loy, A.; Lehner, A.; Lee, N.; Adamczyk, J.; Meier, H.; Ernst, J.; Schleifer, K.H.; Wagner, M. Oligonucleotide microarray for 16S rRNA gene-based detection of all recognized lineages of sulfate-reducing prokaryotes in the environment. Appl. Environ. Microbiol. **2002**, *68*, 5064–5081.
28. Evans, G.M.; Furlong, J.C. *Environmental Biotechnology: Theory and Applications*; John Wiley & Sons, Ltd.: Chichester, 2003.
29. Rittman, B.; McCarty, P.L. *Environmental Biotechnology: Principles and Applications*; McGraw-Hill: Boston, 2002.
30. Armenante, P.M. Bioreactors. In *Biotreatment of Industrial and Hazardous Wastes*; Levin M.A., Gealt, M.A., Eds.; McGraw-Hill, Inc.: New York, 1993; 65–112.
31. Emerson, D. Microbial oxidation of Fe(II) at circumneutral pH. In *Environmental Microbe–Metal Interactions*; Lovley, D.R., Ed.; ASM Press: Washington, DC, 2000; 31–52.
32. Emerson, D.; Moyer, C.L. Neutrophilic Fe-oxidizing bacteria are abundant at the Loihi Seamount hydrothermal vents and play a major role in Fe oxide deposition. Appl. Environ. Microbiol. **2002**, *68*, 3085–3093.
33. Holden, P.A.; Hersman, L.E.; Firestone, M.K. Water content mediated microaerophilic toluene biodegradation in arid vadose zone materials. Microb. Ecol. **2001**, *42*, 256–266.
34. Gonzalez-Flecha, B.; Demple, B. Homeostatic regulation of intracellular hydrogen peroxide concentration in aerobically growing *Escherichia coli*. J. Bacteriol. **1997**, *179*, 382–388.
35. Eriksson, M.; Sodersten, E.; Yu, Z.; Dalhammar, G.; Mohn, W.W. Degradation of polycyclic aromatic hydrocarbons at low temperature under aerobic and nitrate-reducing conditions in enrichment cultures from northern soils. Appl. Environ. Microbiol. **2003**, *69*, 275–284.
36. Marttinen, S.K.; Kettunen, R.H.; Sormunen, K.M.; Rintala, J.A. Removal of bis(2-ethylhexyl)phthalate at a sewage treatment plant. Water Res. **2003**, *37*, 1385–1393.
37. Otal, E.; Lebrato, J. Anaerobic treatment of polyethylene glycol of different molecular weights. Environ. Technol. **2002**, *23*, 1405–1414.
38. Borch, T.; Ambus, P.; Laturnus, F.; Svensmark, B.; Gron, C. Biodegradation of chlorinated solvents in a water unsaturated topsoil. Chemosphere **2003**, *51*, 143–152.
39. Tchobanoglous, G.; Theisen, H.; Vigil, S.A. *Integrated Solid Waste Management: Engineering Principles and Management Issues*; McGraw-Hill Book Co.: Singapore, 1993.
40. Ito, A.; Takachi, T.; Aizawa, J.; Umita, T. Chemical and biological removal of arsenic from sewage sludge. Water Sci. Technol. **2001**, *44*, 59–64.
41. Xiang, L.; Chan, L.C.; Wong, J.W. Removal of heavy metals from anaerobically digested sewage sludge by isolated indigenous iron-oxidizing bacteria. Chemosphere **2000**, *41*, 283–287.
42. Tamaki, S.; Frankenberger, W.T. Jr. Environmental biochemistry of arsenic. Rev. Environ. Contam. Toxicol. **1992**, *124*, 79–110.
43. Vainshtein, M.; Kuschk, P.; Mattusch, J.; Vatsourina, A.; Wiessner, A. Model experiments on the microbial removal of chromium from contaminated groundwater. Water Res. **2003**, *37*, 1401–1405.
44. Atlas, R.M. Bioaugmentation to enhance microbial bioremediation. In *Biotreatment of Industrial and Hazardous Wastes*; Levin, M.A., Gealt, M.A., Eds.; McGraw-Hill, Inc.: New York, 1993; 19–37.
45. De Mot, R.; Parret, A.H. A novel class of self-sufficient cytochrome P450 monooxygenases in prokaryotes. Trends Microbiol. **2002**, *10*, 502–508.
46. Bass, D.H.; Hastings, N.A.; Brown, R.A. Performance of air sparging systems: a review of case studies. J. Hazard. Mater. **2000**, *72*, 101–119.
47. Zappi, M.; White, K.; Hwang, H.M.; Bajpai, R.; Qasim. The fate of hydrogen peroxide as an oxygen source for bioremediation activities within saturated aquifer systems. J. Air. Waste Manag. Assoc. **2000**, *50*, 1818–1830.
48. Ensley, B.D. Designing pathways for environmental purposes. Curr. Opin. Biotechnol. **1994**, *5*, 249–252.

49. Vasilyeva, G.K.; Kreslavski, V.D.; Oh, B.T.; Shea, P.J. Potential of activated carbon to decrease 2,4,6-trinitrotoluene toxicity and accelerate soil decontamination. Environ. Toxicol. Chem. **2001**, *20*, 965–971.
50. Tay, J.-H.; Ivanov, V.; Pan, S.; Tay, S.T.-L. Specific layers in aerobically grown microbial granules. Lett. Appl. Microbiol. **2002**, *34*, 254–258.
51. Tay, S.T.-L.; Ivanov, V.; Yi, S.; Zhuang, W.-Q.; Tay, J.-H. Presence of anaerobic Bacteroides in aerobically grown microbial granules. Microbial Ecol. **2002**, *44*, 278–285.

5

Treatment of Pharmaceutical Wastes

Sudhir Kumar Gupta and Sunil Kumar Gupta
Indian Institute of Technology, Bombay, India

Yung-Tse Hung
Cleveland State University, Cleveland, Ohio, U.S.A.

5.1 INTRODUCTION

The pharmaceutical industry manufactures biological products, medicinal chemicals, botanical products, and the pharmaceutical products covered by Standard Industrial Classification Code Numbers 2831, 2833, and 2834, as well as other commodities. The industry is characterized by a diversity of products, processes, plant sizes, as well as wastewater quantity and quality. In fact, the pharmaceutical industry represents a range of industries with operations and processes as diverse as its products. Hence, it is almost impossible to describe a "typical" pharmaceutical effluent because of such diversity. The growth of pharmaceutical plants was greatly accelerated during World War II by the enormous demands of the armed forces for life-saving products. Manufacture of the new products, particularly the antibiotics that were developed during World War II and later periods, exacerbated the wastewater treatment problems resulting from this industry. Industrialization in the last few decades has given rise to the discharge of liquid, solid, and gaseous emissions into natural systems and consequent degradation of the environment [1]. This in turn has led to an increase in various kinds of diseases, which has necessitated the production of a wide array of pharmaceuticals in many countries. Wastewater treatment and disposal problems have also increased as a result. From 1999 to 2000, the U.S. Geological Survey conducted the first nationwide reconnaissance of the occurrence of pharmaceuticals, hormones, and other organic wastewater contaminants (OWC) in a network of 139 streams across 30 states. The study concluded that OWC were present in 80% of the streams sampled. The most frequently detected compounds were basically of pharmaceutical origin, that is, coprostanol (fecal steroid), cholesterol (plant and animal steroids), *N,N*-diethyltoluamide (insect repellant), caffeine (stimulant), triclosan (antimicrobial disinfectant), and so on [2].

5.2 CATEGORIZATION OF THE PHARMACEUTICAL INDUSTRY

Bulk pharmaceuticals are manufactured using a variety of processes including chemical synthesis, fermentation, extraction, and other complex methods. Moreover, the pharmaceutical industry produces many products using different kinds of raw material as well as processes;

Table 1 Classes of Pharmaceutical Products and Typical Examples [3]

Classes	Subclasses with typical examples
Medicinal	Antibiotics (e.g., penicillins, tetracyclines)
	Vitamins (e.g., B, E, C, A)
	Anti-infective agents (e.g., sulphonamides)
	Central depressants and stimulants (e.g., analgesics, antipyretics, barbiturates)
	Gastro-intestinal agents and therapeutic nutrients
	Hormones and substitutes
	Autonomic drugs
	Antihistamines
	Dermatological agents–local anesthetics (e.g., salicylic acid)
	Expectorants and mucolytic agents
	Renal acting and endema reducing agents
Biologicals	Serums/vaccines/toxoids/antigens
Botanicals	Morphine/reserpine/quinine/curare
	Various alkaloids, codeine, caffeine, etc.

hence it is difficult to generalize its classification. In spite of extreme varieties of processes, raw materials, final products, and uniqueness of plants, a first cut has been made to divide the industry into categories having roughly similar processes, waste disposal problems, and treatment methods. Based on the processes involved in manufacturing, pharmaceutical industries can be subdivided into the following five major subcategories:

1. Fermentation plants;
2. Synthesized organic chemicals plants;
3. Fermentation/synthesized organic chemicals plants (generally moderate to large plants);
4. Biological production plants (production of vaccines–antitoxins);
5. Drug mixing, formulation, and preparation plants (tablets, capsules, solutions, etc.).

Fermentation plants employ fermentation processes to produce medicinal chemicals (fine chemicals). In contrast, synthesized organic chemical plants produce medicinal chemicals by organic synthesis processes. Most plants are actually combinations of these two processes, yielding a third subcategory of fermentation/synthesized organic chemicals plants. Biological production plants produce vaccines and antitoxins. The fifth category comprises drug mixing, formulation, and preparation plants, which produce pharmaceutical preparations in a final form such as tablets, capsules, ointments, and so on.

Another attempt was made to classify the industry based on production of final product. The Kline Guide in 1974 defined the various classes of bulk pharmaceutical final products. Based on that, the NFIC–Denever (recently renamed NEIC, National Enforcement Investigation Center), Washington, D.C., classified the pharmaceutical industry into three major categories as depicted in Table 1 [3].

5.3 PROCESS DESCRIPTION AND WASTE CHARACTERISTICS

Pharmaceutical waste is one of the major complex and toxic industrial wastes [4]. As mentioned earlier, the pharmaceutical industry employs various processes and a wide variety of raw materials to produce an array of final products needed to fulfill national demands. As a result, a number of waste streams with different characteristics and volume are generated, which vary by plant, time, and even season, in order to fulfill the demands of some specific drugs. It has been reported that because of the seasonal use of many products, production within a given pharmaceutical plant often varies throughout the year, which changes the characteristics of wastewater by season [5]. Hence, it is difficult to generalize the characteristics of the effluent discharged from these industries.

Fermentation plants generally produce extremely strong and highly organic wastes, whereas synthetic organic chemical plants produce wastes that are strong, difficult to treat, and frequently inhibitory to biological systems. The production of antitoxins and vaccines by biological plants generates wastewater containing very high BOD (biochemical oxygen demand), COD (chemical oxygen demand), TS (total solids), colloidal solids, toxicity, and odor. The waste load from drug formulating processes is very low compared to the subcategory 1, 2, 3, bulk pharmaceutical manufacturing plants [3]. Characteristics of the waste produced and the process description of various types of pharmaceutical industries are described in the following sections.

5.3.1 Fermentation Plants

These plants use fermentation techniques to produce various pharmaceuticals. A detailed description of the fermentation process including formulation of typical broths, fermentation chemistry, and manufacturing steps of various medicines are given in the NEIC report [6]. Major unit operations involved in the fermentation process are generally comprised of seed production, fermentation (growth), and chemical adjustment of broths, evaporation, filtration, and drying. The waste generated in this process is called spent fermentation broth, which represents the leftover contents of the fermentation tank after the active pharmaceutical ingredients have been extracted. This broth may contain considerable levels of solvents and mycelium, which is the filamentous or vegetative mass of fungi or bacteria responsible for fermentation. One commercial ketone solvent has been reported as having a BOD of approximately 2 kg/L or some 9000 times stronger than untreated domestic sewage. One thousand gallons of this solvent was calculated as equivalent in BOD to the sewage coming from a city of 77,000 people. Similarly, amyl acetate, another common solvent, is reported as having a BOD of about 1 kg/L and acetone shows a BOD of about 400,000 mg/L [7–9]. The nature and composition of a typical spent fermentation broth are depicted in Table 2 [3].

5.3.2 Synthetic Organic Chemical Plants

These plants use the synthesis of various organic chemicals (raw materials) for the production of a wide array of pharmaceuticals. Major unit operations in synthesized organic chemical plants generally include chemical reactions in vessels, solvent extraction, crystallization, filtration, and drying. The waste streams generated from these plants typically consist of cooling waters, condensed steam still bottoms, mother liquors, crystal end product washes, and solvents resulting from the process [10]. The waste produced in this process is strong, difficult to treat, and frequently inhibitory to biological systems. They also contain a wide array of various chemical components prevailing at relatively high concentration produced from the production of chemical intermediates within the plant. Bioassay results on the composite waste from a plant in India approximated 0.3% when expressed as a 48 hour TLm. A typical example of untreated synthetic organic chemical waste for a pharmaceutical plant located in India is given in Table 3

Table 2 Characteristics of a Typical Spent Fermentation Broth [3]

Composition	
Total solids	1–5%
The total solids comprise	
Protein	15–40%
Fat	1–2%
Fibers	1–6%
Ash	5–35%
Carbohydrates	5–27%
Steroids, antibiotics	Present
Vitamin content of the solids	Thiamine, Riboflavin, Pyridoxin, HCl, Folic acid at 4–2,000 μg/g
Ammonia N	100–250 mg/L
BOD	5,000–20,000 mg/L
pH	3–7

BOD, biochemical oxygen demand.

[11]. Various types of waste streams were generated from this plant depending upon the manufacturing process. Waste was segregated into various waste streams such as strong process waste, dilute process waste, service water, and composite waste [12]. The strength and magnitude of various waste streams generated at the Squibb, Inc. synthetic penicillin and antifungal plant in Humaco, Puerto Rico, are given in Table 4.

Many other researchers have segregated the waste generated from a synthetic organic chemical pharmaceutical plant located in Hyderabad, India, into different wastewater streams such as floor washing, also known as condensate waste, acid waste, and alkaline waste [13–15]. This plant is one of the largest of its kind in Asia and is involved in the production of various drugs, such as antipyretics, antitubercular drugs (isonicotinic acid hydrazide), antihelminthic, sulfa drugs, vitamins, and so on. Tables 5 to 8 present the characteristics of each waste stream generated from a synthetic drug plant at Hyderabad, along with the characteristics of the combined waste streams. Wastewater from this plant exhibited considerable BOD variation among the various waste streams generated from the plant. The BOD of the condensate waste

Table 3 Characteristics of Untreated Synthetic Drug Waste [11]

Parameter	Concentration range (mg/L)
p-amino phenol, *p*-nitrophenolate, *p*-nitrochlorobenzene	150–200
Amino-nitrozo, amino-benzene, antipyrene sulfate	170–200
Chlorinated solvents	600–700
Various alcohols	2,500–3,000
Benzene, toluene	400–700
Sulfanilic acid	800–1,000
Sulfa drugs	400–700
Analogous substances	150–200
Calcium chloride	600–700
Sodium chloride	1,500–2,500
Ammonium sulfate	15,000–20,000
Calcium sulfate	800–21,000
Sodium sulfate	800–10,000

Table 4 Characteristics of Synthetic Organic Chemicals, Wastewater at Squibb, Inc., Humaco [12]

Waste	Flow, g/day		BOD (mg/L)	COD (mg/L)	BOD load (lb/day)		COD load (lb/day)	
	Avg.	Max.			Avg.	Max.	Avg.	Max.
Strong process	11,800	17,400	480,000	687,000	47,300	74,200	67,600	105,800
Dilute process	33,800	37,400	640	890	180	190	250	280
Service water	35,300	–	–	–	–	–	–	–
Composite	80,900	–	70,365	109,585	47,500	–	67,900	–

BOD, biochemical oxygen demand; COD, chemical oxygen demand.

was found to be very low compared to other wastes. Acidic waste contributed 50% of the total waste flow at 600 m^3/day and had a pH of 0.6. The combined waste had a pH of 0.8 (including acidic waste stream), whereas the pH of the waste without acidic waste stream was 9.3. The BOD to COD ratio of alkaline, condensate and combined wastewater was around 0.5–0.6, while for the acidic waste alone it was around 0.4, indicating that all these wastewaters are biologically treatable. The combined wastewater had average TOC, COD, and BOD values of 2109 mg/L, 4377 mg/L, and 2221 mg/L. Heavy metal concentration of the wastewater was found to be well below the limits according to IS-3306 (1974). Most of the solids present were in a dissolved form, with practically no suspended solids. The wastewater contained sufficient nitrogen, but was lacking in phosphorus, which is an essential nutrient for biological treatment. The 48-hour TL_m values for alkaline and condensate wastes showed 0.73–2.1% (v/v) and 0.9% (v/v),

Table 5 Characteristics of Alkaline Waste Stream of a Synthetic Drug Plant at Hyderabad [13,15]

Parameters	Ranges (max. to min.)	
	From Ref. [15]	From Ref. [13]
Flow (m^3/day)	1,400–1,920 (1,710)	1,710
pH	4.1–7.5	2.3–11.2
Total alkalinity as $CaCO_3$	1,279–2,140	624–5630
Total solids	1.29–2.55%	11825–23265 mg/L
Total volatile solids	13.1–32.6% of TS	1,457–2,389 mg/L
Total nitrogen (mg/L)	284–1,036 (TKN)	266–669
Total phosphorus (mg/L)	14–42	10–64.8
BOD_5 at 20°C (mg/L)	2,874–4,300	2,980–3,780
COD (mg/L)	5,426–7,848	5,480–7,465
BOD : COD	–	0.506–0.587
BOD : N : P	–	100 : (8.9–17.7) : (0.265–1.82)
Suspended solids (mg/L)	–	11–126
Chlorides as Cl^- (mg/L)	–	2,900–4,500

TS, total solids; TKN, total Kjeldhal nitrogen; BOD, biochemical oxygen demand; COD, chemical oxygen demand.

Table 6 Characteristics of Condensate Waste Stream of a Synthetic Drug Plant at Hyderabad [13,15]

	Ranges (max. to min.)	
Parameters	From Ref. [15]	From Ref. [13]
Flow (m^3/day)	1,570–2,225 (1,990)	1,570–2,225 (1,990)
pH	2.1–7.3	7–7.8
Total alkalinity as $CaCO_3$	498–603	424–520
Total solids	0.31–1.22%	2,742–4,150 mg/L
Total volatile solids	13.6–37.2% of TS	363–800 mg/L
Total nitrogen (mg/L)	120–240 (TKN)	120–131
Total phosphorus (mg/L)	2.8–5	3.1–28.8
BOD_5 at 20°C (mg/L)	1,275–1,600	754–1,385
COD (mg/L)	2,530–3,809	1,604–2,500
BOD : COD	–	0.4–0.688
BOD : N : P	–	100 : (10.9–16.71) : (0.28–3.82)
Suspended solids (mg/L)	–	39–200
Chlorides as Cl^- (mg/L)	–	700–790

TS, total solids; TKN, total Kjeldhal nitrogen; BOD, biochemical oxygen demand; COD, chemical oxygen demand.

respectively. Table 9 gives the characteristics of a typical pharmaceutical industry wastewater located at Bombay producing various types of allopathic medicines [16].

5.3.3 Fermentation/Synthetic Organic Chemical Plants

These plants employ fermentation techniques as well as synthesis of organic chemicals in the manufacturing of various pharmaceuticals. Typically, they are operated on a batch basis via fermentation and organic synthesis, depending upon specific requirements of

Table 7 Characteristics of an Acid Waste Stream of a Synthetic Drug Plant at Hyderabad [13]

Parameters	Ranges (max. to min.)
Flow (m^3/day)	435
pH	0.4–0.65
BOD_5 at 20°C (mg/L)	2,920–3,260
COD (mg/L)	7,190–9,674
BOD/COD ratio	0.34–0.41
Total solids (mg/L)	18,650–23,880
Total volatile solids (mg/L)	15,767–20,891
Suspended solids	Traces
Total nitrogen (mg/L)	352
Total phosphorus (mg/L)	9.4
Total acidity as $CaCO_3$	29,850–48,050
Chlorides as Cl^- (mg/L)	6,500
Sulfate as SO_4^{2-} (mg/L)	15,000

BOD, biochemical oxygen demand; COD, chemical oxygen demand.

Table 8 Characteristics of Combined Wastewater[a] of a Synthetic Drug Plant at Hyderabad [15]

Parameters	Range	Standard deviation
pH	2.9–7.6	–
BOD_5 at 20°C (mg/L)	1,840–2,835	2,221 ± 301
COD (mg/L)	4,000–5,194	4,377 ± 338
BOD/COD ratio	0.46–0.54	–
Total organic carbon (C) (mg/L)	1,965–2,190	2,109 ± 73
BOD exertion rate (*k*) constant[b]	0.24–0.36	0.28 ± 0.02

[a]Alkaline and condensate wastewater mixed in 1:1 ratio.
[b]BOD, biochemical oxygen demand; COD, chemical oxygen demand.

Table 9 Characteristics of Pharmaceutical Industry Wastewater Producing Allopathic Medicines [16]

Parameter	Range of concentration	Average concentration
pH	6.5–7.0	7
BOD (mg/L)	1,200–1,700	1,500
COD (mg/L)	2,000–3,000	2,700
BOD/COD ratio	0.57–0.6	0.55
Suspended solids (mg/L)	300–400	400
Volatile acids (mg/L)	50–80	60
Alkalinity as $CaCO_3$ (mg/L)	50–100	60
Phenols (mg/L)	65–72	65

various pharmaceuticals. Characteristics of the waste generated vary greatly depending upon the manufacturing process and raw materials used in the production of various medicines.

5.3.4 Biological Production Plants

These plants are mainly involved in the production of antitoxins, antisera, vaccines, serums, toxoids, and antigens. The production of antitoxins, antisera, and vaccines generates wastewaters containing animal manure, animal organs, baby fluid, blood, fats, egg fluid and egg shells, spent grains, biological culture, media, feathers, solvents, antiseptic agents, herbicidal components, sanitary loads, and equipment and floor washings. Overall, 180,000 G/day of waste is generated by biological production plants [17]. The various types of waste generated mainly include:

- waste from test animals;
- pathogenic-infectious waste from laboratory research on animal disease;
- toxic chemical wastes from laboratory research on bacteriological, botanical, and zoological problems;
- waste from antisera/antitoxins production;
- sanitary wastes.

Table 10 gives the characteristics of liquid waste arising in liver and beef extract production from a biological production pharmaceutical plant [18]. These wastes can be very high in BOD, COD, TS, colloidal solids, toxicity, color, and odor. The BOD/COD ratio of the

Table 10 Characteristics of Liquid Waste Arising in Liver and Beef Extract Production from a Biological Production Pharmaceutical Wastewater [18]

Constituents	Range	Mean
pH	5–6.3	5.8
Temperature (°C)	26.5–30	28
BOD_5 (mg/L)	11,400–16,100	14,200
COD (mg/L)	17,100–24,200	21,200
BOD/COD ratio	0.66–0.67	0.67
Total solids (TS) (mg/L)	16,500–21,600	20,000
Volatile solids (VS) (mg/L)	15,900–19,600	19,200
TKN (mg/L)	2,160–2,340	2,200
Crude fat (mg/L)	3,800–4,350	4,200
Volatile fatty acids (VFA) (mg/L)	1,060–1,680	1,460

BOD, biochemical oxygen demand; COD, chemical oxygen demand; TKN, total Kjeldhal nitrogen.

Table 11 Characteristics of Typical Spent Stream of Biologicals Production Plant at Greenfield, IN [20]

Parameter	Value
Flow (G/day)	15,000
pH	7.3–7.6
BOD (mg/L)	1,000–1,700
Total solids (TS) (mg/L)	4,000–8,500
Suspended solids (mg/L)	200–800
Percentage suspended solids	5–10

BOD, biochemical oxygen demand.

waste is around 0.66. The waste contains volatile matter as 95% of TS present in the waste, containing easily degradable biopolymers such as fats and proteins. Table 11 presents the characteristics of spent streams generated from a typical biological production plant, Eli Lilly and Co., at Greenfield, IN [19,20].

5.3.5 Drug Mixing, Formulation, and Preparation Plants

Drug formulating processes consist of mixing (liquids or solids), palletizing, encapsulating, and packaging. Raw materials utilized by a drug formulator and packager may include ingredients such as sugar, corn syrup, cocoa, lactose, calcium, gelatin, talc, diatomaceous, earth, alcohol, wine, glycerin, aspirin, penicillin, and so on. These plants are mainly engaged in the production of pharmaceuticals primarily of a nonprescription type, including medications for arthritis, coughs, colds, hay fever, sinus and bacterial infections, sedatives, digestive aids, and skin sunscreens. Wastewater characteristics of such plants vary by season, depending upon the production of medicines to meet seasonal demands. However, the waste can be characterized as being slightly acidic, of high organic strength (BOD, 750–2000 mg/L), relatively low in suspended solids (200–400 mg/L), and exhibiting a degree of toxicity. During the period when cough and cold medications are prepared, the waste may contain high concentrations of mono- and disaccharides and may be deficient in nitrogen [5]. A drug formulation plant usually operates a single shift, five days a week. Since drug formulating is labor-intensive, sanitary waste constitutes a larger part of total wastes generated, therefore waste loads generated from such plants are very low compared to other subcategories of bulk pharmaceutical manufacturing plants.

5.4 SIGNIFICANT PARAMETERS IN PHARMACEUTICAL WASTEWATER TREATMENT

Significant parameters to be considered in designing a treatment and disposal facility for pharmaceutical wastewater are given in Table 12. Biochemical oxygen demand measurements of the waste have been reported to increase greatly with dilution, indicating the presence of toxic or inhibitory substances in some pharmaceutical effluents. The toxicity impact upon various biological treatments by various antibiotics, bactericidal-type compounds, and other pharmaceuticals has been described in the literature [21–24].

Discharge permits for pharmaceutical manufacturing plants place greater attention on high concentrations of ammonia and organic nitrogen in the waste. Considerable amounts of TKN (total Kjeldhal nitrogen) have been found to still remain in the effluent even after undergoing a high level of conventional biological treatment. It has also been reported that the nitrogen load of treated effluent may sometimes exceed even the BOD load. This generates an oxygen demand, increased chlorine demand, and formation of chloramines during chlorination, which may be toxic to fish life and create other suspected health problems. The regulatory authorities have limited the concentration of unoxidized ammonia nitrogen to 0.02 mg/L in treated effluent.

Certain pharmaceutical waste may be quite resistant to biodegradation by conventional biological treatment. For example, various nitroanilines have been used in synthesized production of sulfanilamide and phenol mercury wastes and show resistance against biological attack. Both ortho and meta nitroaniline were not satisfactorily degraded even after a period of many months [25]. Other priority pollutants such as tri-chloro-methyl-proponal (TCMP) and toluene must be given attention in the treatment of pharmaceutical wastewater. With careful controls, *p*-nitroaniline can be biologically degraded, although the reaction requires many days for acclimatization [25,26].

Table 12 Parameters of Significance for the Pharmaceutical Industry Wastewater [3]

pH	Fecal coliform
Temperature	Manganese
BOD_5, BOD_{Ult}	Phenolics
COD	Chromium
Dissolved oxygen	Aluminum
TOC	Cyanides
Solids (suspended and dissolved)	Zinc
Oil and Grease	Lead
Nitrogen, (NH_4 and organic-N)	Copper
Sulfides	Mercury
Toxicity	Iron

BOD, biochemical oxygen demand; COD, chemical oxygen demand; TOC, total organic carbon.

5.5 WASTE RECOVERY AND CONTROL

Production processes used in the pharmaceutical/fine chemical, cosmetic, textile, rubber, and other industries result in wastewaters containing significant levels of aliphatic solvents. It has been reported that of the 1000 tons per year of EC-defined toxic wastes generated in Ireland, organic solvents contribute 66% of the waste [27]. A survey of the constituents of pharmaceutical wastewater in Ireland has reported that aliphatic solvents contribute a significant proportion of the BOD/COD content of pharmaceutical effluents. Organic solvents are flammable, malodorous, and potentially toxic to aquatic organisms and thus require complete elimination by wastewater treatment systems.

Pretreatment and recovery of various useful byproducts such as solvents, acids, sodium sulfate, fermentation solids, and fermentation beers comprise a very important waste control strategy for pharmaceutical plants. Such an approach not only makes expensive biological treatment unnecessary, but also gives economic returns in recovery of valuable byproducts [19,21,28–33].

In fermentation plants, the spent fermentation broth contains considerable levels of solvents and mycelium. As mentioned earlier, these solvents exhibit very high BOD strength and also some of the solvents are not biologically degradable; hence, if not removed/recovered, the latter places a burden on the biological treatment of the waste and destroys the performance efficiency of biological treatment. Intense recovery of these solvents in fermentation processes is thus recommended as a viable option to reduce flow into pharmaceutical effluents. The mycelium, which poses several operational problems during treatment, can be recovered for use as animal feed supplements. Separate filtration, drying, and recovery of mycelium has been recommended as the best method for its use as animal feed or supplements. Moreover, spent fermentation broth contains high levels of nutrients and protein, which attains a high value when incorporated into animal feeds. Large-scale fermentation solids recovery is practiced at Abbott Labs, North Chicago, IL, and has been conducted at Upjohn Co., Kalamazoo, Michigan, and at Abbott Labs, Barceloneta, Puerto Rico [3].

Spent beers contain a substance toxic to the biological system and exhibit considerable organic strength; hence, it needs to be removed/recovered to avoid the extra burden on the biological treatment. Large-scale recovery of antibiotic spent beers by triple-effect evaporators was carried out at Upjohn Co., Kalamazoo, Michigan, in the 1950s. Biochemical oxygen demand reduction with the triple-effect evaporation system was reported to be 96 to 98% for four different types of antibiotic spent beers. A similar practice had been adopted by pharmaceutical plants Pfizer (Terre Haute, IN) and Lederle Labs (Pearl River, NY) for the recovery of spent beers in the 1950s and 1960s, but these practices have been discontinued due to changing products or other conditions.

From 1972 to 1973, Abbott Labs in North Chicago, IL, recovered beers with a BOD_5 (five-day biological oxygen demand) load potential of 20,000 lb/day or greater. In the process, the spent beers were concentrated by multiple effect evaporators to 30% solids and the resulting syrup sold as a poultry feed additive. Any excess was incinerated in the main plant boilers. Abbott Labs reported that an average overall BOD reduction efficiency of the system up to 96% or more could be achieved.

Recovery of valuable products from penicillin, riboflavin, streptomycin, and vitamin B_{12} fermentation has been recommended as a viable waste control strategy when incorporated into animal feeds or supplements. Penicillin wastes, when recovered for animal feed, are reported to contain valuable growth factors, mycelium, and likewise evaporated spray-dried soluble matter [31,32,34].

Recovery of sodium sulfate from waste is an important waste control strategy within synthetic organic pharmaceutical plants. A sodium sulfate waste recovery system was employed in the Hoffmann–La Roche (Belvidere, NJ) plant, which manufactured synthetic organic pharmaceuticals. In 1972, the company reported 80 tons/day of sodium sulfate recovery [3]. The recovery and subsequent sale of sodium sulfate not only gave an economic return, but also reduced the influent sulfate concentration that may otherwise cause sulfide toxicity in anaerobic treatment of the pharmaceutical effluents.

To use water efficiently, the cooling and jacketing tower water must be segregated from the main waste streams and should be recycled and reused in cooling towers. Scavenging and recovery of high-level ammonia waste streams is recommended as a viable option of ammonia recovery for waste streams containing high concentrations of ammonia nitrogen.

The recovery of alcohol by distillation, concentration of organics, and use of waste activated sludge as a soil conditioner and fertilizer has also been reported [35].

Based on extensive experience in wastewater reduction and recovery experience at Bristol Labs (Syracuse, NY) and at the Upjohn Company (Kalamzoo, Michigan), the following practices have been recommended for waste control and recovery of byproducts in pharmaceutical industries [8,9,36,37]:

1. Install stripping towers for solvent removal (recover solvents wherever possible);
2. Conduct a program of sampling and testing solvents on wastewater flows;
3. Collect and incinerate nonreusable combustible solvents and residues;
4. Remove all mycelium;
5. Carefully program dumping of contaminated or spoiled fermentation batches;
6. Eliminate all possible leakage of process materials;
7. Separate clean waters from contaminated wastewaters;
8. Collect and haul selected high organic wastes to land disposal or equivalent;
9. Recycle seal waters on a vacuumed pump system;
10. Improve housekeeping procedures.

5.6 TREATMENT OF PHARMACEUTICAL WASTEWATER

The pharmaceutical industry employs a wide array of wastewater treatment and disposal methods [3]. Wastes generated from these industries vary not only in composition but also in magnitude (volume) by plant, season, and even time, depending on the raw materials and the processes used in manufacturing of various pharmaceuticals. Hence it is very difficult to specify a particular treatment system for such a diversified pharmaceutical industry. Many alternative treatment processes are available to deal with the wide array of waste produced from this industry, but they are specific to the type of industry and associated wastes. Available treatment processes include the activated sludge process, trickling filtration, the powdered activated carbon-fed activated sludge process, and the anaerobic hybrid reactor. An incomplete listing of other treatments includes incineration, anaerobic filters, spray irrigation, oxidation ponds, sludge stabilization, and deep well injection. Based upon extensive experience with waste treatment across the industry, a listing of the available treatments and disposals is summarized as follows [3]:

- Separate filtration of mycelium, drying and recovery of fermentation broth and mycelium for use as animal feed supplements.
- Solvent recovery at centralized facilities or at individual sectors, reuse and/or incineration of collected solvents.
- Special recovery and subsequent sale of sodium sulfate.
- Cooling towers for reuse of cooling and jacketing waters.

- Scavenging and recovery of high-level ammonia waste streams.
- Elimination of barometric condensers.
- Extensive holding and equalization of wastewater prior to main treatment.
- Extensive neutralization and pH adjustment.
- The activated sludge process including multiple-stage, extended aeration, the Unox pure oxygen system, aerated ponds, and other variations.
- The trickling filter process, including conventional rate filters, multiple-stage, high-rate systems, and bio-oxidation roughing towers.
- Treatment of selected waste streams by activated carbon, ion exchange, electro-membranes, chemical coagulation, sand, and dual and multimedia filtration.
- Spray irrigation of fermentation beers and other pharmaceutical wastes.
- Collection of biological, synthetic organic, and pathogenic waste for incineration or disposal by separate means such as steam cooking and sterilization of pathogenic wastes.
- Multiple effects evaporation–steam and/or oil, multiple hearth and rotary kiln incineration, and other special thermal oxidation systems.
- Incineration of mycelium and excess biological sludge. Incineration system may also receive pathogenic wastes, unrecoverable solvents, fermentation broths or syrups, semi-solid and solid wastes, and so on. The system can be further integrated with the burning of odorous air streams.
- Acid cracking at low pH.
- Excess biological sludge can be handled by flotation, thickening, vacuum filtration, centrifugation, degasification, aerobic and/or anaerobic digestion, lagooning, drying, converting to useable product, incineration, land spreading, crop irrigation, composting, or land filling.
- Chlorination, pasteurization, and other equivalent means of disinfecting final effluents. Disinfection is generally utilized inside vaccine-antitoxins production facilities, and in some cases dechlorination may be required.
- Extensive air stream cleaning and treatment systems.
- Municipal waste treatment.

The treatment options cited above are very specific to the type of waste. To have a clear understanding of the various unit operations used in the treatment and disposal of various types of wastes produced in the pharmaceutical industry, the treatment processes can be divided into the following three categories and subcategories:

1. physicochemical treatment process;
2. biological treatment process:
   (i) aerobic treatment,
   (ii) anaerobic treatment,
   (iii) two-stage biological treatment,
   (iv) combined treatment with other waste;
3. integrated treatment and disposal facility for a particular plant wastewater.

5.6.1 Physicochemical Treatment

Physicochemical treatment of pharmaceutical wastewater includes screening, equalization, neutralization/pH adjustment, coagulation/flocculation, sedimentation, adsorption, and ozone and hydrogen peroxide treatment. Detailed descriptions of the various physicochemical treatment processes are described in the following sections.

Extensive Holding and Equalization of Waste

As explained earlier, waste produced from the pharmaceutical industry varies in composition and magnitude depending upon various factors, that is, raw materials, manufacturing processes, process modifications, specific demand of seasonal medicines, and so on. Such variation in the quality and quantity of the wastewater may cause shock as well as underloading to the various treatment systems, which leads to malfunctioning or even failure of treatment processes, particularly biological treatment. To avoid these operational problems, extensive holding and equalization of wastewater is extremely important. Use of an equalization basin has been reported effectively to control shock loading on further treatment units treating the pharmaceutical waste [5]. The retention time and capacity of the holding tank in such cases is designed based on the degree of variability in composition and magnitude of the wastewater.

Neutralization/pH Adjustment

Wastewater generated from the pharmaceutical industry varies greatly in pH, ranging from acidic to alkaline. For example, the pH of an alkaline waste stream from a synthetic organic pharmaceutical plant ranges from 9 to 10, whereas a pH of 0.8 has been reported for acidic waste streams [13,15]. Nevertheless, almost all types of waste streams produced from the pharmaceutical industry are either alkaline or acidic, and require neutralization before biological treatment. Thus, neutralization/pH adjustment of the waste prior to the biological system is a very important treatment unit for the biological treatment of pharmaceutical wastewater. The pH of the wastewater in this unit is adjusted by adding alkali or acid depending upon the requirement of the raw wastewater.

Coagulation/Flocculation

Coagulation and flocculation of the wastewater are carried out for the removal of suspended and colloidal impurities. The application of such treatment units greatly depends upon the suspended and colloidal impurities present in the raw wastewater. Coagulation and flocculation of pharmaceutical wastewater have been reported to be less effective at a pharmaceutical plant in Bombay that produces allopathic medicines [16]. The effects of various coagulants such as $FeSO_4$, $FeCl_3$, and alum on suspended solids and COD removal efficiency were evaluated. The wastewater used in the study contained an average BOD of 1500 mg/L; COD, 2700 mg/L; phenol, 65 mg/L, and SS (suspended solids), 400 mg/L (Table 9). It was found that at the optimum doses of $FeSO_4$ (500 mg/L), $FeCl_3$, (500 mg/L), and alum (250 mg/L), the COD and SS removal efficiency was 24–28% and 70%, respectively. The study indicates that high doses of the coagulants were required, but the COD removal efficiency was marginal. Based on the above results, it was concluded that physicochemical treatment of effluent from this type of plant prior to biological treatment is neither effective nor economical [16]. A similar observation was made in a coagulation study of wastewater from the Alexandria Company for Pharmaceuticals and Chemical Industries (ACPCI) [38].

Air Stripping

Air stripping of pharmaceutical wastewater is a partial treatment used in particular for the removal of volatile organics from wastewater. M/S Hindustan Dorr Oliver, Bombay, in 1977 studied the effect of air stripping on the treatment of pharmaceutical wastewater and reported that a COD removal efficiency up to 30–45% can be achieved by air stripping. It was found that adding caustic soda did not appreciably increase the air stripping efficiency.

Ozone/Hydrogen Peroxide Treatment

Pharmaceutical wastewater contains various kinds of recalcitrant organics such as toluene, phenols, nitrophenols, nitroaniline, trichloromethyl propanol (TCMP), and other pollutants that exhibit resistance against biodegradation. Since these pollutants cannot be easily removed by biological treatment, biologically treated effluent exhibits a considerable oxygen demand, that is, BOD and COD, in the effluent. It has also been reported that activated carbon adsorption may not always be successful in removing such recalcitrant organics [39,40]. Economic constraints may also prohibit the treatment of pharmaceutical wastewater by activated carbon adsorption [41]. In such cases, ozone/hydrogen peroxide treatment may appear to be a proven technology for treating such pollutants from pharmaceutical wastewater.

The removal of organic 1,1,1-trichloro-2-methyl-2-propanol (TCMP), a common preservative found in pharmaceutical effluent, by ozone and hydrogen peroxide treatment has been studied [39]. Oxidation of TCMP was quite effective when it was contained in pure aqueous solutions, but almost nil when the same quantity of TCMP was present in pharmaceutical wastewater. Competitive ozonation of other organic solutes present inhibits the degradation of TCMP in pharmaceutical wastewater. Hence it has been concluded that for effective removal of TCMP by ozone/hydrogen peroxide, biological pretreatment of the wastewater for the removal of other biodegradable organics is crucial. It has been concluded that biological pretreatment of pharmaceutical wastewater before ozonation/hydrogen peroxide treatment should be utilized in order to increase the level of treatment.

5.6.2 Biological Treatment

The biological treatment of pharmaceutical wastewater includes both aerobic and anaerobic treatment systems. Aerobic treatment systems have traditionally been employed, including the activated sludge process, extended aeration activated sludge process, activated sludge process with granular activated carbon, or natural or genetically engineered microorganisms and aerobic fixed growth system, such as trickling filters and rotating biological contactors. Anaerobic treatment includes membrane reactors, continuously stirred tank reactors (anaerobic digestion), upflow filters (anaerobic filters), fluidized bed reactors, and upflow anaerobic sludge blanket reactors. Anaerobic hybrid reactors, which are a combination of suspended growth and attached growth systems, have recently become popular. Pharmaceutical/fine chemical wastewater presents difficult substrates for biological treatment due to their varying content of a wide range of organic chemicals, both natural and xenobiotic, which may not be readily metabolized by the microbial associations present in the bioreactors. Various processes dealing with the biological treatment of pharmaceutical wastewater are summarized in subsequent sections.

Activated Sludge Process

The activated sludge process has been found to be the most efficient treatment for various categories of pharmaceutical wastewater [14,15,19,42–46]. It has also been reported that this process can be successfully employed for the removal of tert-butanol, a common solvent in pharmaceutical wastewater that cannot be degraded by anaerobic treatment [44]. At a volumetric loading rate of 1.05 kg COD/m^3 day, HRT (hydraulic retention time) of 17 hours, and mixed liquor dissolved oxygen concentration of 1 mg/dm^3, the tert-butanol can be completely removed by the activated sludge process.

The activated sludge process has been successfully employed for the treatment of a wide variety of pharmaceutical wastewaters. The American Cynamid Company operated an activated sludge treatment plant to treat wastewater generated from the manufacture of a large variety of chemicals [19]. The activated sludge process has also been successfully employed for the treatment of wastewater in the chemical and pharmaceutical industries [42]. M/S Hindustan Dorr Oliver of Bombay studied the performance of the activated sludge process for the treatment of wastewater from its plant in 1977, and concluded that at an MLSS (mixed liquor suspended solids) concentration of 1800–2200 mg/L and aeration period of 24 hours, a COD removal efficiency of 50–83% can be achieved.

The performance of the activated sludge process for the treatment of wastewater from a synthetic drug factory, has been reported [14,15,45]. One of the biggest plants of its kind in Asia, M/S Indian Drugs and Pharmaceutical Ltd., Hyderabad, went into production in 1966 to make sulfa drugs such as sulfanilamides: antipyretics (phenacetin), B-group vitamins, antitubercular drugs (isonicotinic acid hydrazide) and antihelminthics, and so on.

When the performance of the activated sludge process was first studied for the treatment of simulated pharmaceutical wastewater, it was found that the wastewater was biologically treatable and that this process can be successfully employed for treating wastewater from pharmaceutical plants [45]. Based on Mohanrao's [14] recommendation, the performance of the activated sludge process for the treatment of actual waste streams generated from this plant, that is, alkaline waste, condensate waste, and a mixture of the two along with domestic sewage (1 : 2 : 1) as evaluated. Characteristics of various types of wastes used in the study are depicted in Table 13. The study demonstrated that condensate waste, as well as mixture, could be treated successfully, yielding an effluent BOD of less than 10 mg/L. However, the BOD removal efficiency of the system for the alkaline waste alone was found to be only 70%. The settleability of the activated sludge in all three units was found to be excellent, yielding a sludge volume index 23 and 45. The study indicated that biological treatability of the waste remained the same, although the actual waste was about 10 times diluted compared with the synthetic waste.

In 1984, the performance of a completely mixed activated sludge process for the treatment of combined wastewater was again evaluated. It was found that the activated sludge process was amenable for the treatment of combined wastewater from the plant, concluding that segregation and giving separate treatment for various waste streams of the plant would not be beneficial. The study was conducted at various sludge loading rates (0.14–0.16, 0.17–0.19, and 0.20–0.26 kg BOD/kg MLVSS (mixed liquor volatile suspended solids) per day and indicated that for the lower two loadings, effluent BOD was less than 50 mg/L, while for the other two higher loading

Table 13 Characteristics of Alkaline and Condensate Wastes Generated from a Synthetic Drug Plant at Hyderabad [14]

	Alkaline waste			Condensate waste		
Parameters	Min.	Max.	Avg.	Min.	Max.	Avg.
pH	8.6	9.4	–	7.0	7.6	–
BOD (mg/L)	1025	1345	1204	155	490	257
COD (mg/L)	2475	3420	2827	413	850	572
COD/BOD	2.41	2.54	2.3	2.66	1.73	2.2
Total solids (%)	0.53	0.66	0.63	0.12	0.14	0.13
Volatile solids (% of TS)	29.3	67.7	51.0	36.6	50.6	45.3
Total nitrogen (mg/L)	–	–	560	–	–	56
Total phosphorus (mg/L)	–	–	Nil	–	–	Nil

TS, total solids; BOD, biochemical oxygen demand; COD, chemical oxygen demand.

effluents BOD was less than 100 mg/L. The average TOC, COD, and BOD reductions were around 80, 80, and 99% respectively. The settleability of the activated sludge was found to be excellent with an SVI of 65–72 [15].

A similar study was conducted at Merck & Co. (Stonewall Plant, Elkton, Virginia) to assess the feasibility of the activated sludge process for treating wastewater generated from this plant. This plant is one of the six Merck Chemical Manufacturing Division facilities operated on a batch basis for fermentation and organic synthesis and has been in operation since 1941. A bench-scale study revealed that a food to microorganism (F/M) ratio from 0.15 to 0.25, MLVSS of 3500 mg/L, HRT 4 days, and minimum DO (dissolved oxygen) concentration of 3 mg/L were essential for meeting the proposed effluent limits and maintaining a viable and good settling sludge in the activated sludge process [46]. Based on these design criteria, a pilot plant and full-scale system were designed and studied. The old treatment plant consisted of an equalization basin, neutralization, primary sedimentation, roughing biofilter, activated sludge system, and rock trickling filter with final clarifiers. In the proposed study, the old activated sludge system, rock filter, and final clarifier were replaced with a new single-stage, nitrification-activated sludge system. A schematic diagram of the pilot plant is presented in Figure 1. The study demonstrated that BOD_5 removal efficiencies of the pilot and bench-scale plant were 94 and 98%, respectively. The TKN and NH_4-N removal were found to be 65 and 59%, respectively. It has also been observed that system operation was stable and efficient at F/M ratios ranging from 0.19 to 0.30, but prolonged operation at an F/M ratio less than 0.15 led to an episode of filamentous bulking.

The performance of the activated sludge process has been evaluated for the treatment of ACPCI (Alexandria Company for Pharmaceutical and Chemical Industry) effluent. These drug formulation and preparation-type plants are mainly involved in the production of a wide variety of pharmaceuticals, including analgesics, anthelmintics, antibiotics, cardiacs, chemotherapeutics, urologics, and vitamins. A study indicated that significant dispersed biosolids were found in the treated effluent when applying aeration for 6 hours. However, extending the aeration to 9–12 hours and maintaining the MLSS at levels higher than 2500 mg/L improved sludge

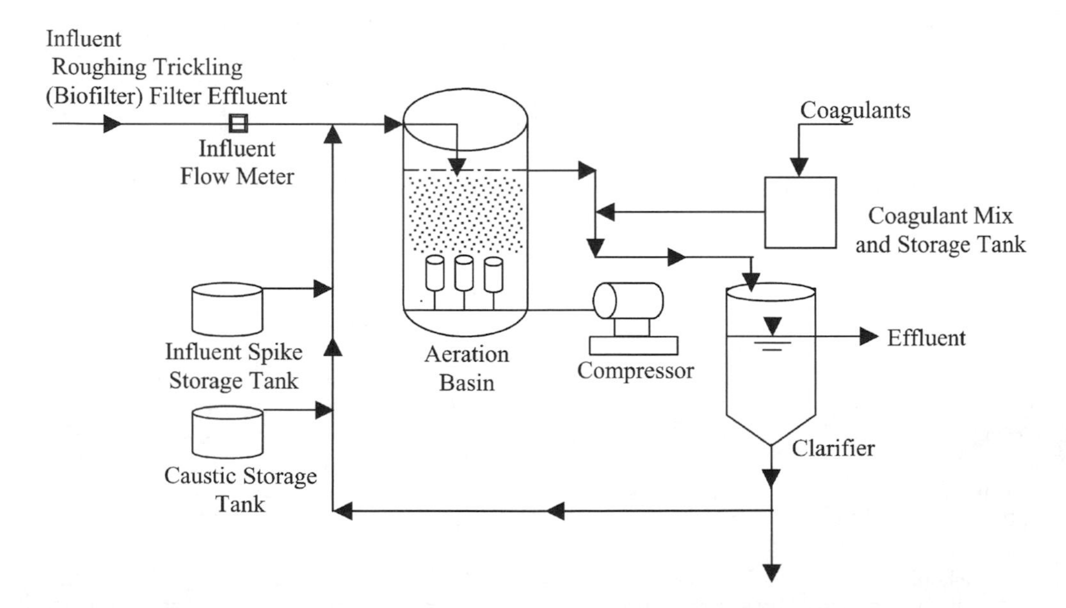

Figure 1 Schematic of the pilot plant at Merck and Co. Stonewall Plant in Elkton, VA.

settling and produced effluent with low SS. The study concluded that the activated sludge process is capable of producing effluent with BOD and SS values within the limits of the Egyptian standards. However, sand filtration was needed for polishing the treated effluent [38].

Powdered Activated Carbon Activated Sludge Process

Various researchers [47,48] have investigated the effect of powdered activated carbon (PAC) on the performance of the activated sludge process for the treatment of pharmaceutical wastewater. Various treatment units such as the activated sludge process (ASP), PAC-ASP, granular activated carbon (GAC), and a resin column were studied and compared in removing priority pollutants from a pharmaceutical plant's wastewater [47]. The wastewater generated from the plant contained 0-nitroaniline (0-NA), 2-nitrophenol (2-NP), 4-nitrophenol (4-NP), 1,1,2-trichloroethane (TCE), 1,1-dichloroethylene (DCE), phenol, various metals, and other organics. Characteristics of the wastewater collected from the holding pond are given in Table 14. The study concluded that there are treatment processes available that can successfully remove the priority pollutants from pharmaceutical wastewater. The treatment systems, ASP, PAC-ASP, and GAC, were all quite efficient in removing phenol, 2-NP and 4-NP, while the resin column was found unable to treat phenol. However, 2-NP and 4-NP can be treated to a certain extent (72 and 65%, respectively). The author further concluded that 1,1,2-dichloroethane and 1,1-dichloroethane can be treated successfully by all four treatment systems, but the efficiency of the resin column and GAC exceeded the other two systems. In terms of TOC removal, ASP and PAC-ASP were found to be more efficient than either GAC or the resin column. However, the performance of the PAC-fed ASP was found to be most efficient. In terms of color removal, PAC, GAC, and the resin process were more efficient than ASP, whereas in terms of arsenic removal, GAC and resin column were found most efficient. The performance summary of various treatment systems is given in Table 15. In general, it may be concluded that the addition of PAC in the ASP produced a better effluent than the ASP.

Addition of PAC to the activated sludge process increases the soluble chemical oxygen demand (SCOD) removal from the pharmaceutical wastewater but no measurable effect in terms

Table 14 Characteristics of Wastewater from a Typical Pharmaceutical Industry [47]

Parameters	Average	Ranges (min.–max.)
Color	4,648	1,800–6,600
TSS (mg/L)	234	47–2,700
VSS (mg/L)	152	17–1,910
TOC (mg/L)	387	205–630
Arsenic (mg/L)	5.82	4–12
o-Nitraniline (ONA) (μg/L)	12,427	3,200–30,500
Phenol (μg/L)	1,034	<10 to 3,700
2-NP (μg/L)	1,271	<10 to 2,900
4-NP (μg/L)	635	<10 to 2,300
TCE (μg/L)	4,080	620–6,550
DCE (μg/L)	291	<10 to 1,060

TSS, total suspended solids; VSS, volatile suspended solids; 4-NP, 4-nitrophenol; 2-NP, 2-nitrophenol; TCE, 1,1,2-trichloroethane; DCE, 1,1-dichloroethylene; TOC, total organic carbon.

Table 15 Performance Efficiency of Various Systems for the Treatment of Pharmaceutical Wastewater [47]

	Removal efficiency (%)			
Parameter	ASP	PAC-ASP	GAC	Resin column
Color	46.3	94.9	96.9	92
TOC	72.4	89.7	43.9	15
Phenol	95.8	>99	95.4	Nil
2-Nitrophenol	93.8	>99.2	99.1	72.3
4-Nitrophenol	89.4	96.5	96.5	65.8
o-Nitraniline	58.6	94.1	99.9	96.7
Arsenic	20.6	42.8	73.9	62.5
1,1,2-trichloroethane	94.2	96.4	99.4	99.8
1,1-dichloroethylene	94.5	>96.6	95.5	96.6

ASP, activated sludge process; PAC-ASP, powdered activated carbon activated sludge process; GAC, granular activated carbon; TOC, total organic carbon.

of soluble-carbonaceous biochemical oxygen demand (S-CBOD) was observed [48]. Moreover, addition of PAC increased the sludge settleability, but the MLSS settling rate remained at a very low level (0.01 to 0.05 cm/min) and resulted in a viscous floating MLSS layer at the surface of the activated sludge unit and clarifier. This study concluded that a PAC-fed ASP cannot be recommended as a viable option for this plant wastewater until the cause of the viscous floating MLSS layer is identified and adequate safeguards against its occurrence are demonstrated. The relationship to estimate the dose of activated carbon required for producing a desired quality of the effluent is given in Eq. (1).

$$\frac{X}{M} = 3.7 \times 10^{-7} C_{\mathrm{e}}^{2.1} \tag{1}$$

where X is the amount of SCOD removal attributed to the PAC (mg/L), M is the PAC dose to the influent (mg/L), and C_{e} is the equilibrium effluent SCOD concentration (mg/L).

Extended Aeration

The performance of the ASP has been found to be more efficient when operating on an extended aeration basis. The design parameters of the process were evaluated for the treatment of combined wastewater from a pharmaceutical and chemical company in North Cairo that produced drugs, diuretics, laboratory chemicals, and so on [49]. The study revealed that at an extended aeration period of 20 hours, COD and BOD removal efficiency ranges of 89–95% and 88–98%, respectively, can be achieved. The COD and BOD values of the treated effluent were found to be 74 mg/L and 43 mg/L, respectively.

In contrast, the performance of an extended aeration system for the treatment of pharmaceutical wastewater at Lincoln, Nebraska, was poor. At an organic loading of 30 kg BOD/day and a detention period of 25 hours, the percentage BOD reduction ranged from 30 to 70%. The degree of treatment provided was quite variable and insufficient to produce a satisfactory effluent. The pilot plant study performed at various feeding rates of 1.5, 2.4, 3.0, 3.6, and 4.8 L/12 hours indicated that at feeding rate of 4.8 L/12 hours, the sludge volume index was 645 and suspended solids were being carried over in the effluent.

Oxidation Ditch

The performance of an oxidation ditch for treating pharmaceutical wastewater has been evaluated and described by many researchers [16,50]. Treatability of wastewater from a typical pharmaceutical industry at Bombay producing various types of allopathic medicines was studied in an oxidation ditch at HRTs ranging from 1 to 3 days, corresponding to an SRT (solid retention time) of 8–16 days. The average MLVSS concentration in the reactor varied from 3000 to 4800 mg/L during the investigation period. The study indicated that on average about 86–91% of influent COD and 50% of phenols could be removed by this process [16].

A pilot-scale oxidation ditch was evaluated for the treatment of pharmaceutical wastewater at a Baroda unit. The treatment system was comprised of neutralization followed by clarifier and oxidation ditch. Primary treatment of the wastewater using neutralization with lime followed by sedimentation in a clarifier demonstrated SS and BOD removal of 30–41% and 28–57%, respectively. The effluent from the clarifier was further treated in an oxidation ditch operating on an extended aeration basis. It was found that at loading of 0.1–0.5 lb BOD/lb MLSS/day, an MLSS concentration of 3000–4000 mg/L, and aeration period of 22 hours, a BOD removal up to 70–80% could be achieved. The high COD of treated effluent indicated the presence of organic constituents resistant to biodegradation. Considering the high COD/BOD ratio of the wastewater, it has been suggested that the biological treatment should be supplemented with chemical treatment for this type of plant wastewater [50].

Aerated Lagoon

The performance studies of aerated lagoons carried out by many researchers [14,51] have demonstrated that lagoons are capable of successfully treating wastewater containing diversified fine chemicals and pharmaceutical intermediates.

A laboratory-scale study of alkaline and condensate waste streams from a synthetic drug factory at Hyderabad demonstrated that an aerated lagoon is capable of treating the wastewater from this industry [14]. The BOD removal rate K of the system was found to be 0.18/day and 0.155/day based on the soluble and total BOD, respectively. Based on the laboratory studies, a flow sheet (Fig. 2) for the treatment of waste was developed and recommended to the factory.

Trickling Filter

The performance of a trickling filter has been studied by many researchers [14,38,49,51–53] and it was found that a high-rate trickling filter was capable of treating wastewater containing diversified fine chemicals and pharmaceutical intermediates to a level of effluent BOD less than 100 mg/L [51]. A similar conclusion was made in the performance study of a trickling filter for the treatment of wastewater from chemical and pharmaceutical units [53].

It has also been reported that wastewater from a pharmaceutical plant manufacturing antibiotics, vitamins, and sulfa drugs can be treated by using a trickling filter [52]. One study evaluated the efficiency of a sand bed filter for the treatment of acidic waste streams from a synthetic organic pharmaceutical plant at Hyderabad. The acidic waste stream was neutralized to a pH of 7.0 and treated separately through a sand bed filter. The sand bed filter was efficient in treating the acidic waste stream to a level proposed for its discharge to municipal sewer [14].

The efficiency of the biological filter (trickling filter) for treatment of combined wastewater from a pharmaceutical and chemical company in North Cairo has been evaluated. The treatment system consisted of a biological filter followed by sedimentation. The degree of treatment was found quite variable. The COD and BOD removal efficiencies of the trickling filter at an average OLR (organic loading rate) of 26.8 g BOD/m^2 day were found to be 43–88%

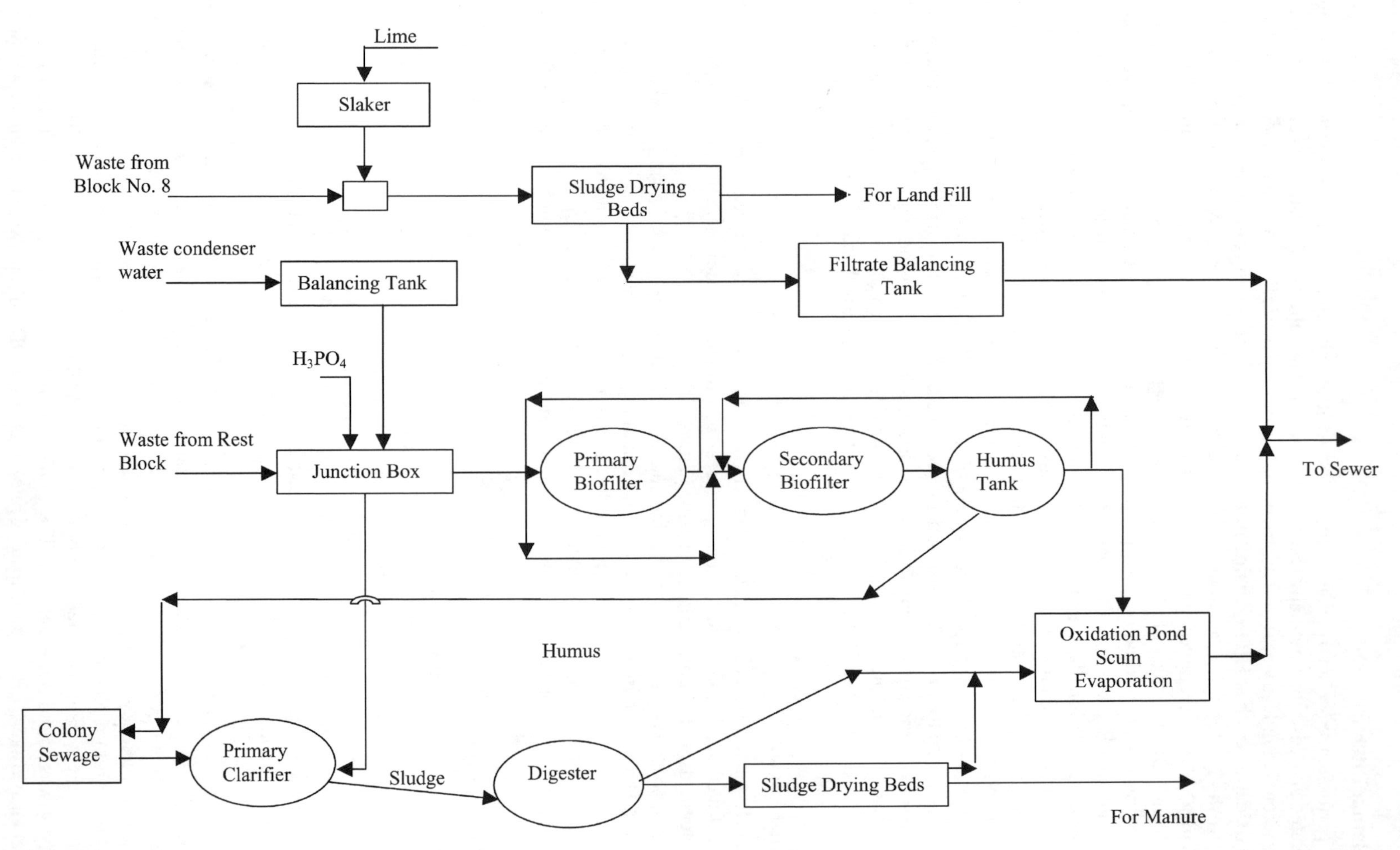

Figure 2 Flow sheet for treatment of synthetic drug waste.

and 58–87%, respectively. The study revealed that a biological filter alone was unable to produce effluents to a level complying with the national standards regulating wastewater disposal into the surface water [49].

Similar conclusions were made in the treatment of ACPCI effluent using a biofilter. The low performance efficiency and presence of dispersed biosolids in the effluent have made the trickling filter unsuitable for the treatment of this plant wastewater [38].

Anaerobic Filter

The anaerobic filter has been reported to be a promising technology for the treatment of wide varieties of pharmaceutical wastewater [4,10,54–59]. The performance of the anaerobic filter was first studied at a pharmaceutical plant in Springfield, Missouri [54]. The characteristics of the waste fed into the reactor are given in Table 16. The treatability study revealed that at an HRT of two days, an OLR ranging from 0.37 to 3.52 kg COD/m^3 day, and influent COD concentration ranging from 1000 to 16,000 mg/L, COD removal efficiencies of 93.7 to 97.8% can be achieved. Moreover, the problem of sludge recycling and sludge disposal in the case of the anaerobic filter can be reduced to a great extent due to the much smaller biomass yield, that is, 0.027 g VSS (volatile suspended solids)/g COD removed. The shock loading study revealed that shock increase in organic loading did not result in a failure of the capability of the filter to treat the waste. This is a distinct feature of anaerobic filters, especially when dealing with pharmaceutical wastewater, which is supposed to cause shock loading due to frequent variation in composition as well as in magnitude of the waste load. In contrast, it has been reported that the

Table 16 Physical and Chemical Characteristics of Pharmaceutical Waste in Springfield, MO [54]

Parameters	Range
pH	7.5–10.1
COD (mg/L)	15,950–16,130
SS (mg/L)	28–32
TS (mg/L)	432–565
Alkalinity (mg/L as $CaCO_3$)	412–540
Nitrogen (mg/L)	
Ammonia	0–11.8
Organic	33.3–34.2
Phosphorus (mg/L)	
Ortho	0.4–0.5
Total	0.9–0.95
Heavy metals (mg/L)	
Lead	0.005–0.007
Copper	0.140
Zinc	0.018–0.11
Manganese	0.020–0.22
Iron	0.05–0.56
Cadmium	0.020–0.01
Calcium	9.7–58.7
Magnesium	7.5–14.7

COD, chemical oxygen demand; SS, suspended solids; TS, total solids.

anaerobic filter fed with pharmaceutical wastewater containing high ammonia nitrogen could not withstand a three-fold increase in OLR [55]. It has been further concluded that the amber color of the untreated waste can be removed through treatment, but due to poor degradability of the odor-producing toluene, the effluent maintained the tell-tale odor of toluene, indicating that it passed through the filter with little or no treatment.

The suitability of the anaerobic filter for treatment of wastewater from a chemically synthesizing pharmaceutical industry has been studied [10]. Characteristics of the strong waste stream used in the study are given in Table 17. The study revealed that at an HRT of 48 hours and COD concentration of 1000 mg/L, waste can be treated at least to a level of treatment generally occurring when employing aerobic treatment. Moreover, methane-rich biogas is generated in this treatment, which can be utilized later as an energy source. Thus the use of an anaerobic filter system would be a net energy producer rather than an energy consumer as in the case of current aerobic systems. In addition, the effluent from this system was found to contain far less color than the effluent from the existing system.

The performance of an anaerobic mesophilic fixed film reactor (AMFFR) and an anaerobic thermophilic fixed film reactor (ATFFR) for the treatment of pharmaceutical wastewater of a typical pharmaceutical plant at Mumbai was studied and compared [56]. The study revealed that at an OLR of 0.51 kg/m^3 day and HRT of 4.7 days, the COD removal efficiency of mesophilic was superior (97%) to the thermophilic reactor (89%). The effect of organic loading and reactor height on the performance of anaerobic mesophilic (30°C) and thermophilic (55°C) fixed film reactors have demonstrated that the AMFFR can take a load of several orders of magnitude higher, with higher removal efficiency compared to the ATFFR for pharmaceutical wastewater [56]. Wastewater used in the study was collected from an equalization tank of the pharmaceutical industry treatment plant at Bombay. The characteristics of the wastewater are given in Table 18. The start-up study has indicated that a starting-up period for the AMFFR (four months) was far less than the starting-up period for the ATFFR (six months). The gas production and methane percentage were also found to be higher in the AMFFR compared to the ATFFR. The effective height of the reactor was found to be in the range of 30–90 cm. Other researchers [10,54,55,58,59] have reported a similar effective height range of 15–90 cm. They have

Table 17 Characteristics of a Concentrated Waste Stream of Synthesized Organic Chemicals—Type Pharmaceutical Industry [10]

Parameters	Sample 1 (28-02-76)	Sample 2 (20-04-76)	Sample 3 (10-10-76)	Sample 4 (20-11-76)
pH	3.6	3.5	2.2	1.6
BOD_5 (mg/L)	Varies	–	–	–
COD (mg/L)	514,900	533,000	89,000	62,530
TS (mg/L)	37,740	38,520	13,090	5,190
TDS (mg/L)	37,650	38,420	13,030	5,180
TVSS (mg/L)	18,880	19,070	5,180	2,090
Dissolved volatile solids (mg/L)	18,800	18,980	5,120	2,080
TKN (mg/L)	19.3	25.8	23.0	33.6
NH_4^-N (mg/L)	BDL	BDL	BDL	BDL
SO_4^{2-} (mg/L)	–	–	75.0	183
Total phosphorous (mg/L)	BDL	BDL	BDL	BDL

BOD, biochemical oxygen demand; COD, chemical oxygen demand; TS, total solids; TDS, total dissolved solids; TVSS, total volatile suspended solids; TKN, total Kjeldhal nitrogen; BDL, below detectable limit.

Table 18 Characteristics of Wastewater from a Typical Pharmaceutical Industry at Bombay [56]

Parameters	Concentration range	Average
pH	5.5–9.2	7.2
COD (mg/L)	1,200–7,000	2,500
TSS (mg/L)	30–55	40
Total alkalinity as $CaCO_3$ (mg/L)	70–1,500	750
TVA (mg/L)	70–2,000	750
NH_4^+-N (mg/L)	80–500	200
PO_4^{3-}P (mg/L)	3.5–35	16
SO_4^{2-} (mg/L)	100–700	300
Chloride (mg/L)	500–1,200	900
Sulfide (mg/L)	2–8	5
Cobalt (mg/L)	0–0.6	0.2
Potassium (mg/L)	5–25	18
Lead (mg/L)	0.05–0.9	0.35
Iron (mg/L)	0.2–0.9	0.45
Zinc (mg/L)	0.05–0.15	0.09
Chromium (mg/L)	0.1–0.6	0.3
Mercury (mg/L)	0.15–0.50	0.25
Copper (mg/L)	0–0.10	0.1
Cadmium (mg/L)	0.07–0.25	0.10
Sodium (mg/L)	200–3,000	2,000
Manganese (mg/L)	0.1–0.4	0.2
Silicon (mg/L)	5–50	25
Magnesium (mg/L)	5–60	40
Tin (mg/L)	0.1–1.5	0.6
Aluminum (mg/L)	0.05–0.20	0.10
Barium (mg/L)	0.1–0.3	0.16
Arsenic (mg/L)	0.1–0.5	0.25
Bismuth (mg/L)	0.09–0.3	0.15
Antimony (mg/L)	0.50–3.0	1.4
Selenium (mg/L)	0.1–0.95	0.38

TVA, total volatile acid; COD, chemical oxygen demand; TSS, total suspended solids.

reported that rapid change in most of the characteristics occurs only in the lower portion of the reactor.

Two-Stage Biological System

The two-stage biological system generally provides a better quality of effluent than the single-stage biological system for the treatment of pharmaceutical wastewater. It has been reported that a single-stage biological system such as activated sludge process and trickling filter alone is not capable of treating the wastewater to the effluent limit proposed for its safe discharge to inland surface water [49]. However, the combined treatment using a two-stage aerobic treatment system is efficient in treating wastewater to a level complying with national regulatory standards. A performance study of a two-stage biological system for the treatment of pharmaceutical wastewater generated from Dorsey Laboratories Plant (drug mixing and formulation type plant) at Lincoln, Nebraska, was carried out and the following conclusions drawn:

- Shock organic and hydraulic loading created serious operational problems in the system. Bulking sludge and the inability to return solids from the clarifier to the aeration unit further complicated plant operation.
- Microscopic observations of the sludge flock showed the presence of filamentous organisms, *Sphaerotilus natans*, in high concentrations. The presence of these organisms was expected to be due to deficiency of the nitrogen in the wastewater.

To overcome the problem of sludge bulking, nitrogen was supplemented in the wastewater as ammonium sulfate, but operational problems continued even after nitrogen was added. Hence, to avoid shock loading on the treatment, the effluent treatment plant (ETP) was expanded. The expanded treatment system (Fig. 3) consists of a communicator, basket screen, equalization basin, biological tower, activated sludge process, disinfection, and filtration. The study indicated that the equalization basin and biological tower effectively controlled shock loading on the activated sludge process. Overall, BOD and COD removal of 96 and 88%, respectively, may be achieved by employing a two-stage biological system [5]. It has also been found that a two-stage biological system generally provides a high degree of treatment. However, bulking sludge causes severe operational problems in the extended aeration system and sand filter.

A two-stage biological treatment system consisting of anaerobic digestion followed by an activated sludge process was developed for the treatment of liquid waste arising from a liver and beef extract production plant. Being rich in proteins and fats, the waste had the following characteristics: pH, 5.8; COD, 21,200 mg/L; BOD, 14,200 mg/L; and TS, 20,000 mg/L. The treatability study of the waste in anaerobic digestion revealed that at an optimum organic loading rate of 0.7 kg COD/m^3 day and an HRT of 30 days, a COD and BOD removal efficiency of 89 and 91% can be achieved [18]. The effluent from anaerobic digestion still contains a COD of 2300 mg/L and BOD of 1200 mg/L. The effluent from anaerobic digestion was settled in a primary settling tank. At an optimum retention time of 60 minutes in the settling tank, the percentage COD and BOD removal increased to 94 and 95%, respectively. The effluent from the settling tank was then subjected to the activated sludge process. At an optimum HRT of 4 days, the COD and BOD removal increased to 96 and 97%, respectively. The effluent from the activated sludge process was settled for 1 hour in a secondary settling tank, which gave an increase in COD and BOD removal to 98 and 99%, respectively. The study therefore revealed that the combination of anaerobic–aerobic treatment resulted in an overall COD and BOD reduction of 98 and 99%, respectively. The final effluent had a COD of 290 mg/L and BOD of 50 mg/L, meeting the effluent standard for land irrigation.

The performance of two-stage biological systems was examined for the treatment of wastewater from a pharmaceutical and chemical company in North Cairo. A combined treatment using an extended aeration system (20 hour aeration) or a fixed film reactor (trickling filter) followed by an activated sludge process (11 hour detention time) was found efficient in treating the wastewater to a level complying with national regulatory standards. From a construction cost point of view, the extended aeration system followed by activated sludge process would be more economical than the fixed film reactor followed by activated sludge process. The flow diagrams of the two recommended alternative treatment processes for the treatment of this plant wastewater are depicted in Figure 4 and Figure 5, respectively [49].

Anaerobic treatment of high-strength wastewater containing high sulfate poses several unique problems. The conversion of sulfate to sulfide inhibits methanogenesis in anaerobic treatment processes and thus reduces the overall performance efficiency of the system. Treatment of high sulfate pharmaceutical wastewater via an anaerobic baffled reactor coupled

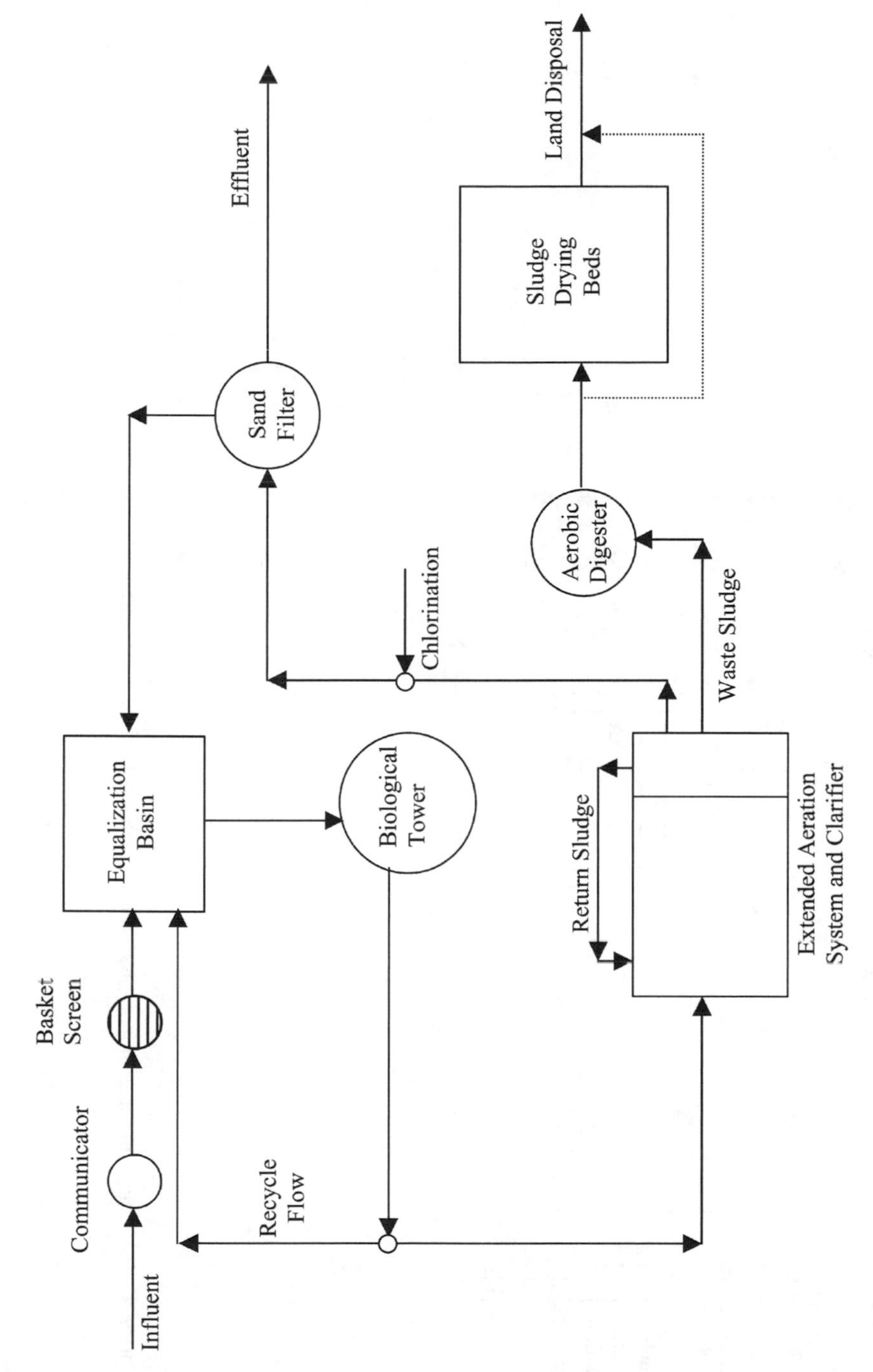

Figure 3 Flow diagram of wastewater treatment plant at Dorsey Laboratory.

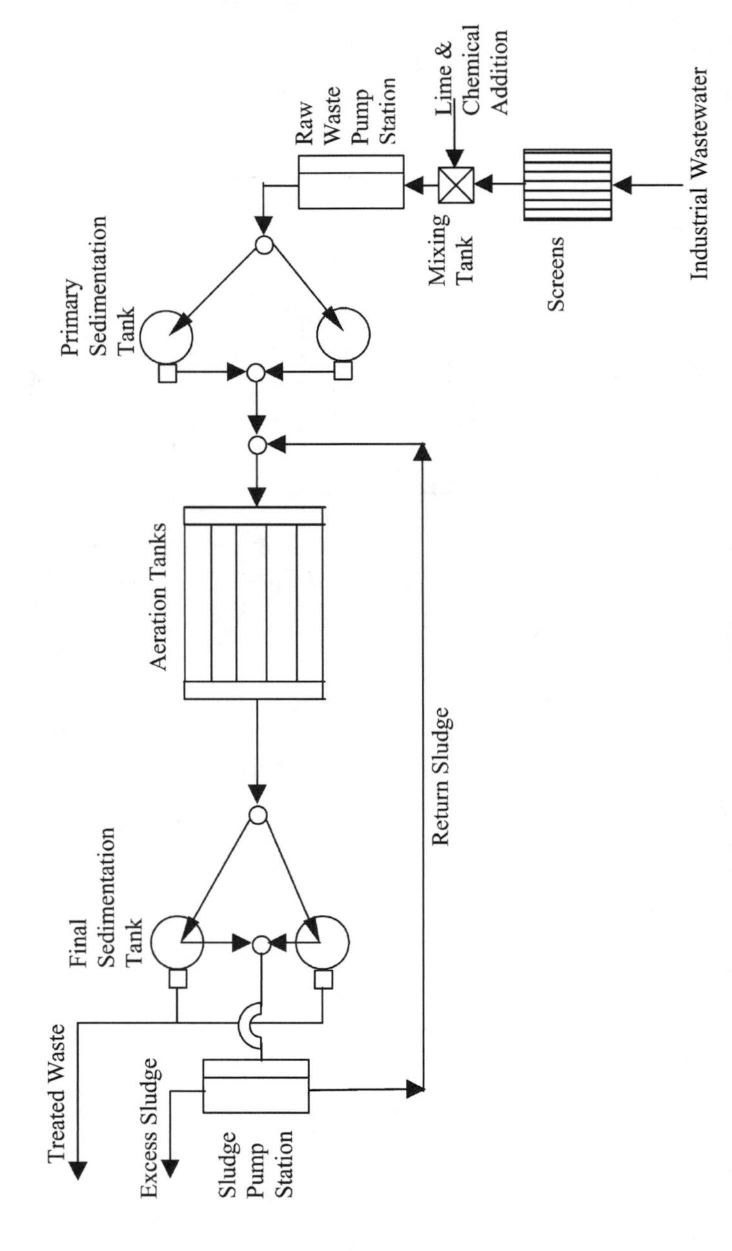

Figure 4 Flow diagram for treatment process using activated sludge, extended aeration.

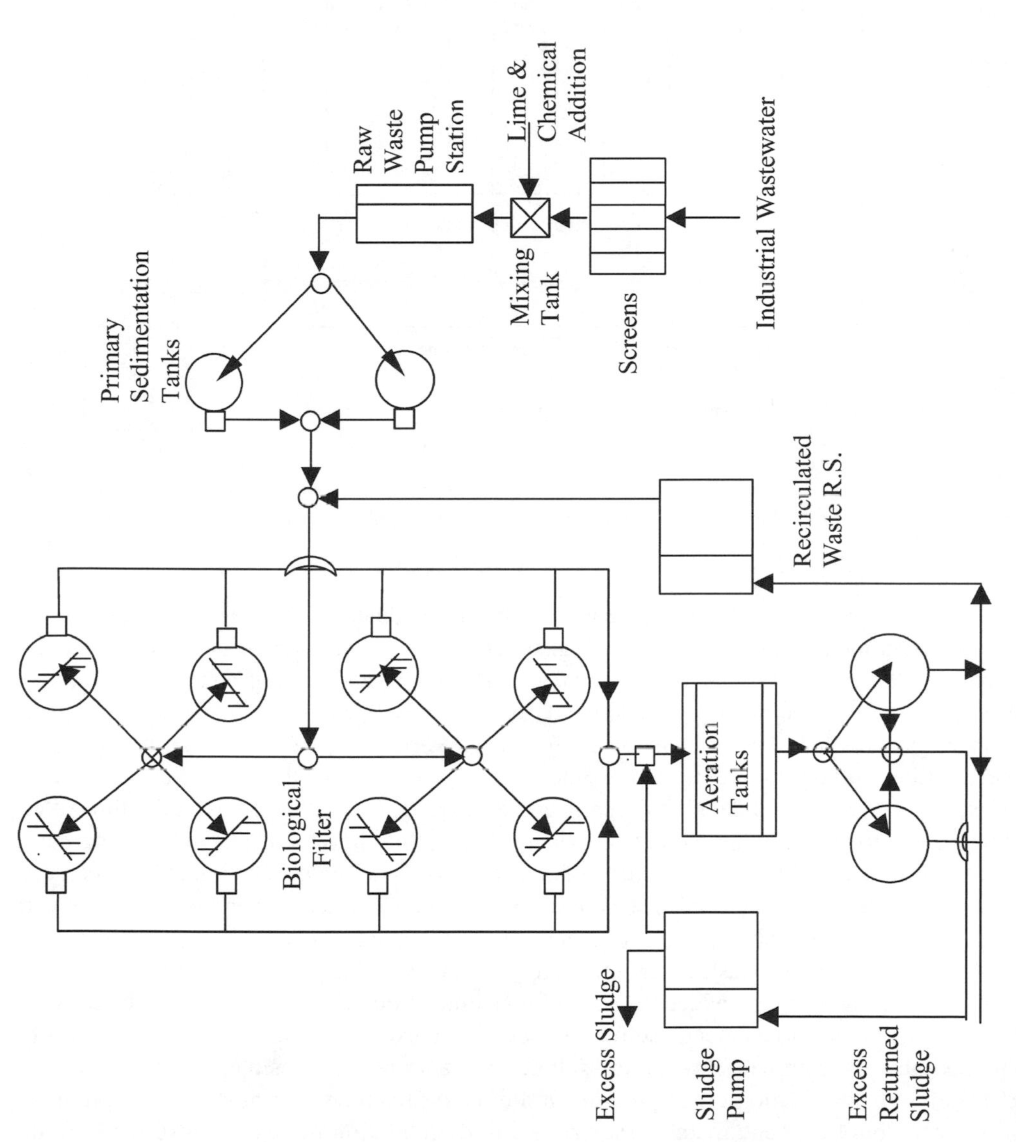

Figure 5 Flow diagram for treatment process using biological filters followed by activated sludge process.

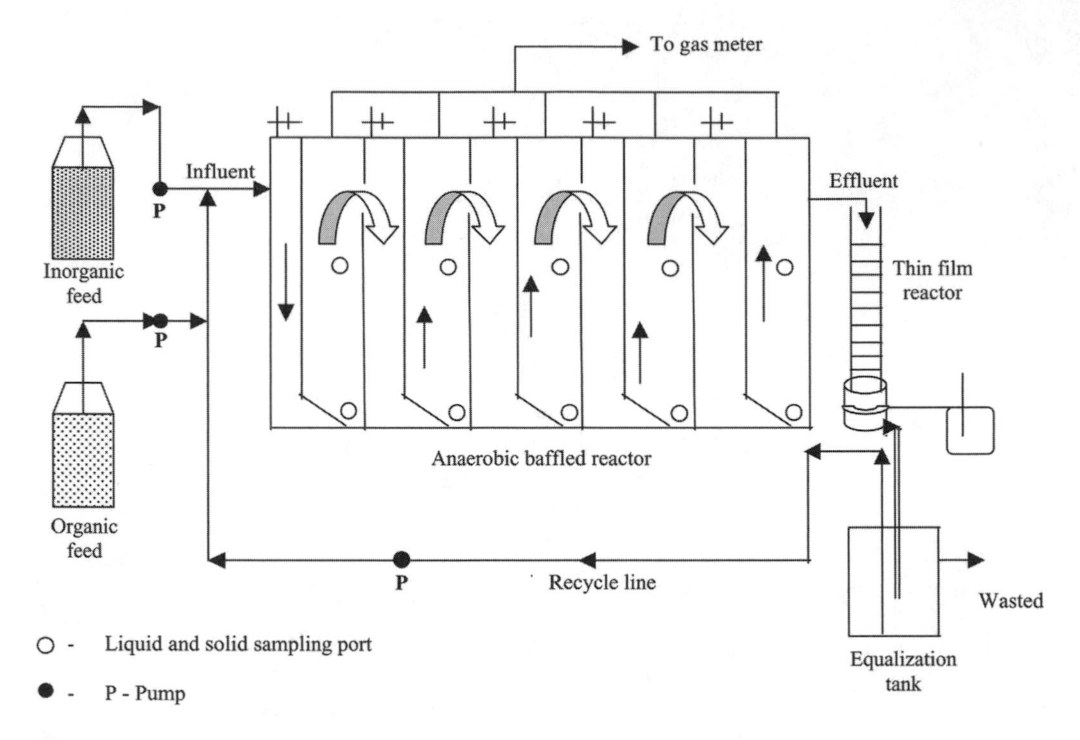

Figure 6 Schematic of anaerobic baffled reactor followed by thin film sulfide oxidizing reactor.

with biological sulfide oxidation was carried out and evaluated. The schematic view of the combined treatment system is given in Fig. 6. The wastewater used in the study contained isopropyl acetate, sulfate, and cellular product. The COD and sulfate concentration of the wastewater were 40,000 mg/L and 5000 mg/L, respectively. Treatment of the wastewater using an anaerobic baffled reactor alone was found effective at 10% dilution but at higher concentration, sulfide inhibition reduced the efficiency of both COD conversion and sulfate conversion. To reduce sulfide inhibition, the treated effluent was subjected to a thin film sulfide oxidizing reactor to facilitate biological oxidation of sulfide into elemental sulfur. The study indicated that at an influent concentration of 40% and HRT of 1 day, COD removal efficiencies greater than 50% can be achieved. The conversion of influent sulfate was greater than 95% with effluent sulfide concentration less than 20 mg/L [60]. Coupled anaerobic/aerobic treatment of high sulfate-containing wastewater effectively alleviated the sulfide inhibition of both methanogenesis and sulfate reduction. A thin film sulfide oxidizing reactor was also effective in converting the sulfide to elemental sulfur without adding excess oxygen, which made recycling of treated anaerobic effluent through the sulfide oxidizing reactor feasible. This indicates that biological sulfide oxidation could provide an alternative method to remove sulfide produced during anaerobic treatment, thereby alleviating sulfide inhibition by removing sulfur from the wastewater stream.

Anaerobic Hybrid Reactor

The anaerobic hybrid reactor is generally a combination of suspended growth and attached growth systems. Recently, this technology has become popular in the treatment of industrial wastewater, in particular in cases of high-strength wastewater. It has been reported that this reactor design presents a viable alternative to continuously stirred reactors, anaerobic filters, and anaerobic fluidized bed reactors for the high-rate treatment of pharmaceutical wastewater containing C_3 and C_4 aliphatic alcohol and other solvents [44]. The suitability of an anaerobic hybrid reactor for the treatment of synthetic pharmaceutical wastewater containing target solvents C_3 and C_4, tert-butanol, sec-butanol, and ethyl acetate was assessed at various organic loadings and varying influent concentrations. The study indicated that isopropanal, isobutanol, and sec-butanol can be almost fully degraded by using the anaerobic hybrid reactor. At OLR ranging from 3.5 to 4.5 kg COD/m^3 day and HRT of 2 days, the reactor achieved total and soluble COD removal efficiencies of 97 and 99%, respectively. However, the reactor was unable to degrade the tert-butanol, resulting in a decrease in soluble COD removal efficiency to 58%. A bacterial enrichment study with the tert-butanol as a sole substrate indicated that this is poorly degradable in anaerobic conditions. The observed recalcitrance of the tert-butanol in the present case contrasts with the findings of earlier researchers, who have listed these solvents as being amenable to anaerobic digestion [61,62]. Degradation of tert-butanol in the activated sludge process has been evaluated, and it was found that aerobic posttreatment/polishing of the anaerobically treated effluent of pharmaceutical wastewater is essential for removing the residual solvent [43]. The addition of a trace metals cocktail in the feed did not affect steady-state reactor performance, but was found beneficial in handling the influent compositional changes. Moreover, the methanogenic activity of the granular sludge fed with trace metals was found significantly higher than the granular sludge of the reference anaerobic hybrid reactor.

Combined Waste Treatment with Other Industrial Waste

The possibility of treatment of pharmaceutical wastewater combined with other industrial waste has been explored and evaluated [63]. One study carried out nitrification of high-strength nitrogenous wastewater (a concentrated stream from a urea plant) in a continuously stirred tank reactor. Pharmaceutical wastewater was used as an organic carbon source to maintain a COD/TKN ratio of 1. The reactor was operated at an HRT of 1.5–2.1 days and solid retention time (SRT) ranging from 10–62.5 days. Characteristics of the wastewater from the urea plant, pharmaceutical wastewater, and combined wastewater are depicted in Table 19. The study concluded that pharmaceutical wastewater may be used as a co-substrate to supply energy for nitrification of high-strength nitrogenous wastewater. Such treatment alternatives establish the advantages of a dual mechanism of treatment, that is, nitrification as well as oxidation of organic pollutants.

Table 19 Characteristics of Urea Plant, Pharmaceutical Plant, and Combined Wastewater [63]

Parameter	Urea plant	Pharmaceutical plant	Combined wastewater
pH	11.0–12.5	5.0–8.0	7.0–9.0
COD (mg/L)	3,520–4,850	1,100–5,500	1,010–1,290
Alkalinity as $CaCO_3$ (g/L)	1.005–1.010	0.30–2.0	4.4–5.45
PO_4^{3-}-P (mg P/L)	0.7–1.0	2.8–14.4	22.8
NH_4^{-}N (mg N/L)	38,000–45,000	30–50	500–550
Urea-N (mg N/L)	1,860–2,380	–	500

COD, chemical oxygen demand.

5.6.3 Integrated Treatment and Disposal Facilities for Specific Pharmaceutical Waste

The above-cited studies demonstrate the performance of a particular unit system for the treatment of specific type of waste stream. A particular unit system alone may not be able to treat the wastewater to a level of effluent standard prescribed for its safe disposal. Hence a number of pretreatments, such as screening, sedimentation, equalization, and neutralization, and post-treatment units such as secondary sedimentation, sludge thickening, digestion and disposal, disinfection, and so on, are extremely important for complete treatment. The effluent treatment and disposal facilities adopted by various types of pharmaceutical industries are described in the following sections.

Treatment of Synthetic Organic Bulk Pharmaceutical Waste

The Hoffman–La Roche plant in Belvedere, NJ, manufactures synthetic organic bulk pharmaceuticals, including dry vitamin powders, sulfa drugs, vitamin C, riboflavin, aromatics, and sodium sulfate salts. An integrated sodium sulfate recovery system was employed in this plant to recover sodium sulfate. The plant's waste control and treatment system includes screening, preclarifier, equalization with aeration (1 day detention time), pH adjustment/neutralization, flocculator-clarifier, activated sludge process, secondary settler, two oxidation ponds in series, sludge thickening, aerobic sludge digestion, sludge drying beds, and final chlorination. The treatment plant was initially designed for a design flow of 1 MGD (million gallons per day) with BOD_5 and TSS removal efficiency of the system at 97.4 and 98%, respectively. Effluent from this plant had a BOD_5 of 50 mg/L and TSS of 20 mg/L. In 1973, the raw waste load at the plant increased from 1 MGD to 1.6 MGD with BOD load of 30,000 lb · BOD/day or more, together with 8400 lb/day of TSS. By late 1973, the effluent load was about twice the design specification. Although data on the performance of the treatment plant for the current waste loads (1973, 1974) were lacking, the author has indicated a typical removal of BOD, COD, and TSS of 97.5, 90, and 90%, respectively.

Treatment of Fermentation/Synthetic Organic Bulk Pharmaceutical Waste

Pfizer, Inc. (Terre Haute, IN) is a fermentation/synthesized organic bulk pharmaceutical type plant mainly involved in the manufacture of streptomycin, terramycin, two undefined antibiotics, fumaric acid, benzoic acid, and so on. This plant employs a five-stage biological system with a retention time of process waste varying from 45 to 65 days. The treatment plant consists of a primary clarifier, two extended aeration (activated sludge) basins in series (12 days detention), secondary settling tank, two clari-digesters in parallel, two standard rate trickling filters in parallel, a high-rate bio-oxidation tower, final clarifier, two aerated stabilization ponds in series, stabilization pond, chlorination, aerobic sludge digester, sludge stabilization pond, land/crop application of stabilized sludges, and holding pond for spent cooling waters (1 day detention). The plant was designed for combined waste of 1.3 MGD of process waste and 5 MGD of spent cooling water flow. In 1972, Pfizer reported average BOD and TSS removal of 98 and 97.5%, respectively. From 1973 to 1974, the BOD and TSS removal were reported to be 99.1 and 97.8%, respectively. The treated effluent contained a BOD of 10–15 mg/L and TSS of 20–30 mg/L. The Pfizer system was capable of giving 50% phosphorous reduction. The TKN, NH_4-N, and organic nitrogen removal were reported to be 75, 67, and 81%, respectively.

A similar plant, Clinton Laboratories (Clinton, IN), is mainly involved in producing a cephalosporin-type antibiotic. Major products include monensin sodium, keflex, and kefzol. The waste generated in this plant includes mycelia, general trash, concentrated chemical wastes, diluted chemical wastes, water process waste, sanitary sewage, and a clear water stream.

The control and treatment system in this plant mainly relies on the chemical destruction of waste rather than biological processes. The plant generates a raw waste load as high as 400,000 lb · BOD/day. From 1973 to 1974, the company reported a total waste flow of 3.5–4.3 MGD containing a BOD of 1710–1960 lb/day, COD of 3700–4000 lb/day, and TSS of 1040–1250 lb/day. The treatment system included the following units:

- concentration of waste streams to minimum volume;
- oversized strippers for solvent recovery;
- stripper system for waste preconditioning;
- Carver–Greenfield multistage, oil dehydration, steam evaporator system (fermentation waste);
- two John Zink thermal oxidation incineration systems (chemical wastes);
- Bartlett–Snow rotary kiln incinerator (plant trash and mycelium);
- small biological treatment plant (sanitary wastes);
- cooling water towers;
- scrubbing of air effluents from incinerators and waste heat boiler on Carver–Greenfield.

Both concentrated and dilute waste were sent to a pair of John Zink thermal oxidizers equipped with adjustable venturi scrubbers for removal of particulates prior to stack discharge. Water process waste originating primarily from fermentation sectors was sent to the Carver–Greenfield evaporation system. The evaporator utilized a multistep oil dehydration process and was equipped with a centrifuge, waste heat boiler, and a venturi scrubber. The Clinton Laboratory reported an overall BOD and COD reduction of 90 and 99%, respectively, depending upon the configuration used.

Treatment of Fermentation, Organic Synthesis Processing, and Chemical Finishing and Packaging Type Bulk Pharmaceutical Waste

Abbott Labs (Chicago, IL) has extensive fermentation, organic synthesis processing, and chemical finishing and packaging facilities and is engaged mainly in production of antibiotics, that is, erythromycin and penicillin, and hundreds of medicinal and fine chemicals. Characteristics of various types of wastes generated from this plant are depicted in Table 20. The typical units involved in the Abbott treatment works are as follows:

- waste screening and neutralization;
- two equalization basins (1.5 day detention);
- six activated sludge basins (100,000 gallon);
- degasification chambers for mixed liquors;

Table 20 Characteristics of the Abbott Laboratory Wastewater [3]

Parameters	Fermentation waste	Chemical waste	Combined waste
Flow (MGD)	0.312	0.262	0.575
pH	6.7	5.4	6.1
BOD (mg/L)	3620	2520	3120
TSS (mg/L)	1660	510	1140
TDS (mg/L)	3590	5690	4620

MGD, million gallons per day; BOD, biochemical oxygen demand; TSS, total suspended solids; TDS, total dissolved solids.

- two final settlers in parallel;
- pasteurization of final process effluent;
- chlorination of process plus cooling flows;
- evaporation/drying of spent fermentation broths;
- enclosure of treatment works;
- centrifuging of excess biological sludge;
- ducting of various odorous air streams to main plant boilers;
- incineration of sludge and odorous air streams in main boilers;
- recovery of select waste streams high in ammonia for bulk fertilizer sales;
- connection to municipal AWT plant.

Process waste averaging 0.6–0.7 MGD was sent to the activated sludge treatment system. Cooling water flows of 14–15 MGD were sent for chlorination before final discharge. This plant also employed a spent fermentation beer recovery system integrated with an expansive incinerator ducting system. Exhaust air from the drying of spent beer was collected into a specially designed duct system. This also collected the odorous stream from the fermentors, exhaust from degassing chambers, and exhaust from the enclosed activated sludge tank and sludge holding tanks. The combined air stream was then carried to the main plant boilers and incinerated therein. Treated effluent characteristics are given in Table 21. In 1972, overall BOD and TOC reductions were reported to be 94.6 and 86%, respectively. In 1973, the average BOD and TOC reductions were reported as 96.7 and 98%, respectively. The annual costs of the Abbott treatment works were U.S. $1.2 million, which was equivalent to U.S. $4.50–5.5 per 1000 gallons of process waste. In view of the state effluent limits of 4 mg/L BOD and 5 mg/L TSS for discharges into Lake Michigan by 1975, the treated effluent is scheduled for connection to the regional municipal AWT plant [29,30,33,64].

A treatment plant including the following units was recommended for handling the wastewater from drug formulation and packaging type bulk pharmaceutical waste [3]:

- possible separate handling of process and sanitary wastes;
- screening;
- equalization (2 days detention or more) with auxiliary aeration;
- activated sludge, multichamber (approximately 24 hour detention);
- secondary settling;

Table 21 Characteristics of Treated Effluent from Abbott Laboratory Works and 1972 Effluent Standards [3]

Parameters	Treated effluent plus cooling water flow	1972 state standard
Flow (MGD)	15	–
pH	7.5	–
BOD (mg/L)	16	20
TSS (mg/L)	20	25
TDS (mg/L)	400	750
Phenolics (mg/L)	0.02	0.30
Mercury (mg/L)	0.0003	0.0005
Coliforms/100 mL	11	400

MGD, million gallons per day; BOD, biochemical oxygen demand; TSS, total suspended solids; TDS, total dissolved solids.

- sludge thickening;
- aerobic digestion of excess sludges with residues to landfill;
- chlorination of final effluent.

A similar system with minor modifications should be fairly adaptable to biological production type pharmaceutical plants.

5.7 OPERATIONAL PROBLEMS AND REMEDIAL MEASURES

Much research has focused on bulking of the sludge in the aerobic treatment of pharmaceutical wastewater [46,65–67]. The filamentous organism *Sphaerotilus natans* has been reported to be responsible for sludge bulking. The growth of these filamentous organisms was coupled with a deficiency of nitrogen in the wastewater and shock organic and hydraulic loading applied in the system. Another researcher identified the Type 021N microorganism as being responsible for sludge bulking [46]. Three microorganisms, Type 0092, *Microtrix parvicella*, and Type 0041, were also identified to be responsible for sludge bulking. It has been further noted that another factor responsible for the bulking of sludge is influent wastewater variability. Subsequently it has been concluded that all three organisms are correlated with filamentous bulking at low organic loading [66]. To deal with the problem of sludge bulking, the addition of nitrogen was recommended, but even after doing so, operational problems continued and the decision was made to expand the treatment facility to avoid shock organic and hydraulic loading in the reactor. It was further observed that the addition of PAC in the activated sludge process resulted in some improvement in sludge settleability; however, the MLSS settling rate remained at a very low level (0.01–0.05 cm/min). The study demonstrated that due to nitrification, the pH decreased, causing a viscous floating layer of MLSS formed on the surface of the aeration basin and clarifier that resulted in significant reductions in the MLSS and PAC concentration in the system.

Chlorination of mixed liquor has been recommended to address the problem of sludge bulking. It was expected that chlorination of the mixed liquor at dosages ranging from 3 to 7.5 lb Cl_2/1000 lb MLSS could control the problem of sludge bulking; however, chlorination had in fact severely affected the treatment process and stopped nitrification. To resolve this problem, it was suggested that the plant should always operate at an F/M ratio above 0.15 to avoid filamentous growth, and that any increase in filaments should be treated before intense chlorination [46]. Another study recommended that sludge bulking be controlled by operating the system at a dissolved oxygen (DO) concentration of MLSS greater than 3 mg/L. An optimal dissolved oxygen control strategy for an activated sludge system in treatment of pharmaceutical wastewater is described by Brandel [68].

Temperature has been shown to affect the performance of the activated sludge process [46]. Pilot plant results indicated that system efficiency was excellent as long as the aeration basin temperature was less than 38°C, whereas at temperatures exceeding 38°C, BOD_5 removal efficiency decreased considerably, accompanied with the cessation of nitrification. High temperatures resulted in killing of the nitrifiers and inhibited carbonaceous removal. Hence, a heat exchanger in the influent line has been suggested to bring down the wastewater temperature.

5.8 ENVIRONMENTAL PROTECTION AGENCY EFFLUENT LIMITATIONS FOR THE PHARMACEUTICAL INDUSTRY

The EPA has developed effluent limitations in terms of percentage reductions of raw waste loads or effluent concentration as shown in Table 22 [3]. Additional parameters that should receive

Table 22 EPA Effluent Limitations for Pharmaceutical Plants [3]

Parameter	Limit
Average daily based on max. monthly raw waste load	
BOD_5	92–95%
COD	80–82%
TSS	82.5%
Ammonia N	70–75%
pH	6–9
Fecal coliforms	Average, 200/100 mL Max. daily, 400/100 mL
For daily limitations = 2 to 3× average daily levels given above, suggested limits for metals, trace ions	
Iron, Zinc	1.0–1.5 mg/L
Mn, Cu	0.5–1 mg/L
Phenolics, total Cr	0.25–0.5 mg/L
Aluminum	1.0–2.0 mg/L
Sulfide (approx.)	0.5 mg/L
Lead	0.1–0.25 mg/L
Mercury (total plant)	45.36 g/day

BOD, biochemical oxygen demand; COD, chemical oxygen demand; TSS, total suspended solids.

attention at many bulk manufacturing plants include copper, cyanides, tin, cadmium, nickel, arsenic, chlorinated hydrocarbons, and pesticides.

In India, domestic and industrial wastewaters are required to meet the standards set out in the Environment (Protection) Third Amendment Rules (1993) and Water (Prevention and Control of Pollution) Act (1974). The tolerance limits for the disposal of industrial effluents into inland surface water are given in Table 23 [69].

5.9 SUMMARY AND CONCLUSIONS

The information included in this chapter on pharmaceutical wastewater encompasses only a fragment of the research in this area. Owing to extreme variability of pharmaceutical wastewater characteristics, treatability studies should be conducted on a case-by-case basis to identify and confirm the required design parameters. As discussed earlier, physico-chemical treatment such as air stripping and coagulation was not found effective and beneficial for this wastewater, but in many cases, sedimentation has been found effective. The treatability study of almost all kinds of waste streams has indicated that waste is biologically treatable. Hence, a combination of physical, chemical, and biological processes seem to be feasible for the treatment of pharmaceutical wastewater. A two-stage biological system or a combination of aerobic and anaerobic processes proved effective for some pharmaceutical wastewater. Keeping in mind the varying characteristics of pharmaceutical wastewater, the shock loading capacity of the treatment units must also be given much attention in identifying and evaluating the technical feasibility of the processes. After identifying the technical feasibility of the processes, the final selection should be made based on economic analysis.

Table 23 Schedule VI of Environment (Protection) Third Amendment Rules (1993) [69]

Serial No.	Parameters	Standards[a] Inland surface water	Public sewers	Land for irrigation	Marine coastal areas
1	Color and odor	b	b	b	b
2	Suspended solids (mg/L), max	100	600	200	(a) For process waste water, 100 (b) For cooling water effluent, 10% above total suspended matter of influent
3	Particle size of suspended solids	Shall pass 850 micron IS sieve	–	–	(a) Floatable solids, solids max. 3 mm (b) Settleable solids, max. 856 microns
4	pH value	5.5–9.0	5.5–9.0	5.5–9.0	5.5–9.0
5	Temperature	Should not exceed 5°C above the receiving water temperature	–	–	Should not exceed 5°C above the receiving water temperature
6	Oil and grease (mg/L), max	10	20	10	20
7	Total residual chlorine (mg/L), max	1	–	–	1
8	Ammonical nitrogen (as N) (mg/L), max	50	50	–	50
9	Total Kjeldahl nitrogen (as N) (mg/L), max	100	–	–	100
10	Free ammonia (as NH_3) (mg/L), max	5	–	–	5
11	Nitrate nitrogen	10	–	–	20
12	BOD_5 (mg/L), max	30	350	100	100
13	COD (mg/L), max	250	–	–	250
14	Arsenic (as As) (mg/L)	0.2	0.2	0.2	0.2
15	Mercury (as Hg) (mg/L), max	0.01	0.01	–	0.01
16	Lead (as Pb) (mg/L), max	0.1	0.1	–	2
17	Cadmium (as Cd) (mg/L), max	2	1	–	2
18	Hexavalent chromium (as Cr^{6+}) (mg/L), max	0.1	2	–	1
19	Total chromium (as Cr) (mg/L), max	2	2	–	2
20	Copper (as Cu) (mg/L), max	3	3	–	2

(continues)

Table 23 Continued

Serial No.	Parameters	Standards[a]			
		Inland surface water	Public sewers	Land for irrigation	Marine coastal areas
21	Zinc (as Zn) (mg/L), max	5	15	–	15
22	Selenium (as Se) (mg/L), max	0.05	0.05	–	0.05
23	Nickel (as N_i) (mg/L), max	3	3	–	5
24	Cyanide (as CN) (mg/L), max	0.2	2	0.2	0.2
25	Fluoride (as F) (mg/L), max	2	15	–	15
26	Dissolved phosphates (as P) (mg/L), max	5	–	–	–
27	Sulfide (as S), (mg/L)	2	–	–	5
28	Phenolic compounds (as C_6H_5OH) (mg/L), max	1	5	–	5
29	Radioactive materials				
	(a) Alpha emitters (micro-Curie, mg/L), max	10^{-7}	10^{-7}	10^{-8}	10^{-7}
	(b) Beta emitters (micro-Curie, mg/L), max	10^{-6}	10^{-6}	10^{-7}	10^{-6}
30	Bio-assay test after 96 hours in 100% effluent	90% survival of fish	90% survival of fish	90% survival of fish	90% survival of fish
31	Manganese (as Mn) (mg/L), max	2	2	2	2
32	Iron (as Fe) (mg/L), max	3	3	3	3
33	Vanadium (as V) (mg/L), max	0.2	0.2	–	0.2

[a]These standards shall be applicable for industries, operations, or processes other than those industries, operations, or process for which standards have been specified in Schedule I.
[b]All efforts should be made to remove color and unpleasant odor as for as practicable.

Based on extensive study and experience in treatment of pharmaceutical wastewater, the following specific conclusions may be drawn:

- Pretreatment of pharmaceutical industry wastewater such as air stripping and coagulation is not beneficial; however, sedimentation of treated effluent was found effective in further reduction of SS and COD of the effluent. Hence, the pretreatment of pharmaceutical wastewater is not advisable.
- In many cases, anaerobic filter treatment was found to successfully treat pharmaceutical industry wastewater. This can be an excellent alternative for conventional aerobic treatment, which is energy intensive and requires the disposal of sludge. The anaerobic filter, on the other hand, can produce energy in the form of biogas and does not require

sludge disposal. Moreover, the anaerobic filter is more resistant and capable of handling shock loading as compared to the aerobic system.

- All waste streams, with the exception of acid waste streams of a synthetic drug factory, must be treated collectively rather than treated separately, as the performance efficiency of combined waste has been proved to be better than that of waste treated separately. Moreover, the segregation of acid waste streams could result in the following benefits:
  - recovery of useful acids from the waste;
  - the volume of the waste needing neutralization has been reduced to 50% and has eliminated the necessity of adjusting the pH of the combined waste for biological treatment;
  - the burden on the biological treatment has been reduced.
- The problem of sludge bulking in the case of the activated sludge process can be controlled in the following ways:
  - chlorination of the mixed liquor;
  - operating the system at min^{m} DO concentration of 3 mg/L;
  - operating the system at higher organic loading.
- Treatment processes such as ASP, PAC-ASP, GAC, and resin columns can successfully remove priority pollutants from pharmaceutical wastewater.
- In general, the trickling filter and activated sludge were found to satisfactorily cope with the needs of wastewater treatment for the pharmaceutical industry.
- Addition of PAC in the activated sludge process was found beneficial in improving the effluent quality, but it cannot be recommended until the problem of viscous layer formation is solved.

5.10 DESIGN EXAMPLES

Example 1

A synthetic organic chemicals plant discharges mainly two types of waste streams, namely strong process waste and dilute process waste. The flow and BOD_5 of the waste streams are given in the following table.

Type of wastes	Flow (GPD)	BOD_5 (mg/L)
Strong process waste	11,800	480,000
Dilute process waste	33,800	640

GPD, gallons per day; BOD, biochemical oxygen demand.

In addition, the plant discharges 35,300 GPD service wastewater. If the total BOD load of the composite waste is 47,500 lb/day, estimate (i) the BOD_5 of the composite waste and domestic waste; and (ii) the BOD load of the each stream and their contribution to the total BOD load of the plant.

Solution

Determine the BOD_5 of the wastes. The first step is to find out the total flow of the composite waste by summing the flow of the various waste streams of the plant.

$$\text{Total flow of the composite waste} = 11{,}800 + 33{,}800 + 35{,}300 = 80{,}900\,\text{GPD}$$

$$\text{BOD}_5 \text{ of the composite waste} = \frac{\text{Total BOD load (lb/day)} \times 453.6\ \text{(g/lb)} \times 1000\ \text{(mg/g)}}{\text{Flow (GPD)} \times 3.785\ \text{(L/gal)}}$$

$$= \frac{47{,}500 \times 453.6 \times 1000}{80{,}900 \times 3.785}$$

$$= 70{,}364.28\,\text{mg/L}$$

$$\text{BOD load of the strong process waste} = \frac{\text{Flow (GPD)} \times 3.785\ \text{(L/gal)} \times \text{BOD}_5\ \text{(mg/L)}}{10^3\ \text{(mg/g)} \times 453.6\ \text{(g/lb)}}$$

$$= \frac{11{,}800 \times 3.785 \times 480{,}000}{1000 \times 453.6}$$

$$= 47{,}262.43\,\text{lb/day}$$

$$\text{BOD load of the dilute process waste} = \frac{\text{Flow (GPD)} \times 3.785\ \text{(L/gal)} \times \text{BOD}_5\ \text{(mg/L)}}{10^3\ \text{(mg/g)} \times 453.6\ \text{(g/lb)}}$$

$$= \frac{33{,}800 \times 3.785 \times 640}{1000 \times 453.6}$$

$$= 180.50\,\text{lb/day}$$

$$\text{BOD load due to domestic waste} = 47{,}500 - (47{,}262.43 + 180.5)$$

$$= 57.05\,\text{lb/day}$$

$$\text{BOD}_5 \text{ of the domestic waste, mg/L} = \frac{\text{BOD load (lb/day)} \times 453.6\ \text{(g/lb)} \times 1000\ \text{(mg/g)}}{\text{Flow (GPD)} \times 3.785\ \text{(L/gal)}}$$

$$= \frac{57.07 \times 453.6 \times 1000}{35{,}300 \times 3.785}$$

$$= 193.75\,\text{mg/L}$$

Comment

The total BOD load of the plant is mainly due to strong process waste. Segregation of strong process waste can result in significant reduction in total BOD load of the plant.

Example 2

The five-days' BOD at 20°C and flow of the various types of waste streams generated from a synthetic drug plant are given in the following table.

Type of wastes	Flow (m^3/day)	BOD_5 (mg/L)
Alkaline waste stream	1710	3500
Condensate waste stream	1990	1275
Acid waste stream	435	3090

BOD, biochemical oxygen demand.

Estimate the BOD_5 and subsequent BOD load of the composite waste. If the acid waste stream has to be segregated for the recovery of acids then (i) find out the BOD_5 of the combined waste excluding acid waste; and (ii) comment on the effect of segregation in BOD loading of the plant.

Solution

$$\text{BOD load of the alkaline waste} = \text{Flow (m}^3\text{/day)} \times BOD_5\ (\text{g/m}^3) \times 10^{-3}\ (\text{kg/g})$$
$$= 1710 \times 3500 \times 10^{-3} = 5985\ \text{kg BOD/day}$$

$$\text{Similarly, BOD load of the condensate waste} = 1990 \times 1275 \times 10^{-3}$$
$$= 3834.61\ \text{kg BOD/day}$$

$$\text{BOD load of the acid waste} = 435 \times 3090 \times 10^{-3} = 1344.15\ \text{kg BOD/day}$$

$$\text{Total BOD load of the composite waste} = 5985 + 3834.61 + 1344.15$$
$$= 11{,}163.76\ \text{kg BOD/day}$$

$$\text{Total flow of the composite waste} = 1710 + 1990 + 435 = 4135\ \text{m}^3$$

$$BOD_5 \text{ of the composite waste} = \frac{\text{BOD load (kg/day)} \times 10^6\ (\text{mg/kg})}{\text{Flow (m}^3\text{/day)} \times 10^3\ (\text{L/m}^3)}$$
$$= \frac{11{,}163.76 \times 10^6}{4135 \times 10^3}$$
$$= 2699.82\ \text{mg/L}$$

$$\text{BOD load of alkaline and condensate waste} = 5985 + 3834.61$$
$$= 9819.61\ \text{kg BOD/day}$$

$$\text{Total flow of the alkaline and condensate waste} = 1710 + 1990$$
$$= 3700\ \text{m}^3$$

$$BOD_5 \text{ of combined (alkaline and condensate)waste} = \frac{9819.61 \times 10^6}{3700 \times 10^3}$$
$$= 2653.95\ \text{mg/L}$$

Comment

Segregation of the acid waste stream has resulted in significant reduction in total BOD load of the plant, but the BOD_5 of the composite waste remains almost the same. Hence the acid waste stream can be segregated from the main stream without affecting the treatability of the waste.

Example 3

A primary sedimentation tank has been designed for the pretreatment of 0.312 MGD of fermentation waste generated from the pharmaceutical industry. The raw waste SS concentration is 1660 mg/L. At a detention time of 2 hours the effluent SS concentration is reduced to 260 mg/L. Determine (i) the SS removal efficiency of the sedimentation tank; and (ii) the quantity of sludge generated per day. Assume the specific gravity of sludge (S_{sl}) is 1.03, which contains 6% solids.

Solution

(A) SS removal efficiency of the tank can be obtained as follows:

$$\text{SS removal efficiency} = \frac{(1660 - 260) \times 100}{1660}$$

$$= 84.34\%$$

(B) Determine the mass of dry solids removed per day.

$$W_s = 0.312\ (\text{MGD}) \times 10^6\ (\text{gal/M}) \times 3.785\ (\text{L/gal}) \times (1660 - 260)\ (\text{mg/L}) \times 10^{-6}\ (\text{kg/mg}) = 1653.29\,\text{kg/day}$$

(C) Determine the volume of sludge produced per day.

$$V_{sl} = \frac{W_s}{\rho_w \times S_{sl} \times P_s}$$

where W_s is the mass of dry solids removed per day, ρ_w is the density of the water, and P_s is the percentage of sludge solids.

$$V_{sl} = \frac{1653.29\ (\text{kg/day})}{1000 \times 1.03 \times 0.06}$$

$$= 26.752\,\text{m}^3\ \ (7067.96\,\text{GPD})$$

Example 4

Physicochemical treatment of a typical pharmaceutical plant generating 33,800 GPD wastewater has indicated that at optimum doses of $FeSO_4$ (500 mg/L), $FeCl_3$ (500 mg/L), and alum (250 mg/L), COD and SS removal of the effluent of 25 and 70%, respectively, can be achieved. Determine the quantities of various chemicals required per day. If 49% strength alum is to be used and 30 days supply is to be stored at the treatment facility, estimate the storage capacity required for the alum.

Solution

(A) The quantities of the various chemicals required per day can be obtained as follows:

$$\text{Quantity of FeSO}_4\text{ required per day} = 500\ (\text{mg/L}) \times 10^{-6}\ (\text{mg/kg}) \times 33{,}800\ (\text{GPD}) \times 3.785\ (\text{L/gal}) = 63.97\,\text{kg/day}$$

$$\text{Quantity of FeCl}_3\text{ required per day} = 500\ (\text{mg/L}) \times 10^{-6}\ (\text{mg/kg}) \times 33{,}800\ (\text{GPD}) \times 3.785\ (\text{L/gal}) = 63.97\,\text{kg/day}$$

$$\text{Quantity of alum required per day} = 250\ (\text{mg/L}) \times 10^{-6}\ (\text{mg/kg}) \times 33{,}800\ (\text{GPD}) \times 3.785\ (\text{L/gal}) = 31.98\,\text{kg/day}$$

(B) Determine the weight of alum per m^3 of 49% liquid alum.

$$\text{Weight per m}^3 = 0.49 \times 80\ (\text{lb/ft}^3) \times 16.0185\ (\text{kg/m}^3 \cdot \text{lb/ft}^3) = 627.925\,\text{kg/m}^3$$

(C) Determine the storage capacity required for 30 days.

$$\text{Storage capacity} = 31.98\ (\text{kg/day}) \times 30\ (\text{days})/627.925\ (\text{kg/m}^3)$$
$$= 1.527\ \text{m}^3\ (1527\ \text{L})$$

Example 5

Estimate the quantity of sludge produced in a chemical precipitation of 1710 m^3/day of pharmaceutical wastewater with SS concentration 560 mg/L. The addition of the $FeSO_4$ (500 mg/L), $FeCl_3$ (500 mg/L), and lime (600 mg/L) increases the SS removal efficiency of the primary sedimentation tank from 60 to 70%. Comment on the chemical precipitation process on the basis of sludge production. Assume $CaCO_3$ solubility = 15 mg/L, specific gravity of sludge = 1.03, and moisture content of the sludge = 95%.

Solution

(A) Determine the mass of SS removed per day without chemical addition.

$$M_{ss1} = 1710\ (\text{m}^3/\text{day}) \times 10^3\ (\text{L/m}^3) \times 0.6 \times 560\ (\text{mg/L}) \times 10^{-6}\ (\text{kg/mg})$$
$$= 574.56\ \text{kg/day}$$

(B) Determine the mass of SS removed per day after chemical addition.

$$M_{ss1} = 1710\ (\text{m}^3/\text{day}) \times 10^3\ (\text{L/m}^3) \times 0.7 \times 560\ (\text{mg/L}) \times 10^{-6}\ (\text{kg/mg})$$
$$= 670.320\ \text{kg/day}$$

(C) Determine the volume of the sludge without chemical addition.

$$V_{sl} = \frac{W_s}{\rho_w \times S_{sl} \times P_s}$$
$$= \frac{574.56\ (\text{kg/day})}{1000 \times 1.03 \times (1 - 0.95)}$$
$$= 11.16\ \text{m}^3/\text{day}\ (2948.48\ \text{GPD})$$

(D) Determine the quantity of sludge produced due to addition of chemicals. This can be calculated from the stochiometry of the chemical reactions taking place with the addition of these chemicals. The chemical reactions taking place are described below. When $FeSO_4$ and lime are added:

$$\underset{(278)}{FeSO_4 \cdot 7H_2O} + \underset{(100)}{Ca(HCO_3)_2} \Longleftrightarrow \underset{(178)}{Fe(HCO_3)_2} + \underset{(136)}{CaSO_4} + \underset{(7 \times 18)}{7H_2O}$$

$$\underset{(178)}{Fe(HCO_3)_2} + \underset{(2 \times 56)}{2Ca(OH)_2} \Longleftrightarrow \underset{(2 \times 100)}{2CaCO_3} + \underset{(89.9)}{Fe(OH)_2} + \underset{(2 \times 18)}{2H_2O}$$

$$\underset{(4 \times 89.9)}{4Fe(OH)_2} + \underset{(32)}{O_2} + \underset{(2 \times 18)}{2H_2O} \Longleftrightarrow \underset{(4 \times 106.9)}{4Fe(OH)_3}$$

$$\underset{(56)}{Ca(OH)_2} + \underset{(44)}{H_2CO_3} \Longleftrightarrow \underset{(100)}{CaCO_3} + \underset{(2 \times 18)}{2H_2O}$$

$$\underset{(56)}{Ca(OH)_2} + \underset{(100)}{Ca(HCO_3)_2} \Longleftrightarrow \underset{(2 \times 100)}{2CaCO_3} + \underset{(2 \times 18)}{2H_2O}$$

When $FeCl_3$ is also added:

$$\underset{(2\times 162)}{2FeCl_3} + \underset{(3\times 56)}{3Ca(OH)_2} \Longleftrightarrow \underset{(2\times 106.9)}{2Fe(OH)_3} + \underset{(3\times 111)}{3CaCl_2}$$

The addition of $FeSO_4$ mainly produces precipitable flocs of $CaCO_3$ and $Fe(OH)_3$. The quantity of $CaCO_3$ precipitated by addition of 500 mg/L of $FeSO_4$ can be estimated as:

$$\begin{aligned}\text{Quantity of } CaCO_3 &= 500\ (\text{mg/L}) \times (200/278)\\ &= 359.71\ \text{mg/L}\end{aligned}$$

Similarly, the quantity of $Fe(OH)_3$ precipitated by addition of 500 mg/L of $FeSO_4$

$$\begin{aligned}&= 500\ (\text{mg/L}) \times (106.9/278)\\ &= 192.27\ \text{mg/L}\end{aligned}$$

Amount of lime consumed in formation of $Fe(OH)_3$ flocs

$$\begin{aligned}&= 192.27\ (\text{mg/L}) \times (56/106.9)\\ &= 100.72\ \text{mg/L}\end{aligned}$$

Similarly, the quantity of $Fe(OH)_3$ precipitated by addition of 500 mg/L of $FeCl_3$

$$\begin{aligned}&= 500\ (\text{mg/L}) \times (106.9/162)\\ &= 329.94\ \text{mg/L}\end{aligned}$$

Amount of lime consumed in formation of $Fe(OH)_3$ flocs by addition of $FeCl_3$

$$\begin{aligned}&= 329.94\ (\text{mg/L}) \times (3 \times 56/2 \times 106.9)\\ &= 259.26\ \text{mg/L}\end{aligned}$$

Total amount of lime remaining

$$\begin{aligned}&= 600 - (100.72 + 259.26)\\ &= 240.02\ \text{mg/L}\end{aligned}$$

Amount of $CaCO_3$ precipitated by addition of lime

$$\begin{aligned}&= 240.02 \times (3 \times 100/2 \times 56)\\ &= 642.91\ \text{mg/L}\end{aligned}$$

Determine the total amount of $CaCO_3$ precipitated per day

$$\begin{aligned}&= 1710\ (\text{m}^3/\text{day}) \times 10^3\ (\text{L/day})\\ &\qquad \times (359.71 + 642.91 - 15)\ (\text{mg/L}) \times 10^{-6}\ (\text{kg/mg})\\ &= 1688.83\ \text{kg/day}\end{aligned}$$

Similarly, the total amount of $Fe(OH)_3$ precipitated per day

$$= 1710\ (m^3/day) \times 10^3\ (L/day) \times (192.27 + 329.94)\ (mg/L) \times 10^{-6}\ (kg/mg)$$
$$= 892.98\ kg/day$$

Total volume of sludge on dry basis per day

$$= 670.32 + 1688.83 + 892.98$$
$$= 3252.13\ mg/L$$

Hence the volume of the sludge produced per day with chemical addition:

$$V_{sl} = \frac{W_s}{\rho_w \times S_{sl} \times P_s}$$
$$= \frac{3252.13\ (kg/day)}{1000 \times 1.03 \times (1 - 0.95)}$$
$$= 63.15\ m^3/day\ (16{,}684.28\ GPD)$$

Comment
The problem of sludge disposal increased to a greater extent in the case of chemical precipitation than in the sedimentation without the chemical.

Example 6

Estimate the food–microorganism ratio (F/M) and sludge age (solid retention time) of an activated sludge process designed to reduce the BOD_5 of the spent stream generated from a biological production plant from 1500 mg/L to 50 mg/L. The wastewater flow is $Q = 15{,}000$ GPD, aeration tank volume = 45 m^3, MLVSS = 3000 mg/L, and net biomass yield coefficient (Y_n) = 0.28 kg/kg. Also compute the performance efficiency of the plant.

Solution

$$\text{Total substrate removed (kg BOD/day)} = Q(\text{GPD}) \times 3.785\ (L/gal) \times (S_i - S_e)\ (mg/L) \times 10^{-6}\ (kg/mg)$$
$$= 15{,}000 \times 3.785 \times (1500 - 50) \times 10^{-6}$$
$$= 82.32\ \text{kg BOD/day}$$

$$\text{Total MLVSS (kg MLVSS)} = \text{MLVSS (mg/L)} \times 10^{-6}\ (kg/mg) \times V\ (m^3) \times 10^3\ (L/m^3)$$

$$\text{Total MLVSS} = 3000 \times 10^{-6} \times 45 \times 10^3 = 135\ \text{kg MLVSS}$$

$$\text{Total substrate applied per day} = 15{,}000 \times 3.785 \times 1500 \times 10^{-6}$$
$$= 85.16\ \text{kg BOD/day}$$

$$\text{F/M ratio (day}^{-1}) = \frac{\text{Total substrate applied (kg BOD/day)}}{\text{Total MLVSS (kg MLVSS)}}$$
$$= (85.16/135) = 0.63\ \text{day}^{-1}$$

$$\text{Net MLVSS produced (kg VSS/day)} = Y_n\ (kg/kg) \times \text{total substrate removed (kg/day)}$$
$$= 0.28 \times 82.32$$
$$= 23.05\ \text{kg VSS/day}$$

$$\text{Sludge age (solid retention time)} \ (\theta_c) = \frac{\text{Total MLVSS}}{\text{Net VSS produced per day}}$$
$$= \frac{135}{23.05} = 5.86 \text{ day}$$

$$\text{BOD removal efficiency} = \frac{(S_i - S_e) \times 100}{S_i}$$
$$= \frac{(1500 - 50) \times 100}{1500}$$
$$= 96.67\%$$

Example 7

Design a complete-mix activated sludge process for the treatment of 1710 m^3/day of settled condensate wastewater with BOD_5, 1500 mg/L generated from a synthetic organic chemical type of pharmaceutical industry. Assume the following conditions are applicable:

1. Effluent contains 25 mg/L biological solids, of which 65% is biodegradable;
2. MLSS concentration in the reactor = 5000 mg/L;
3. MLVSS $(X) = 0.8 \times$ MLSS;
4. Solid retention time, $\theta_c = 5$ days;
5. $BOD_5 = 0.68 \ BOD_L$ (ultimate biological oxygen demand);
6. Return sludge concentration = 1%;
7. Effluent $BOD_5 = 50$ mg/L;
8. Maximum yield coefficient, $Y = 0.6$ mg/mg;
9. Decay constant, $K_d = 0.07 \ \text{day}^{-1}$.

Solution

(A) Determine the influent soluble BOD_5 escaping the treatment:

(i) BOD_L of the biodegradable effluent solid

$$= 25 \ (\text{mg/L}) \times 0.65 \times 1.42 \ (\text{mg } O_2 \text{ consumed/mg cell oxidized})$$
$$= 23.075 \text{ mg/L}$$

(ii) BOD_5 of the effluent SS $= 23.075 \ (\text{mg/L}) \times 0.68$

$$= 15.69 \text{ mg/L (say 15.7 mg/L)}$$

(iii) Influent soluble BOD_5 escaping the treatment

$$= 50 - 15.7$$
$$= 34.3 \text{ mg/L}$$

(B) Efficiency of the process:

(i) Process efficiency based on soluble BOD_5

$$E_s = \frac{(1500 - 34.3) \times 100}{1500}$$
$$= 97.71\%$$

(ii) Similarly, overall plant efficiency of the system

$$E_s = \frac{(1500 - 50) \times 100}{1500}$$
$$= 96.67\%$$

(C) Determine the capacity of the aeration basin:

$$V \text{ (volume)} = \frac{Y\theta_c Q(S_i - S)}{X(1 + K_d\theta_c)}$$

where Y = maximum yield coefficient (mg/mg), θ_c = mean cell residence time (day), Q = flow (m^3/day), S_i = substrate concentration in the influent (mg/L), S = substrate concentration in effluent (mg/L), X = mass concentration of microorganism in reactor (mg/L), and K_d = endogenous decay coefficient (day^{-1}). On substituting the values, the above equation results in:

$$V = \frac{0.6 \text{ (mg/mg)} \times 5 \text{ (day)} \times 1710 \text{ (m}^3\text{/day)(1500} - 50\text{) (mg/L)}}{0.8 \times 5000 \text{ (mg/L)} \times [1 + 0.07 \text{ (day}^{-1}\text{)} \times 5 \text{ (day)]}}$$
$$= 1377.5 \text{ m}^3$$

Check for the F/M ratio and OLR:

$$\text{HRT } (\theta) = V/Q$$
$$\theta = 1377.5 \text{ (m}^3\text{)/1710 (m}^3\text{/day)}$$
$$= 0.805 \text{ day (19.33 hours)}$$
$$\text{F/M ratio} = (S_i/\theta X)$$
$$= \frac{1500 \text{ mg/L}}{0.805 \text{ (day)} \times 0.8 \times 5000 \text{ (mg/L)}}$$
$$= 0.466 \text{ day}^{-1}$$

$$\text{Amount of BOD}_5 \text{ consumed} = (1500 - 34.3) \text{ (mg/L)} \times 10^{-6} \text{ (kg/mg)} \times 1710 \text{ (m}^3\text{/day)} \times 10^3 \text{ (L/m}^3\text{)}$$
$$= 2506.35 \text{ kg BOD/day}$$

$$\text{OLR} = \frac{(1500 - 34.3) \text{ (mg/L)} \times 10^{-6} \text{ (kg/mg)} \times 1710 \text{ (m}^3\text{/day)} \times 10^3 \text{ (L/m}^3\text{)}}{1377.5 \text{ (m}^3\text{)}}$$
$$= (2506.35/1377.5)$$
$$= 1.82 \text{ kg BOD/m}^3\text{day}$$

(D) Sludge recycling

The recycling ratio (r) can be computed as follows:

$$r = \frac{X}{(X_r - X)}$$

where X_r = MLVSS in the recycled effluent

$$= \frac{0.8 \times 5000 \text{ (mg/L)}}{0.8 \times (10{,}000 - 5000) \text{ (mg/L)}}$$

$$= 0.5$$

Hence the recycling flow = 0.5 Q = 0.5 × 1710 (m^3/day) = 855 m^3/day.

(E) Sludge production

(i) Net VSS production = XV/θ_c

$$= \frac{0.8 \times 5000 \text{ (mg/L)} \times 10^{-6} \text{ (kg/mg)} \times 1377.5 \text{ (m}^3\text{)} \times 10^3 \text{ (L/m}^3\text{)}}{5 \text{ (days)}}$$

$$= 1102 \text{ kg/day}$$

(ii) Net SS production = 1102 (kg/day)/0.8

$$= 1377.5 \text{ kg/day}$$

(iii) Volume of the sludge produced

$$= \frac{1377.5 \text{ (kg/day)}}{1000 \text{ (kg/m}^3\text{)} \times 1.03 \times 0.01}$$

$$= 133.73 \text{ m}^3\text{/day}$$

(iv) VSS production per kg BOD$_r$ (biological oxygen demand removed)

$$= \frac{1102 \text{ (kg/day)} \times 10^6 \text{ (mg/kg)}}{(1500 - 34.3) \text{ (mg/L)} \times 1710 \text{ (m}^3\text{/day)} \times 10^3 \text{ (L/m}^3\text{)}}$$

$$= 0.44 \text{ mg/mg}$$

(F) Oxygen requirement

(i) Theoretical O_2 required = (BOD$_L$ removed) − (BOD$_L$ of solids leaving)

$$= 1.47 \, (1500 - 34.3) \text{ (mg/L)} \times 1710 \text{ (m}^3\text{)} \times 10^3 \text{ (L/m}^3\text{)}$$

$$\times 10^{-6} \text{ (kg/mg)} - 1.42 \times 1102 \text{ (kg/day)}$$

$$= 2119.49 \text{ kg/day}$$

(ii) Theoretical air requirement assuming that air contains 23.2% oxygen by weight and density of air = 1.201 kg/m^3

$$= \frac{2119.49 \text{ (kg/day)}}{0.232 \times 1.201 \text{ (kg/m}^3\text{)}}$$

$$= 7606.47 \text{ m}^3\text{/day}$$

(iii) Actual air requirement at an 8% transfer efficiency

$$= 7606.47 \text{ (m}^3\text{/day)}/0.08$$

$$= 95{,}084.65 \text{ m}^3\text{/day}$$

(iv) Check for the air requirement per unit volume

$$= 95{,}084.65\ (m^3/day)/1710\ (m^3/day)$$

$$= 55.60\ m^3/m^3$$

(v) Air requirement per kg of BOD_5 removed

$$= 95{,}084.65\ (m^3/day)/2506.35\ (kg/day)$$

$$= 37.94\ m^3/kg\ BOD_5\ \text{removed}$$

(G) Power requirement assuming the aerators are designed to give 2 kgO_2/kWh and the field efficiency is 70%.

$$\text{Power required} = \frac{2119.49\ (kg/day)}{2\ (kg/kWh) \times 0.7 \times 24\ (h/day)}$$

$$= 63.08\ (kW) \times 1.3410\ (hp/kW)$$

$$= 84.59\ hp\ (\text{say } 85\ hp)$$

Example 8

1710 m^3/day of alkaline waste stream with $BOD_5 = 3500$ mg/L is treated in an extended aeration system. The BOD removal efficiency of the system is 96%. If the volume of the aeration basin is 1780 m^3, estimate (i) detention time (hydraulic retention time, HRT) and organic loading rate (OLR). Also compute the BOD_5 of the treated effluent.

Solution

$$HRT(\theta)\ (day) = \frac{\text{Volume of the tank}, V(m^3)}{\text{Flow}, Q\ (m^3/day)}$$

$$= 1780/1710 = 1.04\ day = 24.98\ hours$$

$$OLR\ (kg\ BOD/m^3 day) = \frac{Q\ (m^3/day) \times 10^3\ (L/m^3) \times E \times S_i\ (mg/L) \times 10^{-6}\ (kg/mg)}{V\ (m^3)}$$

$$= (1710 \times 10^3 \times 0.96 \times 3500 \times 10^{-6})/1780$$

$$= 3.23\ kg\ BOD/m^3\ day$$

$$\%\ \text{BOD removal efficiency} = \frac{(S_i - S_e) \times 100}{S_i}$$

$$96 = \frac{(3500 - S_e) \times 100}{3500}$$

$$S_e = 140\ mg/L$$

Example 9

An extended aeration activated sludge process is designed to treat 2000 m^3/day of condensate waste generated from a synthetic organic chemical plant. The system is operating at an organic loading rate of 1.2 kg COD/m^3 day. If the BOD_5 of influent raw waste and treated effluent is 1275 mg/L and 76.5 mg/L, respectively, determine the HRT and performance efficiency of the system.

Solution

$$\text{OLR (kg BOD/m}^3\text{day)} = \frac{Q\ (\text{m}^3/\text{day}) \times 10^3\ (\text{L/m}^3) \times (S_\text{i} - S_\text{e})\ (\text{mg/L}) \times 10^{-6}\ (\text{kg/mg})}{V\ (\text{m}^3)}$$

$$1.2 = [2000 \times 10^3 \times (1275 - 76.5) \times 10^{-6}]/V$$

$$V = 1997.5\,\text{m}^3$$

$$\text{HRT}, \theta\ (\text{day}) = \frac{V\ (\text{m}^3)}{Q\ (\text{m}^3/\text{day})}$$

$$= 1997.5/2000 = 0.99\,\text{day, say 1 day (24 hours)}$$

$$\%\ \text{BOD removal efficiency} = \frac{(S_\text{i} - S_\text{e}) \times 100}{S_\text{i}}$$

$$= \frac{(1275 - 76.5) \times 100}{1275}$$

$$= 94\%$$

Example 10

Design an extended aeration process for the treatment of 1275 m^3/day of pharmaceutical wastewater with a BOD_5 of 3500 mg/L. Assume the following conditions are applicable:

- Effluent contains 20 mg/L biological solids of which 70% is biodegradable;
- MLSS concentration in the reactor = 6000 mg/L;
- MLVSS = 0.75 × MLVSS;
- Solid retention time, $\theta_c = 12$ days;
- $BOD_5 = 0.68\ BOD_L$;
- Return sludge concentration = 2%;
- Effluent BOD_5 = 30 mg/L;
- $Y = 0.65$ mg/mg;
- Decay constant, $K_d = 0.075\ \text{day}^{-1}$.

Solution

(A) Determine the influent soluble BOD_5 escaping the treatment:

(i) BOD_L of the biodegradable effluent solid

$$= 20\ (\text{mg/L}) \times 0.70 \times 1.42\ (\text{mg O}_2\ \text{consumed/mg cell oxidized})$$

$$= 19.88\,\text{mg/L}$$

(ii) BOD_5 of the effluent SS

$$= 19.88\ (\text{mg/L}) \times 0.68$$
$$= 13.52\ \text{mg/L (say 13.5 mg/L)}$$

(iii) Influent soluble BOD_5 escaping the treatment

$$= 30 - 13.5$$
$$= 16.5\ \text{mg/L}$$

(B) Efficiency of the process:

(i) Process efficiency based on soluble BOD_5

$$E_s = \frac{(3500 - 16.5) \times 100}{3500}$$
$$= 99.5\%$$

(ii) Similarly, overall plant efficiency of the system

$$E_s = \frac{(3500 - 30) \times 100}{3500}$$
$$= 99.1\%$$

(C) Determine the capacity of the aeration basin:

$$V = \frac{Y\theta_c Q(S_i - S)}{X(1 + K_d\theta_c)}$$
$$= \frac{0.65\ (\text{mg/mg}) \times 12\ (\text{day}) \times 1275\ (\text{m}^3/\text{day}) \times (3500 - 30)\ (\text{mg/L})}{0.75 \times 6000\ (\text{mg/L}) \times [1 + 0.075\ (\text{day}^{-1}) \times 12\ (\text{day})]}$$
$$= 4036.15\ \text{m}^3\ (\text{say } 4050\ \text{m}^3)$$

Check for the F/M ratio and OLR and HRT:

$$\text{HRT}(\theta) = V/Q$$
$$\theta = 4050\ (\text{m}^3)/1275\ (\text{m}^3/\text{day})$$
$$= 3.18\ \text{day}$$

$$\text{F/M ratio} = (S_i/\theta X)$$
$$= \frac{3500\ (\text{mg/L})}{3.18\ (\text{day}) \times 0.75 \times 6000\ (\text{mg/L})}$$
$$= 0.24\ \text{day}^{-1}$$

$$\text{Amount of BOD}_5 \text{ removed} = (3500 - 16.5)\ (\text{mg/L}) \times 10^{-6}\ (\text{kg/mg}) \times 1275\ (\text{m}^3/\text{day}) \times 10^3\ (\text{L/m}^3)$$
$$= 4441.46\ \text{kg/day}$$

$$\text{OLR} = \frac{(3500 - 16.5)\ (\text{mg/L}) \times 10^{-6}\ (\text{kg/mg}) \times 1275\ (\text{m}^3/\text{day}) \times 10^3\ (\text{L/m}^3)}{4050\ (\text{m}^3)}$$
$$= (4441.46/4050)$$
$$= 1.10\ \text{kg BOD/m}^3\text{day}$$

(D) Sludge recycling

The recycling ratio (r) can be computed as follows:

$$r = X/(X_r - X)$$
$$= \frac{0.75 \times 6000 \text{ (mg/L)}}{0.75 \times (20{,}000 - 6000) \text{ (mg/L)}}$$
$$= 0.43$$

Hence the recycling flow $= 0.43\ Q = 0.43 \times 1275\ (\text{m}^3/\text{day}) = 548.25\ \text{m}^3/\text{day}$

(E) Sludge production

(i) Net VSS production $= XV/\theta_c$

$$= \frac{0.75 \times 6000 \text{ (mg/L)} \times 10^{-6} \text{ (kg/mg)} \times 4050 \text{ (m}^3) \times 10^3 \text{ (L/m}^3)}{12 \text{ (day)}}$$
$$= 1518.75 \text{ kg/day}$$

(ii) Net SS production $= 1518.75\ (\text{kg/day})/0.75$

$$= 2025 \text{ kg/day}$$

(iii) Volume of the sludge produced $= \frac{2025 \text{ (kg/day)}}{1000 \text{ (kg/m}^3) \times 1.03 \times 0.02}$

$$= 98.3 \text{ m}^3/\text{day}$$

(iv) VSS production per kg BOD_r

$$= \frac{1518.75 \text{ (kg/day)} \times 10^6 \text{ (mg/kg)}}{(3500 - 16.5) \text{ (mg/L)} \times 1275 \text{ (m}^3/\text{day)} \times 10^3 \text{ (L/m}^3)}$$
$$= 0.34 \text{ mg/mg}$$

(F) Oxygen requirement

(i) Theoretical O_2 required $= (\text{BOD}_L \text{ removed}) - (\text{BOD}_L \text{ of solids leaving})$

$$= 1.47\ (3500 - 16.5) \text{ (mg/L)} \times 1275 \text{ (m}^3) \times 10^3 \text{ (L/m}^3) \times 10^{-6} \text{ (kg/mg)} - 1.42 \times 1518.75 \text{ (kg/day)}$$
$$= 4372.32 \text{ kg/day}$$

(ii) Theoretical air requirement assuming that air contains 23.2% oxygen by weight and density of air $= 1.201\ \text{kg/m}^3$

$$= \frac{4372.32 \text{ (kg/day)}}{0.232 \times 1.201 \text{ (kg/m}^3)}$$
$$= 15{,}692.11 \text{ m}^3/\text{day}$$

(iii) Actual air requirement at an 8% transfer efficiency

$$= 15{,}692.11\ (\text{m}^3/\text{day})/0.08$$
$$= 196{,}151.41\ \text{m}^3/\text{day}$$

(iv) Check for the air requirement per unit volume

$$= 196{,}151.41\ (\text{m}^3/\text{day})/1275\ (\text{m}^3/\text{day})$$
$$= 153.84\ \text{m}^3/\text{m}^3$$

(v) Air requirement per kg of BOD_5 removed

$$= 196{,}151.41\ (\text{m}^3/\text{day})/4441.46\ (\text{kg/day})$$
$$= 44.16\ \text{m}^3/\text{kg}\ BOD_5\ \text{removed}$$

(G) Power requirement assuming the aerators are designed to give 2 kgO_2/kWh and the field efficiency is 70%.

$$\text{Power required} = \frac{4372.32\ (\text{kg/day})}{2\ (\text{kg/kWh}) \times 0.7 \times 24\ (\text{h/day})}$$
$$= 130.13\ (\text{kW}) \times 1.3410\ (\text{hp/kW})$$
$$= 174.5\ \text{hp (say 175 hp)}$$

Example 11

A powdered activated carbon fed activated sludge process is designed to treat 15,000 GPD of pharmaceutical wastewater. The SCOD (soluble chemical oxygen demand) of the treated effluent is 590 mg/L. Determine the dose of PAC (powdered activated carbon) required for further reduction of effluent SCOD from 590 mg/L to 200 mg/L. Use the Freundlich equation [48]: $X/M = (3.7 \times 10^{-6})\ C_e^{2.1}$ to determine the dose of powdered activated carbon.

Solution

(A) SCOD concentration at equilibrium $C_e = 200$ mg/L.
(B) Amount of SCOD removal attributed to the PAC, X (mg/L) = 590 − 200 = 390 mg/L.
(C) The dose of activated carbon (M) can be determined by the Freundlich equation:

$$\frac{X}{M} = (3.7 \times 10^{-6})C_e^{2.1}$$

$$M = X/3.7 \times 10^{-6} \times C_e^{2.1}$$

$$M = 390/3.7 \times 10^{-6} \times 200^{2.1}$$
$$M = 1551\ \text{mg/L (1.55 g/L)}$$

(D) The dose of PAC per unit SCOD removed:

$$\frac{X}{M} = \frac{1551\ (\text{mg/L})}{390\ (\text{mg/L})}$$
$$= 3.98\ \text{mg PAC/mg SCOD}_r$$

Example 12

The result of a pilot plant study of PAC-fed activated sludge process is given in the following table.

PAC dose (mg/L)	Effluent SCOD in control reactor (mg/L)	Effluent SCOD in PAC-fed reactor (mg/L)
208	825	459
827	825	265
496	670	314
1520	583	194

PAC, powdered activated carbon; SCOD, soluble chemical oxygen demand.

Using the Freundlich equation $(X/M) = kC_e^{1/n}$, find the values of constants k and n.

Solution

(A) The first step is to estimate the values of X/M against the equilibrium SCOD concentration. The SCOD removal attributed to PAC can be calculated by subtracting the effluent SCOD in the PAC-fed reactor from the effluent SCOD of the control reactor. The estimated values of X and X/M are given in the following table.

PAC dose (mg/L) (M)	Effluent SCOD (mg/L) C_e	SCOD removal by PAC (mg/L) X	Ratio (X/M)	log C_e	log(X/M)
208	459	366	1.76	2.66	0.25
827	265	560	0.68	2.42	−0.17
496	314	356	0.72	2.50	−0.14
1520	194	389	0.20	2.29	−0.70

(B) The second step is to plot the log(X/M) values against the various values of the log C_e as shown in Figure 7. By taking the log of both sides of the Freundlich equation we get a straight line whose intercept gives the value K and slope gives the value of $1/n$. The log of the Freundlich equation results in the following equation:

$$\log\left(\frac{X}{M}\right) = \log K + \left(\frac{1}{n}\right)\log C_e$$

The values of log(X/M) and log C_e have been calculated and given in the table above and plotted as shown in Figure 7. From the graph, the slope of the line gives a value of $1/n = 2.4218$, hence $n = 0.41$ and the intercept gives the value of $\log K = -6.1665$, hence $K = 6.81 \times 10^{-7}$.

Example 13

An aerated lagoon is to be designed to treat 15,000 GPD of spent stream generated from a biological production plant. The depth of the lagoon is restricted to 3.3 m and the HRT of the

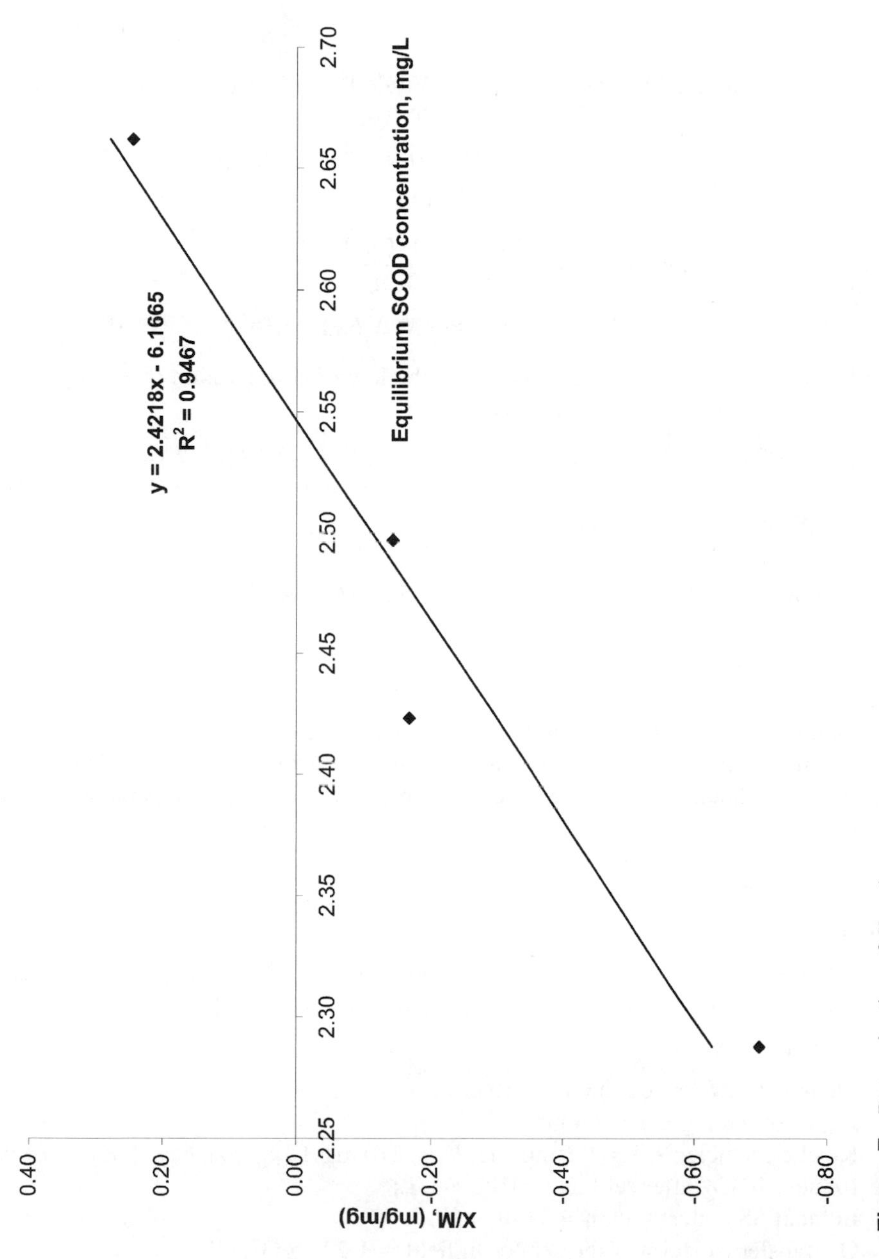

Figure 7 Determination of *K* and *n*.

lagoon is 4 days. Determine the surface area of the lagoon. If the wastewater enters the lagoon at a temperature of 65°C and the mean ambient temperature is 30°C, estimate the lagoon temperature assuming complete mixing condition and exchange coefficient $f = 0.5$ m/day. Also comment on the effect of temperature on process efficiency.

Solution

$$\text{Volume of the aerated lagoon } (m^3) = \text{flow}, V(\text{GPD}) \times 3.785\ (\text{L/gal}) \times 10^{-3}\ (m^3/\text{L}) \times \text{HRT}, \theta(\text{day})$$

$$= 15{,}000 \times 3.785 \times 10^{-3} \times 4$$

$$= 227.1\ m^3$$

$$\text{Surface area of the lagoon } (m^2) = \frac{\text{Volume, } V\ (m^3)}{\text{Depth, } D(m)}$$

$$= (227.1/3.3) = 68.82\ m^2$$

The lagoon temperature can be obtained by the law of conservation of energy:

Total heat gain = Total heat loss

$$Q\ (\text{GPD}) \times 3.785\ (\text{L/gal}) \times 10^3\ (m^3/\text{L}) \times (T_i - T_e)\ (°\text{C}) = f\ (\text{m/day}) \times A\ (m^2) \times (T_w - T_a)\ (°\text{C})$$

For complete mixing condition:

$$T_e = T_w$$

$$15{,}000 \times 3.785 \times 10^3 \times (65 - T_w) = 0.5 \times 68.82 \times (T_w - 30)$$

$$T_w = 65°\text{C}$$

Comment

The temperature of the aerated lagoon of more than 38°C has been reported to decrease the system efficiency. At this high temperature nitrifiers cannot survive. Hence, a heat exchanger on the influent line must be provided to reduce the high temperature of the raw wastewater.

Example 14

Design a flow-through aerated lagoon to treat 0.575 MGD of composite wastewater generated from a pharmaceutical plant. Assume that the following conditions and requirements apply.

1. Mean cell residence time, $\theta_c = 10$ days;
2. Depth of the lagoon = 3.3 m;
3. Kinetic coefficients: $Y = 0.6$ mg/mg, $K_s = 210$ mg/L, $k = 4.6\ \text{day}^{-1}$, $K_d = 0.1\ \text{day}^{-1}$;
4. Influent BOD_5 after settling = 3100 mg/L;
5. Influent SS concentration = 1140 mg/L;
6. O_2 transfer capacity of the aerator in field = 1.22 kg O_2/kWh;
7. Power requirement for mixing = 0.6 hp/1000 m^3.

Solution

(A) Determine the size of the aerated lagoon based on θ_c (solid retention time):

(i) Volume of wastewater generated per day

$$= 0.575\ (\text{MGD}) \times (3.785 \times 10^3)\ (\text{m}^3/\text{Mgal})$$

$$= 2176.37\ \text{m}^3/\text{day}$$

(ii) Volume of the aerated lagoon required $= 2176.37\ (\text{m}^3/\text{day}) \times 10\ (\text{day})$

$$= 21{,}763.75\ \text{m}^3$$

(iii) Surface area of the lagoon $= 21{,}763.75\ (\text{m}^3)/3.3\ (\text{m})$

$$= 6595.07\ \text{m}^2$$

(B) Determine the soluble effluent BOD_5 using the kinetic data:

$$S = \frac{K_s(1 + \theta K_d)}{\theta(Yk - K_d) - 1}$$

$$= \frac{210\ (\text{mg/L})[1 + 10\ (\text{day}) \times 0.1\ (\text{day}^{-1})]}{10\ (\text{day})[0.60\ (\text{mg/mg}) \times 4.6\ (\text{day}^{-1}) - 0.1\ (\text{day}^{-1})] - 1}$$

$$= 16.41\ \text{mg/L}$$

(C) Determine the O_2 requirement:

(i) Estimate the concentration of biological solids produced

$$= \frac{Y(S_i - S)}{1 + K_d\theta_c}$$

$$= \frac{0.6\ (\text{mg/mg})\ (3100 - 16.41)\ (\text{mg/L})}{1 + 0.1\ (\text{day}^{-1}) \times 10\ (\text{day})}$$

$$= 925.08\ \text{mg/L}$$

(ii) Estimate the SS concentration in the lagoon before settling

$$= 925.08 + 1140$$

$$= 2065.08\ \text{mg/L}\ (2065.08\ \text{g/m}^3)$$

(iii) Estimate the amount of solids wasted per day

$$= 925.08\ (\text{g/m}^3) \times 10^{-3}(\text{kg/g}) \times 2176.37\ (\text{m}^3/\text{day})$$
$$= 2013.32\ \text{kg/day}$$

(iv) Estimate the amount of O_2 required $= 1.47\ Q\ (S_i - S) - 1.42\ P_x$

$$= 1.47 \times 2176.37\ (\text{m}^3/\text{day}) \times (3100 - 16.41)\ (\text{mg/L})$$
$$\times\ 10^{-3}\ (\text{kg/mg} \cdot \text{L/m}^3) - 1.42 \times 2013.32\ (\text{kg/day})$$
$$= 7006.33\ \text{kg/day}$$

(D) Determine the power required to meet the O_2 requirement:

$$\text{Power} = \frac{7006.33\ (\text{kg/day})}{(1.22\ \text{kgO}_2/\text{kWh}) \times 24\ (\text{h/day})}$$

$$= 239.29\ \text{kW}$$

$$\text{Power} = 239.29\ (\text{kW}) \times 1.3410\ (\text{hp/kW})$$
$$= 320.89\ \text{hp}$$

(E) Determine the power required for mixing:

(i) Lagoon volume $= 21{,}763.75\ (\text{m}^3) \times 35.3147\ (\text{ft}^3/\text{m}^3)$

$$= 768{,}580.3\ \text{ft}^3$$

(ii) Power required for mixing

$$= 768{,}580.3\ (\text{ft}^3) \times 0.6 \left(\frac{\text{hp}}{1000\ \text{ft}^3}\right)$$

$$= 461.15\ \text{hp}$$

(F) Determine the horse-power rating of the aerator:

Horse power rating = 461.15 hp to fulfill the requirement of both mixing and O_2 supply

Example 15

Design a trickling filter to treat 33,800 GPD of pharmaceutical wastewater using the empirical method of Ten States for the data given below:

1. Influent BOD_5 of the raw wastewater, $S_i = 6000$ mg/L;
2. Efficiency of the filter, $E = 90\%$;
3. Depth of filter is restricted to 1.8 m.

Solution

(A) Determine the recirculation ratio required to give 90% efficiency:

$$E = \frac{(1 + R/Q)}{1.5 + (R/Q)}$$

$$0.90 = \frac{(1 + R/Q)}{1.5 + (R/Q)}$$

$$R/Q = 3.5$$

$$R = 3.5 \times 33{,}800\ (\text{GPD})$$
$$= 118{,}300\ \text{GPD}$$

(B) Determine the filter volume required by providing the maximum organic loading rate 1.2 kg/m^3 day:

$$V = [33{,}800 \text{ (GPD)} \times 3.785 \text{ (L/gal)} \times 0.9 \times 6000 \text{ (mg/L)} \times 10^{-6} \text{ (kg/mg)}]/1.2 \text{ (kg/m}^3 \text{ day)}$$
$$= 575.70 \text{ m}^3$$

(C) Determine the size of filter required:

$$\text{Surface area required} = 575.70 \text{ (m}^3)/1.8 \text{ (m)}$$

$$(\pi/4)D^2 = 319.83 \text{ (m}^2)$$
$$D = 20.18 \text{ m (say 20.2 m)}$$

(D) Hydraulic loading including the recirculation

$$= \frac{(33{,}800 + 3.5 \times 33{,}800) \text{ (GPD)} \times (3.785 \times 10^{-3}) \text{ (m}^3/\text{gal)}}{(\pi/4) \times (20.2)^2 \text{ (m}^2)}$$

$$= 1.80 \text{ m}^3/\text{m}^2 \text{ day}$$

Example 16

Design a UASB (upflow anaerobic sludge blanket) reactor for treatment of 435 m^3/day of wastewater generated from a typical pharmaceutical plant. The COD removal efficiency of the reactor at HRT of 2 days and organic loading of 3.52 kg/m^3 is 94%. Assume the following design data are applicable:

1. Influent COD = 7000 mg/L;
2. Methane yield = 0.35 m^3/kg COD removed;
3. Solubility of methane = 0.028 m^3/m^3 effluent;
4. Biomass yield = 0.027 mg/mg;
5. MLVSS of the sludge bed = 70 kg/m^3;
6. MLVSS of the sludge blanket = 4 kg/m^3;
7. Depth of the sludge bed = 1.5 m;
8. Depth of the sludge blanket = 3.5 m.

Solution

(A) Determine the size of the reactor:

(i) Volume of the reactor, $V = Q \times \theta$

$$V = 435 \text{ (m}^3/\text{day)} \times 2 \text{ (day)}$$

$$= 870 \text{ m}^3$$

(ii) Depth of the reactor, $H = 1.5 + 3.5$

$$H = 5 \text{ m}$$

(iii) Surface area required, $A = V/H$

$$A = 870\ (\text{m}^3)/5\ (\text{m})$$
$$= 174\,\text{m}^2$$

(iv) Diameter of reactor, $D = (4 \times A/\pi)^{1/2}$

$$D = (4 \times 174/\pi)^{1/2}$$
$$= 14.88\,\text{m (say 14.9 m)}$$

(B) Determine the organic loading rate

$$= (Q \times S_\text{i} \times E/V)$$
$$= 435\ (\text{m}^3/\text{day}) \times 7000\ (\text{mg/L}) \times 10^{-3}\ (\text{kg/mg} \cdot \text{L/m}^3) \times 0.94/870\ (\text{m}^3)$$
$$= 3.29\,\text{kg/m}^3\ \text{day} < 3.52\,\text{kg/m}^3\ \text{day, hence OK}$$

(C) Determine the upflow velocity, $v = H/\theta$

$$v = 5\ (\text{m})/[2\ (\text{day}) \times 24\ (\text{h/day})]$$
$$= 0.1\,\text{m/h}$$

(D) Determine the SRT:

(i) Total COD removed per day $= 435\ (\text{m}^3/\text{day}) \times 7000\ (\text{mg/L}) \times 10^{-3}\ (\text{kg/mg} \cdot \text{L/m}^3) \times 0.94$

$$= 2862.3\,\text{kg/day}$$

(ii) Biomass produced per day $= 0.027\ (\text{mg/mg}) \times 2862.3\ (\text{kg/day})$

$$= 77.28\,\text{kg/day}$$

(iii) Total biomass in the reactor

$$= \text{Biomass in the sludge bed} + \text{biomass in the sludge blanket}$$
$$= 70\ (\text{kg/m}^3) \times \pi/4 \times (14.9)^2\ (\text{m}^2) \times 1.5\ (\text{m}) + 4\ (\text{kg/m}^3) \times \pi/4 \times (14.9)^2\ (\text{m}^2) \times 3.5\ (\text{m})$$
$$= 20{,}749.58\,\text{kg}$$

(iv) SRT = Total biomass in the reactor/biomass produced per day

$$= 20{,}749.58\ (\text{kg})/77.28\ (\text{kg/day})$$
$$= 268.5\,\text{day}$$

(E) $$\text{F/M ratio} = \frac{435\ (\text{m}^3/\text{day}) \times 7000\ (\text{mg/L}) \times 10^{-3}\ (\text{kg/mg} \cdot \text{L/m}^3)}{20{,}749.58\ (\text{kg})}$$
$$= 0.15\,\text{day}^{-1}$$

(F) Methane production

(i) Total quantity of methane generated $= 0.35\ (\text{m}^3/\text{kg COD}_\text{r}) \times 2862.3\ (\text{kg COD}_\text{r}/\text{day})$

$$= 1001.8\,\text{m}^3/\text{day}$$

(ii) Methane leaving as dissolved in the effluent $= 0.028\ (m^3/m^3) \times 435\ (m^3/day)$

$= 12.18\ m^3/day$

(iii) Usable methane $= 1001.8 - 12.18$

$= 989.62\ m^3/day$

(G) Specific gas production

(i) Specific gas production per m^3 per m^3 of reactor per day

$= 1001.8\ (m^3/day)/870\ (m^3)$

$= 1.15\ m^3/m^3/day$

(ii) Specific gas production per m^3 per m^3 of effluent

$= 1001.8\ (m^3/day)/435\ (m^3/day)$

$= 2.3\ m^3/m^3$

(H) Energy equivalent of biogas

$= 989.62\ m^3/day \times 10{,}000\ (kcal/m^3)$

$= 989.62 \times 10^4\ kcal/day$

$= 989.62 \times 10^4\ (kcal/day) \times 1.1633 \times 10^{-3}\ (kWh/kcal)$

$= 11{,}512.25\ kWh/day$

(I) Coal equivalent of biogas $= 989.62 \times 10^4\ (kcal/day)/4000\ (kcal/kg)$

$= 2474.05\ kg/day$

$= 2.47\ tonnes/day$

Example 17

Determine the size of the UASB reactor for the treatment of $1275\ m^3/day$ of wastewater generated from a typical pharmaceutical plant with COD of 16,000 mg/L. The COD removal efficiency at an HRT of 4.7 days is 97%. If the following data and conditions are applicable, estimate (i) OLR and upflow velocity and (ii) methane yield and energy equivalent.

1. Overall depth = 4 m;
2. Percentage of methane in biogas = 65%;
3. Specific biogas production rate $= 7.64\ m^3/m^3$ effluent.

Solution

(A) Determine the size of the reactor:

(i) Volume of the reactor, $V = Q \times \theta$

$V = 1275\ (m^3/day) \times 4.7\ (day)$

$= 5992.5\ m^3$

(ii) Surface area required, $A = V/H$

$$A = 5992.5\ (\text{m}^3)/4\ (\text{m})$$
$$= 1498.125\ \text{m}^2$$

(iii) Diameter of reactor, $D = (4 \times A/\pi)^{1/2}$

$$D = (4 \times 1498.125/\pi)^{1/2}$$
$$= 43.67\ \text{m}$$

Because the diameter required is much more, therefore three parallel units must be designed with area of each unit as follows:

$$\text{Area} = 1498.125\ (\text{m}^2)/3$$
$$= 499.375\ \text{m}^2$$

(iv) Diameter of each unit $= (4 \times 499.375/\pi)^{1/2}$

$$= 25.21\ \text{m}$$

(B) Determine the OLR:

(i) Total COD removed/day $= 1275\ (\text{m}^3/\text{day}) \times 16{,}000\ (\text{mg/L}) \times 10^{-3}\ (\text{kg/mg} \cdot \text{L/m}^3) \times 0.97$

$$= 19{,}788\ \text{kg/day}$$

(ii) COD removed per unit $= 19{,}788\ (\text{kg/day})/3$

$$= 6596\ \text{kg/day}$$

(iii) $\text{OLR} = 6596\ (\text{kg/day})/499.375\ (\text{m}^2)$

$$= 13.21\ \text{kg COD/m}^3\ \text{day}$$

(C) Determine the upflow velocity $= H/\theta$

$$= 4\ (\text{m})/[4.7\ (\text{day}) \times 24\ (\text{h/day})]$$
$$= 0.035\ \text{m/h}$$

(D) Determine the methane yield:

(i) Total quantity of methane generated

$$= 7.64\ (\text{m}^3/\text{m}^3\ \text{effluent}) \times 1275\ (\text{m}^3/\text{day}) \times 0.65$$
$$= 6331.65\ \text{m}^3/\text{day}$$

(ii) Methane yield = total methane generated/total COD removed

$$= 6331.65\ (\text{m}^3/\text{day})/19{,}788\ (\text{kg/day})$$
$$= 0.32\ \text{m}^3/\text{kg COD}_\text{r}$$

(E) Determine the energy equivalent:

(i) Methane leaving as dissolved in the effluent

$$= 0.028\ (\text{m}^3/\text{m}^3) \times 1275\ (\text{m}^3/\text{day})$$
$$= 35.7\ \text{m}^3/\text{day}$$

(ii) Usable methane $= 6331.65 - 35.7$

$$= 6295.95\ \text{m}^3/\text{day}$$

(iii) Energy equivalent of biogas

$$= 6295.95\ (\text{m}^3/\text{day}) \times 10{,}000\ (\text{kcal/m}^3)$$
$$= 6295.95 \times 10^4\ \text{kcal/day}$$
$$= 6295.95 \times 10^4\ (\text{kcal/day}) \times 1.1633 \times 10^{-3}\ (\text{kWh/kcal})$$
$$= 73{,}240.79\ \text{kWh/day}$$

5.11 DISCUSSION TOPICS AND PROBLEMS

1. The BOD_5 and flow of various types of waste streams generated from Abbott Laboratory (a typical pharmaceutical plant) are given in the following table.

Type of waste stream	Flow (MGD)	BOD_5 (mg/L)
Chemical waste	0.262	2520
Fermentation waste	0.312	3620

In addition, the plant generates 470 m^3/day of domestic wastewater with $BOD_5 = 675$ mg/L. Calculate (a) BOD_5 of the composite waste and (b) total BOD load of the plant.
[*Answers*: (a) = 2684.09 mg/L; (b) 7091.18 kg BOD/day]
2. A synthetic organic chemical plant generates mainly two types of waste streams, i.e., strong process waste and dilute process waste. The BOD_5 of the 45,000 GPD of combined waste generated from the plant is 75,000 mg/L. If the BOD_5 and flow of the dilute process waste are 1200 mg/L and 33,800 GPD, respectively, estimate (a) BOD_5 of the strong process waste and (b) the BOD load of each waste stream and their contribution to the total BOD load of the plant.
[*Answer*: (a) = 297,717.73 mg/L; (b) BOD load of strong and dilute process waste = 12,620.85 kg/day (98.8%) and 153.52 kg/day (1.2%), respectively]
3. A primary settling tank is designed for the pretreatment of 0.575 MGD of wastewater with SS concentration of 1140 mg/L generated from a typical pharmaceutical plant. If the SS removal efficiency of the sedimentation tank is 60%, find (a) the effluent SS concentration and (b) the quantity of sludge generated. Assume the specific gravity of the sludge is 1.03 and that the sludge contains 5% solids.
[*Answer*: (a) = 456 mg/L; (b) 7636.89 GPD]
4. A typical pharmaceutical industry generates 15,000 GPD of wastewater with SS concentration 800 mg/L. The addition of alum (200 mg/L) and $FeCl_3$ (150 mg/L) reduces the SS concentration of effluent from 800 mg/L to 50 g/L. Determine the quantity of sludge generated per week. Assume the specific gravity of the sludge is 1.04 and that the sludge contains 3% solids.
[*Answer*: 3014.5 GPD]
5. The BOD removal efficiency of an activated sludge process treating 2000 m^3/day of condensate waste generated from a synthetic organic chemical plant is 94%. If the organic loading rate and BOD_5 of the raw waste are 3.171 kg/m^3 day

and 1275 mg/L, estimate (a) BOD_5 of the treated effluent and (b) hydraulic retention time.
[*Answer*: (a) 76.5 mg/L; (b) 9.12 h]

6. An activated sludge process is designed to treat 1950 m^3/day of pharmaceutical wastewater with BOD_5 concentration of 3250 mg/L. If the performance efficiency of the process based on BOD_5 removal is 85%, determine (a) the organic loading rate (OLR) and BOD_5 of the treated effluent, assuming the following data and conditions are applicable:
   (i) Aeration tank volume = 1500 m^3;
   (ii) Depth of the aeration tank = 2.5 m;
   (iii) MLVSS = 6000 mg/L.
   Also compute (b) the hydraulic loading rate and F/M ratio.
   [*Answer*: (a) OLR = 3.59 kg/m^3 day and BOD_5 = 487.5 mg/L; (b) HLR = 3.25 $m^3/m^2 \cdot$ day and F/M ratio = 0.7 day^{-1}]
7. Determine the F/M ratio and solid retention time of an extended aeration system designed for the treatment of 33,800 GPD of pharmaceutical wastewater. The BOD_5 of the raw wastewater and treated effluent are 5000 mg/L and 560 mg/L, respectively. Assume the following data and conditions are applicable:
   (i) HRT = 5 days;
   (ii) MLSS = 5600 mg/L;
   (iii) MLVSS/MLSS = 0.75;
   (iv) $Y_n = 0.45$ mg/mg.
   [*Answer*: F/M = 0.24 day^{-1} and SRT = 10.51 days]
8. The PAC-fed activated sludge process is designed to treat the alkaline waste stream generated from a synthetic organic chemical plant. The influent BOD_5 of the alkaline waste is 1275 mg/L, which can be treated to a BOD_5 of 275 mg/L by the activated sludge process. The addition of PAC at a dose of 500 mg/L gives a further reduction of effluent BOD_5 from 275 mg/L to 150 mg/L. Determine the constant K of the Freundlich equation given below $(X/M) = KC_e^{2.2}$. Also comment on the efficiency of the system before and after addition of PAC.
   [*Answer*: $K = 4.08 \times 10^{-6}$; performance efficiency of the system can be increased by approximately 10% by addition of PAC]
9. An aerated lagoon is designed to treat 435 m^3/day of acid waste stream with a BOD_5 of 3500 mg/L generated from a synthetic organic chemical plant. The depth of lagoon is restricted to 4 m and organic loading rate is 0.7 kg/m^3 day. Estimate (a) the surface area and hydraulic loading rate. If the performance efficiency of the lagoon is 97%, determine (b) the BOD_5 of the treated effluent.
   [*Answer*: (a) $A = 527.44$ m^2 and HLR = 0.82 m^3/m^2 day; (b) 105 mg/L]
10. An aerated lagoon is designed to treat 0.575 MGD of composite waste (including chemical and fermentation waste) with a BOD_5 of 3150 mg/L. The depth and HRT of the lagoon are restricted to 3.5 m and 5 days, respectively. Find (a) the surface area of the lagoon. If the temperature of composite waste entering into the lagoon is 60°C and mean ambient temperature is 15°C during winter, estimate (b) the lagoon temperature assuming complete mixing condition and exchange coefficient $f = 0.54$ m/day. Also comment (c) on the effect of wastewater temperature in the process efficiency of the lagoon.
    [*Answer*: (a) $A = 3109.10$ m^2; (b) $T_w = 40.4$°C; (c) the temperature of waste 60°C will result in the temperature in aerated lagoon being >38°C, which is found to reduce the process efficiency]

11. A trickling filter is designed to treat 435 m^3/day of acid waste stream generated from a synthetic organic chemical plant. The BOD_5 of the acid waste before and after the primary sedimentation is 3250 mg/L and 2850 mg/L, respectively. The efficiency of the filter at a recirculation ratio of 4.5 is 92%. If the depth of filter is restricted to 1.6 m and the value of the constant in Eckenfelder's equation is $n = 0.5$, determine the value of constant K_f assuming the hydraulic loading rate = 17.5 $m^3/m^2 \cdot$ day.
    [*Answer*: $K_f = 3.12$ $m^{1/2} \cdot day^{1/2}$]
12. A pharmaceutical wastewater with BOD_5 of 3000 mg/L is to be treated by a trickling filter. Design the filter for 15,000 GPD of wastewater to give the desired effluent BOD_5 of 50 mg/L. Use the NRC (U.S. National Research Council) equation for the design of the filter. The following data and conditions are applicable:
    (i) Depth of filter = 1.7 m;
    (ii) Recirculation ratio = 2 : 1;
    (iii) Wastewater temperature = 20°C;
    (iv) Assume efficiencies of the two-stage filters are equal: $E_1 = E_2$.
    [*Answer*: (a) $E_1 = E_2 = 87\%$; (b) diameter of 1st stage filter $D_1 = 23$ m and 2nd stage filter $D_2 = 63.95$ m]
13. A UASB reactor is designed to treat 1275 m^3/day of wastewater with a BOD concentration of 2000 mg/L generated from a typical pharmaceutical industry. At an HRT of 1.5 days, the COD and BOD removal efficiencies of the reactor are 80 and 95%, respectively. Determine (a) the size of the reactor; (b) the total quantity of methane produced; and (c) the coal equivalent and energy equivalent. Assume that the following data and conditions are applicable:
    (i) Depth of the reactor is restricted to 4.5 m;
    (ii) Biogas yield = 0.6 m^3/kg COD_r;
    (iii) Methane content of biogas = 70%;
    (iv) Solubility of methane = 0.028 m^3/m^3 effluent;
    (v) Calorific value of methane = 10,000 kcal/m^3;
    (vi) Calorific value of coal = 4000 kcal/kg.
    [*Answer*: (a) Diameter = 23.26 m; (b) 1102.24 m^3/day; (c) 2.67 tons/day and 12,410 kWh/day]
14. The COD removal efficiency of a UASB reactor treating pharmaceutical wastewater is 96% at an organic loading rate of 0.5 kg COD/m^3/day. If the plant generates 33,800 GPD wastewater with a COD concentration of 1000 mg/L and the depth of reactor is restricted to 3 m, estimate (a) the size of the UASB reactor; (b) the HRT; and (c) the specific gas production rate assuming a methane yield of 0.3 m^3/kg COD_r.
    [*Answer*: (a) Diameter = 10.21 m; (b) 1.92 days; (c) 0.29 m^3/m^3 effluent and 0.15 m^3/m^3/day]

NOMENCLATURE

ACPCI	Alexandra Company for Pharmaceutical and Chemical Industry
AMFFR	anaerobic mesophilic fixed film reactor
ASP	activated sludge process
ATFFR	anaerobic thermophilic fixed film reactor

BOD	biochemical oxygen demand (mg/L)
BOD_5	5-day biochemical oxygen demand (mg/L)
BOD_r	biochemical oxygen demand removed (kg/day)
BOD_L	ultimate biochemical oxygen demand (mg/L)
COD	chemical oxygen demand (mg/L)
DCE	dichloroethylene
DO	dissolved oxygen (mg/L)
DOC	dissolved organic carbon (mg/L)
F/M	food to microorganism ratio (day^{-1})
GAC	granular activated carbon
gal	gallon
h	hour
HLR	hydraulic loading rate (m^3/m^3 day)
HRT	hydraulic retention time (day)
kg/day	kilogram per day
L/gal	liter/gallon
mg/L	milligram per liter
MGD	million gallons per day
MLSS	mixed liquor suspended solids (mg/L)
MLVSS	mixed liquor volatile suspended solids (mg/L)
NA	nitroaniline
NEIC	National Enforcement Investigation Center
NP	nitrophenol
OLR	organic loading rate (kg COD/m^3 day)
OWC	organic wastewater contaminants
PAC	powdered activated carbon
SCOD	soluble chemical oxygen demand (mg/L)
S-CBOD	soluble carbonaceous oxygen demand (mg/L)
SRT	solid retention time (day)
SS	suspended solids (mg/L)
SVI	sludge volume index
TCMP	tri-chloromethyl-propanol
TDS	total dissolved solids (mg/L)
TKN	total Kjeldahl nitrogen (mg/L)
TOC	total organic carbon (mg/L)
TS	total solids (mg/L)
UASB	upflow anaerobic sludge blanket
VSS	volatile suspended solids (mg/L)
V/V	volume/volume

REFERENCES

1. Mehta, G.; Prabhu, S.M.; Kantawala, D. Industrial wastewater treatment–The Indian experience. J. Indian Assoc. Environ. Management **1995**, *22*, 276–287.
2. Kolpin, D.W.; Furlong, E.T.; Meyer, M.T.; Thurman, E.M.; Zaugg, S.D.; Barber, L.B.; Buxton, H.T. Pharmaceuticals, hormones and other organic wastewater contaminants in U.S. streams, 1999–2000: A national reconnaissance. Environ. Sci. Technol. **2002**, *36*, 1202–1211.

3. Struzeski, E.J. Status of wastes handling and waste treatment across the pharmaceutical industry and 1977 effluent limitations. *Proceedings of the 35th Industrial Waste Conference*, Purdue University, West Lafayette, IN, 1980, 1095–1108.
4. Seif, H.A.A.; Joshi, S.G.; Gupta, S.K. Effect of organic load and reactor height on the performance of anaerobic mesophilic and thermophilic fixed film reactors in the treatment of pharmaceutical wastewater. Environ. Technol. **1992**, *13*, 1161–1168.
5. Andersen, D.R. Pharmaceutical wastewater treatment: A case study. *Proceedings of the 35th Industrial Waste Conference*, Purdue University, West Lafayette, IN, 1980, 456–462.
6. Struzeski, E.J. *Waste Treatment and Disposal Methods for the Pharmaceutical Industry*. NEIC-Report EPA 330/1-75-001; U.S. Environmental Protection Agency, Office of Enforcement, National Field Investigation Center: Denver, CO, 1975.
7. Molof, A.H.; Zaleiko, N. Parameter of disposal of waste from pharmaceutical industry. Ann. N.Y. Acad. Sci. **1965**, *130*, 851–857.
8. Syracuse, N.Y.; Mann, U.T. Effects of penicillin waste in Ley Creek Sewage Treatment Plant. Sewage Ind. Waste **1951**, *23*, 1457–1460.
9. Gallagher, A., *et al.* Pharmaceutical waste disposal. Sewage Ind. Waste **1954**, *26* (11), 1355–1362.
10. Seeler, T.A.; Jennet, J.C. Treatment of wastewater from a chemically synthesized pharmaceutical manufacturing process with the anaerobic filter. *Proceedings of the 33rd Industrial Waste Conference*, Purdue University, West Lafayette, IN, 1978, 687–695.
11. Patil, D.M.; Shrinivasen, T.K.; Seth, G.K.; Murthy, Y.S. Treatment and disposal of synthetic drug wastes. Environ. Health (India) **1962**, *4*, 96–105.
12. Lawson, J.R.; Woldman, M.L.; Eggerman, P.P. Squibb solves its pharmaceutical wastewater problems in Puerto Rico. Chem. Engng. Progress Symposium Series No. 107, 1971, "Water-1970", 1970, 401–404.
13. Murthy, Y.S.; Subbiah, V.; Rao, D.S.; Reddy, R.C.; Kumar, L.S.; Elyas, S.I.; Rama Rao, K.G.; Gadgill, J.S.; Deshmukh, S.B. Treatment and disposal of wastewater from synthetic drugs plant (I.D.P.L.), Hyderabad, Part I – Wastewater characteristics. Indian J. Environ. Health **1984**, *26* (1), 7–19.
14. Deshmukh, S.B.; Subrahmanyanm, P.V.R.; Mohanrao, G.J. Studies on the treatment of wastes from a synthetic drug plant. Indian. J. Environ. Health **1973**, *15*, 2.
15. Deshmukh, S.B.; Gadgil, J.S.; Subrahmanyanm, P.V.R. Treatment and disposal of wastewater from synthetic drugs plant (I.D.P.L.), Hyderabad, Part II – Biological treatability. Indian J. Environ. Health **1984**, *26*, 20–28.
16. Mayabhate, S.P.; Gupta, S.K.; Joshi, S.G. Biological treatment of pharmaceutical wastewater. Water Air Soil. Poll. **1988**, *38*, 189–197.
17. Howe, R.H.L.; Nicoles, R.A. Waste treatment for veterinary and plant science research and production at Eli Lilly Greenfield Laboratories. *Proceedings of the 14th Industrial Waste Conference*, Purdue University, West Lafayette, IN, 1959; 647–655.
18. Yeole, T.Y.; Gadre, R.V.; Ranade, D.R. Biological treatment of a pharmaceutical waste. Indian J. Environ. Health **1996**, *38* (2), 95–99.
19. Howe, R.H.L. Handling wastes from the billion dollar pharmaceuticals Industry. Waste Engng. **1960**, *31*, 728–753.
20. Howe, R.W.L.; Coates, D.G. Antitoxin and vaccine wastes treated at Eli Lilly plant. Waste Engng. **1955**, *26*, 235.
21. Lines, G. Liquid wastes from the fermentation industries. J. Water Pollut. Conf. **1968**, 655.
22. Genetelli, G.J.; Heukelekian, H.; Hunter, J.V. *Use of Research Techniques for Determination of Design Parameters*. Journal Series, New Jersey Agricultural Experiment Station, Rutgers – The State University of New Jersey, Department of Environmental Sciences: New Brunswick, NJ, 1967.
23. Nedved, T.K.; Bergmann, D.E.; Camens, A.A. Pharmaceutical laboratory BOD studies, a matter of philosophy. *2nd Symposium of Hazardous Chemicals and Disposal*, Indianapolis, IN, 1971.
24. Genetelli, E.J.; Heukelekian, H.; Hunter, J.V. A rational approach to design for a complex chemical waste. *Proceedings of the 5th Texas Water Pollution Association, Industrial Water and Waste Conference*, 1965; 372–396.

25. Young, J.C.; Affleck, S.B. Long-term biodegradability tests of organic industrial wastes. *Proceeding of the 29th Industrial Waste Conference*, Purdue University, West Lafayette, IN, 1974; 26.
26. Howe, R.H.L. Biological degradation of waste containing certain toxic chemical compounds. *Proceedings of the 16th Industrial Waste Conference*, Purdue University, West Lafayette, IN, 1961; 262–276.
27. O'Flaherty, T. The chemical industry. In *Environment and Development in Ireland*; Feehan, J., Ed.; Environmental Institute, University College: Dublin, 1991; 136–142.
28. Colovos, G.C.; Tinklenberg, N. Land disposal of pharmaceutical manufacturing wastes. Biotech. Bioeng. **1962**, *IV*, 153–160.
29. Quane, D.E.; Stumpf, M.R. Coal-fired boilers burn waste sludge and odors in an integrated pollution control system. *46th Annual Water Pollution Control Federation Conference*, Cleveland, OH, 1973.
30. Barker, W.G.; Schwarz, D. Engineering processes for waste control. Civil Engng. Progress **1961**, *65*, 58–61.
31. Blaine, R.K.; Van Lanen, J.H. Application of waste-to-product ratios in fermentation industries. Biotech. Bioeng. **1962**, *IV*, 129–138.
32. Edmondson, K.H. Disposal of antibiotic spent beers by triple effect evaporation. *Proceedings of the 8th Industrial Waste Conference*, Purdue University, West Lafayette, IN, 1953; 46–58.
33. Barker, W.G.; Stumpf, H.R.; Schwarz, D. Unconventional high performance activated sludge treatment of pharmaceutical wastewater. *Proceedings of the 28th Industrial Waste Conference*, Purdue University, West Lafayette, IN, 1973.
34. Jackson, C.J. Fermentation wastes disposal in Great Britain. *Proceedings of the 21st Industrial Waste Conference*, Purdue University, West Lafayette, IN, 1966; 19–32.
35. McCallum, D., *et al.* Wastewater management in the pharmaceutical industry. *3rd International Conference on Effluent Treatment from Biochemical Industry*, Wheatland: Watford, England, 1980.
36. Paradiso, S.J. What to do with wastes when volumes overtake capacity. Industry and Power **1955**, *69*, 35–39.
37. Paradiso, S.J. Disposal of fine chemical wastes. *Proceeding of the 10th Purdue Industrial Waste Conference*, Purdue University, West Lafayette, IN, 1955; 49–60.
38. Hamza, A. Evaluation of treatability of the pharmaceutical wastewater by biological methods. In *Current Practices in Environmental Engineering*; Hamaza, A., Ed.; International Book Traders, Delhi, India. 1984; 37–44.
39. Gulyas, H.; Hemmerling, L.; Sekoulov, I. Moglichkeiten zur weitergehenden Entfernung organischer Inhaltsstoffe aus Abwassern der Altolaufbereitung. Z. Wasser Abwasser Forsch. **1991**, *24*, 253–257.
40. Delaine, J.; Gough, D. An evaluation of process for treatment of pharmaceutical effluents. *3rd International Conference on Effluent Treatment from Biochemical Industry*, Wheatland: Watford, England, 1980.
41. Gulyas, H.; von Bismarck, R.; Hemmerling, L. Treatment of industrial wastewaters with ozone/hydrogen peroxide. Water Sci. Technol. **1995**, *32* (7), 127–134.
42. Dryden, F.E.; Barrett, P.A.; Kissinger, J.C.; Eckenfelder, Jr., W.W. High rate activated sludge treatment of fine chemical wastes. Sewage Ind. Wastes **1956**, *29*, 193.
43. Henary, M.P. *Biological Treatment of Pharmaceutical Wastewater*. Ph.D. thesis, National University of Ireland, 1994.
44. Henary, M.P.; Donlon, B.P.; Lens, P.N.; Colleran, E.M. Use of anaerobic hybrid reactor for treatment of synthetic pharmaceutical wastewaters containing organic solvents. J. Chem. Technol. Biot. **1996**, *66*, 251–264.
45. Mohanrao, G.J.; Subramanyam, P.V.; Deshmukh, S.B.; Saroja, S. Water treatment at a synthetic drug factory in India. J. Water Pollut. Conf. **1970**, *42* (8), 1530–1543.
46. Donahue, R.T. Single stage nitrification activated sludge pilot plant study on a bulk pharmaceutical manufacturing wastewater. *Proceeding of the 38th Industrial Waste Conference*, Purdue University, West Lafayette, IN, 1984; 173–180.
47. Kincannon, D.F. Performance comparison of activated sludge, PAC activated sludge, granular activated carbon and a resin column for removing the priority pollutants from a pharmaceutical

wastewater. *Proceeding of the 35th Industrial Waste Conference*, Purdue University, West Lafayette, IN, 1980; 476–483.

48. Osantowski, R.A.; Dempsey, C.R.; Dostal, K.A. Enhanced COD removal from pharmaceutical wastewater using powdered activated carbon addition to an activated sludge system. *Proceeding of the 35th Industrial Waste Conference*, Purdue University, West Lafayette, IN, 1980; 719–727.
49. El-Gohary, F.A.; Abou-Elela, S.I.; Aly, H.I. Evaluation of biological technologies for the wastewater treatment in the pharmaceutical industry. Water Sci. Technol. **1995**, *32*, 13–20.
50. Gopalan, R., *et al.* Treatment and disposal of effluents from pharmaceutical and dyestuff industries in Baroda. *Proceedings of the Symposium on Environmental Pollution*, Central Public Health Engineering Research Institute, Zonal Laboratory, 1973; 88–94.
51. Vogler, J.F. Chemical and antibiotics waste treatment at William Island, West Virgina; Present chemical wastes treatment. Sewage Ind. Wastes **1952**, *24*, 485.
52. Reimers, A.E.; Rinace, U.S.; Poese, L.E. Trickling filter studies on fine chemicals plant waste. Sewage Ind. Wastes **1954**, *26*, 51.
53. Liontas, J.A. High rate filters treat mixed wastes at Sherp and Dohme. Sewage Ind. Wastes **1954**, *26*, 310.
54. Jennet, J.C.; Dennis, N.D. Anaerobic filter treatment of pharmaceutical waste. J. Water Pollut. Conf. **1975**, *47* (1), 104–121.
55. Elliot, S.F.; Jennet, J.C.; Rgand, M.C. Anaerobic treatment of synthesized organic chemical pharmaceutical wastes. *Proceedings of the 33rd Industrial Waste Conference*, Purdue University, West Lafayette, IN, 1978; 507–514.
56. Seif, H.A.A. *Comparative Study on Treatment of Pharmaceutical Wastewater by Anaerobic Mesophilic and Thermophilic Fixed Film Reactors*. Ph.D. thesis, CESE, IIT, Bombay, India, 1990.
57. Seif, H.A.A.; Joshi, S.G.; Gupta, S.K. Treatment of pharmaceutical wastewater by anaerobic mesophilic and thermophilic fixed film reactors. *First Symposium on Hazard Assessment and Control of Environmental Contaminants in Water*, Kyoto, Japan, 1991; 630–637.
58. Khan, A.N.; Siddiqui, R.H. Wastewater treatment by anaerobic contact filter. Indian J. Environ. Health **1976**, *18* (4), 282.
59. Young, J.C.; Dhab, M.F. The effect of media design on the performance of fixed bed anaerobic reactors. Water Sci. Technol. **1983**, *15*, 369.
60. Fox, P.; Venkatasubbiah, V. Coupled anaerobic/aerobic treatment of high-sulfate wastewater with sulfate reduction and biological sulfide oxidation. Water Sci. Technol. **1996**, *34* (5/6), 359–366.
61. Mormile, M.R.; Liu, S.; Sulfita, J.M. Anaerobic biodegradation of gasoline oxygenates: extrapolation of information to multiple sites and redox conditions. Environ. Sci. Technol. **1994**, *28*, 1727–1732.
62. Speece, R.E. Anaerobic biotechnology for industrial wastewater treatment. Environ. Sci. Technol. **1983**, *17*, 417–427.
63. Gupta, S.K.; Sharma, R. Biological oxidation of high strength nitrogenous wastewater. Water Res. **1996**, *30* (3), 593–600.
64. Otto, R.; Barker, W.G.; Schwarz, D.; Tjarksen, B. Laboratory testing of pharmaceutical wastes for biological control. Biotech. Bioeng. **1962**, *IV*, 139–145.
65. Storm, P.F. *Review of Bulking Episode at the Merck and Co., Inc., Elkton, Virginia Wastewater Treatment Pilot Plant*, Merck & Co., Inc.: Elkton, VA, 1981.
66. Storm, P.F.; Jenkins, D. Identification and significance of filamentous microorganism in activated sludge. *54th Annual Conference of the Water Pollution Control Federation*, Detroit, Michigan, 1981.
67. Jenkins, D. The control of activated sludge bulking. *52nd Annual Conference, California Water Pollution Control Association*, California, 1980.
68. Brandel, J.S. Pharmaceutical company's aeration system saves energy. Ind. Wastes **1980**, *26* (2), 16–19.
69. Arceivala, S.J., Ed. *Wastewater Treatment for Pollution Control*; Tata McGraw-Hill Publishing Company Limited: New Delhi, India, 1998.
70. Hindustan D.O. *Studies on Pharmaceutical Wastewater*; Park-Davies, Hindustan Dorr Oliver: Bombay, India.

6
Treatment of Oilfield and Refinery Wastes

Joseph M. Wong
Black & Veatch, Concord, California, U.S.A.

Yung-Tse Hung
Cleveland State University, Cleveland, Ohio, U.S.A.

The petroleum industry, one of the world's largest industries, has four major branches [1]. The production branch explores for oil and brings it to the surface in oilfields. The transportation branch sends crude oil to refineries and delivers the refined products to consumers. The refining branch processes crude oil into useful products. The marketing branch sells and distributes the petroleum products to consumers. The subject of this chapter is the treatment of liquid wastes from the production and refining branches.

6.1 OIL PRODUCTION

Each year more than 30 billion barrels of crude oil are produced in the world. The average worldwide and U.S. production rates are 83 million and 5.9 million barrels per day (bpd), respectively. Saudi Arabia produced the most crude in 1999, at more than 7.5 million bpd, followed by the former Soviet Union countries, at more than 7.3 million bpd (data taken from Oil & Gas J., December 18, 2000).

Oil production starts with petroleum exploration. Oil geologists study rock formations on and below the Earth's surface to determine where petroleum might be found. The next step is preparing and drilling an oil well. After completing the well, which means bringing the well into production, petroleum is recovered in much the same way as underground water is obtained.

6.1.1 Oil Drilling

There are three well-established methods of drilling [1]. The first oil crews used a technique called cable-tool drilling, which is still used for boring shallow holes in hard rock formation. Today, most U.S. crews use the faster and more accurate method of rotary drilling. On sites where the well must be drilled at an angle, crews use the directional drilling technique. Directional drilling is often used in offshore operations because many wells can be drilled directionally from one platform. Petroleum engineers are also testing a variety of drilling methods to increase the depth of oil wells and reduce the cost of drilling operations.

Cable-tool drilling works in much the same way as a chisel is used to cut wood or stone [1]. A steel cable repeatedly drops and raises a heavy cutting tool called a bit. Bits may be as long as 8 feet (2.4 m) with a diameter of 4 to 12.5 inches (10–31.8 cm). Each time the bit drops, it drives deeper and deeper into the earth. The sharp edges of the bit break up the soil and rock into small particles. From time to time, the workers pull out the cable and drill bit and pour water into the hole. They then scoop up the water and particles at the bottom of the hole with a long steel tool known as a bailer.

The rotary drilling method works like a carpenter's drill boring through wood [1]. The bit on a rotary drill is attached to the end of a series of connected pipes called the drill pipe. The drill pipe is rotated by a turntable on the floor of the derrick. The pipe is lowered into the ground. As the pipe turns, the bit bores through layers of soil and rock. The drilling crew attaches additional lengths of pipe as the hole becomes deeper.

The drill pipe is lowered and raised by a hoisting mechanism called the draw works, which operates somewhat like a fishing rod. Steel cable is unwound from the hoisting drum, then threaded through two sets of pulleys (blocks) – the crown block, at the top of the rig, and the traveling block, which hangs inside the derrick. The workers attach the upper end of the drill pipe to the traveling block with a giant hook. They can then lower the pipe into the hole or lift it out by turning the hoisting drum in one direction or the other.

During rotary drilling, a fluid called drilling mud is pumped down the drill pipe. It flows out of the openings in the bit and then back up between the pipe and the wall of the hole to just below the derrick. This constantly circulating fluid cools and cleans the bit and carries cuttings (pieces of soil and rock) to the surface. Thus, the crew can drill continuously without having to bail out the cuttings from the bottom of the well. The drilling mud also coats the sides of the hole, which helps prevent leaks and cave-ins. In addition, the pressure of the mud on the well reduces the risk of blowouts and gushers.

In cable-tool drilling and most rotary drilling, the well hole is drilled straight down from the derrick floor. In directional drilling, the hole is drilled at an angle using special devices called turbodrills and electrodrills. The motors that power these drills lie directly above the bit and rotate only the lowermost section of the drill pipe. Such drills enable drillers to guide the bit along a slanted path. Drillers may also use tools known as whipstocks to drill at an angle. A whipstock is a long steel wedge grooved like a shoehorn. The wedge is placed in the hole with pointed end upward. The drilling path is slanted as the bit travels along the groove of the whipstock.

6.1.2 Recovering Petroleum

Petroleum is recovered in two ways [1]. If natural energy provides most of the energy to bring the fluid to the surface, the recovery is called primary recovery. If artificial means are used, the process is called enhanced recovery.

In primary recovery the natural energy comes mainly from gas and water in reservoir rocks. The gas may be dissolved in the oil or separated at the top of it in the form of a gas cap. Water, which is heavier than oil, collects below the petroleum. Depending on the source, the energy in the reservoir is called solution-gas drive, gas-cap drive, or water drive. In solution-gas drive, the gas expands and moves toward the opening, carrying some of the liquid with it. In gas-cap drive, gas is trapped in a cap above the oil as well as dissolved in it. As oil is produced from the reservoir, the gas cap expands and drives the oil toward the well. In water drive, water in a reservoir is held in place mainly by underground pressure. If the volume of water is sufficiently large, the reduction of pressure that occurs during oil production causes the water to expand. The water then displaces the petroleum, forcing it to flow into the well.

Enhanced recovery can include a variety of methods designed to increase the amount of oil that flows into a producing well. Secondary recovery consists of replacing the natural energy in a reservoir. Water flooding is the most widely used method, which involves injecting water into the reservoir to cause the oil to flow into the well. Tertiary recovery includes a number of experimental methods of bringing more oil to the surface. These methods may include steam injection or burning some of the petroleum in the reservoir. The heat makes the oil thinner, enabling it to flow more freely into the well.

Oil leaving the producing well is a mixture of liquid petroleum, natural gas, and formation water. Some production may contain as much as 90% produced water [2]. This water must be separated from the oil, as pipeline specifications stipulate maximum water content from as low as 1% to 4%. The initial water–oil separation vessel in a modern treating plant is called a free-water-knockout [2]. Free water, defined as that which separates within five minutes, is drawn off to holding to be clarified prior to reinjection or discharge. Natural gas is also withdrawn from the free-water-knockout and piped to storage. The remaining oil usually contains emulsified water and must be further processed to break the emulsion, usually assisted by heat, electrical energy, or both. The demulsified crude oil flows to a stock tank for pipeline shipment to a refinery.

6.2 OIL REFINING

After crude oil is separated from natural gas, it is transported to refineries and processed into useful products. Refineries range in size from small plants that process about 150 barrels of crude oil per day to giant complexes with a capacity of more than 600,000 bpd [1]. As of January 1, 2002, there are 732 operating refineries in the world and 143 operating refineries in the United States. The worldwide and U.S. crude capacities are 81.2 and 16.6 million bpd, respectively [3]. Table 1 shows the distribution and crude capacities of operating refineries in the United States [3].

A petroleum refinery is a complex combination of interdependent operations engaged in separating crude molecular constituents, molecular cracking, molecular rebuilding, and solvent finishing to produce petroleum-derived products. Figure 1 shows an overall flow diagram for a generalized refinery production scheme [4].

In its 1977 survey, the U.S. Environmental Protection Agency (USEPA) identified over 150 separate processes being used in refineries [5]. A refinery may employ any number or a combination of these processes, depending upon the type of crude processed, the type of product being produced, and the characteristics of the particular refinery. The refining processes can generally be classified as separation, conversion, and chemical treatment processes [1].

Separation processes separate crude oil into some of its fractions. Fractional distillation, solvent extraction, and crystallization are some of the major separation processes.

Conversion processes convert less useful fractions into those that are in greater demand. Cracking and combining processes belong to the class of conversion processes. Cracking processes include thermal cracking and catalytic cracking, which convert heavy fractions into lighter ones. During cracking, hydrogenation may be used to further increase the yield of useful products. Combining processes do the reverse of cracking – they form more complex fractions from simple gaseous hydrocarbons. The major combining processes include polymerization, alkylation, and reforming.

Chemical treatment processes are used to remove impurities from the fractions. The method of treatment depends on the type of crude oil and on the intended use of the petroleum product. Treatment with hydrogen is a widely used method of removing sulfur compounds. Blending with other products or additives may be carried out to achieve certain special properties.

Table 1 Survey of Operating Refineries in the United States (State Capacities as of January 1, 2002)

State	No. of refineries	Crude capacity (b/cd)[a]
Alabama	3	148,225
Alaska	6	373,500
Arkansas	3	67,700
California	20	1,975,100
Colorado	2	88,000
Delaware	1	175,000
Georgia	1	6,000
Hawaii	2	149,000
Illinois	5	940,550
Indiana	2	433,500
Kansas	3	278,500
Kentucky	2	227,500
Louisiana	20	2,703,780
Michigan	1	74,000
Minnesota	2	360,000
Mississippi	2	318,000
Montana	4	175,100
New Jersey	3	557,000
New Mexico	3	97,600
North Dakota	1	58,000
Ohio	4	530,000
Oklahoma	5	438,858
Pennsylvania	5	761,700
Tennessee	1	175,000
Texas	25	4,440,500
Utah	5	160,500
Virginia	1	58,600
Washington	5	618,520
West Virginia	1	11,500
Wisconsin	1	33,250
Wyoming	4	130,000
Total	**143**	**16,564,483**

[a]b/cd = barrels per calendar day.
Source: *Oil & Gas J.*, Dec. 24, 2001.

In addition to these major processes, there are other auxiliary activities that are critical to the operation in a refinery. These auxiliary operations and the major refining processes are briefly described below, along with their wastewater sources [5].

6.2.1 Crude Oil and Product Storage

Crude oil, intermediate, and finished products are stored in tanks of varying size to provide adequate supplies of crude oils for primary fractionation runs of economical duration; to equalize process flows and provide feedstocks for intermediate processing units; and to store final products prior to shipment in adjustment to market demands. Generally, operating schedules permit sufficient detention time for settling of water and suspended materials.

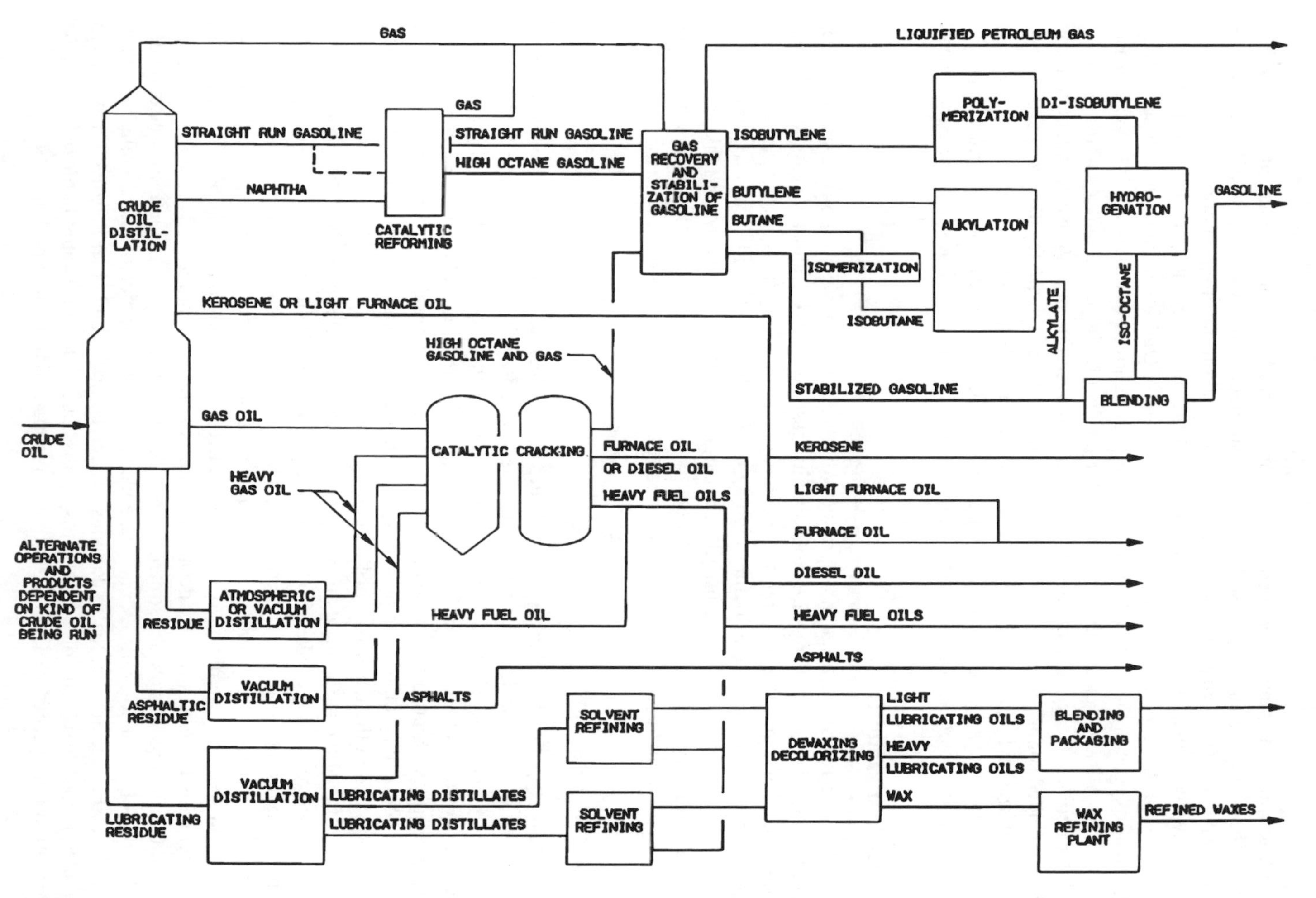

Figure 1 Generalized flowchart for petroleum refining. Crude oil is separated into different fractions and processed into many different products in a refinery. (From Ref. 4.)

Wastewater pollutants associated with storage of crude oil and products are mainly free oil, emulsified oil, and suspended solids. During storage, water and suspended solids in the crude oil separate. The water layer accumulates below the oil, forming a bottom sludge. When the water layer is drawn off, emulsified oil present at the oil–water interface is often lost to the sewers. This waste is high in chemical oxygen demand (COD) levels and, to a lesser extent, biochemical oxygen demand (BOD). Bottom sludge is removed at infrequent intervals. Waste also results from leaks, spills, salt filters (used for product drying), and tank cleaning.

Intermediate storage is frequently the source of polysulfide-bearing wastewaters and iron sulfide suspended solids. Finished product storage can produce high-BOD, alkaline wastewaters, as well as tetraethyl lead. Tank cleaning can contribute large amounts of oil, COD, and suspended solids and a minor amount of BOD. Leaks, spills, and open or poorly ventilated tanks can also be a source of air pollution through evaporation of hydrocarbons into the atmosphere.

6.2.2 Ballast Water Storage

Tankers that ship intermediate and final products discharge ballast water (approximately 30% of the cargo capacity is generally required to maintain vessel stability). Ballast waters have organic contaminants that range from water-soluble alcohol to residual fuels. Brackish water and sediments are also present, contributing high COD and dissolved solids loads to the refinery wastewater. These wastewaters are usually discharged to either a ballast water tank or holding ponds at the refinery. In some cases, the ballast water is discharged directly to the wastewater treatment system, and potentially constitutes a shock load to the treatment system.

6.2.3 Crude Desalting

Common to all types of desalting are an emulsifier and settling tank. Salts can be separated from oil by one of two methods. In the first method, water wash desalting in the presence of chemicals is followed by heating and gravity separation. In the second method, water wash desalting is followed by water–oil separation in a high-voltage electrostatic field acting to agglomerate dispersed droplets. A process flow schematic of electrostatic desalting is shown in Figure 2. Wastewater containing removed impurities is discharged to the wastewater system, and desalted crude oil flows from the upper part of the holding tank.

Much of the bottom sediment and water content in crude oil is a result of the “load-on-top” procedure used on many tankers. This procedure can result in one or more cargo tanks containing mixtures of seawater and crude oil, which cannot be separated by decantation while at sea, and are consequently retained in the crude oil storage at the refinery. Although much of the water and sediment are removed from the crude oil by settling during storage, a significant quantity remains to be removed by desalting before the crude is refined.

The continuous wastewater stream from a desalter contains emulsified oil (occasionally free oil), ammonia, phenol, sulfides, and suspended solids, all of which produce a relatively high BOD and COD concentration. It also contains enough chlorides and other dissolved materials to contribute to the dissolved solids problems in discharges to freshwater bodies. Finally, its temperature often exceeds 95°C (200°F), thus it is a potential thermal pollutant.

6.2.4 Crude Oil Fractionation

Fractionation is the basic refining process for separating crude petroleum into intermediate fractions of specified boiling point ranges. The various subprocesses include prefractionation and atmospheric fractionation, vacuum fractionation, and three-stage crude distillation.

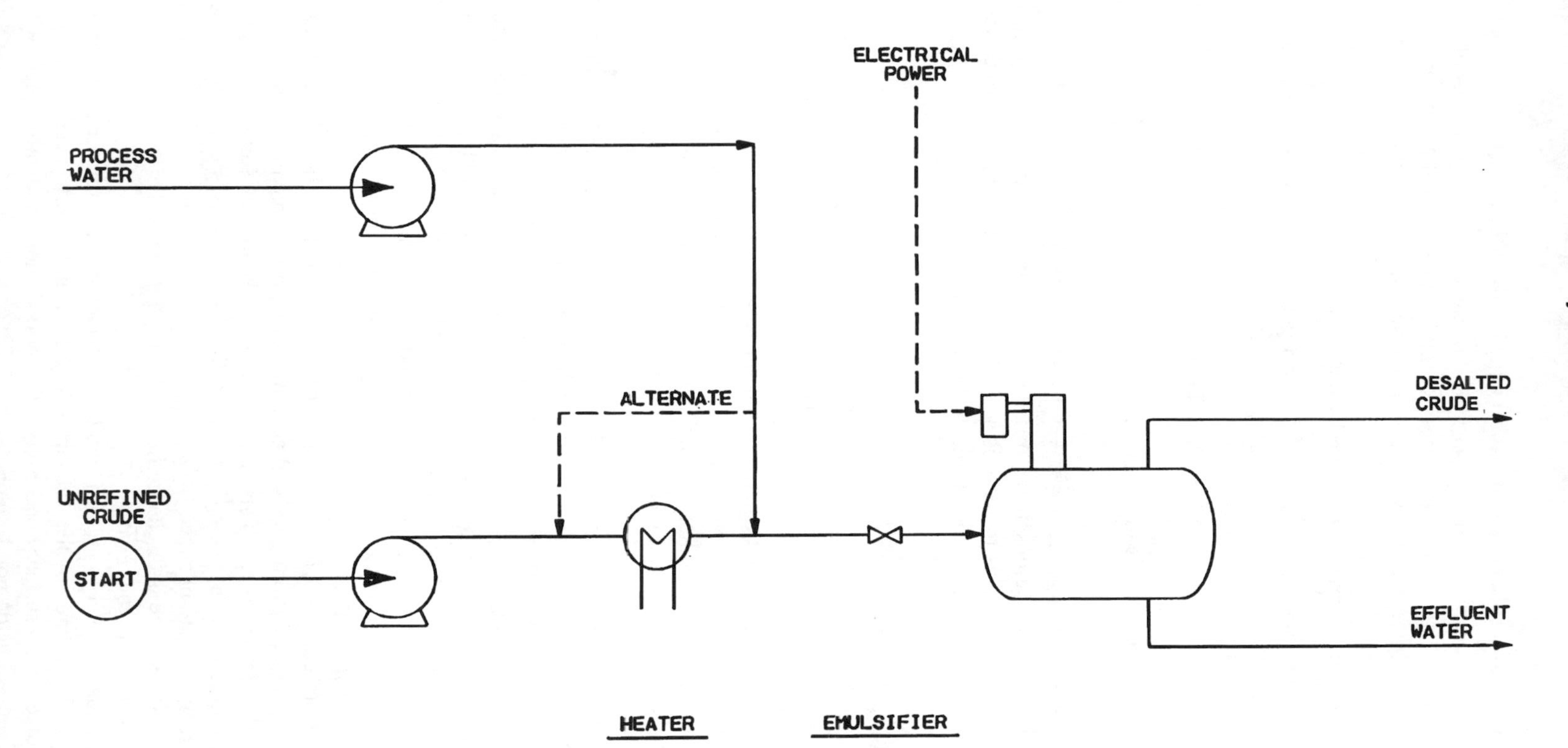

Figure 2 Crude desalting (electrostatic desalting). A high-voltage electrostatic field acts to agglomerate dispersed oil droplets for water–oil separation after water wash desalting. (From Ref. 5.)

Prefractionation and Atmospheric Distillation (Topping or Skimming)

Prefractionation is an optional distillation process to separate economic quantities of very light distillates from the crude oil. Lower temperatures and higher pressures are used than in atmospheric distillation. Some process water can be carried over to the prefractionation tower from the desalting process.

Atmospheric distillation breaks the heated crude oil as follows:

1. Light overhead (gaseous) products (C_5 and lighter) are separated, as in the case of prefractionation.
2. Sidestream distillate cuts of kerosene, heating oil, and gas oil can be separated in a single tower or in a series of topping towers, each tower yielding a successively heavier product stream.
3. Residual or reduced crude oil remains for further refining.

Vacuum Fractionation

The asphaltic residuum from atmospheric distillation amounts to roughly one-third (U.S. average) of the crude charged. This material is sent to vacuum stills, which recover additional heavy gas oil and deasphalting feedstock from the bottoms residue.

Three-Stage Crude Distillation

Three-stage crude distillation, representing only one of many possible combinations of equipment, is shown schematically in Fig. 3. The process consists of (1) an atmospheric fractionating stage, which produces lighter oils; (2) an initial vacuum stage, which produces well-fractionated, lube oil base stocks plus residue for subsequent propane deasphalting; and (3) a second vacuum stage, which fractionates surplus atmospheric bottoms not applicable for lube production, plus surplus initial vacuum stage residuum not required for deasphalting. This stage adds the capability of removing catalytic cracking stock from surplus bottoms to the distillation unit.

Crude oil is first heated in a simple heat exchanger, then in a direct-fired crude charge heater. Combined liquid and vapor effluent flow from the heater to the atmospheric fractionating tower, where the vaporized distillate is fractionated into gasoline overhead product and as many as four liquid sidestream products: naphtha, kerosene, and light and heavy diesel oil. Part of the reduced crude from the bottom of the atmospheric tower is pumped through a direct-fired heater to the vacuum lube fractionator. Bottoms are combined and charged to a third direct-fired heater. In the tower, the distillate is subsequently condensed and withdrawn as two sidestreams. The two sidestreams are combined to form catalytic cracking feedstocks, and an asphalt base stock is withdrawn from the tower bottom.

Wastewater from crude oil fractionation generally comes from three sources. The first source is the water drawn off from overhead accumulators prior to recirculation or transfer of hydrocarbons to other fractionators. This waste is a major source of sulfides and ammonia, especially when sour crudes are being processed. It also contains significant amounts of oil, chlorides, mercaptans, and phenols.

The second waste source is discharge from oil sampling lines. This should be separable, but it may form emulsions in the sewer.

A third waste source is very stable oil emulsions formed in the barometric condensers used to create the reduced pressures in the vacuum distillation units. However, when barometric condensers are replaced with surface condensers, oil vapors do not come into contact with water and consequently emulsions do not develop.

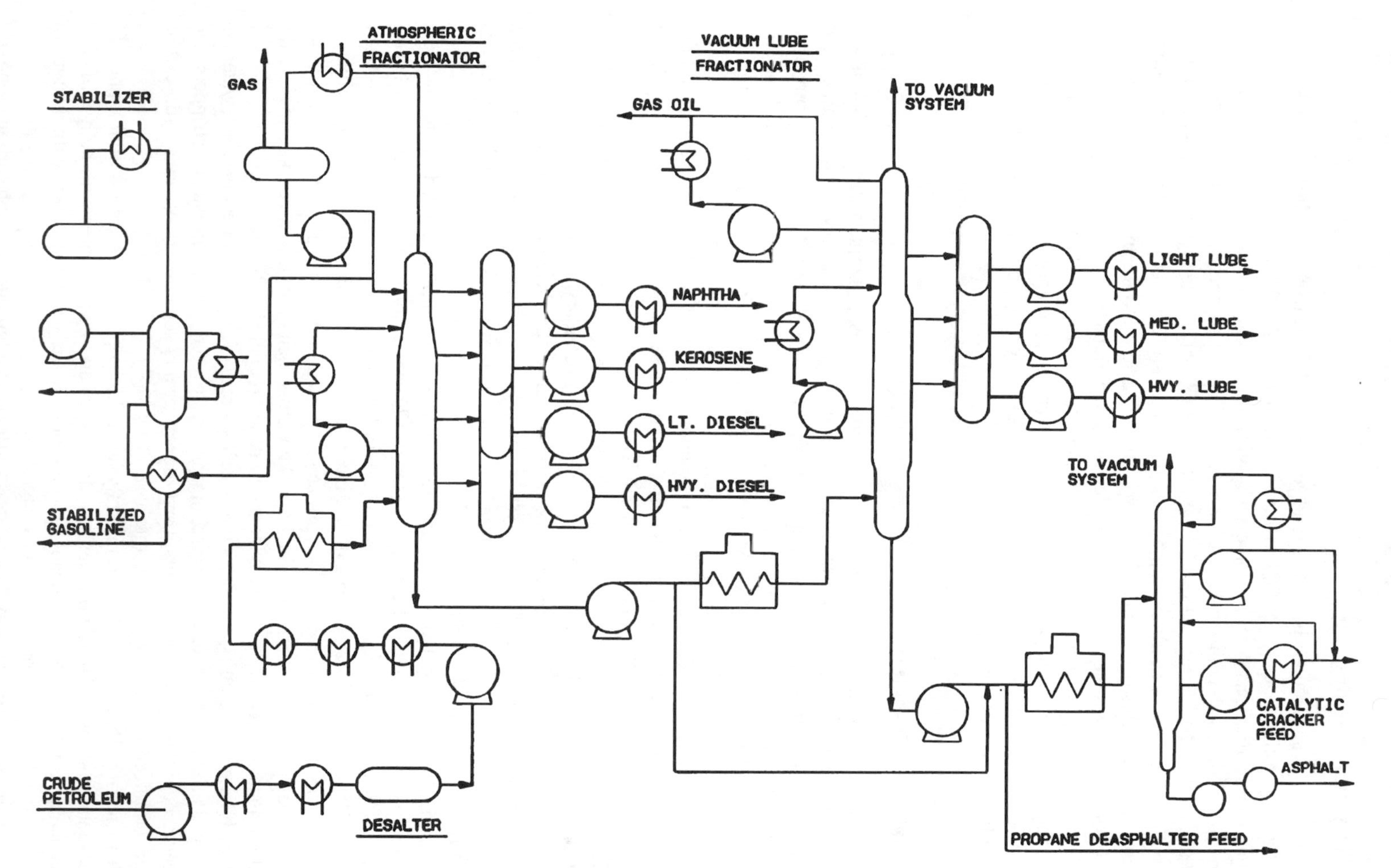

Figure 3 Crude fractionation (crude distillation, three stages). An atmospheric fractionating stage produces lighter oils. An initial vacuum stage produces lube oils. A second vacuum stage fractionates bottoms from the other stages to produce asphalt and catalytic cracker feed. (From Ref. 5.)

6.2.5 Thermal Cracking

Thermal cracking can include visbreaking and coking, in addition to regular thermal cracking. In each of these operations, heavy gas oil fractions (from vacuum stills) are broken down into lower molecular weight fractions such as domestic heating oils, catalytic cracking stock, and other fractions by heating, but without the use of catalyst. Typical thermal cracking conditions are 480–600°C (900–1100°F), and 41.6–69.1 atm (600–1000 psig). The high pressures result from the formation of light hydrocarbons in the cracking reaction (olefins, or unsaturated compounds, are always formed in this chemical conversion). There is also a certain amount of heavy fuel oil and coke formed by polymerization and condensation reactions.

The major source of wastewater in thermal cracking is the overhead accumulator on the fractionator, where water is separated from the hydrocarbon vapor and sent to the sewer system. This water usually contains various oils and fractions and may be high in BOD, COD, ammonia, phenol, sulfides, and alkalinity.

6.2.6 Catalytic Cracking

Catalytic cracking, like thermal cracking, breaks heavy fractions, principally gas oils, into lower molecular weight fractions. The use of catalyst permits operations at lower temperatures and pressures than with thermal cracking, and inhibits the formation of undesirable products. Catalytic cracking is probably the key process in the production of large volumes of high-octane gasoline stocks; furnace oils and other useful middle molecular weight distillates are also produced.

Fluidized catalytic processes, in which the finely powdered catalyst is handled as a fluid, have largely replaced the fixed-bed and moving-bed processes, which use a beaded or pelleted catalyst. A schematic flow diagram of fluid catalytic cracking (FCC) is shown in Fig. 4.

The FCC process involves at least four types of reactions: (1) thermal decomposition; (2) primary catalytic reactions at the catalyst surface; (3) secondary catalytic reactions between the primary products; and (4) removal of polymerization products from further reactions by adsorption onto the surface of the catalyst as coke. This last reaction is the key to catalytic cracking because it permits decomposition reactions to move closer to completion than is possible in simple thermal cracking.

Cracking catalysts include synthetic and natural silica-alumina, treated bentonite clay, fuller's earth, aluminum hydrosilicates, and bauxite. These catalysts are in the form of beads, pellets, and powder, and are used in a fixed, moving, or fluidized bed. The catalyst is usually heated and lifted into the reactor area by the incoming oil feed which, in turn, is immediately vaporized upon contact. Vapors from the reactors pass upward through a cyclone separator which removes most of the entrained catalyst. The vapors then enter the fractionator, where the desired products are removed and heavier fractions are recycled to the reactor.

Catalytic cracking units are one of the largest sources of sour and phenolic wastewaters in a refinery. Pollutants from catalytic cracking generally come from the steam strippers and overhead accumulators on fractionators, used to recover and separate the various hydrocarbon fractions produced in the catalytic reactors.

The major pollutants resulting from catalytic cracking operations are oil, sulfides, phenols, cyanides, and ammonia. These pollutants produce an alkaline wastewater with high BOD and COD concentrations. Sulfide and phenol concentrations in the wastewater vary with the type of crude oil being processed, but at times are significant.

Regeneration of spent catalyst in the steam stripper may produce enough carbon monoxide and fine catalyst particles to constitute an air pollution problem.

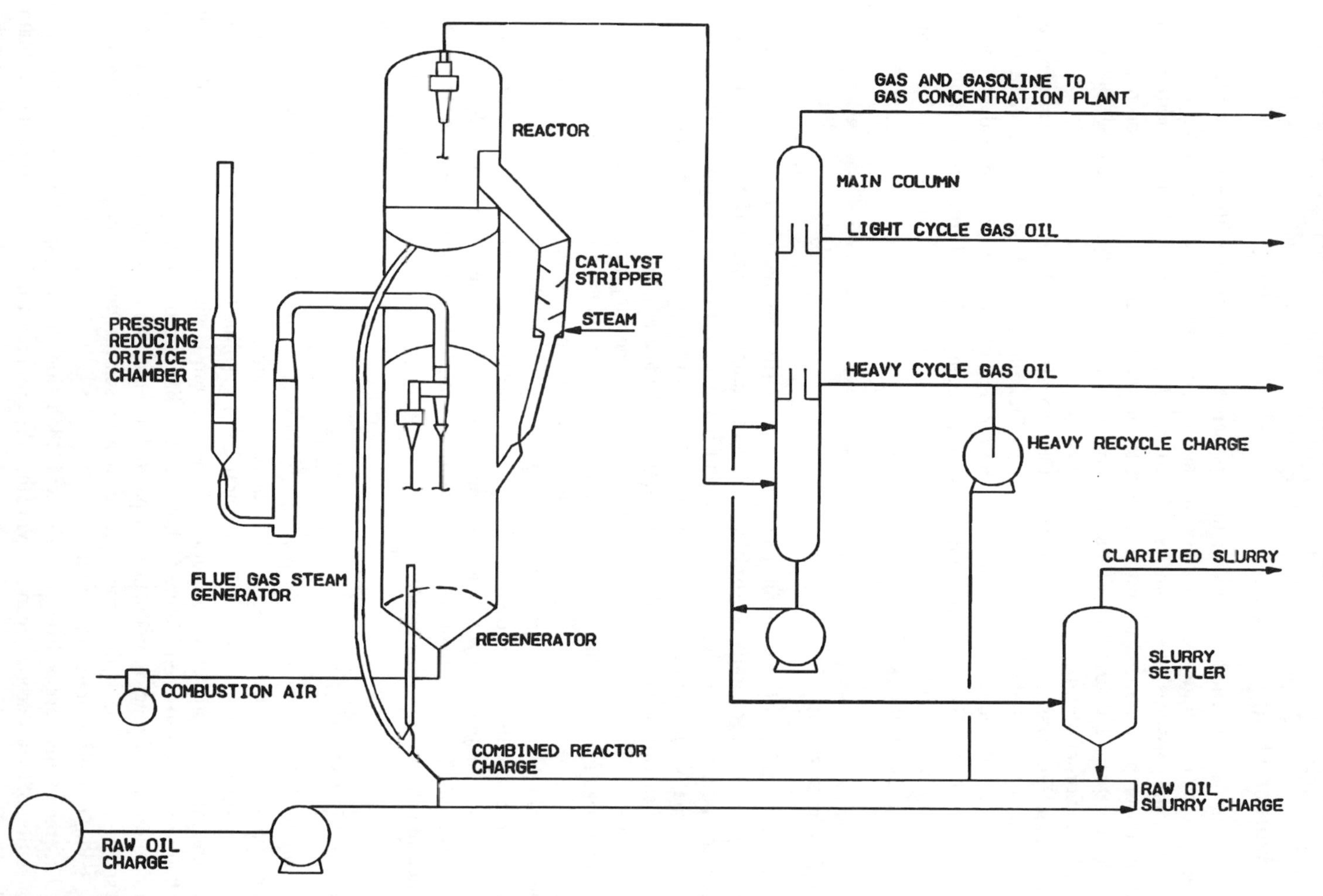

Figure 4 Catalytic cracking (fluid catalytic cracking). Heavy fraction gas oils are cracked (broken down) into lower molecular weight fractions in the presence of finely powdered catalyst, handled as a fluid. (From Ref. 5.)

6.2.7 Hydrocracking

This process is basically catalytic cracking in the presence of hydrogen, with lower temperatures and higher pressures than FCC. Hydrocracking temperatures range from 200 to 425°C (400–800°F) and pressures range from 7.8 to 137.0 atm (100–2000 psig). Actual conditions and hydrogen consumption depend upon the feedstock and the degree of hydrogenation required. The molecular weight distribution of the products is similar to catalyst cracking, but with reduced formation of olefins.

At least one wastewater stream from the process should be high in sulfides, as hydrocracking reduces the sulfur content of the material being cracked. Most of the sulfides are in the gas products that are sent to a treating unit for removal or recovery of sulfur and ammonia. However, some of the H_2S dissolves in the wastewater being collected from the separator and fractionator following the hydrocracking reactor. This water is probably high in sulfides and may contain significant quantities of phenols and ammonia.

6.2.8 Polymerization

Polymerization units convert olefin feedstocks (primarily propylene) into higher octane polymers. These units generally consist of a feed treatment unit (to remove H_2S, mercaptans, and nitrogen compounds), a catalytic reactor, an acid removal section, and a gas stabilizer. The catalyst is usually phosphoric acid, although sulfuric acid is used in some older methods. The catalytic reaction occurs at 150–224°C (300–435°F) and at a pressure of 11.2–137.0 atm (150–2000 psig). The temperature and pressure vary with the subprocess used.

Polymerization is a rather dirty process in terms of pounds of pollutants per barrel of charge, but because of the small polymerization capacity in most refineries, the total waste production from the process is small. Even though the process makes use of acid catalysts, the waste stream is alkaline because the acid catalyst in most subprocesses is recycled, and any remaining acid is removed by caustic washing. Most of the waste material comes from the pretreatment of feedstock, which removes sulfides, mercaptans, and ammonia from the feedstock in caustic and acid wastes.

6.2.9 Alkylation

Alkylation is the reaction of an isoparaffin (usually isobutane) and an olefin (propylene, butylene, amylenes) in the presence of a catalyst at carefully controlled temperatures and pressures to produce a high-octane alkylate for use as a gasoline blending component. Propane and butane are also produced. Sulfuric acid is the most widely used catalyst, although hydrofluoric acid is also used. Figure 5 shows a flow diagram of the alkylation process using sulfuric acid [6]. The reactor products are separated in a catalyst recovery unit, from which the catalyst is recycled. The hydrocarbon stream then passes through a caustic and water wash before going to the fractionation section.

The major discharges from sulfuric acid alkylation are the spent caustics from the neutralization of hydrocarbon streams leaving the alkylation reactor. These wastewaters contain dissolved and suspended solids, sulfides, oils, and other contaminants. Water drawn off from the overhead accumulators contains varying amounts of oil, sulfides, and other contaminants, but is not a major source of waste. Most refineries process the waste sulfuric acid stream from the reactor to recover clean acids, use it to neutralize other waste streams, or sell it.

Hydrofluoric acid (HF) alkylation units have small acid rerun units to purify the acid for reuse. HF units do not have a spent acid or spent caustic waste stream. Any leaks or spills that

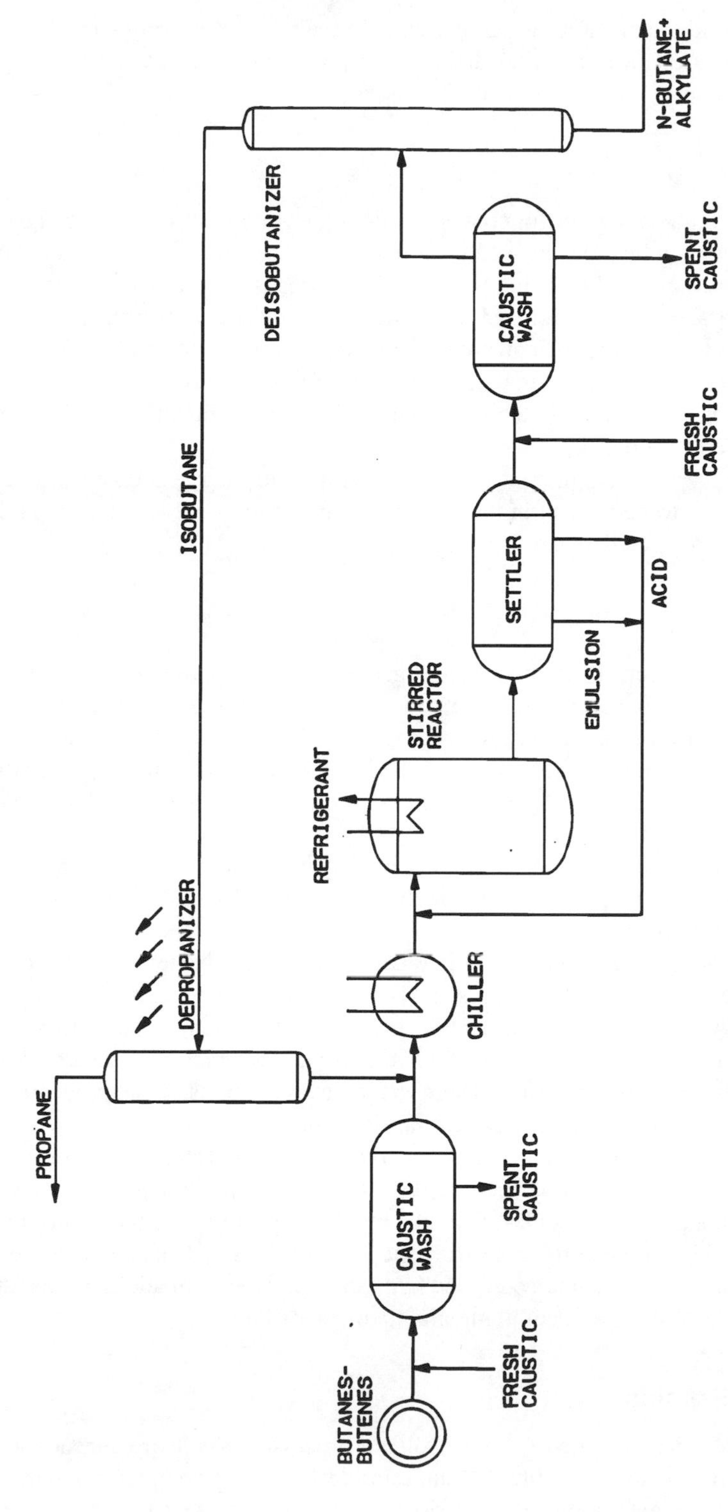

Figure 5 Alkylation process using sulfuric acid. Butanes and butenes react in the presence of a catalyst (sulfuric acid) to form an alkylate for use as a gasoline blending component. Propane and butane are also produced. (From Ref. 6.)

involve loss of fluorides constitute a serious and difficult pollution problem. Formation of fluorosilicates has caused line plugging and similar problems. The major sources of waste materials are the overhead accumulators on the fractionator.

6.2.10 Isomerization

Isomerization is a process technique for converting light gasoline stocks into their higher octane isomers. The greatest application has been, indirectly, in the conversion of isobutane from normal butane for use as feedstock for the alkylation process. In a typical subprocess, the desulfurized feedstock is first fractionated to separate isoparaffins from normal paraffins. The normal paraffins are then heated, compressed, and passed through the catalytic hydrogenation reactor, which isomerizes the *n*-paraffin to its respective high-octane isomer. After separation of hydrogen, the liquids are sent to a stabilizer, where motor fuel blending stock or synthetic isomers are removed as products.

Isomerization wastewaters present no major pollutant discharge problems. Sulfides and ammonia are not likely to be present in the effluent. Isomerization wastewaters should also be low in phenolics and oxygen demand.

6.2.11 Reforming

Reforming converts low-octane naphtha, heavy gasoline, and naphthene-rich stocks to high-octane gasoline blending stock, aromatics for petrochemical use, and isobutane. Hydrogen is a significant byproduct. Reforming is a mild decomposing process, as some reduction occurs in molecular size and boiling range of the feedstock. Feedstocks are usually hydrotreated to remove sulfur and nitrogen compounds prior to charging to the reformer, because the platinum catalysts widely used are readily poisoned.

The predominant reaction during reforming is dehydrogenation of naphthenes. Important secondary reactions are isomerization and dehydrocyclization of paraffins. All three reactions result in high-octane products.

One subprocess may be divided into three parts: the reactor heater section, in which the charge plus recycle gas is heated and passed over the catalyst in a series of reactions; the separator drum, in which the reactor effluent is separated into gas and liquid streams, the gas being compressed for recycle; and the stabilizer section, in which the separated liquid is stabilized to the desired vapor pressure. There are many variations in subprocesses, but the essential and frequently only difference is in catalyst involved.

Reforming is a relatively clean process. The volume of wastewater flow is small, and none of the wastewater streams has high concentrations of significant pollutants. The wastewater is alkaline, and the major pollutant is sulfide from the overhead accumulator on the stripping tower used to remove light hydrocarbon fractions from the reactor effluent. The overhead accumulator catches any water that may be contained in the hydrocarbon vapors. In addition to sulfides, the wastewater contains small amounts of ammonia, mercaptans, and oil.

6.2.12 Solvent Refining

Refineries employ a wide spectrum of contact solvent processes, which are dependent upon the differential solubilities of the desirable and undesirable feedstock components. The principal steps are countercurrent extraction, separation of solvent and product by heating and fractionation, and solvent recovery. Naphthenics, aromatics, unsaturated hydrocarbons, and sulfur and other inorganics are separated, with the solvent extract yielding high-purity products. Many of the solvent processes may produce process wastewaters that contain small amounts of the solvents employed. However, these are usually minimized because of the economic incentives for reuse of the solvents.

The major solvent refining processes include solvent deasphalting, solvent dewaxing, lube oil solvent refining, aromatic extraction, and butadiene extraction. These processes are briefly described below.

Solvent deasphalting is carried out primarily to recover lube or catalytic cracking feedstocks from asphaltic residuals, with asphalt as a byproduct. Propane deasphalting is the predominant technique. The vacuum fractionation residual is mixed in a fixed proportion with a solvent in which asphalt is not soluble. The solvent is recovered from the oil via steam stripping and fractionation, and is reused. The asphalt produced by this method is normally blended into fuel oil or other asphaltic residuals.

Solvent dewaxing removes wax from lubricating oil stocks, promoting crystallization of the wax. Solvents include furfural, phenol, cresylic acid-propane (DuoSol), liquid sulfur dioxide (Eleleanu process), B,B-dichloroethyl ether, methyl ethyl ketone, nitrobenzene, and sulfurbenzene. The process yields de-oiled waxes, wax-free lubricating oils, aromatics, and recovered solvents.

Lube oil solvent refining includes a collection of subprocesses improving the quality of lubricating oil stock. The raffinate or refined lube oils obtain improved viscosity, color, oxidation resistance, and temperature characteristics. A particular solvent is selected to obtain the desired quality raffinate. The solvents include furfural, phenol, sulfur dioxide, and propane.

Aromatic extraction removes benzene, toluene, and xylene (BTX) that are formed as byproducts in the reforming process. The reformed products are fractionated to give a BTX concentrate cut, which, in turn, is extracted from the napthalene and the paraffinics with a glycol base solvent.

Butadiene extraction accounts for some 15% of the U.S. supply of butadiene, which is extracted from the C4 cuts from the high-temperature petroleum cracking processes. Furfural or cuprous ammonia acetate is commonly used for the solvent extraction.

The major potential pollutants from the various solvent refining subprocesses are the solvents themselves. Many of the solvents, such as phenol, glycol, and amines, can produce a high BOD. Under ideal conditions the solvents are continually recirculated with no losses to the sewer. Unfortunately, some solvent is always lost through pump seals, flange leaks, and other sources. The main source of wastewater is from the bottom of fractionation towers. Oil and solvent are the major wastewater constituents.

6.2.13 Hydrotreating

Hydrotreating processes are used to saturate olefins, and to remove sulfur and nitrogen compounds, odor, color, gum-forming materials, and others by catalytic action in the presence of hydrogen, from either straight-run or cracked petroleum fractions. In most subprocesses, the feedstock is mixed with hydrogen, heated, and charged to the catalytic reactor. The reactor products are cooled, and the hydrogen, impurities, and high-grade product separated. The principal difference between the many subprocesses is the catalyst; the process flow is similar for essentially all subprocesses. Figure 6 shows a flow diagram of the hydrotreating process [2].

Hydrotreating reduces the sulfur content of product streams from sour crudes by 90% or more. Nitrogen removal requires more severe operating conditions, but generally 80% reductions or better are accomplished.

The primary variables influencing hydrotreating are hydrogen partial pressure, process temperature, and contact time. Higher hydrogen pressure gives a better removal of undesirable

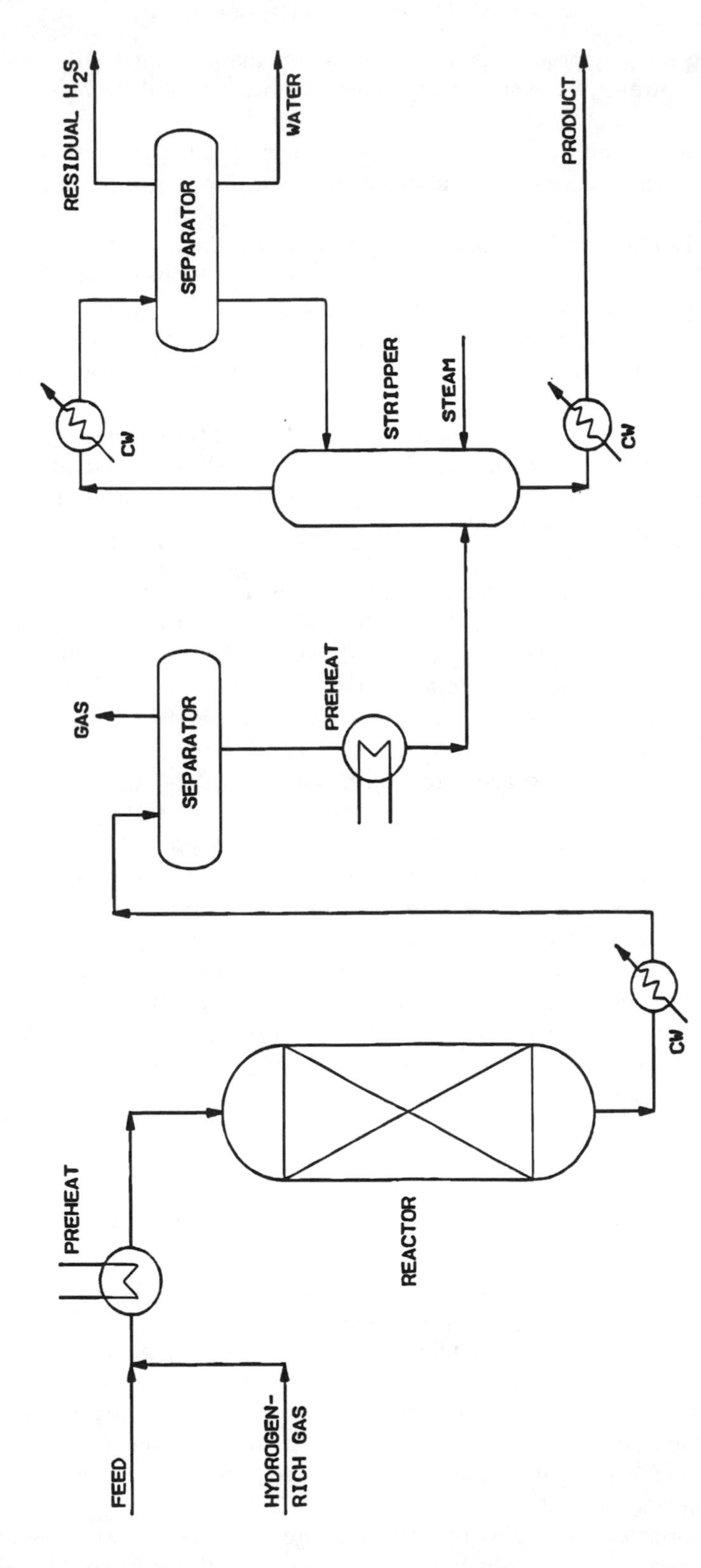

Figure 6 Hydrotreating process. Hydrogen reacts with hydrocarbon feed to remove sulfur from the stream. The formed hydrogen sulfide is steam-stripped from the product. (From Ref. 2.)

materials and a better rate of hydrogenation. Make-up hydrogen requirements are generally great enough to require a hydrogen production unit. Excessive temperatures increase the formation of coke, and the contact time is set to give adequate treatment without excessive hydrogen usage or undue coke formation. For the various hydrotreating processes, the pressures range from 7.8 to 205 atm (100 to 3000 psig). Temperatures range from less than 177°C (350°F) to as high as 450°C (850°F); most processing is carried out in the range 315–427°C (600–800°F). Hydrogen consumption is usually less than 5.67 cubic meters (200 scf) per barrel of charge.

The principal hydrotreating subprocesses used are as follows:

- pretreatment of catalytic reformer feedstock;
- naphtha desulfurization;
- lube oil polishing;
- pretreatment of catalytic cracking feedstock;
- heavy gas-oil and residual desulfurization;
- naphtha saturation.

The strength and quantity of wastewaters generated by hydrotreating depends upon the subprocess used and feedstock. Ammonia and sulfides are the primary contaminants, but phenols may also be present if the feedstock boiling range is sufficiently high.

6.2.14 Grease Manufacturing

Grease manufacturing processes require accurate measurements of feed components, intimate mixing, and rapid heating and cooling, together with milling, dehydration, and polishing in batch reactions. The feed components include soap and petroleum oils with inorganic clays and other additives.

Grease is primarily a soap and lube oil mixture. The properties of grease are determined in large part by the properties of the soap component. For example, sodium soap grease is water soluble and not suitable for water contact service. A calcium soap grease, on the other hand, can be used in water service. The soap may be purchased as a raw material or may be manufactured on site as an auxiliary process.

Only small volumes of wastewater are discharged from a grease manufacturing process. A small amount of oil is lost to the wastewater system through leaks in pumps. The largest waste loading occurs when the batch units are washed, resulting in soap and oil discharges to the sewer system.

6.2.15 Asphalt Production

Asphalt feedstock (flux) is contacted with hot air at 200–280°C (400–550°F) to obtain desirable asphalt product. Both batch and continuous processes are in operation at present, but the batch process is more prevalent because of its versatility. Nonrecoverable catalytic compounds include copper sulfate, zinc chloride, ferric chloride, aluminum chloride, phosphorus pentoxide, and others. The catalyst does not normally contaminate the process water effluent.

Wastewaters from asphalt blowing contain high concentrations of oil and have high oxygen demand. Small quantities of phenols may also be present.

6.2.16 Drying and Sweetening

Drying and sweetening is a broad class of processes used to remove sulfur compounds, water, and other impurities from gasoline, kerosene, jet fuels, domestic heating oils, and other middle distillate products. Sweetening is the removal of hydrogen sulfide, mercaptans, and thiophenes, which impart a foul odor and decrease the tetraethyl lead susceptibility of gasoline. The major sweetening operations are oxidation of mercaptans or disulfides, removal of mercaptans, and destruction and removal of all sulfur compounds. Drying is accomplished by salt filters or absorptive clay beds. Electric fields are sometimes used to facilitate separation of the product.

The most common waste stream from drying and sweetening operations is spent caustic. The spent caustic is characterized as phenolic or sulfidic, depending on which is present in the largest concentration; this in turn is mainly determined by the product stream being treated. Phenolic spent caustics contain phenol, cresols, xylenols, sulfur compounds, and neutral oils. Sulfidic spent caustics are rich in sulfides, but do not contain any phenols. These spent caustics have very high BOD and COD. The phenolic caustic streams are usually sold for the recovery of phenolic materials.

Other waste streams from the process result from water washing of the treated product and regeneration of the treating solution such as sodium plumbite (Na_2PbO_2) in doctor sweetening. These waste streams contain small amounts of oil and the treating material, such as sodium plumbite (or copper from copper chloride sweetening).

The treating of sour gases produces a purified gas stream and an acid gas stream rich in hydrogen sulfide. The H_2S-rich stream can be flared, burned as fuel, or processed for recovery of elemental sulfur.

6.2.17 Lube Oil Finishing

Solvent-refined and dewaxed lube oil stocks can be further refined by clay or acid treatment to remove color-forming and other undesirable materials. Continuous contact filtration, in which an oil–clay slurry is heated and the oil removed by vacuum filtration, is the most widely used subprocess.

Acid treatment of lubricating oils produces acid-bearing wastes occurring as rinse waters, sludges, and discharges from sampling, leaks, and shutdowns. The waste streams are also high in dissolved and suspended solids, sulfates, sulfonates, and stable oil emulsions.

Handling of acid sludge can create additional problems. Some refineries burn the acid sludge as fuel, which produces large volumes of sulfur dioxide that can cause air pollution problems. Other refineries neutralize the sludge with alkaline wastes and discharge it to the sewer, resulting in both organic and inorganic pollution. The best method of disposal is probably processing to recover the sulfuric acid, but this also produces a wastewater stream containing acid, sulfur compounds, and emulsified oil.

Clay treatment results in only small quantities of wastewater being discharged to the sewer. Clay, free oil, and emulsified oil are the major waste constituents. However, the operation of clay recovery kilns involves potential air pollution problems of hydrocarbon and particulate emissions. Spent clays are usually disposed of by landfill.

6.2.18 Blending and Packaging

Blending is the final step in producing finished petroleum products to meet quality specifications and market demands. The largest volume operation is the blending of various gasoline stocks (including alkylates and other high-octane components) and antiknock (tetraethyl lead), antirust, anti-icing, and other additives. Diesel fuels, lube oils, and waxes involve blending of various components and additives. Packaging at refineries is generally highly automated and restricted to high-volume consumer-oriented products such as motor oils.

These are relatively clean processes because care is taken to avoid loss of product through spillage. The primary source of waste material is from the washing of railroad tank cars or tankers prior to loading finished products. These wash waters are high in emulsified oil.

Tetraethyl lead was the major additive blended into gasolines in the past, and it must be carefully handled because of its high toxicity if it is still used. Sludges from finished gasoline storage tanks can contain large amounts of lead if tetraethyl lead is still used and should not be washed into the wastewater system.

6.2.19 Hydrogen Manufacture

The rapid growth of hydrotreating and hydrocracking has increased the demand for hydrogen beyond the level of byproduct hydrogen available from reforming and other refinery processes. The most widely used process for the manufacture of hydrogen in the refinery is steam reforming, which utilizes refinery gases as a charge stock. The charge is purified to remove sulfur compounds that would temporarily deactivate the catalysts.

The desulfurized feedstock is mixed with superheated steam and charged to the hydrogen furnace. On the catalyst, the hydrocarbons are converted to hydrogen, carbon monoxide, and carbon dioxide. The furnace supplies the heat needed to maintain the reaction temperature.

The gases from the furnace are cooled by the addition of condensate and steam, and then passed through a converter containing a high or low temperature shift catalyst, depending on the degree of carbon monoxide conversion desired. Carbon dioxide and hydrogen are produced by the reaction of the carbon monoxide with steam.

The gas mixture from the converter is cooled and passed to a hydrogen purifying system where carbon dioxide is absorbed into amine solutions and later driven off to the atmosphere by heating the rich amine solution in the reactivator.

Because some refining processes require a minimum of carbon dioxide in the product gas, the oxides are reacted with hydrogen in a methanation step. This reaction takes place in the methanator over a nickel catalyst at elevated temperatures.

Hydrocarbon impurities in the product hydrogen usually are not detrimental to the processes where this hydrogen will be used. Thus, a small amount of hydrocarbon is tolerable in the effluent gas.

The hydrogen manufacture process is relatively clean. In the steam reforming subprocess a potential waste source is the desulfurization unit, which is required for feedstock that has not already been desulfurized. This waste stream contains oil, sulfur compounds, and phenol. In the partial oxidation subprocess, free carbon is removed by a water wash. Carbon dioxide is discharged to the atmosphere at several points in the subprocess.

6.2.20 Utility Functions

Utility functions such as the supply of steam and cooling water generally are set up to service several processes. Boiler feed water is prepared and steam is generated in a single boiler house. Noncontact steam used for surface heating is circulated through a closed loop, whereby various quantities are made available for the specific requirements of the different processes. The condensate is nearly always recycled to the boiler house, where a certain portion is discharged as blowdown.

Steam is also used as a diluent, stripping medium, or source of vacuum through the use of steam jet ejectors. This steam actually contacts the hydrocarbons in the manufacturing processes and is a source of contact process wastewater when condensed.

Noncontact cooling water is normally supplied to several processes from the utilities area. The system is either a loop that utilizes one or more evaporative cooling towers, or a once-through system with direct discharge.

Cooling towers work by moving a predetermined flow of ambient air through the tower with large fans. A small amount of the water is evaporated by the air; thus, through latent heat transfer, the remainder of the circulated water is cooled.

Wastewater streams from the utility functions include boiler and cooling tower blowdowns and waste brine and sludge produced by demineralizing and other water treatment systems. The quantity and quality of the wastewater streams depend on the design of the systems and the water source. These streams usually contain high dissolved and suspended solids concentrations and treatment chemicals from the boiler and cooling tower. The blowdown streams also have elevated temperatures.

6.3 WASTE SOURCES AND CHARACTERISTICS

Wastes generated from oil fields include produced water, drilling muds and cuttings, and tank bottom sludges. These wastes are associated with the drilling, recovery, and storage of crude oil. Wastes from petroleum refineries generally include process wastewater, wastewater from utility operations, contaminated storm water, sanitary waste, and miscellaneous contaminated streams. These waste streams are usually discharged to a central wastewater treatment system; some of these streams, such as sour water, are pretreated first.

6.3.1 Oil Field Wastes

The most important wastes from oil fields are produced water and drilling muds. The characteristics of these waste streams are discussed below.

Produced water is the water brought to the surface with the oil from a production well. It is estimated that for every barrel of oil produced, on average 2–3 barrels of water are produced, ranging from a negligible amount up to values over 100 barrels of water per barrel of oil [7]. Once on the surface, the water and oil are separated. The oil is prepared for distribution, leaving the water to be disposed of by some means.

Produced water is typically saline. A great deal of data exist regarding the quality of the inorganic components of the produced water [8]. Table 2 is a summary of this information. To date very little information has been published regarding the concentration of the traditional pollutant parameters in the produced water. Table 3 presents the ranges of various water quality

Table 2 Inorganic Components in Oilfield Produced Water

Constituent	Concentration (mg/L)
Sodium	12,000–150,000
Potassium	30–4,000
Calcium	1,000–120,000
Magnesium	500–25,000
Chloride	20,000–250,000
Bromide	50–5,000
Iodide	1–300
Bicarbonate	0–3,600

Source: From Ref. 9.

Table 3 Produced Water Quality

Parameter	Range (mg/L)
Biochemical oxygen demand	50–1,400
Chemical oxygen demand	450–5,900
Phenols	0.7–7.6
Oil and grease	15–290
Ammonia nitrogen	4–206
Total suspended solids	35–300
Sulfides	0.2–800
pH	6.7–9.0

Source: From Ref. 14.

parameters measured in produced water from over 30 individual wells in several California oilfields [9]. Work done by Chevron showed that typical produced waters from the U.S. west coast and the Gulf of Mexico, after oil removal, had compositions ranging from 20,000 to 135,000 mg/L total dissolved solids, 45 to 130 mg/L ammonia (as N), and 0.1 to 3.0 mg/L phenols [10].

Drilling muds are fluids that are pumped into the bore holes to aid in the drilling process. Most are water based and contain barite, lignite, chrome lignosulfate, and sodium hydroxide [11], but oil-based drilling muds are still used for economic and safety reasons [12]. Used muds can be removed by vacuum trucks, pumped down the well annulus, or allowed to dewater in pits, which are then covered with soil or disposed of by land farming.

The main components of pollution concern in drilling muds include (1) oil itself, especially in oil fluids, (2) salts, and (3) soluble trace elements consisting of zinc, lead, copper, cadmium, nickel, mercury, arsenic, barium, and chromium associated with low grades of barite [13]. Owing to its variability, very little information has been published regarding the concentration of pollutants in spent drilling mud. Copa and Dietrich [14] conducted a wet air oxidation experiment on a sample of spent drilling mud taken from a storage lagoon. The material was a concentrated mud, having a suspended solids concentration of approximately 500 g/L. The original drilling mud contained emulsifying agents and oils, which inhibited dewaterability. The characteristics of the diluted (4 : 1) spent drilling mud are shown in Table 4.

Table 4 Characteristics of Spent Drilling Mud

Analyses	Concentration (diluted 4 : 1)
COD, soluble (mg/L)	5,720
BOD, soluble (mg/L)	2,625
TOC, soluble (mg/L)	2,010
Total solids (g/L)	113.4
Ash (g/L)	107.5
Suspended solids (g/L)	103.7
Suspended ash (g/L)	100.9
Specific filtration resistance ($cm^2/g \times 10^{-7}$)	155

Source: From Ref. 14.

6.3.2 Refinery Wastewater

The sources of wastewater generation in petroleum refineries have been discussed previously in this chapter. Table 5 presents a qualitative evaluation of wastewater flow and characteristics by fundamental refinery processes [5]. The trend of the industry has been to reduce wastewater production by improving the management of the wastewater systems. Table 6 shows wastewater loadings and volumes per unit fundamental process throughput in older, typical, and newer technologies [15]. Table 7 shows typical wastewater characteristics associated with several refinery processes [16].

In addition to those from the fundamental processes, wastewaters are also generated from other auxiliary operations in refineries. Figure 7 shows the various sources of wastewater and their primary pollutants in a refinery [17].

In the USEPA study to develop effluent limitation guidelines [7], refinery operations were grouped together to produce five subcategories based on raw waste load, product mix, refinery processes, and wastewater generation characteristics. These subcategories are described below.

1. *Topping* Includes topping, catalytic reforming, asphalt production, or lube oil manufacturing processes, but excludes any facility with cracking or thermal operations.
2. *Cracking* Includes topping and cracking.
3. *Petrochemical* Includes topping, cracking, and petrochemical operations.
4. *Lube* Includes topping, cracking, and lube oil manufacturing processes.
5. *Integrated* Includes topping, cracking, lube oil manufacturing processes, and petrochemical operations.

The term petrochemical operations means the production of second-generation petrochemicals (alcohols, ketones, cumene, styrene, and so on) or first-generation petrochemicals and isomerization products (BTX, olefins, cyclohexane, and so on) when 15% or more of refinery production is as first-generation petrochemicals and isomerization products.

All five subcategories of refineries generate wastewaters containing similar constituents. However, the concentrations and loading of the constituents (raw waste load) vary among the categories. The raw waste loads, and their variabilities, for the five petroleum refining subcategories are presented in Tables 8 to 12 [7].

In addition to the conventional pollutant constituents, USEPA made a survey of the presence of the 126 toxic pollutants listed as "priority pollutants" in refinery operations in 1977 [5]. The survey responses indicated that 71 toxic pollutants were purchased as raw or intermediate materials; 19 of these were purchased by single refineries. At least 10% of all refineries purchase the following toxic pollutants: benzene, carbon tetrachloride, l,l,l-trichloroethane, phenol, toluene, zinc and its compounds, chromium and its compounds, copper and its compounds, and lead and its compounds. Zinc and chromium are purchased by 28% of all refineries, and lead is purchased by nearly 48% of all plants.

Forty-five priority pollutants are manufactured as final or intermediate materials; 15 of these are manufactured at single refineries. Benzene, ethylbenzene, phenol, and toluene are manufactured by at least 10% of all refineries. Of all refineries, 8% manufacture cyanides, while more than 20% manufacture benzene and toluene. Hence, priority pollutants are expected to be present in refinery wastewaters. The EPA's short-term and long-term sampling programs conducted later detected and quantified 22 to 28 priority pollutants in refinery effluent samples [5].

Table 5 Qualitative Evaluation of Wastewater Flow and Characteristics by Fundamental Refinery Processes

Production processes	Flow	BOD	COD	Phenol	Sulfide	Oil	Emulsified oil	pH	Temperature	Ammonia	Chloride	Acidity	Alkalinity	Suspended solids
Crude oil and product storage	XX	X	XXX	X		XXX	XX	O	O	O		O		XX
Crude desalting	XX	XX	XX	X	XXX	X	XXX	X	XXX	XX	XXX	O	X	XXX
Crude distillation	XXX	X	X	XX	XXX	XX	XXX	X	XX	XXX	X	O	X	X
Thermal cracking	X	X	X	X	X	X		XX	XX	X	X	O	XX	X
Catalytic cracking	XXX	XX	XX	XXX	XXX	X	X	XXX	XX	XXX	X	O	XXX	X
Hydrocracking	X			XX	XX				XX	XX				
Polymerization	X	X	X	O	X	X	O	X	X	X	X	X	O	X
Alkylation	XX	X	X	O	XX	X	O	XX	X	X	XX	XX	O	XX
Isomerization	X													
Reforming	X	O	O	X	X	X	O	O	X	X	O	O	O	O
Solvent refining	X		X	X	O		X	X	O			O	X	
Asphalt blowing	XXX	XXX	XXX	X		XXX								
Dewaxing	X	XXX	XXX	X	O	X	O							
Hydrotreating	X	X	X		XX	O	O	XX		XX	O	O	X	O
Drying and sweetening	XXX	XXX	X	XX	O	O	X	XX	O	X	O	X	X	XX

XXX = major contribution; XX = moderate contribution; X = minor contribution; O = insignificant; Blank = no data.
BOD, biochemical oxygen demand; COD, chemical oxygen demand.
Source: From Ref. 5.

Table 6 Waste Loadings and Volumes Per Unit of Fundamental Process Throughput in Older, Typical, and Newer Technologies

	Older technology				Typical technology				Newer technology			
Fundamental process	Flow (gal/bbl)	BOD (lb/bbl)	Phenol (lb/bbl)	Sulfides (lb/bbl)	Flow (gal/bbl)	BOD (lb/bbl)	Phenol (lb/bbl)	Sulfides (lb/bbl)	Flow (gal/bbl)	BOD (lb/bbl)	Phenol (lb/bbl)	Sulfides (lb/bbl)
Crude oil and product storage	4	0.001	—	—	4	0.001	—	—	4	0.001	—	—
Crude desalting	2	0.002	0.20	0.002	2	0.002	0.10	0.002	2	0.002	0.05	0.002
Crude fractionation	100	0.020	3.0	0.001	50	0.0002	1.0	0.001	10	0.0002	1.0	0.001
Thermal cracking	66	0.001	7.0	0.002	2	0.001	0.2	0.001	1.5	0.001	0.2	0.001
Catalytic cracking	85	0.062	50.0	0.03	30	0.010	20	0.003	5	0.010	5	0.003
Hydrocracking		Not in this technology				Not in this technology			5	—	—	—
Reforming	9	T	0.7	T	6	T	0.7	0.001	6	T	0.7	0.001
Polymerization	300	0.003	1.4	0.22	140	0.003	0.4	0.010		Not in this technology		
Alkylation	173	0.001	0.1	0.005	60	0.001	0.1	0.010	20	0.001	0.1	0.020
Isomerization		Not in this technology				Not in this technology			—	—	—	—

Solvent refining	8	—	3	T	8	—	3	T	8	—	3	T
Dewaxing	24	0.52	2	T	23	0.50	1.5	T	20	0.25	1.5	T
Hydrotreating	1	0.002	0.6	0.007	1	0.002	0.01	0.002	8	0.002	0.01	0.002
Deasphalting	—	—	—	—	—	—	—	—	—	—	—	—
Drying and sweetening	100	0.10	10	—	40	0.05	10	—	40	0.05	10	—
Wax finishing	—	—	—	—	—	—	—	—	—	—	—	—
Grease manufacturing	—	—	—	—	—	—	—	—	—	—	—	—
Lube oil finishing	—	—	—	—	—	—	—	—	—	—	—	—
Hydrogen manufacture	Not in this technology				Not in this technology				—	—	—	—
Blending and packaging	—	—	—	—	—	—	—	—	—	—	—	—

T = trace; — = data not available for reasonable estimate; BOD, biochemical oxygen demand.
gal/bbl = gallons of wastewater per barrel of oil processed.
lb/bbl = pounds of contaminant per barrel of oil processed.
Source: From Ref. 15.

Table 7 Typical Waste Characteristics

	Spent caustic stream			
Characteristic	Benzene sulfonation scrubbing	Orthophenylphenol washing	Alkylate washing	Polymerization
Alkalinity (mg/L)	33,800	18,400	46,250	209,330
BOD (mg/L)	53,600	18,400	256	8,440
COD (mg/L)	112,000	67,600	3,230	50,350
pH	13.2	9–12	12.8	12.7
Phenols (mg/L)	8.3	5,500	50	22.2
NaOH (wt %)	1	0.2–0.5		
Na_2SO_4 (wt %)	1.5–2.5			
Sulfates (mg/L)	3,760	2,440		
Sulfides (mg/L)			2	3,060
Sulfites (mg/L)	7,100	4,720		
Total solids (mg/L)	90,300	40,800		
	Process waste			
Characteristic	Crude Desalting	Catalytic cracking	Naphtha cracking	Sour condensates from distillation cracking, etc.
Ammonia (mg/L)	80			135–6,550
BOD (mg/L)	60–610	230–440		500–1,000
COD (mg/L)	124–470	500–2,800	53–180	500–2,000
Oil (mg/L)	20–516	200–2,600	160	100–1,000
pH (mg/L)	7.2–9.1			4.5–9.5
Phenols (mg/L)	10–25	20–26	6–10	100–1,000
Salt (as NaCl) (wt %)	0.4–25			
Sulfides (mg/L)	0–13			390–8,250 (H_2S)
	Acid waste			
Characteristic	Acid wash: alkylation	Acid wash: phenol still bottoms	Acid wash: orthophenylphenol	Sulfite wash: liquid OP-phenol distillation
Acidity (mg/L)	1,105–12,325		24,120	675
BOD (mg/L)	31	20,800	13,600	105,000
COD (mg/L)	1,251	248,000	23,400	689,000
Dissolved solids (mg/L)		340,500	81,300	176,800
Oil (mg/L)	131.5			
pH	0.6–1.9	1.0	1.1	3.8
Phenols (mg/L)		3,800	1,500	16,400
Sulfate (mg/L)			54,700	
Sulfite (mg/L)		34,800	2,920	74,000
Total solids (mg/L)		403,200	81,600	176,900

BOD, biochemical oxygen demand; COD, chemical oxygen demand.
Source: From Ref. 16,

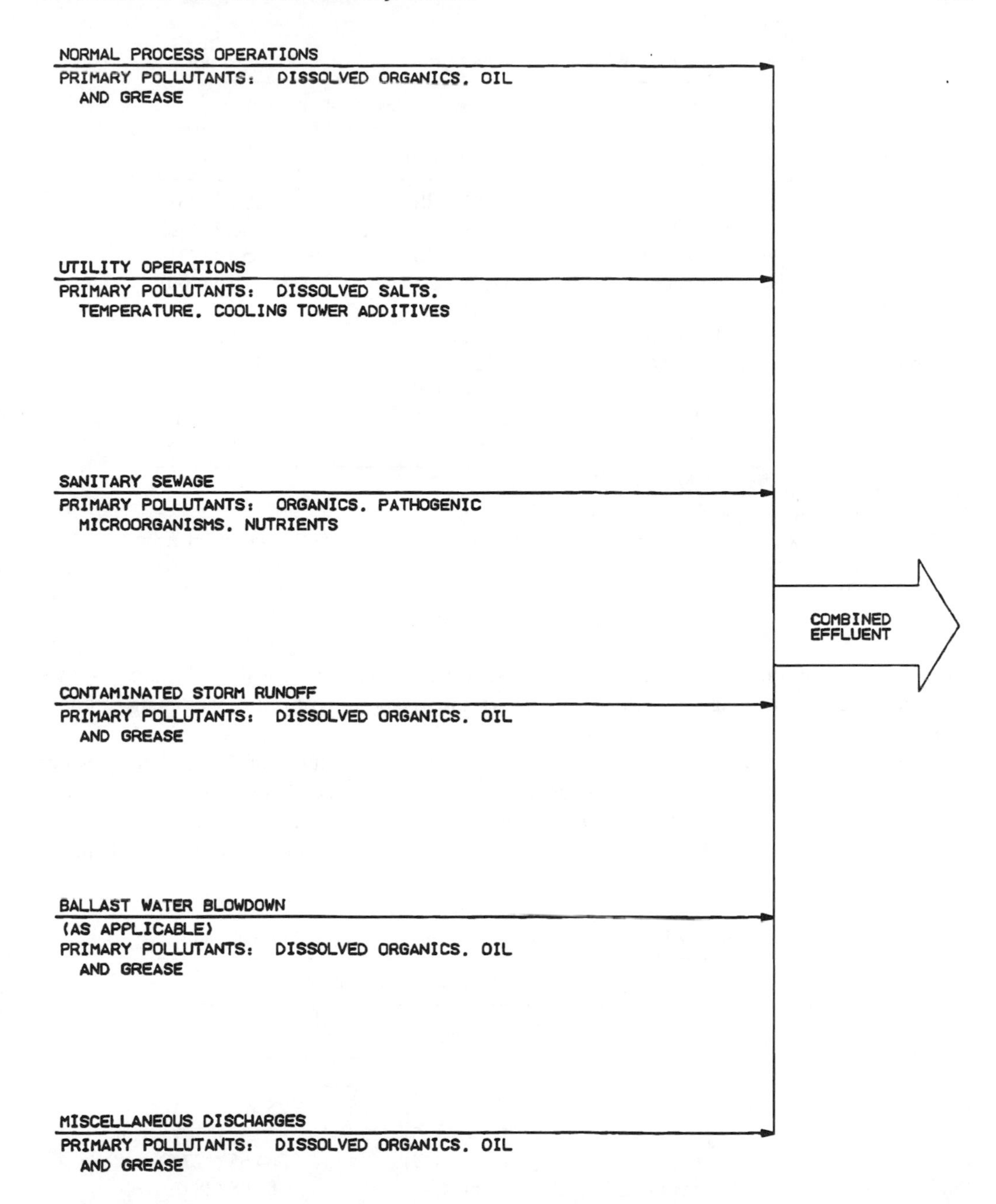

Figure 7 Components of pollutants by source. These principal pollutants are present in waste streams from each refinery operations/sources. (From Ref. 17.)

6.3.3 Refinery Solid and Hazardous Wastes

According to a USEPA survey, many of the more than 150 separate processes used in petroleum refineries generate large quantities of hazardous wastes. Typical wastes generated from refinery processes include bottom sediments and water from crude storage tanks, spent amines, spent acids and caustics, spent clays, spent glycol, catalyst fines, spent Streford solution and sulfur,

Table 8 Topping Subcategory Raw Waste Load Effluent from Refinery API Separator[a]

	Probability of occurrence, percent less than or equal to		
Parameters	10	50 (median)	90
BOD_5	1.29 (0.45)	3.43 (1.2)	217.36 (76)
COD	3.43 (1.2)	37.18 (13)	486.2 (170)
TOC	1.09 (0.38)	8.01 (2.8)	65.78 (23)
TSS	0.74 (0.26)	11.73 (4.1)	286 (100)
O&G	1.03 (0.36)	8.29 (2.9)	88.66 (31)
Phenols	0.001 (0.0004)	0.034 (0.012)	1.06 (0.37)
Ammonia	0.077 (0.027)	1.20 (0.42)	19.45 (6.8)
Sulfides	0.002 (0.00065)	0.054 (0.019)	1.52 (0.53)
Chromium	0.0002 (0.00007)	0.007 (0.0025)	0.29 (0.1)
Flow[b]	8.00 (2.8)	66.64 (23.3)	557.7 (195)

[a]Values represent kg/1000 m^3(lb/1000 bbl) of feedstock throughput.
[b]1000 m^3/1000^3 feedstock throughput (gallons/bbl).
BOD, biochemical oxygen demand; COD, chemical oxygen demand; TOC, total organic carbon; TSS, total suspended solids; O&G, oil and grease.
Source: From Ref. 7.

coking fines, slop oil, and storage tank bottoms. Most are hazardous wastes. Figure 8 shows a refinery schematic diagram indicating representative sources of solid wastes in refinery systems [18].

Also, the plant's utility systems often contribute to the volume of waste. Utility water systems generate raw water treatment sludge, lime softening sludge, demineralizer regenerants, and cooling tower sludge. These wastes may or may not be hazardous, depending on characteristics such as pH and metal concentrations. Figure 9 shows a refinery schematic

Table 9 Cracking Subcategory Raw Waste Load Effluent from Refinery API Separator[a]

	Probability of occurrence, percent less than or equal to		
Parameters	10	50 (median)	90
BOD_5	14.3 (5.0)	72.93 (25.5)	466.18 (163)
COD	27.74 (9.7)	217.36 (76.0)	2516.8 (880)
TOC	5.43 (1.9)	41.47 (14.5)	320.32 (112)
O&G	2.86 (1.0)	31.17 (10.9)	364.65 (127.5)
Phenols	0.19 (0.068)	4.00 (1.4)	80.08 (28.0)
TSS	0.94 (0.33)	18.16 (6.35)	360.36 (126.0)
Sulfur	0.01 (0.0035)	0.94 (0.33)	39.47 (13.8)
Chromium	0.0008 (0.00028)	0.25 (0.088)	4.15 (1.45)
Ammonia	2.35 (0.82)	28.31 (9.9)	174.46 (61.0)
Flow[b]	3.29 (1.15)	92.95 (32.5)	2745.6 (960.0)

[a]Values represent kg/1000 m^3 (lb/1000 bbl) of feedstock throughput.
[b]1000 m^3/1000 m^3 feedstock throughput (gallons/bbl).
BOD, biochemical oxygen demand; COD, chemical oxygen demand; TOC, total organic carbon; TSS, total suspended solids; O&G, oil and grease.
Source: From Ref. 7.

Table 10 Petrochemical Subcategory Raw Waste Load Effluent from Refinery API Separator[a]

	Probability of occurrence, percent less than or equal to		
Parameters	10	50 (median)	90
BOD_5	40.90 (14.3)	171.6 (60)	715 (250)
COD	200.2 (70)	463.32 (162)	1086.8 (380)
TOC	48.62 (17)	148.72 (52)	457.6 (160)
TSS	6.29 (2.2)	48.62 (17)	371.8 (130)
O&G	12.01 (4.2)	52.91 (18.5)	234.52 (82)
Phenols	2.55 (0.89)	7.72 (2.7)	23.74 (8.3)
Ammonia	5.43 (1.9)	34.32 (12)	205.92 (72)
Sulfides	0.009 (0.003)	0.86 (0.3)	91.52 (32)
Chromium	0.014 (0.005)	0.234 (0.085)	3.86 (1.35)
Flow[b]	26.60 (9.3)	108.68 (38)	443.3 (155)

[a]Values represent kg/1000 m^3 (lb/1000 bbl) of feedstock throughput.
[b]1000 m^3/1000 m^3 feedstock throughput (gallons/bbl).
BOD, biochemical oxygen demand; COD, chemical oxygen demand; TOC, total organic carbon; TSS, total suspended solids; O&G, oil and grease.
Source: From Ref. 7.

diagram indicating representative sources of solid waste in utility water systems [18]. Wastes generated from wastewater treatment systems include API/CPI separator sludge, dissolved-air flotation or induced-air flotation system floats, pond and tank sediments, and biosolids. Of these, only the biosolids from the biological wastewater treatment system may be nonhazardous. Figure 10 shows a refinery schematic diagram indicating representative sources of solids waste in wastewater treatment systems [18].

Table 11 Lube Subcategory Raw Waste Load Effluent from Refinery API Separator[a]

	Probability of occurrence, percent less than or equal to		
Parameters	10	50 (median)	90
BOD_5	62.92 (22)	217.36 (76)	757.9 (265)
COD	165.88 (58)	543.4 (190)	2288 (800)
TOC	31.46 (11)	108.68 (38)	386.1 (135)
TSS	17.16 (6)	71.5 (25)	311.74 (109)
O&G	23.74 (8.3)	120.12 (42)	600.6 (210)
Phenols	4.58 (1.6)	8.29 (2.9)	51.91 (18.5)
Ammonia	6.5 (2.3)	24.1 (8.5)	96.2 (34)
Sulfides	0.00001 (0.000005)	0.014 (0.005)	20.02 (7.0)
Chromium	0.002 (0.0006)	0.046 (0.016)	1.23 (0.43)
Flow[b]	68.64 (24)	117.26 (41)	772.2 (270)

[a]Values represent kg/1000 m^3 (lb/1000 bbl) of feedstock throughput.
[b]1000 m^3/1000 m^3 feedstock throughput (gallons/bbl).
BOD, biochemical oxygen demand; COD, chemical oxygen demand; TOC, total organic carbon; TSS, total suspended solids; O&G, oil and grease.
Source: From Ref. 7.

Table 12 Integrated Subcategory Raw Waste Load Effluent from Refinery API Separator[a]

	Probability of occurrence, percent less than or equal to		
Parameters	10	50 (median)	90
BOD_5	63.49 (22.2)	197.34 (69.0)	614.9 (215)
COD	72.93 (25.5)	328.9 (115)	1487.2 (520)
TOC	28.6 (10.0)	139.0 (48.6)	677.82 (237)
O&G	20.88 (7.3)	74.93 (26.2)	268.84 (94.0)
Phenol	0.61 (0.215)	3.78 (132)	22.60 (7.9)
TSS	15.16 (5.3)	58.06 (20.3)	225.94 (79.0)
Sulfur	0.52 (0.182)	2.00 (0.70)	7.87 (2.75)
Chromium	0.12 (0.043)	0.49 (0.17)	121.55 (42.5)
Ammonia	3.43 (1.20)	20.50 (7.15)	121.55 (42.5)
Flow[b]	40.04 (14.0)	234.52 (82.0)	1372.8 (480)

[a]Values given represent kg/1000 m^3 (lb/1000 bbl) of feedstock throughput.
[b]1000 m^3/1000 m^3 feedstock throughput (gallons/bbl).
BOD, biochemical oxygen demand; COD, chemical oxygen demand; TOC, total organic carbon; TSS, total suspended solids; O&G, oil and grease.
Source: From Ref. 7.

The amount and type of wastes generated in a refinery depend on a variety of factors such as crude capacity, number of refining processes, crude source, and operating procedures. A 130,000 bpd integrated refinery on the West Coast generates about 50,000 tons per year of hazardous waste (including recycled streams and unfiltered sludges). The major wastes are wastewater treatment plant sludge, spent caustics, Stretford solution and sulfur, and spent catalysts [19]. A much simpler 50,000 bpd refinery generates only 400 tons per year of hazardous waste. Major wastes in this refinery are wastewater treatment plant sludge (dewatered by pressure filtration), spent catalysts, and spent clay filter media [19].

6.4 ENVIRONMENTAL REGULATIONS

Three categories of regulatory limitations apply to wastewater discharge from industrial facilities such as oilfields and petroleum refineries [20]. The first category includes effluent limitations, which are designed to control those industry-specific wastewater constituents deemed significant from the standpoints of water quality impact and treatability in conventional treatment systems. In the United States, these limitations are the EPA Effluent Guidelines, issued under Public Law 92-500.

The second category includes pretreatment discharge requirements established both by the EPA and certain municipalities that treat combined industrial and domestic wastes in their publicly owned treatment works. These standards have not been updated by USEPA as of 2003.

The third category includes effluent limitations associated with maintaining or establishing desirable water uses in certain bodies of effluent-receiving waters, that is, water-quality-limiting segments as defined in Public Law 92-500. This last category became the overriding category in many locations in the United States when the EPA published its final surface water toxics control rule on June 2, 1989 [21]. These three categories of effluent limitations are discussed below.

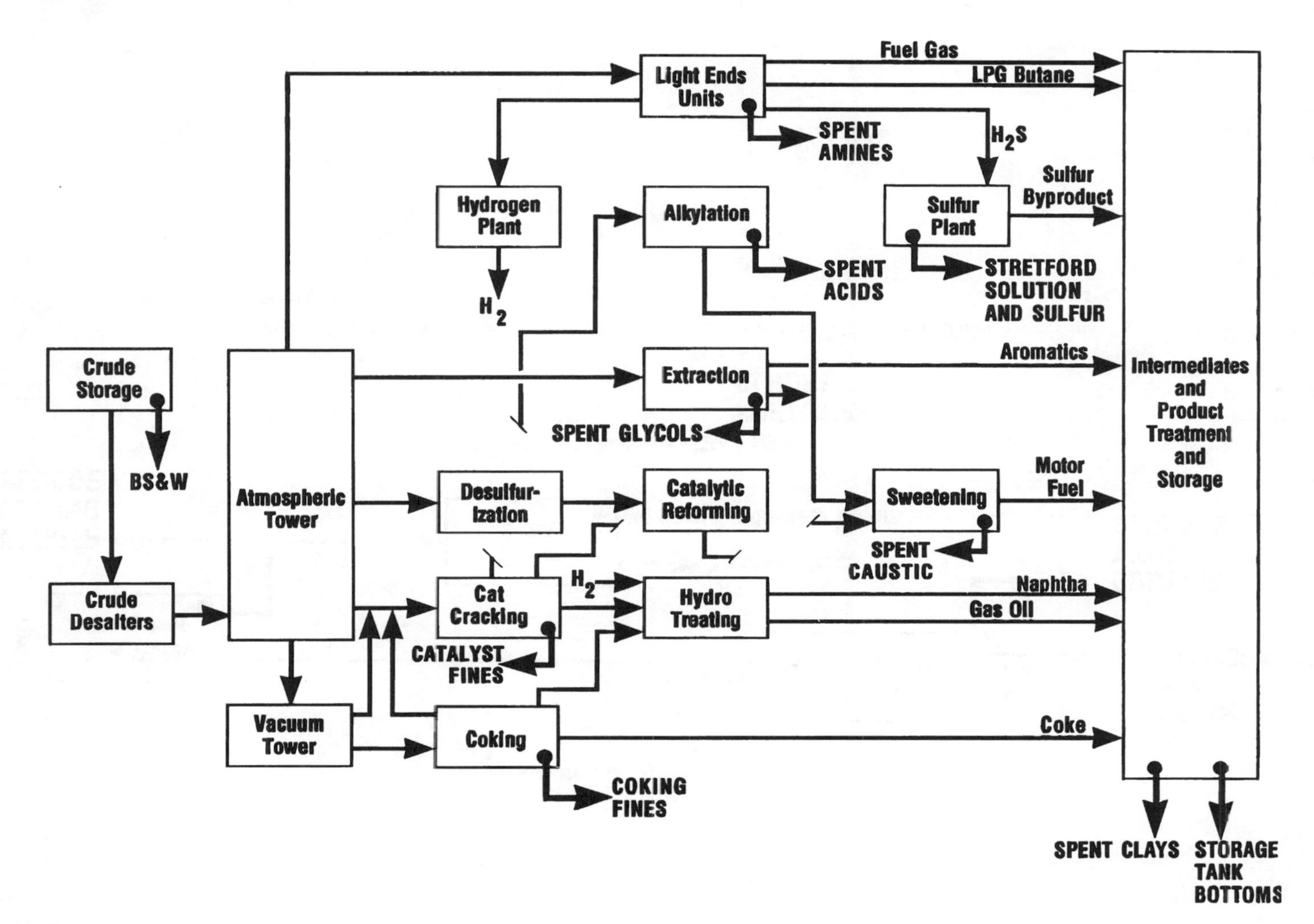

Figure 8 Refinery schematic diagram indicating representative sources of solid waste in refinery system. Most solid wastes from refineries are considered hazardous wastes in the United States. (From Ref. 18.)

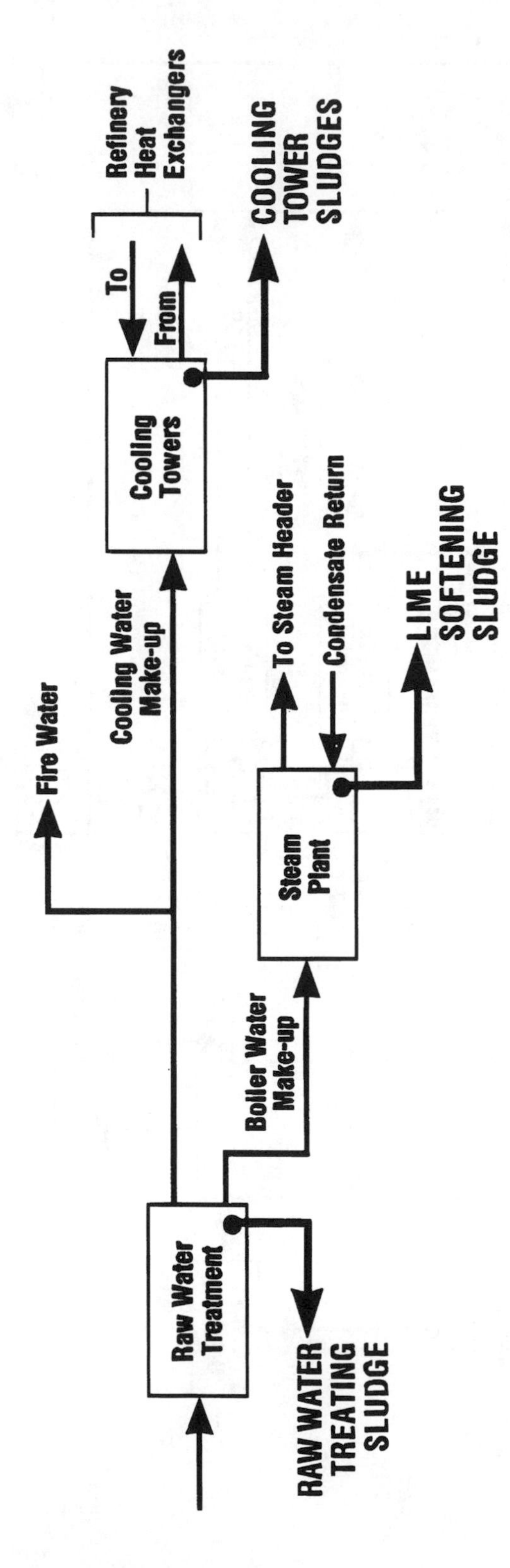

Figure 9 Refinery schematic diagram indicating representative sources of solid waste in utility water system. These wastes may not be classified as hazardous in the United States. (From Ref. 18.)

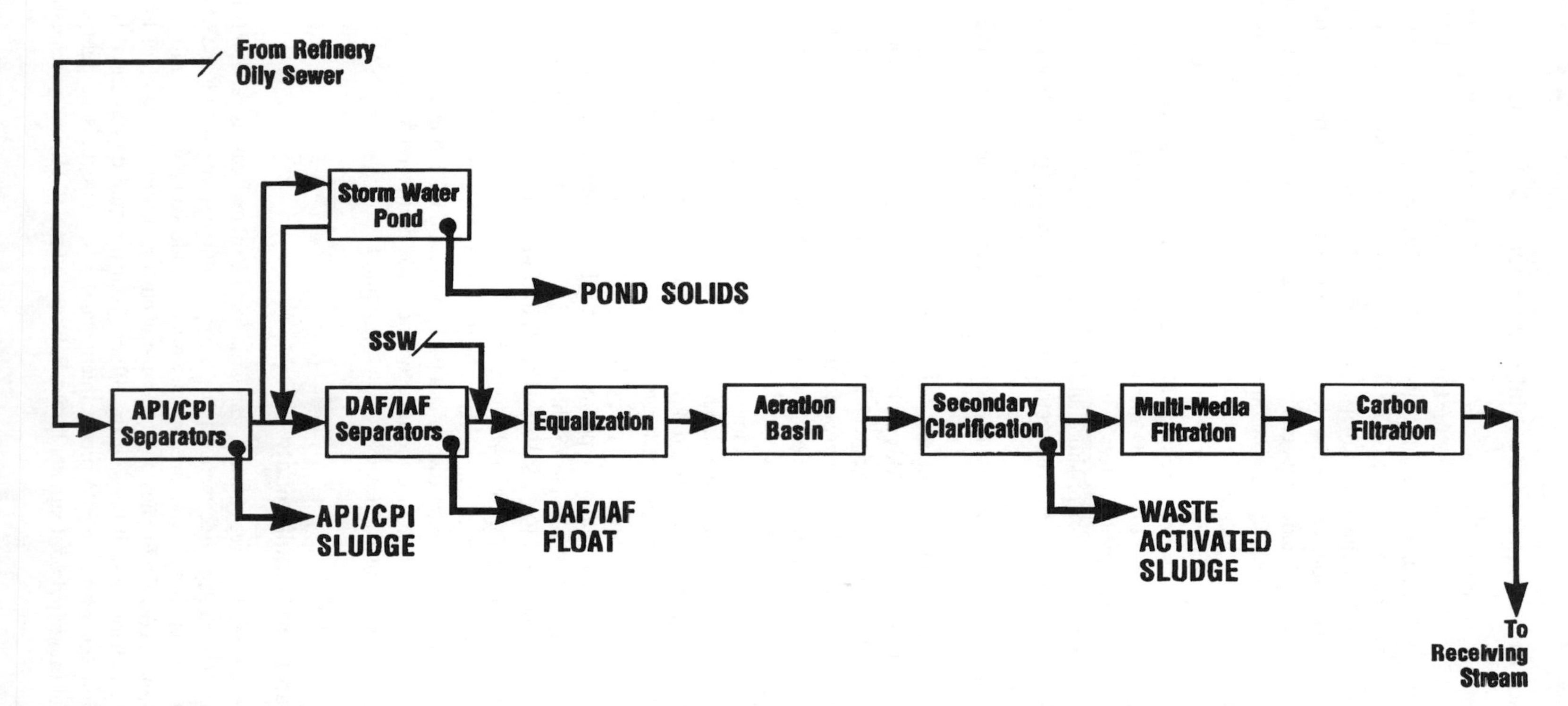

Figure 10 Refinery schematic diagram indicating representative sources of solid waste in wastewater treatment system. All wastes except waste activated sludge are classified as hazardous wastes because of their oil contents. (From Ref. 18.)

6.4.1 Effluent Guidelines for Industrial Point Source Categories

USEPA has established effluent limitations on wastewater constituents for various industrial categories. The EPA effluent limitations for Oil and Gas Extraction Point Source Category are under 40 CFR Part 435 (Code of Federal Register, 1988). The regulations differentiate between offshore, onshore, and coastal facilities. The limitation for onshore oil and gas facilities is no discharge of wastewater pollutants into navigable waters from any source associated with production, field exploration, drilling, well completion, or well treatment (produced water, drilling muds, drill cuttings, and produced sand). Owing to a challenge in court (API vs. EPA, 1981), the limitation was suspended for facilities located in the Santa Maria Basin of California.

For onshore facilities located in the continental United States and west of the 98th meridian for which the produced water has a use in agriculture or wildlife propagation when discharged into navigable waters, discharge of produced water is allowed if its oil and grease (O&G) concentration does not exceed 35 mg/L. Other wastes from these onshore facilities are not to be discharged to navigable waters.

The effluent limitations for offshore and coastal oil and gas facilities are identical. The main criteria for discharge are O&G concentrations. For produced water, the effluent limitations are 72 mg/L of O&G maximum for anyone day and 48 mg/L of O&G average for 30 consecutive days. For other industrial wastes from these facilities, the effluent limitations are no discharge of free oil.

The EPA promulgated Effluent Guidelines and Standards for the Petroleum Refining Industry under 40 CFR Part 419 on May 9, 1974, and published the most recent update to Part 419 on August 12, 1985 (Federal Register). Standards for direct dischargers are mass-limited, not concentration-limited, and are expressed in pounds per 1000 barrels of feedstock. The standards are further subdivided into five subcategories within the petroleum refining category, as described earlier in this chapter. The standards for each subcategory may in turn be modified by "size" and "process" factors. For example, in the topping subcategory, a plant of less than 24,000 bpsd of feedstock would have a size factor of 1.02 applied to the effluent limitations, and a plant of 150,000 bpsd or greater would have a size factor of 1.57 applied.

The EPA has established four different control technologies for the petroleum refining industry: best practicable control technology (BPT), best available technology economically achievable (BAT), best conventional pollutant control technology (BCT), and new source performance standards (NSPS). Table 13 shows the BPT and NSPS standards that must be met by the various subcategories (40 CFR Part 419). The limitations for BPT actually incorporate those of both BAT and BCT for this industry.

In addition to these effluent standards, the EPA has also established separate BPT, BAT, BCT, and NSPS standards for ballast water and BPT, BAT, and BCT standards for contaminated storm water (40 CFR Part 419). Once-through cooling water is allowed for direct discharge if the total organic carbon concentration does not exceed 5 mg/L.

6.4.2 Pretreatment Requirements

Presently there are no EPA pretreatment standards for the oil and gas extraction (oilfield) point source category. The EPA pretreatment standards for discharge from existing and new petroleum refining facilities to publicly owned treatment works include 100 mg/L each for oil and grease (O&G) and ammonia (as N). For new facilities a total chromium concentration of 1 mg/L for the cooling tower discharge part of the refinery effluent is also required (40 CPR Part 419).

In addition to meeting the EPA pretreatment standards, indirect dischargers are required to meet individual municipal pretreatment limits. Publicly owned treatment works establish limits

Table 13 Effluent Standards for Five Subcategories of the Petroleum Refining Point Source Category

	Effluent limitation (daily average for 30 consecutive days, in lbs/1000 bbl of feedstock)									
	Topping		Cracking		Petroleum		Lube		Integrated	
Parameters	BPT	NS	BPT	NS	BPT	NS	BPT	NS	BPT	NS
BOD_5	4.25	2.2	5.5	3.1	6.5	4.1	9.1	6.5	10.2	7.8
TSS	3.6	1.9	4.4	2.5	5.25	3.3	8.0	5.3	8.4	6.3
COD	31.3	11.2	38.4	21.0	38.4	24.0	66.0	45.0	70.0	54.0
O&G	1.3	0.70	1.6	0.3	2.1	1.3	3.0	2.0	3.2	2.4
Phenolic compounds	0.027	0.16	0.036	0.020	0.0425	0.027	0.065	0.043	0.068	0.051
Ammonia as N	0.45	0.45	3.0	3.0	3.8	3.8	3.8	3.8	3.8	3.8
Sulfide	0.024	0.012	0.029	0.017	0.035	0.022	0.053	0.035	0.056	0.042
Total chromium	0.071	0.037	0.088	0.049	0.107	0.068	0.160	0.105	0.17	0.13
Hexavalent chromium	0.0044	0.0025	0.0056	0.0032	0.0072	0.0044	0.160	0.0072	0.011	0.008

Note: pH (within the range of 6.0 to 9.0); BPT incorporates BAT and BCT; BPT, best practicable control technology; NS, new source performance standards; BOD, biochemical oxygen demand; TSS, total suspended solids; COD, chemical oxygen demand; O&G, oil and grease; BAT, best available technology economically achievable; BCT, best conventional pollutant control technology.

Source: From Ref. 40 CFR, Part 419, 1988.

to control pollutants that could be deleterious to conventional biological treatment systems or that could cause the municipality to violate receiving water standards. Table 14 shows the industrial effluent limits established by the City of San Jose, CA (San Jose Municipal Code, 1988). This city has also adopted an effluent toxicity requirement for industrial dischargers using the public sewer. Discharges are not to exceed a median threshold limit of 50%.

6.4.3 Water Quality Based Limitations

In the United States, as control of conventional pollutants has been significantly achieved, increased emphasis is being placed on reduction of toxic pollutants. The EPA has developed a water quality based approach to achieve desired water quality where treatment control based discharge limits have proved to be insufficient [22]. The procedure for establishing effluent limitations for point sources discharging to a water quality based segment generally involves the use of some type of mathematical model or allocation procedure to apportion the allowable

Table 14 Industrial Waste Pretreatment Limits for a Publicly Owned Treatment Works

Toxic substance	Max. allowable concentration (mg/L)
Aldehyde	5.0
Antimony	5.0
Arsenic	1.0
Barium	5.0
Beryllium	1.0
Boron	1.0
Cadmium	0.7
Chlorinated hydrocarbons, including but not limited to pesticides, herbicides, algaecides	Trace
Chromium, total	1.0
Copper	2.7
Cyanides	1.0
Fluorides	10.0
Formaldehydes	5.0
Lead	0.4
Manganese	0.5
Mercury	0.010
Methyl ethyl ketone and other water insoluble ketones	5.0
Nickel	2.6
Phenol and derivatives	30.0
Selenium	2.0
Silver	0.7
Sulfides	1.0
Toluene	5.0
Xylene	5.0
Zinc	2.6
pH, su	5.0 to 10.5

SU = standard unit

Source: From City of San Jose, CA, Municipal Code, 1988.

loading of a particular toxicant to each discharge in the segment. These allocations are generally made by the state regulatory agency [20].

State and regional regulatory agencies may also establish general effluent limitations for a particular water body to control the total discharge of toxic pollutants. Table 15 shows the discharge limits for toxic pollutants established by the San Francisco Bay Regional Water Quality Board (1986).

This agency has also adopted biomonitoring and toxicity requirements for municipal and industrial dischargers. Biomonitoring, or whole-effluent toxicity testing, has become a requirement for many discharges in the United States. As of 1988, more than 6000 discharge permits have toxicity limits to protect against chronic toxicity [22]. When a discharge exceeds the toxicity limits, the discharger must conduct a toxicity identification evaluation (TIE) and a toxicity reduction evaluation (TRE). A TRE is a site-specific investigation of the effluent to identify the causative toxicants that may be eliminated or reduced, or treatment methods that can reduce effluent toxicity.

6.5 CONTROL AND TREATMENT TECHNIQUES FOR OILFIELD WASTES

Major waste liquids arising from oil and gas production include produced water and drilling fluids and muds. These waste streams are handled and disposed of separately.

6.5.1 Produced Water Treatment and Disposal

Produced water (brine) disposal practices may be divided into the broad categories of surface discharge, subsurface discharge, evaporation, and reuse. Approximately 30 states produce some amount of oil or gas, and brine handling practises vary considerably because of variations in climate, geology, brine quantity and quality, and regulatory framework [23].

Table 15 Effluent Limitations for Selected Toxic Pollutants for Discharge to Surface Waters (All Values in μg/L)

	Daily average	
	Shallow water	Deep water
Arsenic	20	200
Cadmium	10	30
Chromium (VI)	11	110
Copper	20	200
Cyanide	25	25
Lead	5.6	56
Mercury	1	1
Nickel	7.1	71
Silver	2.3	23
Zinc	58	580
Phenols	500	500
PAHs	15	150

PAHs = Polynuclear aromatic hydrocarbons

Source: From Water Quality Control Plan, San Francisco Bay Basin, 1986.

Surface Discharge

Because onshore oil and gas facilities are not allowed to discharge wastes to navigable waters, surface discharge is only practiced at coastal facilities. In some states indirect surface discharge is practiced by simple dilution through an existing municipal or industrial wastewater treatment facility [23].

The main pollutant of concern for brine discharge is oil and grease (O&G). However, other pollutants may be important if they violate state-set water quality criteria for local water bodies. Michalczyk et al. [9] suggested a typical production water treatment system to meet the criteria of the California Ocean Plan. As shown in Fig. 11, treatment processes include equalization, oil removal by flotation, pH adjustment, and activated sludge. Experimental results obtained by Michalczyk *et al.* indicate that biological treatment effectively reduces BOD/COD and phenol in oilfield produced waters to acceptable levels, but nitrification can be inhibited by inorganic or biologically refractory organic compounds. Wang et al. [24] reported the use of hydrocarbon deterioration bacteria with gas lift processing to treat oily produced water. With oil content above 300 mg/L, COD of 250–480 mg/L, the treated water has 10 mg/L of oil and less than 120 mg/L of COD. A special group of bacteria named WS3 were selected for treatment testing after an elaborate screening process.

Palmer et al. [10] reported the results of two pilot field studies of treating oilfield produced water by biodisks in southern California. The TDS concentration of the produced water was 20,000 mg/L. The results indicate that dissolved organics and ammonium compounds can be

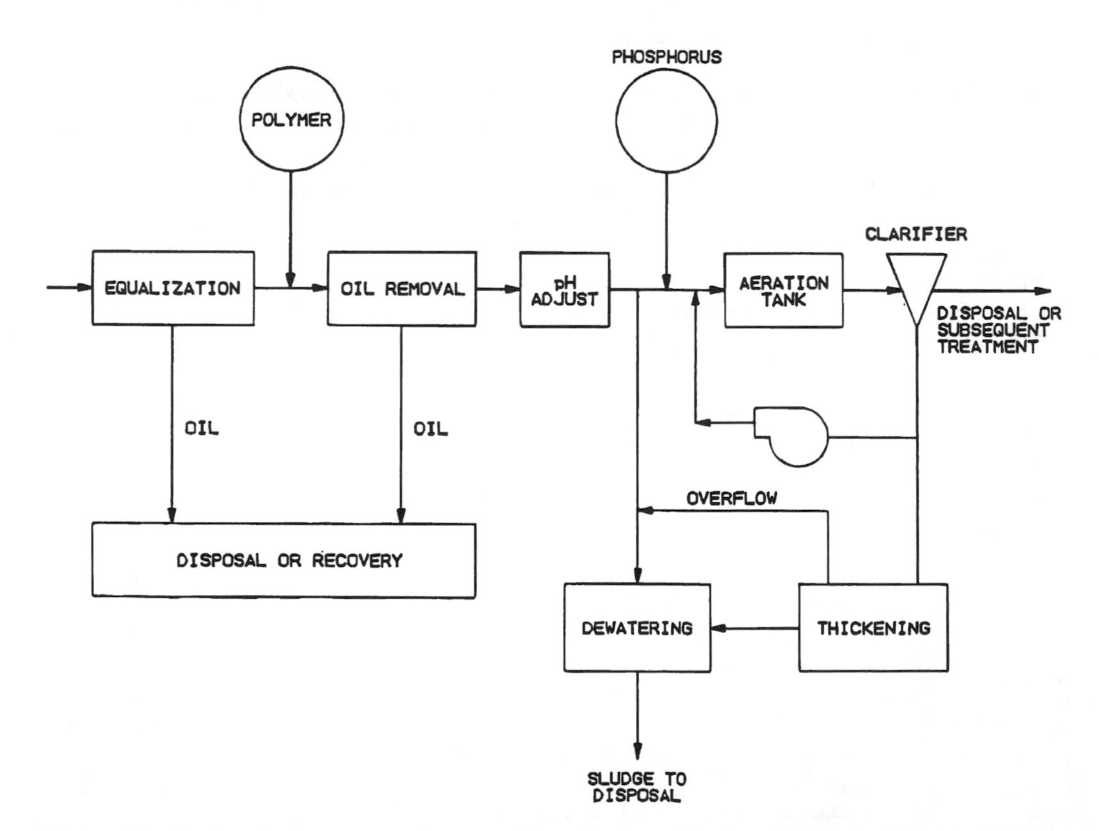

Figure 11 Produced water treatment system. Treatment is mainly for oil and organics removal. (From Ref. 9.)

removed by biological oxidation in a biodisk unit to meet California Ocean Plan criteria. Earlier, Beyer et al. [25] demonstrated the feasibility of biological oxidation by aerated lagoons to remove dissolved compounds such as ammonia and phenols from produced waters. Ali et al. [26] conducted laboratory and field tests to successfully demonstrate that a two-stage filtration process can effectively reduce oil and grease content in offshore discharged produced water. The first stage (Crudesorb) removes dispersed oil and grease droplets, and the second stage (polymeric resin) removes dissolved hydrocarbons, aliphatic carboxylic acids, cyclic carboxylic acids, aromatic carboxylic acids, and phenolic compounds.

Subsurface Discharge

Disposal of brine in subsurface wells is probably the most widely used control method, especially in the western and southern oil and gas producing states [23]. For this to be an effective disposal option, two conditions must be met: the natural aquifer must be naturally saline and must not leak to freshwater aquifers, and the reinjection pressure must not exceed the fracture pressure of the formation [9]. Produced water is usually pretreated to prevent equipment from being corroded and to prevent plugging of the sand at the base of the well. Pretreatment may include the removal of oils and floating material, suspended solids, biological growth, dissolved gases, precipitable ions, acidity, or alkalinity [27]. A typical system is shown in Fig. 12.

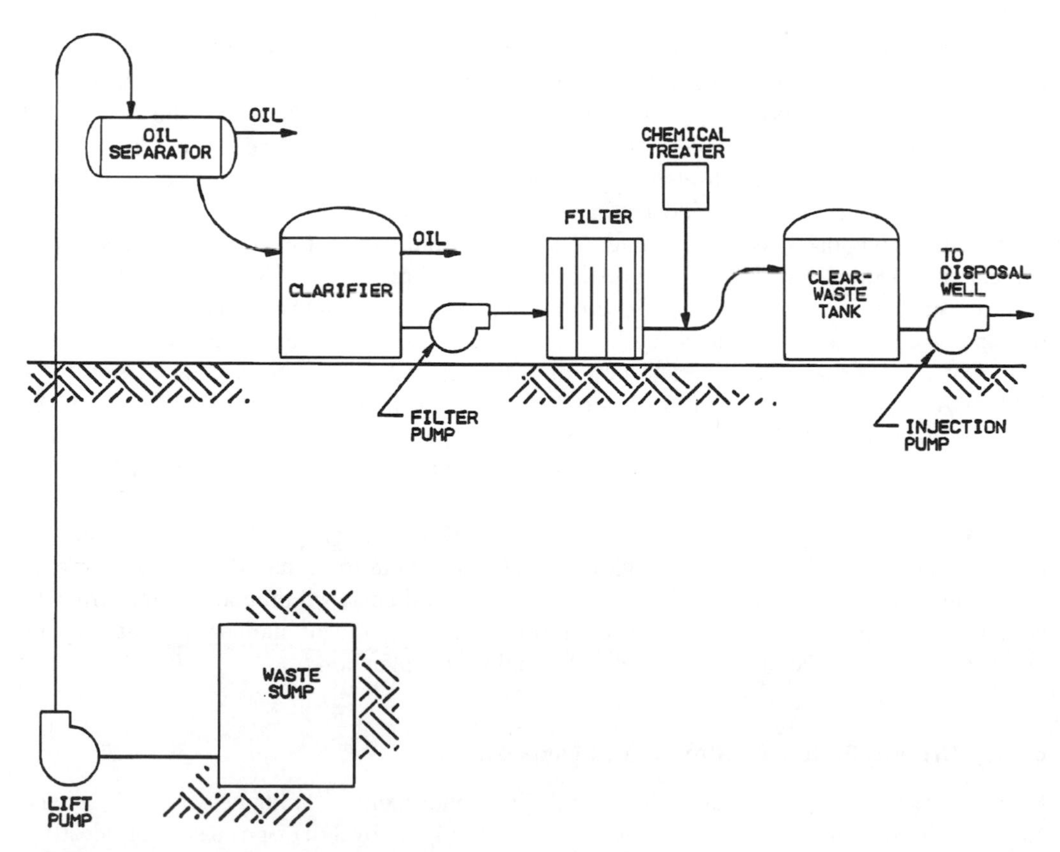

Figure 12 Typical subsurface waste disposal system. Waste is treated for oil removal, filtered, and chemically treated before subterranean injection. (From Ref. 27.)

In the United States, injection wells are classified into three categories: Class 1 wells are used to inject hazardous wastes; Class 2 wells are used to inject fluids brought to the surface in connection with the production of oil and gas or for disposal of salt water; Class 3 covers solution mining wells [28]. Class 1 wells are heavily regulated by the EPA. However, tougher rules for casing and cementing are being considered for Class 2 wells. After conducting a random sample of Class 2 wells in four states in 1987 and 1988, the General Accounting Office (GAO) claimed that federal and state regulations are not preventing brine injection wells from contaminating U.S. drinking water aquifers [29]. The GAO recommended that the EPA require all existing injection wells to be checked for leakage and require state agencies to examine permit applications for new injection wells more closely.

Evaporation

The use of open pits or ponds for evaporation of brine is widely practiced in southwestern states where evaporation exceeds precipitation [23]. For example, about 75% of all oil and gas waste fluids are disposed of by evaporation pits in New Mexico [30]. Evaporation ponds require large land areas, and they may contaminate groundwater. Today regulators view evaporation pits with disfavor because faulty pond design and operation have allowed salts to migrate into usable groundwater reservoirs [9].

Reuse

The most desirable disposal method is to reuse the produced water. Produced water can be treated and reinjected into the subsurface reservoir to cause the oil to flow into the well to increase yields (water flooding). The produced water is usually filtered to remove oil and suspended solids before injection. In a steam flooding project in the Far East, produced water was treated and used as feedwater to generate steam for enhanced oil recovery [31]. The treatment processes included induced air flotation, filtration, softening, and deaeration. Treatment technologies for reclaiming oilfield-produced water for beneficial reuse were evaluated by Doran et al. [32]. The investigators selected precipitating softening and high-pH reverse osmosis (RO) for pilot testing based on a literature review and benchscale softening tests that indicated hardness, boron, and silica removal could be simultaneously optimized. The results of a pilot study were used to perform a conceptual design and cost estimation for a 7000 m^3/day (44,000 bpd) treatment facility for converting produced water to drinking water or other reuse quality [32]. Depending on reuse quality requirements, the capital cost ranged from \$3.1 million to \$12.3 million, and the operating and maintenance (O&M) cost from \$0.28/m^3 to \$1.45/m^3 of water recovered.

Another potential use of the brine is for highway application. Sodium and calcium chlorides have been widely used in highway applications both for winter deicing and for road stabilization and dust control. Sack et al. [23] sampled and analyzed produced brines from 13 counties representing 8 different geological formations. A significant number of West Virginia brines were found to be of suitable quality for highway application.

6.5.2 Drilling Fluids Treatment and Disposal

Potential treatment and disposal methods for drilling fluids include (1) fluid ejection, (2) pit and solids encapsulation, (3) injection into safe formations, (4) removal to disposal sites off location, (5) incineration, (6) microorganism processing, and (7) distillation, liquid extraction, and chemical fixation [13].

Direct Ejection of Fluids

This disposal method only applies to water-based drilling fluids. The fluids may be spread directly over adjacent agricultural or forest land after adjustment of pH and ion content. Treatment may include coagulation, flocculation, filtration, and pH adjustment before spreading. A major consideration is chloride ion content. With higher chlorides, some transport of the fluid to a better disposal site may be necessary.

Pit and Solids Encapsulation

Pit encapsulation means constructing a reserve pit to contain the fluids and sealing it at the end of drilling. Normal procedures involve slurry trenching for sidewalls, plowing in organic-treated bentonite for a bottom base, placing a synthetic liner in this excavation, and covering the liner with additional soil containing some bentonite for puncture protection. The pit is then filled with waste drilling fluid. At well completion, the fluid is allowed to evaporate. When the fluid is substantially dewatered, the pit is covered with a top layer of soil containing organic-treated bentonite. The location of the burial site is recorded.

Solids encapsulation is removing solids from a fluid by some form of polymer coating procedure. The coated solids are buried. One novel treatment method adds a microorganism "cocktail", along with nutrients, to a fluid containing suspended solids and high chloride ion content [13]. The microorganisms utilize chloride during growth and coagulate the solids. After sufficient aging, clean water can be pumped off, leaving the coagulated solids residue, which is buried. Chloride ion concentration is normally below 200 mg/L following aging.

Pumping into Safe Formation

Deep well injection of spent fluids is another possible alternative. The criteria for injection of drilling fluids are similar to those for injection of produced water discussed earlier in this chapter.

Removal to Designated Disposal Sites

The fluids can be hauled by vacuum trucks to an approved disposal site for such wastes. There are different classes of disposal sites. If regulatory agencies require that a fluid be disposed in a hazardous waste "secure" landfill, the cost would be very high.

Incineration

Incineration offers the complete destruction of oil and organic materials. However, it is very expensive and may cause air pollution. Incineration would be used when other less costly options are unavailable.

Microorganism Processing

Biological treatment may be used to degrade the oil and grease fractions in drilling fluids prior to solids separation. Marks et al. [33] conducted batch treatment tests for drilling fluids and production sludges and demonstrated that biological treatment is feasible. However, more biokinetic tests are required for further evaluation.

Distillation, Liquid Extraction, and Chemical Treatment

Several emerging processes may be applicable for treatment of oily drilling muds prior to disposal. One process being tested in Europe involves the use of an electric distillation kiln to break down solids-laden oil-based drilling muds [13]. Another process uses critical fluid to extract oil and organics from oily sludges so that they can be landfilled [34]. Copa and Dietrich [14] treated a sample of spent drilling mud with wet air oxidation. The COD content was reduced by 45 to 64% and the dewaterability of the mud was improved.

Chemical fixation is another possible process to handle drilling fluids. A typical process uses a mixture of potassium or sodium silicate with portland cement to turn a drilling fluid into a soil-like solid that may be left in place, used as a landfill, or even used as a construction material [13].

6.6 IN-PLANT CONTROL AND TREATMENT TECHNIQUES FOR REFINERY WASTES

The management of wastes from refineries includes in-plant source control, pretreatment, and end-of-pipe treatment. In-plant source control reduces the overall pollutant load that must be treated in an end-of-pipe treatment system. Pretreatment reduces or eliminates a particular pollutant before it is diluted in the main wastewater stream, and may provide an opportunity for material recovery. End-of-pipe treatment is the final stage for meeting regulatory discharge requirements and protection of stream water quality. These techniques are discussed in more detail in the following sections.

6.6.1 In-Plant Source Control

Source control means different things to different people. Here it means knowing the sources and the amounts of water and contaminants and continuously monitoring them, then reducing the amounts by in-plant operating and equipment changes.

There are many ways in a refinery to reduce the amount of wastewater flows and contaminants. These can include good housekeeping, process modifications, and recycle–reuse.

Good Housekeeping

Good housekeeping can play an important role in reducing unnecessary flows that must be treated downstream. Good housekeeping practices include minimizing waste when sampling product lines; shutting off washdown hoses when not in use; having a good maintenance program to keep the refinery as leakproof as possible; and individually treating waste streams with special characteristics, such as spent cleaning solutions [35].

Many more things can be done; here are just a few. The use of dry cleaning, without chemicals, aids in reducing water discharges to the sewer. Using vacuum trucks to clean up spills, then charging this recovered material to slop oil tanks, reduces the discharge of both oil and water to the wastewater system. Process units should be curbed to prevent the contamination of clean runoff with oily storm runoff and to prevent spills from spreading widely. Sewers should be flushed regularly to prevent the buildup of material, eliminating sudden surges of pollutants during heavy rains. Collection vessels should be provided whenever maintenance is performed on liquid processing units, to prevent accidental discharges to the sewers.

Housekeeping practices within a refinery can have substantial impact on the loads discharged to the wastewater facilities. Knowlton [36] reported how source control by good housekeeping helped a Chevron refinery meet new NPDES permit requirements. Good housekeeping practises to reduce wastewater loads require judicious planning, organization, and operational philosophy. They also require good communication and education for all personnel involved. A refinery newsletter is a good tool to communicate and educate refinery personnel on pollution control issues.

Process Modifications

Many new and modified refineries incorporate reduced water use and pollutant loading into their process and equipment design. Modifications include:

1. Substitution of improved catalysts, which require less regeneration and thus lower wastewater loads.
2. Replacement of barometric condensers with surface condensers or air fan coolers, reducing a major source of oil–water emulsion.
3. Substitution of air cooling devices for water cooling systems.
4. Use of hydrocracking and hydrotreating processes that produce lower wastewater loadings than existing processes.
5. Improved drying, sweetening, and finishing procedures to minimize spent caustics and acids, water washes, and filter solids requiring disposal.

Wastewater Recycle–Reuse

Wastewater reuse is a good way to reduce overall pollutant loadings. However, water quality is critical in water reuse. The contaminants present must be compatible with the reuse. For example, reuse waters with high solids content are not satisfactory for crude unit desalting. Stripped foul water containing low H_2S and ammonia and high concentrations of phenols has essentially no solids. It is suitable for crude unit desalter wash water if the phenols extracted by the crude are subsequently converted by hydroprocessing units into nonphenolic compounds [36]. Some other examples include:

1. Use of recycling cooling towers to replace a once-through cooling system.
2. Reuse of cooling tower blowdown as seal water on high-temperature pumps, where mechanical seals are not practicable.
3. Use of stripped sour water as low-pressure boiler makeup.
4. Reuse of wastewater treatment plant effluent as cooling water, as scrubber water, or as plant makeup water.
5. Putting high-pressure water in cokers through a gravity separator to remove floating oil and settleable coke fines.

6.6.2 Segregation and Pretreatment

The first step in good pretreatment practice is the segregation of major wastewater streams. This frequently simplifies waste treating problems as well as reducing treatment facility costs. Treatment at the source is also helpful in recovering byproducts that otherwise would not be economically recovered from combined wastes downstream [35]. Four major pretreatment processes that are applicable to individual process effluents or groups of effluents within a refinery are sour water stripping, spent caustics treatment, ballast water separation, and slop oil recovery. These are discussed below.

Sour Water Stripping

Many processes in a refinery use steam as a stripping medium in distillation and as a diluent to reduce the hydrocarbon partial pressure in catalytic or thermal cracking [37]. The steam is eventually condensed as a liquid effluent commonly referred to as sour or foul water. The two most prevalent pollutants found in sour water are H_2S and NH_3 resulting from the destruction of organic sulfur and nitrogen compounds during desulfurization, denitrification, and hydrotreating. Phenols and cyanides also may be present in sour water.

The purpose of sour water pretreatment is to remove sulfides (H_2S, ammonium sulfide, and polysulfides) before the waste enters the sewer. The sour water can be treated by stripping with steam or flue gas, air oxidation to convert sulfides to thiosulfates, or vaporization and incineration.

Sour water strippers are designed primarily for the removal of sulfides and can be expected to achieve 85–99% removal. If acid is not required to enhance sulfide stripping, ammonia will also be stripped, the percentage varying widely with stripping pH and temperature. Depending on pH, temperature, and contaminant partial pressure, phenols and cyanides can also be stripped with removal as high as 30%.

There are many different types of strippers, but most of them involve the downward flow of sour water through a trayed or packed tower while an ascending flow of stripping steam or gas removes the pollutants. The stripping medium can be steam, flue gas, fuel gas, or any inert gas. Owing to its higher efficiency, the majority of installed refinery sour water strippers employ steam as both a heating medium and a stripping gas [37]. Some of the steam strippers are provided with overhead condensers to remove the stripping steam from the overhead H_2S and NH_3. The condensed steam is recycled or refluxed back to the stripper. The results of a 1972 survey by the American Petroleum Institute suggested that, overall, refluxed strippers remove a greater percentage of H_2S and NH_3 than nonrefluxed strippers [5].

The operating conditions of sour water strippers vary from 0.1 to 3.5 atm (1–50 psig) and from 38 to 132°C (100–270°F). The sour water may or may not be acidified with mineral acid prior to stripping. H_2S is much easier to remove than NH_3. In pure water at 100°F, for example, the Henry's Law coefficient for NH_3 is 38,000 ppm/psia, whereas that for H_2S is 184 ppm/psia [37]. To remove 90% of the NH_3, a temperature of 110°C (230°F) or higher is usually employed, but 90% or more of the H_2S can be removed at 100°F.

Two-stage strippers are installed in some refineries to enhance the separate recovery of sulfide and ammonia. Acidification with a mineral acid is used to fix the NH_3 in the first stage and allow more efficient H_2S removal. In the second stage the pH is readjusted by adding caustics for efficient NH_3 removal. One example is the Chevron WWT process, which is essentially two-stage stripping with ammonia purification, so that the H_2S and NH_3 are separated. The H_2S goes to a conventional Claus sulfur plant and the NH_3 can be used as a fertilizer [20]. Figure 13 shows a schematic flow diagram of the Chevron WWT process.

Another way to treat sour water is air oxidation under elevated temperature and pressure. Compressed air is injected into the stream followed by sufficient steam to raise the reaction temperature to at least 88°C (190°F). Reaction pressure of 3.7 to 7 atm (50–100 psig) is required. Oxidation proceeds rapidly and converts practically all the sulfides to thiosulfates, and about 10% of the thiosulfates to sulfate [38]. Air oxidation, however, is much less effective than stripping in reducing the oxygen demand of sour waters, as the remaining thiosulfates can later be oxidized to sulfates by aquatic microorganisms. Air oxidation is sometimes carried out after sour water stripping as a sulfide polishing step.

Stripping of sour water is normally carried out to remove sulfides, hence the effluent may contain 50 to 100 ppm of NH_3, or even considerably more, depending on the influent ammonia

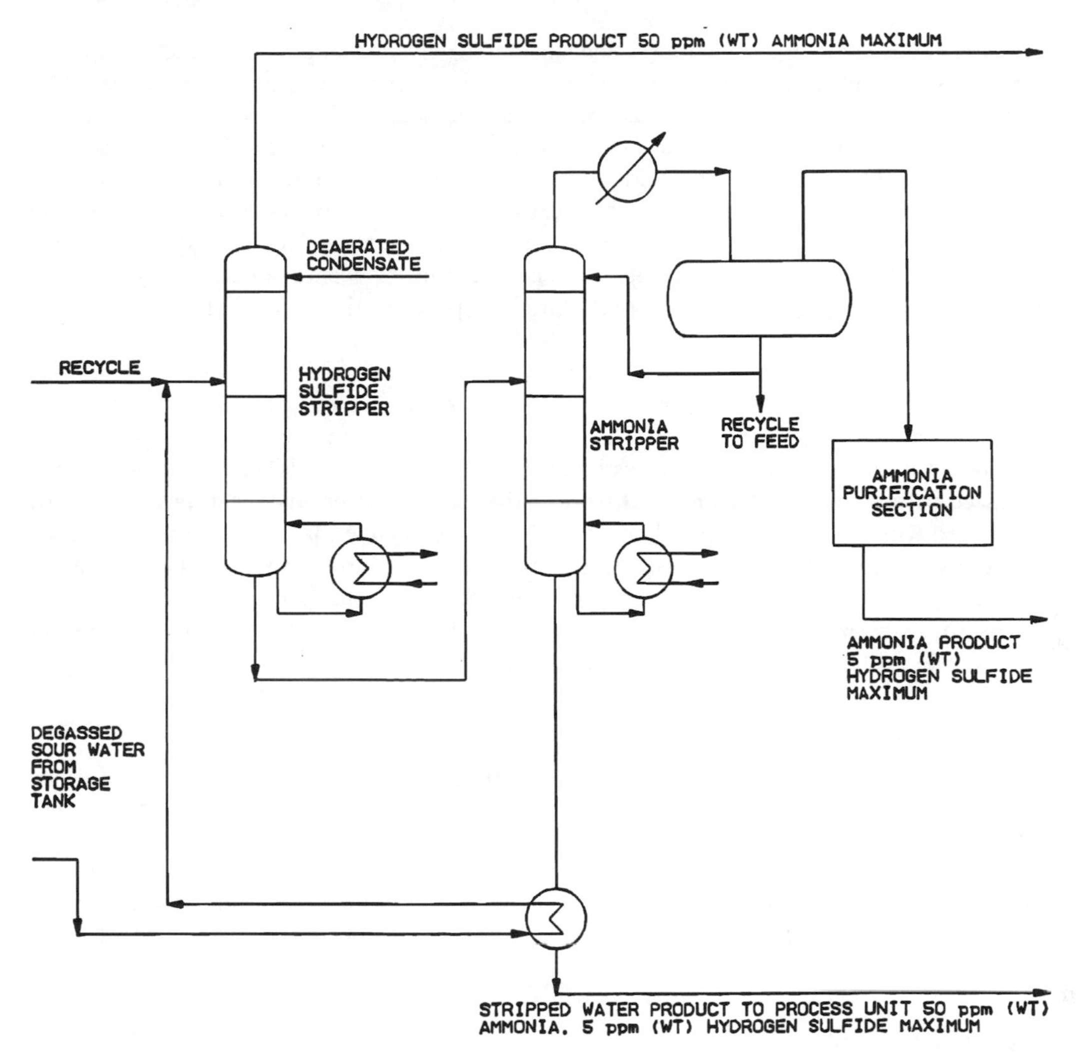

Figure 13 Chevron WWT process. Acid is used in first stage to enhance hydrogen sulfide removal. Caustic is used in second stage to enhance ammonia removal. (From Ref. 20.)

concentration. Values of NH_3 have been reported to be as low as 1 ppm, but generally the effluent NH_3 concentration is held to approximately 50 ppm to provide nutrient nitrogen for the refinery biological waste treatment system. Because of more stringent effluent requirements for NH_3, many refineries seek to improve the sour water stripping systems for NH_3 removal. This can be done by (1) increasing the number of trays, (2) increasing the steam rate, (3) increasing tower height, and (4) adding a second column in series. All these methods are now available to the refining industry [5].

Spent Caustics Treatment

Caustics are widely used in petroleum refineries. Typical uses are to neutralize and to extract acidic materials that may occur naturally in crude oil, acidic reaction products that may be produced by various chemical treating processes, and acidic materials formed during thermal and catalytic cracking such as H_2S, phenolics, and organic acids.

Spent caustics may therefore contain sulfides, mercaptides, sulfates, sulfonates, phenolates, naphthenates, and other similar organic and inorganic compounds [38]. Spent caustics can also be classified as phenolic and sulfidic [37]. Sulfidic spent caustics are rich in sulfides, contain no phenols, and can be oxidized with air. Phenolic spent caustics are rich in phenols and must be neutralized with acid to release and remove the phenols.

At least four companies process spent caustics to market the phenolics and the sodium hyposulfite. However, the market is limited and most of the spent caustics are very dilute, so the cost of shipping the water makes this operation uneconomic. Concentration can be increased by recycling spent caustics at the treater or recycling the spent caustics found in the water bottoms of intermediate product tanks [39].

Some refineries neutralize the caustic with spent sulfuric acid from other refining processes, and charge it to the sour water stripper. This removes the H_2S. The bottoms from the sour water stripper go to the desalter, where the phenolics can be extracted by the crude oil.

Spent caustics usually originate as batch dumps, and the batches may be combined and equalized before being treated and discharged to the refinery sewer. Spent caustics can also be neutralized with flue gas to form carbonates. Sulfides, mercaptides, phenolates, and other basic salts are converted by the flue gas (reaction time 16–24 hours) stripping. Phenols can be removed, then used as a fuel or sold. H_2S and mercaptans are usually stripped and burned in a heater. Some sulfur is recovered from stripper gases. The treated solution contains mixtures of carbonates, sulfates, sulfites, thiosulfates, and some phenolic compounds.

Ballast Water Separation

Ballast water normally is not discharged directly to the refinery sewer system because of the intermittent high-volume discharges [38]. The potentially high contents of salt, oil, and organics in ballast water would upset the treatment facilities if not controlled. Ballast water may also be treated separately by heating, settling, and at times filtration. The settling tank can also be provided with a steam coil for heating the tank contents to help break emulsions, and an air coil to provide agitation. The recovered oil, which may be considerable, is generally sent to the slop oil system.

Slop Oil Treatment

Separator skimmings, which are generally referred to as slop oil, require treatment before they can be reused because they contain an excess amount of solids and water. Solids and water contents of about 1% generally interfere with processing [38].

In most cases slop oils are easily treated by heating to 88°C (190°F) for 12 to 14 hours. At the end of settling, three definite layers exist: a top layer of clean oil; a middle layer of secondary emulsion; and a bottom layer of water containing soluble components, suspended solids, and oil. It may be advantageous or even necessary to use acid or specific chemical demulsifiers to break slop oil emulsions. The water layer has high BOD and COD contents, but also low pH (after acid treatment), and must be treated before it can be discharged. Slop oil can also be successfully treated by centrifugation or by precoat filtration using diatomaceous earth.

6.6.3 End-of-Pipe Treatment

Conventional refinery wastewater treatment technology is mainly concerned with removing oil, organics, and suspended solids before discharge. However, because of new stringent discharge requirements for specific toxic constituents as well as whole-effluent toxicity, specific advanced treatment processes are becoming a necessity for many refineries. This section describes the conventional treatment processes used in refineries. Specific advanced treatment processes are described in the next section.

Conventional refinery wastewater treatment processes can be categorized into primary, intermediate, secondary, and tertiary treatment processes [17]. Primary processes include API separators and parallel or corrugated plate interceptors (CPI) to remove free oil. Intermediate processes include dissolved air flotation (DAF) or induced air flotation (IAF) and equalization. Secondary processes include biological treatment processes in their different forms or combinations. These can include activated sludge, trickling filters, aerated lagoons, stabilization, and rotating biological contactors (RBC). Tertiary treatment processes include filtration and granular activated carbon (GAC) adsorption. Activated sludge enhanced with powdered activated carbon (PACT®), a combination of secondary and tertiary processes, is discussed in the next section.

API Separators

The API separator is a widely used gravity separator for removal of free oil from refinery wastewater. It can be installed either in the central wastewater treatment plant or as an upstream pretreatment process to remove gross quantities of free oil and solids.

The process involves removal of materials less dense than water (such as oil) and suspended materials that are more dense than water by settling. The API separator does not separate substances in solution, nor does it break emulsions. The effectiveness of a separator depends on the temperature of the water, the density and size of the oil globules, and the amounts and characteristics of the suspended materials. The susceptibility to separation (STS) test is normally used as a guide to determine what portion of the influent to a separator is amenable to gravity separation [38]. In terms of globule size, an API separator is effective down to globule diameters of 0.015 cm (15 microns).

The API has long been active in the study of oil–water separators. Its design recommendations are clearly and adequately set forth in the API manual [40]. The basic design of an API separator is a long rectangular basin, with enough detention time for most of the oil to float to the surface and be removed. Most API separators are divided into more than one bay to maintain laminar flow within the separator, making the separator more effective. They are usually equipped with scrapers to move the oil to the downstream end of the separator where it is collected in a slotted pipe or on a drum. On their return to the upstream end, the scrapers travel along the bottom moving the solids to a collection trough. Sludge can be dewatered and either incinerated or disposed of in hazardous waste landfills. To control volatile organic compound emissions to the atmosphere, U.S. refineries are required to install covers for oil–water separators (40 CFR Part 60).

Because of the limitations in gravity separator design, the lower limit of free oil in API separator effluent is usually around 50 mg/L. Removal of other contaminants in an API separator is highly variable. Table 16 shows typical removal efficiencies of oil separator units for several contaminants [17]. Chemical oxygen demand removal efficiencies range from 16 to 45%, and suspended solids removal ranges from 33 to 68%.

Parallel and Corrugated Plate Separators

Parallel and corrugated plate separators are improved types of oil–water separators with tilted plates installed at an angle of 45°. This increases the collection area many times while decreasing the overall size of the unit accordingly. As the water flows through the separator, the oil droplets coalesce on the underside of the plates and travel upward to where the oil is collected.

Table 16 Typical Efficiencies of Oil Separation Units

Oil content					
Influent (mg/L)	Effluent (mg/L)	Oil (percent removed)	Type of separator	COD (percent removed)	SS (percent removed)
300	40	87	Parallel plate	–	–
220	49	78	API	45	–
108	20	82	Circular	–	–
108	50	54	Circular	16	–
98	44	55	API	–	–
100	40	60	API	–	–
42	20	52	API	–	–
2,000	746	63	API	22	33
1,250	170	87	API	–	68
1,400	270	81	API	–	35

COD, chemical oxygen demand; SS, suspended solids.
Source: From Ref. 17.

Because of the coalescing action, these separators can separate oil droplets as small as 0.006 mm (6 microns) in diameter and produce effluent-free oil concentrations as low as 10 mg/L [27].

There is a broad range of applications for tilted-plate separators. As little space is required, they can be installed to polish the effluent from existing API separators that are either overloaded or improperly designed, or they can be installed parallel with existing separators, reducing the hydraulic load and enhancing the oil removal capacity of the system.

Dissolved Air Flotation

Dissolved air flotation (DAF) is a process commonly used in refineries to enhance oil and suspended solids from gravity-separator effluent. In some refineries it is used as a secondary clarifier for activated sludge systems and as a sludge thickener. The process involves pressurizing the influent or recycled wastewater at 3–5 atm (40–70 psig) then releasing the pressure, which creates minute bubbles that float the suspended and oily particulates to the surface. The float solids are removed by a mechanical surface collector.

If a significant portion of the oil is emulsified, chemical addition with rapid-mix and flocculation chambers are a part of the flotation unit, breaking the emulsion and enhancing the separation. Chemicals normally used include salts of iron and aluminum and polyelectrolytes.

Dissolved air flotation in combination with flocculation can reduce oil content in refinery wastewater to levels approaching oil solubility [40]. According to Katz [41], DAF plus chemical aids for flocculation can be expected to reduce BOD and COD by 30–50% and to reduce total oil to the range 5–25 mg/L. Table 17 shows some data for oil removal from refinery wastewater [27]. Removal efficiencies range from 70 to 90%. The accepted design overflow rates for DAF units are between 60 and 120 L/min per square meter (1.5–3.0 gpm/sq ft) [17].

Dissolved air flotation equipment is available from a number of manufacturers. Packaged units of steel construction are available with capacities to 7.6 cu m/min (2000 gpm). The essential elements of the DAF system are the pressurizing pump, air injection facilities, pressurization tank or contact vessel, back-pressure regulating device, and the flotation chamber [40].

Three principal variations in the process design of DAF systems are full-flow, split-flow, and recycle operation (Fig. 14). Full-flow operation consists of pressurizing the entire waste

Table 17 Oil Removal by Dissolved Air Flotation in Refineries

Coagulant dosage (mg/L)	Oil concentration (mg/L)		Removal (percent)
	Influent	Effluent	
0	125	35	72
100 (alum)	100	10	90
130 (alum)	580	68	88
0	170	52	70

Source: From Ref. 27.

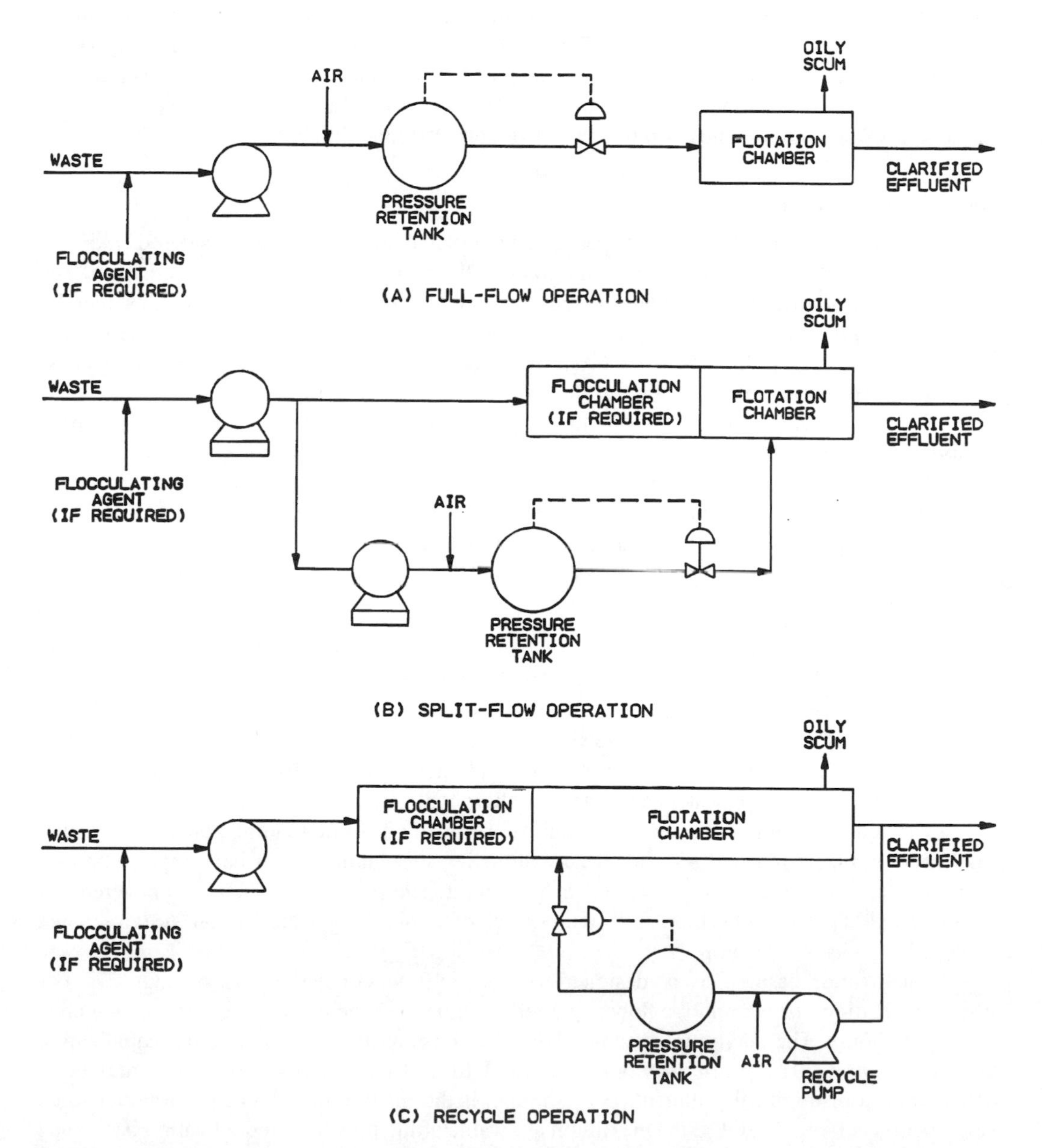

Figure 14 Variations in dissolved air flotation (DAF) design. (A) Full-flow operation; (B) split-flow operation; (C) recycle operation. (From Ref. 40.)

stream, followed by release of pressure and bubble formation at the inlet to the flotation chamber. Split operation consists of pressurizing only part of the waste flow and diverting the remainder directly into the flocculation or flotation chamber. Recycle operation consists of pressurizing a recycle stream of the clarified effluent. The recycled stream usually amounts to between 20 and 50% of the oily wastewater flow. The pressure is released and the bubble-containing recycle stream is mixed with the unit influent. Each of these variations has its advantages and disadvantages.

A relatively new design of a high-rate DAF unit uses a shallow bed system (Supracell) with only 3 minutes of retention time and operated at an overflow rate of 140 Lpm/sq m (3.5 gpm/sq ft) [42]. This unit has been used for industrial and municipal wastewater treatment and offers lower capital cost and headroom requirements. It was installed at a petrochemical complex in Texas as a secondary clarifier to improve the operation and the capacity of an existing activated sludge system [43]. In recent years, nitrogen has replaced air in covered DAF systems because of the potential for explosion. These systems are called dissolved nitrogen flotation (DNF) systems. The operations of DAF and DNF are similar.

Induced Air Flotation

The induced air flotation (IAF) system operates on the same principles as a pressurized DAF unit [27]. The air, however, is typically mixed into the effluent by a rotor-disperser mechanism. The rotor, which is submerged in the liquid, forces the liquid through the disperser openings, thereby creating a negative pressure. This pulls the air downward into the liquid, causing the desired gas–liquid contact. The liquid moves through the flotation cell(s), and the float skimmings pass over the overflow weirs on each side of the unit.

The advantages of an IAF unit include significantly lower capital cost and smaller space requirements than a DAF unit. On the other hand, it has a higher connected power requirement than a DAF unit. It also has a higher volume of float skimmings: the normal range is 3–7% of the incoming flow for IAF units and less than 1% for DAF units [27]. Induced air flotation units have been used in petroleum refineries and oilfields for removing free oil and suspended solids.

Equalization

The purpose of equalization is to dampen out surges in flows and loadings to maintain optimum conditions for subsequent treatment processes. This is especially necessary for a biological treatment plant, as high concentrations of certain materials will upset or completely kill the bacteria in the biotreater. Many wastewater discharges within refinery complexes are from washdowns, tank cleanings, batch operations, and inadvertent spills, necessitating a basin capable of receiving these waters and allowing their controlled release [17].

The equalization step in a refinery usually consists of one or more large ponds or tanks that may contain mixers to stir the wastes. Many refineries are planning to or have replaced ponds with steel tanks because of the requirements for groundwater protection. The use of covered and vented tanks also provides more positive control of odors from equalization systems when anaerobic conditions develop.

Equalization basins may be designed to equalize flow, concentrations, or both [27]. For flow equalization, the cumulative flow is plotted vs. time over the equalization period, which is usually 24 hours. The maximum volume above the constant-discharge line is the equalization volume required. The basin may also be sized to restrict the discharge to a maximum concentration of a critical pollutant. For example, if the maximum effluent from an activated sludge unit is 20 mg/L BOD_5, the maximum allowable effluent from the equalization basin may be computed and thereby provide a basis for sizing the unit. Novotny and England [44] suggested a formula for computing the required equalization time for the case of near-constant wastewater flow and a normal statistical distribution of wastewater composite analyses.

Activated Sludge

Activated sludge is the most common biological treatment process because of the high rate and degree of organic stabilization possible. It is widely used in treating refinery wastewater [5].

Activated sludge is an aerobic biological treatment process in which high concentrations (1500–3000 mg/L) of newly grown and recycled microorganisms are suspended uniformly throughout a holding tank to which raw wastewaters are added. Oxygen is introduced by mechanical aerators, diffused air systems, or other means. The organic materials in the waste are removed by the microbiological growth and stabilized by biochemical synthesis and oxidation reactions. The term activated stems from the fact that the microbial sludge is a floc that is highly active in adsorbing colloidal and suspended waste matter from the aqueous stream [25].

The basic activated sludge system consists of an aeration tank followed by a sedimentation tank. The microbial floc removed in the sedimentation tank is recycled to the aeration tank to maintain a high concentration of active microorganisms. Although the microorganisms remove almost all of the organic matter from the waste being treated, much of the converted organic matter remains in the system in the form of microbial cells. Because of their oxygen demand, these cells must be removed from the treated wastewater before discharge. Thus, final sedimentation and recirculation of biological solids are important elements in an activated sludge system.

Although refinery wastewaters are generally highly amenable to activated sludge treatments, the exact treatability of a refinery-petrochemical installation is a function of many factors such as the classification of the refinery, the type of crude charge, the age of the facility and nature of its collection system, the relative effluent volume attributed to utility blowdown, and the degree of in-plant control. For these reasons, activated sludge facilities vary from one installation to another [17]. Treatability studies using bench- and pilot-scale trials therefore are used to formulate the basic design criteria and predict treated effluent quality.

The three basic types of activated sludge systems are conventional, contact stabilization, and extended aeration systems [40]. Other types include high-purity oxygen systems and sequencing batch reactors, but these are not commonly used in refineries.

The conventional activated sludge system allows for absorption, flocculation, and synthesis in a single step. It usually employs long, rectangular aeration tanks that approximate plug flow conditions, or crossflow aeration tanks that approach complete mixing. The oxygen utilization rate is high in the beginning of the aeration tank, but decreases with aeration time or distance down the tank. Where complete treatment is required, the oxygen utilization rate approaches the endogenous value toward the end of the aeration tank. The conventional process can operate over a wide loading range, which is limited by flocculation, settling, and separation requirements of the microbial flocs.

The contact stabilization system provides for removal of the organics from the wastewater by contact with activated sludge (absorption) and transfer to a separate aeration tank for oxidation and synthesis. This process is applicable to wastes containing a large proportion of the BOD in suspended or colloidal form. The influent is first contacted with the activated sludge in an aeration basin for a relatively short retention period (15–30 minutes). This contact basin removes the suspended or colloidal content from the stream by absorption on the sludge floc. The mixed liquor flows to the settler-thickener where the clarified effluent overflows and the thickened sludge flows to a stabilization basin. A small part of the thickened sludge is discarded as waste. The recycled sludge is aerated in the stabilization basin for 1–5 hours. During this period, the adsorbed organics undergo synthesis and endogenous respiration and the sludge becomes stabilized. This process results in savings in total basin area as only the recycled sludge, not the whole waste stream, is subjected to long-time aeration. However, if the oxygen demand of the influent is due mostly to dissolved rather than suspended contaminants, the short retention period in the contact basin may not produce a satisfactory effluent [25].

The extended aeration system is one in which the synthesized cells undergo autooxidation, resulting in a minimum of solids disposal. Extended aeration is reaction-defined rather than a hydraulically-defined mode and can be designed as a plug flow or a complete mix system. Design parameters include a food/microorganism ratio (F/M) of 0.05–0.15, a sludge age of 15–35 days, and mixed-liquor suspended solids (MLSS) concentrations of 3000–5000 mg/L [27]. This process has low cell growth rates, low sludge yields, and high oxygen requirements compared with the conventional activated sludge process. The advantages are high-quality effluent and less sludge production.

The extended aeration process can be sensitive to sudden increases in flow due to resultant high-MLSS loadings on the final clarifier, but is relatively insensitive to shock loads in concentrations due to the buffering effect of the large biomass volume. Because of the long sludge age, nitrification can be incorporated into the design of the extended aeration process. Extended aeration in the form of loop-reactor or ditch systems has been used significantly in wastewater treatment during recent years.

Other variations of activated sludge such as deep shaft high-rate activated sludge and sequencing batch reactor (SBR) have been used for refinery wastewater treatment. A refined deep shaft process has been installed and in operation at the Chevron refinery in Burnaby, British Columbia, Canada, since 1996 [45]. In the course of a recent wastewater treatment upgrade, a BP refinery on the eastern Australian coast converted an existing lagoon to an SBR system [46].

The design organic load for most activated sludge systems ranges from 0.1 to 1.0 lb BOD_5/(day)(lb MLSS) [17]. Higher loadings can be imposed, but generally at the expense of poorer efficiency and higher organic levels in the treated effluent. Table 18 shows the performance of typical activated sludge systems in refineries based on loading and retention time [40].

One particularly important parameter for the influent to an activated sludge system in a refinery is oil and grease, which can lower floc density to a level where the sludge-settling properties are destroyed. A study conducted for USEPA [47] indicated that an activated sludge

Table 18 Performance of Typical Activated Sludge Systems

Type of waste	Sludge loading (lbs of BOD/day/lbs of sludge)	Detention time (hours)	Percentage of BOD reduction	Comment
Refinery	0.6	4–5	90–95	99% phenol removal
Refinery	0.3–0.4	3–4	90–95	98% phenol removal
Refinery	0.1–0.2	18–22	88–92	Minimal sludge production
Petrochemical	0.65–0.76	8–10	95–97	Sludge bulks for long periods

BOD, biochemical oxygen demand.
Source: From Ref. 40.

system will perform satisfactorily with continuous loading of hexane extractables of 0.1 lb/lb MLSS. It was recommended that the influent to the biological system should contain less than 75 mg/L hexane extractables and preferably less than 50 mg/L.

Aerated Lagoon

Aerated lagoons are low-rate biological systems in which a flow-through basin allows microorganisms in contact with the wastewater to reduce organic constituents biochemically. Unlike activated sludge there is no solids recycle. Retention times are usually between 3 and 10 days [38]. Oxygenation and mixing can be carried out with mechanical or diffused aeration units and through induced surface aeration. Depths of 3–4.3 m (10–14 ft) are used to accommodate the aeration equipment and minimize area requirements [40].

Aerated lagoons have been extensively used in refineries to treat wastewaters because of their ease of operation and maintenance, ability to equalize wastewater, and ability to dissipate heat when desirable. However, because of their inherent limitations, they are usually used upstream from waste stabilization ponds or as an interim treatment process that can be converted to an activated sludge system. BOD_5 reductions in completely mixed aerated lagoons may range from 40 to 60%, with little or no reduction in suspended solids [38]. Because of more stringent effluent discharge requirements, many lagoons have been converted to other more effective processes such as SBR, as discussed previously in the activated sludge section [46].

Waste Stabilization Ponds

A stabilization pond is a simple pond in which aeration is not mechanically enhanced. Its shallow depth allows the pond to function aerobically without mechanical aerators. Algae in the pond produce oxygen through photosynthesis, which is then used by the bacteria to oxidize the wastes. Because of the low loadings, little biological sludge is produced and the pond is fairly resistant to upsets due to shock loadings.

The stabilization pond is practical where land is plentiful and cheap. It has a large surface area and a shallow depth, usually not exceeding 2 m (6 ft). Stabilization ponds have a long retention, ranging from 11 to 110 days [38], depending on the land available as well as the design requirement.

Stabilization ponds have been successfully used in the treatment of refinery and petrochemical wastewaters. They are used either as the major treatment step or as a polishing process after other treatment processes. In the United States, because land is generally quite expensive, the use of waste stabilization ponds is limited [17].

Trickling Filter

The trickling filter is a packed bed of medium covered with biological slime growth through which the wastewater is passed. Wastewater is sprinkled onto the medium through a rotating distribution system above the bed. As the wastewater passes through the slime, organics and oxygen diffuse into the microbial mass where they are oxidized to carbon dioxide, water, and metabolic byproducts [17]. The trickling filter is followed by a clarifier to settle sloughed-off slimes. Recycle flow may be taken either before or after clarification.

Conventional trickling filters contain 6 to 10 cm (2.5 to 4 in) rocks and vary in depth from 1 to 2.5 m (3–8 ft). Hydraulic loadings are 20 Lpm/sq m (0.5 gpm/sq ft) or less. Plastic packings are employed in depths up to 12 m (40 ft), with hydraulic loadings as high as 240 Lpm/sq m (6.0 gpm/sq ft) [40]. Trickling filters are fixed reactors and are simple to operate. However, the reaction rate for treating soluble industrial wastewaters is relatively low, hence they are not economically attractive for high treatment efficiency (85% BOD reduction) of such wastewaters [27].

The petroleum industry uses them mostly as roughing devices to reduce the loading on activated sludge systems. In some cases, trickling filters are used to pretreat steam-stripped sour water before mixing it with other refinery wastewater streams for secondary treatment [48].

Rotating Biological Contactors

Rotating biological contactors (RBCs) have attracted widespread attention in the United States since 1969 [5]. RBCs generally consist of rows of plastic discs mounted on horizontal shafts that turn slowly keeping the disc about 40% immersed in a shallow tank containing wastewater as shown in Fig. 15. A 1 to 4 mm layer of slime biomass is developed on the media. This is equivalent to 2500–10,000 mg/L of MLSS in a mixed system [27]. Single RBC units are up to 3.7 m (12 ft) in diameter and 7.6 m (25 ft) long, containing up to 9300 square meters (100,000 sq ft) of surface in one section.

The RBC is a combination of fixed film reactor and mechanical aerator. The fixed film reactor is the disc upon which microorganisms attach themselves and grow. Aeration occurs while a section of disc is above water level. Microorganisms produce a film on the surface of the disc, remove organic matter from the wastewater, and accumulate on each disc. Excess biomass is stripped and returned to the wastewater stream by the shearing action of water against rotating discs. Waste biomass is held in suspension by the mixing action of the discs, and carried out of the reactor for removal in a clarifier. Treatment efficiency can be improved by increasing the number of RBCs in series, and by temperature control, sludge recycle, and chemical addition.

Advantages of RBCs include the ability to sustain shock loads because of high microorganism concentrations, ease of expansion because of modular design, and low power consumption, which may be particularly attractive for industrial application. Full-scale RBC installations in refineries have performances in removal of oxygen-demanding pollutants comparable to activated sludge systems [5].

Filtration

The use of filtration to polish biological treatment system effluent has become more popular in recent years because of more stringent discharge requirements. The 1977 EPA survey of petroleum refineries indicated that 27 of 259 plants used filtration as part of the existing treatment scheme and 16 others planned to install filtration systems in the near future [5]. Filtration can improve effluent quality by removing oil, suspended solids, and associated BOD and COD, and carryover metals that have already been precipitated and flocculated. Improved effluent filtration in one recent instance helped a Colorado refinery to meet the newly adopted discharge toxicity requirements [49].

Granular-medium filters are the predominant type of filtration systems used in refineries. The medium can be sand, dual medium of anthracite and sand, or multimedium of anthracite, sand, and garnet. As the water passes down through a filter, the suspended matter is caught in the pores. When the pressure drop through the filter becomes excessive, the flow is reversed to remove the collected solids. The backwash cycle occurs approximately once a day, depending on the loading, and usually lasts for 10 to 15 minutes. The normal surface loading rate is between 80 and 200 Lpm/sq m (2–5 gpm/sq ft). Coagulants such as iron and aluminum salts and polyelectrolytes can enhance suspended solids removal.

Several advanced filtration systems are finding applications in treating refinery wastewaters. Examples include the HydroClear filter (Zimpro, Rothschild, WI) and the Dynasand filter (Parkson Corporation, Fort Lauderdale, FL). The HydroClear filter employs a single sand medium (0.35–0.45 mm) with an air mix (pulsation) for solids suspension and regeneration of the filter surface. Filter operation enables periodic regeneration of the medium

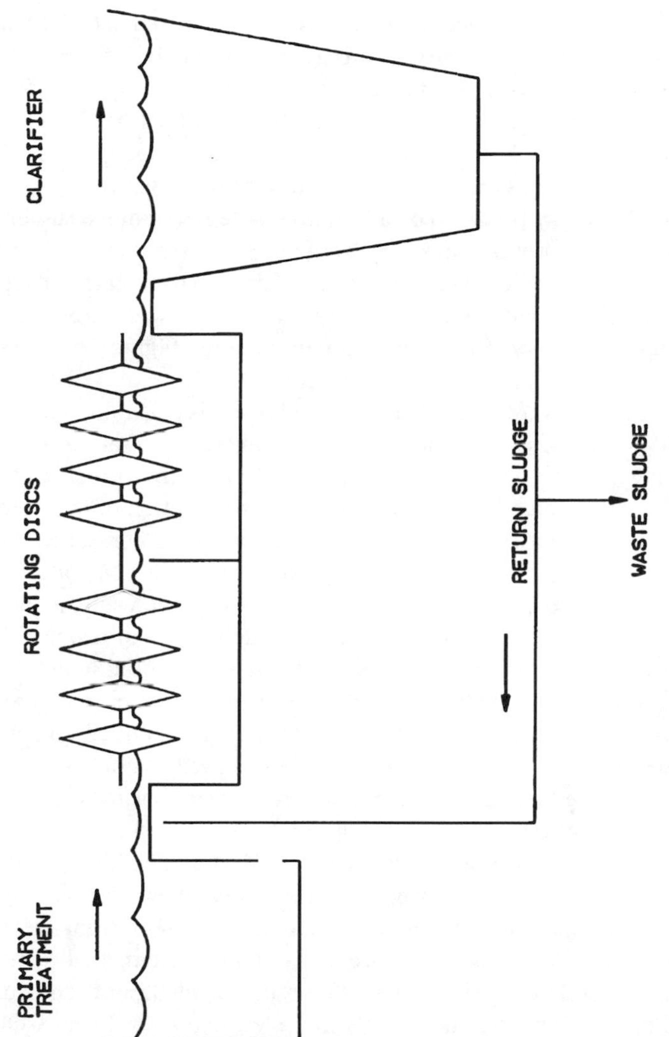

Figure 15 Rotating biological contactors. Plastic discs rotate slowly in a shallow tank. About 60% of each disk is above water surface for aeration. (From Ref. 5.)

surface without backwashing. The Dynasand filter is a continuous self-cleaning upflow deep-bed granular-medium filter [27]. The filter medium is cleaned continuously by recycling the sand internally through an airlift pipe and sand washer. The regenerated sand is redistributed on top of the bed, allowing for a continuous uninterrupted flow of filtered water and reject water.

Activated Carbon Adsorption

Activated carbon adsorption is most often employed for removal of organic constituents from water and wastewater. Granular activated carbon (GAC) or powdered activated carbon (PAC) may be used. Granular activated carbon columns can be used for secondary treatment of industrial wastewaters or for tertiary treatment to remove residual organics from biological treatment effluent. The primary use of PAC in wastewater treatment has been in the PACT® process (Zimpro), in which PAC is added to the activated sludge process for enhanced performance. This process is discussed in the next section of this chapter.

A GAC system is generally preceded by a filtration system to remove suspended solids to minimize plugging of the adsorption sites (pores). Filtered water flows to a bank of GAC columns arranged in series or parallel. As the water flows through the columns the pollutants are adsorbed onto the carbon, gradually filling the pores. The exhausted carbon is removed for regeneration in a furnace or disposed of in appropriate landfills. Figure 16 shows the process flowsheet for a GAC system with a regeneration system.

The adsorption of organics from the liquid to a solid phase is generally assumed to occur in three stages [50]. The first is the movement of the contaminant (adsorbate or solute) through a film surface surrounding the solid phase (adsorbant). The second is the diffusion of the adsorbate within the pores of the activated carbon. The final stage is the sorption of the material onto the surface of the sorbing medium. The overall rate of adsorption is controlled by the rate of diffusion of the solute molecules within the capillary pores of the carbon particles [27].

Adsorption can be divided into two types. Chemical adsorption results in the formation of a monomolecular layer of the adsorbate on the surface through forces of residual valence of the surface molecules. Physical adsorption results from molecular condensation in the capillaries of the solid. In general, substances of the highest molecular weight are most easily adsorbed [27].

Currently the use of full-scale GAC systems in the U.S. petroleum refining industry is very limited. Some refineries used GAC as the secondary treatment process but have discontinued the operations. Two examples are the Atlantic Richfield (Arco) system near Wilmington, CA, and the British Petroleum (BP) system in Marcus Hook, PA [17].

The Arco GAC system was designed to treat 50 MGD of combined storm runoff and process water during periods of rainfall when the treatment plant of Los Angeles County Sanitation District (LACSD) cannot accommodate the storm runoff from the refinery. The GAC system included 12 adsorber cells, a carbon handling system, and a multiple-hearth regeneration system. The design was based on COD removal of 85% at an average influent concentration of 250 mg/L. The operating results indicated that the effluent COD was in the range of the predicted level when the influent concentration did not exceed the design basis. However, the carbon consumption rate ranged from 0.30 to 0.35 kg COD removed per kg of carbon, rather than the 1.75 kg COD/kg carbon predicted. The system is no longer in operation primarily because the treatment requirements imposed by LACSD have been changed.

The BP refinery used a filtration/GAC system to treat API separator effluent before discharge. It consisted of three parallel adsorbers each containing 42,000 kg of carbon in beds 14 m (45 ft) deep. The design contact time was 40 minutes and theoretical carbon capacity was 0.3 kg TOC/kg carbon. The regeneration facility was a 1.5 m diameter, multiple-hearth furnace. After several years of operation, BP abandoned the GAC system and installed a biological

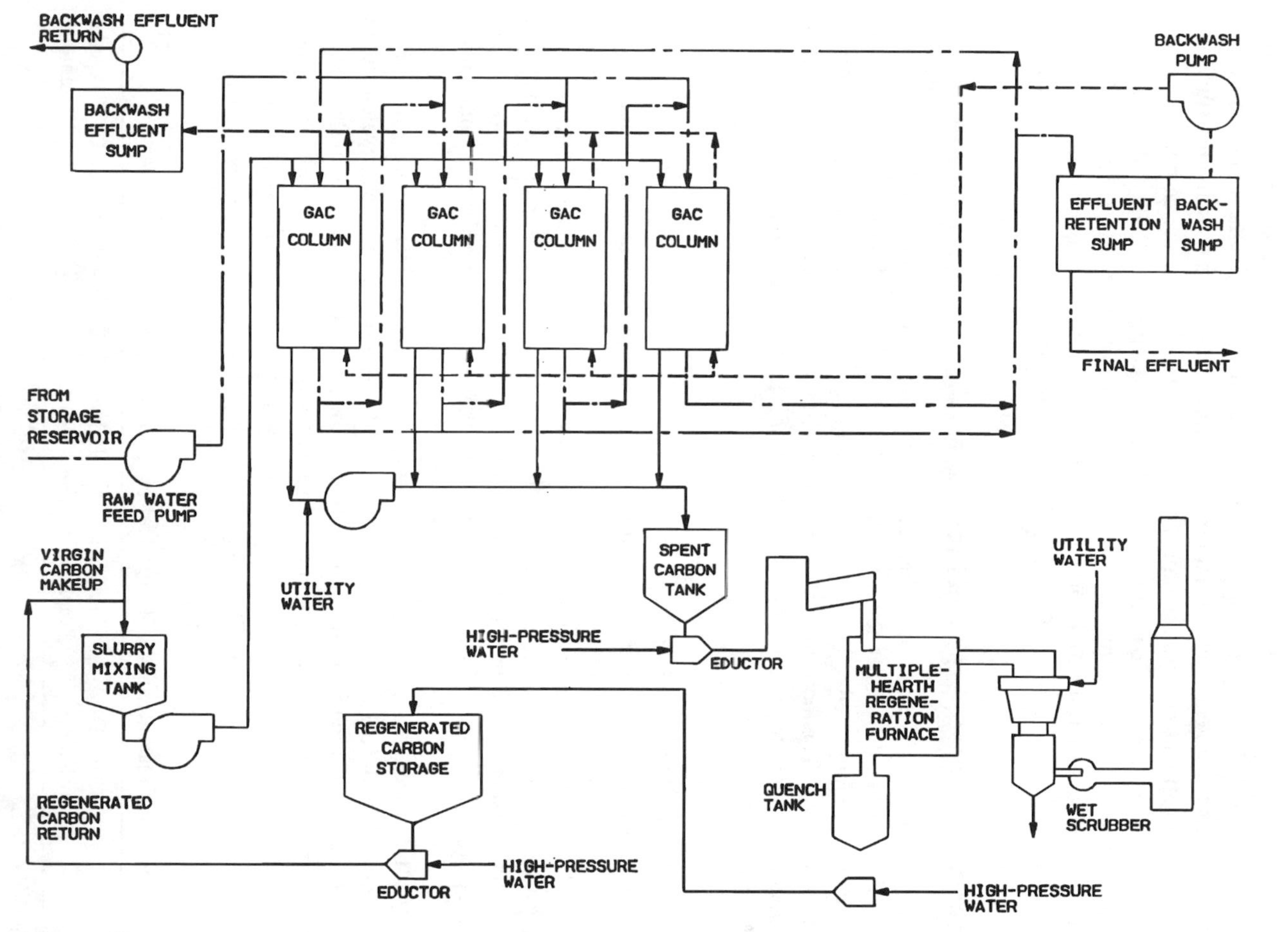

Figure 16 Process flowsheet of a GAC system with regeneration. In this complete GAC adsorption and regeneration system, four GAC columns can be operated in parallel or in series. Spent carbon is transferred to a multiple-hearth furnace for thermal regeneration. Regenerated carbon is mixed with virgin makeup and pumped back to the GAC columns. The GAC columns are backwashed periodically. (From Ref. 27.)

treatment system for secondary treatment because of operational problems including inadequate pretreatment of the API separator effluent in terms of O&G and soluble organics removal, buildup of anaerobic biological growths and oily materials in the carbon media, and a 40% decrease in adsorptive capacity of the regenerated carbon.

The use of GAC systems to follow biological treatment processes is a more promising application. Adding GAC as a polishing process may be necessary in the future in certain refineries to meet more stringent discharge requirements for toxic constituents. In pilot studies of GAC as a tertiary treatment process for refinery and petrochemical plants, carbon adsorption following biological treatment was particularly effective in reducing both BOD and COD to low levels; Table 19 shows the results for COD removal in some of these studies [51]. Activated carbon also removes a variety of toxic organic compounds from water and wastewater [52]. More discussions of GAC for control of whole effluent toxicity are presented in the next section.

6.6.4 Specific Advanced Treatment Processes

Many refineries in the United States are being required to control whole-effluent toxicity as well as specific toxic constituents to meet new wastewater discharge limits. There can be a variety of toxic constituents that may need to be controlled, depending on waste characteristics and local water quality objectives. The more common constituents in refinery wastewater include cyanide and heavy metals. The treatment processes for control of whole-effluent toxicity, cyanide, and heavy metals are discussed below.

Control of Whole Effluent Toxicity

Any treatment process that can remove the toxicity-causing constituents can reduce whole-effluent toxicity of a discharge. If the primary cause of effluent toxicity can be identified through the TIE or TRE procedures, specific treatment processes can be incorporated into the existing treatment system to control the toxicity. However, for a complex wastewater such as that from refinery and petrochemical facilities, the cause of toxicity may not be easily identified. The toxicity can be caused by a combination of constituents that exhibit synergistic or antagonistic effects.

The PACT® process has great potential for controlling whole-effluent toxicity in refinery wastewater. It involves the addition of powdered activated carbon (PAC) to the activated sludge process for enhanced performance [53]. Figure 17 shows the process flow diagram of the PACT® process. The addition of PAC has several process advantages: decreasing variability in effluent quality, removing nondegradable organics by adsorption, reducing inhibitions in industrial wastewater treatment (e.g., nitrification), and removing refractory priority pollutants [27].

Table 19 Carbon Pilot Plant Results for Petrochemical and Refining Wastewaters

Type of wastewater	Design Q (MGD)	Process application	Influent COD (mg/L)	Effluent COD (mg/L)	Percent removal
Petrochemical	3	Tertiary	150	49	67
Refinery	26	Tertiary	100	41	59
Refinery	28	Tertiary	300	50	83
Refinery	8	Tertiary	100	40	60
Petrochemical	29	Tertiary	150	48	68

Source: From Ref. 51.

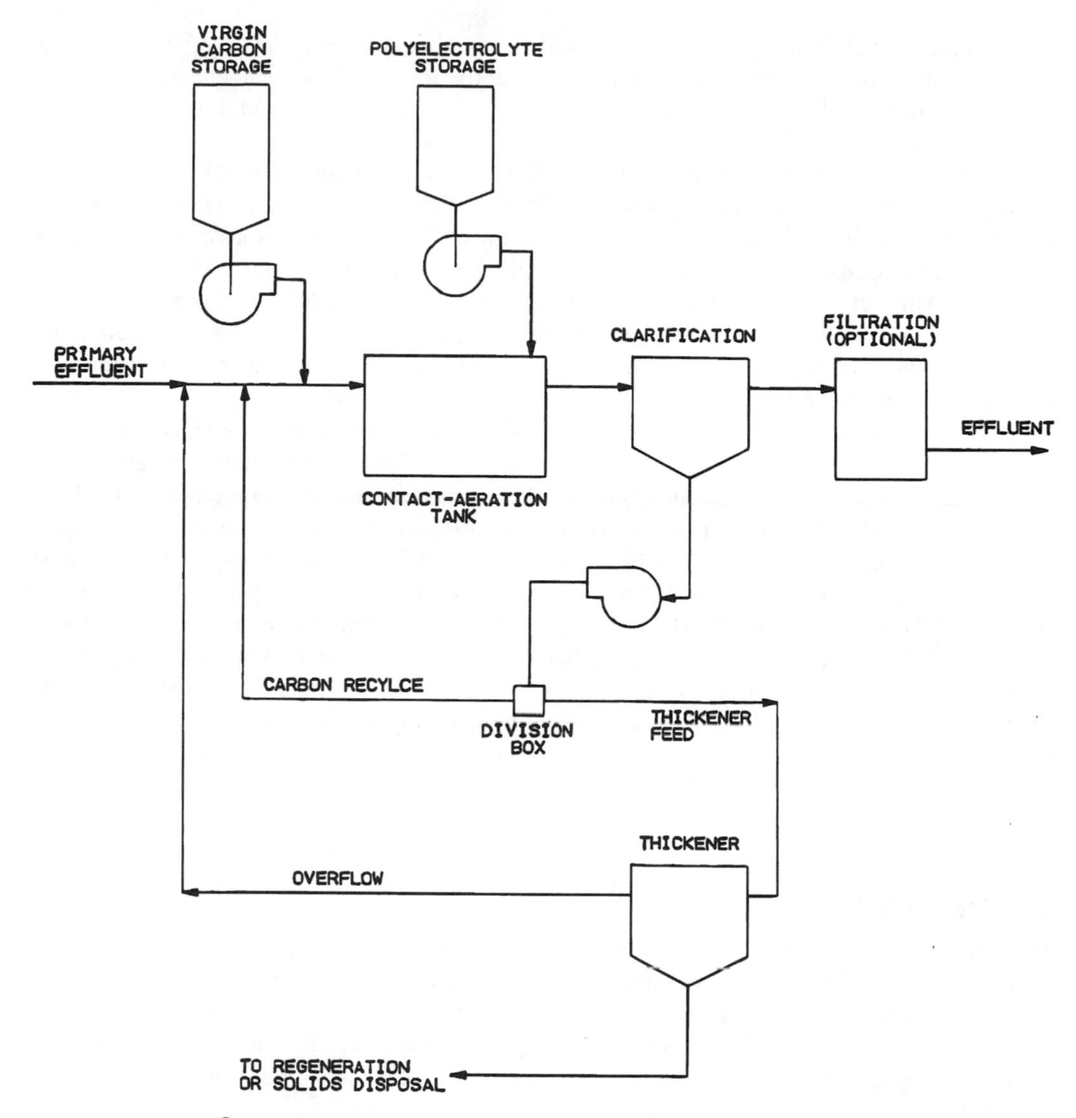

Figure 17 PACT® wastewater treatment system general process diagram. Powdered activated carbon is added to the aeration tank influent in an activated sludge system. Polyelectrolyte is added to enhance flocculation of carbon fines and microorganisms. Filtration may or may not be required. (From Ref. 27.)

Several studies have added PAC to petroleum refinery activated sludge systems. Rizzo [54] reported on a plant test in which carbon was added to an extended aeration treatment system at the Sun Oil Refinery in Corpus Christi, TX. Test results showed that even very small carbon dosages (9–24 mg/L) significantly improved removal of BOD, COD, and total suspended solids, as well as producing uniform effluent quality, a clearer effluent and eliminating foam. Grieves *et al.* [55] reported on a pilot plant study at the Amoco refinery in Texas City where PAC was added to the activated sludge process in 10 gal (37.9 L) pilot plant aerators. Significant amounts of soluble organic carbon (53%), soluble COD (60%), NH_3-N (98%), and phenolics were removed after 50 mg/L of PAC was added. The amounts removed increased with increasing carbon dosage.

Thibault et al. [56] reported on a field-scale test with aerator PAC levels of 1000 mg/L or more in an Exxon refinery. They found significantly improved effluent quality and noted improvement in shock loading resistance leading to process stability. An additional 10% of TOC and COD was removed.

Wong and Maroney [48] reported on a pilot-plant comparison of PACT® and extended aeration for toxicity reduction in wastewater from a West Coast refinery. The average PAC dosage used was approximately 70 mg/L in the influent. Flow-through bioassays were used to monitor the toxicity of the treated effluent. Although both PACT® and extended aeration performed similarly in COD removal, only the PACT® system yielded an effluent meeting the discharge requirements for whole-effluent toxicity. A full-scale PACT® system installed at this refinery has been operating satisfactorily. Similar results in toxicity reduction have been reported for wastewaters from other industries [57].

Butterworth [58] has presented case histories of how refineries have used GAC to achieve compliance with NPDES permit requirements for toxicity. There are five major refineries in the San Francisco Bay Area as of 2003. Because of the stringent toxicity requirements for direct discharge to the Bay, four of the major refineries have installed GAC systems to polish secondary treatment plant effluent prior to discharge (Chevron Texaco, Valero, Tesoro, and Shell Equilon). The one exception is the ConocoPhilips Refinery in Rodeo, which has a PACT® system for organics and toxicity removal. These GAC systems are designed mainly to reduce toxicity rather than COD. The toxicity of treated refinery effluent is believed to be caused mainly by naphthenic acid [59]. The spent GAC from the refineries is regenerated offsite by a contractor. The cost of GAC treatment in these refineries has been lower than anticipated because COD removal is not critical for meeting toxicity requirements, and thus the GAC beds can last much longer between regenerations.

Cyanide Control

Historically, refinery cyanide control was not a concern because cyanide levels in refinery effluent were usually much lower than those in wastewaters from metal finishing and plating industries. Regulatory agencies have now established new and more stringent cyanide effluent limits for most wastewater discharges. One example is the cyanide effluent limit of 0.025 mg/L (as total cyanide) in the San Francisco Bay imposed by the California Water Resources Control Board [60].

Fluid catalytic cracking (FCC) and coker units generate most of the cyanides in refineries [61]. Cracking organic nitrogen compounds liberates cyanide and other nitrogen compounds, such as ammonia and thiocyanates. Figure 18 shows a simplified FCC/coker gas fractionation system and the path the waste stream containing cyanide follows in a typical refinery [62]. The FCC/coker reactor gases, including cyanide and NH_3, go overhead on the fractionation column, where water is injected into the overhead line for corrosion control. This water is collected in an accumulator and pumped to a steam stripper along with other sour water to remove NH_3 and H_2S. Part of the cyanide is also removed. The remaining cyanide goes to the wastewater treatment system where simple cyanide is biodegraded and complexed cyanide may pass through the treatment plant and be discharged.

Because the complexed cyanide species that pass through biological treatment plants are usually very stable, common cyanide removal methods such as chlorination and precipitation do not reduce the effluent cyanide concentrations to below detection limits. Wong and Maroney [62] identified four potential end-of-pipe treatment processes to remove cyanide to very low levels: (1) a cyanide-selective ion exchange resin Amberlite IRA-958 developed by Rohm and

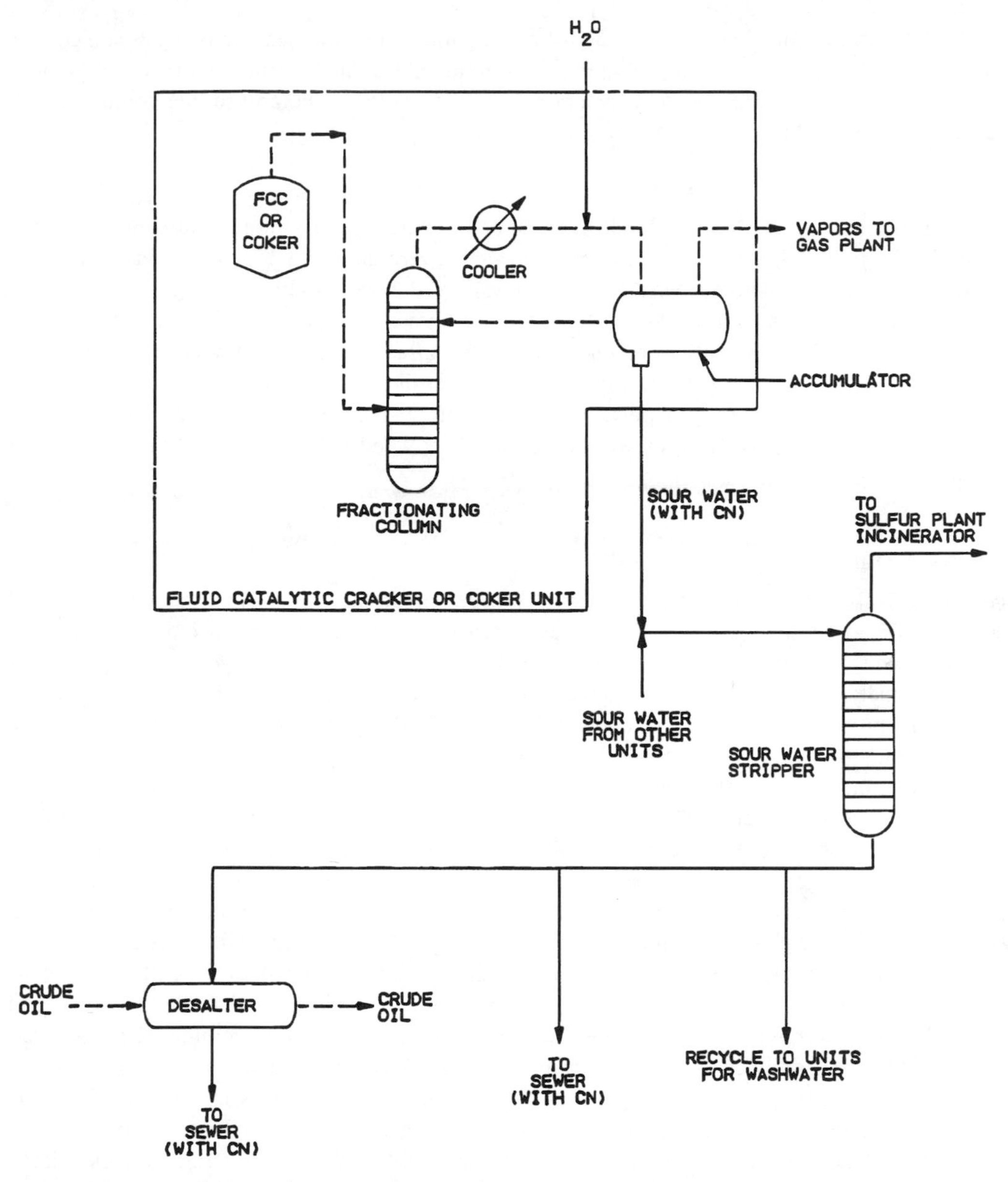

Figure 18 Cyanide generation and disposal in a typical refinery. Cyanide and other gases are formed in FCC or coker units during cracking of organics and go overhead on the fractionating column. Wash water dissolves these gases and becomes sour water. Part of the cyanide is removed by the sour water stripper and the rest goes to the sewer and eventually to the wastewater treatment system. (From Ref. 48.)

Haas Company (Philadelphia, PA), (2) reverse osmosis (RO), (3) adsorption/oxidation with PAC and copper, and (4) ultraviolet irradiation/ozonation (UV/O_3).

All these processes are very expensive for the purpose of removing a small amount of cyanide. The adsorption/oxidation process with PAC and copper could be easily incorporated into existing biological treatment systems; however, the concern of copper toxicity in the final effluent makes this process undesirable.

The most economical cyanide control method in a refinery appears to be upstream control using polysulfides. Sodium and ammonium polysulfide (APS) have been used to inhibit cyanide-induced corrosion in FCC and coker fractionation systems [63]. The polysulfide combines with cyanide, forming thiocyanate according to the reaction

$$CN^- + S_x^{-2} \rightarrow SCN^- + S_{x-1}$$

The thiocyanate is readily biodegradable and is innocuous in refinery effluent. Knowlton *et al.* [63] reported that one large refinery generated several hundred pounds per day of cyanide in its FCC and coker units. When APS solution was used to thoroughly scrub gases produced in the FCC and coker units, the cyanide content in the final effluent was consistently less than the detection limit. The polysulfide treatment method is effective at high temperatures and when the cyanide is still in the free form. However, careful design and operation control are critical to the success of implementing a polysulfide treatment system. Some refineries have reported severe fouling and plugging in the sour water strippers when APS was used [64].

Heavy Metals Removal

Heavy metals such as copper, zinc, lead, nickel, silver, arsenic, selenium, cadmium and chromium may originate from many sources within a refinery and may, in specific cases, require end-of-pipe treatment. Some agencies have set discharge limits that are beyond the capability of common metals removal processes such as lime precipitation and clarification to achieve. Other treatment processes such as iron coprecipitation and adsorption, ion exchange, and reverse osmosis may be required to achieve these low effluent concentrations [52].

The iron coprecipitation and adsorption process involves adding an iron salt such as ferric chloride to the wastewater. The iron hydrolyzes, forming an amorphous iron hydroxide floc. Metals adsorb onto the floc and are removed by clarification or filtration. Cationic metals (e.g., cadmium, zinc) are best adsorbed at high pH and anionic metals (e.g., arsenic, selenium) adsorb better at low pH. The process can remove metals to very low levels, in the ppb range [65]. It operates within the physiological pH range (6–9) and produces relatively few waste solids. This process can be incorporated at one or more points in an existing treatment plant if metals removal is required. The success of the process may depend on the forms of the metallic species and the extent of interference by organics in the wastewater. The iron coprecipitation process has been used successfully in several San Francisco Bay Area refineries to remove selenium to below 50 ppb in treated effluent. Based on bench- and pilot-scale tests in a refinery, a ferric chloride dosage of 50 mg/L as Fe was necessary to achieve the required 50 ppb selenium at all times [66]. The iron coprecipitation system in the Shell Equilon Refinery generates a large amount of iron sludge for disposal. An outside contractor uses an onsite belt filter press system to dewater the iron sludge before its offsite disposal as hazardous waste in California (Glaze, D.E., 2002, personal communication).

Ion exchange can be used to remove soluble heavy metals to very low levels [67]. Because it can remove all ionic species in water and thus chemical regeneration cost is high, its use has been more common for treatment of water or wastewater with low dissolved solids. Pretreatment is required to prevent excessive resin fouling. There are many ion-specific resins for removal of different metals [68]. However, several different resins are needed when different metals must be removed. One significant use of ion exchange wastewater treatment is for chromate-containing blowdown from recirculating cooling water systems [52]. With proper pH adjustment (to pH 4.0–5.0), the chromate is removed even in the presence of several hundred mg/L sulfate and chloride [69].

Reverse osmosis can remove dissolved metals to very low levels. It can also remove a variety of pollutants such as cyanide and residual organics from refinery wastewater. However, because it is an expensive process, it would be competitive only if removal of total dissolved solids is also required. It also requires extensive pretreatment to prevent membrane fouling and deterioration [52]. The pretreatment processes may include filtration to remove suspended solids, pH adjustment, softening, and activated carbon treatment to remove organics and chlorine. A major drawback of the RO process is the handling and disposal of the reject stream, which can amount to 20–30% of the influent flow.

6.6.5 Treatment Modifications Due to Newer Regulations

Since 1990, several new or revised U.S. environmental regulations, which significantly affect refinery wastewater treatment systems, have been promulgated. The most important ones include the revised Toxicity Characteristics (TC) rule, the Primary Sludge rule, and the Benzene NESHAP (National Emissions Standards for Hazardous Air Pollutants) rule. These regulations and their impacts on refinery wastewater facilities are briefly discussed below.

Revised TC Rule

The TC rule was revised to include 26 organic chemicals, including benzene and cresols. It broadened the definition of a characteristically toxic hazardous waste to include a large number of wastes that were previously not included. This rule came into effect on September 25, 1990. The presence of benzene, for example, renders a wastewater hazardous when benzene concentrations are greater than 0.5 mg/L [70]. Refinery waste streams typically contain benzene. The greatest impact of the revised TC rule is on ponds, lagoons, and impoundments that have been managing wastewater that was not previously considered to be hazardous [71]. These units become RCRA regulated surface impoundments if they receive TC hazardous wastewater, and would have had to be retrofitted with two liners and leachate collection by March 19, 1994. Because these units are usually very large, they are very costly to retrofit. Several alternatives are available to retrofitting these units. Some refineries replaced the ponds and lagoons with above-ground tanks. Other plants have installed air or steam stripping facilities to remove benzene before the wastewater enters these surface impounds. And yet others installed high-rate biological treatment systems to biodegrade benzene so that they can continue to use the ponds without retrofitting. This is economically feasible because benzene can be easily biodegraded.

One of the Bay Area refineries has installed a second above-ground biological treatment system to treat waste streams with higher benzene concentrations (Glaze, D.E., 2002, personal communication). The existing biotreater is pond-based with DAF clarifiers. Figure 19 shows a block flow diagram of the revised effluent treatment system [72]. The process train includes conventional refinery treatment processes, two different biological treatment systems, an iron coprecipitation system for selenium removal, and GAC for toxicity reduction.

Primary Sludge Listing

The Primary Sludge rule, effective May 2, 1991, lists primary petroleum refinery sludge, designated F037 and F038, as hazardous wastes [70]. It governs all sludges generated from the separation of oil/water/solids during the storage or primary treatment of process wastewaters and oily cooling waters. These include API separator sludge, DAF floats, and sludges from all surface impoundments prior to biological treatment. Surface impoundments that receive or generate these wastes must comply with minimum technology requirements (MTRs) within four years of the promulgation date. Examples of these MTRs are double liners, leachate collection,

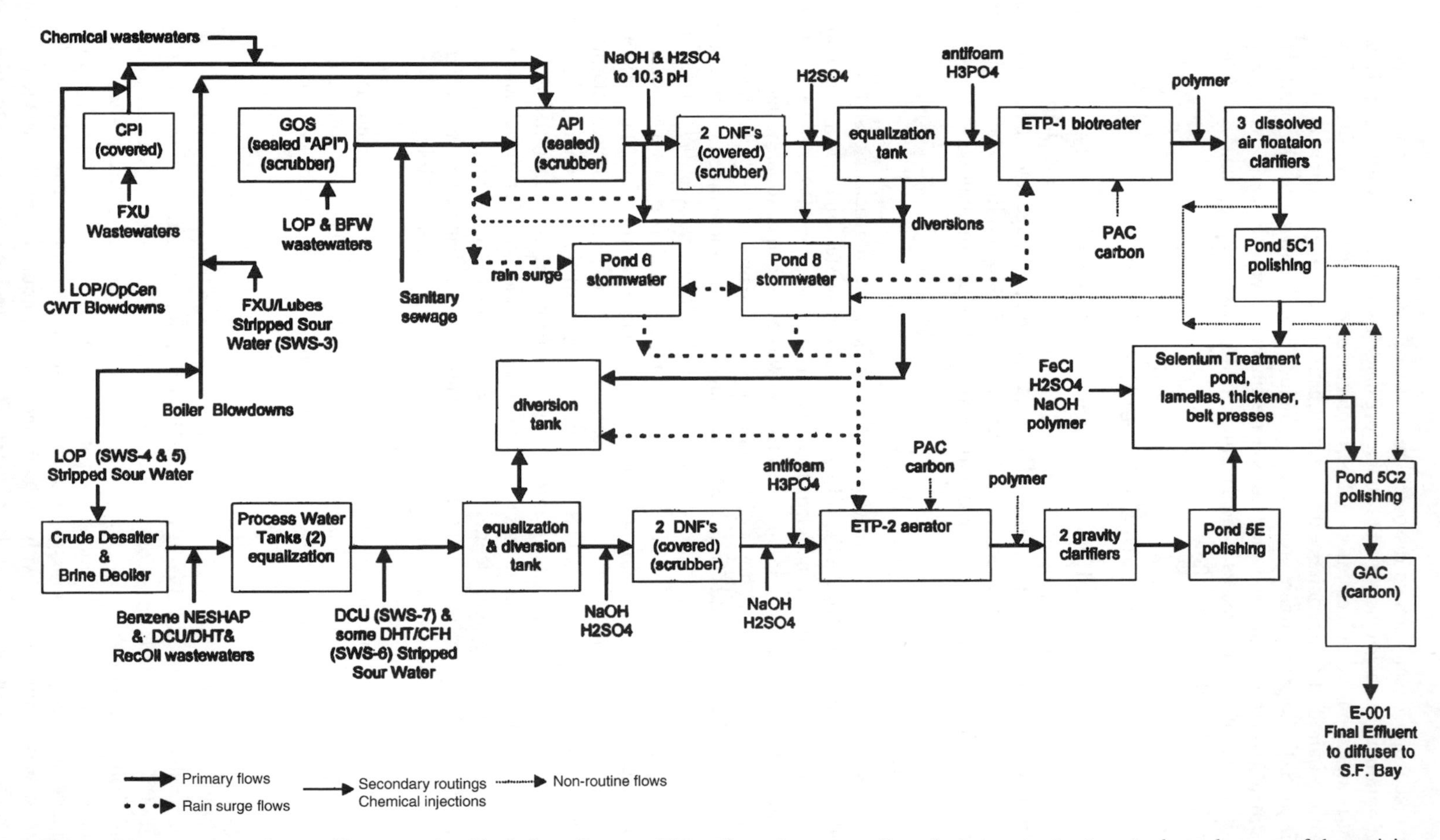

Figure 19 Bay area refinery effluent treating block flow diagram. This refinery has a complicated wastewater treatment scheme because of the toxicity characteristics rule to separate streams with higher benzene concentrations for treatment in aboveground biotreater. (From Ref. 72.)

and groundwater monitoring. Most refiners chose to reconfigure their wastewater collection and treatment systems by replacing impoundments with above-ground tanks, and by lining or enclosing process wastewater conveyance ditches. About 25 U.S. refineries practice sludge coking to dispose of oily, indigenous sludges [70]. In this process, the sludge is injected into coke drums during the quench cycle.

Benzene NESHAP

USEPA issued the NESHAP for benzene waste operations March 7, 1990, under the Clean Air Act. The compliance date was May 1992. It affects not only equipment leaks but also emissions of benzene in wastewater streams. Facilities with greater than 10 tonnes/year benzene in wastewater streams are affected. They must identify wastewater streams containing greater than 10 mg/L benzene and divert them to units that will reduce benzene to acceptable levels, that is, below 10 mg/L or by 98%. This rule affected most major refineries and olefins plants. Mobil Corp. spent $10 million on a benzene recovery project at its Chalmette, LA, refinery. The refinery uses vacuum steam stripping to decrease benzene emissions by about 10 tonnes/year. One Gulf Coast petrochemical plant has also spent $10 million on a wastewater stripping facility, which reduced benzene levels from several thousand mg/L to less than 5 mg/L [70].

On March 5, 1992, USEPA delayed the effective date of the NESHAP until it clarified some confusing points raised by members of the petroleum industry. The final clarifying amendments to the benzene NESHAP were issued by USEPA on January 7, 1993 [72a].

6.6.6 Treatment for Recycle/Reuse and Zero Discharge

Petroleum refineries require a reliable supply of fresh water for steam generation, process cooling, product manufacturing, and other purposes. Because fresh water is becoming more valuable in many parts of the world, many locations have undertaken to reclaim and reuse waters for cooling, steam generation, and process use [73]. Bresnahan [74] presented two case studies that illustrate some of the technical challenges that were encountered when reusing water in refining and petrochemical complexes. One case was use of reclaimed municipal wastewater for most of the cooling towers at Mobil's Torrance Refinery in Los Angeles County, CA, which began in 1995. After working with chemical suppliers to formulate an appropriate treatment program together with optimization and continuous improvements, the reuse program has been operating successfully.

Another case involves the 300,000 bpd Chevron refinery in the San Francisco Bay Area. It is the largest user of potable water in the area [75]. Nearly half of the refinery's water demand (23,000 m^3/day) is used as makeup water in the cooling towers. The water utility identified the potential water reuse for this application in 1979. A pilot plant testing program was completed in 1987, which demonstrated that using lime/soda ash softening treatment on secondary effluent would produce a consistently high-quality reclaimed water for use as makeup water in the refinery's cooling towers. A full-scale plant (23,000 m^3/day) was completed in 1995. Figure 20 shows a process flow schematic of the reclamation plant. Secondary effluent from the WWTP is stored in a 6400 m^3 equalization tank. The influent is pumped to two 17 m diameter solids contact clarifiers after chemical treatment with lime/soda ash. The clarifier overflow is pH adjusted and filtered by four deep-bed, continuous-backwash sand filtration units. The filter effluent is disinfected by sodium hypochlorite for 90 minutes before being pumped to the refinery. The sludge from the clarifiers is thickened in two 10.7 m diameter thickeners and dewatered by a plate and frame filter press with 1.5 m plates.

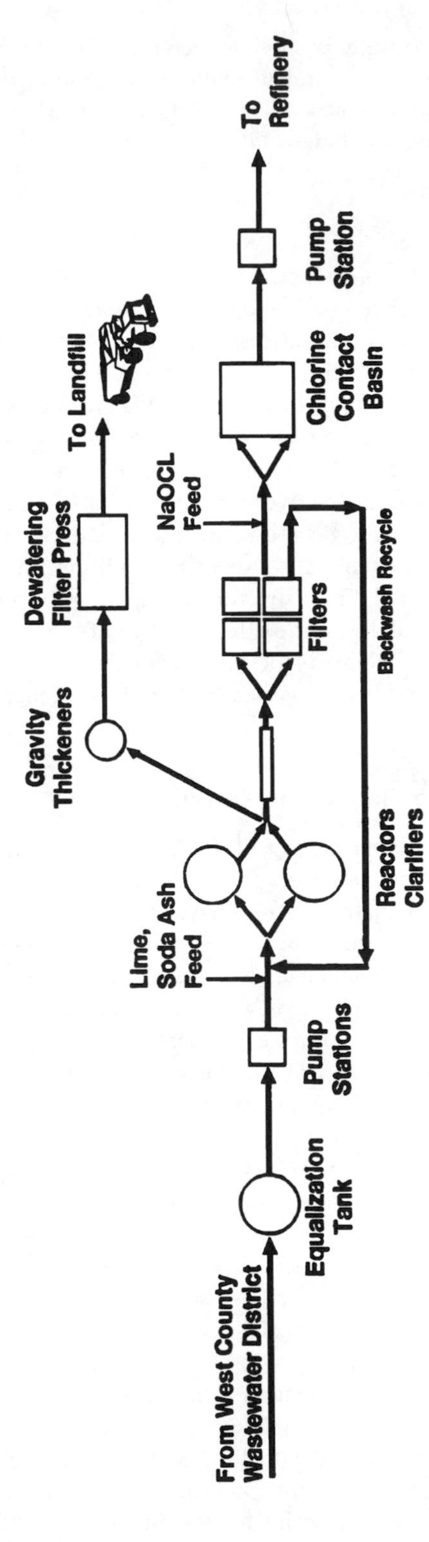

Figure 20 EBMUD North Richmond Reclamation Plant process flow schematic. The tertiary treated wastewater is reused in the Richmond Chevron refinery as cooling tower makeup. (From Ref. 75.)

The El Segundo, CA, Chevron refinery takes a further step in water reuse [76]. It receives 16,300 m^3/day of reclaimed water to feed its boilers. Microfiltration (MF) and RO are used to treat secondary effluent from the Hyperion Wastewater Treatment Plant to provide low-pressure boiler feedwater while a second pass RO is used to produce high-pressure boiler feedwater.

The concept of water reuse and zero liquid discharge in petroleum refineries has been proposed and debated for many years [77]. The principal drawback for zero liquid discharge is the generation of large amount of solid waste, mostly salt from the wastewater. It is this problem that caused USEPA to back off from zero liquid discharge in the 1970s, and it remains the primary deterrent today. However, there are two refineries in Mexico that have recently gone to zero discharge [78]. Wastewater from the refineries and nearby municipalities are treated with biological, physical/chemical processes, RO, brine concentrator evaporator and crystallizer to maximize water recycle to the refineries, minimize water makeup from the river and to attain zero liquid discharge. Figure 21 shows a process schematic diagram of the refinery wastewater recycle/zero liquid discharge system.

6.7 POLLUTION PREVENTION/HAZARDOUS WASTE MINIMIZATION

Refineries generate a large amount of hazardous wastes. As a result, they have been hit hard by environmental regulations and unfavorable public opinion, and Congress mandated in 1984 that refineries minimize waste [79]. In California, refiners turned to waste minimization, or pollution prevention, en masse in 1991 when the state's Source Reduction and Hazardous Waste

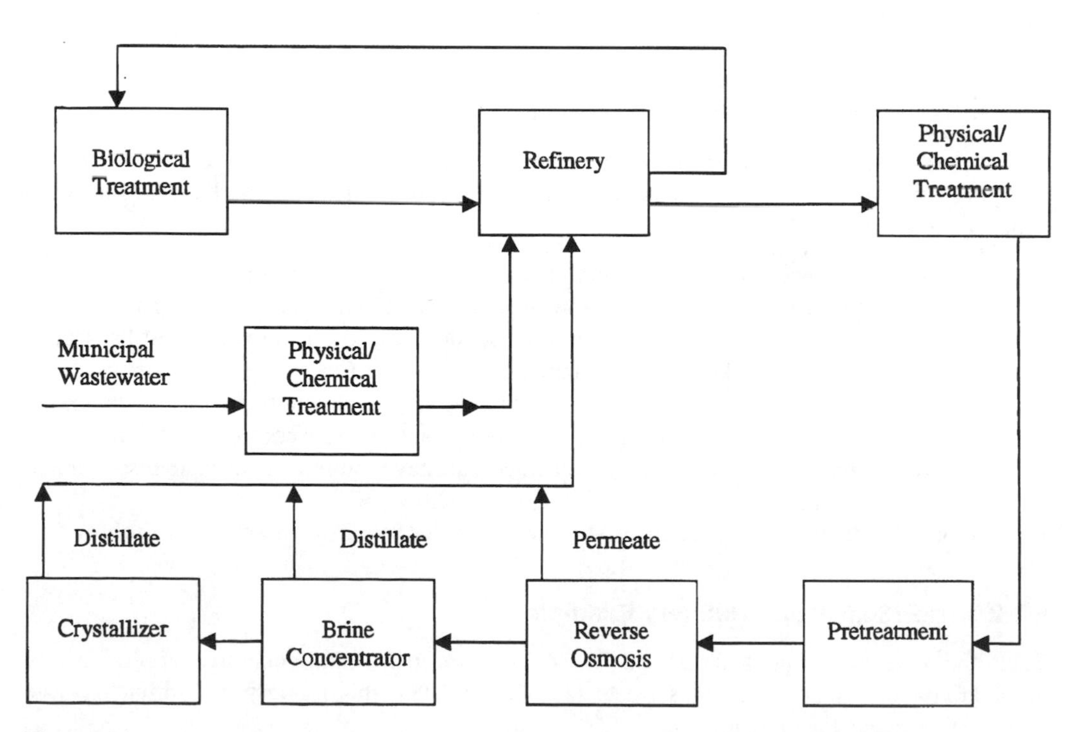

Figure 21 Refinery wastewater recycle/zero liquid discharge scheme. Pretreatment and reverse osmosis are used to recycle water, and brine concentrator and crystallizer are used to treat the rejects to achieve zero liquid discharge. (From Ref. 78.)

Management Review Act of 1989, commonly referred to as Senate Bill (SB) 14, went into effect. Inspired by USEPA and California regulations, other states have pursued similarly restrictive paths.

6.7.1 Pollution Prevention Program

A pollution prevention or waste minimization program usually consists of the following [79]:

- conducting a waste survey;
- screening waste streams for minimization opportunities;
- developing minimization options;
- screening minimization options;
- evaluating high-priority options;
- scheduling and implementing desirable options;
- evaluating and reviewing program performance periodically.

For a waste minimization program to succeed, refinery managers must provide the necessary staff and other resources to accomplish their goals. A team committed to the tasks is usually assembled. Because a refinery is a complex facility and there are numerous emission sources and waste streams to take into account, the team should consider and give the highest priority to:

- *Pollution prevention hierarchy* – The order of preference (highest to lowest) is source reduction, recycling, treatment, and secure land disposal;
- *Reduction of waste volume* – Volume reduction will usually reduce cost for handling, treatment, and disposal;
- Ease of implementation;
- Proven performance;
- Safety and health risks to employees and the public.

Some waste minimization approaches are proving to be more successful than others. Studying several refineries for waste minimization opportunities led to these eye-opening conclusions [79]:

- Housekeeping is the most cost-effective way to minimize waste;
- Solids that enter the refinery's wastewater treatment system automatically are classified as hazardous waste. Therefore, refiners can lower the volume of hazardous waste generated by keeping nonhazardous waste out of the treatment system;
- Raw materials (e.g., crude source) substitution is difficult because the choice of material is dictated by economics, availability, and the process units at the refinery;
- Process modifications can be implemented but may require considerable research and development;
- In-plant and offsite reuse of wastes plays a major role in waste minimization.

6.7.2 The 130,000 bpd Refinery Example

Take the example of a 130,000 bpd West Coast refinery that generates approximately 50,000 tons per year of hazardous waste [19]. Since 1984, this refinery has initiated waste management practices to handle:

- *Spent caustic*. At the end of 1990, 100% of the spent caustic was recycled onsite or offsite. The alkylation/dimersol and fluid catalytic cracking unit (FCCU) spent caustic is recycled to neutralize acidic wastewater. The virgin light-ends spent caustic is transported offsite for reuse at a paper manufacturing facility. The alkylation unit propane spent caustic is used in three other alkylation unit caustic washes. A Minalk Treating System was installed at the FCCU to replace an existing Merox Treating System to convert mercaptans to disulfide. This replacement reduces FCCU spent caustic generation by 1700 tons/year.

- *Stretford solution*. Since 1987, 100% of the waste Stretford solution has been shipped to a metals reclamation facility. Vanadium is reclaimed as vanadium pentoxide.
- *Wastewater treatment sludge*. In 1987, the refinery started a program to recycle this sludge to the coker within the refinery. At the end of 1990, 60% of the sludge was recycled to the coker. The remaining 40% was dewatered onsite at a belt filter press and then landfilled offsite or incinerated. Since 1986, the refinery has paved five plant areas to reduce the amount of dirt and debris washing into the sewer.
- *Catalysts, desiccants, and catalyst inerts*. In 1988, the refinery began to recycle nonhazardous catalysts, desiccants, and catalyst fines. It recycles electrostatic precipitator fines, Claus catalyst, and catalyst support inerts for use in cement manufacture. Two other catalysts, zinc oxide and iron chromate from the hydrogen plant, are reprocessed at smelters to recover the metals.

California SB 14 regulations required the refinery to further evaluate source reduction opportunities. The following are some of the measures identified by the refinery for further evaluation and implementation:

- Modification of the coker silo area to reduce dirt and debris to the wastewater treatment system;
- Reuse of the waste Stretford sulfur stream at a sulfuric acid manufacturer;
- Use of a transportable treatment unit to oxidize thiosulfate salts in the Stretford solution to allow them to be recycled in the Stretford process;
- Installation of sulfur de-entrainment devices in the Claus Plant sulfur condensers to allow them to be recycled in the Stretford process;
- Installation of asphalt lips around sewer to inhibit entry of dirt and debris;
- Evaluation of increasing the amount of wastewater treatment sludge recycled to the coker.

REFERENCES

1. Bass, D.M. Petroleum. In *The World Book Encyclopedia*; World Book, Inc.: Chicago, IL, 1989; 330–350.
2. Kemmer, F.N. *The Nalco Water Handbook*; McGraw-Hill: New York, 1979.
3. Stell, J. Survey of Operating Refineries in the U.S. Oil & Gas J. **2001**, December 24.
4. Shreve, R.N. *Chemical Process Industries*, 3rd Ed.; McGraw-Hill: New York, 1967.
5. U.S. Environmental Protection Agency. *Development Document for Effluent Limitations, Guidelines and Standards for the Petroleum Refining Point Source Category,* EPA-440/1-82-014, 1982.
6. Hengstebeck, R.J. *Petroleum Processing*; McGraw-Hill: New York, 1959.
7. U.S. Environmental Protection Agency. *Brine Disposal Treatment Practices Relating to the Oil Production Industry*, EPA-660/2-74-037, 1974.
8. U.S. Environmental Protection Agency. *Development Document for Interim Final Effluent Guidelines and Proposed New Source Performance Standards for the Oil and Gas Extraction Point Source Category*, EPA 440/1-76-005-a-Group II, 1976.

9. Michalczyk, B.L.; Pollock, T.E.; White, H.R. Treatment of oil field production waters. *Proceedings of the Industrial Wastes Symposium, 57th Annual WPCF Conference*, New Orleans, Louisiana, 1984, 457–471.
10. Palmer, L.L.; Beyer, A.H.; Stock, J. Biological oxidation of dissolved compounds in oilfield produced water by a field pilot biodisk. J. Petrol. Tech. **1981**, *June*, 1136.
11. Bates, M.H. Land farming of reserve pit fluids and sludges: Fates of selected contaminants. Wat. Res. **1988**, *22*, 793.
12. Vielvoye, R. Cleaning up cuttings. Oil & Gas J. **1989**, *Sept. 25*, 46.
13. Nesbitt, L.E.; Sanders, J.A. Drilling fluid disposal. J. Petrol. Tech. **1981**, *Dec.*, 2377–2381.
14. Copa, W.M.; Dietrich, M.J. Wet oxidation of oils, oil refinery sludges, and spent drilling muds. *Proceedings of Oil Waste Management Alternatives Symposium*, California Dept. of Health Services: Oakland, California, April, 1988.
15. U.S. Department of the Interior. *The Cost of Clean Water, Volume III, Industrial Waste Profile No. 5–Petroleum Refining*; Federal Water Pollution Control Administration, 1967.
16. Gloyna, E.F.; Ford, D.L. *The Characteristics and Pollutional Problems Associated With Petrochemical Wastes*; Federal Water Pollution Control Administration, Ada, OK, 1970.
17. Ford, D.L. Water pollution control in the petroleum industry. In *Industrial Wastewater Management Handbook*; Azad, H.S., Ed.; McGraw-Hill: New York, 1976.
18. Bryant, J.S.; Moores, C.W. Disposal of hazardous wastes from petroleum refineries. *Proceedings, 45th Purdue Industrial Waste Conference*, West Lafayette, IN, 1990: Lewis Publishers, Inc.
19. Wong, J.M. Hazardous waste minimization (SB 14) in California petroleum refineries. *Proceedings of 50th Purdue Industrial Waste Conference*, Lewis Publishers, Inc., Chelsea, MI, 1993.
20. Sittig, M. *Petroleum Refining Industry Energy Saving and Environmental Control*; Noyes Data Corp.: Park Ridge, NJ, 1978.
21. Truitt, R. POTWs feel the heat of toxics control. Water/Engng. & Mgnt. **1989**, *Sept.*, 14–15.
22. Thomas, N.A. Use of biomonitoring to control toxics in the United States. Wat. Sci. Technol. **1988**, *20*(10), 101–108.
23. Sack, W.A.; Eck, R.W.; Romano, C.R. Recovery of waste brines for highway applications. *Proceedings of 40th Purdue Industrial Waste Conference*, West Lafayette, IN, 1985.
24. Wang, W.D.; Li, X.M.; Chen, Y.; Zhang, S.T.; Jiang, Y. The technology of microbial treating drained water of oil field. *Proceedings, SPE Asia Pacific Improved Oil Recovery Conference*, Kuala Lumpur, Malaysia, 8–9 October, 2001.
25. Beyer, A.H.; Palmer, L.L.; Stock, J. Biological oxidation of dissolved compounds in oilfield-produced water by a pilot aerated lagoon. J. Petrol. Technol. **1979**, *Feb.*, 241–245.
26. Ali, S.A.; Henry, L.R.; Darlington, J.W.; Occapinti, J. New filtration process cuts contaminants from offshore produced water. Oil & Gas J. **1998**, *Nov.*, 73.
27. Eckenfelder, W.W., Jr. *Industrial Water Pollution Control*, 2nd Ed.; McGraw-Hill: New York, 1989.
28. McNally, R. Tougher rules challenge future for injection wells. Petrol. Eng. Int. **1987**, *July*, 28–30.
29. Anonymous. GAO finds brine still contaminates aquifers. Oil and Gas J. **1989**, *Oct. 16*, 38.
30. Waite, B.A.; Blauvelt, S.C.; Moody, J.L. *Oil and Gas Well Pollution Abatement Project*, ME No. 81495; Moody and Associates, Inc.: Meadville, PA, 1983.
31. Chen, J.C.T.; Stephenson, R.L. Cost effective treatment of oil field produced wastewater for 'Wet Stream' generation – a case history. *Proceedings, 39th Purdue Industrial Waste Conference*, West Lafayette, IN, 1984.
32. Doran, G.F.; Carini, F.H.; Fruth, D.A.; Drago, J.A.; Leong, L.Y.C. Evaluation of technologies to treat oil field produced water to drinking water or reuse quality. *Proceedings, 68th SPE Annual Western Regional Meeting*, Bakersfield, CA, 38830, 1998.
33. Marks, R.E.; Field, S.D.; Wojtanowicz, A. Biodegradation of oilfield production pit sludges. *Proceedings, 42nd Purdue Industrial Waste Conference*, West Lafayette, IN, 1987.
34. Moses, J.; Abrishamian, R. Case study: SITE program puts critical fluid solvent extraction to the test. Hazardous Waste Management Magazine **1988**, *Jan/Feb.*

35. U.S. Environmental Protection Agency. *Development Document for Proposed Effluent Limitations Guidelines and New Source Performance Standards for the Petroleum Refining Point Source Category*, Report EPA-440/1-73/014, 1973.
36. Knowlton, H.E. Source control in petroleum refineries. *Proceedings, National Petroleum Refiners Association Annual Meeting*, San Antonio, TX, March 19–21, 1978.
37. Beychok, M.R. *Aqueous Wastes from Petroleum and Petrochemical Plants*; John Wiley & Sons: London, 1967.
38. Hackman, E.E., III. *Toxic Organic Chemicals Destruction and Waste Treatment*; Noyes Data Corp.: Park Ridge, NJ, 1978.
39. Knowlton, H.E. Control refinery odors and effluent quality to meet environmental regulations. *Proceedings, National Petroleum Refiners Association Annual Meeting*, San Antonio, TX, March 24–26, 1985.
40. American Petroleum Institute. *Manual on Disposal of Refinery Wastes*; American Petroleum Institute: Washington, DC, 1969.
41. Katz, W.J. Adsorption – secret of success in separating solid by air flotation. *Wastes Eng.* **1959**, *July.*
42. Krofta, M.; Wang, L.K. Flotation technology and secondary clarification. *TAPPI J.* **1987**, *70*(4).
43. Krofta, M.; Guss, D.; Wang, L.K. Development of low-cost flotation technology and systems for wastewater treatment. *Proceedings, 42nd Purdue Industrial Waste Conference*, West Lafayette, IN, 1987.
44. Novotny, V.; England, A.J. Water Res. **1974**, *8*, 325.
45. Anonymous. A refined process. Water Qual. Int. **1998**, *1–2*, 37.
46. Hudson, N.; Doyle, J.; Lant, P.; Roach, N.; de Bruyn, B.; Staib, C. Sequencing batch reactor technology: The key to a BP refinery (Bulwer Island) upgraded environmental protection system: A low cost lagoon based retrofit. Water Sci. Technol. **2001**, *43*(3), 339.
47. U.S. Environmental Protection Agency. *The Impact of Oily Materials on Activated Sludge Systems*, Hydroscience, Inc., EPA Project No. 12050 DSH, March 1971.
48. Wong, J.M.; Maroney, P.M. Pilot plant comparison of extended aeration and PACT® for toxicity reduction in refinery wastewater. *Proceedings, 44th Purdue Industrial Waste Conference*, West Lafayette, IN, 1989.
49. Brown and Caldwell Consulting Engineers. Confidential report; Brown and Caldwell: Denver, CO, October 1989.
50. Adams, C.E., Jr.; Ford, D.L.; Eckenfelder, W.W., Jr. *Development of Design and Operational Criteria for Wastewater Treatment*; Enviro Press, Inc.: Nashville, Tennessee, 1981.
51. Ford, D.L.; Manning, F.S. Treatment of petroleum refinery wastewater. In *Carbon Adsorption Handbook*; Cheremisinoff, P.N., Ellerbusch, F., Ed.; Ann Arbor Science: Ann Arbor, MI, 1978.
52. Patterson, J.W. *Industrial Wastewater Treatment Technology*, 2nd Ed.; Butterworth: Boston, 1985.
53. Hutton, D.G.; Robertaccio, F.L. Waste water treatment process. U.S. Patent 3,904,518, September 9, 1975.
54. Rizzo, J.A. Case history: Use of powdered activated carbon in an activated sludge system. *First Open Forum on Petroleum Refinery Wastewaters*, Tulsa, OK, 1976.
55. Grieves, C.G.; Stenstrom, M.K.; Walk, J.D.; Grutsch, J.F. Effluent quality improvement by powdered activated carbon in refinery activated sludge processes. *API Refining Department, 42nd Midyear Meeting*, Chicago, IL, May 11, 1977.
56. Thibault, G.T.; Tracy, K.D.; Wilkinson, J.B. Evaluation of powdered activated carbon treatment for improving activated sludge performance. *API Refining Department, 42nd Midyear Meeting*, Chicago, IL, May 11, 1977.
57. Zimpro, Inc. CIBA-GEIGY meeting tough bioassay test. Reactor **1986**, *June*, 13–14.
58. Butterworth, S.L. Granular activated carbon as a toxicity reduction technology for wastewater treatment. Proc. Am. Chem. Soc. Spring Natl. Meet. Fuel Chem. Div. **1996**, *41*(1), 466.
59. Wong, D.C.L.; van Compernolle, R.; Nowlin, J.G.; O'Neal, D.L.; Johnson, G.M. Use of supercritical fluid extraction and fast ion bombardment mass spectrometry to identify toxic chemicals from a refinery effluent adsorbed onto granular activated carbon. Chemosphere **1996**, *32*, 621.

60. California Water Resources Control Board. *Water Quality Control Plan for the San Francisco Bay Basin*, Region 2, 1986.
61. Prather, B.; Berkemeyer, R. Cyanide sources in petroleum refineries. *Proceedings, 30th Purdue Industrial Waste Conference*, West Lafayette, IN, 1975.
62. Wong, J.M.; Maroney, P.M. Cyanide control in petroleum refineries. *Proceedings, 44th Purdue Industrial Waste Conference*, West Lafayette, IN, 1975.
63. Knowlton, H.E.; Coombs, J.; Allen, E. Chevron process reduces FCC/coker corrosion and saves energy. Oil Gas J. **1980**, *April 14.*
64. Kunz, R.; Casey, J.; Huff, J. Refinery cyanide: A regulatory dilemma. Hydrocarbon Proc. **1978**, *October.*
65. Merrill, D.T.; Manzione, M.A.; Peterson, J.J.; Parker, D.S.; Chow, W.; Hobbs, A.D. Field evaluation of arsenic and selenium removal by iron coprecipitation. J. WPCF **1986**, *58*, 18–26.
66. Nurdogan, Y.; Schroeder, R.P.; Meyer, C.L. Selenium removal from petroleum refinery wastewater. *Proceedings, 49th Purdue Industrial Waste Conference*, West Lafayette, IN, 1994.
67. U.S. Environmental Protection Agency (1981). Summary Report: Control and Treatment Technology for Metal Finishing Industry; Ion Exchange, EPA 625/8-81-007.
68. Peters, R.W.; Ku, Y.; Bhattacharyya, D. Evaluation of recent treatment techniques for removal of heavy metals from industrial wastewaters. AIChE Symposium Series **1985**, *81*(243).
69. Anderson, R.E. Some examples of the concentration of trace heavy metals with ion exchange resins. *Proceedings, Traces of Heavy Metals in Water-Removal Processes and Monitoring*, USEPA 902/9-74-001, 1974.
70. Rhodes, A.K. Recent and pending regulations push refiners to the limit. Oil Gas J. **1991**, *Dec. 16*, 39–46.
71. American Petroleum Institute. *Applying the Revised Toxicity Characteristic to the Petroleum Industry*; API: Washington, DC, 1991.
72a. Wong, J.M. Advanced Wastewater treatment for refineries. Proc. Petroleum Refinery/Petrochemical Wastewater Treatment and River Basin/Water Quality Management Workshop, Kaohsiung, Taiwan, ROC, December 4–7, 2001.
72. *Environmental Reporter*; The Bureau of National Affairs, Inc.: Washington, DC, January 15, 1993.
73. Wong, J.M. Petrochemicals. Water Environ. Res. **1998**, *71*(5), 828.
74. Bresnahan, W.T. Water reuse in oil refineries. *Proceedings, Natl. Assoc. Corrosion Eng. Corrosion '96 Conf.*, Houston, TX, 1996.
75. Wong, J.M. Water conservation and reuse in the petrochemical industries. *Proc. Industrial Water Conservation Workshop*, Taipei, Taiwan, ROC, April 11–12, 1996.
76. Anonymous. Effluent recycling plant expands to provide boiler feedwater to Chevron refinery. Civil Eng. **2000**, *70*(7), 14.
77. Diepolder, P. Is 'Zero Discharge' Realistic? Hydrocarbon Proc. **1992**, *October.*
78. Heimbigner, B. Water and wastewater treatment in petroleum refineries. Presented at the Plock Refinery, Poland, June, 1999.
79. Wong, J.M. Pollution prevention/waste minimization in california petroleum refineries. OCEESA J. **2002**, *19*, 1.

7

Treatment of Soap and Detergent Industry Wastes

Constantine Yapijakis
The Cooper Union, New York, New York, U.S.A.

Lawrence K. Wang
Lenox Institute of Water Technology and Krofta Engineering Corporation, Lenox, Massachusetts and Zorex Corporation, Newtonville, New York, U.S.A.

7.1 INTRODUCTION

Natural soap was one of the earliest chemicals produced by man. Historically, its first use as a cleaning compound dates back to Ancient Egypt [1–4]. In modern times, the soap and detergent industry, although a major one, produces relatively small volumes of liquid wastes directly. However, it causes great public concern when its products are discharged after use in homes, service establishments, and factories [5–22].

A number of soap substitutes were developed for the first time during World War I, but the large-scale production of synthetic surface-active agents (surfactants) became commercially feasible only after World War II. Since the early 1950s, surfactants have replaced soap in cleaning and laundry formulations in virtually all countries with an industrialized society. Over the past 40 years, the total world production of synthetic detergents increased about 50-fold, but this expansion in use has not been paralleled by a significant increase in the detectable amounts of surfactants in soils or natural water bodies to which waste surfactants have been discharged [4]. This is due to the fact that the biological degradation of these compounds has primarily been taking place in the environment or in treatment plants.

Water pollution resulting from the production or use of detergents represents a typical case of the problems that followed the very rapid evolution of industrialization that contributed to the improvement of quality of life after World War II. Prior to that time, this problem did not exist. The continuing increase in consumption of detergents (in particular, their domestic use) and the tremendous increase in production of surfactants are the origin of a type of pollution whose most significant impact is the formation of toxic or nuisance foams in rivers, lakes, and treatment plants.

7.1.1 Classification of Surfactants

Soaps and detergents are formulated products designed to meet various cost and performance standards. The formulated products contain many components, such as surfactants to tie up unwanted materials (commercial detergents usually contain only 10–30% surfactants), builders or polyphosphate salts to improve surfactant processes and remove calcium and magnesium ions, and bleaches to increase reflectance of visible light. They also contain various additives designed to remove stains (enzymes), prevent soil re-deposition, regulate foam, reduce washing machine corrosion, brighten colors, give an agreeable odor, prevent caking, and help processing of the formulated detergent [18].

The classification of surfactants in common usage depends on their electrolytic dissociation, which allows the determination of the nature of the hydrophilic polar group, for example, anionic, cationic, nonionic, and amphoteric. As reported by Greek [18], the total 1988 U.S. production of surfactants consisted of 62% anionic, 10% cationic, 27% nonionic, and 1% amphoteric.

Anionic Surfactants

Anionic surfactants produce a negatively charged surfactant ion in aqueous solution, usually derived from a sulfate, carboxylate, or sulfonate grouping. The usual types of these compounds are carboxylic acids and derivatives (largely based on natural oils), sulfonic acid derivatives (alkylbenzene sulfonates LAS or ABS and other sulfonates), and sulfuric acid esters and salts (largely sulfated alcohols and ethers). Alkyl sulfates are readily biodegradable, often disappearing within 24 hours in river water or sewage plants [23]. Because of their instability in acidic conditions, they were to a considerable extent replaced by ABS and LAS, which have been the most widely used of the surfactants because of their excellent cleaning properties, chemical stability, and low cost. Their biodegradation has been the subject of numerous investigations [24].

Cationic Surfactants

Cationic surfactants produce a positively charged surfactant ion in solution and are mainly quaternary nitrogen compounds such as amines and derivatives and quaternary ammonium salts. Owing to their poor cleaning properties, they are little used as detergents; rather their use is a result of their bacteriocidal qualities. Relatively little is known about the mechanisms of biodegradation of these compounds.

Nonionic Surfactants

Nonionic surfactants are mainly carboxylic acid amides and esters and their derivatives, and ethers (alkoxylated alcohols), and they have been gradually replacing ABS in detergent formulations (especially as an increasingly popular active ingredient of automatic washing machine formulations) since the 1960s. Therefore, their removal in wastewater treatment is of great significance, but although it is known that they readily biodegrade, many facts about their metabolism are unclear [25]. In nonionic surfactants, both the hydrophilic and hydrophobic groups are organic, so the cumulative effect of the multiple weak organic hydrophils is the cause of their surface-active qualities. These products are effective in hard water and are very low foamers.

Amphoteric Surfactants

As previously mentioned, amphoteric surfactants presently represent a minor fraction of the total surfactants production with only specialty uses. They are compounds with both anionic and cationic properties in aqueous solutions, depending on the pH of the system in which they work. The main types of these compounds are essentially analogs of linear alkane sulfonates, which provide numerous points for the initiation of biodegradation, and pyridinium compounds that also have a positively charged N-atom (but in the ring) and they are very resistant to biodegradation [26].

7.1.2 Sources of Detergents in Waters and Wastewaters

The concentrations of detergent that actually find their way into wastewaters and surface water bodies have quite diverse origins: (a) Soaps and detergents, as well as their component compounds, are introduced into wastewaters and water bodies at the point of their manufacture, at storage facilities and distribution warehouses, and at points of accidental spills on their routes of transportation (the origin of pollution is dealt with in this chapter). (b) The additional industrial origin of detergent pollution notably results from the use of surfactants in various industries, such as textiles, cosmetics, leather tanning and products, paper, metals, dyes and paints, production of domestic soaps and detergents, and from the use of detergents in commercial/industrial laundries and dry cleaners. (c) The contribution from agricultural activities is due to the surface runoff transporting of surfactants that are included in the formulation of insecticides and fungicides [27]. (d) The origin with the most rapid growth since the 1950s comprises the wastewaters from urban areas and it is due to the increased domestic usage of detergents and, equally important, their use in cleaning public spaces, sidewalks, and street surfaces.

7.1.3 Problem and Biodegradation

Notable improvements in washing and cleaning resulted from the introduction and increasing use of synthetic detergents. However, this also caused difficulties in sewage treatment and led to a new form of pollution, the main visible effect of which was the formation of objectionable quantities of foam on rivers. Although biodegradation of surfactants in soils and natural waters was inferred by the observation that they did not accumulate in the environment, there was widespread concern that their much higher concentrations in the effluents from large industrial areas would have significant local impacts. In agreement with public authorities, the manufacturers fairly quickly introduced products of a different type.

The surface-active agents in these new products are biodegradable (called "soft" in contrast to the former "hard" ones). They are to a great extent eliminated by normal sewage treatment, and the self-purification occurring in water courses also has some beneficial effects [28]. However, the introduction of biodegradable products has not solved all the problems connected to surfactants (i.e., sludge digestion, toxicity, and interference with oxygen transfer), but it has made a significant improvement. Studies of surfactant biodegradation have shown that the molecular architecture of the surfactant largely determines its biological characteristics [4]. Nevertheless, one of the later most pressing environmental problems was not the effects of the surfactants themselves, but the eutrophication of natural water bodies by the polyphosphate builders that go into detergent formulations. This led many local authorities to enact restrictions in or even prohibition of the use of phosphate detergents.

7.2 IMPACTS OF DETERGENT PRODUCTION AND USE

Surfactants retain their foaming properties in natural waters in concentrations as low as 1 mg/L, and although such concentrations are nontoxic to humans [24], the presence of surfactants in drinking water is esthetically undesirable. More important, however, is the generation of large volumes of foam in activated sludge plants and below weirs and dams on rivers.

7.2.1 Impacts in Rivers

The principal factors that influence the formation and stability of foams in rivers [27] are the presence of ABS-type detergents, the concentration of more or less degraded proteins and colloidal particles, the presence and concentration of mineral salts, and the temperature and pH of the water. Additional very important factors are the biochemical oxygen demand (BOD) of the water, which under given conditions represents the quantity of biodegradable material, the time of travel and the conditions influencing the reactions of the compounds presumed responsible for foaming, between the point of discharge and the location of foam appearance, and last but not least, the concentration of calcium ion, which is the main constituent of hardness in most natural waters and merits particular attention with regard to foam development.

The minimum concentrations of ABS or other detergents above which foam formation occurs vary considerably, depending on the water medium, that is, river or sewage, and its level of pollution (mineral or organic). Therefore, it is not merely the concentration of detergents that controls foam formation, but rather their combined action with other substances present in the waters. Various studies have shown [27] that the concentration of detergents measured in the foams is quite significantly higher, up to three orders of magnitude, than that measured at the same time in solution in the river waters.

The formation of foam also constitutes trouble and worries for river navigation. For instance, in the areas of dams and river locks, the turbulence caused by the intensive traffic of barges and by the incessant opening and closing of the lock gates results in foam formation that may cover entire boats and leave a sticky deposit on the decks of barges and piers. This renders them extremely slippery and may be the cause of injuries. Also, when winds are strong, masses of foam are detached and transported to great distances in the neighboring areas, causing problems in automobile traffic by deposition on car windshields and by rendering the road surfaces slippery. Finally, masses of foam floating on river waters represent an esthetically objectionable nuisance and a problem for the tourism industry.

7.2.2 Impacts on Public Health

For a long time, detergents were utilized in laboratories for the isolation, through concentration in the foam, of mycobacteria such as the bacillus of Koch (tuberculosis), as reported in the annals of the Pasteur Institute [27]. This phenomenon of extraction by foam points to the danger existing in river waters where numerous such microorganisms may be present due to sewage pollution. The foam transported by wind could possibly serve as the source of a disease epidemic. In fact, this problem limits itself to the mycobacteria and viruses (such as those of hepatitis and polio), which are the only microorganisms able to resist the disinfecting power of detergents. Therefore, waterborne epidemics could also be spread through airborne detergent foams.

7.2.3 Impacts on Biodegradation of Organics

Surfactant concentrations in polluted natural water bodies interfere with the self-purification process in several ways. First, certain detergents such as ABS are refractory or difficult to biodegrade and even toxic or inhibitory to microorganisms, and influence the BOD exhibited by organic pollution in surface waters. On the other hand, readily biodegradable detergents could impose an extreme short-term burden on the self-purification capacity of a water course, possibly introducing anaerobic conditions.

Surfactant concentrations also exert a negative influence on the bio-oxidation of certain substances, as evidenced in studies with even readily biodegradable substances [7]. It should be noted that this protection of substances from bio-oxidation is only temporary and it slowly reduces until its virtual disappearance in about a week for most substances. This phenomenon serves to retard the self-purification process in organically polluted rivers, even in the presence of high concentrations of dissolved oxygen.

An additional way in which detergent concentrations interfere with the self-purification process in polluted rivers consists of their negative action on the oxygen rate of transfer and dissolution into waters. According to Gameson [16], the presence of surfactants in a water course could reduce its re-aeration capacity by as much as 40%, depending on other parameters such as turbulence. In relatively calm waters such as estuaries, under certain conditions, the reduction of re-aeration could be as much as 70%. It is the anionic surfactants, especially the ABS, that have the overall greatest negative impact on the natural self-purification mechanisms of rivers.

7.2.4 Impacts on Wastewater Treatment Processes

Despite the initial apprehension over the possible extent of impacts of surfactants on the physicochemical or biological treatment processes of municipal and industrial wastewaters, it soon became evident that no major interference occurred. As mentioned previously, the greatest problem proved to be the layers of foam that not only hindered normal sewage plant operation, but when wind-blown into urban areas, also aided the probable transmission of fecal pathogens present in sewage.

The first unit process in a sewage treatment plant is primary sedimentation, which depends on simple settling of solids partially assisted by flocculation of the finer particles. The stability, nonflocculating property, of a fine particle dispersion could be influenced by the surface tension of the liquid or by the solid/liquid interface tension – hence, by the presence of surfactants. Depending on the conditions, primarily the size of the particles in suspension, a given concentration of detergents could either decrease (finer particles) or increase (larger particles) the rate of sedimentation [23]. The synergistic or antagonistic action of certain inorganic salts, which are included in the formulation of commercial detergent products, is also influential.

The effect of surfactants on wastewater oils and greases depends on the nature of the latter, as well as on the structure of the lipophilic group of the detergent that assists solubilization. As is the case, emulsification could be more or less complete. This results in a more or less significant impact on the efficiency of physical treatment designed for their removal. On the other hand, the emulsifying surfactants play a role in protecting the oil and grease molecules from attacking bacteria in a biological unit process.

In water treatment plants, the coagulation/flocculation process was found early to be affected by the presence of surfactants in the raw water supply. In general, the anionic detergents stabilize colloidal particle suspensions or turbidity solids, which, in most cases, are negatively charged. Langelier [29] reported problems with water clarification due to surfactants, although according to Nichols and Koepp [30] and Todd [31] concentrations of surfactants on the order of 4–5 ppm interfered with flocculation. The floc, instead of settling to the bottom, floats to the surface of sedimentation tanks. Other studies, such as those conducted by Smith *et al.* [32] and Cohen [10], indicated that this interference could be not so much due to the surfactants themselves, but to the additives included in their formulation, that is, phosphate complexes. Such interference was observed both for alum and ferric sulfate coagulant, but the use of certain organic polymer flocculants was shown to overcome this problem.

Concentrations of detergents, such as those generally found in municipal wastewaters, have been shown to insignificantly impact on the treatment efficiency of biological sewage treatment plants [33]. Studies indicated that significant impacts on efficiency can be observed only for considerable concentrations of detergents, such as those that could possibly be found in undiluted industrial wastewaters, on the order of 30 ppm and above. As previously mentioned, it is through their influence of water aeration that the surfactants impact the organics' biodegradation process. As little as 0.1 mg/L of surfactant reduces to nearly half the oxygen absorption rate in a river, but in sewage aeration units the system could be easily designed to compensate. This is achieved through the use of the alpha and beta factors in the design equation of an aeration system.

Surfactants are only partially biodegraded in a sewage treatment plant, so that a considerable proportion may be discharged into surface water bodies with the final effluent. The shorter the overall detention time of the treatment plant, the higher the surfactant concentration in the discharged effluent. By the early 1960s, the concentration of surfactants in the final effluents from sewage treatment plants was in the 5–10 ppm range, and while dilution occurs at the site of discharge, the resulting values of concentration were well above the threshold for foaming. In more recent times, with the advent of more readily biodegradable surfactants, foaming within treatment plants and in natural water bodies is a much more rare and limited phenomenon.

Finally, according to Prat and Giraud [27], the process of anaerobic sludge digestion, commonly used to further stabilize biological sludge prior to disposal and to produce methane gas, is not affected by concentrations of surfactants in the treated sludge up to 500 ppm or when it does not contain too high an amount of phosphates. These levels of concentration are not found in municipal or industrial effluents, but within the biological treatment processes a large part of the detergents is passed to the sludge solids. By this, it could presumably build up to concentrations (especially of ABS surfactants) that may affect somewhat the sludge digestion process, that is, methane gas production. Also, it seems that anaerobic digestion [34] does not decompose surfactants and, therefore, their accumulation could pose problems with the use of the final sludge product as a fertilizer.

The phenomena related to surface tension in groundwater interfere with the mechanisms of water flow in the soil. The presence of detergents in wastewaters discharged on soil for groundwater recharge or filtered through sand beds would cause an increase in headloss and leave a deposit of surfactant film on the filter media, thereby affecting permeability. Surfactants, especially those resistant to biodegradation, constitute a pollutant that tends to accumulate in groundwater and has been found to remain in the soil for a few years without appreciable decomposition. Because surfactants modify the permeability of soil, their presence could possibly facilitate the penetration of other pollutants, that is, chemicals or microorganisms, to depths where they would not have reached due to the filtering action of the soil, thereby increasing groundwater pollution [35].

7.2.5 Impacts on Drinking Water

From all the aforementioned, it is obvious that detergents find their way into drinking water supplies in various ways. As far as imparting odor to drinking water, only heavy doses of anionic surfactants yield an unpleasant odor [36], and someone has to have a very sensitive nose to smell detergent doses of 50 mg/L or less. On the other hand, it seems that the impact of detergent doses on the sense of taste of various individuals varies considerably. As reported by Cohen [10], the U.S. Public Health Service conducted a series of taste tests which showed that although 50% of the people in the test group detected a concentration of 60 mg/L of ABS in drinking water, only 5% of them detected a concentration of 16 mg/L. Because tests like this have been conducted using commercial detergent formulations, most probably the observed taste is not due to the surfactants but rather to the additives or perfumes added to the products. However, the actual limit for detergents in drinking water in the United States is a concentration of only 0.5 mg/L, less than even the most sensitive palates can discern.

7.2.6 Toxicity of Detergents

There is an upper limit of surfactant concentration in natural waters above which the existence of aquatic life, particularly higher animal life, is endangered. Trout are particularly sensitive to concentrations as low as 1 ppm and show symptoms similar to asphyxia [4]. On the other hand, numerous studies, which extended over a period of months and required test animals to drink significantly high doses of surfactants, showed absolutely no apparent ill effects due to digested detergents. Also, there are no instances in which the trace amounts of detergents present in drinking water were directly connected to adverse effects on human health.

River pollution from anionic surfactants, the primarily toxic ones, is of two types: (a) acute toxic pollution due to, for example, an accidental spill from a container of full-strength surfactant products, and (b) chronic pollution due to the daily discharges of municipal and industrial wastewaters. The international literature contains the result of numerous studies that have established dosages for both types of pollutional toxicity due to detergents, for most types of aquatic life such as species of fish.

7.3 CURRENT PERSPECTIVE AND FUTURE OUTLOOK

This section summarizes the main points of a recent product report [18], which presented the new products of the detergent industry and its proposed direction in the foreseeable future.

If recent product innovations sell successfully in test markets in the United States and other countries, rapid growth could begin again for the entire soap and detergent industry and especially for individual sectors of that industry. Among these new products are formulations that combine bleaching materials and other components, and detergents and fabric softeners sold in concentrated forms. These concentrated materials, so well accepted in Japan, are now becoming commercially significant in Western Europe. Their more widespread use will allow the industry to store and transport significantly smaller volumes of detergents, with the consequent reduction of environmental risks from housecleaning and spills. Some components of detergents such as enzymes will very likely grow in use, although the use of phosphates employed as builders will continue to drop for environmental reasons. Consumers shift to liquid formulations in areas where phosphate materials are banned from detergents, because they perceive that the liquid detergents perform better than powdered ones without phosphates.

In fuel markets, detergent formulations such as gasoline additives that limit the buildup of deposits in car engines and fuel injectors will very likely grow fast from a small base, with the likelihood of an increase in spills and discharges from this industrial source. Soap, on the other hand, has now become a small part (17%) of the total output of surfactants, whereas the anionic forms (which include soaps) accounted for 62% of total U.S. production in 1988. Liquid detergents (many of the LAS type), which are generally higher in surfactant concentrations than powdered ones, will continue to increase in production volume, therefore creating greater surfactant pollution problems due to housecleaning and spills. (Also, a powdered detergent spill creates less of a problem, as it is easier to just scoop up or vacuum.)

Changes in the use of builders resulting from environmental concerns have been pushing surfactant production demand. Outright legal bans or consumer pressures on the use of inorganic phosphates and other materials as builders generally have led formulators to raise the contents of surfactants in detergents. Builders provide several functions, most important of which are to aid the detergency action and to tie up and remove calcium and magnesium from the wash water, dirt, and the fabric or other material being cleaned. Besides sodium and potassium phosphates, other builders that may be used in various detergent formulations are citric acid and derivatives, zeolites, and other alkalis. Citric acid causes caking and is not used in powdered detergents, but it finds considerable use in liquid detergents. In some detergent formulations, larger and larger amounts of soda ash (sodium carbonate) are replacing inert ingredients due to its functionality as a builder, an agglomerating aid, a carrier for surfactants, and a source of alkalinity.

Incorporating bleaching agents into detergent formulations for home laundry has accelerated, because its performance allows users to curtail the need to store as well as add (as a second step) bleaching material. Because U.S. home laundry requires shorter wash times and lower temperatures than European home laundry, chlorine bleaches (mainly sodium hypochlorite) have long dominated the U.S. market. Institutional and industrial laundry bleaching, when done, has also favored chlorine bleaches (often chlorinated isocyanurates) because of their rapid action. Other kinds of bleaching agents used in the detergent markets are largely sodium perborates and percarbonates other than hydrogen peroxide itself.

The peroxygen bleaches are forecast to grow rapidly, for both environmental and technical reasons, as regulatory pressures drive the institutional and industrial market away from chlorine bleaches and toward the peroxygen ones. The Clean Water Act amendments are requiring lower levels of trihalomethanes (products of reaction of organics and chlorine) in wastewaters. Expensive systems may be needed to clean up effluents, or the industrial users of chlorine bleaches will have to pay higher and higher surcharges to municipalities for handling chlorine-containing wastewaters that are put into sewers. Current and expected changes in bleaching materials for various segments of the detergent industry are but part of sweeping changes to come due to environmental concerns and responses to efforts to improve the world environment.

Both detergent manufacturers and their suppliers will make greater efforts to develop more "environmentally friendly" products. BASF, for example, has developed a new biodegradable stabilizer for perborate bleach, which is now being evaluated for use in detergents. The existing detergent material, such as LAS and its precursor linear alkylbenzene, known to be nontoxic and environmentally safe as well as effective, will continue to be widely used. It will be difficult, however, to gain approval for new materials to be used in detergent formulations until their environmental performance has been shown to meet existing guidelines. Some countries, for example, tend to favor a formal regulation or law (i.e., the EEC countries) prohibiting the manufacture, importation, or use of detergents that are not satisfactorily biodegradable [28].

7.4 INDUSTRIAL OPERATION AND WASTEWATER

The soap and detergent industry is a basic chemical manufacturing industry in which essentially both the mixing and chemical reactions of raw materials are involved in production. Also, short- and long-term chemicals storage and warehousing, as well as loading/unloading and transportation of chemicals, are involved in the operation.

7.4.1 Manufacture and Formulation

This industry produces liquid and solid cleaning agents for domestic and industrial use, including laundry, dishwashing, bar soaps, specialty cleaners, and industrial cleaning products. It can be broadly divided (Fig. 1) into two categories: (a) soap manufacture that is based on the processing of natural fat; and (b) detergent manufacture that is based on the processing of

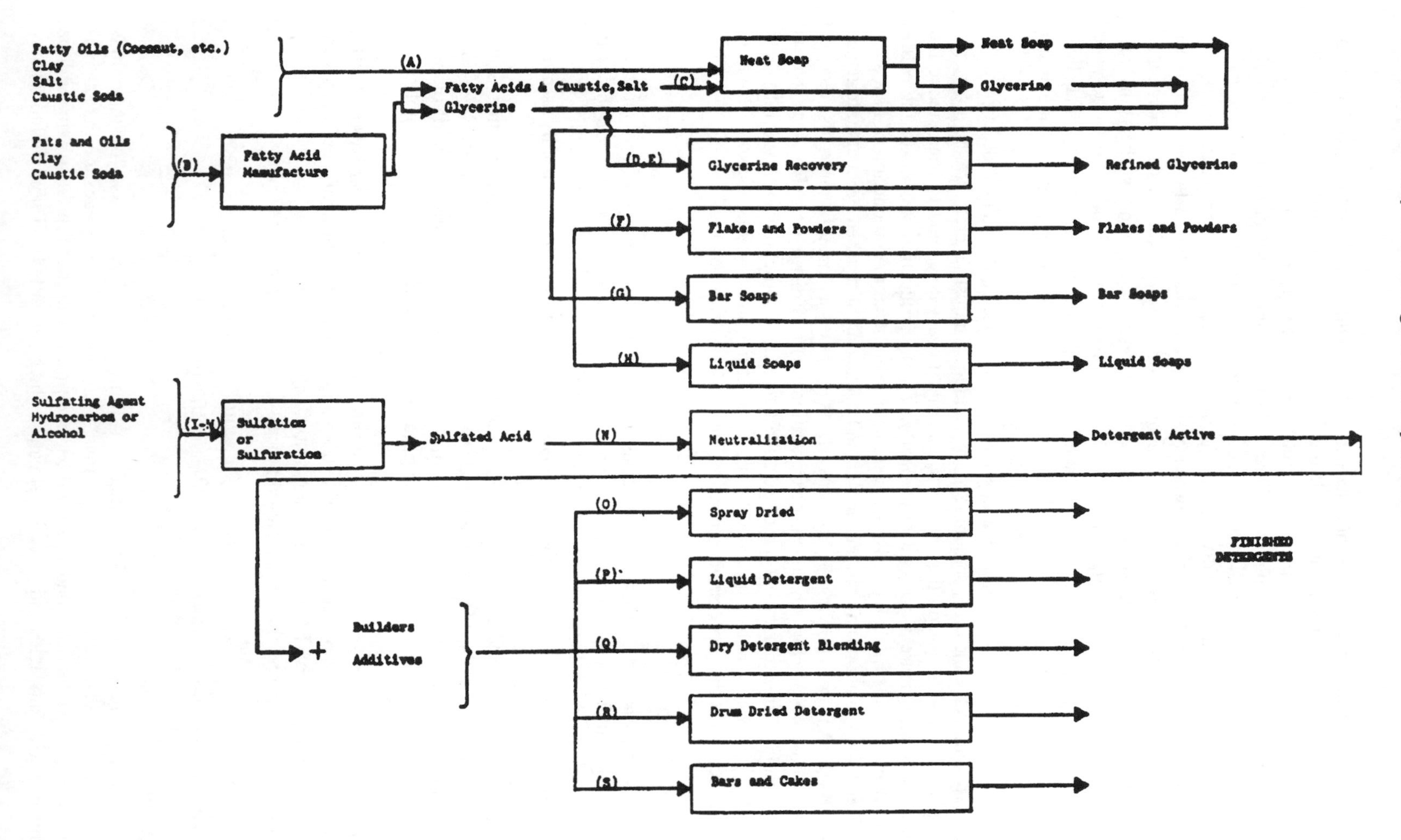

Figure 1 Flow diagram of soap and detergent manufacture (from Ref. 13).

petrochemicals. The information presented here includes establishments primarily involved in the production of soap, synthetic organic detergents, inorganic alkaline detergents, or any combinations of these, and plants producing crude and refined glycerine from vegetable and animal fats and oils. Types of facilities not discussed here include plants primarily involved in the production of shampoo or shaving creams/soaps, whether from soap or surfactants, and of synthetic glycerine as well as specialty cleaners, polishing and sanitation preparations.

Numerous processing steps exist between basic raw materials for surfactants and other components that are used to improve performance and desirability, and the finished marketable products of the soap and detergent industry. Inorganic and organic compounds such as ethylene, propylene, benzene, natural fatty oils, ammonia, phosphate rock, trona, chlorine, peroxides, and silicates are among the various basic raw materials being used by the industry. The final formulation of the industry's numerous marketable products involves both simple mixing of and chemical reactions among compounds such as the above.

The categorization system of the various main production streams and their descriptions is taken from federal guidelines [13] pertaining to state and local industrial pretreatment programs. It will be used in the discussion that ensues to identify process flows and to characterize the resulting raw waste. Figure 1 shows a flow diagram for the production streams of the entire industry. Manufacturing of soap consists of two major operations: the production of neat soap (65–70% hot soap solution) and the preparation and packaging of finished products into flakes and powders (F), bar soaps (G), and liquid soaps (H). Many neat soap manufacturers also recover glycerine as a byproduct for subsequent concentration (D) and distillation (E). Neat soap is generally produced in either of two processes: the batch kettle process (A) or the fatty acid neutralization process, which is preceded by the fat splitting process (B, C). (Note, letters in parentheses represent the processes described in the following sections.)

Batch Kettle Process (A)

This process consists of the following operations: (a) receiving and storage of raw materials, (b) fat refining and bleaching, and (c) soap boiling. The major wastewater sources, as shown in the process flow diagram (Fig. 2), are the washouts of both the storage and refining tanks, as well as from leaks and spills of fats and oils around these tanks. These streams are usually skimmed for fat recovery prior to discharge to the sewer.

The fat refining and bleaching operation is carried out to remove impurities that would cause color and odor in the finished soap. The wastewater from this source has a high soap concentration, treatment chemicals, fatty impurities, emulsified fats, and sulfuric acid solutions of fatty acids. Where steam is used for heating, the condensate may contain low-molecular-weight fatty acids, which are highly odorous, partially soluble materials.

The soap boiling process produces two concentrated waste streams: sewer lyes that result from the reclaiming of scrap soap and the brine from Nigre processing. Both of these wastes are low volume, high pH, with BOD values up to 45,000 mg/L.

Soap manufacture by the neutralization process is a two-step process:

fat + water → fatty acid + glycerine (*fat splitting*) (B)

fatty acid + caustic → soap (*fatty acid neutralization*) (C)

Fat Splitting (B)

The manufacture of fatty acid from fat is called fat splitting (B), and the process flow diagram is shown in Fig. 3. Washouts from the storage, transfer, and pretreatment stages are the same as those for process (A). Process condensate and barometric condensate from fat splitting will be contaminated with fatty acids and glycerine streams, which are settled and skimmed to recover

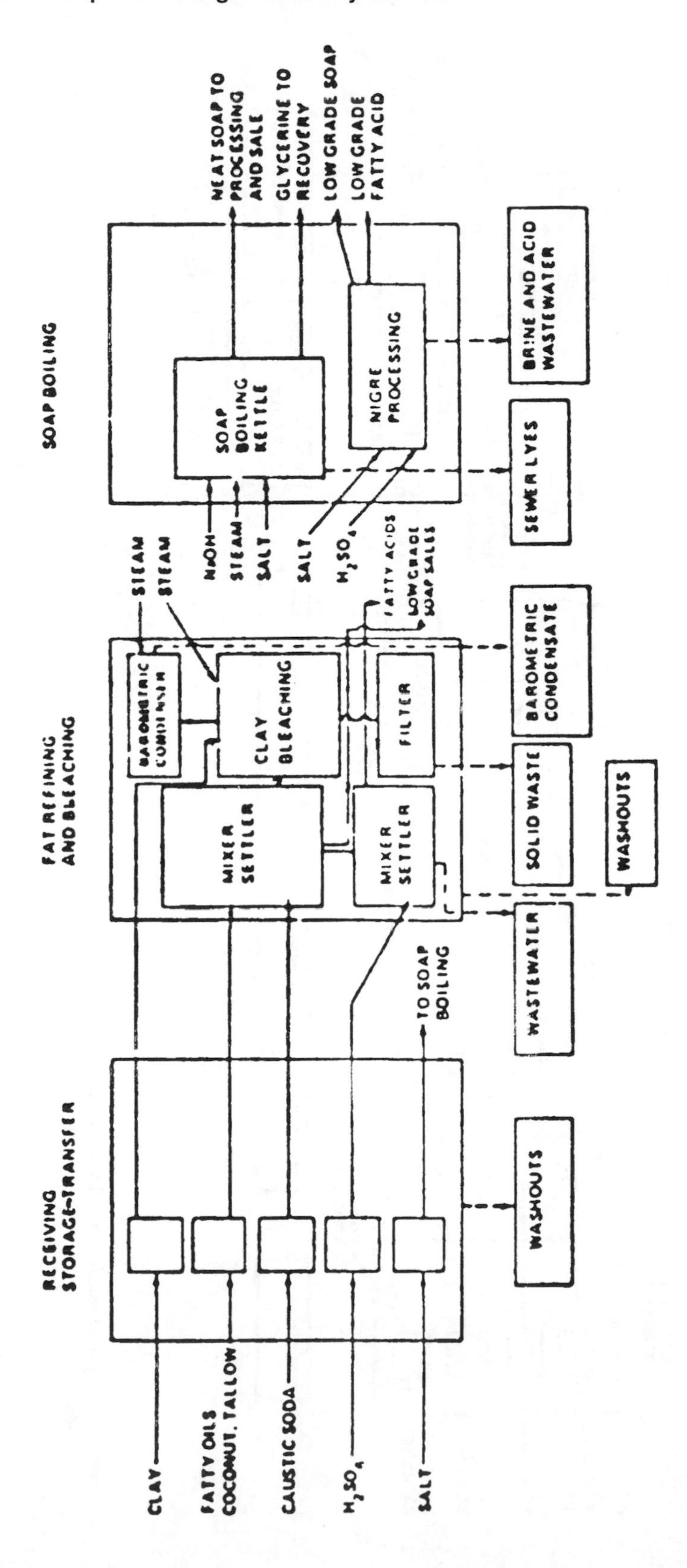

Figure 2 Soap manufacture by batch kettle (A) (from Ref. 13).

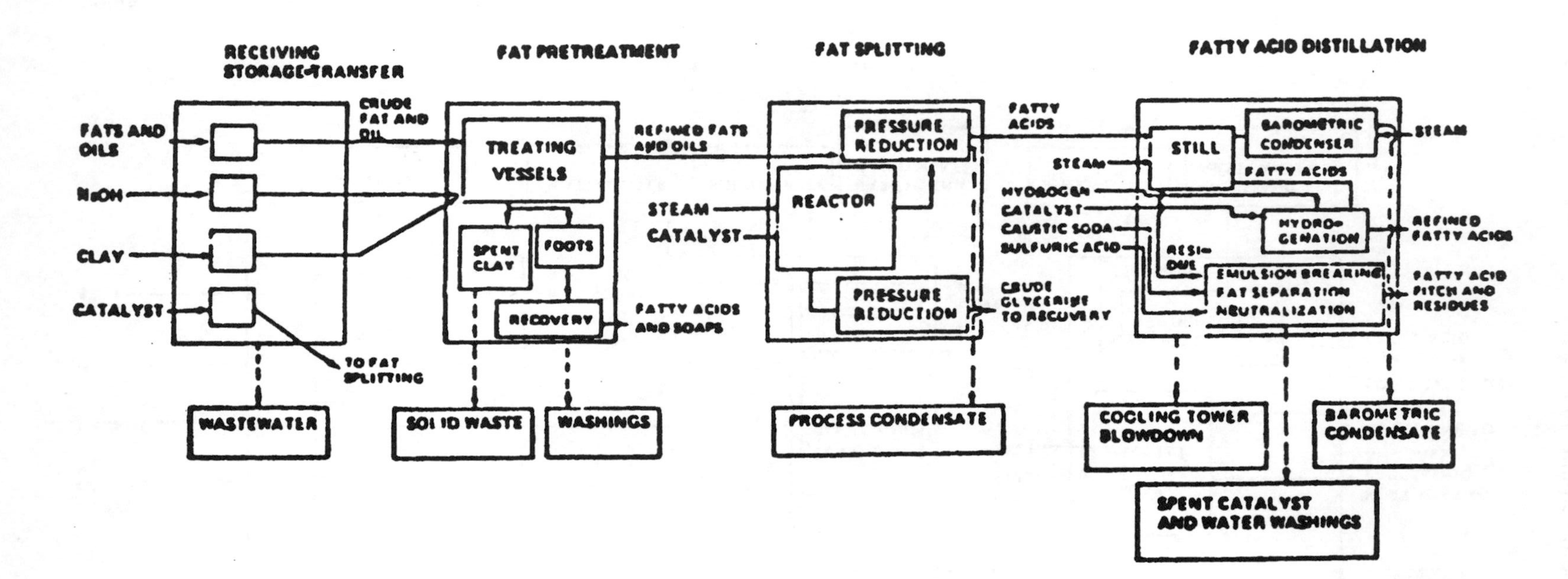

Figure 3 Fatty acid manufacture by fat splitting (B) (from Ref. 13).

the insoluble fatty acids that are processed for sale. The water will typically circulate through a cooling tower and be reused. Occasional purges of part of this stream to the sewer release high concentrations of BOD and some grease and oil.

In the fatty acid distillation process, wastewater is generated as a result of an acidification process, which breaks the emulsion. This wastewater is neutralized and sent to the sewer. It will contain salt from the neutralization, zinc and alkaline earth metal salts from the fat splitting catalyst, and emulsified fatty acids and fatty acid polymers.

Fatty Acid Neutralization (C)

Soap making by this method is a faster process than the kettle boil process and generates less wastewater effluent (Fig. 4). Because it is faster, simpler, and cleaner than the kettle boil process, it is the preferred process among larger as well as small manufacturers.

Often, sodium carbonate is used in place of caustic. When liquid soaps (at room temperature) are desired, the more soluble potassium soaps are made by substituting potassium hydroxide for the sodium hydroxide (lye). This process is relatively simple and high-purity raw materials are converted to soap with essentially no byproducts. Leaks, spills, storm runoff, and washouts are absent. There is only one wastewater of consequence: the sewer lyes from reclaiming of scrap. The sewer lyes contain the excess caustic soda and salt added to grain out the soap. Also, they contain some dirt and paper not removed in the strainer.

Glycerine Recovery Process (D, E)

A process flow diagram for the glycerine recovery process uses the glycerine byproducts from kettle boiling (A) and fat splitting (B). The process consists of three steps (Fig. 5): (a) pretreatment to remove impurities, (b) concentration of glycerine by evaporation, and (c) distillation to a finished product of 98% purity.

There are three wastewaters of consequence from this process: two barometric condensates, one from evaporation and one from distillation, plus the glycerine foots or still bottoms. Contaminants from the condensates are essentially glycerine with a little entrained salt. In the distillation process, the glycerine foots or still bottoms leave a glassy dark brown amorphous solid rich in salt that is disposed of in the wastewater stream. It contains glycerine, glycerine polymers, and salt. The organics will contribute to BOD, COD (chemical oxygen demand), and dissolved solids. The sodium chloride will also contribute to dissolved solids. Little or no suspended solids, oil, and grease or pH effect should be seen.

Glycerine can also be purified by the use of ion-exchange resins to remove sodium chloride salt, followed by evaporation of the water. This process puts additional salts into the wastewater but results in less organic contamination.

7.4.2 Production of Finished Soaps and Process Wastes

The production of finished soaps utilizes the neat soap produced in processes A and C to prepare and package finished soap. These finished products are soap flakes and powders (F), bar soaps (G), and liquid soap (H). See Figures 6, 7, and 8 for their respective flow diagrams.

Flakes and Powders (F)

Neat soap may or may not be blended with other products before flaking or powdering. Neat soap is sometimes filtered to remove gel particles and run into a reactor (crutcher) for mixing with builders. After thorough mixing, the finished formulation is run through various mechanical operations to produce flakes and powders. Because all of the evaporated moisture goes to the atmosphere, there is no wastewater effluent.

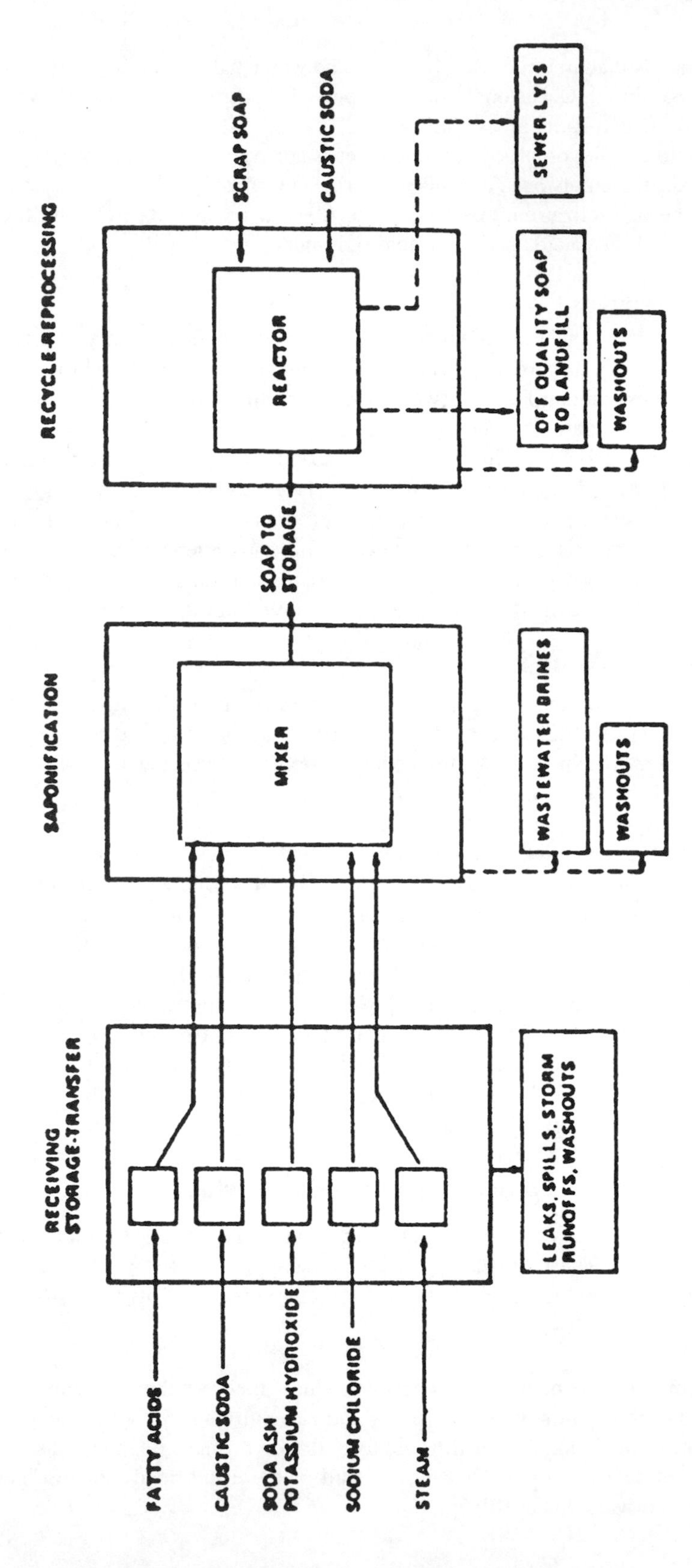

Figure 4 Soap from fatty acid neutralization (C) (from Ref. 13).

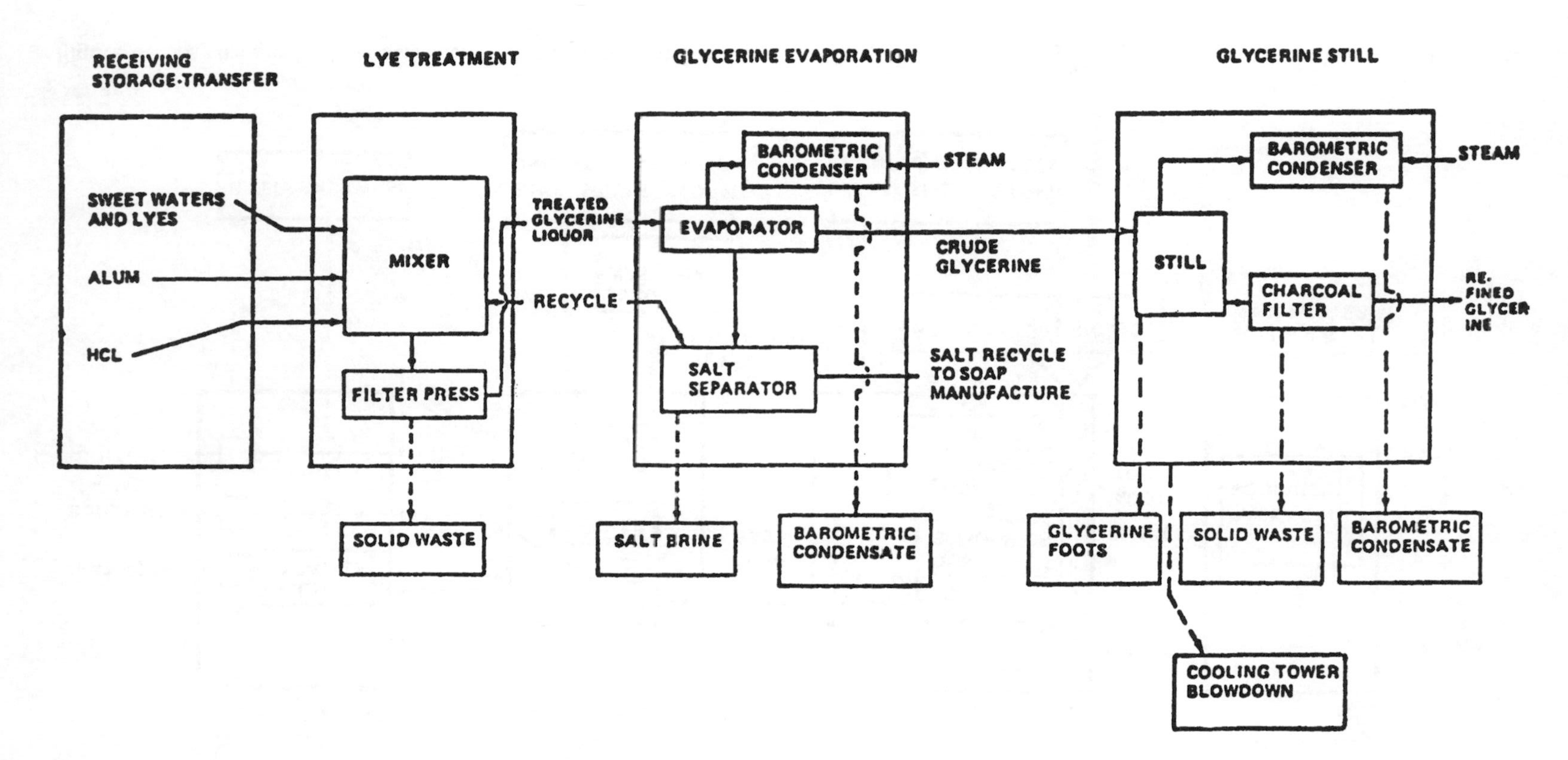

Figure 5 Glycerine recovery process flow diagram (D, E) (from Ref. 13).

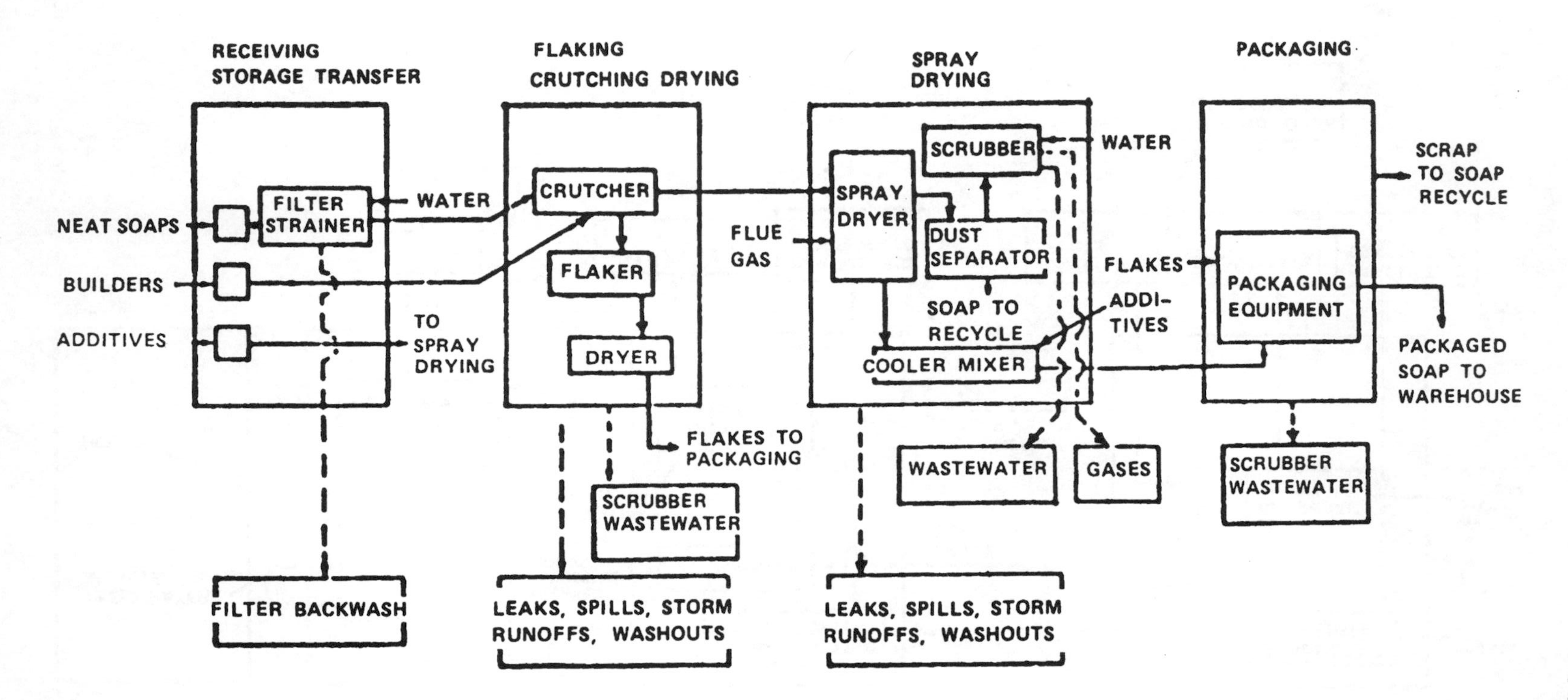

Figure 6 Soap flake and powder manufacture (F) (from Ref. 13).

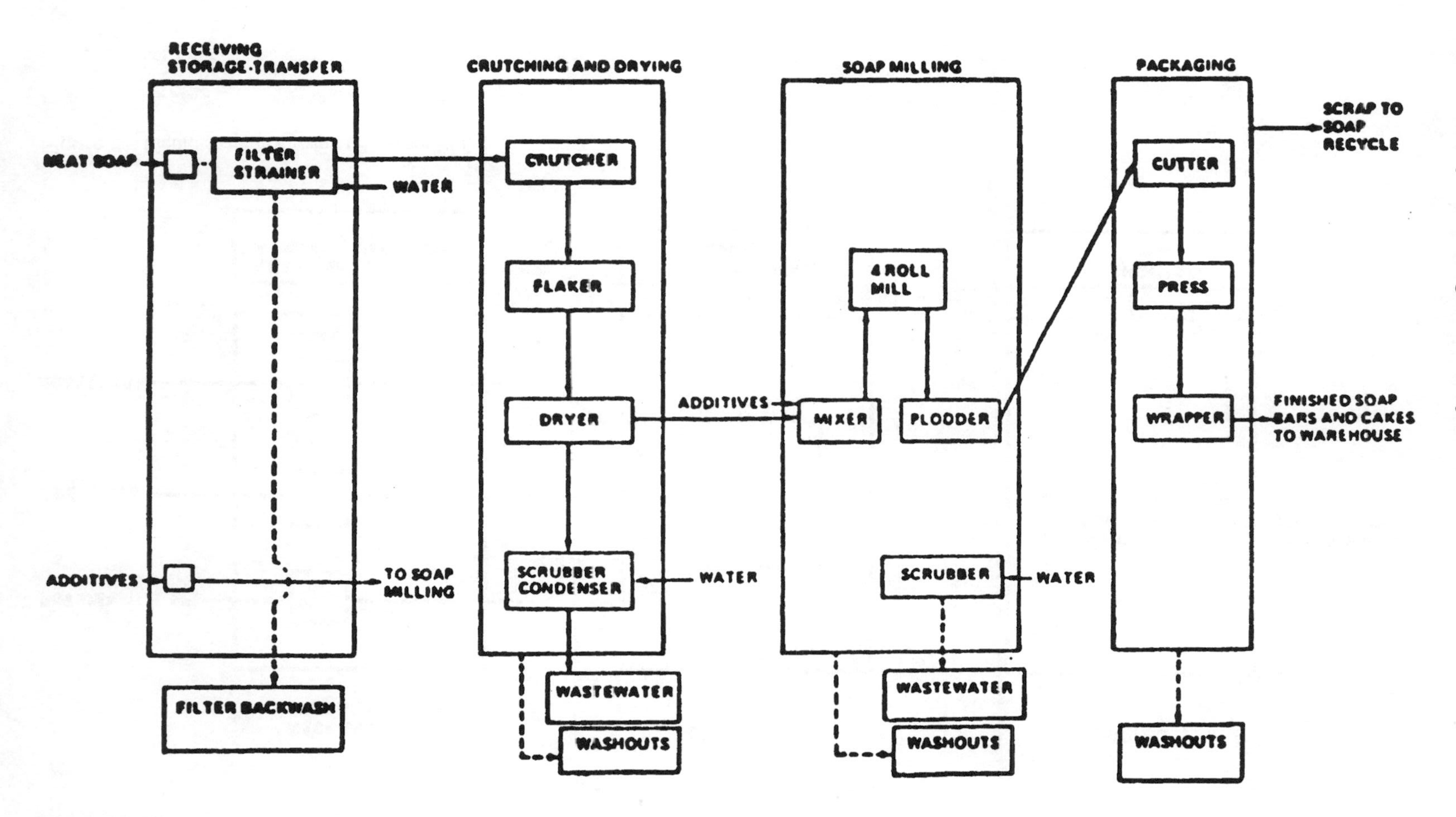

Figure 7 Bar soap manufacture (G) (from Ref. 13).

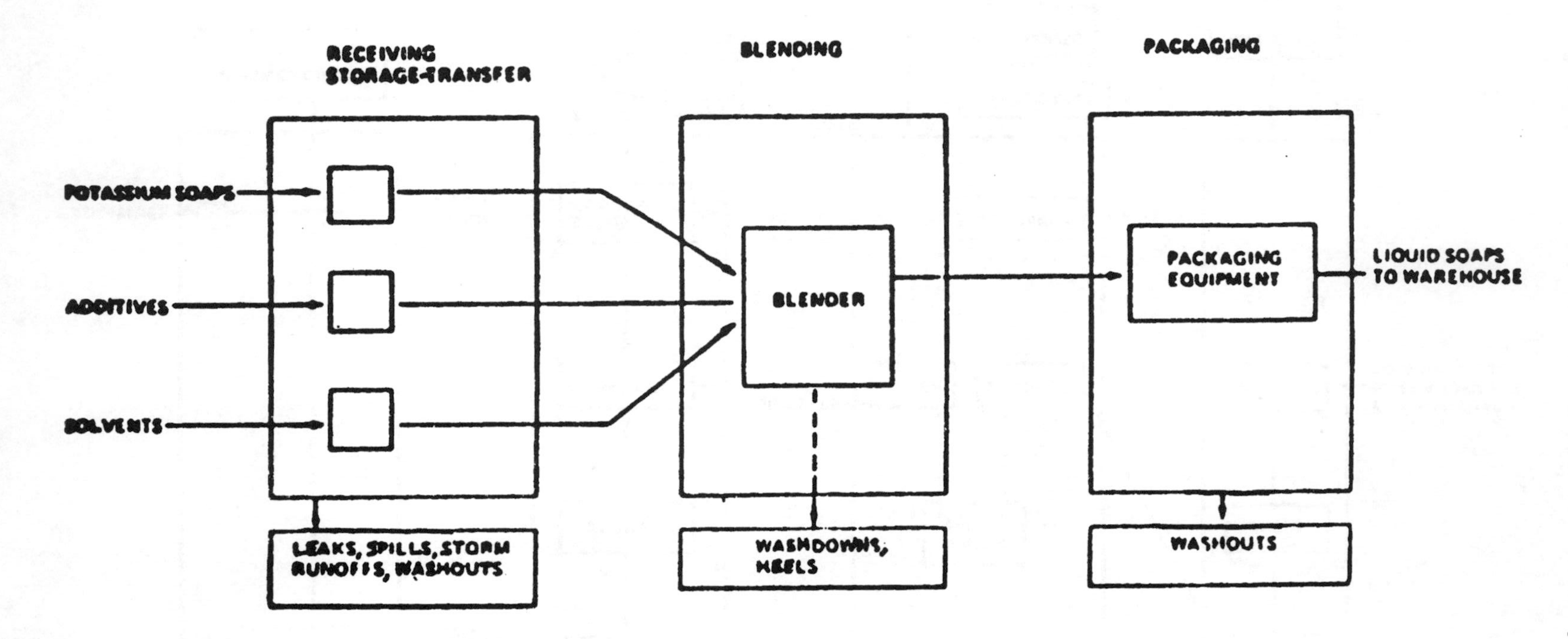

Figure 8 Liquid soap processing (H) (from Ref. 13).

Some operations will include a scrap soap reboil to recover reclaimed soap. The soap reboil is salted out for soap recovery and the salt water is recycled. After frequent recycling, the salt water becomes so contaminated that it must be discharged to the sewer. Occasional washdown of the crutcher may be needed. The tower is usually cleaned down dry. There is also some gland water that flows over the pump shaft, picking up any minor leaks. This will contribute a very small, but finite, effluent loading.

There are a number of possible effluents shown on the flow diagram for process F (Fig. 6). However, a survey of the industry showed that most operating plants either recycled any wastewater to extinction or used dry clean-up processes. Occasionally, water will be used for clean-up.

Bar Soaps (G)

The procedure for bar soap manufacture (O) will vary significantly from plant to plant, depending on the particular clientele served. A typical flow diagram for process O is shown in Figure 7. The amount of water used in bar soap manufacture varies greatly. In many cases, the entire bar soap processing operation is carried out without generating a single wastewater stream. The equipment is all cleaned dry, without any washups. In other cases, due to housekeeping requirements associated with the particular bar soap processes, there are one or more wastewater streams from air scrubbers.

The major waste streams in bar soap manufacture are the filter backwash, scrubber waters, or condensate from a vacuum drier, and water from equipment washdown. The main contaminant of all these streams is soap that will contribute primarily BOD and COD to the wastewater.

Liquid Soap (H)

In the making of liquid soap, neat soap (often the potassium soap of fatty acids) is blended in a mixing tank with other ingredients such as alcohols or glycols to produce a finished product, or the pine oil and kerosene for a product with greater solvency and versatility (Fig. 8). The final blended product may be, and often is, filtered to achieve a sparkling clarity before being drummed. In making liquid soap, water is used to wash out the filter press and other equipment. According to manufacturers, there are very few effluent leaks. Spills can be recycled or handled dry. Washout between batches is usually unnecessary or can be recycled to extinction.

7.4.3 Detergent Manufacture and Waste Streams

Detergents, as mentioned previously, can be formulated with a variety of organic and inorganic chemicals, depending on the cleaning characteristics desired. A finished, packaged detergent customarily consists of two main components: the active ingredient or surfactant, and the builder. The processes discussed in the following will include the manufacture and processing of the surfactant as well as the preparation of the finished, marketable detergent. The production of the surfactant (Fig. 1) is generally a two-step process: (a) sulfation or sulfonation, and (b) neutralization.

7.4.4 Surfactant Manufacture and Waste Streams

Oleum Sulfonation/Sulfation (I)

One of the most important active ingredients of detergents is the sulfate or sulfonate compounds made via the oleum route. A process flow diagram is shown in Figure 9. In most cases, the sulfonation/sulfation is carried out continuously in a reactor where the oleum (a solution of sulfur trioxide in sulfuric acid) is brought into contact with the hydrocarbon or alcohol and a

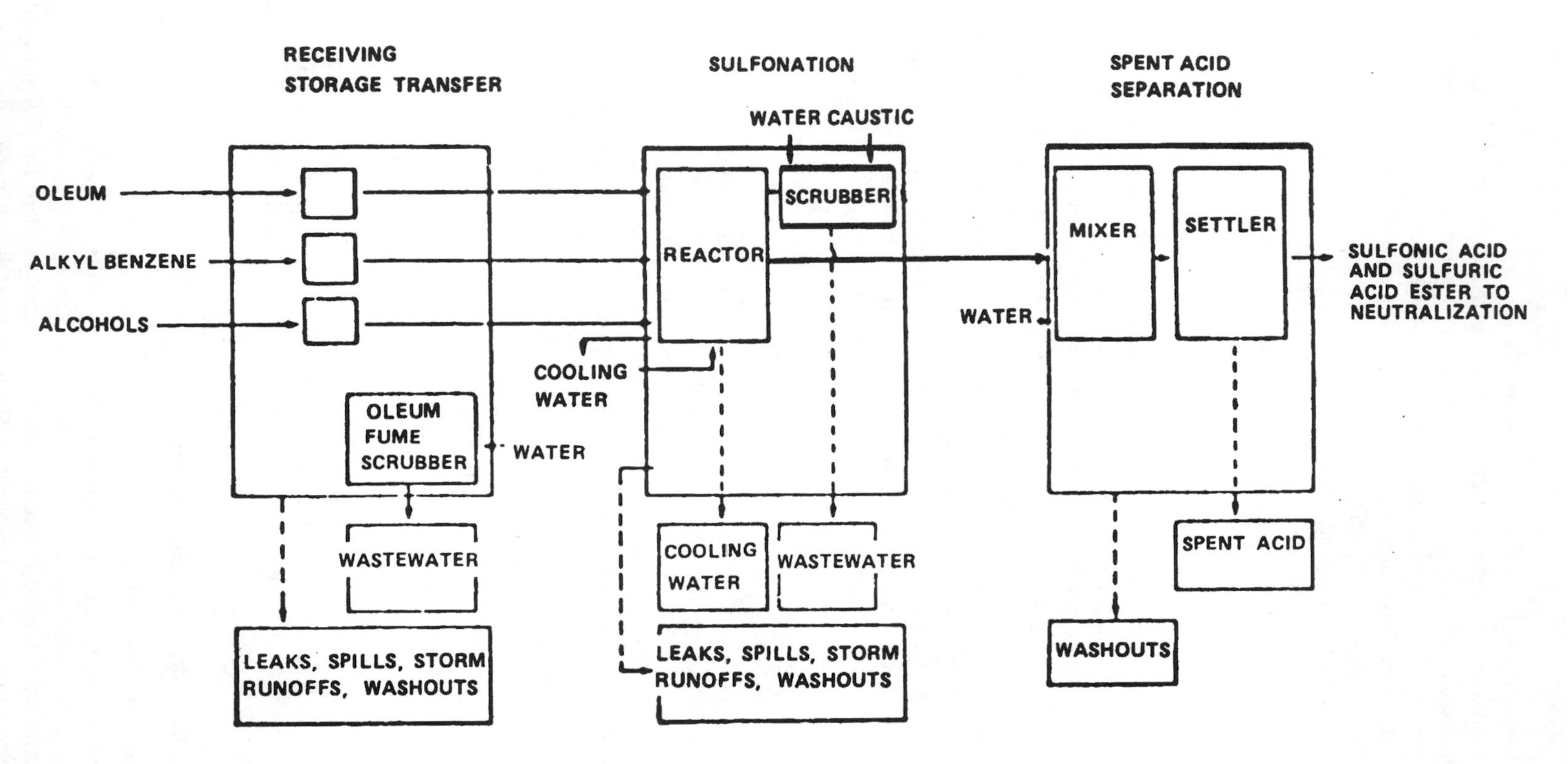

Figure 9 Oleum sulfation and sulfonation (batch and continuous) (I) (from Ref. 13).

rapid reaction ensues. The stream is then mixed with water, where the surfactant separates and is then sent to a settler. The spent acid is drawn off and usually forwarded for reprocessing, and the sulfonated/sulfated materials are sent to be neutralized.

This process is normally operated continuously and performs indefinitely without need of periodic cleanout. A stream of water is generally played over pump shafts to pick up leaks as well as to cool the pumps. Wastewater flow from this source is quite modest, but continual.

Air–SO_3 Sulfation/Sulfonation (J)

This process for surfactant manufacture has many advantages and is used extensively. With SO_3 sulfation, no water is generated in the reaction. A process flow diagram is shown in Figure 10. SO_3 can be generated at the plant by burning sulfur or sulfur dioxide with air instead of obtaining it as a liquid. Because of this reaction's particular tendency to char the product, the reactor system must be cleaned thoroughly on a regular basis. In addition, there are usually several airborne sulfonic acid streams that must be scrubbed, with the wastewater going to the sewer during sulfation.

SO_3 Solvent and Vacuum Sulfonation (K)

Undiluted SO_3 and organic reactant are fed into the vacuum reactor through a mixing nozzle. A process flow diagram is shown in Figure 11. This system produces a high-quality product, but offsetting this is the high operating cost of maintaining the vacuum. Other than occasional washout, the process is essentially free of wastewater generation.

Sulfamic Acid Sulfation (L)

Sulfamic acid is a mild sulfating agent and is used only in very specialized quality areas because of the high reagent price. A process flow diagram is shown in Figure 12. Washouts are the only wastewater effluents from this process as well.

Chlorosulfonic Acid Sulfation (M)

For products requiring high-quality sulfates, chlorosulfonic acid is an excellent corrosive agent that generates hydrochloric acid as a byproduct. A process flow diagram is shown in Figure 13. The effluent washouts are minimal.

Neutralization of Sulfuric Acid Esters and Sulfonic Acids (N)

This step is essential in the manufacture of detergent active ingredients as it converts the sulfonic acids or sulfuric acid esters (products produced by processes I–M) into neutral surfactants. It is a potential source of some oil and grease, but occasional leaks and spills around the pump and valves are the only expected source of wastewater contamination. A process flow diagram is shown in Figure 14.

7.4.5 Detergent Formulation and Process Wastes

Spray-Dried Detergents (O)

In this segment of the processing, the neutralized sulfonates and/or sulfates are first blended with builders and additives in the crutcher. The slurry is then pumped to the top of a spray tower of about 4.5–6.1 m (15–20 ft) in diameter by 45–61 m (150–200 ft) in height, where nozzles spray out detergent slurry. A large volume of hot air enters the bottom of the tower and rises to

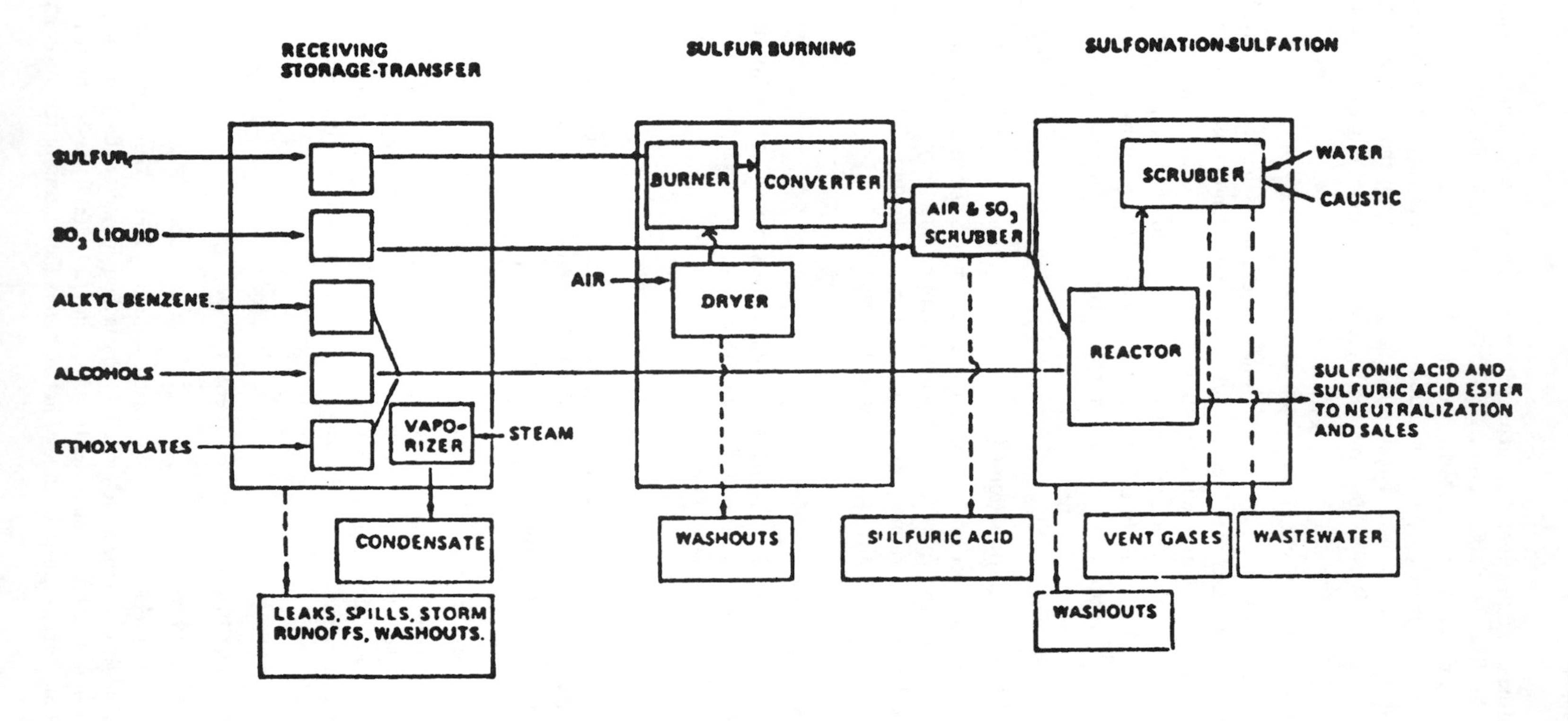

Figure 10 Air–SO_3 sulfation and sulfonation (batch and continuous) (J) (from Ref. 13).

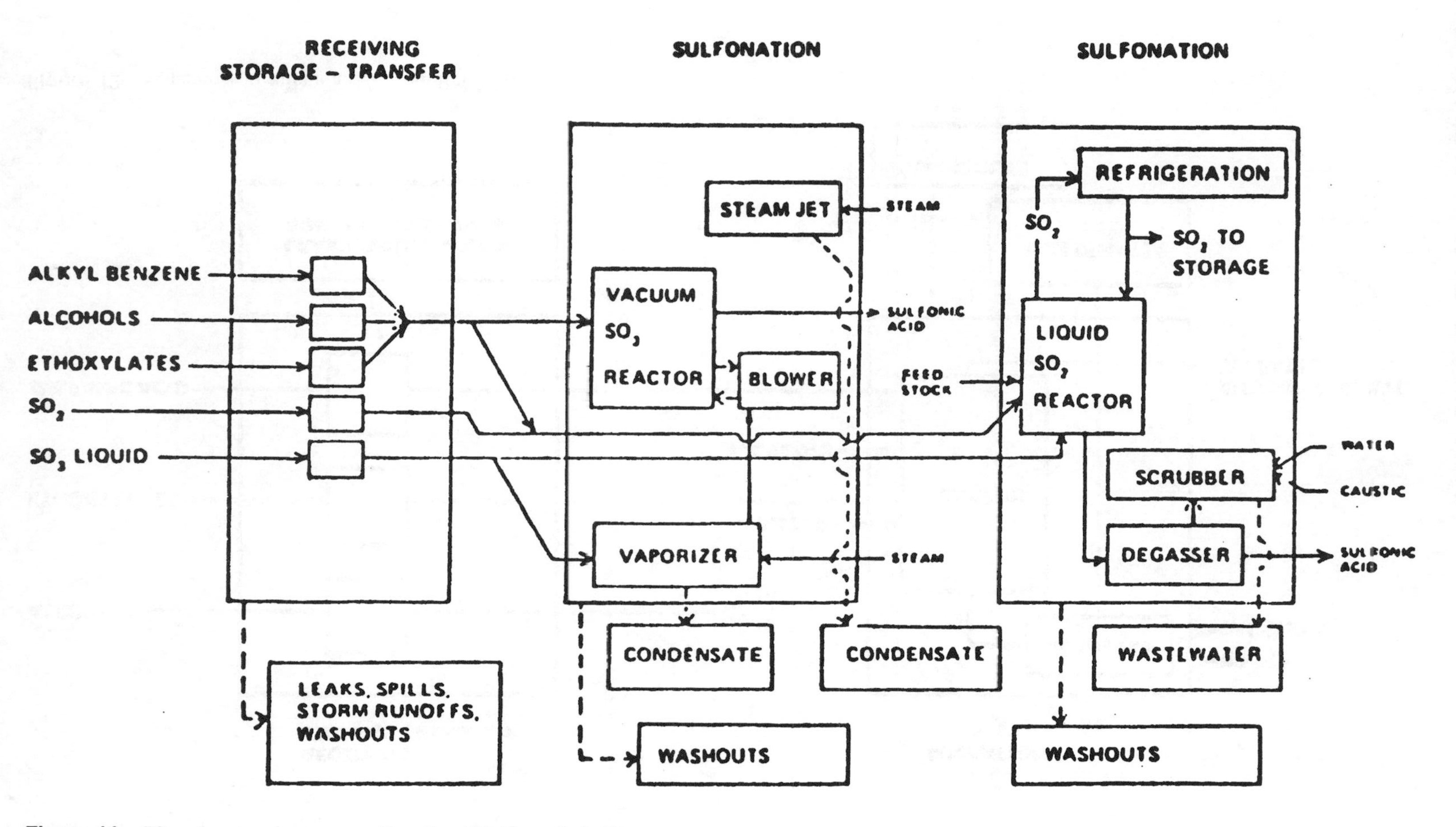

Figure 11 SO_3 solvent and vacuum sulfonation (K) (from Ref. 13).

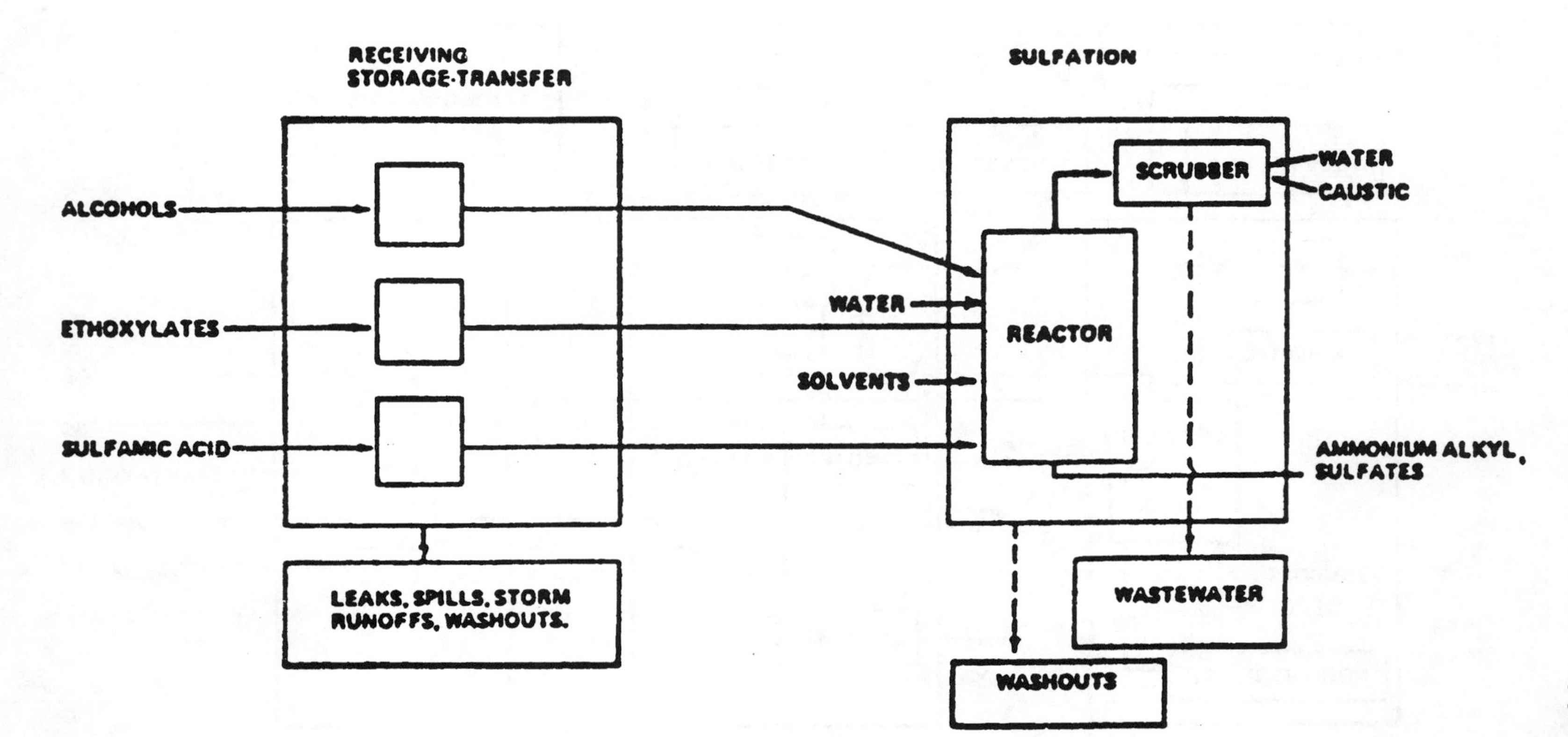

Figure 12 Sulfamic acid sulfation (L) (from Ref. 13).

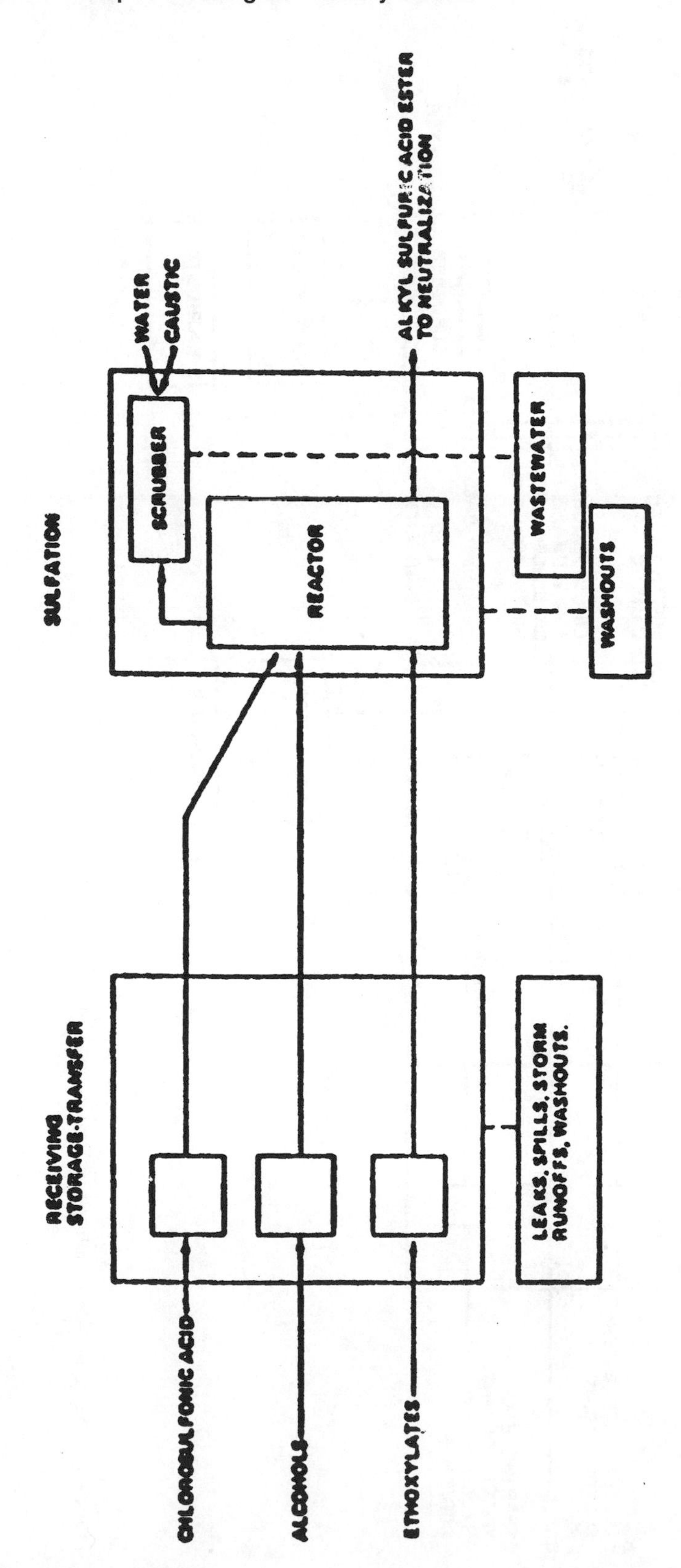

Figure 13 Chlorosulfonic acid sulfation (M) (from Ref. 13).

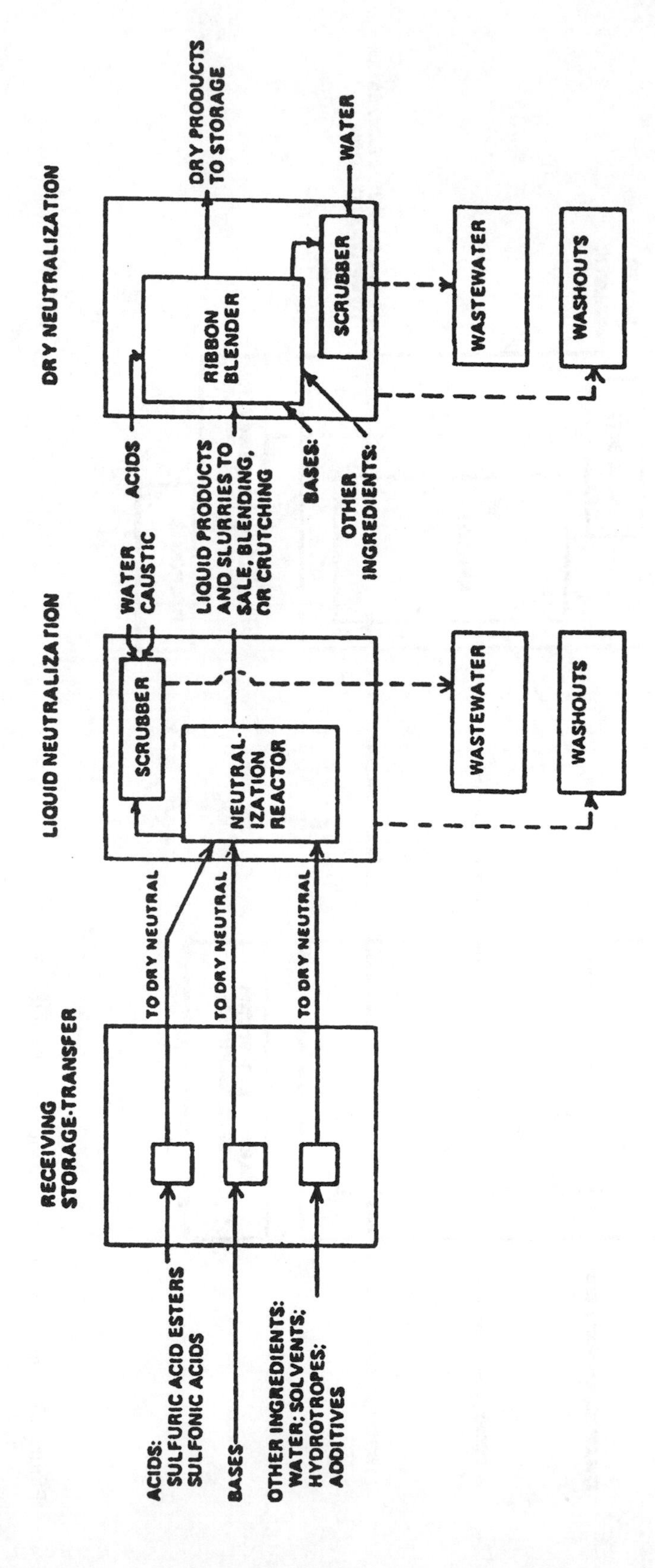

Figure 14 Neutralization of sulfuric acid esters and sulfonic acids (N) (from Ref. 13).

meet the falling detergent. The design preparation of this step will determine the detergent particle's shape, size, and density, which in turn determine its solubility rate in the washing process.

The air coming from the tower will be carrying dust particles that must be scrubbed, thus generating a wastewater stream. The spray towers are periodically shut down and cleaned. The tower walls are scraped and thoroughly washed down. The final step is mandatory because the manufacturers must be careful to avoid contamination to the subsequent formulation.

Wastewater streams are rather numerous, as seen in the flow diagram of Figure 15. They include many washouts of equipment from the crutchers to the spray tower itself. One wastewater flow that has high loadings is that of the air scrubber, which cleans and cools the hot gases exiting from this tower. All the plants recycle some of the wastewater generated, while some of the plants recycle all the flow generated. Owing to increasingly stringent air quality requirements, it can be expected that fewer plants will be able to maintain a complete recycle system of all water flows in the spray tower area. After the powder comes from the spray tower, it is further blended and then packaged.

Liquid Detergents (P)

Detergent actives are pumped into mixing tanks where they are blended with numerous ingredients, ranging from perfumes to dyes. A process flow diagram is shown in Figure 16. From here, the fully formulated liquid detergent is run down to the filling line for filling, capping, labeling, and so on. Whenever the filling line is to change to a different product, the filling system must be thoroughly cleaned out to avoid cross contamination.

Dry Detergent Blending (Q)

Fully dried surfactant materials are blended with additives in dry mixers. Normal operation will see many succeeding batches of detergent mixed in the same equipment without anything but dry cleaning. However, when a change in formulation occurs, the equipment must be completely washed down and a modest amount of wastewater is generated. A process flow diagram is shown in Figure 17.

Drum-Dried Detergent (R)

This process is one method of converting liquid slurry to a powder and should be essentially free of the generation of wastewater discharge other than occasional washdown. A process flow diagram is shown in Figure 18.

Detergent Bars and Cakes (S)

Detergent bars are either 100% synthetic detergent or a blend of detergent and soap. They are blended in essentially the same manner as conventional soap. Fairly frequent cleanups generate a wastewater stream. A process flow diagram is shown in Figure 19.

7.4.6 Wastewater Characteristics

Wastewaters from the manufacturing, processing, and formulation of organic chemicals such as soaps and detergents cannot be exactly characterized. The wastewater streams are usually expected to contain trace or larger concentrations of all raw materials used in the plant, all intermediate compounds produced during manufacture, all final products, coproducts, and byproducts, and the auxiliary or processing chemicals employed. It is desirable, from the

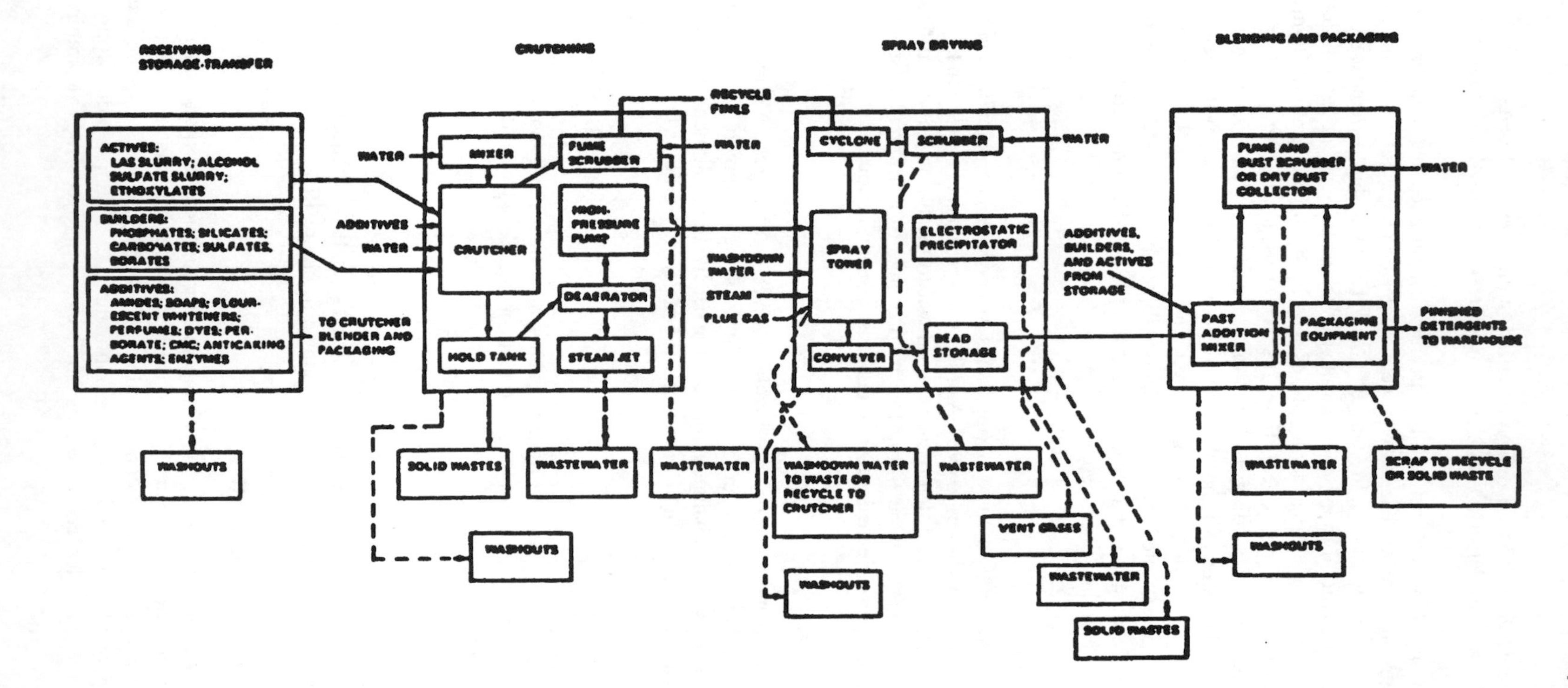

Figure 15 Spray-dried detergent production (O) (from Ref. 13).

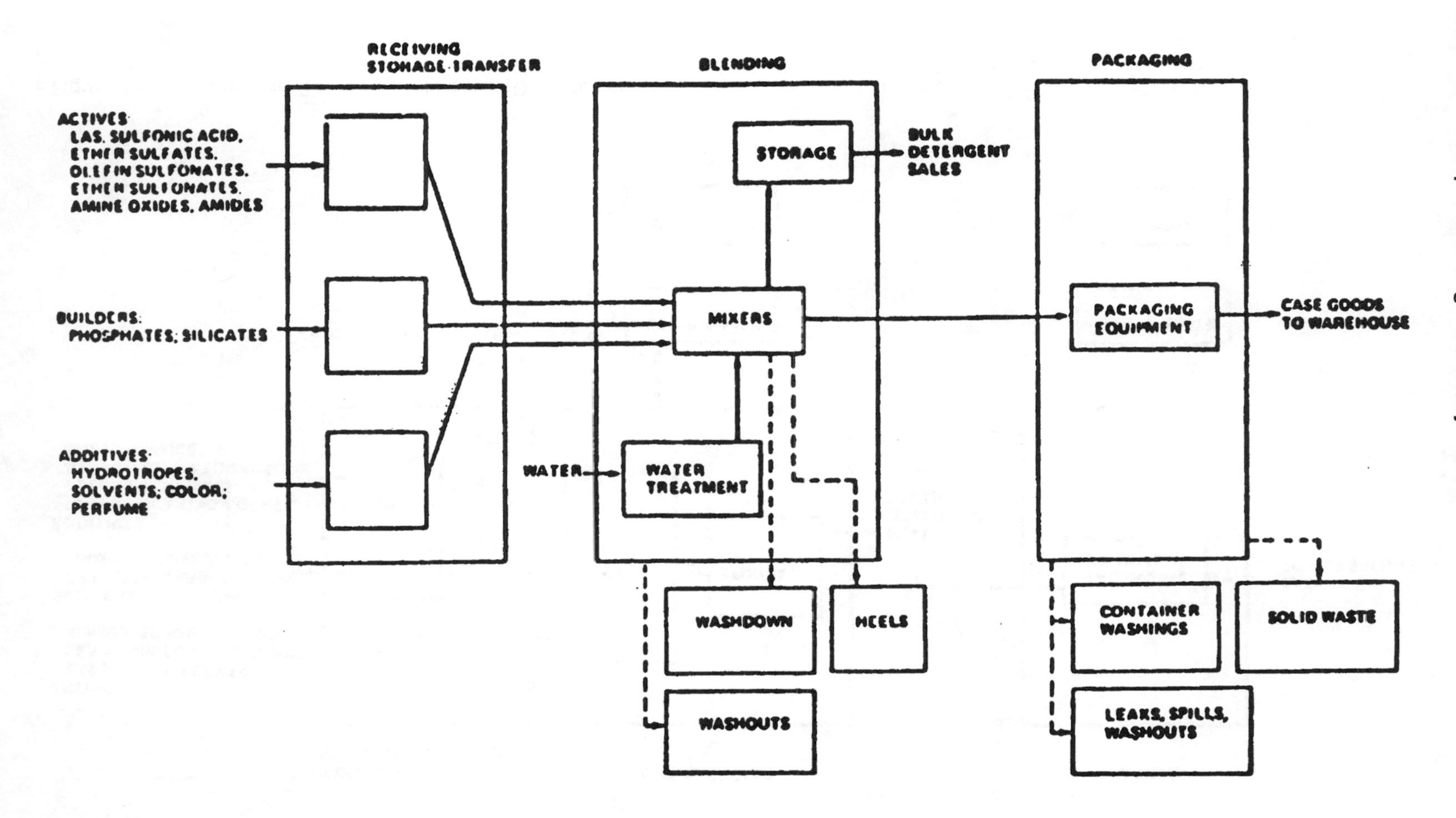

Figure 16 Liquid detergent manufacture (P) (from Ref. 13).

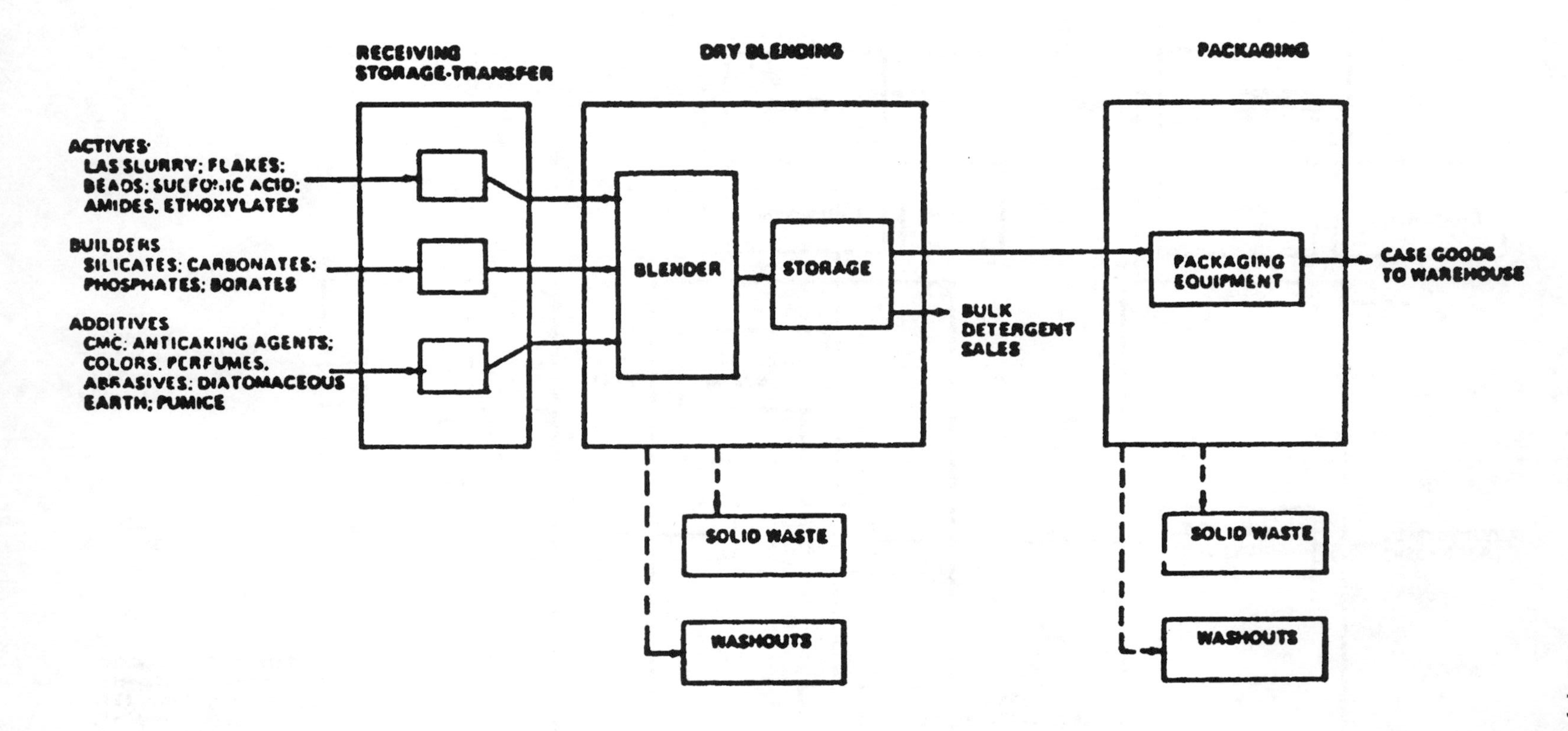

Figure 17 Detergent manufacture by dry blending (Q) (from Ref. 13).

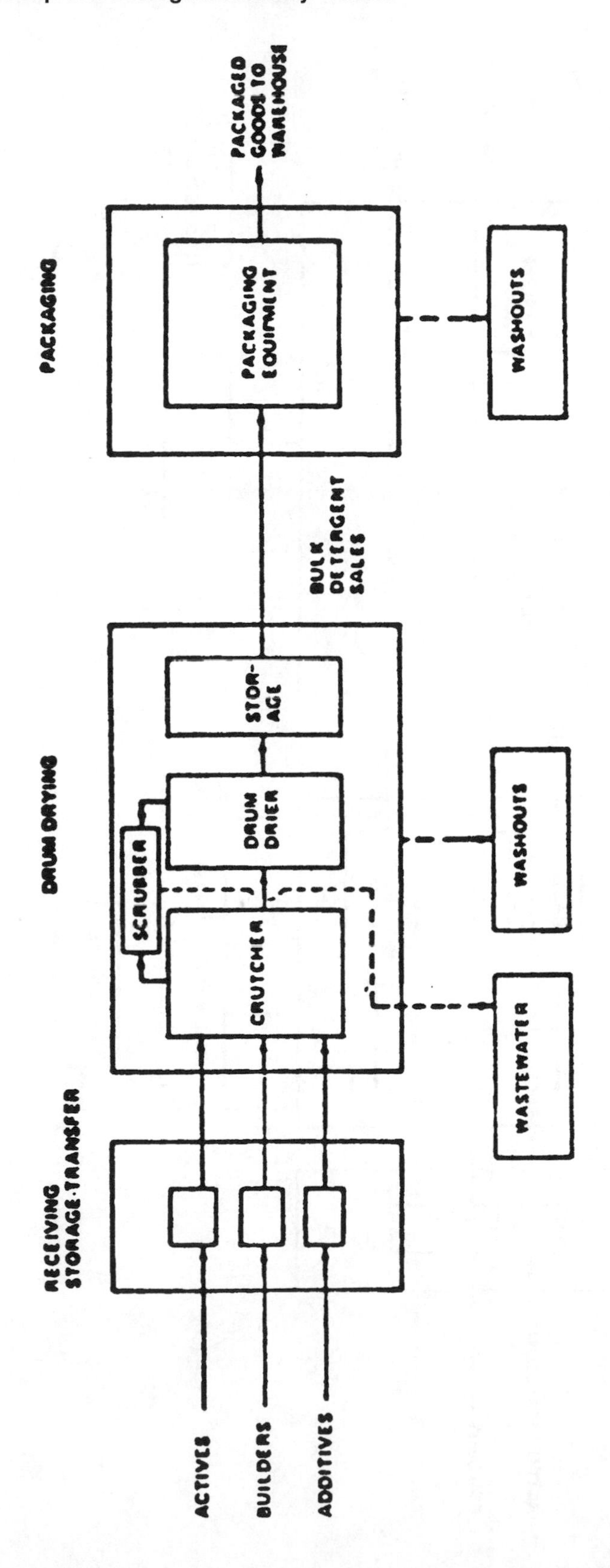

Figure 18 Drum-dried detergent manufacture (R) (from Ref. 13).

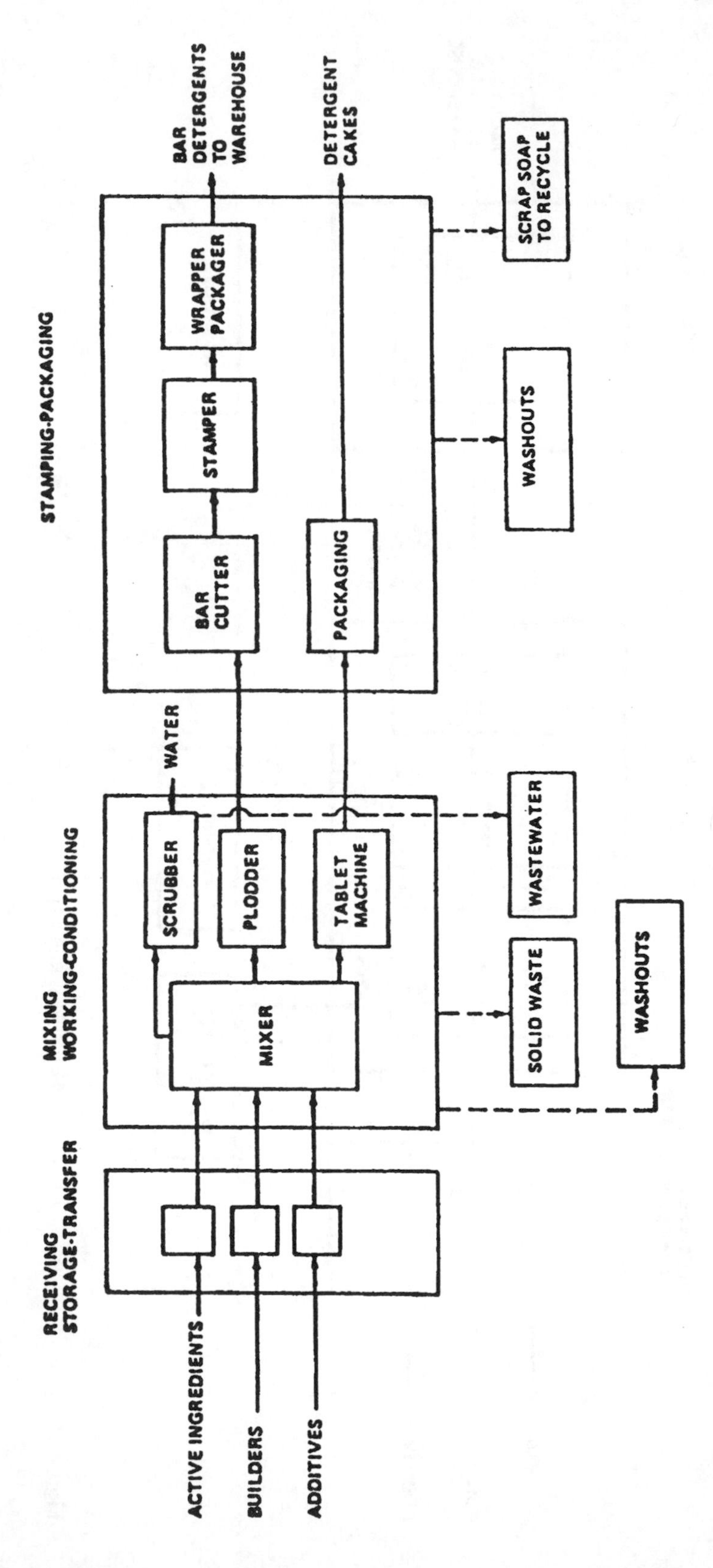

Figure 19 Detergent bar and cake manufacture (S) (from Ref. 13).

viewpoint of economics, that these substances not be lost, but some losses and spills appear unavoidable and some intentional dumping does take place during housecleaning and vessel emptying and preparation operations.

According to a study by the USEPA [12], which presents estimates of industrial wastewater generation as well as related pollution parameter concentrations, the wastewater volume discharged from soap and detergent manufacturing facilities per unit of production ranges from 0.3 to 2.8 gal/lb (2.5–23.4 L/kg) of product. The reported ranges of concentration (mg/L) for BOD, suspended solids, COD, and grease were 500–1200, 400–2100, 400–1800, and about 300, respectively. These data were based on a study of the literature and the field experience of governmental and private organizations. The values represent plant operating experience for several plants consisting of 24 hour composite samples taken at frequent intervals. The ranges for flow and other parameters generally represent variations in the level of plant technology or variations in flow and quality parameters from different subprocesses. In particular, the more advanced and modern the level of production technology, the smaller the volume of wastewater discharged per unit of product. The large variability (up to one order of magnitude) in the ranges is generally due to the heterogeneity of products and processes in the soap and detergent industry.

The federal guidelines [13] for state and local pretreatment programs reported the raw wastewater characteristics (Table 1) in mg/L concentration and the flows and water quality parameters (Table 2) based on the production or 1 ton of product manufactured for the subcategories of the industry. Most soap and detergent manufacturing plants contain two or more of the subcategories shown in Table 3, and their wastewaters are a composite of these individual unit processes.

7.5 U.S. CODE OF FEDERAL REGULATIONS

The information presented in this section has been taken from the U.S. Code of Federal Regulations (40 CFR), containing documents related to the protection of the environment [14], in particular, the regulations contained in Part 417, Soap and Detergent Manufacturing Point Source Category, pertaining to effluent limitations guidelines and pretreatment or performance standards for each of the 19 subcategories shown in Table 3.

The effluent guideline regulations and standards of 40 CFR, Part 417, were promulgated on February 11, 1975. According to the most recent notice in the Federal Register [15] regarding industrial categories and regulations, no review is under way or planned and no revision is proposed for the soap and detergent industry. The effluent guidelines and standards applicable to this industrial category include: (a) the best practicable control technology currently available (BPT); (b) the best available technology economically achievable (BAT); (c) pretreatment standards for existing sources (PSES); (d) standards of performance for new sources (NSPS); and (e) pretreatment standards for new sources (PSNS).

For all 19 subcategories of the soap and detergent manufacture industry, there are no pretreatment standards establishing the quantity and quality of pollutants or pollutant properties that may be discharged to a publicly owned treatment works (POTW) by an existing or new point source. If the major contributing industry is an existing point source discharging pollutants to navigable waters, it will be subject to Section 301 of the Federal Water Pollution Control Act and to the provisions of 40 CFR, Part 128. However, practically all the soap and detergent manufacturing plants in the United States discharge their wastewaters into municipal sewer systems. The effluent limitations guidelines for certain subcategories regarding BPT, BAT, and NSPS are presented in Tables 4–10.

Table 1 Soap and Detergent Industry Raw Wastewater Characteristics

Parameter	Batch kettle (A)	Fat splitting (B)	Fatty acid neutralization (C)	Glycerine concentration (D)	Glycerine distillation (E)	Flakes and powders (F)	Bar soap (G)	Liquid soap (H)
BOD (mg/L)	3600[a]	60–3600[a]	400				1600–3000[a]	
COD (mg/L)	4267[a]	115–6000[a]	1000					
TSS (mg/L)	1600–6420	115–6000	775					
Oil and grease (mg/L)	250[a]	13–760[a]	200[a]					
pH	5–13.5	High	High	Neutral	Neutral	Neutral	Neutral	Neutral
Chlorides (mg/L)	20–47 m[a]							
Zinc (mg/L)		Present						
Nickel (mg/L)		Present						

Parameter	Oleum sulfation and sulfonation (I)	Air sulfation and sulfonation (J)	SO_3 solvent and vacuum (K)	Sulfamic acid sulfation (L)	Chloro-sulfonic (M)	Neutral sulfuric (N)	Spray-dried (O)	Liquid detergent (P)	Dry blend (Q)	Drum-dried (R)	Bars and cakes (S)
BOD (mg/L)	75–2000[a]	380–520				8.5–6 m[a]	48–19 m[a]	65–3400[a]	Neg.		
COD (mg/L)	220–6000[a]	920–1589[a]				245–21 m[a]	150–60 m[a]	640–11 m[a]			
TSS (mg/L)	100–3000										
Oil and grease (mg/L)	100–3000[a]										
pH	1–2[a]	2–7[a]	Low	Low	Low	Low					
Surfactant (mg/L)	250–7000							60–2 m			
Boron (mg/L)	Present	Present	Present	Present	Present	Present	Present	Present	Present	Present	Present

[a] In high levels these parameters may be inhibitory to biological systems; m = thousands; BOD, biochemical oxygen demand; COD, chemical oxygen demand; TSS, total suspended solids.

Source: Ref. 10.

Table 2 Raw Wastewater Characteristics Based on Production

Parameter	Batch kettle (A)	Fat splitting (B)	Fatty acid neutralization (C)	Glycerine concentration (D)	Glycerine distillation (E)	Flakes and powders (F)	Bar soap (G)	Liquid soap (H)
Flow range (L/kkg)[a]	623/2500	3.3M/1924M	258			Neg.		Neg.
Flow type	B	B	B	B	B	B	B	B
BOD (kg/kkg)[b]	6	12	0.1	15	5	0.1	3.4	0.1
COD (kg/kkg)	10	22	0.25	30	10	0.3	5.7	0.3
TSS (kg/kkg)	4	22	0.2	2	2	0.1	5.8	0.1
Oil and grease (kg/kkg)	0.9	2.5	0.05	1	1	0.1	0.4	0.1

Parameter	Oleum sulfation and sulfonation (I)	SO_3 sulfation and sulfonation (J)	SO_3 solvent and vacuum sulfonation (K)	Sulfamic acid sulfation (L)	Chloro-sulfonic (M)	Neutral sulfuric acid esters (N)	Spray-dried (O)	Liquid detergent (P)	Dry blend (Q)	Drum-dried (R)	Bars and cakes (S)
Flow range (L/kkg)[a]	100/2740	249				10/4170	41/2084	625/6250			
Flow type	C	C	B	B	B	B&C	B	B	B	B	B
BOD (kg/kkg)[b]	0.2	3	3	3	3	0.10	0.1–0.8	2–5	0.1	0.1	7
COD (kg/kkg)	0.6	9	9	9	9	0.3	0.3–25	4–7	0.5	0.3	22
TSS (kg/kkg)	0.3	0.3	0.3	0.3	0.3	0.3	0.1–1.0		0.1	0.1	2
Oil and grease (kg/kkg)	0.3	0.5	0.5	0.5	0.5	0.1	Nil–0.3			0.1	0.2
Chloride (kg/kkg)					5						
Surfactant (kg/kkg)	0.7	3	3	3	3	0.2	0.2–1.5	1.3–3.3		0.1	5

[a] L/kkg, L/1000 kg product produced (lower limit/upper limit).
[b] kg/kkg, kg/1000 kg product produced.
B = Batch; C = Continuous; Neg. = Negligible; M = Thousand.
Source: Ref. 13.

Table 3 Soap and Detergent Categorization

Category	Subcategory	Code
Soap manufacture	Batch kettle and continuous	A
	Fatty acid manufacture by fat splitting	B
	Soap from fatty acid neutralization	C
	Glycerine recovery	
	Glycerine concentration	D
	Glycerine distillation	E
	Soap flakes and powders	F
	Bar soaps	G
	Liquid soap	H
Detergent manufacture	Oleum sulfonation and sulfation (batch and continuous)	I
	Air–SO_3 sulfation and sulfonation (batch and continuous)	J
	SO_3 solvent and vacuum sulfonation	K
	Sulfamic acid sulfation	L
	Chlorosulfonic acid sulfation	M
	Neutralization of sulfuric acid esters and sulfonic acids	N
	Spray-dried detergents	O
	Liquid detergent manufacture	P
	Detergent manufacture by dry blending	Q
	Drum-dried detergents	R
	Detergent bars and cakes	S

Source: Ref. 10.

Table 4 Effluent Limitations for Subpart A, Batch Kettle

	Effluent limitations [metric units (kg/1000 kg of anhydrous product)]	
Effluent characteristic	Maximum for any 1 day	Average of daily values for 30 consecutive days shall not exceed
(a) BPT		
BOD_5	1.80	0.60
COD	4.50	1.50
TSS	1.20	0.40
Oil and grease	0.30	0.10
pH	[a]	a
(b) BAT and NSPS		
BOD_5	0.80	0.40
COD	2.10	1.05
TSS	0.80	0.40
Oil and grease	0.10	0.05
pH	[a]	[a]

[a] Within the range 6.0–9.0.

BAT, best available technology economically achievable; NSPS, standards of performance for new sources.

Source: Ref. 14.

Table 5 Effluent Limitations for Subpart C, Soap by Fatty Acid

Effluent characteristic	Effluent limitations [metric units (kg/1000 kg of anhydrous product)]	
	Maximum for any 1 day	Average of daily values for 30 consecutive days shall not exceed
(a) BPT		
BOD_5	0.03	0.01
COD	0.15	0.05
TSS	0.06	0.02
Oil and grease	0.03	0.01
pH	[a]	[a]
(b) BAT		
BOD_5	0.02	0.01
COD	0.10	0.05
TSS	0.04	0.02
Oil and grease	0.02	0.01
pH	[a]	[a]
(c) NSPS		
BOD_5	0.02	0.01
COD	0.10	0.05
TSS	0.04	0.02
Oil and grease	0.02	0.01
pH	[a]	[a]

[a] Within the range 6.0–9.0.
Source: Ref. 14.

7.6 WASTEWATER CONTROL AND TREATMENT

The sources and characteristics of wastewater streams from the various subcategories in soap and detergent manufacturing, as well as some of the possibilities for recycling and treatment, have been discussed in Section 7.4. The pollution control and treatment methods and unit processes used are discussed in more detail in the following sections. The details of the process design criteria for these unit treatment processes can be found in any design handbooks.

7.6.1 In-Plant Control and Recycle

Significant in-plant control of both waste quantity and quality is possible, particularly in the soap manufacturing subcategories where maximum flows may be 100 times the minimum. Considerably less in-plant water conservation and recycle are possible in the detergent industry, where flows per unit of product are smaller.

The largest in-plant modification that can be made is the changing or replacement of the barometric condensers (subcategories A, B, D, and E). The wastewater quantity discharged from these processes can be significantly reduced by recycling the barometric cooling water through fat skimmers, from which valuable fats and oils can be recovered, and then through the cooling towers. The only waste with this type of cooling would be the continuous small blowdown from

Table 6 Effluent Limitations for Subpart D, Glycerine Concentration

Effluent characteristic	Effluent limitations [metric units (kg/1000 kg of anhydrous product)] Maximum for any 1 day	Average of daily values for 30 consecutive days shall not exceed
(a) BPT		
BOD_5	4.50	1.50
COD	13.50	4.50
TSS	0.60	0.20
Oil and grease	0.30	0.10
pH	a	a
(b) BAT		
BOD_5	0.80	0.40
COD	2.40	1.20
TSS	0.20	0.10
Oil and grease	0.08	0.04
pH	a	a
(c) NSPS		
BOD_5	0.80	0.40
COD	2.40	1.20
TSS	0.20	0.10
Oil and grease	0.08	0.04
pH	a	a

[a] Within the range 6.0–9.0.
Source: Ref. 14.

the skimmer. Replacement with surface condensers has been used in several plants to reduce both the waste flow and quantity of organics wasted.

Significant reduction of water usage is possible in the manufacture of liquid detergents (P) by the installation of water recycle piping and tankage and by the use of air rather than water to blowdown filling lines. In the production of bar soaps (G), the volume of discharge and the level of contamination can be reduced materially by installation of an atmospheric flash evaporator ahead of the vacuum drier. Finally, pollutant carryover from distillation columns such as those used in glycerine concentration (D) or fatty acid separation (B) can be reduced by the use of two additional special trays.

In another document [37] presenting techniques adopted by the French for pollution prevention, a new process of detergent manufacturing effluent recycle is described. As shown in Figure 20, the washout effluents from reaction and/or mixing vessels and washwater leaks from the paste preparation and pulverization pump operations are collected and recycled for use in the paste preparation process. The claim has been that pollution generation at such a plant is significantly reduced and, although the savings on water and raw materials are small, the capital and operating costs are less than those for building a wastewater treatment facility.

Besselievre [2] has reported in a review of water reuse and recycling by the industry that soap and detergent manufacturing facilities have shown an average ratio of reused and recycled water to total wastewater effluent of about 2:1. That is, over two-thirds of the generated wastewater stream in an average plant has been reused and recycled. Of this volume, about 66% has been used as cooling water and the remaining 34% for the process or other purposes.

Table 7 Effluent Limitations for Subpart G, Bar Soaps

	Effluent limitations [metric units (kg/1000 kg of anhydrous product)]	
Effluent characteristic	Maximum for any 1 day	Average of daily values for 30 consecutive days shall not exceed
(a) BPT		
BOD_5	1.02	0.34
COD	2.55	0.85
TSS	1.74	0.58
Oil and grease	0.12	0.04
pH	a	a
(b) BAT		
BOD_5	0.40	0.20
COD	1.20	0.60
TSS	0.68	0.34
Oil and grease	0.06	0.03
pH	a	a
(c) NSPS		
BOD_5	0.40	0.20
COD	1.20	0.60
TSS	0.68	0.34
Oil and grease	0.06	0.03
pH	a	a

[a]Within the range 6.0–9.0.
Source: Ref. 14.

7.6.2 Wastewater Treatment Methods

The soap and detergent manufacturing industry makes routine use of various physicochemical and biological pretreatment methods to control the quality of its discharges. A survey of these treatment processes is presented in Table 11 [13], which also shows the usual removal efficiencies of each unit process on the various pollutants of concern. According to Nemerow [38] and Wang and Krofta [39], the origin of major wastes is in washing and purifying soaps and detergents and the resulting major pollutants are high BOD and certain soaps (oily and greasy, alkali, and high-temperature wastes), which are removed primarily through air flotation and skimming, and precipitation with the use of $CaCl_2$ as a coagulant.

Figure 21 presents a composite flow diagram describing a complete treatment train of the unit processes that may be used in a large soap and detergent manufacturing plant to treat its wastes. As a minimum requirement, flow equalization to smooth out peak discharges should be utilized even at a production facility that has a small-volume batch operation. Larger plants with integrated product lines may require additional treatment of their wastewaters for both suspended solids and organic materials' reduction. Coagulation and sedimentation are used by the industry for removing the greater portion of the large solid particles in its waste. On the other hand, sand or mixed-bed filters used after biological treatment can be utilized to eliminate fine particles. One of the biological treatment processes or, alternatively, granular or powdered activated carbon is the usual method employed for the removal of particulate or soluble organics from the waste streams. Finally, as a tertiary step for removing particular ionized pollutants or

Table 8 Effluent Limitations for Subpart H, Liquid Soaps

Effluent characteristic	Effluent limitations [metric units (kg/1000 kg of anhydrous product)]	
	Maximum for any 1 day	Average of daily values for 30 consecutive days shall not exceed
(a) BPT		
BOD_5	0.03	0.01
COD	0.15	0.05
TSS	0.03	0.01
Oil and grease	0.03	0.01
pH	a	a
(b) BAT		
BOD_5	0.02	0.01
COD	0.10	0.05
TSS	0.02	0.01
Oil and grease	0.02	0.01
pH	a	a
(c) NSPS		
BOD_5	0.02	0.01
COD	0.10	0.05
TSS	0.02	0.01
Oil and grease	0.02	0.01
pH	a	a

[a] Within the range 6.0–9.0.
Source: Ref. 14.

total dissolved solids (TDS), a few manufacturing facilities have employed either ion exchange or the reverse osmosis process.

Flotation or Foam Separation

One of the principal applications of vacuum and pressure (air) flotation is in commercial installations with colloidal wastes from soap and detergent factories [20,40–42]. Wastewaters from soap production are collected in traps on skimming tanks, with subsequent recovery floating of fatty acids.

Foam separation or fractionation [40,41,43–45] can be used to extra advantage: not only do surfactants congregate at the air/liquid interfaces, but other colloidal materials and ionized compounds that form a complex with the surfactants tend to also be concentrated by this method. An incidental, but often important, advantage of air flotation processes is the aerobic condition developed, which tends to stabilize the sludge and skimmings so that they are less likely to turn septic. However, disposal means for the foamate can be a serious problem in the use of this procedure [46]. It has been reported that foam separation has been able to remove 70–80% of synthetic detergents, at a wide range of costs [2]. Gibbs [17] reported the successful use of fine bubble flotation and 40 mm detention in treating soap manufacture wastes, where the skimmed sludge was periodically returned to the soap factory for reprocessing. According to Wang [47–49], the dissolved air flotation process is both technically and economically feasible for the removal of detergents and soaps (i.e., surfactants) from water.

Table 9 Effluent Limitations for Subpart I, Oleum Sulfonation

Effluent characteristic	Effluent limitations [metric units (kg/1000 kg of anhydrous product)]	
	Maximum for any 1 day	Average of daily values for 30 consecutive days shall not exceed
(a) BPT		
BOD_5	0.09	0.02
COD	0.40	0.09
TSS	0.15	0.03
Surfactants	0.15	0.03
Oil and grease	0.25	0.07
pH	a	a
(b) BAT		
BOD_5	0.07	0.02
COD	0.27	0.09
TSS	0.09	0.03
Surfactants	0.09	0.03
Oil and grease	0.21	0.07
pH	a	a
(c) NSPS		
BOD_5	0.03	0.01
COD	0.09	0.03
TSS	0.06	0.02
Surfactants	0.03	0.01
Oil and grease	0.12	0.04
pH	a	a

[a] Within the range 6.0–9.0.
Source: Ref. 14.

Activated Carbon Adsorption

Colloidal and soluble organic materials can be removed from solution through adsorption onto granular or powdered activated carbon, such as the particularly troublesome hard surfactants. Refractory substances resistant to biodegradation, such as ABS, are difficult or impossible to remove by conventional biological treatment, and so they are frequently removed by activated carbon adsorption [11]. The activated carbon application is made either in mixed-batch contact tanks with subsequent settling or filtration, or in flow-through GAC columns or contact beds. Obviously, because it is an expensive process, adsorption is being used as a polishing step of pretreated waste effluents. Nevertheless, according to Koziorowski and Kucharski [22] much better results of surfactant removal have been achieved with adsorption than coagulation/settling. Wang [50–52] used both powdered activated carbon (PAC) and coagulation/settling/DAF for successful removal of surfactants.

Coagulation/Flocculation/Settling/Flotation

As mentioned previously in Section 7.2.4, the coagulation/flocculation process was found to be affected by the presence of surfactants in the raw water or wastewater. Such interference was observed for both alum and ferric sulfate coagulant, but the use of certain organic polymer

Table 10 Effluent Limitations for Subpart P, Liquid Detergents

Effluent characteristic	Effluent limitations [metric units (kg/1000 kg of anhydrous product)]	
	Maximum for any 1 day	Average of daily values for 30 consecutive days shall not exceed
(a) BPT[a]		
BOD_5	0.60	0.20
COD	1.80	0.60
TSS	0.015	0.005
Surfactants	0.39	0.13
Oil and grease	0.015	0.005
pH	c	c
(b) BPT[b]		
BOD_5	0.05	
COD	0.15	
TSS	0.002	
Surfactants	0.04	
Oil and grease	0.002	
pH	c	
(c) BAT[a]		
BOD_5	0.10	0.05
COD	0.44	0.22
TSS	0.01	0.005
Surfactants	0.10	0.05
Oil and grease	0.01	0.005
pH	c	c
(d) BAT[b]		
BOD_5	0.02	
COD	0.07	
TSS	0.002	
Surfactants	0.02	
Oil and grease	0.002	
pH	c	
(e) NSPS[a]		
BOD_5	0.10	0.05
COD	0.44	0.22
TSS	0.01	0.005
Surfactants	0.10	0.05
Oil and grease	0.01	0.005
pH	c	c
(f) NSPS[b]		
BOD_5	0.02	
COD	0.07	
TSS	0.002	
Surfactants	0.02	
Oil and grease	0.002	
pH	c	

[a]For normal liquid detergent operations.
[b]For fast turnaround operation of automated fill lines.
[c]Within the range 6.0–9.0.
Source: Ref. 14.

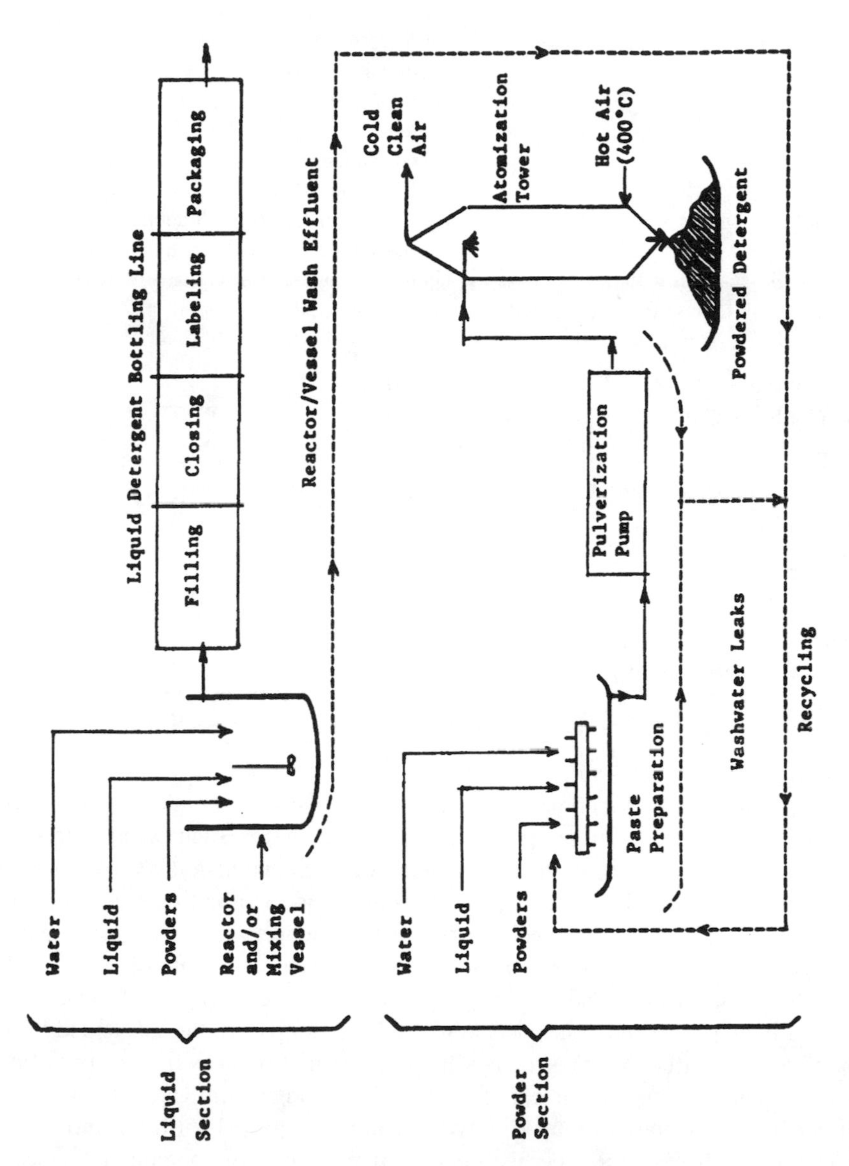

Figure 20 Process modification for wastewater recycling in detergent manufacture (from Ref. 37).

Table 11 Treatment Methods in the Soap and Detergent Industry

Pollutant and method	Efficiency (percentage of pollutant removed)
Oil and grease	
API-type separation	Up to 90% of free oils and greases. Variable on emulsified oil.
Carbon adsorption	Up to 95% of both free and emulsified oils.
Flotation	Without the addition of solid phase, alum, or iron, 70–80% of both free and emulsified oil. With the addition of chemicals, 90%.
Mixed-media filtration	Up to 95% of free oils. Efficiency in removing emulsified oils unknown.
Coagulation/sedimentation with iron, alum, or solid phase (bentonite, etc.)	Up to 95% of free oil. Up to 90% of emulsified oil.
Suspended solids	
Mixed-media filtration	70–80%
Coagulation/sedimentation	50–80%
BOD and COD	
Bioconversions (with final clarifier)	60–95% or more
Carbon adsorption	Up to 90%
Residual suspended solids	
Sand or mixed-media filtration	50–95%
Dissolved solids	
Ion exchange or reverse osmosis	Up to 90%

Source: Ref. 13.

flocculants was shown to overcome this problem. However, chemical coagulation and flocculation for settling may not prove to be very efficient for such wastewaters. Wastes containing emulsified oils can be clarified by coagulation, if the emulsion is broken through the addition of salts such as $CaCl_2$, the coagulant of choice for soap and detergent manufacture wastewaters [11]. Also, lime or other calcium chemicals have been used in the treatment of such wastes whose soapy constituents are precipitated as insoluble calcium soaps of fairly satisfactory flocculating ("hardness" scales) and settling properties. Treatment with $CaCl_2$ can be used to remove practically all grease and suspended solids and a major part of the suspended BOD [19]. Using carbon dioxide (carbonation) as an auxiliary precipitant reduces the amount of calcium chloride required and improves treatment efficiency. The sludge from $CaCl_2$ treatment can be removed either by sedimentation or by dissolved air flotation [39,53–56]. For monitoring and control of chemical coagulation, flocculation, sedimentation and flotation processes, many analytical procedures and testing procedures have been developed [57–64].

Ion Exchange and Exclusion

The ion-exchange process has been used effectively in the field of waste disposal. The use of continuous ion exchange and resin regeneration systems has further improved the economic feasibility of the applications over the fixed-bed systems. One of the reported [1] special

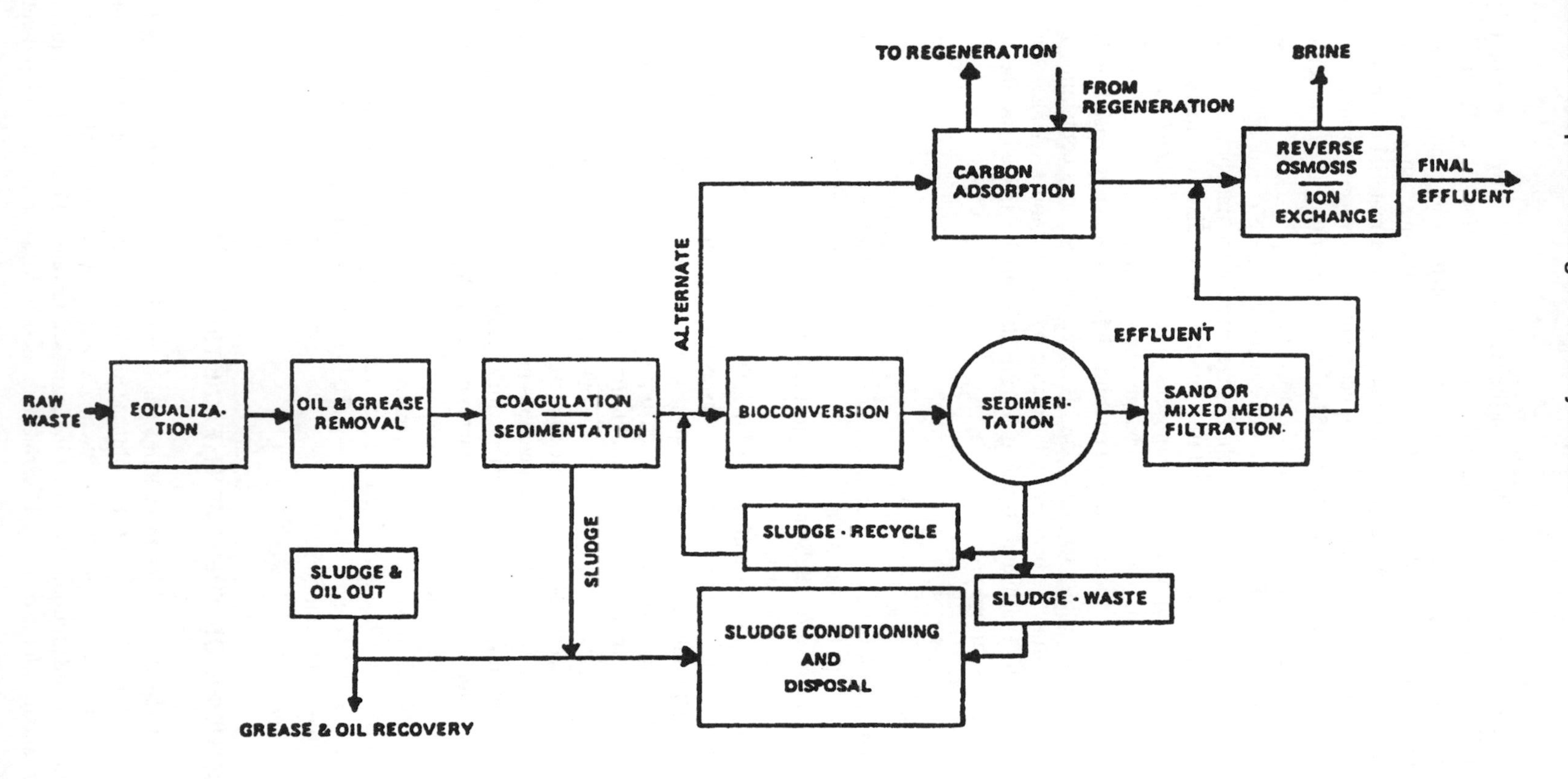

Figure 21 Composite flowsheet of waste treatment in soap and detergent industry (from Ref. 13).

applications of the ion-exchange resins has been the removal of ABS by the use of a Type II porous anion exchanger that is a strong base and depends on a chloride cycle. This resin system is regenerated by removing a great part of the ABS absorbed on the resin beads with the help of a mixture of hydrocarbons (HC) and acetone. Other organic pollutants can also be removed by ion-exchange resins, and the main problem is whether the organic material can be eluted from the resin using normal regeneration or whether it is economically advisable to simply discard the used resin. Wang and Wood [65] and Wang [51,52,66] successfully used the ion-exchange process for the removal of cationic surfactant from water.

The separation of ionic from nonionic substances can be effected by the use of ion exclusion [46]. Ion exchange can be used to purify glycerine for the final product of chemically pure glycerine and reduce losses to waste, but the concentration of dissolved ionizable solids or salts (ash) largely impacts on the overall operating costs. Economically, when the crude or sweet water contains under 1.5% ash, straight ion exchange using a cation and anion mixed bed can be used, whereas for higher percentages of dissolved solids, it is economically feasible to follow the ion exchange with an ion-exclusion system. For instance, waste streams containing 0.2–0.5% ash and 3–5% glycerine may be economically treated by straight ion exchange, while waste streams containing 5–10% ash and 3–5% glycerine have to be treated by the combined ion-exchange and ion-exclusion processes.

Biological Treatment

Regarding biological destruction, as mentioned previously, surfactants are known to cause a great deal of trouble due to foaming and toxicity [103] in municipal treatment plants. The behavior of these substances depends on their type [22], that is, anionic and nonionic detergents increase the amount of activated sludge, whereas cationic detergents reduce it, and also the various compounds decompose to a different degree. The activated sludge process is feasible for the treatment of soap and detergent industry wastes but, in general, not as satisfactory as trickling filters. The turbulence in the aeration tank induces frothing to occur, and also the presence of soaps and detergents reduces the absorption efficiency from air bubbles to liquid aeration by increasing the resistance of the liquid film.

On the other hand, detergent production wastewaters have been treated with appreciable success on fixed-film process units such as trickling filters [2]. Also, processes such as lagoons, oxidation or stabilization ponds, and aerated lagoons have all been used successfully in treating soap and detergent manufacturing wastewaters. Finally, Vath [102] demonstrated that both linear anionic and nonionic ethoxylated surfactants underwent degradation, as shown by a loss of surfactant properties, under anaerobic treatment.

Wang *et al.* [42,67,68] have developed innovative biological process and sequencing batch reactors (SBR) specifically for removal of volatile organic compounds (VOCs) and surfactants. Related analytical procedures [57–64,71–91] available for process monitoring and control are available in the literature.

7.7 CASE STUDIES OF TREATMENT FACILITIES

Soap and detergent manufacture and formulation plants are situated in many areas in the United States and other countries. At most, if not all of these locations, the wastewaters from production and cleanup activities are discharged to municipal sewer systems and treated together with domestic, commercial, institutional, and other industrial wastewaters. Following the precipitous reduction in production and use of “hard” surfactants such as ABS, no discernible problems in operation and treatment efficiency due to the combined treatment of surfactant manufacture wastes at these municipal sewage treatment plants (most of which employ biological processes) have been reported. In fact, there is a significantly larger portion of surfactants and related compounds being discharged to the municipal facilities from user sources. In most cases, the industrial discharge is simply surcharged due to its high-strength BOD concentration.

7.7.1 Colgate–Palmolive Plant

Possibly the most representative treatment facility that handles wastewaters from the production of soaps, detergents, glycerines, and personal care products is Colgate–Palmolive Company's plant at Jeffersonville, IN [3]. The production wastes had received treatment since 1968 [21] in a completely mixed activated sludge plant with a 0.6 MGD design flow and consisting of a 0.5 MG mixed equalization and storage basin, aeration basin, and final clarifier. The treated effluent was discharged to the Ohio River, combined with rain drainage and cooling waters. During operation, it was observed that waste overloads to the plant caused a deterioration of effluent quality and that the system recovered very slowly, particularly from surfactant short-term peaks. In addition, the fact that ABS had been eliminated and more LAS and nonionic surfactants were being produced, as well as the changes in product formulation, may have been the reasons for the Colgate treatment plant's generally less than acceptable effluent quality. (Note that 1 MG = 3785 m^3, 1 MGD = 3785 m^3/day.)

Owing to the fact that the company considered the treatment efficiency in need of more dependable results, in 1972–1973 several chemical pretreatment and biological treatment studies were undertaken in order to modify and improve the existing system. As a result, a modified treatment plant was designed, constructed, and placed in operation. A new 1.5 MG mixed flow and pollutant load equalization basin is provided prior to chemical pretreatment, and a flash mixer with lime addition precedes a flocculator/clarifier unit. Ahead of the pre-existing equalization and aeration basins, the capability for pH adjustment and nutrient supplementation was added. Chemical sludge is wasted to two lagoons where thickening and dewatering (normally 15–30% solids) take place.

The intermediate storage basin helps equalize upsets in the chemical pretreatment system, provides neutralization contact time, and allows for storage of pretreated wastewater to supply to the biological treatment unit whenever a prolonged shutdown of the chemical pretreatment occurs. Such shutdowns are planned for part of the weekend and whenever manufacturing stoppage occurs in order to cut down on costs. According to Brownell [3], waste loads to the pretreatment plant diminish during plantwide vacations and production shutdowns, and bypassing the chemical pretreatment allows for a more constant loading of the aeration basins at those times. In this way, the previously encountered problems in the start-up of the biological treatment unit after shutdowns were reduced.

The pollutant removal efficiency of this plant is normally quite high, with overall MBAS (methylene blue active substances) removals at 98–99% and monthly average overall BOD_5 removals ranging from 88 to 98% (most months averaging about 95%). The reported MBAS removals achieved in the chemical pretreatment units normally averaged 60–80%. Occasional high MBAS concentrations in the effluent from the chemical pretreatment system were controlled through the addition of $FeCl_2$ and an organic polymer that supplemented the regular dose of lime and increased suspended solids' capture. Also, high oil and grease concentrations were occasionally observed after spills of fatty acid, mineral oil, olefin, and tallow, and historically this caused problems with the biological system. In the chemical pretreatment units, adequate oil and grease removals were obtained through the addition of $FeCl_2$. Finally, COD removals in the chemical system were quite consistent and averaged about 50% (COD was about twice the BOD_5).

In the biological step of treatment, removal efficiency for BOD_5 was very good, often averaging over 90%. During normal operating periods, the activated sludge system appeared incapable of treating MBAS levels of over 100 lb/day (45.4 kg/day) without significant undesirable foaming. The BOD_5 loading was normally kept at 0.15–0.18 g/day/g (or lb/day/lb) MLVSS, but it had to be reduced whenever increased foaming occurred. Finally, suspended solids concentrations in the secondary clarifier effluent were occasionally quite high, although the overflow rate averaged only 510 gal/day/ft^2 and as low as 320 gpd/ft^2 (13–20.8 m^3/day/m^2). The use of polymer flocculants considerably improved the effluent turbidity, reducing it by 50–75%, and because higher effluent solids contribute to high effluent BOD_5, it was reduced as well. Therefore, although the Colgate–Palmolive waste treatment plant occasionally experiences operating problems, it generally achieves high levels of pollutant removal efficiencies.

Many analytical procedures have been developed for determination of MBAS [73,75] and COD/DO [61,89–91] concentrations in water and wastewater, in turn, for monitoring the efficiency of treatment processes.

7.7.2 Combined Treatment of Industrial and Municipal Wastes

Most soap and detergent manufacturing facilities, as mentioned previously, discharge their untreated or pretreated wastes into municipal systems. The compositions of these wastewaters vary widely, with some being readily biodegradable and others inhibitory to normal biological treatment processes. In order to allow and surcharge such an effluent to a municipal treatment plant, an evaluation of its treatability is required. Such a detailed assessment of the wastewaters discharged from a factory manufacturing detergents and cleaning materials in the vicinity of Pinxton, England, was reported by Shapland [92]. The average weekly effluent discharged from a small collection and equalization tank was 119 m^3/day (21.8 gpm), which contributes about 4% of the flow to the Pinxton sewage treatment plant.

Monitoring of the diurnal variation in wastewater pollutant strengths on different days showed that no regular diurnal pattern exists and the discharged wastewaters are changeable. In particular, the pH value was observed to vary rapidly over a wide range and, therefore, pH correction in the equalization tank would be a minimum required pretreatment prior to discharge into the sewers in such cases. The increase in organic loading contributed to the Pinxton plant by the detergent factory is much higher than the hydraulic loading, representing an average of 32% BOD increase in the raw influent and 60% BOD increase in the primary settled effluent, but it does not present a problem because the plant is biologically and hydraulically underloaded.

The treatability investigation of combined factory and municipal wastewaters involved laboratory-scale activated sludge plants and rolling tubes (fixed-film) units. The influent feed to these units was settled industrial effluent (with its pH adjusted to 10) mixed in various proportions with settled municipal effluent. The variation of hydraulic loading enabled the rotating tubes to be operated at similar biological loadings. In the activated sludge units, the mixed liquor suspended solids (MLSS) were maintained at about 3000 mg/L, a difficult task since frothing and floe break-up caused solids loss. The overall results showed that more consistent removals were obtained with the fixed-film system, probably due to the loss of solids from the aeration units [93].

At 3 and 6% by vol. industrial waste combination, slight to no biological inhibition was caused either to the fixed-film or activated sludge system. The results of sample analysis from the inhibitory runs showed that in two of the three cases, the possible cause of inhibition was the presence of chloroxylenes and brominated compounds. The third case represented only temporary inhibition, since the rolling tubes provided adequate treatment after a period of acclimation. Finally, the general conclusion reached in the investigation was that the detergent factory effluent may be accepted at 3% by vol. equalized flow to the municipal fixed-film treatment plant, that is, up to 200 m^3/day (36.7 gpm), without any noticeable efficiency reduction.

7.7.3 Treatability of Oily Wastes from Soap Manufacture

McCarty [94] addressed the subject of the treatability of animal and vegetable oils and fats in municipal treatment systems. In general, certain reported treatment difficulties in biological systems are attributed to the presence of fats, oils, and other "grease" components in wastewaters. However, as opposed to mineral-type oils, animal and vegetable oils and fats such as those discharged by soap manufacture plants are readily biodegradable and generally nontoxic, although differences exist as to the difficulties caused depending on the form (floatable or emulsified) and type (hydrocarbons, fatty acids, glycerides, sterols, etc.). In general, shorter-chain-length fatty acids, unsaturated acids, and soluble acids are more readily degraded than longer-chain, saturated, and insoluble ones. The more insoluble and larger fatty acid particles have been found to require greater time for degradation than those with opposite characteristics. It has also been reported that animal and vegetable oils, fats, and fatty acids are metabolized quickly in anaerobic systems and generate the major portion of methane in regular anaerobic sludge digestion.

McCarty [94] also reported on the results of laboratory investigations in the treatability of selected industrial oily wastes from soap manufacturing and food processing by the Procter & Gamble Co. in Cincinnati, OH, when combined with municipal sewage or sludge. The grease content of the industrial wastes was high in all cases, ranging from 13 to 32% of the waste COD, and it was about 2.9 g of COD per gram of grease. It was found that it is possible to treat about equal COD mixtures of the industrial wastes with municipal sewage using the activated sludge process and achieve removal efficiencies similar to those for municipal sewage alone.

The grease components of the industrial wastes were readily degraded by anaerobic treatment, with removal efficiencies ranging from 82 to 92%. Sludges from the anaerobic digestion of an industrial/municipal mixture could be dewatered with generally high doses of chemical conditioning ($FeCl_2$), but these stringent requirements seemed a result of the hard-to-dewater municipal waste sludge. In conclusion, the Procter & Gamble Co. industrial wastes were readily treated when mixed with municipal sewage without significant adverse impacts, given sufficient plant design capacity to handle the combined wastes hydraulically and biologically. Also, there was no problem with the anaerobic digestion of combined wastes, if adequate mixing facilities are provided to prevent the formation of scum layers.

For treatment process control, Wang [85–87] has developed rapid methods for determination of oil and grease and dissolved proteins in the wastewaters.

7.7.4 Removal of Nonionic Surfactants by Adsorption

Nonionic surfactants, as mentioned previously, have been widely adopted due to their characteristics and properties and, in particular, because they do not require the presence of undesirable phosphate or caustic builders in detergent formulation. However, the relatively lesser degree of biodegradability is an important disadvantage of the nonionic surfactants compared to the ionic ones. Adsorption on activated carbon and various types of clay particles is, therefore, one of the processes that has been effective in removing heterodisperse nonionic surfactants – those that utilize a polyhydroxyl alcohol as a lipophilic phase – from wastewaters [6]. In another study by Carberry and Geyer [5], the adsorptive capacity kinetics of polydisperse nonionic surfactants – those that utilize a hydrocarbon species as a lipophilic base – removal by granular activated carbon and clay were investigated. Both clay particulates of different types and various activated carbons were tested and proven efficient in adsorbing nonionic surfactants. Of all the clays and carbons studied, Bentolite-L appeared to be the superior adsorbent (9.95% mol/kg vs. 0.53 mol/kg for Hydrodarco 400), but reaction rate constants for all adsorbents tested appeared to be strikingly similar.

7.7.5 Removal of Anionic Detergents with Inorganic Gels

Inorganic gels exhibiting ion-exchange and sorption characteristics are more stable than synthetic organic resins, which have also been used for the removal of detergents from wastewaters [95]. The sorption efficiency and number of cycles for which inorganic gels can be used without much loss in sorption capacity would compensate the cost involved in their preparation. Zinc and copper ferrocyanide have been shown to possess promising sorption characteristics for cationic and anionic surfactants. Of the two, copper ferrocyanide is a better scavenger for anionic detergents, which have a relatively small rate and degree of biodegradation and their presence in raw water causes problems in coagulation and sedimentation.

The cation-exchange capacity of the copper ferrocyanide gel used was found to be about 2.60 meq/g and its anion-exchange capacity about 0.21 meq/g. In all cases of various doses of gel used and types of anionic surfactants being removed, the tests indicated that a batch contact time of about 12 hours was sufficient for achieving maximum removals. Trials with various fractions of particle size demonstrated that both uptake and desorption (important in material regeneration) were most convenient and maximized on 170–200 BSS mesh size particles. Also, the adsorption of anionic surfactants was found to be maximum at pH 4 and decreased with an increase in pH.

The presence of NaCl and $CaCl_2$ salts (mono and bivalent cations) in solution was shown to increase the adsorption of anionic surfactants in the pH range 4–7, whereas the presence of $AlCl_3$ salt (trivalent cation) caused a greater increase in adsorption in the same pH range. However, at salt concentrations greater than about 0.6 M, the adsorption of the studied anionic surfactants started decreasing. On the other hand, almost complete desorption could be obtained by the use of K_2SO_4 or a mixture of H_2SO_4 and alcohol, both of which were found to be equally effective. In conclusion, although in these studies the sorption capacity of the adsorbent gel was not fully exploited, the anionic detergent uptake on copper ferrocyanide was found to be comparable to fly ash and activated carbon.

7.7.6 Removal of Cationic Surfactants

There are few demonstrated methods for the removal of cationic surfactants from wastewater, as mentioned previously, and ion exchange and ultrafiltration are two of them. Chiang and Etzel [8] developed a procedure for selecting from these the optimum removal process for cationic surfactants from wastewaters. Preliminary batch-test investigations led to the selection of one resin (Rohm & Haas "Amberlite," Amb-200) with the best characteristics possible (i.e., high exchange capacity with a rapid reaction rate, not very fine mesh resin that would cause an excessive pressure drop and other operational problems, macroporous resin that has advantages over the gel structure resins for the exchange of large organic molecules) to be used in optimizing removal factors in the column studies vs. the performance of ultrafiltration membranes (Sepa-97 CA RO/UF selected). The cyclic operation of the ion-exchange (H^+) column consisted of the following stops: backwash, regeneration, rinse, and exhaustion (service).

The ion-exchange tests indicated that the breakthrough capacity or total amount adsorbed by the resin column was greater for low-molecular-weight rather than high-molecular-weight surfactants. Furthermore, the breakthrough capacity for each cationic surfactant was significantly influenced (capacity decreases as the influent concentration increases) by the corresponding relationship of the influent concentration to the surfactant critical micelle concentration (CMC). A NaCl/ethanol/water (10% NaCl plus 50% ethanol) solution was found to be optimum in regenerating the exhausted resin.

In the separation tests with the use of a UF membrane, the rejection efficiency for the C_{16} cationic surfactants was found to be in the range 90–99%, whereas for the C_{12} surfactants it ranged from 72 to 86%, when the feed concentration of each surfactant was greater than its corresponding CMC value. Therefore, UF rejection efficiency seems to be dependent on the respective hydrated micelle diameter and CMC value. In conclusion, the study showed that for cationic surfactants removal, if the feed concentration of a surfactant is higher than its CMC value, then the UF membrane process is found to be the best. However, if the feed concentration of a surfactant is less than its CMC value, then ion exchange is the best process for its removal.

Initial and residual cationic surfactant concentrations in a water or wastewater treatment system can be determined by titration methods, colorimetric methods, or UV method [69–71, 77–79,81]. Additional references for cationic surfactant removal are available elsewhere [44,45,51,65,66].

7.7.7 Adsorption of Anionic Surfactant by Rubber

Removal of anionic surfactants has been studied or reported by many investigators [96–101]. It has been reported [101] that the efficiency of rubber granules, a low-cost adsorbent material, is efficient for the removal of sodium dodecyl sulfate (SDS), which is a representative member of anionic surfactants (AS). Previous studies on the absorption of AS on various adsorbents such as alumina and activated carbon showed 80–90% removals, while the sodium form of type A Zeolite did not have a good efficiency; however, these adsorbing materials are not cost-effective. In this study, a very low-cost scrap rubber in the form of granules (the waste product of tires locally purchased for US$0.20 per kg) was used to remove AS from the water environment. Tires contain 25–30% by weight carbon black as reinforcing filler and hydroxyl and/or carboxyl groups; both the carbon black and carboxyl group are responsible for the high degree of adsorption. In addition to the abundance and low cost of the waste tire rubber, the advantage is the possibility of reusing the exhausted rubber granules as an additive to asphalt as road material.

Earlier, Shalaby and El-Feky [98] had reported successful adsorption of nonionic surfactant from its aqueous solution onto commercial rubber. The average size of sieved adsorbent granules used was 75, 150, and 425 m. It was observed that within 1 hour, with all three sizes, the removal of AS was the same, about 78%. But after 5 hours, the removal was found to be 90% for the 75 m average size, while it was only about 85% for the other two larger sizes (adsorption is a surface phenomenon and as the size decreases, the surface area increases). Tests performed with initial adsorbate (SDS) concentrations of 2, 4, and 6 mg/L and doses of adsorbent varying between 5 and 15 g/L showed a removal efficiency in all cases of 65–75% within 1 hour, which only increased to about 80% after 7 hours. The effect of solution pH on adsorption of AS by rubber granules was also studied over a pH range of 3–13 using an initial AS concentration and an adsorbent dose of 3 mg/L and 10 g/L, respectively. Over a 6 hour contact time, with increase of pH, the removal of AS decreased practically linearly from 86 to 72%, probably due to interference of OH^- ion, which has similar charge to that of AS.

The effect of Ca^{2+} ion, which is very common in waters, was investigated over a range of 0–170 ppm calcium and it was shown that about 80–89% removal of AS occurred throughout this range. Similarly high levels of AS removal (87–93%) were observed for iron concentrations from 20 to 207 ppm, possibly due to formation of insoluble salt with the anionic part of the surfactant causing increased removal. On the other hand, the ionic strength of the solution in the form of NO_3^- concentration ranging from 150 to 1500 ppm was shown to reduce SDS removal efficiency to 71–77%, while the effect of chloride concentration (in the range 15–1200 mg/L) on AS removal by rubber granules was found to be adverse, down to 34–48% of SDS, which might be due to competition for adsorbing sites.

For treatment process control, initial and residual anionic surfactant concentrations in a water treatment system can be determined by titration methods or colorimetric methods [75,76,80,84,90]. The most recent technical information on management and treatment of the soap and detergent industry waste is available from the state of New York [104].

REFERENCES

1. Abrams, I.M.; Lewon, S.M. J. Am. Water Works Assoc. **1962**, *54* (*5*).
2. Besselievre, E.B. *The Treatment of Industrial Wastes*; McGraw-Hill: New York, NY, 1969.
3. Brownell, R.P. Chemical-biological treatment of surfactant wastewater. *Proceedings of the 30th Industrial Waste Conference*, Purdue University, Lafayette, IN, 1975, Vol. 30, 1085.
4. Callely, A.G. *Treatment of Industrial Effluents*; Haisted Press: New York, NY, 1976.
5. Carberry, J.B.; Geyer, A.T. Adsorption of non-ionic surfactants by activated carbon and clay. *Proceedings of the 32nd Industrial Waste Conference*, Purdue University, Lafayette, IN, 1977, Vol. 32, 867.
6. Carberry, J.B. Clay adsorption treatment of non-ionic surfactants in wastewater. J. Water Poll. Control Fed. **1977**, *49*, 452.
7. Chambon, M.; Giraud, A. *Bull. Ac. Nt. Medecine (France)* **1960**, *144*, 623–628.
8. Chiang, P.C.; Etzel, J.E. Procedure for selecting the optimum removal process for cationic surfactants. In *Toxic and Hazardous Waste*; LaGrega, Hendrian, Eds.; Butterworth: Boston, MA, 1983.
9. Cohen, J.M. *Taste and Odor of ABS*; US Dept. of Health, Education and Welfare Dept.: Cincinnati, OH, 1962.
10. Cohen, J.M. J. Am. Water Works Assoc. **1959**, *51*, 1255–1266.
11. Eckenfelder, W.W. *Industrial Water Pollution Control*; McGraw-Hill: New York, NY, 1989.
12. USEPA. *Development Document on Guidelines for Soap and Detergent Manufacturing*, EPA-440/1-74-018a; US Government Printing Office: Washington, DC, Construction Grants Program, 1974.
13. USEPA. *Federal Guidelines on State and Local Pretreatment Programs*, EPA-430/9-76-017c; US Government Printing Office: Washington, DC, Construction Grants Program, 1977; 8-13-1–8-13-25.
14. Federal Register. *Code of Federal Regulations*; US Government Printing Office: Washington, DC, 1987; CFR 40, Part 417, 362–412.
15. Federal Register. *Notices, Appendix A, Master Chart of Industrial Categories and Regulations*; US Government Printing Office: Washington, DC, 1990; Jan. 2, *55* (*1*), 102–103.
16. Gameson, A.L.H. J. Inst. Water Engrs (UK) **1955**, *9*, 571.
17. Gibbs, F.S. The removal of fatty acids and soaps from soap manufacturing wastewaters. *Proceedings of the 5th Industrial Waste Conference*, Purdue University, Lafayette, IN, 1949, Vol. 5, p. 400.
18. Greek, B.F. Detergent industry ponders products for new decade. Chem. & Eng. News **1990**, *Jan. 29*, 37–60.
19. Gumham, C.F. *Principles of Industrial Waste Treatment*; Wiley: New York, NY, 1955.
20. Gurnham, C.F. (Ed.) *Industrial Wastewater Control*; Academic Press: New York, NY, 1965.

21. Herin, J.L. Development and operation of an aeration waste treatment plant. *Proceedings of the 25th Industrial Waste Conference*, Purdue University, Lafayette, IN, 1970, Vol. 25, p. 420.
22. Koziorowski, B.; Kucharski, J. *Industrial Waste Disposal*; Pergamon Press: Oxford, UK, 1972.
23. Payne, W.J. Pure culture studies of the degradation of detergent compounds. Biotech. Bioeng. **1963**, *5*, 355.
24. Swisher, R.D. *Surfactant Biodegradation*; Marcel Dekker: New York, NY, 1970.
25. Osburn, O.W.; Benedict, III, H. Polyethoxylated alkyl phenols: relationship of structure to biodegradation mechanism. J. Am. Oil Chem. Soc. **1966**, *43*, 141.
26. Wright, K.A.; Cain, R.B. Microbial metabolism of pyridinium compounds. Biochem. J. **1972**, *128*, 543.
27. Prat; Giraud, A. *La Pollution des Eaux par les Detergents*, Report 16602; Scientific Committee of OECD: Paris, France, 1964.
28. OECD. *Pollution par les Detergents*, Report by Expert Group on Biodegradability of Surfactants: Paris, France, 1971.
29. Langelier, W.F. Proc. Am. Soc. Civil Engrs. **1952**, *78* (*118*), February.
30. Nichols, M.S.; Koepp, E. J. Am. Water Works Assoc. **1961**, *53*, 303.
31. Todd, R. Water Sewage Works **1954**, *101*, 80.
32. Smith, R.S. *et al.* J. Am. Water Works Assoc. **1956**, *48*, 55.
33. McGauhey, P.H.; Klein, S.A. *Sewage and Indust. Wastes*, **1959**, *31* (*8*), 877–899.
34. Lawson, R. *Sewage and Indust. Wastes*, **1959**, *31* (*8*), 877–899.
35. Robeck, G. *et al.* J. Am. Water Works Assoc. **1962**, *54*, 75.
36. Renn, E.; Barada, M. J. Am. Water Works Assoc. **1961**, *53*, 129–134.
37. Overcash, M.R. *Techniques for Industrial Pollution Prevention*; Lewis Publishers: MI, 1986.
38. Nemerow, N.L. *Industrial Water Pollution*; Addison-Wesley: Reading, MA, 1978.
39. Wang, L.K.; Krofta, M. *Flotation Engineering*, 3rd Ed.; Lenox Institute of Water Technology: Lenox, MA, 2002; 252p.
40. Wang, L.K.; Kurylko, L.; Hrycyk, O. Removal of Volatile Compounds and Surfactants from Liquid. US Patent No. 5,122,166, June 1992.
41. Hrycyk, O.; Kurylko, L.; Wang, L.K. Removal of Volatile Compounds and Surfactants from Liquid. US Patent No. 5,122,165, June 1992.
42. Wang, L.K.; Kurylko, L.; Wang, M.H.S. *Combined Coarse and Fine Bubbles Separation System*. US Patent No. 5,275,732, January 1994.
43. Wang, L.K. *Continuous Bubble Fractionation Process*. Ph.D. Dissertation; Rutgers University, NJ, USA, 1972; 171 p.
44. Wang, M.H.S.; Granstrom, M.; Wang, L.K. Lignin separation by continuous ion flotation: investigation of physical operational parameters. Water Res. Bull. **1974**, *10* (*2*), 283–294.
45. Wang, M.H.S.; Granstrom, M.; Wang, L.K. Removal of lignin from water by precipitate flotation. J. Environ. Eng. Div., Proc. of ASCE **1974**, *100* (*EE3*), 629–640.
46. Ross, R.D. (Ed.) *Industrial Waste Disposal*; Reinhold: New York, NY, 1968.
47. Wang, L.K. Treatment of various industrial wastewaters by dissolved air flotation. *Proceedings of the N.Y.–N.J. Environmental Expo*. Secaucus, NJ, October, 1990.
48. Wang, L.K. Potable water treatment by dissolved air flotation and filtration. J. AWWA **1982**, *74* (*6*), 304–310.
49. Wang, L.K. *The State-of-the-Art Technologies for Water Treatment and Management*, UNIDO Manual No. 8-8-95; United Nations Industrial Development Organization: Vienna, Austria, 1995; 145 p.
50. Wang, L.K. The adsorption of dissolved organics from industrial effluents onto activated carbon. J. Appl. Chem. Biotechnol. **1975**, *25* (*7*), 491–503.
51. Wang, L.K. Water treatment with multiphase flow reactor and cationic surfactants. J. AWWA **1978**, *70*, 522–528.
52. Wang, L.K. Application and determination of anionic surfactants. Indust. Engng. Chem. **1978**, *17* (*3*), 186–195.
53. Krofta, M.; Wang, L.K. *Design of Dissolved Air Flotation Systems for Industrial Pretreatment and Municipal Wastewater Treatment – Design and Energy Considerations*; AIChE National Conference, Houston, TX, NTIS-PB83-232868, 1983, 30 p.

54. Krofta, M.; Wang, L.K. *Design of Dissolved Air Flotation Systems for Industrial Pretreatment and Municipal Wastewater Treatment – Case History of Practical Applications*; AIChE National Conference, Houston, TX, NTIS-PB83-232850, 1983, 25 p.
55. Krofta, M.; Wang, L.K. Development of a total closed water system for a deinking plant. *AWWA Water Reuse Symposium III* **1984**, *2*, 881–898.
56. Krofta, M.; Wang, L.K. *Flotation and Related Adsorptive Bubble Separation Processes*, 4th Ed.; Lenox Institute of Water Technology: Lenox, MA, 2001; 185 p.
57. Wang, L.K. Polyelectrolyte determination at low concentration. Indust. Engng. Chem., Prod. Res. Devel. **1975**, *14* (*4*), 312–314.
58. Wang, L.K. *A Modified Standard Method for the Determination of Ozone Residual Concentration by Spectrophotometer*, PB84-204684; US Department of Commerce, National Technical Information Service: Springfield, VA, 1983; 16 p.
59. Wang, L.K. *Process Control Using Zeta Potential and Colloid Titration Techniques*, PB87-179099/AS; US Department of Commerce, National Technical Information Service: Springfield, VA, 1984, 126 p.
60. Wang, L.K. *Determination of Polyelectrolytes and Colloidal Charges*, PB86-169307; US Department of Commerce, National Technical Information Service: Springfield, VA, 1984; 47 p.
61. Wang, L.K. *Alternative COD Method for Reduction of Hazardous Waste Production*, PB86-169323; US Department of Commerce, National Technical Information Service: Springfield, VA, 1985; 7 p.
62. Wang, L.K. *Determination of Solids and Water Content of Highly Concentrated Sludge Slurries and Cakes*, PB85-182624/AS; US Department of Commerce, National Technical Information Service: Springfield, VA, 1985, 9 p.
63. Wang, L.K. *Laboratory Simulation and Optimization of Water Treatment Processes*, PB88-168414/AS; US Department of Commerce, National Technical Information Service: Springfield, VA, 1986; 54 p.
64. Chao, L.; Wang, L.K.; Wang, M.H.S. *Use of the Ames Mutagenicity Bioassay as a Water Quality Monitoring Method*, PB88-168422/AS; US Department of Commerce, National Technical Information Service: Springfield, VA, 1986; 25 p.
65. Wang, L.K.; Wood, G.W. *Water Treatment by Disinfection, Flotation and Ion Exchange Process System*, PB82-213349; US Department of Commerce, Nat. Tech. Information Service: Springfield, VA, 1982.
66. Wang, L.K. *Water Treatment by Disinfection, Flotation and Ion Exchange Process System*, PB82-213349; US Department of Commerce, National Technical Information Service: Springfield, VA, 1982; 115 p.
67. Wang, L.K.; Kurylko, L.; Wang, M.H.S. Sequencing Batch Liquid Treatment. US Patent No. 5,354,458, October 1994.
68. Wang, L.K.; Kurylko, L.; Hrycyk, O. Biological Process for Groundwater and Wastewater Treatment. US Patent No. 5,451,320, September 1995.
69. Wang, L.K. Neutralization effect of anionic and cationic surfactants. J. New Engl. Water Works Assoc. **1976**, *90* (*4*), 354–359.
70. Wang, L.K. *Cationic Surfactant Determination Using Alternate Organic Solvent*, PB86-194164/AS; US Department of Commerce, National Technical Information Service: Springfield, VA, 1983; 12 p.
71. Wang, L.K. *The Effects of Cationic Surfactant Concentration on Bubble Dynamics in a Bubble Fractionation Column*, PB86-197845/AS; US Department of Commerce, National Technical Information Service: Springfield, VA, 1984; 47 p.
72. Wang, L.K. A proposed method for the analysis of anionic surfactants. J. Am. Water Works Assoc. **1975**, *67* (*1*), 6–8.
73. Wang, L.K. Modified methylene blue method for estimating the MBAS concentration. J. Am. Water Works Assoc. **1975**, *67* (*1*), 19–21.
74. Wang, L.K. Analysis of LAS, ABS and commercial detergents by two-phase titration. Water Res. Bull. **1975**, *11* (2), 267–277.

75. Wang, L.K. Evaluation of two methylene blue methods for analyzing MBAS concentrations in aqueous solutions. J. Am. Water Works Assoc. **1975**, *67* (*4*), 182–184.
76. Wang, L.K. Determination of anionic surfactants with Azure A and quaternary ammonium salt. Analy. Chem. **1975**, *47* (*8*), 1472–1475.
77. Wang, L.K. Determining cationic surfactant concentration. Indust. Engng. Chem. Prod. Res. Devel. **1975**, *14* (*3*), 210–212.
78. Wang, L.K. A test method for analyzing either anionic or cationic surfactants in industrial water. J. Am. Oil Chemists Soc. **1975**, *52* (*9*), 340–346.
79. Wang, L.K. Rapid colorimetric analysis of cationic and anionic surfactants. J. New Engl. Water Works Assoc. **1975**, *89* (*4*), 301–314.
80. Wang, L.K. Direct two-phase titration method for analyzing anionic nonsoap surfactants in fresh and saline waters. J. Environ. Health **1975**, *38*, 159–163.
81. Wang, L.K. Analyzing cetyldimethylbenzylammonium chloride by using ultraviolet absorbance. Indust. Engng. Chem. Prod. Res. Devel. **1976**, *15* (*1*), 68–70.
82. Wang, L.K. Role of polyelectrolytes in the filtration of colloidal particles from water and wastewater. Separ. Purif. Meth. **1977**, *6* (*1*), 153–187.
83. Wang, L.K. Application and determination of organic polymers. Water, Air Soil Poll. **1978**, *9*, 337–348.
84. Wang, L.K. Application and determination of anionic surfactants. Indust. Engng. Chem. **1978**, *17* (*3*), 186–195.
85. Wang, L.K. *Selected Topics on Water Quality Analysis*, PB87-174066; US Department of Commerce, National Technical Information Service: Springfield, VA, 1982; 189 p.
86. Wang, L.K. *Rapid and Accurate Determination of Oil and Grease by Spectrophotometric Methods*, PB83-180760; US Department of Commerce, National Technical Information Service: Springfield, VA, 1983; 31 p.
87. Wang, L.K. *A New Spectrophotometric Method for Determination of Dissolved Proteins in Low Concentration Range*, PB84-204692; US Department of Commerce, National Technical Information Service: Springfield, VA, 1983; 12 p.
88. Wang, L.K.; DeMichele, E.; Wang, M.H.S. *Simplified Laboratory Procedures for DO Determination*, PB88-168067/AS; US Department of Commerce, National Technical Information Service: Springfield, VA, 1985; 13 p.
89. Wang, L.K.; DeMichele, E.; Wang, M.H.S. *Simplified Laboratory Procedures for COD Determination Using Dichromate Reflux Method*, PB86-193885/AS; US Department of Commerce, National Technical Information Service: Springfield, VA, 1986; 8 p.
90. AWWA, WEF, APHA. *Standard Methods for the Examination of Water and Wastewater*, AWWA/WEF/APHA, 2000.
91. Wang, L.K. *Recent Advances in Water Quality Analysis*, PB88-168406/AS; US Department of Commerce, National Technical Information Service: Springfield, VA, 1986; 100 p.
92. Shapland, K. Industrial effluent treatability. J. Water Poll. Control (UK), **1986**, p. 75.
93. Yapijakis, C. Treatment of soap and detergent industry wastes. In *Handbook of Industrial Waste Treatment*, Chap. 5; Wang, L.K., Wang, M.H.S., Eds.; Marcel Dekker, Inc.: New York, NY, 1992; 229–292.
94. McCarty, P.L. Treatability of oily wastewaters from food processing and soap manufacture. *Proceedings of the 27th Industrial Waste Conference*; Purdue University, Lafayette, IN, 1972; Vol. 27, p. 867.
95. Srivastava, S.K. *et al.* Use of inorganic gels for the removal of anionic detergents. *Proceedings of the 36th Industrial Waste Conference*, Purdue University, Lafayette, IN, 1981; Vol. 36, Sec. 20, p. 1162.
96. Bevia, F.R.; Prats, D.; Rico, C. Elimination of LAS during sewage treatment, drying and compostage of sludge and soil amending processes. In *Organic Contaminants in Wastewater, Sludge and Sediment*; Quaghebeur, D., Temmerman, I., Agelitti, G., Eds.; Elservier Applied Science: London, 1989.
97. Mathru, A.K.; Gupta, B.N. Detergents and cosmetics technology. Indian J. Environ. Protection **1998**, *18* (2), 90–94.

98. Shalaby, M.N.; El-Feky, A.A. Adsorption of nonionic surfactant from its aqueous solution onto commercial rubber. J. Dispersion Sci. Technol. **1999**, *20* (*5*), 1389–1406.
99. Sing, B.P. Separation of hazardous organic pollutants from fluids by selective adsorption. Indian J. Environ. Protection **1994**, *14* (*10*), 748–752.
100. Wayt, H.J.; Wilson, D.J. Soil clean-up by in-situ surfactant flushing II: theory of miscellar solubilization. Separ. Sci. Technol. **1989**, *24*, 905–907.
101. Wang, L.K.; Yapijakis, C.; Li, Y.; Hung, Y.; Lo, H.H. Wastewater treatment in soap and detergent industry. OCEESA J. **2003**, *20* (*2*), 63–66.
102. Vath, C.A. Soap and Chem. Specif. **1964**, *March.*
103. Swisher, R.D. Exposure levels and oral toxicity of surfactants. Arch. Environ. Health **1968**, *17*, 232.
104. NYSDEC. Soap and Detergent Manufacturing Point Source Catergory. 4. CFR. Protection of Environment. Part 417. New York State Department of Environmental Conservation. Albany, New York. Feb. 2004.

8
Treatment of Textile Wastes

Thomas Bechtold and Eduard Burtscher
Leopold Franzens University, Innsbruck, Austria

Yung-Tse Hung
Cleveland State University, Cleveland, Ohio, U.S.A.

8.1 IDENTIFICATION AND CLASSIFICATION OF TEXTILE WASTES

8.1.1 Textile Processes

The production of textiles represents one of the big consumers of high water quality. As a result of various processes, considerable amounts of polluted water are released. Representative magnitudes for water consumption are 100–200 L of water per kilogram of textile product. Considering an annual production of 40 million tons of textile fibers, the release of wasted water can be estimated to exceed 4–8 billion cubic meters per year.

The production of a textile requires several stages of mechanical processing such as spinning, weaving, knitting, and garment production, which seem to be insulated from the wet treatment processes like pretreatment, dyeing, printing, and finishing operations, but there is a strong interrelation between treatment processes in the dry state and consecutive wet treatments.

For a long time the toxicity of released wastewater was mainly determined by the detection of biological effects from pollution, high bulks of foam, or intensively colored rivers near textile plants. Times have changed and the identification and classification of wastewater currently are fixed by communal regulations [1,2].

General regulations define the most important substances to be observed critically by the applicant, and propose general strategies to be applied for minimization of the release of hazardous substances. The proposed set of actions has to be integrated into processes and production steps [3]. Figure 1 gives a general overview of a textile plant and also indicates strategic positions for actions to minimize ecological impact. In this figure, the textile plant is defined as a structure that changes the properties of a textile raw material to obtain a desired product pattern. The activities to treat hazardous wastes can range from legal prohibition to cost-saving recycling of chemicals. Depending on the type of product and treatment, these steps can show extreme variability.

Normally the legal regulations are interpreted as a set of wastewater limits that have to be kept, but in fact the situation is more complex and at present a complex structure of actions has been defined and has described useful strategies to improve an actual situation.

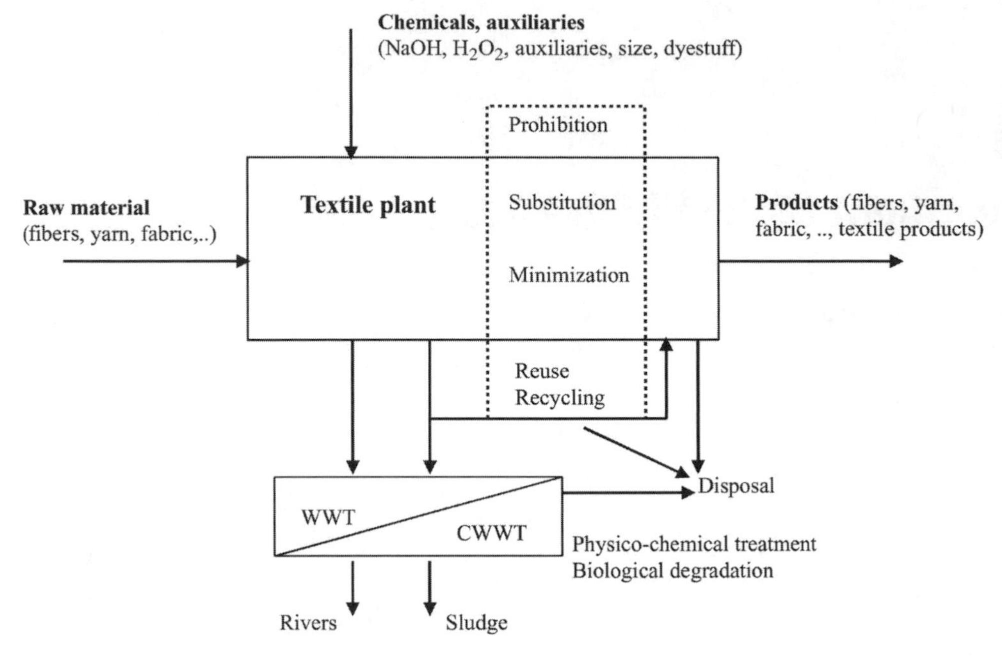

Figure 1 Flow structure of a textile plant (from Refs. 2 and 3).

8.1.2 Strategies to Reach Existing Requirements

Figure 2 shows a general action path recommended to minimize a present problem in the wastewater released from a textile plant [3,4].

Replacement and Minimization

As a first step substances that are known to cause problems in the wastewater have to be replaced by less hazardous chemicals or the process itself should be reconsidered; for example,

- use of high-temperature dyeing (HT-dyeing) processes for polyester fibers (PES) instead of carrier processes;
- replacement of chloro-organic carriers;
- replacement of preservatives containing As, Hg, or Sn organic compounds;
- replacement of alkylphenolethoxylates (APEO) in surfactants [5];
- substitution of "chlorine" bleach for natural fibers by peroxide bleach processes;
- substitution of sizes with poor biodegradability, e.g., carboxymethylcellulose (CMC);
- replacement of "hard" complexing agents like ethylene-diamine-tetra-acetic acid (EDTA), phosphonates.

The implementation of these steps into a dyehouse reduces the chemical load of the released wastewater considerably. In particular the replacement of substances that exhibit high toxicity or very low biodegradability will facilitate the following efficient treatment of the wastewater.

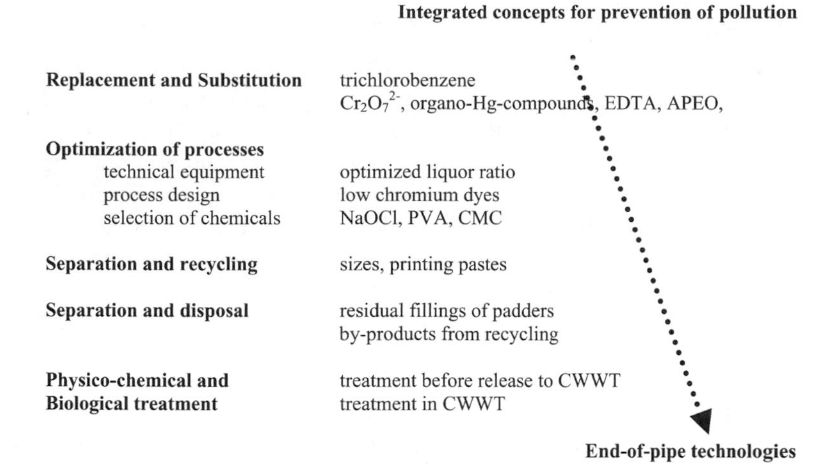

Figure 2 Action path for consideration and improvement of an existing situation (from Refs. 1–3 and 9).

Optimization of Processes

The second general step recommended to improve an existing situation is the optimization of treatment steps with regard to a lowering of the released amounts of hazardous substances [6,7]. In many cases this strategy is more intelligent and less expensive than a concentration of activities on the final treatment of released effluents. Typical examples for possible optimization are:

- reconsideration of dyestuffs and machinery chosen in exhaust dyeing (degree of exhaustion, fixation, liquor ratio);
- optimization of dyes and reducing agent in sulfur dyeing;
- optimization of residual volumes of padders and printing machines;
- optimization of water consumption.

Separation and Recycling

Besides the replacement of substances, the improvement of processes on an optimization of the handling of rather concentrated liquors, for example, used in sizing, caustic treatment like mercerization, dyeing, finishing processes, or in textile printing processes is the next step. As a desired goal, a recycling of a main part of the substances should be attempted. Examples that can be mentioned include the recovery and regeneration of sizes and caustic soda solutions, and the recovery of lanolin from wool washing.

Separation and Treatment for Disposal or Drain

If regeneration is impossible, a separate collection of a certain type of waste and an optimized treatment of the concentrates is more efficient and cheaper than a treatment of the full waste stream. Such treatments will concentrate on a minimization of costs for disposal (e.g., disposal of sludge, printing pastes, chemical products) or reaching existing limits defined for various parameters analyzed in the wastewater, for example, pH value, content of heavy metals, chemical oxygen demand (COD), adsorbable halogenated organic compounds (AOX) [8].

General Wastewater Treatment

In any case the wastewater will finally be fed into rivers, lakes, or the sea; thus some wastewater treatments have to be performed before the textile effluents are released either to the communal wastewater treatment plant (CWWT) or into the rivers, lakes, and so on. Normally physical and (bio-) chemical treatments (e.g., adjustment of pH, temperature, sedimentation, flocculation) are performed in the textile plant, while the following biological treatment (aerobic, anaerobic degradation) is performed either in the textile plant or in a CWWT. The site of the biological treatment is dependent on the location of the textile plant; however, a biological treatment of textile effluents preceding release into surface water is state of the art.

8.1.3 Definitions and Limits

For a long time the treatment of textile effluents has concentrated mainly on two aspects: regeneration of concentrated effluents with regard to savings of chemicals and lowering of chemical costs and treatment of effluents with high toxicity.

Over the last decade the situation has changed and limits for a considerable number of compounds and parameters have been defined to avoid problems with regard to the following:

- biotoxicity (e.g., disturbance of biodegradation processes);
- heavy metal content (accumulation in sludge of CWWT);
- corrosion problems (e.g., sulfate can cause corrosion of concrete tubes);
- total COD/BOD load in the released effluents (capacity of the CWWT).

Table 1 gives an extract of important parameters for wasted water from textile plants, as defined by the Austrian Government [1]. The table contains limits defined for both direct release into surface water (rivers) and for release into a CWWT.

Table 1 can be used as a guide to define "hazardous" wastes from textile plants. Besides the direct toxicity of substances like chlorinated hydrocarbons, organo-Hg compounds, or concentrated alkaline solutions, other parameters have been defined with regard to problems during biodegradation or accumulation in the sludge from CWWT. A particular situation is found with colored effluents, where limits for spectral absorption have been defined. While the toxicity of textile dyes is comparably low, these limits were derived from the visual aspect of the water released from a textile plant because they look "unhealthy."

As a result of these regulations, textile companies have to apply a strategic concept to lower both the daily load released into the wastewater stream and the concentrations of hazardous substances therein. On the basis of the action plan given in Figure 2, a stepwise improvement of the present situation of a plant has to be undertaken.

Owing to the extreme diversity of the textile processes and products, it is impossible to develop a realistic concept for an efficient wastewater treatment without detailed analysis of the particular situation of a textile plant. The more intelligently the applied technical concept has been designed, the lower will be the expected costs for installation and working of the equipment.

In the following sections techniques and technical solutions are given as examples that can be adapted to a certain problem.

To facilitate an overview and to consider the specific differences of textile fibers during pretreatment, dyeing, and finishing, the sections have been focused on the most important types of fibers: wool, cotton, and synthetic fibers. Mixtures of fibers can be seen as systems combining problems of the single fiber types. In Section 8.3 end-of-pipe technologies have been summarized.

Table 1 Representative Limits Defined for Release of Textile Waste Water

Limits for emission	Release into river	Release into CWWT
General parameters		
Temperature (°C)	30	40
Toxicity	<2	No hindrance of biodegradation
Filter residue (mg/L)	30	500
Sediments (mL/L)	<0.3	—
pH	6.5–8.5	6.5–9.5
Color, spectral coefficient of extinction:		
436 nm (yellow) (m^{-1})	7.0	28.0
525 nm (red) (m^{-1})	5.0	24.0
620 nm (blue) (m^{-1})	3.0	20.0
Inorganic parameters (mg/L)		
Aluminum	3	Limited by filter residue
Lead	0.5	0.5
Cadmium	0.1	0.1
Chromium total	0.5	1
Chromium-VI	0.1	0.1
Iron	2	Limited by filter residue
Cobalt	0.5	0.5
Copper	0.5	0.5
Zinc	2	2
Tin	1	1
Free chlorine (as Cl_2)	0.2	0.5
Chlorine total (as Cl_2)	0.4	1
Ammonium (as N)	5	—
Total phosphor (as P)	1	No problems in P elimination
Sulfate (as SO_4)	—	200
Organic parameters (mg/L)		
TOC (total organic carbon as C)	50	$>70\%$ biodegradation
COD (chemical oxygen demand as O_2)	150	$>70\%$ biodegradation
BOD_5 (biological oxygen demand as O_2)	20	—
AOX (adsorbable organic halogen as Cl)	0.5	0.5
Total hydrocarbon	5	15
VOX (volatile organic halogen)	0.1	0.2
Phenol index calculated as phenol	0.1	10
Total anionic and nonionic surfactants	1	No problems in sewer and CWWT

Source: Ref. 1.

8.1.4 IPPC Directive of the European Community

In the legislation of different national governments, some limits were defined especially for wastewater and air. The activities in Europe are covered by the Council Directive 96/61/EC concerning Integrated Pollution Prevention and Control (IPPC) [9]. This means that all environmental media (water, air, energy, ground) and a comprehensive description of the production have to be considered. In addition a broad harmonization of requirements for the approval of industrial plants can be reached.

The classification of a company as an IPPC plant is based on the definition of the work concerning plants for the pretreatment (operations such as washing, bleaching, mercerization) or dyeing of fibers or textiles where the treatment capacity exceeds 10 tons per day. As a firm basis of reference the capacity will be calculated as the potential output a company could have in 24 hours. Capacity means what a plant is designed for and not what is really achieved (actual production). The treatment of fibers and textiles covers fibers, yarns, and fabric in the wider sense of the word, that is, including knitted and woven materials and carpets. As most textiles are treated with continuous working machines with a very high theoretical maximum capacity, a lot of companies have to fulfill the directions for IPPC plants.

To reach the aim of the directive an efficient and progressive state of development is defined by the best available techniques (BAT). In practice, this means precaution against environmental pollution by the use of these techniques, special equipment and better way of production, and an efficient use of energy for prevention of accidents and provisions for a shutdown of a production plant. The term *best available techniques* is defined as the most effective and advanced stage in the development of activities and their methods of operation that indicate the practical suitability of particular techniques for providing in principle the basis for emission limit values designed to prevent and, where it is not practicable, generally to reduce emissions and the impact on the environment as a whole. These available techniques are developed on a scale that allows implementation under economically and technically viable conditions, taking into consideration the costs and advantages when the techniques are used.

In the best available technology reference document (BREF), particular attention is given to the processes of fiber preparation, pretreatment, dyeing, printing, and finishing, but it also includes upstream processes that may have a significant influence on the environmental impact of textile processing. The treatment of all main fiber types as natural fibers (cotton, linen, wool, and silk), man-made fibers derived from natural polymers, such as viscose and celluloseacetate, as well as from synthetic polymers (such as polyester, polyamide, polyacrylnitrile, polyurethane, polypropylene) are described, including blends of these textile substrates. Beside general information about the industrial sector and the industrial processes, the situation in the plants is described by data about current emission and consumption.

A catalogue of emission reduction or other environmentally beneficial techniques that are considered to be most relevant in the determination of BAT (both generally and in specific cases) is given as a pool of possible techniques including both process integrated and end-of-pipe techniques, thus covering pollution prevention and pollution control measures. Techniques presented may apply to the improvement of existing installations, or to new installations, or a combination of both, considering various cost/benefit situations including both lower and higher cost techniques. To obtain a limitation of emission impact, different techniques are proposed corresponding to the basic possibilities for pollution prevention:

- handling of concentrates from various processes such as textile pretreatment, residual dye liquors from semicontinuous and continuous dyeing, residual printing pastes, residual finishing liquors, residues of prepared but not applied dyestuffs, textile auxiliaries, and so on;
- recovery of chemicals such as NaOH, sizing agents, indigo;
- assessment of textile auxiliaries aiming at a reduction of emissions of refractory and toxic compounds to water by substituting harmful substances with less harmful alternatives;

- reduction of releases to air from thermal treatment installations like stenter frames;
- reduction of releases to water by applying process-integrated measures and considering the available options for wastewater treatment; wastewater treatment including pretreatment onsite before discharge to the sewer as well as treatment of effluent onsite in case of discharge to rivers; efficiency of treatment of textile wastewater together with municipal wastewater;
- options for handling and treatment of residues and waste from different sources;
- minimizing of energy consumption used in energy-intensive processes such as pretreatment, fixation of dyes, finishing operation, and drying.

8.2 FIBER-SPECIFIC PROCESSES

The activities described in this section intend to minimize or avoid the release of chemicals into the stream wastewater by substitution, optimization, reuse, and recycling. Besides a lowering of the costs for following up general wastewater treatment, benefits due to minimization of chemical consumption are intended. As there are various specific problems arising from the particular treatment steps applied for different fibers, this section concentrates on the most important problems. Table 2 gives an overview of the annual production of textile fibers [10].

8.2.1 Protein Fibers: Wool

General

The annual production of wool is approximately 1.2 million tons, which corresponds to a share of 2% of the total production of textile fibers. A simplified route for the preparation, dyeing, and finishing of woolen textiles is shown in Figure 3.

Table 2 Annual Production of Textile Fibers 2001

Type of fiber	Mt/year
Man-made fibers	
Synthetics	31.6
Polyester	19.2
Polypropylene	5.8
Polyamide	3.7
Acrylics	2.6
Others	0.3
Cellulosics	2.7
Natural fibers	
Cotton	19.8
Jute	3.1
Ramie	0.2
Linen	0.6
Wool	1.2
Silk	0.1
Total	59.2

Mt, million tons.
Source: Ref. 10.

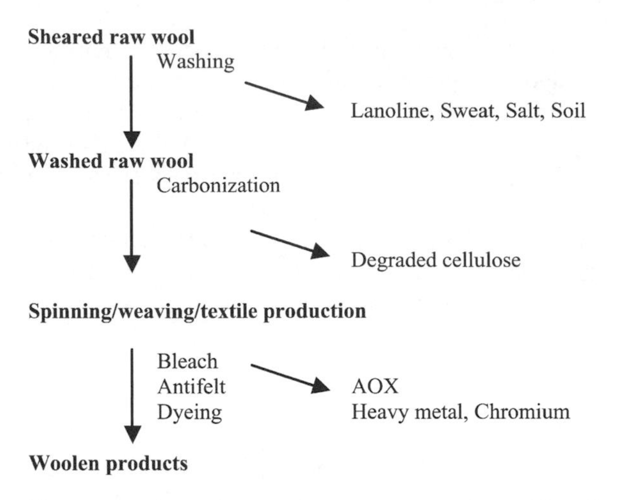

Figure 3 General processing route of woolen textiles (from Ref. 3).

Besides more general strategies of process optimization, three representative steps will be discussed in more detail because of their particular importance with regard to wastewater. The main problem resulting from these steps is given in parentheses:

- washing of raw wool (COD);
- antifelt treatment of wool (AOX);
- dyeing processes (chromium).

Washing of Raw Wool

The high content of impurities in raw wool has to be removed before further processing, for example, in carbonization, spinning, and weaving. As a considerable part of the raw material (approx. 30%) is removed and released into the wastewater, washing of raw wool can cause heavy pollution problems. These difficulties are not due to the toxicity of the released components, but result from the high concentrations and the load of organic material released in the form of dispersed and dissolved substances. Figure 4 gives an overview of a general set of techniques that can be applied to lower the initial COD in the effluent from approximately 80,000 mg/L to a final value of 12,000 mg/L [11,12].

The lanolin extracted from the wool is purified further for use in cosmetics, hand cream, boot-polish, and so on. Part of the permeate from the ultrafiltration is recycled to save fresh water. A particular advantage arises from the fact that the dissolved sweat components exhibit

Table 3 Average Composition of Raw Wool

Component	%
Fiber, protein	58
Wool-fat, lanolin, waxes	14
Soil, plant material (cellulose)	13
Sweat/salt, water soluble	5
Humidity	10

Source: Refs. 3,11,12.

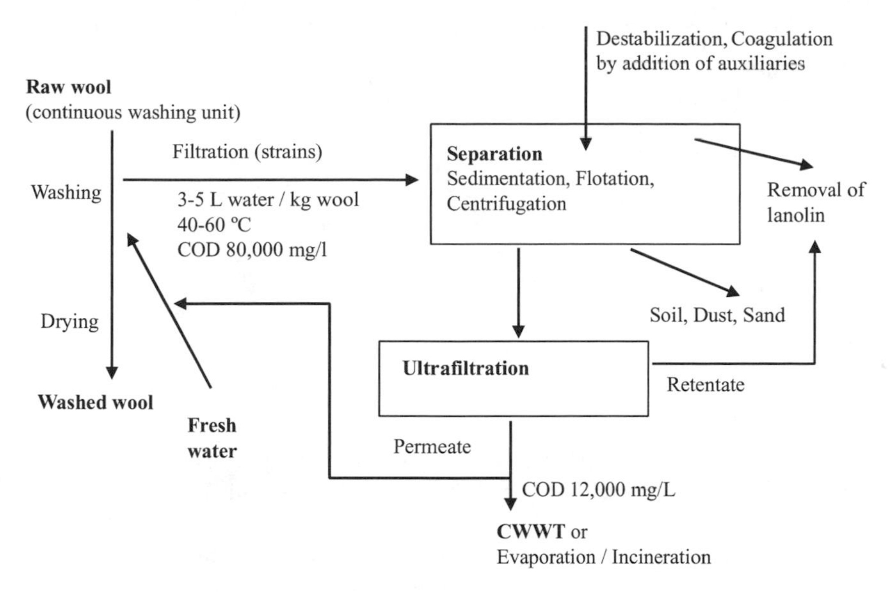

Figure 4 General scheme for the treatment of effluents from wool washing (from Refs. 11–13).

distinct washing properties for raw wool and thus a certain content of dissolved sweat is favorable to improve the washing effect.

Various treatment concepts have been presented in the literature [11–13]. Besides the release of the pre-treated wastewater into the CWWT and aerobic biodegradation, in some cases evaporation of the wastewater and incineration of the residue are performed.

Antifelt Finishing of Wool

The surface of a wool hair is covered by keratin sheds, which cause a distinct tendency to shrinkage and formation of felts. This behavior is usually undesirable and thus an antifelt finishing is the most important treatment during the processing of woolen textiles. One of the most important standard procedures, the Hercosett finish, is based on the oxidative treatment of wool by application of compounds that release chlorine. Examples for applied chemicals are NaOCl, Cl_2 gas, and dichloroisocyanuric acid (DCCA) [14].

Such processes lead to the formation of adsorbable halogenated organic compounds (AOX) in high concentrations. Typical concentrations found in a continuous antifelt treatment are shown in Table 4. The high dissolved organic carbon (DOC) determined in the baths is one of the sources for the formation of high concentrations of chlorinated compounds. The formation of chlorinated products is the result of chemical reactions directly with the fiber, with organic compounds released from the fibers, and with added auxiliaries.

An average size of continuous treatment plant for antifelt treatment of wool releases approximately 140 g/hour AOX. As an optimization of the process is possible only within certain limits, alternative processes for an antifelt treatment have to be chosen to substitute the chlorination process, for example, enzymatic processes, oxidative processes ($KMnO_4$, persulfate), or corona or plasma treatment. In many cases combinations with resin treatments are proposed.

Table 4 Concentrations for AOX Determined in the Chlorination Bath of the Chlorine-Hercosett Process

Parameter	Concentration
AOX	20 mg/L
$CHCl_3$	160–1200 μg/L
CCl_4	25–50 μg/L
DOC	1110 mg/L

Source: Refs. 3,14.

Chromium in Wool Dyeing

A considerable part of the wool dyes contains Cr complexes. The average consumption of dyes used in 1992 is shown in Table 5. At this time approximately 70% of all dyes used contain chromium.

As shown in Table 1 the wastewater limit for chromium is 0.5–1 mg/L and Cr^{VI} is 0.1 mg/L. While conventional 1 : 2 and 1 : 1 dyes permit chromium concentrations in the dyebath at the end of the dyeing process of 3.0–13.0 mg/L Cr, the application of modern dyestuffs and optimized processes permits final concentrations to approximately 1 ppm. By general optimization of the process (e.g., dosage of acid), use of dyes with a high degree of exhaustion, and minimal concentration of free chromium [15], final bath concentrations below 4 ppm can be reached, even for black shades. By application of such procedures the exhaustion of the chromium should reach values of better than 95% of the initial value.

Owing to the low limits for concentrations of chromium the proposed processes for wastewater treatment concentrate on the removal, for example, by flocculation and precipitation, but as a result chromium-containing sludge/precipitate or concentrates are obtained that need further treatment.

8.2.2 Cellulose Fibers: Cotton

General

Cellulose fibers (Co, CV, CMD, CLY) represent the main group of textile fibers used [10]. In this section cotton will be considered as a representative type of fiber because the treatments for other cellulose fibers are similar in many cases, and often milder conditions are applied for other cellulose fibers.

Table 5 Dyestuff Consumption in Wool Dyeing

Dyestuff	%
1 : 2 Metal-complex[a]	35
Chromium dyes[a]	30
Acid dyes	28
1 : 1 Metal complex dyes[a]	4
Reactive dyes	3

[a] Contain Cr or Cr-salts are added.
Source: Refs. 3,15.

Sources for textile effluents that need further treatment are found in all steps of processing. Table 6 shows a list of important parameters and wastes that require further treatment.

Sizing–Desizing

Before weaving, the warp is covered with a layer of polymer to withstand the mechanical stress (abrasion, tension) during weaving. These polymer coatings are so-called sizes. Normally native starch, modified starch like carboxymethyl-starch (CMS), carboxymethyl-cellulose (CMC), polyvinylalcohols (PVA), polyacrylates, and proteins can be used. The amount of added polymer for staple yarns like Co is between 8 and 20% of the weight of the warp. As a result, in many cases the final amount of polymer to be removed in the desizing step is approximately 5–10% of the weight of the fabric.

Sizing is not necessary in the case of knitted material, and much lower amounts are required for filament yarn (2–10% of the weight of the warp). The main problem resulting from the desizing step is the high load in COD found in the polymer-containing effluent. Table 7 summarizes the COD and biological oxygen demand (BOD) values determined for various sizes.

To estimate the COD/BOD load released from a desizing step, Eqs. (1) and (2) can be used:

$$L_{COD} = Cpm \times 10^{-3} \quad (1)$$

$$L_{BOD} = Bpm \times 10^{-3} \quad (2)$$

Table 6 Processing of Cotton: Process Steps and Selected Parameters

Process step	Critical parameter	Component
Desizing	COD $\neq$ BOD	Starch, modified starch, PVA, polyacrylates
Scouring	COD $\neq$ BOD	Organic load released from cotton and added auxiliaries
	Complexing agents	EDTA, phosphonates
	pH	NaOH
Bleach		
Hypochlorite	AOX	Chlorinated compounds
Peroxide	Complexing agents	EDTA, phosphonates
Mercerization	pH	NaOH
Dyeing		
Direct	Salt	NaCl, Na_2SO_4
Reactive	Color	Hydrolyzed dyes
	Salt	NaCl, Na_2SO_4
	pH	NaOH
Vat	pH	NaOH
	Sulfate	Na_2SO_4, Na_2SO_3
Indigo	Color	Indigo
	Salt	Na_2SO_4
Printing	Printing pastes	Concentrated chemical load
	Washwater (COD, BOD, color)	Thickener, dyestuff
Finishing	Filling of padder	Concentrated chemical load

Source: Ref. 3.

Table 7 COD and BOD per Mass of Size Released

Type of size	COD C (mg/g)	BOD B (mg/g)
Starch	900–1000	500–600
CMC	800–1000	50–90
PVA	1700	30–80
Polyacrylate	1350–1650	<50
Galactomannane	1000–1150	400
PES-dispersion	1600–1700	<50
Protein	1200	700–800

Source: Ref. 3.

Desizing of $m = 1000$ kg of goods, which contain 5% of weight starch size ($p = 0.05$) cause a load $L_{COD} = 50$ kg and $L_{BOD} = 30$ kg. Using 10 L of water for desizing of 1 kg of fabric, a total volume of 10,000 L will be required and the load $L_{COD} = 50$ kg will be diluted in this volume. As a result, a COD value of 5000 mg/L can be calculated for the effluent.

Two different paths can be followed to describe the behavior of sizes released in effluents:

- Biodegradation, which refers to the complete biodegradation of sizes like starch. Here high values of COD are coupled to high BOD.
- Bioelimination is detected by BOD, which is rather low BOD, compared to the COD. In such cases the polymer is removed from the waste stream in the WWT/CWWT by flocculation, adsorption, hydrolysis, and, to a certain degree, by biodegradation. Representatives are PVA, CMC, and acrylate sizes [16,17].

The strategies to handle size-containing wastes are dependent on the type of size and particularly on the technique of desizing (Fig. 5). In the case of starch, the desizing step is usually performed by enzymatic degradation, and in some cases oxidative degradation is used. However, the starch is degraded and a reuse is not possible in such cases. The disadvantage of a high COD caused by the released partially degraded starch is accompanied by easy biodegradation, thus the effluents can be treated in a WWT/CWWT with sufficient capacity for biodegradation with no further problems.

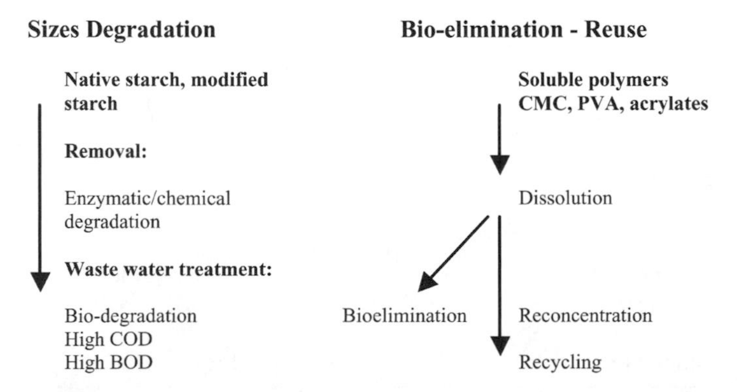

Figure 5 Desizing and treatment of size-containing wastes (from Refs. 18–24).

Water-soluble sizes permit a recycling of the polymer for further weaving processes. Various techniques have been proposed to regenerate sizes released from the fabric. General requirements that have to be considered as fundamentals for possible reuse of sizes are summarized as the following:

- easy and short distance transportation of recovered size to sizing/weaving plant;
- known composition of sizes;
- development of standardized recipes;
- stable composition of recovered size/no degradation.

In practice, a recycling of sizes is hindered for a number of reasons. In many cases various qualities of fabric containing different sizes are treated in a dyehouse and the type of size is often not known. The selection of sizes with regard to easy biodegradation/bioelimination is necessary. When a regeneration is intended a direct interaction between the selection of size, desizing procedure, recycling processes, and the sizing/weaving process have to be considered.

Two general technological strategies have been developed and proposed:

- removal of water soluble sizes by washing;
- reconcentration in the washing machine or by UF/evaporation. Figure 6 gives an overview of these two techniques.

Washing techniques have been proposed for PVA and acrylate sizes [18]. When applying washing techniques the volume of concentrated washwater for each size is limited by the volume actually spent in the following up sizing process (e.g., 900 L in Fig. 6) [19–21]. The use of higher amounts of water would increase the mass of recovered size, but the dilution of the regenerate is too much and hinders a reuse without reconcentration. A typical balance for a full process for acrylate sizes is shown in Figure 7 [22].

The advantage of UF techniques is the higher rate of size recovery, because a reduction of volume is possible. In some cases an evaporation step is used as final concentration step because the viscosity of the sizes increases and the permeate flow is reduced substantially. Problems can result from a change in the composition of the size due to changes in the molecular weight distribution as a result of the cutoff of the UF membrane. Attention has to be paid to avoid biodegradation of the recovered sizes, which changes the properties of the polymer and causes intensive odor of the regenerates.

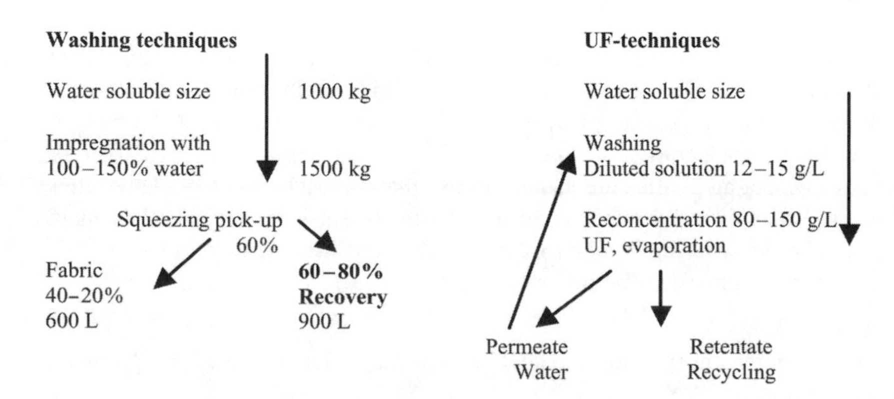

Figure 6 Recycling of sizes (from Refs. 18–24).

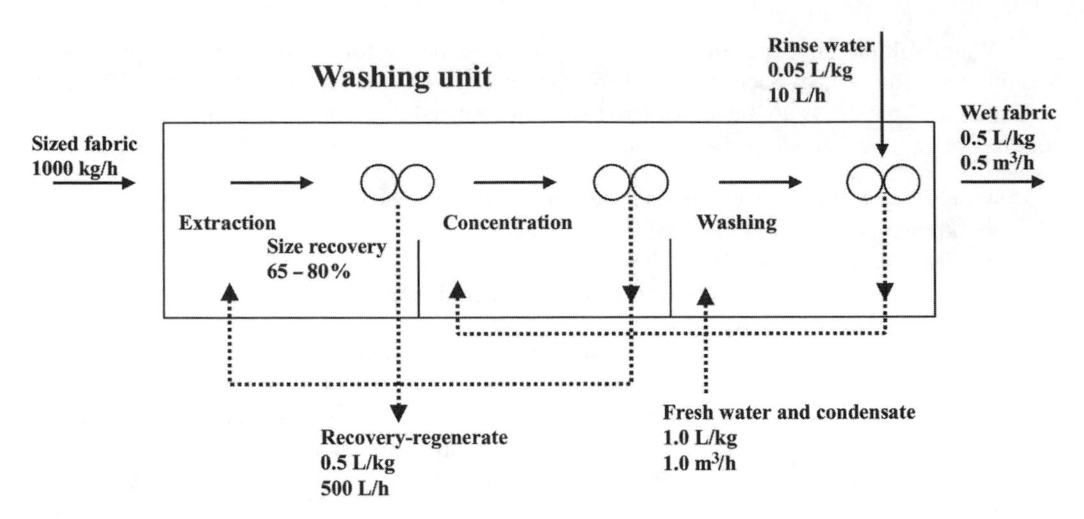

Figure 7 Recovery of sizes by washing techniques (from Refs. 3 and 22).

In general, for a recovery of sizes, the following points have to be defined:

- establishment of continuous and time-stable conditions in sizing/desizing/regeneration/reuse;
- low amount of impurities in the regenerates due to dyes, colored fibers, dust from singing;
- establishment of an organizational structure that is able to handle the recovered products.

In many cases savings due to lowered costs for size and COD in the wastewater exceed the expenses for investment and running costs; thus acceptable data for ROI of less than two years are given in the literature [23,24].

Scouring, Alkaline Pretreatment, and Peroxide Bleach

A central step of pretreatment of natural cellulose fibers like cotton or linen for dyeing and printing is the alkaline scouring and bleach of the fibers. Figure 8 gives an overview for the pretreatment of cotton. Besides the destruction of the natural yellow-gray color of the fibers by the bleach chemicals, a considerable part of the organic compounds is removed from the fibers during the alkaline scouring step [3]. Average values of the compounds present in raw cotton are given in Table 8.

Assuming an average COD for the released compounds of 200 mgO_2/g, a total COD of 20 gO_2 per 1 kg of cotton is transported into the wastewater. In a batch treatment applying a liquor ratio of 1 : 10, 1 kg of cotton is extracted with 10 L of water, thus a COD of 2000 mg O_2/L can be estimated without consideration of the COD resulting from added auxiliaries or complexing agents. At present auxiliaries are usually in use that are easily biodegradable; thus after neutralization no problems should appear during the treatment in a CWWT. The main problem arising from alkaline scouring is therefore due to the considerable load in COD.

A typical recipe for alkaline scouring processes (liquor ratio 1 : 10) is as follows:

2–8 g/L NaOH;
0.3–3 g/L complexing agent (polyphosphate, carbohydrates, polyacrylate, phosphonate, nitrilo-tri-acetic acid (NTA);
0.5–3 g/L surfactant.

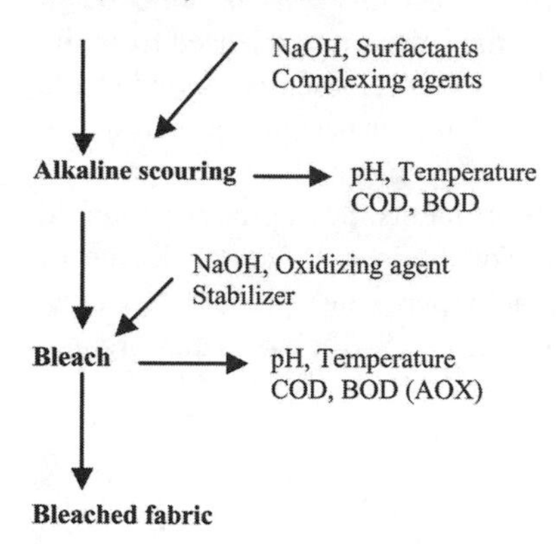

Figure 8 General scheme for the pretreatment of cotton (from Refs. 3 and 27).

The total water consumption of the treatment including the rinsing step is approximately up to 50 L/kg. When the composition of an auxiliary is known, an estimation of the COD can be made by calculation of the oxygen demand for total oxidation. Examples are given below for Na–polyacrylate ($-CH_2-CHCO_2Na-$) and for Na–gluconate. Basing on Eqs. (3) and (4), the oxygen demand for 1 g of compound can be calculated.

$$C_6H_{11}O_7Na + 5.5O_2 \rightleftarrows NaOH + 6CO_2 + 6H_2O \tag{3}$$

$$(-CH_2-CHCO_2Na-) + 3O_2 \rightleftarrows NaOH + 3CO_2 + 2H_2O \tag{4}$$

For the oxidation of 1 g of Na–gluconate, 810 mg of O_2 are required, and for the oxidation of 1 g of Na–polyacrylate, 1020 mgO_2 will be necessary. Technical products are mainly liquid formulations and the actual composition is given very rarely, but on the basis of the content of active compounds and an assumption of the chemical structure, an estimation of the contribution of the auxiliaries to the COD can be made.

The COD contribution of a recipe using 2 g/L of an auxiliary that contains 50% polyacrylate to the total COD in the wastewater will be approximately $COD = 2 \times 0.50 \times 1020 = 1020\ mgO_2/L$.

Generally the treatment of waste water from alkaline scouring/bleaching (peroxide) processes will require an adjustment of pH and temperature, which is normally made by mixing with wastewater from other treatment steps. When surfactants, complexing agents, and so on,

Table 8 Average Composition of Raw Cotton

Component	%
Cellulose	80 90
Hemicellulose, pectin	4–6
Waxes, fat	0.5–1
Proteins	1.5
Minerals (Ca, Mg, K, Na, P)	1–2
Other components	0.5–1
Humidity	6–8

Source: Ref. 3.

with good biodegradability/bioelimination have been selected, the COD load is removed in CWWT without problems. The main load of the COD is due to the substances released from the fibers and added auxiliaries, thus an optimization of the load of COD released is limited to the auxiliaries only; however, these components will represent only a minor part of the total COD.

The application of chlorine bleach on the basis of hypochlorite/chlorite for the preparation of cotton/linen results in considerable formation of AOX in the effluents. Such processes should be replaced by bleach processes on the basis of peroxide. To obtain a sufficient degree of whiteness during the bleach, a two-step bleach (peracetic acid/peroxide) process has been proposed in the literature [25–27]. Such processes avoid the formation of chlorinated organic compounds (AOX).

Mercerization

Depending on conditions applied, the treatment of cotton textiles in concentrated alkaline solutions, for example, 300 g/L NaOH, leads to increased luster, improved dimensional stability, high uptake of dyes, and changes in strength and hand. Usually a continuous treatment process is applied. As a result enormous amounts of concentrated caustic soda solution have to be removed during the washing step. As a typical value approximately 300 g of NaOH are transported per 1 kg of cotton into the following up stabilization/washing baths. In the stabilization step the caustic soda is rapidly removed by washing with diluted caustic soda solutions. The effluents from the stabilization step contain approximately 40–60 g/L NaOH. Figure 9 gives an overview of the steps during mercerization of cotton.

The high costs for the consumed NaOH and the costs for neutralization of the NaOH in wastewater favor the recycling of NaOH by reconcentration procedures. Normally a re-concentration is made up to at least 400 g/L NaOH. Starting from a diluted NaOH containing 50 g/L NaOH, 7.8 L of water has to be removed to obtain 1 L NaOH with 440 g/L. The reconcentration is usually made by reboiling. For this purpose evaporation plants with several evaporation stages are in use. The use of several stages (normally at least three stages) is of

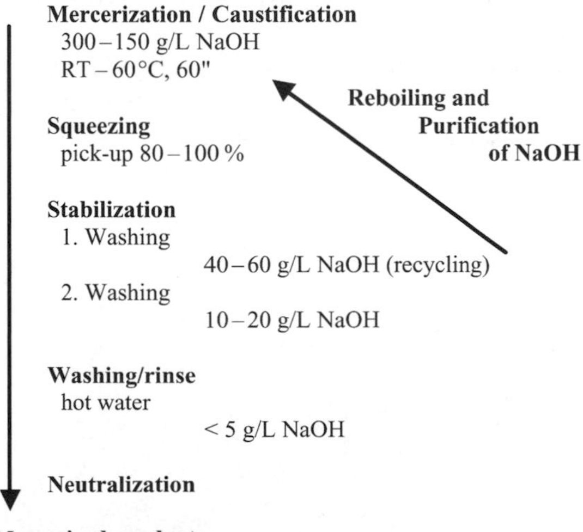

Figure 9 Mercerization of cotton (from Refs. 28–31).

importance to keep the energy consumption of the process within reasonable dimensions. Typical values for energy consumption are 0.2–0.3 kWh/kg of evaporated water. Large amounts of waste energy are released from the condensation of the evaporated water and have to be used in the form of warm water. Care has to be taken to achieve a reuse of the warm water because the degree of heat recovery is essential to obtain an acceptable return on investment (ROI) of the unit [28].

Purification of the reboiled caustic soda is important to remove sizes (raw-mercerization), dyes (mercerization of dyed materials), fibers, and impurities released from the fibers. Important techniques are filtration, centrifugation, flotation processes, and oxidative processes [29–31]. The application of membrane processes for reconcentration is limited to low concentrations of NaOH because of the insufficient chemical stability of the membranes.

The reuse of the diluted caustic soda from the first stabilization compartment in other processes, for example, alkaline scouring, has been recommended. Problems can arise from variations in concentration and impurities present in the reused lye, so the recycling of the diluted NaOH for other treatment processes is not used widely. As the amount of caustic soda that can be reused for other processes is low compared to the amount of NaOH released from the mercerization step, regeneration by evaporation is normally the favored process.

Dyeing of Cellulose Fibers

Dyeing of cellulose textiles can be performed at all stages of textile processing, for example, fibers, yarn, fabric, or garment dyeing. Depending on the desired final properties of the dyed material, various classes of dyes are used, which are collected in gamuts of common application. Important classes of dyes are direct dyes, reactive dyes, and vat dyes, including indigo and sulfur dyes.

Wastewater problems mainly arise from three different sources:

- dyestuff: colored effluents, AOX, heavy metal content (Cu, Ni) [32];
- dispersing agents in dyestuff formulation: COD, poor biodegradability;
- auxiliaries, chemicals added: salt content (NaCl, Na_2SO_4), sulfide, pH value (NaOH, soda, silicates), COD (glucose, hydroxyacetone), N-content (urea).

Direct Dyes. For direct dyes a degree of fixation in the range 70–90% is given in the literature [33–35]. When optimized dyes and processes with a high degree of fixation are implemented into a dyehouse, problems of colored wastewater can be minimized. As heavy metal ions are mainly present in complexed form in the dyestuff, a lowering of the Cu and Ni content in the wastewater goes in parallel with an increase in dyestuff fixation. A similar situation is found with AOX values, which result from the halogen bound in the dyestuff molecules. In dyehouses where chlorine bleach has been substituted by other bleach chemicals, halogens bound in dyes can cause a main contribution to the AOX value found in the wastewater.

Reactive Dyes. The situation with regard to heavy metals (e.g., Cu, Ni from phthalocyanine dyes) and AOX from covalently bound halogen is comparable with direct dyes. Selection of processes with a high fixation of dyestuff yields a considerable decrease in Cu/Ni concentrations and AOX. For the fixation process certain amounts of alkaline are added to the dyebath. As the total amount of alkali used is low compared to the consumption of alkali during mercerization, scouring, and bleach, high pH due to the alkali from reactive dyeing is of minor relevance. Two main problems have to be mentioned in connection with reactive dyeing [36]:

- High load of soluble salt (NaCl, Na_2SO_4). For acceptable exhaustion of dyes, considerable concentrations of salt (up to 50 g/L) are required in exhaust dyeing processes. The release of the used dyebath transports a rather high load of salt into the wastewater stream. When a liquor ratio of 1 : 10 is applied, 10 L of dyebath are used for dyeing of 1 kg of goods, thus at a salt concentration of 50 g/L an amount of 0.5 kg salt is released for dyeing of 1 kg of goods.
- Colored wastewater. The problem of relatively high dyestuff concentrations in wastewater particularly arises when dyestuff exhaustion and fixation proceed only to a limited degree, typically only 70–80%, so that between 30 and 20% of the dye is released with the spent dyebath and the washing baths that follow. Such a situation is observed particularly with reactive dyeing processes where a covalent reaction of the dye with the fiber takes place but some of the reactive groups become hydrolyzed during dyeing and thus some dye remains unfixed in the dyebath. Depending on the general method of dyeing, two different qualities of colored wastewater can be identified (Fig. 10).

Particularly in the case of dyes with a limited degree of fixation the dyestuff content in the wasted water leads to intensively colored wastewater. As the reactive group of the unfixed dyestuff is hydrolyzed into an inactive form, a reuse is not possible. On the basis of an exhaust dyeing with 5% color depth, a liquor ratio of 1 : 10, and a degree of dyestuff fixation of 70–80% corresponding to 3.5–4 g/L of dye are fixed on the goods and 1.5–1 g/L of hydrolyzed dyes are released with the dyebath.

For exhaust dyeing processes a reduction of the liquor ratio leads to significant improvements. When the dyestuff fixation is known for a certain liquor ratio, the lowering of the amount of unfixed dye released into the wasted water can be estimated as a function of the liquor ratio (LR). The amount of dyestuff on the fiber, m_{DF}, can be calculated using Eq. (5), and the total amount of dyestuff in the dyebath, m_D, can be calculated using Eq. (6).

$$m_{DF} = m_F p_F \tag{5}$$

$$m_D = m_F c_D LR \tag{6}$$

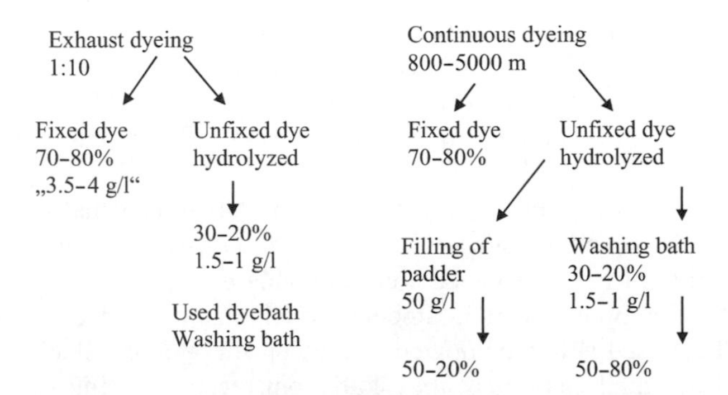

Figure 10 Sources for colored wastes from textile dyeing operations (from Ref. 55).

On the basis of Eqs. (5) and (6), the part of dyestuff released as hydrolyzed dye can be estimated using Eq. (7).

$$L = \frac{m_F c_D LR}{m_F p_F + m_F c_D LR} = \frac{c_D LR}{p_F + c_D LR} \quad (7)$$

When a color depth of 5% (50 g dyestuff per 1 kg of goods) is used as basis for a calculation and a dyestuff fixation of 80% is observed at a liquor ratio of 1 : 10 (10 L of dyebath for 1 kg of goods), then a mass of 40 g dyestuff is fixed on the textile while 10 g remain in the dyebath as hydrolyzed dye. The dyestuff concentration c_D in the used bath is then 1 g/L ($p_F = 0.05$, $LR = 10$, $c_D = 1$ g/L).

While at *LR* 1 : 10 a fixation of 80% is observed, a reduction of *LR* to 1 : 5 lowers the losses of dyestuff to approximately 11%, and a degree of fixation of 89% is expected. These results clearly indicate the importance of a low liquor ratio to optimize the degree of dyestuff fixation.

Another source of highly colored dyebaths is found in continuous dyeing processes where the last filling of the padder required to complete the process at well-defined conditions has to be withdrawn at the end of the padding process. Dyestuff concentrations of 50 g L^{-3} technical dyestuff are quite usual for such dye liquors.

For a dyestuff fixation of 70–80% and a color depth of 5% a concentration of 1.5–1 g/L hydrolyzed dye is expected in the wastewater, when 10 L of washing water is applied per 1 kg of goods. The emission of colored wastewater here can be divided into two different sources, the wastewater from the washing of the dyed material and the residual filling of the padder.

Depending on the length of the dyed piece (800–5000 m) the contribution of the filling of the padder to the total dyestuff concentration in the wasted water is estimated between 50 and 20%.

In general there are two different qualities of colored wasted water:

- The fillings of the padder. High dyestuff concentrations of approximately 50 g/L, high concentration of alkali;
- Spent dyebaths and washing baths. Low concentration of dyestuff, approximately 1 g/L, low concentration of alkali.

Besides an optimization of the dyestuff and the dyeing processes with regard to improved dyebath, exhaustion, the problem of colored wastewater released from dyehouses, has led to numerous technical developments proposed to overcome it.

A large number of techniques have been described in the literature, for example, dyestuff adsorption, oxidative and reductive treatments, electrochemical oxidation or reduction methods, electrochemical treatment with flocculation, membrane separation processes, and biological methods [37–55]. Each of these techniques offers special advantages, but they can also be understood as a source of coupled problems, for example, consumption of chemicals, increased COD, AOX, increased chemical load in the wastewater, and formation of sludge that has to be disposed.

The techniques for decolorization of dye-containing solutions can be applied at different stages:

- Treatment of concentrated dyestuff solutions (e.g., filling of padder), which is an efficient way to handle such concentrates, but as shown in Figure 10 usually only part of the released dyestuff is decolorized by treatment of such baths.

- Treatment of separately collected and reconcentrated baths that initially contain dyestuff concentrations of approximately 1 g/L and are reconcentrated to approximately 10–20 g/L dyestuff by membrane filtration. Such techniques yield considerable amounts of recyclable water, but care has to be taken to avoid any disturbing effect during reuse caused by salt and alkali content in the regenerate. The concentrated dyestuff solution can be treated with similar methods as concentrated dye solutions from fillings of padder.
- Treatment of the total wastewater: this technique will be discussed in Section 8.3, "End-of-pipe Technologies." The general scheme of such treatments is shown in Figure 11.

Vat Dyes. Vat dyes are normally present in their insoluble oxidized form. During their application in the dyeing process the dyestuffs are reduced in alkaline solution by addition of reducing agents, for example, dithionite, hydroxyacetone, or formaldehydsulfoxylates. Vat dyes normally exhibit an excellent degree of fixation; thus, the problem of colored wastewater is of minor relevance. In addition, vat dyes are readily reoxidized in the wastewater into the insoluble oxidized form that precipitates and thus shows lower absorbance. The main problem in the wastewater released form reducing agents which cause certain load in the effluents (XX1). In the case of dithionite, sulfate is formed that can cause corrosion of concrete tubes, and in the case of hydroxyacetone, the COD is increased considerably. A substitution of the nonregenerable reducing agents by electrochemical reduction has been proposed in the literature [56].

Sulfur Dyes. Similar to the vat dyes, sulfur dyes are applied in reduced form. Owing to the lower redox potential of the dyes, reducing agents such as sulfide, polysulfide, glucose, hydroxyacetone, or mixtures of glucose with dithionite are in use. Sulfides should be replaced by other organic reducing agents mentioned above; in such cases the COD is increased but the products are easily biodegradable. In comparison to the vat dyes the degree of fixation is lower with sulfur dyes. As such, dyes are mainly used for dark shades and colored effluents have to be treated with methods similar to the processes mentioned with reactive dyes.

Indigo. Dyeing with indigo for the Denim market (jeans) is unique. Here a nonuniform dyeing through the cross-section of the yarn is the desired type of quality. There is only one dye in use, indigo. For this type of textile the warp is dyed before the weaving process and special techniques are applied on unique dyeing machines specialized to produce indigo-dyed warp yarn [57]. Figure 12 presents a scheme of the dyeing process. After the warp yarn has been wetted and squeezed, it is immersed into the dyebath, which contains the reduced indigo dye (from 1 to 5 g/L), for a few seconds. After mangling to 80–90% expression, the reduced dyestuff on the material is oxidized completely during an air passage that lasts for 60–120 s. The immersion/squeezing/

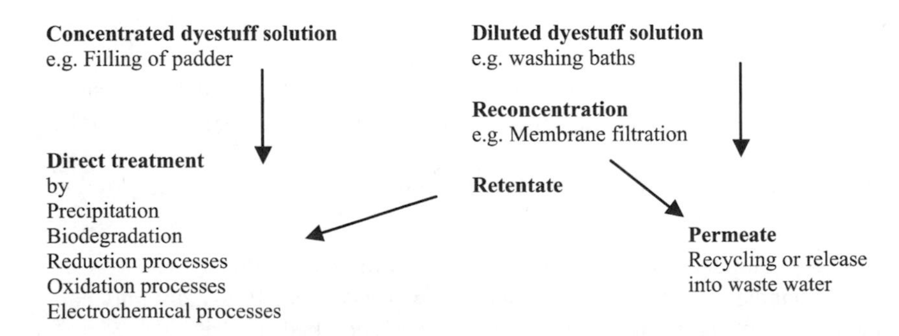

Figure 11 Treatment scheme for colored wasted water (from Ref. 54).

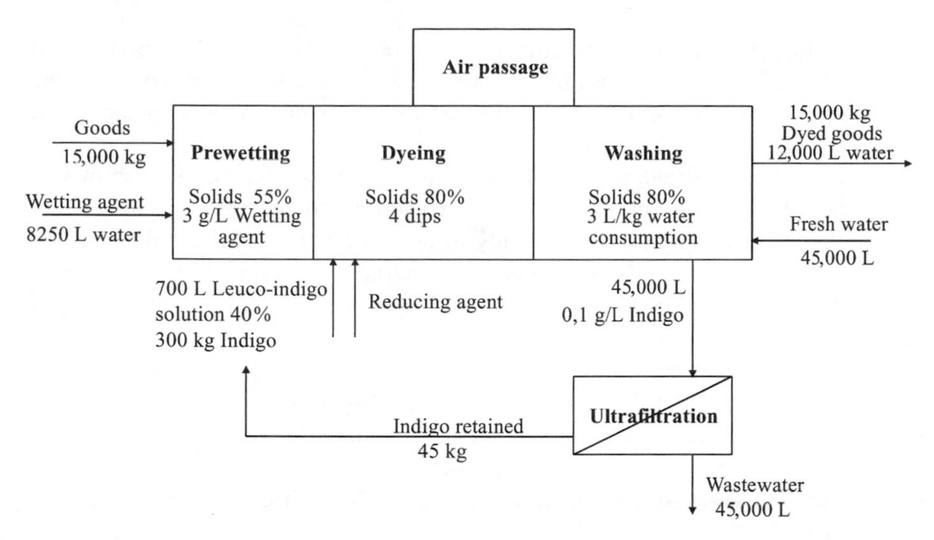

Figure 12 Flow scheme for indigo recovery in continuous yarn dyeing for denim (from Ref. 57).

oxidation cycle is repeated several times and the dyestuff is applied layer by layer. After the last oxidation passage the dyed material is washed and dried. Table 9 presents the typical data describing the production scale of such a dyeing unit. Two main difficulties exist at present:

- the indigo content in the wasted water, which causes colored wasted water;
- the sulfate or COD content in the washing water due to the use of dithionite or hydroxyacetone as reducing agents.

Table 9 Working Conditions and Production Data for a Full-Scale Indigo Dyeing Range

Production rate	
Cotton yarn	15,000 kg/day
	11.9 kg/min
Hours of operation	21 h/day
Warp speed	35 m/min
Depth of shade	2% indigo
Consumption of chemicals	
Reducing agent $Na_2S_2O_4$	50–126 kg/day
	40–100 g/min
Water	3–5 L/kg
	45–74 m^3/day
Composition of dyebath	
Wetting agent	0.5 g/L
Pre-reduced indigo	1–4 g/L
NaOH to maintain pH to	11.5–12.0
Temperature	20–30°C
Redox potential	<-700 mV

Source: Refs. 57–59.

A considerable improvement of the situation could be obtained by the use of prereduced indigo instead of the reduction of the dyestuff in a stock vat [58]. By use of prereduced indigo the sulfate concentration in the wasted water can be lowered to approximately 50% of the initial value.

A recovery of the dispersed indigo from the wastewater can be obtained by use of UF. Owing to the low price of indigo the cost savings due to dyestuff recovery are poor compared to the investment. The problem of sulfate load can only be solved by the use of more expensive organic reducing agents, which can be degraded by anaerobic digestion [59]. Additional improvements are expected from the use of electrochemical methods for the reduction of dyestuff instead of nonregenerable reducing agents [57].

Figure 12 shows a flow scheme of a complete installation including the recycling of the diluted dyebath by ultrafiltration (UF) with regard to the dispersed oxidized indigo. The permeate is used as washing water or released, and after reduction of the dyestuff in a stock vat, the indigo-containing permeate is reused for dyeing processes.

The reuse of purified wastewater from dyeing processes for pretreatment processes has also been studied in detail [60].

8.2.3 Dyeing of Synthetic Fibers

Polyester PES

Polyester fibers represent the most important group of man-made fibers. With an annual production volume of 19.2 Mt, polyester fibers hold second position in world production of textile fibers [10]. Polyester is usually dyed with disperse dyes.

Three techniques are in use at present:

- High temperature (HT) processes. To exceed the glass-transition temperature processes and to achieve sufficient rate of dyeing and leveling, the temperature of the dyebath is elevated to 110–115°C in high-temperature dyeing apparatus. Normally such processes are limited to batch processes and specialized equipment has to be used to stand the high pressure.
- Dyeing with use of carriers. The addition of organic compounds of low molecular weight permits the temperature to be lowered below 100°C for polyester dyeing; thus dyeings can also be performed in normal pressure equipment. The chloro-organic compounds widely used in the 1970s have now been replaced by chlorine-free carriers such as aromatic esters, substituted phenols.
- Thermosol dyeing. The characteristics of low-molecular-weight polyester dyes can be utilized in thermosol dyeing processes. In this continuous dyeing process the material is impregnated with the dispersed dye, dried, and heated to a temperature of approximately 200–210°C. The dyestuff is fixed by sublimation into the fiber.

Generally only low amounts of chemicals are added to the dyebaths and the degree of dyestuff fixation is high, so except for the application of carriers, which has to be considered carefully, and dispersing agents with limited biodegradability, the dyeing of polyester fibers causes minor problems with regard to the release of hazardous wastes [61].

An important innovative technique to replace water as the solvent in dyeing processes is the use of supercritical fluids, for example, supercritical CO_2 for dyeing processes. Successful trials have been conducted in various scales with different fibers and full-scale production has been performed in the case of PES dyeing [62,63]. Besides the handling of high pressure equipment, the development of special dyestuff formulations is required.

Elastomer Fibers: Elastan, Lycra™

An increasing percentage of textiles is now designed with elastic properties, which are obtained by the introduction of elastic fibers into them. The pretreatment of elastomer-containing fibers can be regarded as representative for the pretreatment of other man-made fibers. To improve the behavior of these fibers during spinning, winding, weaving, and knitting, considerable amounts of auxiliaries are added. Typical examples for such compounds are:

- fatty amines;
- polyethylene glycols;
- hydrocarbons;
- silicone compounds.

In particular in the case of elastomer fibers, such compounds (in many cases silicone compounds) add up to 2.5–8% of the weight of the fibers. Besides problems in removing these oily components during pretreatment, for example, washing of the textiles, the compounds are then detected in the wastewater in considerable amounts. As the addition of such auxiliaries is required for technical purposes, an optimization of the situation has to be achieved by direct cooperation between the fiber/yarn/fabric producer and the textile dyehouses.

8.2.4 Dyeing on Standing Dyebath

A method to lower the release of chemicals, auxiliaries, and residual dyestuff in exhaust processes is dyeing on a standing dyebath. In such a technique the exhausted dyebath, which contains the auxiliaries, chemicals (salt), and dyestuff, is reused for the next dyeing after a replenishment of the exhausted dyestuff and lost chemicals. In fact, such techniques are not as widely in use as might be expected because a set of requirements has to be fulfilled to introduce them:

- no accumulation of chemicals (e.g., spent reducing agents in vat dyeing will lead to increasing salt concentration);
- no formation of dyestuff byproducts (hydrolyzed dye in reactive dyeing);
- the run of the dyeing process has to be suited for dyeing on a standing bath (no dosing of chemicals);
- the size of a batch that has to be dyed at the same conditions has to be significant.

Examples for such techniques are found in sulfur dyeing for black shades and in a special form in indigo dyeing for denim, where a continuous replenishment of the dyebath is performed for a long period of production.

8.2.5 Textile Printing Operations

Numerous variations of textile printing processes are found in textile production depending on the type of fiber, applied dyes, desired effect, and fashion.

At present, flat screen printing and rotary screen printing are the main techniques used. Here the dyestuff is dissolved/dispersed in a printing paste containing thickener and chemicals. With every change of color, the filling of the dosing unit and of the screen has to be withdrawn. As such, changes frequently involve considerable amounts of used printing pastes having to be handled. In addition, the equipment (screen, pumps, and containers) have to be cleaned, so a distinct load is released into the wastewater. This amount increases with shorter lengths of printed batch. Table 10 gives two examples for the composition of printing pastes.

Table 10 Composition of 1000 g Printing Pastes for Pigment Printing and Two-Phase Reactive Printing

Pigment printing	Mass (g)	Two-phase reactive printing	Mass (g)
Pigment	5–80	Dyestuff	1–100
Thickener (e.g., polyacrylate)	15–45	Urea	50
Emulsifier (e.g., fattyalcohol-polyglycolethers)	5–10	Alginate thickener	400
Binder (e.g., copolymers from butylacrylate, acrylonitrile, styrol)	60–80	m-Nitro-benzene-sulfonic acid Na-salt	15
Fixation agent (melamine formaldehyde condensation prod.)	5–10	Buffer (e.g., NaH_2PO_4)	2–3
Catalysator (e.g., $MgCl_2$)	0–2		
Softener (fatty acid ester)	5–10		
Anti-foam agent	0–3		
Water	ad 1000	Water	ad 1000

Source: Ref. 3.

A COD of reactive printing pastes of 150,000–200,000 mgO_2/kg for pigment paste values of up to 350,000 O_2/kg is realistic. Additional problems arise from the AOX content (chlorine containing dyestuff) and from heavy metal content resulting from metal ions complexed in the dyes (e.g., Co, Cu, Ni). Attention also has to be given to the use of antimicrobial agents in the printing pastes, which are added to block the microbial growth that results in degradation of the thickener and lowering of the viscosity of the printing paste.

Generally, any release of printing pastes into the wastewater should be avoided, and in many countries such action is forbidden. Figure 13 gives an overview of the possible proceedings to minimize chemical load in the wasted water from the release of printing pastes [64,65].

First the consumption of printing pastes has to be minimized by:

- Minimization of the required volumes to fill the equipment, e.g., printing screen, tubes, pumps, and container. By optimization, a filling of up to 8 kg can be reduced to a consumption less than 2 kg per filling.
- Exact calculation and metering of the consumption of printing paste to avoid excess of pastes.

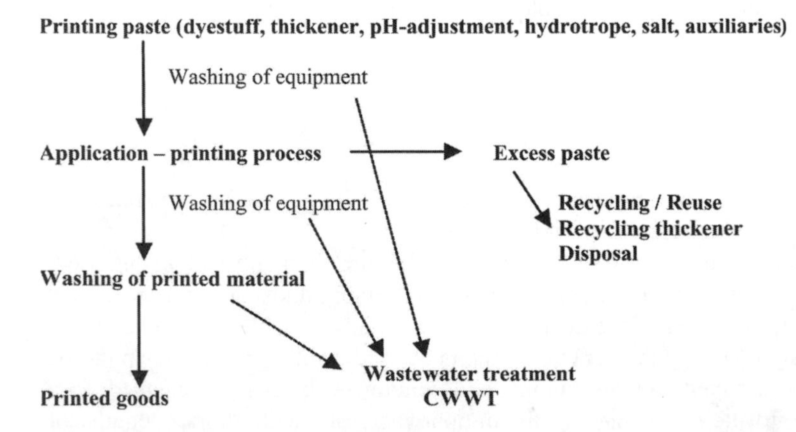

Figure 13 Minimization of chemical load from textile printing (from Ref. 57).

The minimization of the filling of the equipment is of particular importance for the production of short lengths, for example, during sample printing. In particular for the production of very short lengths (e.g., 120 m), a considerable portion of the printing paste is required for the filling of the printing machine. Depending on the coverage factor of a pattern, approximately 55–80% of the paste is used for printing, while 45–20% is spent for the filling of the printing machine, which is considerable with a mass of 5 kg in this example. When a length of 1000 m is produced the portion of paste spent for the filling reduces to 10–3% of the total mass of printing paste [66]. The high consumption of printing pastes for the production of short length samples causes high costs for the production of a collection of new patterns and thus at present digital printing techniques are recommended to substitute for the expensive full-scale production of design samples.

The high content of dissolved compounds and the broad variations in the concentration of dyes and auxiliaries make a direct recycling of pastes difficult. Supported by calculation programs, a certain portion of printing pastes can be added for the preparation of new pastes [67]. In the most simple case, the preparation of pastes for the printing of black color is carried out.

If disposal is necessary, various techniques can be used: drying and incineration, binding in concrete, and anaerobic degradation [64,65].

A recent technique to achieve a reuse of the thickener is the precipitation of the thickener by addition of organic solvent (e.g., methanol). After removal of the dyes and chemicals the thickener can be reused for the preparation of new pastes. The removed chemicals and dyes are collected and discarded [68]. By this method a considerable part of the COD-forming compounds can be recycled and the AOX and heavy metal content in the wastewater from textile printing can be reduced.

The replacement of classical textile printing techniques by digital printing techniques (ink-jet and bubble jet) is in full progress. Present limitations result from the availability of appropriate formulations of inks/dyes and fixation techniques. The comparable low production speed and limitations with regard to the quality of the textile material can be expected to be overcome within the next 5–10 years.

8.2.6 Finishing Processes

A great part of the variation in the final properties of a textile is adjusted for by finishing procedures, for example, wrinkle resistance, soil repellence, hydrophobic properties, flame retardance, and antimicrobial properties [69]. In many cases chemicals are added by padding/squeezing followed by drying/fixation, for example, in a stenter.

Representative groups of chemicals used are:

- urea-formaldehyde resins for crosslinking of cellulose textiles, e.g., dimethylol-dihydroxyethylene-urea (DMDHEU);
- dispersions of polymers (polyacrylesters, polyethylene, silicones);
- fluorocarbon compounds.

The applied products are fixed on the textile by drying/curing, but similar to the pad batch dyeing procedures, the last filling of the padding unit needs additional attention. A release of such concentrated finishing baths can introduce a COD of up to 200,000 mgO_2/L of liquor [70].

In a first attempt the volumes of residual baths have to be optimized and a reorganization of the recipes with regard to feed of residual excess volumes of a finishing bath into similar finishing recipes is recommended [71]. If reuse is not possible, a careful check of recipes with regard to easy biodegradation/bioelimination is necessary.

8.3 END-OF-PIPE TECHNIQUES

8.3.1 First Steps

The application of end-of-pipe technologies as general procedures for the treatment of wastewater has changed from simple procedures to sophisticated concepts, applying a consecutive set of methods that has been adapted to the particular situation of a textile plant [72]. As already discussed in the previous sections, the separation of concentrated wastes and the treatment of small volumes of concentrates are much more efficient compared to a global treatment of mixed wastes.

Numerous techniques and types of equipment have been developed and tested in laboratory tests, on a pilot scale, or in full technical application. The introduction of a technique is always coupled to a general wastewater treatment concept and has to consider the individual situation of a textile producer [73–75].

As a first step, a separation of different types of wastewater into the following groups is recommended:

- Concentrated liquids: fillings of padders (dyeing, finishing), printing pastes, used dyebaths;
- Medium polluted wastes (e.g., washing, rinsing baths);
- Low to zero polluted wastes (e.g., cooling water).

Basic general procedures applied are:

- Collection and mixing of released baths to level pH and temperature maxima in the final wastewater stream;
- Adjustment of pH by neutralization. Cellulose dyeing and finishing companies mainly release alkaline baths, which can be neutralized by introduction of CO_2-containing waste gas from the power/steam generation plant [76].

8.3.2 Overview

According to Schönberger and Kaps [3], the various methods for the treatment of wastewater from textile plants can be divided into the groups given in Table 11.

Table 11 Techniques for Wastewater Treatment

Separation, concentration	Decompositon, degradation	Exchange processes
Membrane techniques: Microfiltration, ultrafiltration (UF), nanofiltration (NF), reverse osmosis (RO)	Oxidation: Aerobic, wet oxidation, ozonation, peroxides (incl. Fenton's reagents), electrochemical oxidation	Ion-exchange
Mechanical Processes: Sedimentation, filtration	Incineration	
Evaporation	Reduction: chemical, electrochemical	
Precipitation, flocculation		
Flotation		
Adsorption		
Formation of inclusion complexes		
Extraction processes		
Stripping		

Source: Ref. 3.

The application of a certain technology for wastewater treatment is dependent on the type of wastewater, thus different technologies have been proposed and are applied at present. Normally a combination of procedures and equipment are applied and a big variety of concepts have been realized. To facilitate an overview of the different techniques, the most important processes are discussed in this section. Full concepts that are specialized to a distinct situation are given in the references [77–82]. Some of the techniques have already been discussed in Section 8.2.

8.3.3 Desizing, Pretreatment

The anaerobic biodegradation of sizes is favorable because the aerobic degradation of size-containing wastewater requires approximately 1 kWh/kg of BOD, while the anaerobic degradation yields 0.5–1.5 kWh/kg of BOD and in addition releases a lower volume of sludge. A general problem for biological treatment steps can be identified with the demand for a rather constant feed of load into the biological system to obtain constant conditions in microbial growth.

Theoretically, polymer-containing wastewater from desizing can be purified for water recycling by removal and reconcentration of the polymer by ultrafiltration or evaporation, but the high costs of investment and additional expenses for the disposal of the concentrate hinder the introduction of such techniques as a general treatment process.

For the degradation of polymers like PVA and carboxy-methyl-cellulose (CMC), low-pressure wet oxidation (5–20 bar, $<200^{\circ}C$) has been proposed [83]. In this process oxygen and a catalyst are used to destroy the organic material by oxidation.

The application of evaporation processes for purification and recycling of wastewater has been used in various concepts. The main problems that have to be considered are:

- energy consumption and heat recovery;
- incrustation and cleaning;
- corrosion;
- treatment of concentrated residues (e.g., incineration, disposal).

In many countries the disposal of the concentrated residues formed is rather complicated because this material has to be handled as hazardous waste.

The removal of fiber/yarn preparation during the pretreatment of knitted material can be identified as an important source of oil, grease, and silicones in wastewater. A general treatment can be performed by means of precipitation, flocculation, membrane filtration, and evaporation.

The removal of these components is required because these components are not biodegraded in the CWWT, but mainly adsorb on the sludge. When the sludge from the CWWT is used as fertilizer for farming, these components are transported to farmland and thus get released there. The reuse of bleach baths after catalase treatment has also been proposed in the literature [84].

8.3.4 Treatment of Wastewater from Dyeing Processes

The wastewater from dyeing processes contains a lot of components in various concentrations, for example, dyestuff, alkali, acid, salt, and auxiliaries [85]. In a first basic step, a separation of the wastewater stream according to the degree of chemical load should be performed.

A treatment of wastewater with low pollution for reuse can be achieved by the combination of:

- adjustment of pH and temperature;
- sedimentation, precipitation [86];

- flocculation ($Fe^{2+/3+}$, Al^{3+}, polyelectrolyte) [87];
- filtration (e.g., sand filter);
- adsorption (e.g., activated carbon) [79,88–93];
- ozone treatment.

In many cases the removal/destruction of the intensive color is the main goal to be achieved. Important techniques are given in the literature [66,94,95]:

Oxidative processes can be based on ozonation, UV treatment, hydrogen peroxide, and Fenton's reagent for the destruction of the chromopore [96–107].

Aerobic biodegradation processes often show unsatisfying results because a number of azo dyes are resistant to aerobic microbiological attack. The main process for removal of dyes in the aerobic part of a CWWT is based on an adsorption of the dyes on the biomass. Further problems in the destruction of chromophores result during the treatment of phthalocyanine dyes, anthraquinoid dyes, and vat and sulfur dyes, which contain rather persistent chromophores.

Reductive Processes. A reductive cleavage of the azo groups can be achieved by direct introduction of the dyes into the anaerobic step of a CWWT, but this method is restricted for heavy-metal-containing dyes, for example, phthalocyanine dyes, because of contamination of the sludge. In many cases the reductive destruction of colored dye baths is performed by the addition of reducing chemicals such as $Na_2S_2O_4$ and Fe^{2+} salts. As such processes generally lead to an increased load in the wastewater, such treatments should be replaced. The formation of aromatic amines as a result of the application of reducing conditions has to be considered in detail for every application.

Precipitation/Flocculation. Various chemicals can be added to textile wastewater to obtain precipitation/flocculation of colored substances:

- Addition of iron salt/$Ca(OH)_2$ is a rather simple and cheap method to form sludge, but the costs for separation and disposal of the sludge must be considered [108].
- Destabilization of the dissolved compounds by addition of iron or aluminum salts and addition of polyelectrolytes to support agglomeration and formation of larger size precipitation.

The removal of precipitate can be achieved by sedimentation, flotation, and filtration. If a recycling of water is intended, additional purification, for example, by adsorption methods, is needed to remove any added metal ions and flocculation auxiliaries.

At present these methods, which are based on the formation of a large amount of sludge containing substances of low/limited biodegradability, should only be used after careful optimization of the process conditions.

Membrane Processes [109,110]. Depending on the desired application, membrane techniques can be divided into:

- micro-, ultrafiltration (e.g., polymers, pressure $p = 1–10$ bar);
- nanofiltration (e.g., organic molecules, $p = 10–40$ bar);
- reverse osmosis (e.g., salt, $p = 10–80$ bar).

In the case of purification of water the permeate is the cleaned water and the removed components are collected in the concentrate [111–113]. Various modules can be used, such as plate-modules, tubes, and capillary modules. For water purification and recycling processes the following aspects have to be considered:

- high permeate flow;
- selectivity;
- stability and life-time of membrane and equipment;

- cleaning of membrane;
- tendency for membrane fouling;
- costs.

Today numerous membrane filtration units for removal of dissolved dyes such as reactive dyes are in full-scale operation. The treatment of the remaining concentrates still remains difficult. At present the following have been proposed in the literature and tested in full-scale operation:

- evaporation, incineration;
- anaerobic degradation;
- electrochemical reduction.

Electrochemical Processes. The reductive cleavage of azo-group-containing dyes has been applied on a full scale for the decolorization of concentrates from batch dyeing. Depending on the color, decolorization of up to 80% of the initial absorbance can be obtained. Mixed processes consist of combinations of electrochemical treatment and precipitation by use of dissolving electrodes [43,49]. Such techniques have been described in the literature and have, in part, also been tested on a full scale. Anodic processes that form chlorine from oxidation of chloride have also been proposed to destroy dyes, but care has to be taken with regard to the chlorine and chlorinated products (AOX) formed [114,115].

A special technique proposed in the literature for the removal of dyes is the inclusion of dye into cave molecules such as crown-ethers/cucurbituril, but developments with regard to regeneration and disposal of the crown ether have to be performed to permit introduction into full-scale application [116].

Adsorption processes and ion-pair extraction processes can also be used to remove color from wastewater [117–119]. The main problem to be solved in adsorption processes is the further treatment of the loaded adsorbents (regeneration, disposal). A similar situation is found in ion-pair extraction, where a concentrated organic phase results from the process and further treatment of this product is required.

Evaporation can also be used to purify wastewater, particularly in the case of heavy-metal-containing wastewater where a removal of the heavy metal ions is achieved, but again the problem of further treatment or disposal of the formed concentrated residue has to be solved [80].

In many cases combinations of the techniques are applied to obtain an optimized process fitting on the individual situation of the textile dyehouse, for example:

- nanofiltraton–oxidation processes;
- nanofiltration–evaporation–oxidation;
- evaporation–oxidation.

Another full-scale process combines catalytic oxidation including biodegradation, adsorption, precipitation/flocculation, and reverse osmosis [120].

8.3.5 Wastewater from Printing and Finishing Processes

The main difference in the wastes from dyeing processes is identified in the presence of thickeners and, in some cases, additional difficulties can arise from the added auxiliaries and hydrotropes (e.g., urea).

As a result, a high COD is found in the effluents and end-of-pipe technologies that form sludge have to face a high amount of precipitate.

In pigment printing the dyestuff pigments are bound to the textile by means of a polymer binder system and no additional washing is performed; however, wastewater is released from the cleaning of the equipment and machinery.

Printing pastes should be recycled whenever possible. Disposal is possible by incineration and biological degradation. Problems can arise in biodegradation from preservatives added to the pastes to avoid microbial growth and in cases of high formaldehyde and heavy metal content.

As a high number of different chemicals are applied in finishing processes, reuse is difficult in many cases. A high number of the used compounds show low biodegradability, so disposal is recommended in many cases. Techniques proposed in the literature include incineration, low-pressure wet oxidation [H_2O_2, Fe salt, NaOH, $Ca(OH)_2$], and precipitation by addition of high concentrations of Na_2SO_4 [121].

8.3.6 General Treatment Procedures

For the treatment of already mixed wastewater, various methods have been proposed and tested in full-scale application; examples are:

- Oxidation processes: oxidation in the presence of carbon particles and coupled precipitation [$FeSO_4$, $Ca(OH)_2$, polyelectrolyte] [37];
- Biological oxidation/degradation including sedimentation;
- Coupling of physical processes (flotation, sedimentation) [82,122];
- Aerobic/anaerobic biological degradation [123–133].

In some cases (particularly reactive dyes) dyes can pass the aerobic, anaerobic degradation step and colored water is observed at the end of the treatment. In such cases a special treatment of the colored wastewater (reduction, adsorption, precipitation) has to be introduced [105,134–137]. In the presence of low concentrations of organic compounds, ozonation can be used as a final "polishing" step.

NOMENCLATURE

AOX	adsorbable halogenated compounds
APEO	alkylphenol-ethoxylates, surfactants
B	factors for BOD from Table 7 (mg/g)
BOD	biological oxygen demand (mg/L)
C	factors for COD from Table 7 (mg/g)
c_D	concentration of hydrolyzed dyestuff in spent dyebath (kg/L)
CLY	lyocell fiber
CMC	carboxymethyl cellulose (size, thickener for printing)
CMD	modal fiber
Co	cotton fiber
COD	chemical oxygen demand (mg/L)
CV	viscose fiber
CWWT	communal wastewater treatment plant (e.g., combination of sedimentation, aerobic treatment, anaerobic treatment, nitrification, and elimination of phosphor)
DOC	dissolved organic carbon (mg/L)
EDTA	ethylene-diamine-tetra-acetic-acid (complexing agent)
LR	liquor ratio as volume of dyebath per mass of goods (L/kg)
L	losses, part of dyestuff released into the wastewater stream (dimensionless)

L_{BOD}	released load in BOD (kgO_2)
L_{COD}	released load in COD (kgO_2)
m	mass of desized fabric (kg)
m_{DF}	mass of dyestuff fixed on the fiber (kg)
m_F	mass of goods (kg)
m_D	mass of dyestuff in spent dyebath (kg)
NF	nanofiltration
NTA	nitrilotriacetic acid (complexing agent)
p	mass of size in fabric (kg/kg)
p_F	fixation of dyestuff in dyed material (kg/kg)
PVA	polyvinyl alcohol (type of size)
RO	reverse osmosis
UF	ultrafiltration

REFERENCES

1. Anon. *Verordnung über die Begrenzung von Abwasseremissionen aus Textilbe-trieben 2/14.*, BGBL 1992/612, *Regulation of the Ministry of Agriculture and Forestry*: Austria, 1992.
2. Müller, K.; Schönberger, H. Der Anhang 38 zur Abwasserverordnung. *Melliand Textil.* **2002**, *83*, 256–261.
3. Schönberger, H.; Kaps, U. *Reduktion der Abwasserbelastung in der Textilindustrie*, Umweltbundesamt; Berlin, 1994.
4. Schramm, W.; Jantschgi, J. Comparative assessment of textile dyeing technologies from a preventive environmental protection point of view. *J. Soc. Dyers Colour* **1999**, *115*, 130–135.
5. Naylor, C.G. Environmental fate and safety of nonylphenol ethoxylates. *Text. Chem. Color.* **1995**, *27*, 29–33.
6. Glober, B.; Lorrain H. Waste minimisation in the dyehouse. *Text. Chem. Color.* **1993**, *25*, 15–20.
7. Shah, II.A.; Sharma, M.A.; Doshi, S.M.; Pillay, G.R. Practical approach towards energy conservation, economy and effluent control in a process-house. *Colourage* **1989**, *15*, 20c.
8. Müller, B. Adsorbable organic halogens in textile effluents. *Rev. Prog. Coloration* **1992**, *22*, 14–21.
9. Anon. *Richtline 96/61/EG des Rates vom 24. Sept. 1996 über die integrierte Vermeidung und Verminderung der Umweltverschmutzung*, Amtsblatt der Europ. Gemeinschaften, L 257/26, 10.10.1996, 26–40.
10. Anon. World production: 59 million tons textile fibres. In *Man-Made Fibre Year Book*; IBP, International Business Press Publisher: Frankfurt a.M., Germany, August 2002; 21.
11. Gibson, M.D.M.; Morgan, W.V.; Robinson, B. Aspekte der Wollwäsche und Abwasseraufbereitung. *Textil Praxis Int.*, **1979**, **34**, 437–444, 701–708.
12. Hoffmann, R.; Timmer, G. Abwasserreinigung der BWK – Ergebnisse nach Optimierung des Gesamtsystems. *Melliand Textil.* **1994**, *75*, 831–837.
13. Hoffmann, R.; Timmer, G. Abwasserreinigung der BWK – eine Lösung für Wollwäschereien und kämmereien. *Melliand Textil.* **1991**, *72*, 562–566.
14. Heiz, H. Chlor/Hercosett-Ausrüstung von Wolle. *Textilveredlung*, **1981**, *16*, 43–53.
15. Duffield, P.A.; Holt, R.R.D.; Smith, J.R. Färben mit geringem Restchromgehalt im Abwasser. *Melliand Textil.* **1991**, *72*, 938–942.
16. Doser, C.; Zschocke, P.; Biedermann, J.; Süssmuth, R.; Trauter, J. Mikrobieller Abbau von Polyacrylsäureschlichten. *Textilveredlung* **1997**, *32*, 245–249.
17. Klein, M.; Zschocke, P.; Süssmuth, R.; Trauter, J. Über den anaeroben biologischen Abbau der Polyvinylalkohole. *Textilveredlung* **1997**, *32*, 241–245.
18. Rüttiger, W.; Schenk, W.; Würz, A. Recycling-Verfahren für Schlichtemittel. *Chem-Ing-Tech.* **1983**, *33*, 490–494.

19. Deschler, O. Schlichte-Rückgewinnung und Wiederverwendung – Resumée nach 5 Jahren Praxis. *Melliand Textil.* **1983**, *64*, 716–720.
20. Hoechst AG. German Patent Application, DAS 2808920, 02 March 1978.
21. BASF AG. German Patent Application, DAS 2543815, 01 October 1975.
22. Rüttiger, W. Abwasserentlastung durch Rückgewinnung und Aufkonzentrierung von Acrylat-Schlichten. *Textil Praxis Int.*, **1983**, *38*, 975–977, 1117–1121.
23. Byazeed, A.; Trauter, J. Untersuchungen von Veränderungen der physikalischen und technologischen Eigenschaften wasserlöslicher Schlichtemittel bei der Ultrafiltration. *Textil Praxis Int.*, **1992**, *47*, 220–229.
24. Trauter, J. Anwendung der Ultrafiltration für das Schlichtemittel- und Indigo-Recycling. *Melliand Textil.* **1993**, *74*, 559–563.
25. Olip, V. Peressigsäure eine Bleichchemikalie. *Melliand Textil.* **1992**, *73*, 819–822.
26. Wurster, P. Die Peressigsäurebleiche – eine Alternative zu Bleichverfahren mit halogenhaltigen Oxidationsmitteln. *Textil Praxis Int.* **1992**, *47*, 960–965.
27. Cheng, K.M. An improvement in effluent disposal with emphasis on cotton pretreatment processes. *Text. Chem. Color.* **1998**, *30* (3), 15–21.
28. Bechtold, T.; Gmeiner, D.; Burtscher, E.; Bösch, I.; Bobleter, O. Flotation of particles suspended in lye by the decomposition of hydrogen peroxide. *Separ. Sci. Technol.* **1989**, *24*, 441–451.
29. Galda, K. Rückgewinnung von Mercerisierlauge. *Melliand Textil.* **1998**, *79*, 38–39.
30. Son, E.J.; Choe, E.K. Nanofiltration membrane technology for caustic soda recovery. *Text. Chem. Color. Am. D.* **2000**, *32*, 46–52.
31. Bechtold, T.; Burtscher, E.; Sejkora, G.; Bobleter, O. Modern methods of lye recovery. *Int. Textil Bulletin* **1985**, *31*, 5–26.
32. Baughman, G.L. Fate of copper in copperized dyes during biological waste treatment. I: Direct dyes. *Textile Chem. Color. Am. D.* **2000**, *32* (1), 51–55.
33. Anon. ATV-Arbeitsgruppe 7.2.23, Abwässer in der Textilindustrie – Arbeitsbericht der ATV-Arbeitsgruppe 7.2.23 "Textilveredlungsindustrie". *Korrespondenz Abwasser* **1989**, *36*, 1074–1084.
34. *Ecological and toxicological association of the dyestuff manufacturing industry (ETAD). E 3022 Environmental hazard and risk assessment in the context of EC directive 79/831/* from 31 May 1991.
35. Beckmann, W.; Sewekow, U. Farbige Abwasser aus der Reaktiv-färberei: Probleme und Wege zur Lösung. *Textil Praxis Int.* **1991**, *46*, 346–348.
36. Kwok, W.Y.; Xin, J.H.; Sin, K.M. Quantitative prediction of the degree of pollution of effluent from reactive dye mixtures. *Color. Technol.* **2002**, *118*, 174–180.
37. Marmagne, O.; Coste, C. Color removal from textile plant effluents. *Am. Dyest. Rep.* **1966**, *85*, 15–21.
38. Tsui, L.S.; Roy, W.R.; Cole M.A. Removal of dissolved textile dyes from wastewater by a compost sorbent. *J. Soc. Dyers Colour.* **2003**, *119*, 14–18.
39. Tokuda, J.; Ohura, R.; Iwasaki, T.; Takeuchi, Y.; Kashiwada, A.; Nango, M. Decoloration of azo dyes by hydrogen catalyzed by water-soluble manganese porphyrins. *Textile Res. J.* **1999**, *69*, 956–960.
40. Gregor, K.H. Aufbereitung und Wiederverwertung von Färbereiabwässern. *Melliand Textil.* **1998**, *79*, 643–646.
41. Perkins, W.S.; Walsh, W.K.; Reed, I.E.; Namboodri, C.G. A demonstration of reuse of spent dyebath water following color removal with ozone. *Text. Chem Color.* **1995**, *28*, 31–37.
42. Yoshida, Y.; Ogata, S.; Nakamatsu, S.; Shimamune, T.; Kikawa, K.; Inoue, H.; Iwakura, C. Decoloration of azo dye using atomic hydrogen permeating through a Pt-modified palladized Pd sheet electrode. *Electrochim. Acta* **1999**, *45*, 409–414.
43. Laschinger, M. Implementing dyehouse wastewater treatment systems. *Am. Dyest. Rep.* **1996**, *85*, 23–27.
44. Kolb, M.; Korger, P.; Funke, B. Entfärbung von textilem Abwasser mit Dithionit. *Melliand Textil.* **1988**, *69*, 286–287.

45. McClung, S.M.; Lemley, A.T. Electrochemical treatment and HPLC analysis of wastewater containing acid dyes. *Text. Chem. Color.* **1994**, *26*, 17–22.
46. Elgal, G.M. Recycling and disposing of dyebath solutions. *Text. Chem. Color.* **1986**, *18*, 15–20.
47. Burtscher, E.; Bechtold, T.; Amann, A.; Turcanu, A.; Schramm, C.; Bobleter, O. Aspekte der Teilstrombehandlung Textiler Abwässer unter besonderer Berücksichtigung der Farbigkeit. *Melliand Textil.* **1993**, *74*, 903–907.
48. Marzinkowski, J.M.; van Clewe, B. Wasserkreislaufführung durch Membranfiltration der farbigen Abwässer. *Melliand Textil.* **1998**, *79*, 174–177.
49. Wilcock, A.; Brewster, M.; Tincher, W. Using electrochemical technology to treat textile wastewater: three case studies. *Am. Dyest. Rep.* **1992**, *81*, 15–22.
50. Diaper, C.; Correia, V.M.; Judd, S.J. The use of membranes for the recycling of water and chemicals from dyehouse effluents: an economic assessment. *J. Soc. Dyers Colour.* **1996**, *112*, 273–281.
51. Knapp, J.S.; Zhang, F.; Tapely, K.N. Decolourisation of Orange II by a wood-rotting fungus. *J. Chem. Technol. Biot.* **1997**, *69*, 289–296.
52. Oxspring, A.A.; McMullan, G.; Smyth, W.F.; Marchant, R. Decolourisation and metabolism of the reactive textile dye, Remazol Black 5, by an immobilized microbial consortium. *Biotechnol. Lett.* **1996**, *18*, 527–530.
53. Nigam, P.; McMullan, G.; Banat, I.M.; Marchant, R. Decolourisation of effluent from the textile industry by a microbial consortium. *Biotechnol. Lett.* **1997**, *18*, 117–120.
54. Bechtold, T.; Burtscher, E.; Turcanu, A. Cathodic decolourisation of textile waste water containing reactive dyes using a multi-cathode electrolyser. *J. Chem Technol. Biot.* **2001**, *76*, 300–311.
55. Bechtold, T.; Mader, J.; Mader, C. Entfärbung von Reaktivfarbstoffen durch kathodische Reduktion. *Melliand Textil.* **2002**, *83*, 361–364.
56. Bechtold, T.; Burtscher, E.; Bobleter, O. Application of electrochemical processes and electroanalytical methods in textile chemistry. *Curr. Topics Electrochem.* **1998**, *6*, 97–110.
57. Bechtold, T.; Burtscher, E.; Kühnel, G.; Bobleter, O. Electrochemical processes in indigo dyeing. *J. Soc. Dyers Colour.* **1997**, *113*, 135–144.
58. DyStar/BASF Technical information Bulletin, TI/T 245e, 1994.
59. Rümmele, W. Lösung des Abwasserproblems in einer Indigofärberei. *Textilveredlung* **1989**, *24*, 96–97.
60. Denter, U.; Schollmeyer, E. Einsatz von mit Alkali- und Erdalkalisalzen belastetem Prozeßwasser in der Vorbehandlung von Textilien. *Textil Praxis Int.* **1991**, *46*, 644–646, 1343–1348.
61. Richter, P. Möglichkeiten zur Abwasserentlastung beim Färben von Polyesterfasern. *Melliand Textil.* **1993**, *74*, 872–875.
62. Bach, E.; Cleve, E.; Schüttken, J.; Schollmeyer, E.; Rucker, J.W. Correlation of solubility data of azo disperse dyes with the dye uptake of poly(ethylene terephthalate) fibres in supercritical carbon dioxide. *Color. Technol.* **2001**, *117*, 13–18.
63. Schmidt, A.; Bach, E.; Schollmeyer, E. Supercritical fluid dyeing of cotton modified with 2,4,6-trichloro-1,3,5-triazine. *J. Soc. Dyers Colour.* **2003**, *119*, 31–36.
64. Schönberger, H. Reduzierung der Abwasserbelastung in Textildruckereien durch produktionsintegrierte Massnahmen. *Textilveredlung* **1994**, *29*, 128–133.
65. Provost, J.R. Effluent improvement by source reduction of chemicals used in textile printing. *J. Soc. Dyers Colour.* **1992**, *108*, 260–264.
66. Pierce, J. Colour in textile effluents – the origins of the problem. *J. Soc. Dyers Colour.* **1994**, *110*, 131–133.
67. Brocks, J. Minimierung der Abwasserbelastung in der Druckerei. *Textil Praxis Int.* **1992**, *47*, 550–554.
68. Marte, W.; Meyer, U. Verdicker-Recyclierung im Textildruck. *Textil-veredlung* **1995**, *30*, 64–68.
69. Roberts, D.L.; Hall, M.E.; Horrocks, A.R. Environmental aspects of flame-retardant textiles – an overview. *Rev. Prog. Color.* **1992**, *22*, 48–57.
70. Puk, R.; Sedlak, D. Ein Konzept zur Minimierung von Restflotten. *Textil Praxis Int.* **1992**, *47*, 238–245.

71. Teichmann, R. Methodik zum Wiedereinsatz von konzentrierten Ausrüstungsmittel-Restflotten. *Textilveredlung* **1997**, *32*, 131–135.
72. Coia-Ahlman, S.; Groff, K.A. Textile wastes. *Res. J. Water Pollut. C.* **1990**, *62*, 473–478.
73. Höhn, W.; Reinigungmöglichkeiten für textile Abwässer im Überblick. *Melliand Textil.* **1998**, *79*, 647–649.
74. Park, J.; Shore, J. Water for the dyehouse: Supply, consumption, recovery and disposal. *J. Soc. Dyers Colour.* **1984**, *100*, 383–399.
75. Athanasopoulos, N. Abwässer bei der Veredlung von Baumwolle und Baumwollmischungen – von der Versuchsanlage zur Betriebsanlage. *Melliand Textil.* **1990**, *71*, 619–628.
76. Anon. Neutralisation alkalischer Abwässer mit Rauchgas. *Chemie-fasern/Textilindustrie* **1976**, *78*, 924.
77. Wilking, A.; Frahne, D. Textilabwasser – Behandlungsverfahren der 90er Jahre. *Melliand Textil.* **1993**, *74*, 897–900.
78. Sharma, M.A. Treatment of cotton textile mills effluent – a case study. *Colourage* **1989**, *36*, 15–21.
79. Schulze-Rettmer, R. Treatment of textile dyeing wastewater by adsorption/bio-oxidation process. *Text. Chem. Color.* **1998**, *30*, 19–23.
80. Richarts, F. Integrales Konzept der Ver- und Entsorgung von Textilveredlungsbetrieben. *Textil Praxis Int.* **1991**, *46*, 567–572.
81. Mukherjee, A.K.; Bhuvanesh, G.; Chowdhury, S.M.S. Separation dyes from cotton dyeing effluent using cationic polyelectrolytes. *Am. Dyest. Rep.* **1999**, *88*, 25–28.
82. Reid, R. On-site colour removal at Courtaulds Textiles. *J. Soc. Dyers Colour.* **1996**, *112*, 140–141.
83. Horak, O. Katalyische Naßoxidation von biologisch schwer abbaubaren Abwasserinhaltsstoffen unter milden Reaktionsbedingungen. *Chem. Ing. Tech.* **1990**, *62*, 555–557.
84. Tzanov, T.; Costa, S.; Guebitz, G.; Cavaco-Paulo, A. Effect of temperature and bath composition on the dyeing of cotton with catalase treated bleaching effluent. *J. Soc. Dyers Colour.* **2001**, *117*, 116–170.
85. Sewekow, U. Färbereiabwässer – behördliche Anforderungen und Problemlösungen. *Melliand Textil.* **1989**, *70*, 589–596.
86. Fiola, R.; Luce, R. Wastewater solutions for the blue jeans processing industry. *Am. Dyest. Rep.* **1998**, *87*, 54–55.
87. Papic, S.; Koprivanac, N.; Bozic, A.L. Removal of reactive dyes from wastewater using Fe(III) coagulant. *J. Soc. Dyers Colour.* **2000**, *116*, 352–359.
88. Netpradit, S.; Thiravetyan, P.; Towprayoon S. Application of "waste" metalhydroxide sludge for adsorption of azo reactive dyes. *Water Res.* **2003**, *37*, 763–772.
89. Kadirvelu, K.; Kavipriya, M.; Karthika, C.; Radhika, M.; Vennilamani N.; Pattabhi S. Utilization of various agricultural wastes for activated carbon preparation and application for the removal of dyes and metal ions from aqueous solutions. *Biores. Technol.* **2003**, *87*, 129–132.
90. Sun, Q.; Yang, L. The adsorption of basic dyes from aqueous solution on modified peat-resin particle. *Water Res.* **2003**, *37*, 1535–1544.
91. Puranik, S.A.; Rathi, A.K.A. Treatment of wastewater pollutant from direct dyes. *Am. Dyest. Rep.* **1999**, *88*, 42–50.
92. Gärtner, R.; Müller, W.; Schulz, G.; Lehr, T. Neue Sorptionsmaterialien auf Basis speziell aufbereiteter Polyamidabfälle als Adsorptivreiniger für Färbereiabwässer. *Melliand Textil.* **1996**, *77*, 67–72.
93. Laszo, J.A. Preparing an ion exchange resin from sugarcane bagasse to remove reactive dye from wastewater. *Text. Chem. Color.* **1996**, *85*, 13–17.
94. Beckmann, W.; Sewekow, U. Farbige Abwasser aus Reaktivfärberei: Probleme und Wege zur Lösung. *Textil Praxis Int.* **1991**, *46*, 445–449.
95. Cooper, P. Removing colour from dyehouse waste waters – a critical review of technology available. *J. Soc. Dyers Colour.* **1993**, *109*, 97–100.
96. Hickman, W.S. Environmental aspects of textile processing. *J. Soc. Dyers Colour.* **1993**, *109*, 32–37.
97. Huang, C.R.; Lin, Y.K.; Shu, H.Y. Wastewater decolorization and toc-reduction by sequential treatment. *Am. Dyest. Rep.* **1994**, *83*, 15–18.

98. Arslan-Alaton, I.; Balcioglu, I.A. Heterogenous photocatalytic treatment of dyebath wastewater in a TFFB reactor. *AATCC Review* **2002**, *2*, 33–36.
99. Hassan, M.M.; Hawkyard, C.J. Reuse of spent dyebath following decolorisation with ozone. *J. Soc. Dyers Colour.* **2002**, *118*, 104–111.
100. Fung, P.C.; Sin, K.M.; Tsui, S.M. Decolorisation and degradation kinetics of reactive dye wastewater by a UV/ultrasonic/peroxide system. *J. Soc. Dyers Colour.* **2000**, *116*, 170–173.
101. Tokuda, J.; Oura, R.; Iwasaki, T.; Takeuchi, Y.; Kashiwada, A.; Nango, M. Decolorisation of azo dyes with hydrogen peroxide catalyzed by manganese protoporphyrins. *J. Soc. Dyers Colour.* **2000**, *116*, 42–47.
102. Perkins, W.S. Oxidative decolorisation of dyes in aqueous medium. *Text. Chem. Color. Am. D.* **1999**, *1*, 33–37.
103. Ferrero, F. Oxidative degradation of dyes and surfactant in the Fenton and photo-Fenton treatment of dyehouse effluents. *J. Soc. Dyers Colour.* **2000**, *116*, 148–153.
104. Yang, Y.; Wyat, D.T.; Bahorsky, M. Decolorisation of dyes using UV/H_2O_2 photochemical oxidation. *Text. Chem. Color.* **1998**, *30*, 27–35.
105. Strickland, A.F.; Perkins, W.S. Decolorisation of continuous dyeing wastewater by ozonation. *Text. Chem. Color.* **1995**, *27*, 11–15.
106. Uygur, A. An overview of oxidative and photooxidative decolorisation treatments of textile waste waters. *J. Soc. Dyers Colour.* **1997**, *113*, 211–217.
107. Namboodri, C.G.; Walsh, W.K. Ultraviolet light/hydrogen peroxide system for decolorizing spent reactive waste water. *Am. Dyest. Rep.* **1996**, *85*, 15–25.
108. Kolb, M.; Funke, B.; Gerber, H.-P.; Peschen, N. Entfärbung von Abwasser aus Textilbetrieben mit Fe(II)/$Ca(OH)_2$. *Korrespondenz Abwasser* **1987**, *34*, 238–241.
109. Wehlmann, U. Reinigen von Abwasser aus der Textilveredlung mit Membranverfahren. *Melliand Textil.* **1997**, *78*, 249–252.
110. Majewska-Nowak, K.; Winnicki, T.; Wisniewski, J. Effect of flow conditions on ultrafiltration efficiency of dye solutions and textile effluents. *Desalination* **1989**, *71*, 127–135.
111. Marzinkowski, J.M.; van Clewe, B. Wasserkreislaufführung durch Membranfiltration der farbigen Abwässer. *Melliand Textil.* **1998**, *79*, 174–177.
112. Diaper, C.; Correia, V.M.; Judd, S.J. The use of membranes for the recycling of water and chemicals from dyehouse effluents: an economic assessment. *J. Soc. Dyers Colour.* **1996**, *112*, 270–280.
113. Schäfer, T.; Trauter, J.; Janitza, J. Aufarbeitung von Färbereiabwässern durch Nanofiltration. *Textilveredlung* **1997**, *32*, 79–83.
114. Vilaseca, M.M.; Gutierrez, M.C.; Crespi, M. Biologische Abbaubarkeit von Abwässern nach elektrochemischer Behandlung. *Melliand Textil.* **2002**, *83*, 558–560.
115. Brincell INC. 2109 West 2300 South, Salt Lake City, Utah.
116. Buschmann, H.-J.; Schollmeyer, E. Selektive Abtrennung von Schwermetallkationen aus Färbereiabwässern. *Melliand Textil.* **1991**, *72*, 543–544.
117. Buschmann, H.-J.; Schollmeyer, E. Die Entfärbung von textilem Abwasser durch Bildung von Farbstoffeinschlußverbindungen. *Textilveredlung* **1998**, *33*, 44–47.
118. Smith, B.; Konce, T.; Hudson, S. Decolorizing dye wastewater using chitosan. *Am. Dyest. Rep.* **1993**, *82*, 18–36.
119. Steenken-Richter, I.; Kermer, W.D. Decolorising textile effluents. *J. Soc. Dyers Colour.* **1992**, *108*, 182–186.
120. Anon. Reinigung und Teil-Wiederverwendung von Maschenwaren-Abwässern mit Hilfe von Kokskohle-Adsorbens. *Textil Praxis Int.* **1986**, *41*, 943–949.
121. Janitza, J.; Koscielski, S. Reinigung und Wiederverwertung von Druckereiabwässern. *Int. Text. Bull.* **1996**, *4/96*, 28–32.
122. Glöckler, R. Optimiertes Textilabwasserreinigungsverfahren. *Melliand Textil.* **1995**, *76*, 1020–1021.
123. Yang, Q.; Yang, M.; Pritsch, K.; Yediler, A.; Hagn, A.; Schloter, M.; Kettrup, A. Decolorization of synthetic dyes and production of manganese-dependent peroxidase by new fungal isolates. *Biotechnol. Lett.* **2003**, *25*, 709–713.

124. Abadulla, E.; Robra, K.; Gübitz, G.M.; Silva, L.M.; Cavaco-Paulo, A. Enzymatic decolorisation of textile dyeing effluents. *Textile Res. J.* **2000**, *70*, 409–414.
125. Churchley, J.H.; Greaves, A.J.; Hutchings, M.G.; Phillips, D.A.S.; Taylor, J.A. A chemometric approach to understanding the bioelimination of anionic, water-soluble dyes by a biomass – Part 3: Direct dyes. *J. Soc. Dyers Colour.* **2000**, *116*, 279–284.
126. Goncalves, I.M.C.; Gomes, A.; Bras, R.; Ferra, M.I.A.; Amorim, M.T.P.; Porter, R.S. Biological treatment of effluent containing textile dyes. *J. Soc. Dyers Colour.* **2000**, *116*, 393–397.
127. Metosh-Dickey, C.; Davis, T.M.; McEntire, C.A.; Christopher, J.; DeLoach, H.; Portier, R.J. COD, color, and sludge reduction using immobilized bioreactor technology. *Textile Chem. Color. Am. D.* **2000**, *32*, 28–31.
128. Moreira, M.T.; Mieglo, I.; Feijoo, G.; Lema, J.M. Evaluation of different fungal strains in the decolourisation of synthetic dyes. *Biotechnol. Lett.* **2000**, *22*, 1499–1503.
129. Willmott, N.; Guthrie, J.; Nelson, G. The biotechnology approach to colour removal from textile effluent. *J. Soc. Dyers Colour.* **1998**, *114*, 38–41.
130. Cao, H.; Hardin, I.R.; Akin D.E.; Optimization of conditions for microbial decolorisation of textile wastewater: starch as a carbon source. *AATCC Rev.*, **2001**, *7*, 37–42.
131. Beckert, M.; Ohmann, U.; Platzer, B.; Bäuerle, U.; Burkert, G. Langzeiterfahrungen beim Reinigen von Abwässern nach dem SB-Verfahren. *Melliand Textil.* **2000**, *81*, 64–68.
132. Feitkenhauer, H.; Meer, U.; Marte, W.; Integration der anaeroben Abwasservorbehandlung in das Wassermanagementkonzept. *Melliand Textil.* **1999**, *80*, 303–306.
133. Gähr, F.; Lehr, T. Verbesserung der anaeroben Abbaubarkeit von Teilströmen aus der Textilveredlung durch Ozonbehandlung. *Textilveredlung* **1997**, *32*, 70–73.
134. Mock, B.; Hamouda, H. Ozone application to color destruction of industrial wastewater – Part I: Experimental. *Am. Dyest. Rep.* **1998**, *87*, 18–22.
135. Streibelt, H.P. Abwasser-Reinigung und -Wiederverwertung in der Textilveredlung. *Chemiefasern/Textilindustrie.* **1986**, *36/88*, 401–402.
136. Küßner, J.; Janitza, J.; Koscielski, S. Entfärbung, Reinigung und Wiederverwertung von farbigen Abwässern der Textilveredlungsindustrie und umweltgerechte Entsorgung der Reinigungsneben-produkte. *Textil Praxis Int.* **1992**, *47*, 736–741.
137. Wragg, P. Waste water recycling – a case study. *J. Soc. Dyers Colour.* **1993**, *109*, 280–282.

9
Treatment of Phosphate Industry Wastes

Constantine Yapijakis
The Cooper Union, New York, New York, U.S.A.

Lawrence K. Wang
Lenox Institute of Water Technology and Krofta Engineering Corporation, Lenox, Massachusetts and Zorex Corporation, Newtonville, New York, U.S.A.

9.1 INTRODUCTION

The phosphate manufacturing and phosphate fertilizer industry includes the production of elemental phosphorus, various phosphorus-derived chemicals, phosphate fertilizer chemicals, and other nonfertilizer phosphate chemicals [1–30]. Chemicals that are derived from phosphorus include phosphoric acid (dry process), phosphorus pentoxide, phosphorus pentasulfide, phosphorus trichloride, phosphorus oxychloride, sodium tripolyphosphate, and calcium phosphates [8]. The nonfertilizer phosphate production part of the industry includes defluorinated phosphate rock, defluorinated phosphoric acid, and sodium phosphate salts. The phosphate fertilizer segment of the industry produces the primary phosphorus nutrient source for the agricultural industry and for other applications of chemical fertilization. Many of these fertilizer products are toxic to aquatic life at certain levels of concentration, and many are also hazardous to human life and health when contact is made in a concentrated form.

9.1.1 Sources of Raw Materials

The basic raw materials used by the phosphorus chemicals, phosphates, and phosphate fertilizer manufacturing industry are mined phosphate rock and phosphoric acid produced by the wet process.

Ten to 15 million years ago, many species of marine life withdrew minute forms of phosphorus dissolved in the oceans, combined with such substances as calcium, limestone, and quartz sand, in order to construct their shells and bodies [30]. When these multitudes of marine organisms died, their shells and bodies (along with sea-life excretions and inorganic precipitates) settled to the ocean bottom where thick layers of such deposits – containing phosphorus among other things – were eventually formed. Land areas that formerly were at the ocean bottom millions of years ago and where such large deposits have been discovered are now being commercially mined for phosphate rock. About 70% of the world supply of phosphate rock comes from such an area around Bartow in central Florida, which was part of the Atlantic Ocean 10 million years ago [1]. Other significant phosphate rock mining and processing operations can be found in Jordan, Algeria, and Morocco [28].

9.1.2 Characteristics of Phosphate Rock Deposits

According to a literature survey conducted by Shahalam [28], the contents of various chemicals found in the natural mined phosphate rocks vary widely, depending on location, as shown in Table 1. For instance, the mineralogical and chemical analyses of low-grade hard phosphate from the different mined beds of phosphate rock in the Rusaifa area of Jordan indicate that the phosphates are of three main types: carbonate, siliceous, and silicate-carbonate. Phosphate deposits in this area exist in four distinct layers, of which the two deepest – first and second (the thickness of bed is about 3 and 3.5 m, respectively, and depth varies from about 20 to 30 m) – appear to be suitable for a currently cost-effective mining operation. A summary of the data from chemical analyses of the ores is shown in Table 2 [28].

Screen tests of the size fraction obtained from rocks mined from these beds, which were crushed through normal crushers of the phosphate processing plant in the area, indicated that the best recovery of phosphate in the first (deepest) bed is obtained from phosphate gains recovered at grain sizes of mesh 10–20 (standard). The high dust (particles of less than 200 mesh) portion of 11.60% by wt. of the ores remains as a potential air pollution source; however, the chemical analyses of these ores showed that crushing to smaller grain sizes tends to increase phosphate recovery. The highest percentage of phosphate from the second bed (next deepest) is also recovered from grain sizes of 10–20 mesh; however, substantial amounts of phosphate are also found in sizes of 40–100 mesh. Currently, the crushing operation usually maintains a maximum grain size between 15 and 30 mesh.

The phosphate rock deposits in the Florida region are in the form of small pebbles embedded in a matrix of phosphatic sands and clays [31]. These deposits are overlain with lime

Table 1 Range of Concentrations of Various Chemicals in Phosphate Ores

Chemical	Range
Fluorine	2.8–5.6%[a]
Sulphur (SO_3)	0.8–7.52%[a]
Carbon (CO_2)	2.07–10.7%[a]
Strontium	180–1683 ppm[b]
Manganese	0.001–0.004%[a]
Barium	0.044–0.40%[a]
Chlorine	0.20–1.42%[a]
Zinc	59–765 ppm[b]
Nickel	7–244 ppm[b]
Cobalt	31–34 ppm[b]
Chromium	12–895 ppm[b]
Copper	18–46 ppm[b]
Vanadium	0.03–0.08%[a]
Cadmium	0.038–1.5 ppm[b]
Uranium	4–8 ppm[b]
P_2O_5	40–55%[c]
Silica	3–34%[c]
Carbon (C)	14–48%[c]

[a] % by wt.
[b] Parts per million.
[c] Kusaifa Rocks only (% by wt.).
Source: Ref. 28.

Table 2 Chemical Analysis of Different Size Fractions of Phosphate in Mining Beds at Rusaifa

Size fractions in mesh	First bed (average chemical composition)			Second bed (average chemical composition)			Fourth bed (average chemical composition)		
	P_2O_5%	$CaCO_3$%	Insoluble%	P_2O_5%	$CaCO_3$%	Insoluble%	P_2O_5%	$CaCO_3$%	Insoluble%
<10	21.35	10.01	36.46	15.73	35.64	23.35	17.99	47.95	2.08
10–20	21.03	10.67	36.94	18.04	32.60	21.26	21.33	36.58	2.25
20–30	21.03	10.46	37.30	19.63	29.23	19.82	26.95	30.81	2.48
30–40	21.82	10.56	36.21	21.80	25.24	17.72	29.38	28.51	2.00
40–60	22.04	10.65	33.70	25.96	22.40	14.78	32.26	20.66	2.49
60–100	24.12	11.27	29.64	26.65	18.13	11.75	30.88	26.76	1.23
100–150	24.32	10.90	28.64	26.76	21.71	13.17	26.55	33.51	1.41
150–200	25.92	11.95	24.30	24.46	23.18	14.45	23.18	40.25	2.52
>200	25.28	12.50	23.42	23.37	28.06	12.99	21.29	40.25	4.24
Average total sample	Chemical					25.19	34.21	2.32	
Mineralogical	47.0	31.0[a]	18.0[b]	27.0[a]	50.0	15.0[b]	55.0	35.0[a]	3.5[b]

[a] Carbonaceous materials.
[b] Silica.
Source: Ref. 28.

rock and nonphosphate sands and can be found at depths varying from a few feet to hundreds of feet, although the current economical mining operations seldom reach beyond 18.3 m (60 ft) of depth.

9.1.3 Mining and Phosphate Rock Processing

Mechanized open-cut mining is used to first strip off the overburden and then to excavate in strips the exposed phosphate rock bed matrix. In the Rusaifa area of Jordan, the stripping ratio of overburden to phosphate rock is about 7 : 1 by wt. [28]. Following crushing and screening of the mined rocks in which the dust (less than 200 mesh) is rejected, they go through "beneficiation" processing. The unit processes involved in this wet treatment of the crushed rocks for the purpose of removing the mud and sand from the phosphate grains include slurrification, wet screening, agitation and hydrocycloning in a two-stage operation, followed by rotating filtration and thickening, with a final step of drying the phosphate rocks and separating the dusts. The beneficiation plant makes use of about 85% of the total volume of process water used in phosphate rock production.

Phosphate rocks from crushing and screening, which contain about 60% tricalcium phosphate, are fed into the beneficiation plant for upgrading by rejection of the larger than 4 mm over-size particles. Two stages of agitation follow the hydrocycloning, the underflow of which (over 270 mesh particles) is fed to rotary filters from which phosphatic cakes results (with 16–18% moisture). The hydrocyclone overflow contains undesirable slimes of silica carbonates and clay materials and is fed to gravity thickeners. The thickener underflow consisting of wastewater and slimes is directly discharged, along with wastewater from dust-removing cyclones in the drying operation, into the nearby river.

In a typical mining operation in Florida, the excavated phosphate rockbed matrix is dumped into a pit where it is slurrified by mixing it with water and subsequently carried to a washer plant [31]. In this operation, the larger particles are separated by the use of screens, shaker tables, and size-separation hydrocyclone units. The next step involves recovery of all particles larger than what is considered dust, that is, 200 mesh, through the use of both clarifiers for hydraulic sizing and a flotation process in which selective coating (using materials such as caustic soda, fuel oil, and a mixture of fatty acids and resins from the manufacture of chemical wood pulp known as tall oil, or resin oil from the flotation clarifier) of phosphate particles takes place after pH adjustment with NaOH.

The phosphate concentration in the tailings is upgraded to a level adequate for commercial exploitation through removal of the nonphosphate sand particles by flotation [32], in which the silica solids are selectively coated with an amine and floated off following a slurry dewatering and sulfuric acid treatment step. The commercial quality, kiln-dried phosphate rock product is sold directly as fertilizer, processed to normal superphosphate or triple superphosphate, or burned in electric furnaces to produce elemental phosphorus or phosphoric acid, as described in Section 9.2.

9.2 INDUSTRIAL OPERATIONS AND WASTEWATERS

The phosphate manufacturing and phosphate fertilizer industry is a basic chemical manufacturing industry, in which essentially both the mixing and chemical reactions of raw materials are involved in production. Also, short- and long-term chemical storage and warehousing, as well as loading/unloading and transportation of chemicals, are involved in the operation. In the case of fertilizer production, only the manufacturing of phosphate fertilizers and mixed and blend fertilizers containing phosphate along with nitrogen and/or potassium is presented here.

Regarding wastewater generation, volumes resulting from the production of phosphorus are several orders of magnitude greater than the wastewaters generated in any of the other product categories. Elemental phosphorus is an important wastewater contaminant common to all segments of the phosphate manufacturing industry, if the phossy water (water containing colloidal phosphorus) is not recycled to the phosphorus production facility for reuse.

9.2.1 Categorization in Phosphate Production

As previously mentioned, the phosphate manufacturing industry is broadly subdivided into two main categories: phosphorus-derived chemicals and other nonfertilizer phosphate chemicals. For the purposes of raw waste characterization and delineation of pretreatment information, the industry is further subdivided into six subcategories. The following categorization system (Table 3) of the various main production streams and their descriptions are taken from the federal guidelines [8] pertaining to state and local industrial pretreatment programs. It will be used in the following discussion to identify process flows and characterize the resulting raw waste. Figure 1 shows a flow diagram for the production streams of the entire phosphate manufacturing industry.

The manufacture of phosphorus-derived chemicals is almost entirely based on the production of elemental phosphorus from mined phosphate rock. Ferrophosphorus, widely used in the metallurgical industries, is a direct byproduct of the phosphorus production process. In the United States, over 85% of elemental phosphorus production is used to manufacture high-grade phosphoric acid by the furnace or dry process as opposed to the wet process that converts phosphate rock directly into low-grade phosphoric acid. The remainder of the elemental phosphorus is either marketed directly or converted into phosphorus chemicals. The furnace-grade phosphoric acid is marketed directly, mostly to the food and fertilizer industries. Finally, phosphoric acid is employed to manufacture sodium tripolyphosphate, which is used in detergents and for water treatment, and calcium phosphate, which is used in foods and animal feeds.

On the other hand, defluorinated phosphate rock is utilized as an animal feed ingredient. Defluorinated phosphoric acid is mainly used in the production of animal foodstuffs and liquid fertilizers. Finally, sodium phosphates, produced from wet process acid as the raw material, are used as intermediates in the production of cleaning compounds.

Table 3 Categorization System in Phosphorous-Derived and Nonfertilizer Phosphate Chemicals Production

Main category	Subcategory	Code
1. Phosphorus-derived chemicals	Phosphorus production	A
	Phosphorus-consuming	B
	Phosphate	C
2. Other nonfertilizer phosphate chemicals	Defluorinated phosphate rock	D
	Defluorinated phosphoric acid	E
	Sodium phosphates	F

Source: Ref. 8.

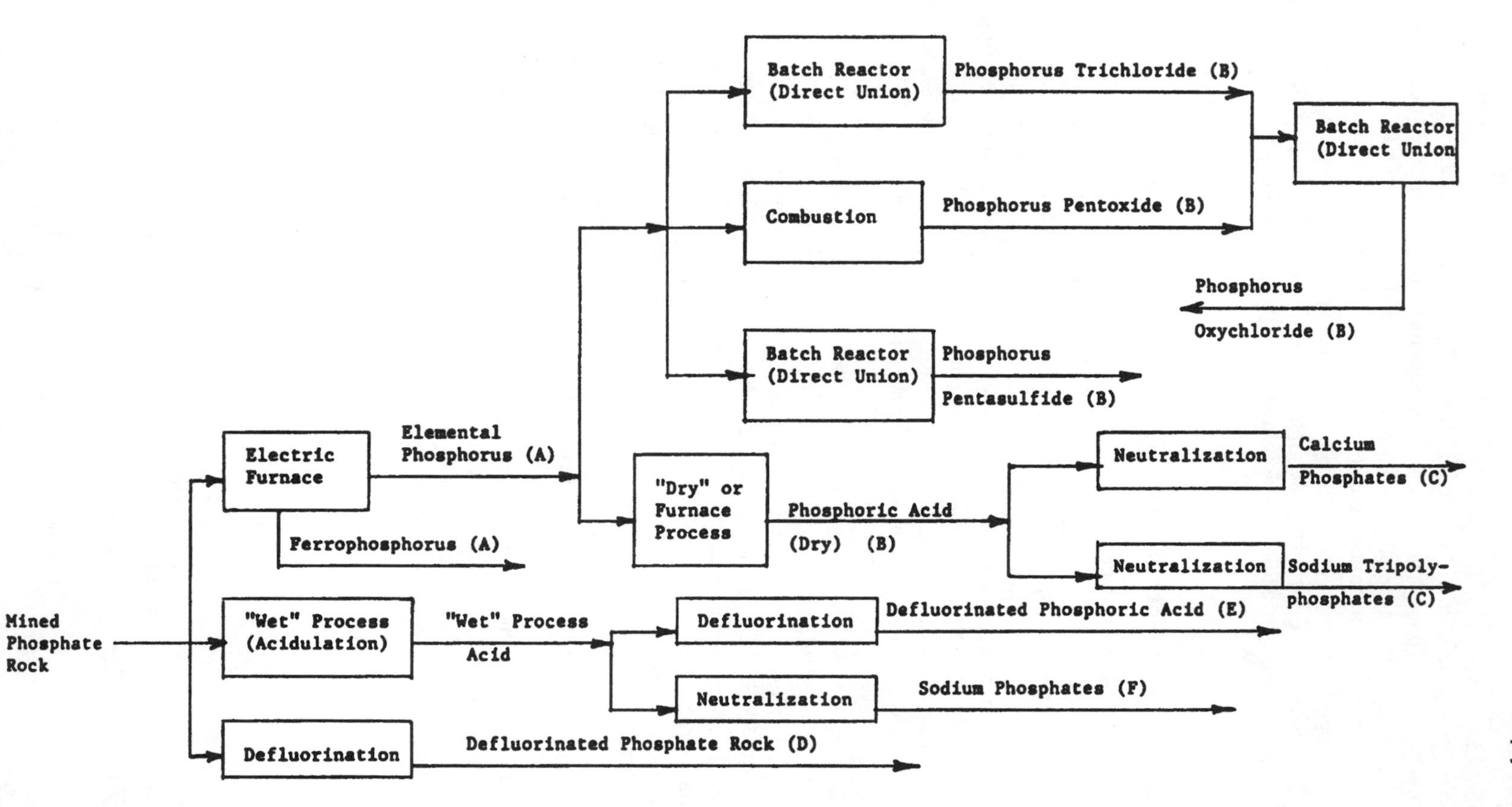

Figure 1 Phosphate manufacturing industry flow diagram (from Ref. 8).

9.2.2 Phosphorus and Phosphate Compounds

Phosphorus Production

Phosphorus is manufactured by the reduction of commercial-quality phosphate rock by coke in an electric furnace, with silica used as a flux. Slag, ferrophosphorus (from iron contained in the phosphate rock), and carbon monoxide are reaction byproducts. The standard process, as shown in Figure 2, consists of three basic parts: phosphate rock preparation, smelting in an electric furnace, and recovery of the resulting phosphorus. Phosphate rock ores are first blended so that the furnace feed is of uniform composition and then pretreated by heat drying, sizing or agglomerating the particles, and heat treatment.

The burden of treated rock, coke, and sand is fed to the furnace (which is extensively water-cooled) by incrementally adding weighed quantities of each material to a common conveyor belt. Slag and ferrophosphorus are tapped periodically, whereas the hot furnace gases (90% CO and 10% phosphorus) pass through an electrostatic precipitator that removes the dust before phosphorus condensation. The phosphorus is condensed by direct impingement of a hot water spray, sometimes enhanced by heat transfer through water-cooled condenser walls. Liquid phosphorus drains into a water sump, where the water maintains a seal from the atmosphere. Liquid phosphorus is stored in steam-heated tanks under a water blanket and transferred into tank cars by pumping or hot water displacement. The tank cars have protective blankets of water and are equipped with steam coils for remelting at the destination.

There are numerous sources of fumes from the furnace operation, such as dust from the raw materials feeding and fumes emitted from electrode penetrations and tapping. These fumes, which consist of dust, phosphorus vapor (immediately oxidized to phosphorus pentaoxide), and carbon monoxide, are collected and scrubbed. Principal wastewater streams consist of calciner scrubber liquor, phosphorus condenser and other phossy water, and slag-quenching water.

Phosphorus Consuming

This subcategory involves phosphoric acid (dry process), phosphorus pentoxide, phosphorus pentasulfide, phosphorus trichloride, and phosphorus oxychloride. In the standard dry process for phosphoric acid production, liquid phosphorus is burned in the air, the resulting gaseous phosphorus pentaoxide is absorbed and hydrated in a water spray, and the mist is collected with an electrostatic precipitator. Regardless of the process variation, phosphoric acid is made with the consumption of water and no aqueous wastes are generated by the process.

Solid anhydrous phosphorus pentaoxide is manufactured by burning liquid phosphorus in an excess of dried air in a combustion chamber and condensing the vapor in a roomlike structure. Condensed phosphorus pentaoxide is mechanically scraped from the walls using moving chains and is discharged from the bottom of the barn with a screw conveyor. Phosphorus pentasulfide is manufactured by directly reacting phosphorus and sulfur, both in liquid form, in a highly exothermic batch operation. Because the reactants and products are highly flammable at the reaction temperature, the reactor is continuously purged with nitrogen and a water seal is used in the vent line.

Phosphorus trichloride is manufactured by loading liquid phosphorus into a jacketed batch reactor. Chlorine is bubbled through the liquid, and phosphorus trichloride is refluxed until all the phosphorus is consumed. Cooling water is used in the reactor jacket and care is taken to avoid an excess of chlorine and the resulting formation of phosphorus pentachloride. Phosphorus oxychloride is manufactured by the reaction of phosphorus trichloride, chlorine, and solid phosphorus pentaoxide in a batch operation. Liquid phosphorus trichloride is loaded to the reactor, solid phosphorus pentoxide added, and chlorine bubbled through the mixture. Steam is

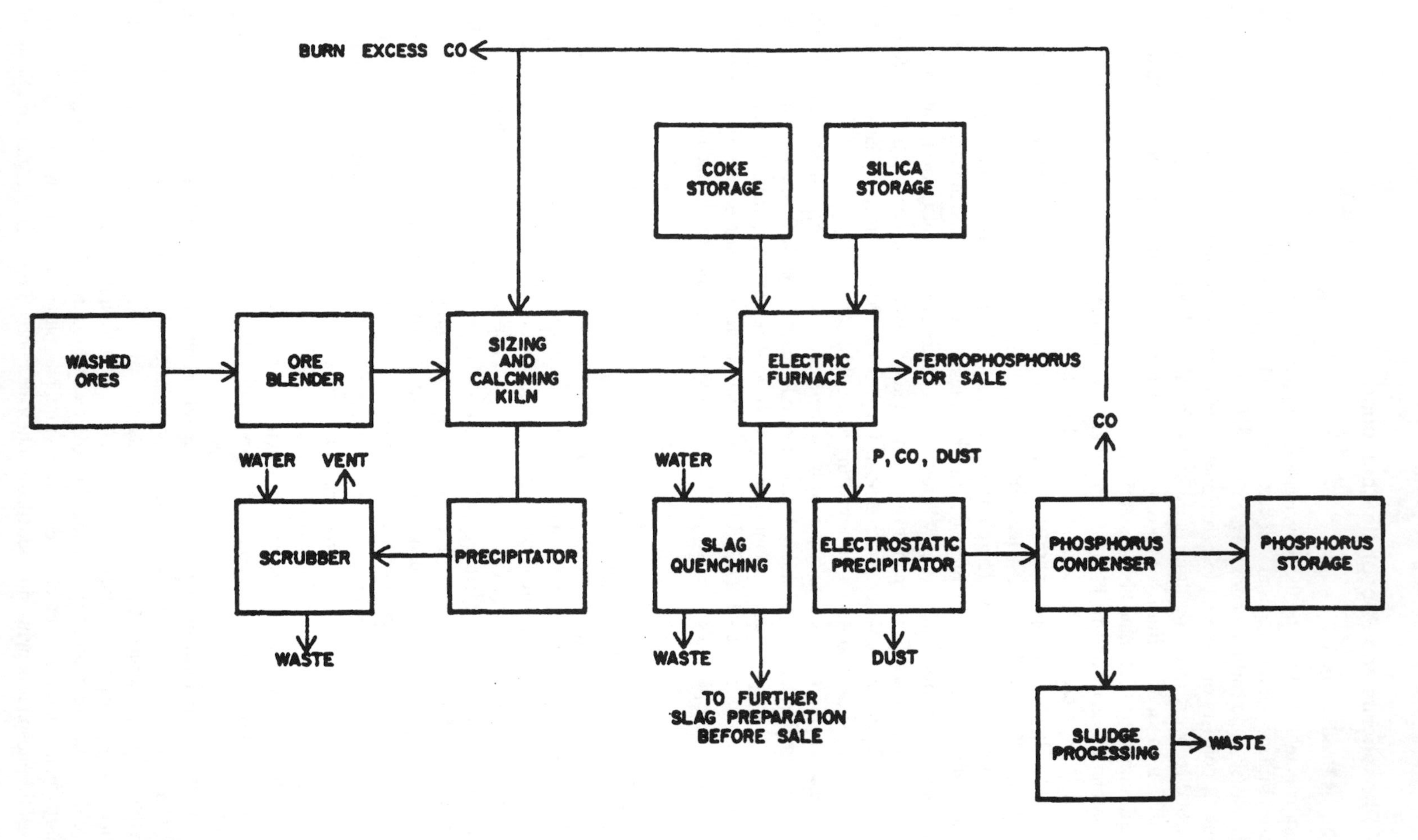

Figure 2 Standard phosphorus process flow diagram (from Ref. 8).

supplied to the reactor jacket, water to the reflux condenser is shut off, and the product is distilled over and collected.

Because phosphorus is transported and stored under a water blanket, phossy water is a raw waste material at phosphorus-consuming plants. Another source of phossy wastewater results when reactor contents (containing phosphorus) are dumped into a sewer line due to operator error, emergency conditions, or inadvertent leaks and spills.

Phosphate

This subcategory involves sodium tripolyphosphate and calcium phosphates. Sodium tripolyphosphate is manufactured by the neutralization of phosphoric acid by soda ash or caustic soda and soda ash, with the subsequent calcining of the dried mono- and disodium phosphate crystals. This product is then slowly cooled or tempered to produce the condensed form of the phosphates.

The nonfertilizer calcium phosphates are manufactured by the neutralization of phosphoric acid with lime. The processes for different calcium phosphates differ substantially in the amount and type of lime and amount of process water used. Relatively pure, food-grade monocalcium phosphate (MCP), dicalcium phosphate (DCP), and tricalcium phosphate (TCP) are manufactured in a stirred batch reactor from furnace-grade acid and lime slurry, as shown in the process flow diagram of Figure 3. Dicalcium phosphate is also manufactured for livestock feed supplement use, with much lower specifications on product purity.

Sodium tripolyphosphate manufacture generates no process wastes. Wastewaters from the manufacture of calcium phosphates are generated from a dewatering of the phosphate slurry and wet scrubbing of the airborne solids during product operations.

Defluorinated Phosphate Rock

The primary raw material for the defluorination process is fluorapatite phosphate rock. Other raw materials used in much smaller amounts, but critical to the process, are sodium-containing reagents, wet process phosphoric acid, and silica. These are fed into either a rotary kiln or a fluidized bed reactor that requires a modular and predried charge. Reaction temperatures are maintained in the 1205–1366°C range, whereas the retention time varies from 30 to 90 min. From the kiln or fluidized bed reactor, the defluorinated product is quickly quenched with air or water, followed by crushing and sizing for storage and shipment. A typical flow diagram for the fluidized bed process is shown in Figure 4.

Wastewaters are generated in the process of scrubbing contaminants from gaseous effluent streams. This water requirement is of significant volume and process conditions normally permit the use of recirculated contaminated water for this service, thereby effectively reducing the discharged wastewater volume. Leaks and spills are routinely collected as part of process efficiency and housekeeping and, in any case, their quantity is minor and normally periodic.

Defluorinated Phosphoric Acid

One method used in order to defluorinate wet process phosphoric acid is vacuum evaporation. The concentration of 54% P_2O_5 acid to a 68–72% P_2O_5 strength is performed in vessels that use high-pressure (30.6–37.4 atm or 450–550 psig) steam or an externally heated Dowtherm solution as the heat energy source for evaporation of water from the acid. Fluorine removal from the acid occurs concurrently with the water vapor loss. A typical process flow diagram for vacuum-type evaporation is shown in Figure 5.

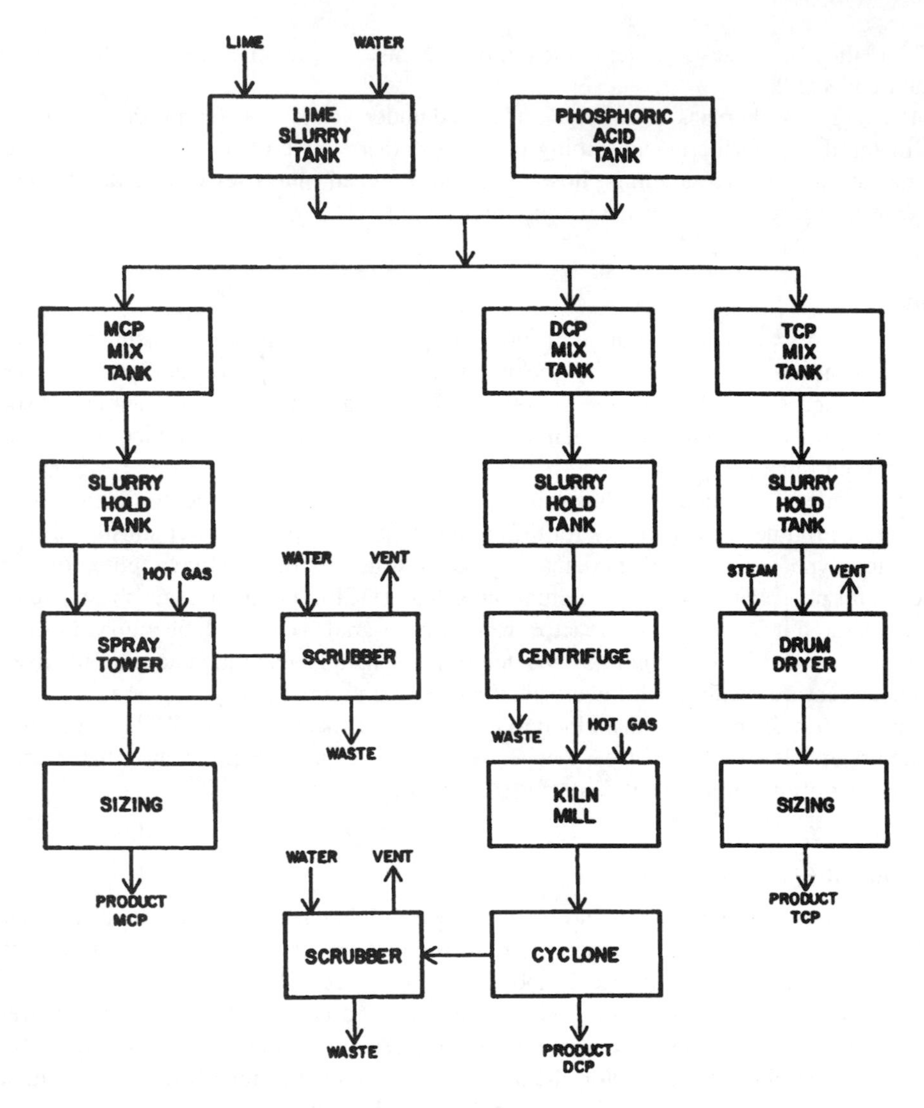

Figure 3 Standard process for food-grade calcium phosphates (from Ref. 8).

A second method of phosphoric acid defluorination entails the direct contact of hot combustion gases (from fuel oil or gas burners) with the acid by bubbling them through the acid. Evaporated and defluorinated product acid is sent to an acid cooler, while the gaseous effluents from the evaporation chamber flow to a series of gas scrubbing and absorption units. Finally, aeration can also be used for defluorinating phosphoric acid. In this process, diatomaceous silica or spray-dried silica gel is mixed with commercial 54% P_2O_5 phosphoric acid. Hydrogen fluoride in the impure phosphoric acid is converted to fluosilicic acid, which in turn breaks down to SiF_4 and is stripped from the heated mixture by simple aeration.

The major wastewater source in the defluorination processes is the wet scrubbing of contaminants from the gaseous effluent streams. However, process conditions normally permit the use of recirculated contaminated water for this service, thereby effectively reducing the discharged wastewater volume.

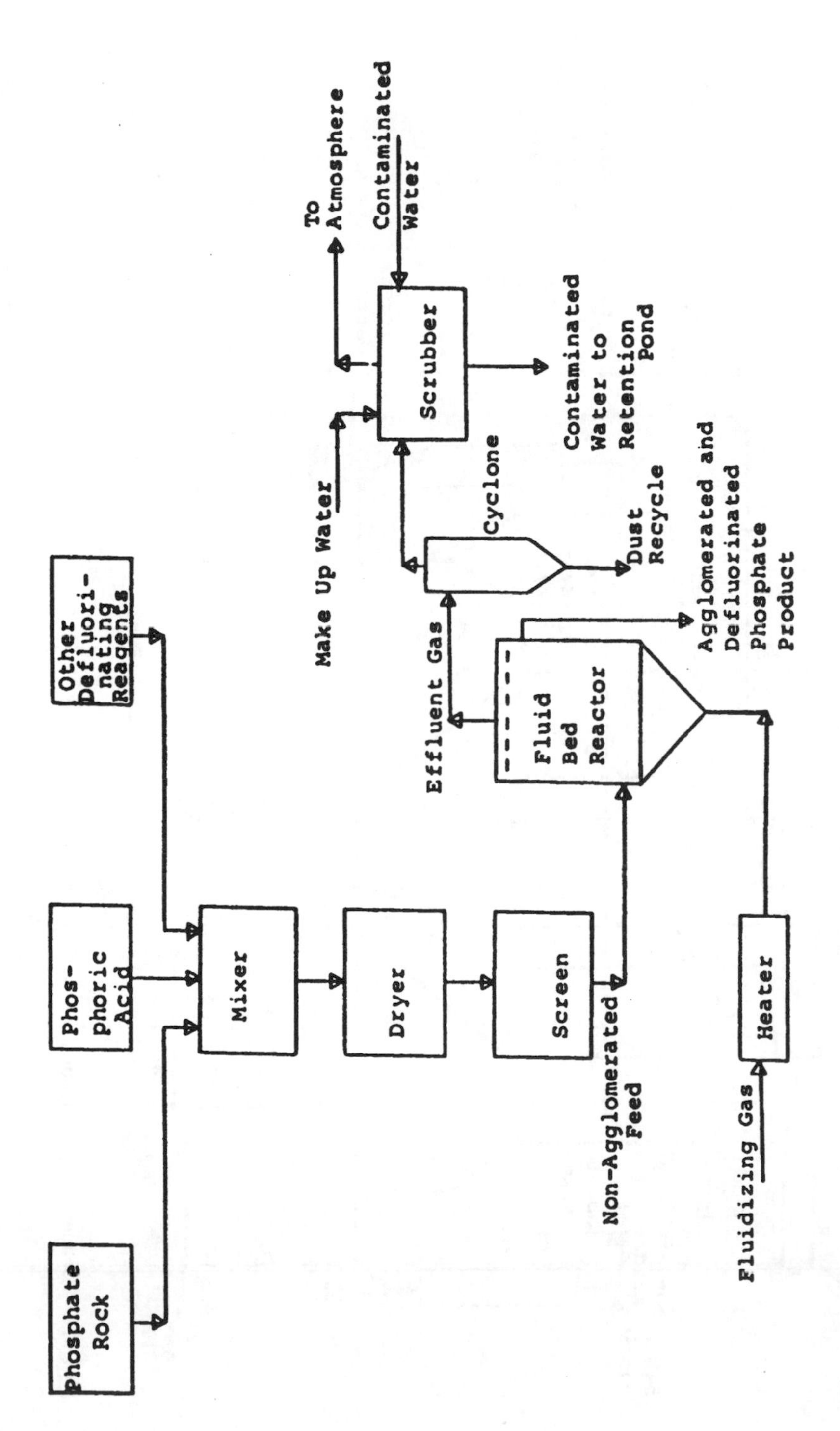

Figure 4 Defluorinated phosphate rock fluid bed process (from Ref. 8).

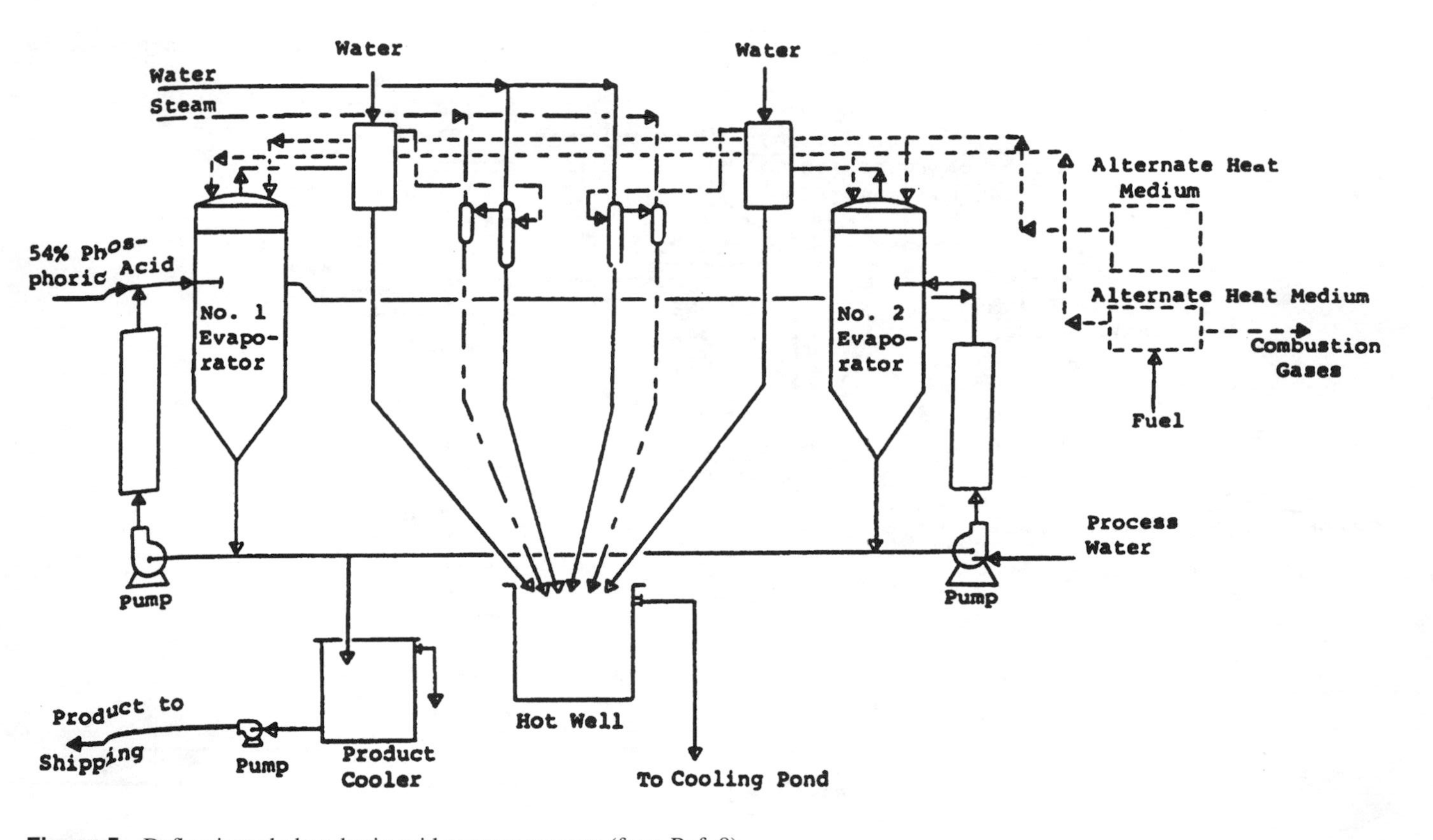

Figure 5 Defluorinated phosphoric acid vacuum process (from Ref. 8).

Sodium Phosphates

In the manufacture of sodium phosphates, the removal of contaminants from the wet process acid takes place in a series of separate neutralization steps. The first step involves the removal of fluosilicates with recycled sodium phosphate liquor. The next step precipitates the minor quantities of arsenic present by adding sodium sulfide to the solution, while barium carbonate is added to remove the excess sulfate. The partially neutralized acid still contains iron and aluminum phosphates, and some residual fluorine.

A second neutralization is carried out with soda ash to a pH level of about 4. Special heating, agitation, and retention are next employed to adequately condition the slurry so that filtration separation of the contaminants can be accomplished. The remaining solution is sufficiently pure for the production of monosodium phosphate, which can be further converted into other compounds such as sodium metaphosphate, disodium phosphate, and trisodium phosphate. A typical process flow diagram is shown in Figure 6. Wastewater effluents from these processes originate from leaks and spills, filtration backwashes, and gas scrubber wastewaters.

9.2.3 Categorization in Phosphate Fertilizer Production

The fertilizer industry comprises nitrogen-based, phosphate-based, and potassium-based fertilizer manufacturing, as well as combinations of these nutrients in mixed and blend fertilizer formulations. Only the phosphate-based fertilizer industry is discussed here and, therefore, the categorization mainly involves two broad divisions: (a) the phosphate fertilizer industry (A) and (b) the mixed and blend fertilizer industry (G) in which one of the components is a phosphate compound. The following categorization system of the various separate processes and their production streams and descriptions is taken from the federal guidelines [8] pertaining to state and local industrial pretreatment programs. It will be used in the discussion that ensues to identify process flows and characterize the resulting raw waste. Figure 7 shows a flow diagram for the production streams of the entire phosphate and nitrogen fertilizer manufacturing industry.

9.2.4 Phosphate and Mixed and Blend Fertilizer Manufacture

Phosphate Fertilizer (A)

The phosphate fertilizer industry is defined as eight separate processes: phosphate rock grinding, wet process phosphoric acid, phosphoric acid concentration, phosphoric acid clarification, normal superphosphate, triple superphosphate, ammonium phosphate, and sulfuric acid. Practically all phosphate manufacturers combine the various effluents into a large recycle water system. It is only when the quantity of recycle water increases beyond the capacity to contain it that effluent treatment is necessary.

Phosphate Rock Grinding. Phosphate rock is mined and mechanically ground to provide the optimum particle size required for phosphoric acid production. There are no liquid waste effluents.

Wet Process Phosphoric Acid. A production process flow diagram is shown in Figure 8. Insoluble phosphate rock is changed to water-soluble phosphoric acid by solubilizing the phosphate rock with an acid, generally sulfuric or nitric. The phosphoric acid produced from the nitric acid process is blended with other ingredients to produce a fertilizer, whereas the phosphoric acid produced from the sulfuric acid process must be concentrated before further use. Minor quantities of fluorine, iron, aluminum, silica, and uranium are usually the most serious waste effluent problems.

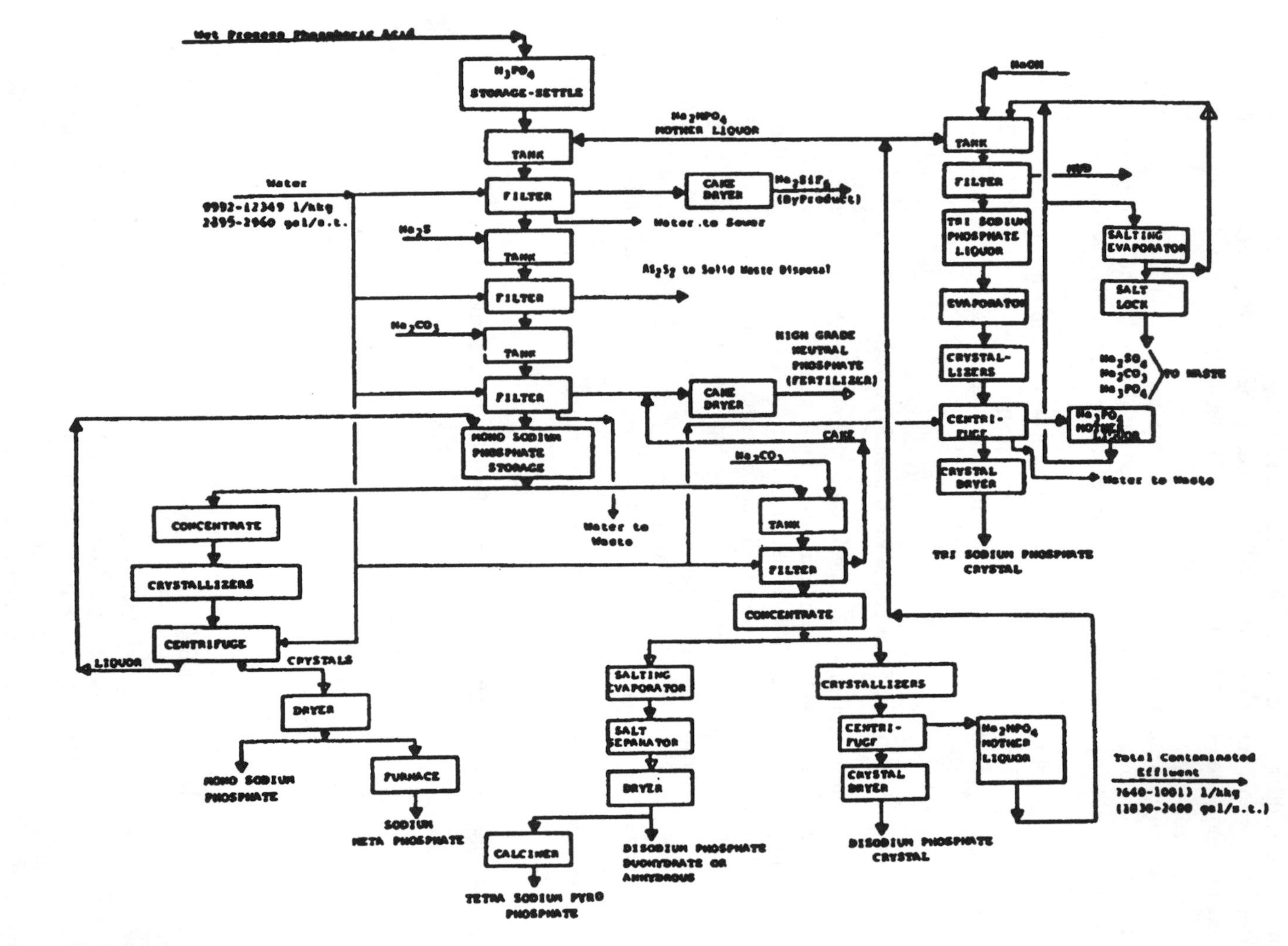

Figure 6 Sodium phosphate process from wet process (from Ref. 8).

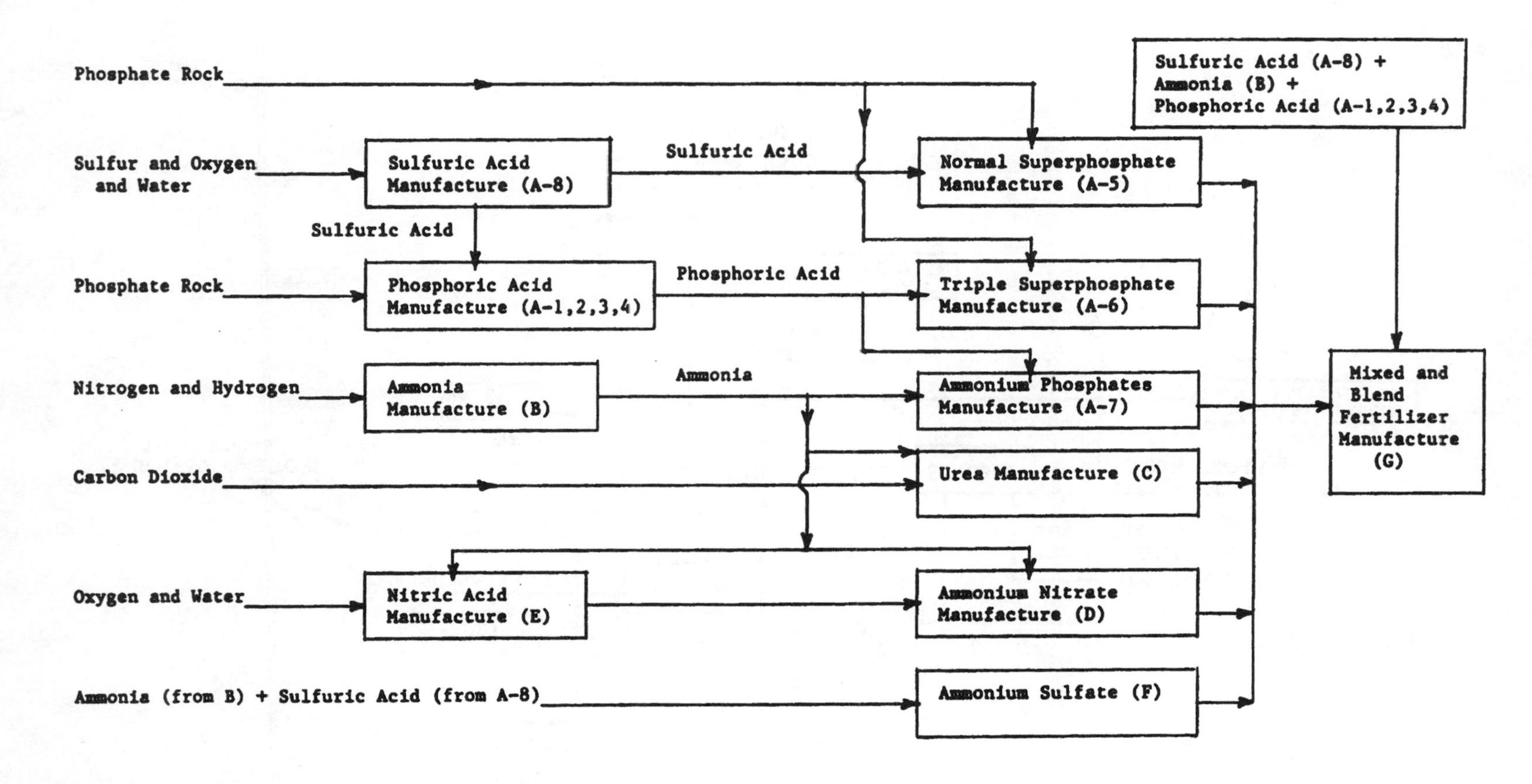

Figure 7 Flow diagram of fertilizer products manufacturing (from Ref. 8).

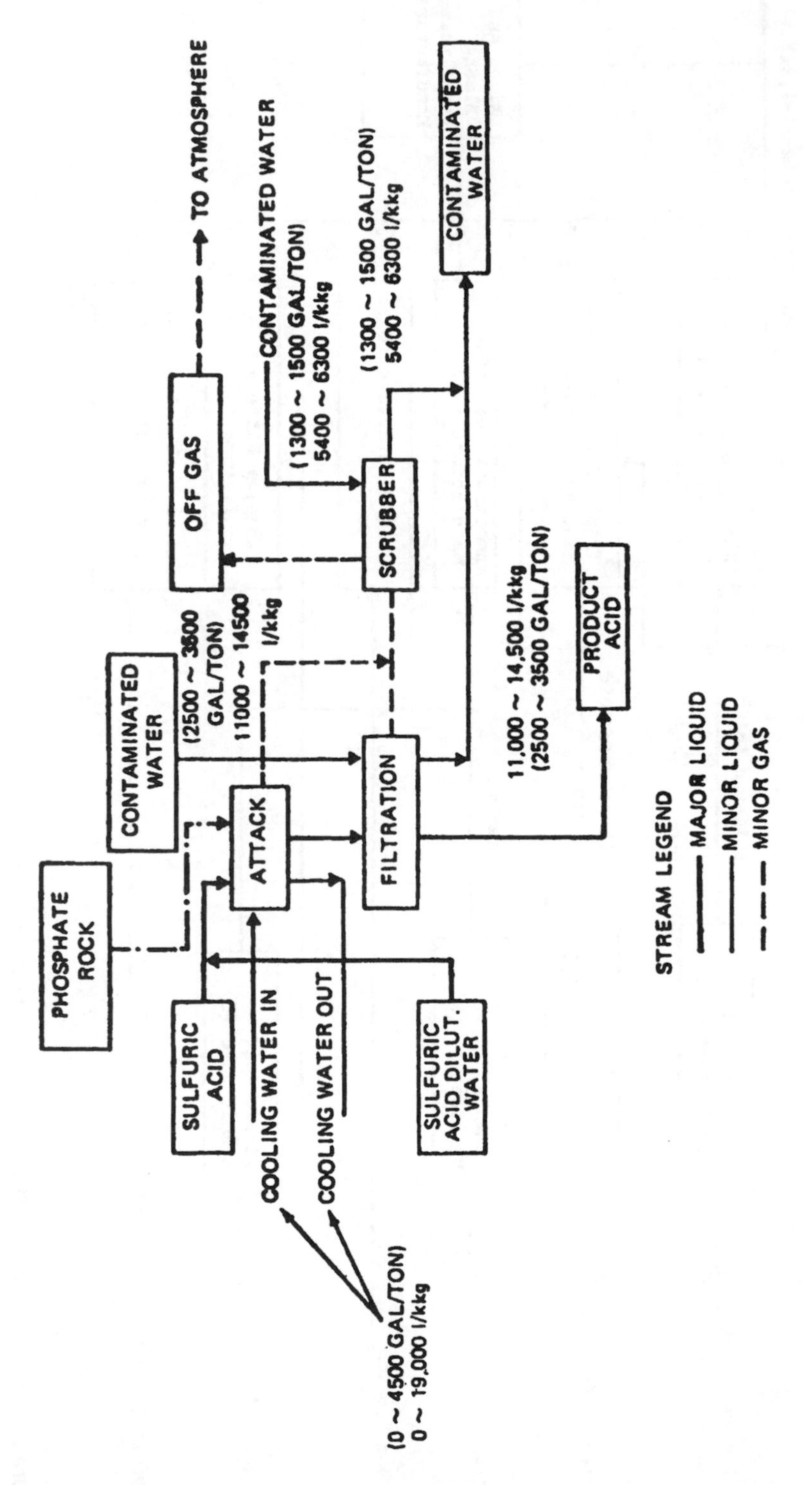

Figure 8 Wet process phosphoric acid (H_2SO_4) acidulation (from Ref. 8).

Phosphoric Acid Concentration. Phosphoric acid produced with sulfuric acid cannot be used for processing due to its very low concentration. It is therefore concentrated to 50–54% by evaporation. Waste streams contain fluorine and phosphoric acid.

Phosphoric Acid Clarification. When the phosphoric acid has been concentrated, iron and aluminum phosphates, gypsum and fluorosilicates become insoluble and can pose problems during acid storage. They are therefore removed by clarification and/or centrifugation.

Normal Superphosphate. Normal superphosphate is produced by the reaction between ground phosphate rock and sulfuric acid, followed by three to eight weeks of curing time. Obnoxious gases are generated by this process.

Triple Superphosphate (TSP). Triple superphosphate is produced by the reaction between ground phosphate rock and phosphoric acid by one of two processes. One utilizes concentrated phosphoric acid and generates obnoxious gases. The dilute phosphoric acid process permits the ready collection of dusts and obnoxious gases generated.

Ammonium Phosphate. Ammonium phosphate, a concentrated water-soluble plant food, is produced by reacting ammonia and phosphoric acid. The resultant slurry is dried, stored, and shipped to marketing.

Sulfuric Acid. Essentially all sulfuric acid manufactured in this industry is produced by the "contact" process, in which SO_2 and oxygen contact each other on the surface of a catalyst (vanadium pentaoxide) to form SO_3 gas. Sulfur trioxide gas is added to water to form sulfuric acid. The sulfur dioxide used in the process is produced by burning elemental sulfur in a furnace.

In addition, the process is designed to capture a high percentage of the energy released by the exothermic reactions occurring in the oxidation of sulfur to sulfur trioxide. This energy is used to produce steam, which is then utilized for other plant unit operations or converted into electrical energy. It is the raw water treatment necessary to condition water for this steam production that generates essentially all the wastewater effluent from this process.

Mixed and Blend Fertilizer (G)

Mixed Fertilizer. The raw materials used to produce mixed fertilizers include inorganic acids, solutions, double nutrient fertilizers, and all types of straight fertilizers. The choice of raw materials depends on the specific nitrogen-phosphate-potassium (N-P-K) formulation to be produced and on the cost of the different materials from which they can be made.

The mixed fertilizer process involves the controlled addition of both dry and liquid raw materials to a granulator, which is normally a rotary drum, but pug mills are also used. Raw materials, plus some recycled product material, are mixed to form an essentially homogeneous granular product. Wet granules from the granulator are discharged into a rotary drier, where the excess water is evaporated and dried granules from the drier are then sized on vibrating screens. Over- and undersized granules are separated for use as recycle material in the granulator. Commercial-product-size granules are cooled and then conveyed to storage or shipping.

Blend Fertilizer. Raw materials used to produce blend fertilizers are a combination of granular dry straight and mixed fertilizer materials with an essentially identical particle size. Although many materials can be utilized, the five most commonly used in this process are ammonium nitrate, urea, triple superphosphate, diammonium phosphate, and potash. These raw materials are stored in a multicompartmented bin and withdrawn in the precise quantities needed to produce the nitrogen-phosphorus-potassium (N-P-K) formulation desired. Raw material addition is normally done by batch weighing, and the combination of batch-weighed and granular raw materials is then conveyed to a mechanical blender for mixing. From the blender, the product is conveyed to storage or shipping.

9.2.5 Wastewater Characteristics and Sources

Wastewaters from the manufacturing, processing, and formulation of inorganic chemicals such as phosphorus compounds, phosphates, and phosphate fertilizers cannot be exactly characterized. The wastewater streams are usually expected to contain trace or large concentrations of all raw materials used in the plant; all intermediate compounds produced during manufacture; all final products, coproducts, and byproducts; and the auxiliary or processing chemicals employed. It is desirable from the viewpoint of economics that these substances not be lost, but some losses and spills appear unavoidable and some intentional dumping does take place during housecleaning, vessel emptying, and preparation operations.

The federal guidelines [8] for state and local pretreatment programs reported the raw wastewater characteristics (Table 4) in mg/L concentration, and flows and quality parameters (Table 5) based on the production of 1 ton of the product manufactured, for each of the six subcategories of the phosphate manufacturing industry. Few fertilizer plants discharge wastewaters to municipal treatment systems. Most use ponds for the collection and storage of wastewaters, pH control, chemical treatment, and settling of suspended solids. Whenever available retention pond capacities in the phosphate fertilizer industry are exceeded, the wastewater overflows are treated and discharged to nearby surface water bodies. The federal guidelines [8] reported the range of wastewater characteristics (Table 6) in mg/L concentrations for typical retention ponds used by the phosphate fertilizer industry.

The specific types of wastewater sources in the phosphate fertilizer industry are (a) water treatment plant wastes from raw water filtration, clarification, softening and deionization, which principally consist of only the impurities removed from the raw water (such as carbonates, hydroxides, bicarbonates, and silica) plus minor quantities of treatment chemicals; (b) closed-loop cooling tower blowdown, the quality of which varies with the makeup of water impurities and inhibitor chemicals used (note: the only cooling water contamination from process liquids is through mechanical leaks in heat exchanger equipment, and Table 7 shows the normal range of contaminants that may be found in cooling water blowdown systems [26]); (c) boiler blowdown, which is similar to cooling tower blowdown but the quality differs as shown in Table 8 [26]; (d) contaminated water or gypsum pond water, which is the impounded and reused water that accumulates sizable concentrations of many cations and anions, but mainly fluorine and phosphorus concentrations of 8500 mg/L F and in excess of 5000 mg/L P are not unusual; concentrations of radium 226 in recycled gypsum pond water are 60–100 picocuries/L, and its acidity reaches extremely high levels (pH 1–2); (e) wastewater from spills and leaks that, when possible, is reintroduced directly to the process or into the contaminated water system; and (f) nonpoint-source discharges that originate from the dry fertilizer dust covering the general plant area and then dissolve in rainwater and snowmelt, which become contaminated.

In the specific case of wastewater generated from the condenser water bleedoff in the production of elemental phosphorus from phosphate rock in an electric furnace, Yapijakis [33] reported that the flow varies from 10 to 100 gpm (2.3–23 m^3/hour), depending on the particular installation. The most important contaminants in this waste are elemental phosphorus, which is colloidally dispersed and may ignite if allowed to dry out, and fluorine, which is also present in the furnace gases. The general characteristics of this type of wastewater (if no soda ash or ammonia were added to the condenser water) are given in Table 9.

As previously mentioned, fertilizer manufacturing may create problems within all environmental media, that is, air pollution, water pollution, and solid wastes disposal difficulties. In particular, the liquid waste effluents generated from phosphate and mixed and blend fertilizer production streams originate from a variety of sources and may be summarized [17,27] as follows: (a) ammonia-bearing wastes from ammonia production; (b) ammonium salts such as

Table 4 Phosphate Manufacturing Industry Raw Waste Characteristics

Parameter (mg/L except pH)	Phosphorus production (A)	Phosphorus-consuming (B)	Phosphate (C)	Defluorinated phosphate rock (D)	Defluorinated phosphoric acid (E)	Sodium phosphate (F)
Flow type	C	B	B	B	B	B
BOD_5			–	3	15	31
SS	100		24,000–54,000	16	30	416
						460
TDS			1900–7000[a]	2,250[a]	28,780[a]	1640[a]
COD				48	306	55
pH (unit)				1.65[a]	1.29[a]	7.8
Phosphorus	21					
Phosphate	59		7000[a]			
Sulfate	260			350	4770	240
F	126			1930	967	15
HCl		0–800				
H_2SO_3		0–34				
$H_3PO_3 + H_3PO_4$		17–500				
HF, H_2SiF_6, H_2SiO_3			1900[a]			
Chloride				101	65	90
Calcium				40	1700[a]	95
Magnesium				12	106	
Aluminum				58	260	
Iron				8[a]	180[a]	
Arsenic				0.38[a]	0.83[a]	
Zinc				5.2[a]	5.3[a]	
Total acidity	128					
Total phosphorus				600	5590[a]	250

[a] In high levels, these parameters may be inhibitory to biological systems.
B = batch process.
C = continuous process.
BOD, biochemical oxygen demand; SS, suspended solids; TDS, total dissolved solids; COD, chemical oxygen demand.
Source: Ref. 8.

Table 5 Phosphate Manufacturing Industry Raw Waste Characteristics Based on Production

Parameter (kg/kkg except pH and flow)	Phosphorus production (A)	Phosphorus-consuming (B)	Phosphate (C)	Defluorinated phosphate rock (D)	Defluorinated phosphoric acid (E)	Sodium phosphate (F)
Flow range (L/kkg)	425,000	38,000	10,920	45,890	18,020–70,510	7640–10,020
Flow type	C	B	B	B	B	B
BOD_5					0.27–1.06	0.2–0.3
SS	42.5		22.5–50	0.73	0.54–2.11	3.5–4.6
TDS			4.0–14.6	103	519–2031	12.5–16.40
COD				2.2	5.5–21.5	0.4–0.52
pH (unit)				1.65	1.29	7.8
Phosphorus	9					
Phosphate	25		15			
Sulfate	111			16	86–336	1.8–2.36
F	53.5			88	17.4–68.1	0.1–0.13
HCl		0–3				
H_2SO_3		0–1.0				
$H_3PO_3 + H_3PO_4$		0.5–2.5				
HF, H_2SiF_6, H_2SiO_3			12			
Chloride				4.6	1.17–4.58	0.68–0.90
Calcium				1.8	30.6–120	0.72–0.94
Magnesium				0.6	1.9–7.43	
Aluminum				2.7	4.7–18.39	
Iron				0.37	3.2–12.52	
Arsenic				0.02	0.02–0.08	
Zinc				0.24	0.09–0.35	
Total acidity	54.5					
Total phosphorus				27.5	101–395	1.91–2.51

BOD, biochemical oxygen demand; SS, suspended solids; TDS, total dissolved solids; COD, chemical oxygen demand.
Source: Ref. 8.

Table 6 Raw Wastewater Characteristics of Phosphate Fertilizer Industry Retention Ponds

Quality parameter	Phosphate (A)
Suspended solids (mg/L)	800–1200
pH (unit)	1–2
Ammonia (mg/L)	450–500
Sulfate (mg/L)	4000
Chloride (mg/L)	58
Total phosphate (mg/L)	3–5M
Fluoride (mg/L)	6–8.5M
Aluminum (mg/L)	110
Iron (mg/L)	85
Radium 226 (picocuries/L)	60–100

M = thousand.
Source: Ref. 8.

ammonium phosphate; (c) phosphates and fluoride wastes from phosphate and superphosphate production; (d) acidic spillages from sulfuric acid and phosphoric acid production; (e) spent solutions from the regeneration of ion-exchange units; (f) phosphate, chromate, copper sulfate, and zinc wastes from cooling tower blowdown; (g) salts of metals such as iron, copper, manganese, molybdenum, and cobalt; (h) sludge discharged from clarifiers and backwash water from sand filters; and (i) scrubber wastes from gas purification processes.

Considerable variation, therefore, is observed in quantities and wastewater characteristics at different plants. According to a UNIDO report [34], the most important factors that contribute to excessive in-plant materials losses and, therefore, probable subsequent pollution are the age of the facilities (low efficiency, poor process control), the state of maintenance and repair (especially of control equipment), variations in feedstock and difficulties in adjusting processes to cope, and an operational management philosophy such as consideration for pollution control and prevention of materials loss. Because of process cooling requirements, fertilizer manufacturing facilities may have an overall large water demand, with the wastewater effluent discharge largely dependent on the extent of in-plant recirculation [17]. Facilities designed on

Table 7 Range of Concentrations of Contaminants in Cooling Water

Cooling water contaminant	Concentration (mg/L)
Chromate	0–250
Sulfate	500–3000
Chloride	35 160
Phosphate	10–50
Zinc	0–30
TDS	500–10,000
TSS	0–50
Biocides	0–100

TDS, total dissolved solids; TSS, total suspended solids.
Source: Ref. 26.

Table 8 Range of Concentrations of Contaminants in Boiler Blowdown Waste

Boiler blowdown contaminant	Concentration (mg/L)
Phosphate	5–50
Sulfite	0–100
TDS	500–3500
Zinc	0–10
Alkalinity	50–700
Hardness	50–500
Silica (SiO_2)	25–80

TDS, total dissolved solids.
Source: Ref. 26.

a once-through process cooling stream generally discharge from 1000 to over 10,000 m^3/hour wastewater effluents that are primarily cooling water.

9.3 IMPACTS OF PHOSPHATE INDUSTRY POLLUTION

The possibility of the phosphate industry adversely affecting streams did not arise until 1927, when the flotation process was perfected for increasing the recovery of fine-grain pebble phosphate [12]. A modern phosphate mining and processing facility typically has a 30,000 gpm (1892 L/s) water supply demand and requires large areas for clear water reservoirs, slime settling basins, and tailings sand storage. With the help of such facilities, the discharge of wastes into nearby surface water bodies is largely prevented, unless heavy rainfall inputs generate volumes that exceed available storage capacity.

According to research results reported by Fuller [12], the removal of semicolloidal matter in settling areas or ponds seems to be one of the primary problems concerning water pollution control. The results of DO and BOD surveys indicated that receiving streams were actually

Table 9 Range of Concentrations of Contaminants in Condenser Waste from Electric Furnace Production of Phosphorous

Quality parameter	Concentration or value
pH	1.5–2.0
Temperature	120–150°F
Elemental phosphorus	400–2500 Mg/L
Total suspended solids	1000–5000 Mg/L
Fluorine	500–2000 Mg/L
Silica	300–700 Mg/L
P_2O_5	600–900 Mg/L
Reducing substances (as I_2)	40–50 Mg/L
Ionic charge of particles	Predominantly positive (+)

Source: Ref. 15.

improved in this respect by the effluents from phosphate operations. On the other hand, no detrimental effects on fish were found, but there is the possibility of destruction of fish food (aquatic microorganisms and plankton) under certain conditions.

The wastewater characteristics vary from one production facility to the next, and even the particular flow magnitude and location of discharge will significantly influence its aquatic environmental impact. The degree to which a receiving surface water body dilutes a wastewater effluent at the point of discharge is important, as are the minor contaminants that may occasionally have significant impacts. Fertilizer manufacturing wastes, in general, affect water quality primarily through the contribution of nitrogen and phosphorus, whose impacts have been extensively documented in the literature. Significant levels of phosphates assist in inducing eutrophication, and in many receiving waters they may be more important (growth-limiting agent) than nitrogenous compounds. Under such circumstances, programs to control eutrophication have generally attempted to reduce phosphate concentrations in order to prevent excessive algal and macrophyte growth [14].

In addition to the above major contaminants, pollution from the discharge of fertilizer manufacturing wastes may be caused by such secondary pollutants as oil and grease, hexavalent chromium, arsenic, and fluoride. As reported by Beg et al. [3], in certain cases, the presence of one or more of these pollutants may have adverse impacts on the quality of a receiving water, due primarily to toxic properties, or can be inhibitory to the nitrification process. Finally, oil and grease concentrations may have a significant detrimental effect on the oxygen transfer characteristics of the receiving surface water body.

The manufacture of phosphate fertilizers also generates great volumes of solid wastes known as phosphogypsum, which creates serious difficulties, especially in large production facilities [18]. The disposal of phosphogypsum wastes requires large areas impervious to the infiltration of effluents, because they usually contain fluorine and phosphorus compounds that would have a harmful impact on the quality of a receiving water. Dumping of phosphogypsum in the sea would be acceptable only at coastal areas of deep oceans with strong currents that guarantee thorough mixing and high dilution.

9.4 U.S. CODE OF FEDERAL REGULATIONS

The information presented here has been taken from the U.S. Code of Federal Regulations, 40 CFR, containing documents related to the protection of the environment [9]. In particular, the regulations contained in Part 418, Fertilizer Manufacturing Point Source Category (Subpart A, Phosphate Subcategory, and Subpart O, Mixed and Blend Fertilizer Production Subcategory), and Part 422, Phosphate Manufacturing Point Source Category, pertain to effluent limitations guidelines and pretreatment or performance standards for each of the six subcategories shown in Table 3.

9.4.1 Phosphate Fertilizer Manufacture

The effluent guideline regulations and standards of 40 CFR, Part 418, were promulgated on July 29, 1987. According to the most recent notice in the Federal Register [10] regarding industrial categories and regulations, no review is under way or planned and no revision proposed for the fertilizer manufacturing industry. The effluent guidelines and standards applicable to this industrial category are (a) the best practicable control technology currently available (BPT); (b) the best available technology economically achievable (BAT); (c) the best conventional pollutant control technology (BCT); (d) new source performance standards (NSPS); and (e) new source pretreatment standards for new sources (NSPS).

The provisions of 40 CFR, Part 418, Subpart A, Phosphate Subcategory, are applicable to discharges resulting from the manufacture of sulfuric acid by sulfur burning, wet process phosphoric acid, normal superphosphate, triple superphosphate, and ammonium phosphate. The limitations applied to process wastewater, which establish the quantity of pollutants or pollutant properties that may be discharged by a point source into a surface water body after the application of various types of control technologies, are shown in Table 10. The total suspended solids limitation is waived for process wastewater from a calcium sulfate (phosphogypsum) storage pile runoff facility, operated separately or in combination with a water recirculation system, which is chemically treated and then clarified or settled to meet the other pollutant limitations. The concentrations of pollutants discharged in contaminated nonprocess wastewater, that is, any water including precipitation runoff that comes into incidental contact with any raw material, intermediate or finished product, byproduct, or waste product by means of precipitation, accidental spills, or leaks and other nonprocess discharges, should not exceed the values given in Table 10.

The provisions of Subpart G, Mixed and Blend Fertilizer Production Subcategory, are applicable to discharges resulting from the production of mixed fertilizer and blend fertilizer (or compound fertilizers), such as nitrogen/phosphorus (NP) or nitrogen/phosphorus/potassium (NPK) balanced fertilizers of a range of formulations. The plant processes involved in fertilizer compounding comprise mainly blending and granulation plants, with in-built flexibility to

Table 10 Effluent Limitations (mg/L) for Subpart A, Phosphate Fertilizer

Effluent characteristic	Maximum for any 1 day	Average of daily values for 30 consecutive days shall not exceed
(a) BPT		
Total phosphorus (as P)	105	35
Fluoride	75	25
TSS	150	50
(b) BAT		
Total phosphorus (as P)	105	35
Fluoride	75	25
(c) BCT		
TSS	150	50
(d) NSPS		
Total phosphorus (as P)	105	35
Fluoride	75	25
TSS	150	50
(e) Contaminated nonprocess wastewater		
Total phosphorus (as P)	105	35
Fluoride	75	25

BPT, best practicable control technology currently available; BAT, best available technology economically available; BCT, best conventional pollutant control technology; NSPS, standards of performance for new sources.
Source: Ref. 9.

produce NPK grades in varying proportions [22]. According to Subpart O, "mixed fertilizer" means a mixture of wet and/or dry straight fertilizer materials, mixed fertilizer materials, fillers, and additives prepared through chemical reaction to a given formulation, whereas "blend fertilizer" means a mixture of dry, straight, and mixed fertilizer materials. The effluent limitations guidelines for BPT, BCT, and BAT, and the standards of performance for new sources, allow no discharge of process wastewater pollutants to navigable waters. Finally, the pretreatment standards establishing the quantity of pollutants that may be discharged to publicly owned treatment works (POTW) by a new source are given in Table 11.

9.4.2 Phosphate Manufacturing

The effluent guideline regulations and standards of 40 CFR, Part 422, were promulgated on July 9, 1986. According to the most recent notice in the Federal Register [10] regarding industrial categories and regulations, no review is under way or planned and no revision proposed for the phosphate manufacturing industry. The effluent guidelines and standards applicable to this industrial category are (a) the best practicable control technology currently available (BPT); (b) the best conventional pollutant control technology (BCT); (c) the best available technology economically achievable (BAT); and (d) new source performance standards (NSPS).

The provisions of 40 CFR, Part 422, Phosphate Manufacturing, are applicable to discharges of pollutants resulting from the production of the chemicals described by the six subcategories shown in Table 3. The effluent limitations guidelines for Subpart D, Defluorinated Phosphate Rock Subcategory, are shown in Table 12, and the limitations for contaminated nonprocess wastewater do not include a value for total suspended solids (TSS). Tables 13 and 14 show the effluent limitations guidelines for Subpart E, Defluorinated Phosphoric Acid, and Subpart F, Sodium Phosphate Subcategories, respectively, and again the limitations for contaminated nonprocess wastewater do not include a value for TSS. As can be seen, only for Subpart F are the effluent limitations given as kilograms of pollutant per ton of product (or lb/1000 lb).

9.4.3 Effluent Standards in Other Countries

The control of wastewater discharges from the phosphate and phosphate fertilizer industry in various countries differs significantly, as is the case with effluents from other industries. The discharges may be regulated on the basis of the receiving medium, that is, whether the disposal is

Table 11 Effluent Limitations (mg/L except for pH) for Subpart G, Mixed and Blend Fertilizer

Effluent characteristic	Average of daily values for 30 consecutive days shall not exceed
BOD_5	–
TSS	–
pH	–
NH_3—N	30
NO_3—N	30
Total P	35

BOD, biochemical oxygen demand; TSS, total suspended solids.
Source: Ref. 9.

Table 12 Effluent Limitations (mg/L except for pH) for Subpart D, Defluorinated Phosphate Rock

Effluent characteristic	Maximum for any 1 day	Average of daily values for 30 consecutive days shall not exceed
(a) BPT and NSPS		
Total phosphorus (as P)	105	35
Fluoride (as F)	75	25
TSS	150	50
pH	a	a
(b) BPT and BAT for nonprocess wastewater, and BAT for process wastewater		
Total phosphorus (as P)	105	35
Fluoride (as F)	75	25
pH	a	a
(c) BCT		
TSS	150	50
pH	a	a

[a] Within the range 6.0–9.5.
BPT, best practicable control technology currently available; BAT, best available technology economically available; BCT, best conventional pollutant control technology; NSPS, standards of performance for new sources.
Source: Ref. 9.

to land, municipal sewer system, inland surface water bodies, or coastal areas. Consideration may be given to environmental, socio-economic, and water-quality requirements and objectives, as well as to an assessment of the nature and impacts of the specific industrial effluents, which leads to an approach of either specific industry subcategories or classification of waters, or on a case-by-case basis. To a more limited extent than in the United States, the Indian Central Board for Prevention and Control of Water Pollution established a fertilizer industry subcommittee that adopted suitable effluent standards, proposed effective pollution control measures, and established subcategories for the fertilizer industry [11].

Pollution control legislation and standards in many countries are based on the adoption of systems of water classification. This approach of environmental management can make use of either a broad system of classifications with a limited number of subcategories or, as in Japan, a detailed system of subcategories such as river groups for various uses, lakes, and coastal waters. Within such a framework, specific cases of discharge standards could also be considered under circumstances of serious localized environmental impacts. Other countries, such as the United Kingdom and Finland, have a more flexible approach and review discharge standards for fertilizer plants on a case-by-case basis, with no established uniform guidelines [17]. The assessment of each case is based on the nature and volume of the discharge, the characteristics of receiving waters, and the available pollution control technology.

9.5 WASTEWATER CONTROL AND TREATMENT

The sources and characteristics of wastewater streams from the various subcategories in phosphate and phosphate fertilizer manufacturing, as well as some of the possibilities for

Table 13 Effluent Limitations (mg/L except for pH) for Subpart E, Defluorinated Phosphoric Acid

Effluent characteristic	Maximum for any 1 day	Average of daily values for 30 consecutive days shall not exceed
(a) BPT and NSPS		
Total phosphorus (as P)	105	35
Fluoride (as F)	75	25
TSS	150	50
pH	a	a
(b) BAT for process and nonprocess wastewater, and BPT for nonprocess wastewater		
Total phosphorus (as P)	105	35
Fluoride (as F)	75	25
(c) BCT		
TSS	150	50
pH	a	a

[a] Within the range 6.0–9.5.
BPT, best practicable control technology currently available; BAT, best available technology economically available; BCT, best conventional pollutant control technology; NSPS, new source performance standards.
Source: Ref. 9.

recycling and treatment, were discussed in Section 9.2. The pollution control and treatment methods and unit processes used are discussed in more detail in the following. The details of the process design criteria for these unit treatment processes can be found in any design handbook.

9.5.1 In-Plant Control, Recycle, and Process Modification

The primary consideration for in-plant control of pollutants that enter waste streams through random accidental occurrences, such as leaks, spills, and process upsets, is establishing loss prevention and recovery systems. In the case of fertilizer manufacture, a significant portion of contaminants may be separated at the source from process wastes by dedicated recovery systems, improved plant operations, retention of spilled liquids, and the installation of localized interceptors of leaks such as oil drip trays for pumps and compressors [17]. Also, certain treatment systems installed (i.e., ion-exchange, oil recovery, and hydrolyzer-stripper systems) may, in effect, be recovery systems for direct or indirect reuse of effluent constituents. Finally, the use of effluent gas scrubbers to improve in-plant operations by preventing gaseous product losses may also prevent the airborne deposition of various pollutants within the general plant area, from where they end up as surface drainage runoff contaminants.

Cooling Water

Cooling water constitutes a major portion of the total in-plant wastes in fertilizer manufacturing and it includes water coming into direct contact with the gases processed (largest percentage) and water that has no such contact. The latter stream can be readily used in a closed-cycle system, but sometimes the direct contact cooling water is also recycled (after treatment to

Table 14 Effluent Limitations (mg/L except for pH) for Subpart F, Sodium Phosphates

Effluent characteristic	Maximum for any 1 day	Average of daily values for 30 consecutive days shall not exceed
(a) BPT		
TSS	0.50	0.25
Total phosphorus (as P)	0.80	0.40
Fluoride (as F)	0.30	0.15
pH	a	a
(b) BAT		
Total phosphorus (as P)	0.56	0.28
Fluoride (as F)	0.21	0.11
(c) NSPS		
TSS	0.35	0.18
Total phosphorus (as P)	0.56	0.28
Fluoride (as F)	0.21	0.11
pH	a	a
(d) BCT		
TSS	0.35	0.18
pH	a	a

[a] Within the range 6.0–9.5.
BPT, best practicable control technology currently available; BAT, best available technology economically available; BCT, best conventional pollutant control technology; NSPS, new source performance standards.
Source: Ref. 9.

remove dissolved gases and other contaminants, and clarification). By recycling, the amount of these wastewaters can be reduced by 80–90%, with a corresponding reduction in gas content and suspended solids in the wastes discharged to sewers or surface water [18,35,36]. Wang [35,36], Caswell [37], and Hallett [38] introduce new technologies and methodologies for cooling water treatment and recycle.

Phosphate Manufacturing

Significant in-plant control of both waste quantity and quality is possible for most subcategories of the phosphate manufacturing industry. Important control measures include stringent in-process abatement, good housekeeping practices, containment provisions, and segregation practices [8]. In the phosphorus chemicals industry (subcategories A, B, and C in Table 3), plant effluent can be segregated into noncontact cooling water, process water, and auxiliary streams comprising ion-exchange regenerants, cooling tower blowdowns, boiler blowdowns, leaks, and washings. Many plants have accomplished the desired segregation of these streams, often by a painstaking rerouting of the sewer lines. The use of once-through scrubber waste should be discouraged; however, there are plants that recycle the scrubber water from a sump, thus satisfying the scrubber water flow rate demands on the basis of mass transfer considerations while retaining control of water usage.

The containment of phossy water from phosphorus transfer and storage operations is an important control measure in the phosphorus-consuming sub-category B. Although displaced phossy water is normally shipped back to the phosphorus-producing facility, the usual practice in phosphorus storage tanks is to maintain a water blanket over the phosphorus for safety reasons.

This practice is undesirable because the addition of makeup water often results in the discharge of phossy water, unless an auxiliary tank collects phossy water overflows from the storage tanks, thus ensuring zero discharge. A closed-loop system is then possible if the phossy water from the auxiliary tank is reused as makeup for the main phosphorus tank.

Another special problem in phosphorus-consuming subcategory B is the inadvertent spills of elemental phosphorus into the plant sewer pipes. Provision should be made for collecting, segregating, and bypassing such spills, and a recommended control measure is the installation of a trap of sufficient volume just downstream of reaction vessels. In the phosphates subcategory C, an important area of concern is the pickup by stormwater of dust originating from the handling, storing, conveying, sizing, packaging, and shipping of finely divided solid products. Airborne dust can be minimized through air pollution abatement practices, and stormwater pickup could be further controlled through strict dust cleanup programs.

In the defluorinated phosphate rock (D) and defluorinated phosphoric acid (F) subcategories, water used in scrubbing contaminants from gaseous effluent streams constitutes a significant part of the process water requirements. In both subcategories, process conditions do permit the use of contaminated water for this service. Some special precautions are essential at a plant producing sodium phosphates (F), where all meta-, tetra-, pyro-, and polyphosphate wastewater in spills should be diverted to the reuse pond. These phosphates do not precipitate satisfactorily in the lime treatment process and interfere with the removal of fluoride and suspended solids. Because unlined ponds are the most common treatment facility in the phosphate manufacturing industry, the prevention of pond failure is vitally important. Failures of these ponds sometimes occur because they are unlined and may be improperly designed for containment in times of heavy rainfall. Design criteria for ponds and dikes should be based on anticipated maximum rainfall and drainage requirements. Failure to put in toe drainage in dikes is a major problem, and massive contamination from dike failure is a major concern for industries utilizing ponds.

The following are possible process modifications and plant arrangements [7] that could help reduce wastewater volumes, contaminant quantities, and treatment costs:

1. In ammonium phosphate production and mixed and blend fertilizer manufacturing, one possibility is the integration of an ammonia process condensate steam stripping column into the condensate-boiler feedwater systems of an ammonia plant, with or without further stripper bottoms treatment depending on the boiler quality makeup needed.
2. Contaminated wastewater collection systems designed so that common contaminant streams can be segregated and treated in minor quantities for improved efficiencies and reduced treatment costs.
3. In ammonium phosphate and mixed and blend fertilizer (G) production, another possibility is to design for a lower-pressure steam level (i.e., 42–62 atm) in the ammonia plant to make process condensate recovery easier and less costly.
4. When possible, the installation of air-cooled vapor condensers and heat exchangers would minimize cooling water circulation and subsequent blowdown.

In a recent document [25] presenting techniques adopted by the French for pollution prevention, a new process modification for steam segregation and recycle in phosphoric acid production is described. As shown in Figure 9, raw water from the sludge/fluorine separation system is recycled to the heat-exchange system of the sulfuric acid dilution unit and the wastewater used in plaster manufacture. Furthermore, decanted supernatant from the phosphogypsum deposit pond is recycled for treatment in the water filtration unit. The claim was that this process modification permits an important reduction in pollution by

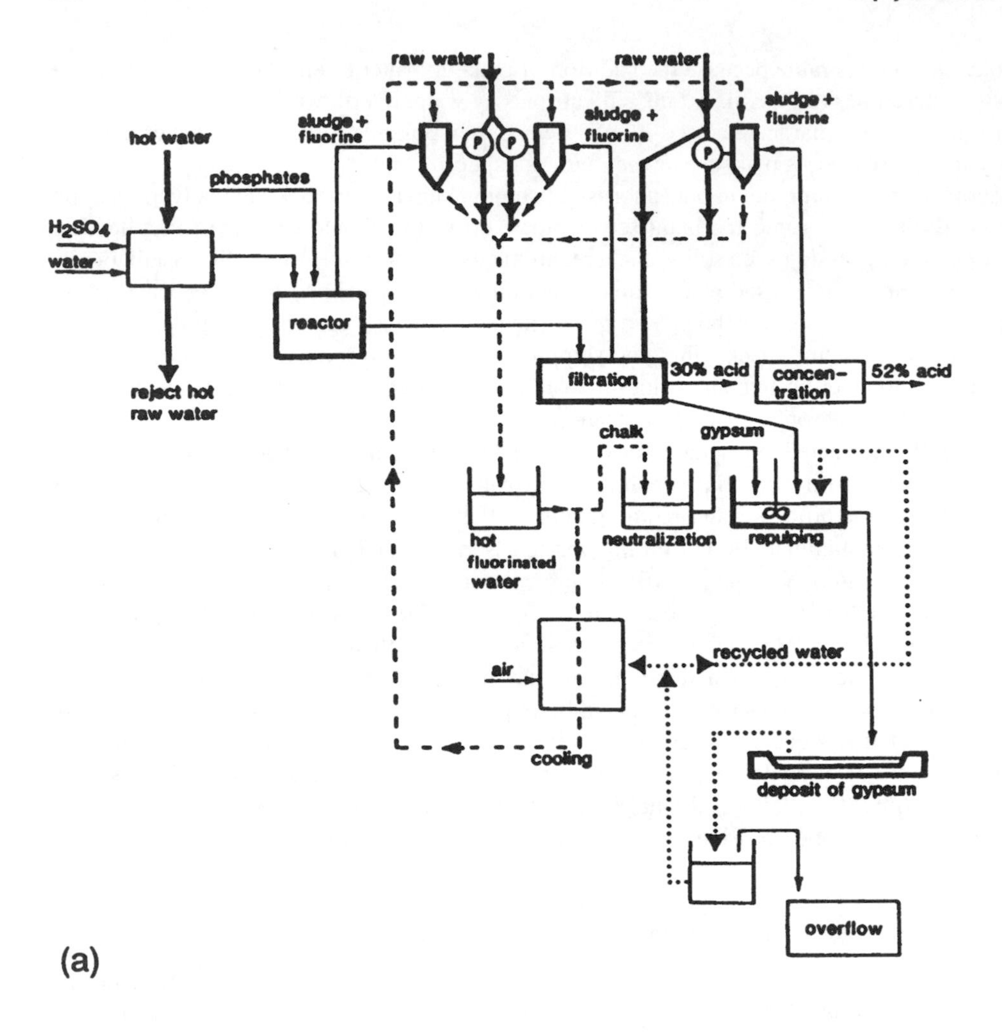

Figure 9 (a) The old process flow diagram and (b) the new process modification for steam segregation and recycle in phosphoric acid production (from Ref. 25).

fluorine, and that it makes the treatment of effluents easier and in some cases allows specific recycling. Finally, the new process produced a small reduction in water consumption, either by recycle or discharging a small volume of polluted process water downstream, and required no particular equipment and very few alterations in the mainstream lines of the old process.

9.5.2 Wastewater Treatment Methods

Phosphate Manufacturing

Nemerow [23] summarized the major characteristics of wastes from phosphate and phosphorus compounds production (i.e., clays, slimes and tall oils, low pH, high suspended solids, phosphorus, silica, and fluoride) and suggested the major treatment and disposal methods such as lagooning, mechanical clarification, coagulation, and settling of refined wastewaters. The

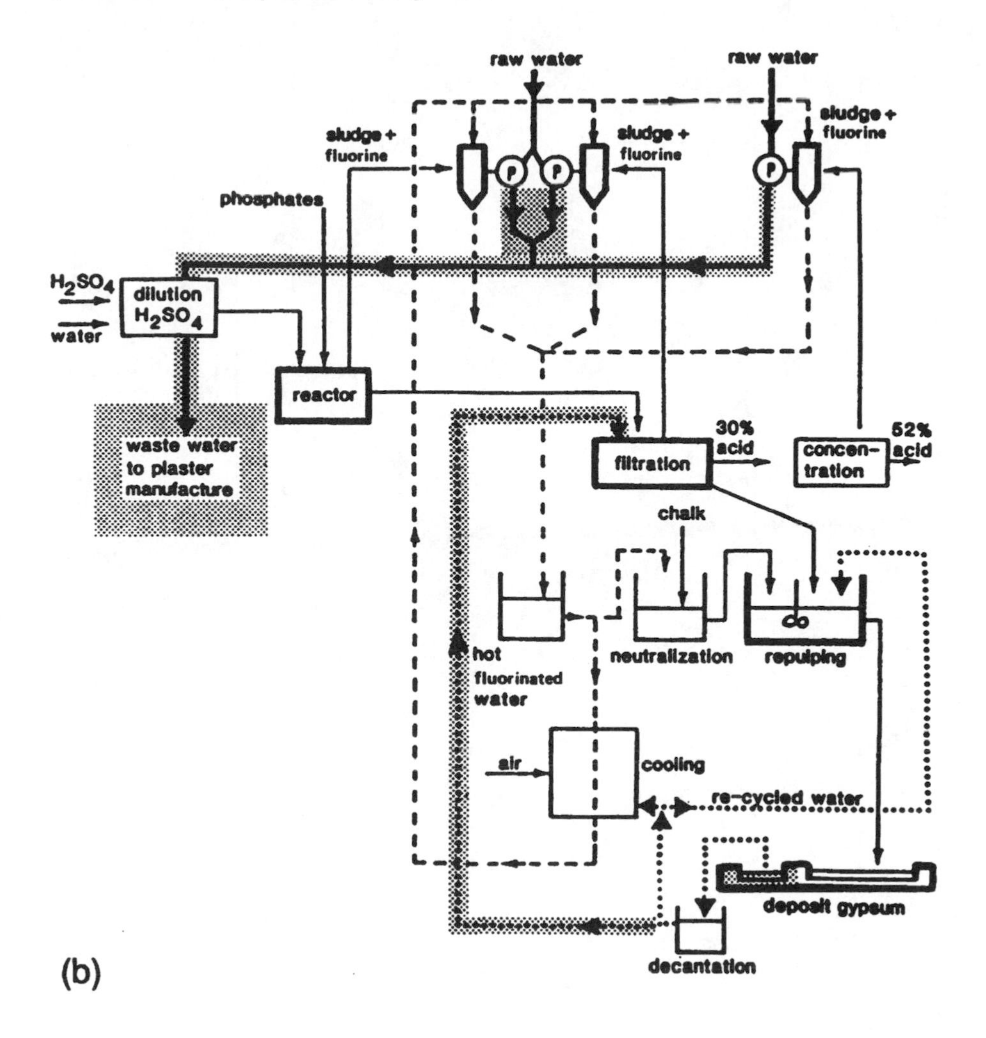

Figure 9 Continued.

various wastewater treatment practices for each of the six subcategories (Table 3) of the phosphate manufacturing industry were summarized by the USEPA [8] as shown in Table 15. The percent removal efficiencies indicated in this table pertain to the raw waste loads of process effluents from each of these six subcategories. As can be seen from Table 15, the predominant method of removal of primary pollutants such as TDS, TSS, total phosphate, phosphorus, fluoride, sulfate, and for pH adjustment or neutralization is lime treatment followed by sedimentation.

Phosphate Fertilizer Production

Contaminated water from the phosphate fertilizer subcategory A is collected in gypsum ponds and treated for pH adjustment and control of phosphorus and fluorides. Treatment is achieved by "double liming" or a two-stage neutralization procedure, in which phosphates and fluorides precipitate out [7]. The first treatment stage provides sufficient neutralization to raise the pH from 1 to 2 to a pH level of at least 8. The resultant effectiveness of the treatment depends on the

Table 15 Phosphate Manufacturing Industry Wastewater Treatment Practices and Unit Removal Efficiencies (%) and Effluent pH (unit)

Pollutant and method	Phosphorus production (A)	Phosphorus-consuming (B)	Phosphate (C)	Defluorinated phosphate rock (D)	Defluorinated phosphoric acid (E)	Sodium phosphate (F)
TDS						
Lime treatment and sedimentation[a]	99		99			
TSS						
Lime treatment and sedimentation[a]	99		99			
Flocculation, clarification, and dewatering		92				
Total phosphate						
Lime treatment and sedimentation[a]	97	73–97	97			
Phosphorus						
Lime treatment and sedimentation[a]				90	99	88
Flocculation, clarification, and dewatering		92				
Sulfate						
Lime treatment and sedimentation[a]	98		98			
Fluoride						
Lime treatment and sedimentation[a]	99		99	98	96	0
pH (effluent level)						
Lime treatment and sedimentation[a] (neutralization)				6–8[b]	6–8[b]	6–8[b]

[a] Preceded by recycle of phossy water and evaporation of some process water in subcategories A, B, and C.
TDS, total dissolved solids; TSS, total suspended solids.
[b] pH values in units.
Source: Ref. 8.

point of mixing of lime addition and on the constancy of pH control. Flurosilisic acid reacts with lime and precipitates calcium fluoride in this step of the treatment.

The wastewater is again treated with a second lime addition to raise the pH level from 8 to at least 9 (where phosphate removal rates of 95% may be achieved), although two-stage dosing to pH 11 may be employed. Concentrations of phosphorus and fluoride with a magnitude of 6500 and 9000 mg/L, respectively, can be reduced to 5–500 mg/L P and 30–60 mg/L F. Soluble orthophosphate and lime react to form an insoluble precipitate, calcium hydroxy apatite [17]. Sludges formed by lime addition to phosphate wastes from phosphate manufacturing or fertilizer production are generally compact and possess good settling and dewatering characteristics, and removal rates of 80–90% for both phosphate and fluoride may be readily achieved [13].

The seepage collection of contaminated water from phosphogypsum ponds and re-impoundment is accomplished by the construction of a seepage collection ditch around the perimeter of the diked storage area and the erection of a secondary dike surrounding the first [8]. The base of these dikes is usually natural soil from the immediate area, and these combined earth/gypsum dikes tend to have continuous seepage through them (Fig. 10). The seepage collection ditch between the two dikes needs to be of sufficient depth and size to not only collect contaminated water seepage, but also to permit collection of seeping surface runoff from the immediate outer perimeter of the seepage ditch. This is accomplished by the erection of the small secondary dike, which also serves as a backup or reserve dike in the event of a failure of the primary major dike.

The sulfuric acid plant has boiler blowdown and cooling tower blowdown waste streams, which are uncontaminated. However, accidental spills of acid can and do occur, and when they do, the spills contaminate the blowdown streams. Therefore, neutralization facilities should be supplied for the blowdown waste streams (Table 15), which involves the installation of a reliable pH or conductivity continuous-monitoring unit on the plant effluent stream. The second part of the system is a retaining area through which non-contaminated effluent normally flows. The detection and alarm system, when activated, causes a plant shutdown that allows location of the failure and initiation of necessary repairs. Such a system, therefore, provides the continuous protection of natural drainage waters, as well as the means to correct a process disruption.

Mixed fertilizer (subcategory G) treatment technology consists of a closed-loop contaminated water system, which includes a retention pond to settle suspended solids. The water is then recycled back to the system. There are no liquid waste streams associated with the blend fertilizer (subcategory G) process, except when liquid air scrubbers are used to prevent air pollution. Dry removals of air pollutants prevent a wastewater stream from being formed.

Phosphate and Fluoride Removal

Phosphates may be removed from wastewaters by the use of chemical precipitation as insoluble calcium phosphate, aluminum phosphate, and iron phosphate [5]. The liming process has been discussed previously, lime being typically added as a slurry, and the system used is designed as either a single- or two-stage one. Polyelectrolytes have been employed in some plants to improve overall settling. Clarifier/flocculators or sludge-blanket clarifiers are used in a number of facilities [11]. Alternatively, the dissolved air flotation (DAF) process is both technically and economically feasible for phosphate and fluoride removal, according to Wang *et al.* [39–43] and Krofta and Wang [44–48]. Both conventional biological sequencing batch reactors (SBR) and innovative physicochemical sequencing batch reactors (PC-SBR) have been proven to be highly efficient for phosphate and fluoride precipitation and removal [32,43].

A number of aluminum compounds, such as alum and sodium aluminate, have also been used by Layer and Wang [20] as phosphate precipitants at an optimum pH range of 5.5–6.5, as

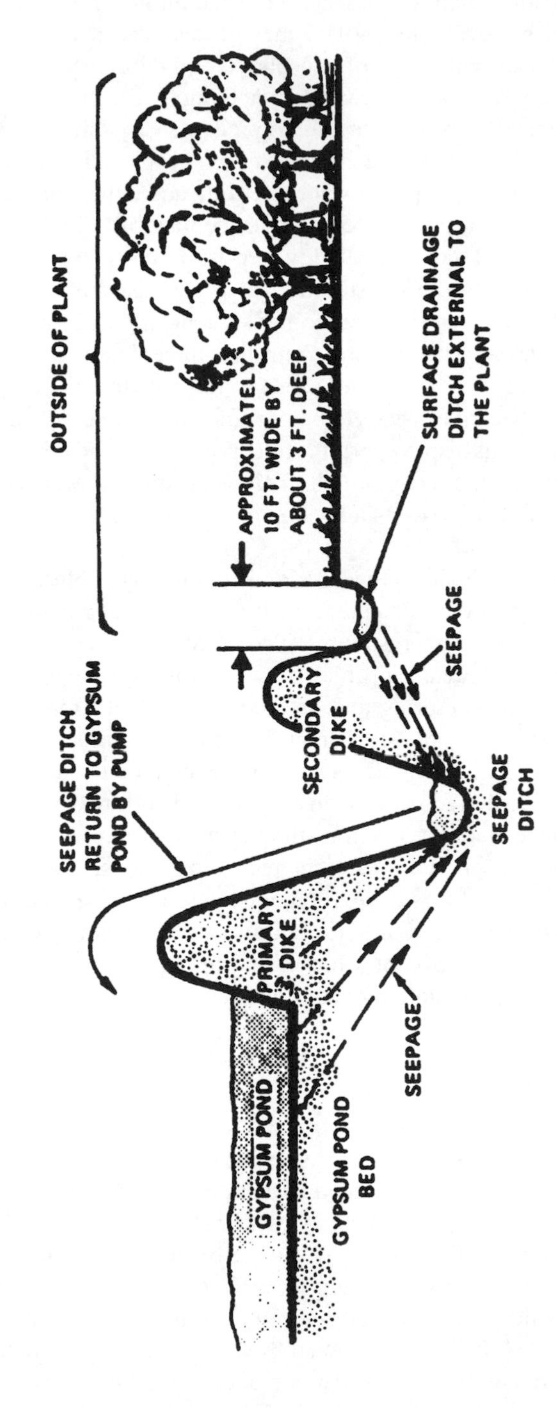

Figure 10 Phosphogypsum pond water seepage control (from Ref. 7).

have iron compounds such as ferrous sulfate, ferric sulfate, ferric chloride, and spent pickle liquor [17]. The optimum pH range for the ferric salts is 4.5–5, and for the ferrous salts it is 7–8, although both aluminum and iron salts have a tendency to form hydroxyl and phosphate complexes. As reported by Ghokas [13], sludge solids produced by aluminum and iron salts precipitation of phosphates are generally less settleable and more voluminous than those produced by lime treatment.

Removal of Other Contaminants

Compressor houses, tank farm areas, cooling water, and loading or unloading bays may be sources of oil in wastewater discharges from phosphate and fertilizer manufacturing. Oil concentrations may range from 100–900 mg/L and can be removed by such units as coke filters or recovered by the use of separators (usually operating in series to recover high oil levels). Chromates and dichromates are present in cooling tower blowdown, at levels of about 10 mg/L, because they are used in cooling water for corrosion inhibition. They may be removed (over 90%) from cooling tower discharges through chemical reduction (i.e., the use of sulfuric acid for lowering pH to less than 4 and addition of a reducing agent) and precipitation as the hydroxide by the addition of lime or NaOH. Chromates may also be recovered and reused from cooling tower blowdown by the use of ion exchange, at recovery levels in excess of 99%, employing a special weak base anion resin that is regenerated with caustic soda.

Phosphoric Acid Production

The use of the electric furnace process (Fig. 11) and acidulation of phosphate-bearing rock is made commercially to produce phosphoric acid. In the first method, elemental phosphorus is first produced from phosphate ore, coke, and silica in an electric furnace, and then the phosphorus is burned with air to form P_2O_5, which is cooled and reacted with water to form orthophosphoric acid (Section 9.2.2). Extremely high acid mist loadings from the acid plant are common, and there are five types of mist-collection equipment generally used: packed towers, electrostatic precipitators, venturi scrubbers, fiber mist eliminators, and wire mesh contactors [21]. Choosing one of these control equipments depends on the required contaminant removal efficiency, the required materials of construction, the pressure loss allowed through the device, and capital and operating costs of the installation (with very high removal efficiencies being the primary factor). The venturi scrubber is widely used for mist collection and is particularly applicable to acid

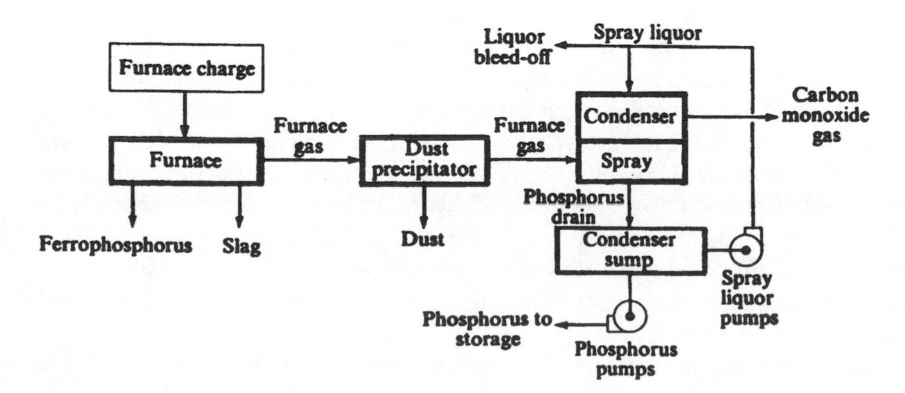

Figure 11 Flow diagram of electric furnace process for phosphorous production (from Ref. 15).

plants burning sludge. The sludge burned is an emulsion of phosphorus, water, and solids carried out in the gas stream from the phosphorus electric furnace as dust or volatilized materials. Impurities vary from 15 to 20% and the venturi scrubber can efficiently collect the acid mist and fine dust discharged in the exhaust from the hydrator.

Wet process phosphoric acid is made by reacting pulverized, beneficiated phosphate ore with sulfuric acid to form calcium sulfate (gypsum) and dilute phosphoric acid (see Section 9.2.4). The insoluble calcium sulfate and other solids are removed by filtration, and the weak (32%) phosphoric is then concentrated in evaporators to acid containing about 55% P_2O_5. Mist and gaseous emissions from the gypsum filter, the phosphoric acid concentrator, and the acidulation off-gas are controlled with scrubbers or other equipment. The preparation of the phosphate ore generates dust from drying and grinding operations, and this is generally controlled with a combination of dry cyclones and wet scrubbers [21]. The material collected by the cyclones is recycled, and the scrubber water discharged to the waste phosphogypsum ponds. Most frequently, simple towers and wet cyclonic scrubbers are used, but at some plants the dry cyclone is followed by an electrostatic precipitator.

9.6 CASE STUDIES OF TREATMENT FACILITIES

Phosphate production and phosphate fertilizer manufacturing facilities are situated in many areas in the United States (primarily in Florida and California) and in other countries such as Algeria, Jordan, and Morocco, as previously mentioned. The wastewaters from production and cleanup activities and surface runoff in most of these locations are stored, treated, and recycled, and the excess overflows are discharged into natural water systems. In those facilities where wastewater from production and cleanup activities and drainage are discharged into municipal water systems and treated together with domestic, commercial, institutional, and other industrial wastewaters, a degree of pretreatment is required to meet federal guidelines or local ordinances such as those presented in Section 9.4. For instance, according to the USEPA [6], the pretreatment unit operations required for the phosphate fertilizer industry comprise solids separation and neutralization, and it may be achieved by either a suspended biological process, a fixed-film biological process, or an independent physicochemical system.

9.6.1 Pebble Phosphate Mining Industry

In one of the earlier reports on the phosphate mining and manufacturing industry in Florida and its water pollution control efforts, Wakefield [31] gave the following generalized account. Because of the huge volumes of water being used for washing, hydraulic sizing, flotation, and concentration of phosphate ores (i.e., one of the main mines of a larger company requires about 60 MGD or 2.63 m^3/s), and since makeup water is not readily available and excess wastewater constitutes a major disposal problem, the recovery and reuse of water have always been of great importance. Waste products from the mining and processing operation consist of large quantities of nonphosphating sands and clays, together with unrecovered phosphatic materials less than 300 mesh in size, and they are pumped into huge lagoons. Easily settled sands fill the near-end, leaving the rest to be gradually filled with "slimes" (a semicolloidal water suspension), while a thin layer of virtually clear water at the surface of the lagoon flows over spillways and is returned to the washers for reuse.

The above ideal wastewater management, however, is infeasible during wet weather (rains of 3 in./day or 7.6 cm/day are frequent in the tropical climate of Florida), because more water goes into the lagoons than can be used by the washers, and the excess volume must be discharged into nearby surface waters. In particular in small streams, this results in highly turbid waters due to the fact that the larger of the slowly consolidating slimes is near the surface of the lagoons in most cases. Furthermore, there are no core walls (for cost reasons) in the large earthen dams forming the settling lagoons; rather, it is usual to depend on the slimes to seal their inner face and prevent excessive seepage. The entire operation, therefore, involves a delicate balance of slime input and weir discharge to accomplish the objective of a maximum of water reuse with a minimum of danger of dam failure and a minimum of turbid discharge to the stream. Such dams have failed very often and the pollution effect of the volume of, for example, a 100 acre (0.4 km^2) pond with 25–30 ft (7.6–9.1 m) of consolidated slime being discharged into a stream with a mean flow of possibly 300 cfs (8.5 m^3/s) has been devastating.

There are, however, other much more frequent situations when effluents of higher or lower turbidity are discharged to streams to protect dam structures or as excess flow when it rains heavily. These discharges of turbid wastewater volumes may be due to underdesigning the settling lagoons or because the wastewater slimes are of a more completely colloidal nature and do not clarify too well. They usually continue over extended periods and cause noticeable stream turbidities, although nothing approaching those encountered after a lagoon dam failure. When streams contain appreciable turbidity due to phosphate industry effluents, local sports fishermen claim that it ruins fishing (but mostly they just prefer fishing in clear waters), while industry managers have demonstrated that fish are not affected by stocking mined-out pits and settling ponds with bass and other species. Finally, a positive effect of discharging moderate levels of turbidity noted in a water treatment plant located downstream is a decrease in chemical costs, undoubtedly due to greater turbidity in the raw water that aids coagulation of color and other impurities.

9.6.2 Phosphate Industry Waste Disposal

In one of the earliest extensive studies and reports on the disposal of wastes from the phosphate mining and processing industry in Florida, Specht [29] reviewed the waste treatment and disposal practices in the various phases of phosphate and phosphorus manufacturing. Regarding waste disposal from mining and beneficiation operations, he reported the use of specially constructed settling areas for the clay and quartz sand separated in the washing and flotation processes and also for the clarification of water to be reused in the process or discharged into streams. As mining processes, the mined-out areas are then used as supplementary settling lagoons, with the wastewater circulating through them using specially made cuts, similar to the slow movement through settling areas that are frequently divided into compartments. During the dry season, as mentioned previously, very little (if any) water is wasted to the streams, but sometimes an estimated maximum of 10% of the total amount of water used is wasted at some facilities during the rainy season. This may represent a significant volume, given the large quantities of water needed in phosphate mining (2000–8000 gpm or 7.6–30 m^3/min) and at the recovery plants (4000–50,000 gpm or 15–190 m^3/min), depending on the size of the plant and its method of treatment.

Occasionally, the phosphate "slime" is difficult to settle in the lagoons because of its true colloidal nature, and the use of calcium sulfate or other electrolytes can promote coagulation, agglomeration, and settling of the particles. Usually an addition of calcium sulfate is unnecessary, because it is present in the wastewater from the sand-flotation process. Generally, it has been shown [33] that the clear effluent from the phosphate mining and beneficiation operation is not deleterious to fish life, but the occurrence of a dam break may result in adverse effects [19].

In superphosphate production, fluoride vapors are removed from the mixing vessel, den or barn, and elevators under negative pressure and passed through water sprays or suitable scrubbers. A multiple-step scrubber is required to remove all the fluorides from the gases and vapors, and the scrubbing water containing the recovered fluorosilisic acid and insoluble silicon hydroxide is recycled to concentrate the acid to 18–25%. The hydrated silica is removed from the acid by filtration and is washed with fresh water and then deposited in settling areas or dumps. In triple superphosphate (also known as double, treble, multiple, or concentrated superphosphate) manufacturing, the calcium sulfate cake from the phosphoric acid production is transferred into settling areas after being washed, where the solid material is retained. The clarified water that contains dissolved calcium sulfate, dilute phosphoric acid, and some fluorosilisic acid is either recycled for use in the plant or treated in a two-step process to remove the soluble fluorides, as described in Section 9.5.2. Water from plant washing and the evaporators may also be added to the wastes sent to the calcium sulfate settling area.

According to Specht [30], in the two-step process to remove fluorides and phosphoric acid, water entering the first step may contain about 1700 mg/L F and 5000 mg/L P_2O_5, and it is treated with lime slurry or ground limestone to a pH of 3.2–3.8. Insoluble calcium fluorides settle out and the fluoride concentration is lowered to about 50 mg/L F, whereas the P_2O_5 content is reduced only slightly. The clarified supernatant is transferred to another collection area where lime slurry is added to bring the solution to pH 7, and the resultant precipitate of P is removed by settling. The final clear water, which contains only 3–5 mg/L F and practically no P_2O_5, is either returned to the plant for reuse or discharged to surface waters. The two-step process is required to reduce fluorides in the water below 25 mg/L F, because a single-step treatment to pH 7 lowers the fluoride content only to 25–40 mg/L F. In the process where the triple phosphate is to be granulated or nodulized, the material is transferred directly from the reaction mixer to a rotary dryer, and the fluorides in the dryer gases are scrubbed with water.

In making defluorinated phosphate by heating phosphate rock, one method of fluoride recovery consists of absorption in a tower of lump limestone at temperatures above the dew point of the stack gas, where the reaction product separates from the limestone lumps in the form of fines. A second method of recovery consists of passing the gases through a series of water sprays in three separate spray chambers, of which the first one is used primarily as a cooling chamber for the hot exit gases of the furnace. In the second chamber, the acidic water is recycled to bring its concentration to about 5% equivalence of hydrofluoric acid in the effluent, by withdrawing acid and adding fresh water to the system. In the final chamber, scrubbing is supplemented by adding finely ground limestone blown into the chamber with the entering gases. Hydrochloric acid is sometimes formed as a byproduct from the fluoride recovery in the spray chambers and this is neutralized with NaOH and lime slurry before being transferred to settling areas.

9.6.3 Ammonium Phosphate Fertilizer and Phosphoric Acid Plant

The fertilizer industry is plagued with a tremendous problem concerning waste disposal and dust because of the very nature of production that involves large volumes of dusty material. Jones and Olmsted [16] described the waste disposal problems and pollution control efforts at such a plant, Northwest Cooperative Mills, in St. Paul, Minnesota. Two types of problems are associated with waste from the manufacture of ammonium phosphate: wastes from combining ammonia and phosphoric acid and the subsequent drying and cooling of the products, and wastes from the handling of the finished product arising primarily from the bagging of the product prior to shipping. Because the ammonia process has to be "forced" by introducing excess amounts of ammonia than the phosphoric acid is capable of absorbing, there is high ammonia content in the exhaust air stream from the ammoniator. Because it is neither economically sound nor environmentally acceptable to exhaust this to the atmosphere, an acid scrubber is employed to recover the ammonia without condensing with it the steam that nearly saturates the exhausted air stream.

Drying and cooling the products of ammonium phosphate production are conventionally achieved in a rotary drum, and a means must be provided to remove the dust particles from the air streams to be exhausted to the atmosphere. At the Minnesota plant, a high-efficiency dry cyclone recovery system followed by a wet scrubber was designed. In this way, material recovered from the dry collector (and recycled to the process) pays for the dry system and minimizes the load and disposal problem in the wet scrubber, because it eliminates the need for a system to recover the wet waste material that is discharged to the gypsum disposal pond for settling.

The remaining problem of removing dust from discharges to the ambient air originating from the bagging and shipping operations is the one most neglected in the fertilizer industry, causing complaints from neighbors. Jones and Olmsted [16] reported the installation of an elaborate, relatively expensive, system of suction pickups at each transfer point of the products in the entire bagging system and shipping platform areas. The collected dust streams are passed over a positive cloth media collector before discharge to the atmosphere, and the system recovers sufficient products to cover only operating expenses.

The filtered gypsum cake from phosphoric acid production, slurried with water to about 30%, is pumped to the settling lagoons, from where the clarified water is recycled to process. To provide a startup area, approximately two acres (or 8100 m^2) of the disposal area were blacktopped to seal the soil surface against seepage, and the gypsum collected in this area was worked outward to provide a seal for enlarging the settling area. A dike-separated section of the disposal area was designated as a collection basin for all drainage waters at the plant site. From this basin, after the settling of suspended solid impurities, these waters are discharged to surface waters under supervision from a continuous monitoring and alarm system that guards against accidental contamination from any other source.

Air streams from the digestion system, vacuum cooler, concentrator, and other areas where fluorine is evolved are connected to a highly efficient absorption system, providing extremely high volumes of water relative to the stream. The effluent from this absorption system forms part of the recycled water and is eventually discharged as part of the product used for fertilizer manufacture. The Minnesota plant requires a constant recirculating water load in excess of 3000 gpm (11.4 m^3/min), but multiple use and recycle reduce makeup requirements to less than 400 gpm (1.5 m^3/min) or a mere 13% of total water use.

9.6.4 Rusaifa Phosphate Mining and Processing Plant

As reported by Shahalam *et al.* [28], Jordan stands third in the region following Morocco and Algeria with respect to the mining of phosphate rock and production of phosphates, and the Jordanian phosphate industry is bound to grow, with time creating additional environmental problems. The paper presented the results of a study assessing the phosphate deposits and pollution resulting from a phosphate processing operation in Rusaifa, Jordan. The beneficiation plant uses about 85% of the total process water, and the overflow from the hydrocyclones (rejected as slimes of silica carbonates and clay materials) is taken to a gravity thickener, the underflow (about 0.93 m^3/min or 245 gpm) from which is directly discharged to a nearby holding pond as wastewater and slimes (sludge containing about 25% solids by wt.). As shown in Table 16, the sludge contains significant amounts of fluorine, sulfate, P_2O_5, and organics, and it is unsuitable for direct or indirect discharge into natural waters, in accordance with U.S. 40 CFR

Table 16 Chemical Analysis of the Characteristics of Underflow Sludge from the Gravity Thickener

Parameter	Contents
pH	7.75
Solids	About 10% by sludge volume
P_2O_5	24.4% of dry solid
Total orthophosphate	0.30 mg/L
SiO_2 (inorganic)	14.16% of dry solid
Total SiO_2	16.83% of dry solid
Organics	1.00% of dry solid
Total chlorine	0.09% of dry solid
CaO	40.43% of dry solid
Fluorine	2.96% of dry solid
$CaSO_4$	0.32% of dry solid

Source: Ref. 28.

(see Section 9.4). The water portion of this thick sludge partially evaporates and the remainder percolates into the ground through the bottom of the holding pond.

The second major wastewater discharge (about 1.2 m^3/min or 318 gpm) is from the bottom of the scrubbers used after a dry dust collector cyclone to reduce the dust concentration in the effluent air stream from the phosphate dryer. The underflow of the scrubber contains a high concentration of dust and mud, laden with tiny phosphate organic and inorganic silica and clay particles, and is disposed of in a nearby stream. The main pollutant in the flow to the stream is P_2O_5 at a concentration of nearly 1200 mg/L and it remains mostly as suspended particles.

Finally, most of the P_2O_5 content in the solid wastes generated from Rusaifa Mining is mainly from overburdens and exists as solid particles. Because it does not dissolve in water readily, the secondary pollution potential with respect to P_2O_5 is practically nonexistent. However, loose overburdens in piles, when carried off by rainwater, may create problems for nearby stream(s) due to the high suspended solids concentration resulting in stormwater run-off. Also, loose overburdens blown by strong winds may cause airborne dust problems in neighboring areas; therefore, a planned land reclamation of the mined-out areas is the best approach to minimizing potential pollution from the solid wastes.

9.6.5 Furnace Wastes from Phosphorus Manufacture

The electric furnace process (Fig. 11) for the conversion of phosphate rock into phosphorus was described by Horton *et al.* [15] in a paper that also presented the results of a pilot plant study of treating the wastes produced. The process, as well as the handling of the various waste streams for pollution control, are discussed in Section 9.5.2. In processing the phosphate, the major source of wastewater is the condenser water bleedoff from the reduction furnace, the flow of which varies from 10 to 100 gpm (2.3–22.7 m^3/hour) and its quality characteristics are presented in Table 7.

Phossy water, a waste product in the production of elemental phosphorus by the electric furnace process, contains from 1000 to 5000 mg/L suspended solids that include 400–2500 mg/L of elemental phosphorus, distributed as liquid colloidal particles. These particles are usually positively charged, although this varies depending on the operation of the electrostatic precipitators. Furthermore, the chemical equilibrium between the fluoride and flurosilicate ions introduces an important source of variation in suspended solids that is a pH function. Commonly used coagulants such as alum or ferric chloride were unsatisfactory for wastewater clarification because of the positively charged particles, as were inorganic polyelectrolytes despite their improved performance. However, high-molecular-weight protein molecules at a suitable pH level (which varies for each protein) produced excellent coagulation and were highly successful in clarifying phossy water.

Horton *et al.* [15] investigated or attempted such potential treatment and disposal methods as lagooning, oxidizing, settling (with or without prior chemical coagulation), filtering, and centrifuging and concluded that the best solution appears to be coagulation and settling. The pilot units installed to evaluate this optimum system are shown in Figure 12, together with a summary of the experimental results. The proper pH for optimum coagulation with proteins alone was obscured at higher pH levels by the formation of silica, which tended to encrust the pipelines. It was found that the addition of clay, as a weighting agent with the coagulant, eliminated the scale problem without decreasing the settling rates. Finally, it was concluded that in the pilot plant it was possible to obtain a 40-fold concentration of suspended solids (or 25% solids) by a simple coagulation and sedimentation process.

9.6.6 Phosphate Fertilizer Industry in Eastern Europe

Koziorowski and Kucharski [18] presented a survey of fertilizer industry experience in Eastern European countries and compared it with the United States and Western European equivalents. For instance, they stated that HF and silicofluoric acid are evolved during the process of

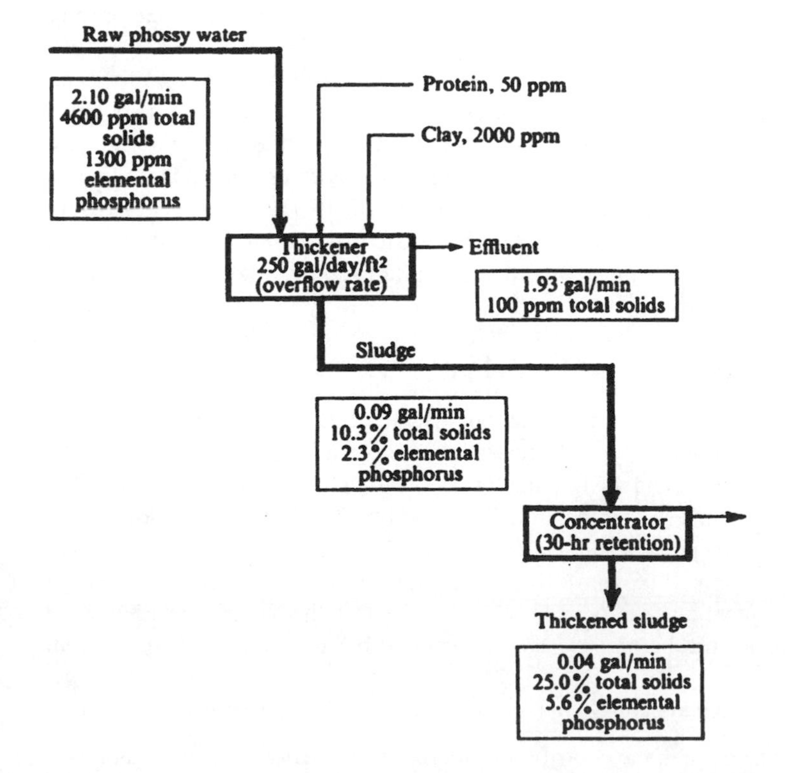

Figure 12 Summary of materials balance in a pilot plant for recovery of phosphorus from phossy water (from Ref. 15). 1 ppm = 1 mg/L.

dissolving phosphorite in the manufacture of normal superphosphate. These constituents are removed from the acidic and highly toxic gases by washing with water or brine in closed condenser equipment, and the wastes from this process are usually slightly acidic, clear, and colorless.

The production of superphosphate is often combined with the manufacture of sodium fluorosilicate and then the amount of wastes is larger.

Czechoslovakian experiments have shown that for every 1000 tons of superphosphate (20% P_2O_5 content) produced, 133 m^3 (4700 ft^3) of postcrystallization liquor from the crystallization of sodium fluorosilicate and 67 m^3 (2370 ft^3) of washings are discharged. The liquor contains 20–25 g/L NaCl, 25–35 g/L HCl, 10–15 g/L silisic acid, and 8–10 g/L sodium silicofluoride. For this waste, the most significant aspect of treatment is recovering the sodium silicofluoride from the brine used to absorb fluorine compounds from the gaseous waste streams, and this is relatively easy to accomplish since it settles nearly ten times as fast as silicic acid gel and, therefore, it is separated by sedimentation. The silicofluoride recovered is a valuable byproduct that, following filtration, washing with water, and drying, is used as a flux in enamel shops, glass works, and other applications.

In Czechoslovakian phosphate fertilizer plants, the superphosphate production wastewaters are further treated by neutralization on crushed limestone beds contained in special tanks that are followed by settling tanks for clarification of the wastewater. The beds have from three to five layers (with a minimum bed height of 0.35–1.60 m), treat a range of acidity of wastes from 438 to 890 meq/L, and are designed for a hydraulic load ranging from 0.13 to 0.52 cm^3/cm^2 s (1.9–7.7 gpm/ft^2) at operating temperatures of 20–28°C. This experience agrees with results reported from Polish plants, the limestone used contains 56% CaO, and it was found in practice that coarse particles of 3–5 mm give better results because less material is carried away. In the former USSR, superphosphate wastes were neutralized with powdered limestone or milk of lime.

The neutralized wastes leaving the settling tank contain primarily dissolved sodium and calcium chlorides. As previously mentioned, the manufacture of phosphate fertilizers also yields large quantities of phosphogypsum, which often contains significant amounts of fluorine and phosphorus compounds and requires large areas for dumping. It has been estimated that for each ton of phosphorite processed, a wet process phosphoric acid plant yields 1.4–1.6 tons of gypsum containing about 30% water and 66% calcium sulfate. This waste material has been used for the production of building materials such as plasterboard, ammonium sulfate, but primarily sulfuric acid and cement.

9.6.7 Phosphoric Acid and N-P-K Fertilizer Plant

According to the literature [3,17,33], the heterogeneous nature of fertilizer production plants precludes the possibility of presenting a "typical" case study of such a facility. Nevertheless, the wastewater flows, the characteristics, and the treatment systems for a phosphoric acid and N-P-K fertilizer plant were parts of a large fertilizer manufacturing facility. The full facility additionally included an ammonia plant, a urea plant, a sulfuric acid plant, and a nitric acid plant. The typical effluent flows were 183 m^3/hour (806 gpm) from the phosphoric plant and 4.4 m^3/hour (20 gpm) from the water treatment plant associated with it, whereas in the N-P-K plant they were 420 m^3/hour (1850 gpm) from the barometric condenser and 108 m^3/hour (476 gpm) from other effluent sources.

These wastewater effluents had quality characteristics that could be described as follows:

1. In the phosphoric acid plant, the contributing sources of effluent are the cooling tower bleedoff and the scrubber liquor solution that contains concentrations ranging for phosphate from 160 to 200 mg/L and for fluoride from 225 to 7000 mg/L.

2. In the water treatment plant, the wastewater effluent is slightly acidic in nature.
3. In the N-P-K plant, the barometric condenser effluent has a pH range of 5.5–8, and concentrations of ammonia-nitrogen of about 250 mg/L, fluoride of about 10 mg/L, and trace levels of phosphate.
4. The N-P-K plant's other effluents contain concentrations of ammonia-nitrogen of about 2000 mg/L, fluoride of about 350 mg/L, and phosphate of about 3000 mg/L.

The wastewater treatment systems utilized for the phosphoric acid and N-P-K plant effluents are shown in Figure 13. As can be seen, the cooling tower bleedoff and scrubber liquor from the phosphoric acid plant are treated together with N-P-K plant effluents by a two-stage lime slurry addition to precipitate out the phosphates and fluorides, reducing them to levels of less than 10 mg/L. The treated effluent pH is adjusted to 5.5–7 using sulfuric acid, and it is discharged to a river, while the precipitated slurry containing the phosphates and fluorides is disposed of in lagoons. As can be seen in the right-hand side of Figure 13, the effluent of the barometric condenser has its pH adjusted to 11 by adding lime to remove residual ammonia-nitrogen, and subsequently waste steam is introduced to remove free ammonia, and the final effluent is mixed with the water treatment plant effluent prior to discharge in a river.

9.6.8 Environmentally Balanced Industrial Complexes

Unlike common industrial parks where factories are selected simply on the basis of their willingness to share the real estate, environmentally balanced industrial complexes (EBIC) are a selective collection of compatible industrial plants located together in a complex so as to minimize environmental impacts and industrial production costs [24,33]. These objectives are accomplished by utilizing the waste materials of one plant as the raw materials for another with a minimum of transportation, storage, and raw materials preparation costs. It is obvious that when an industry neither needs to treat its wastes, nor is required to import, store, and pretreat its raw materials, its overall production costs must be reduced significantly. Additionally, any material reuse costs in an EBIC will be difficult to identify and more easily absorbed into reasonable production costs.

Such EBICs are especially appropriate for large, water-consuming, and waste-producing industries whose wastes are usually detrimental to the environment, if discharged, but they are also amenable to reuse by close association with satellite industrial plants using wastes from and

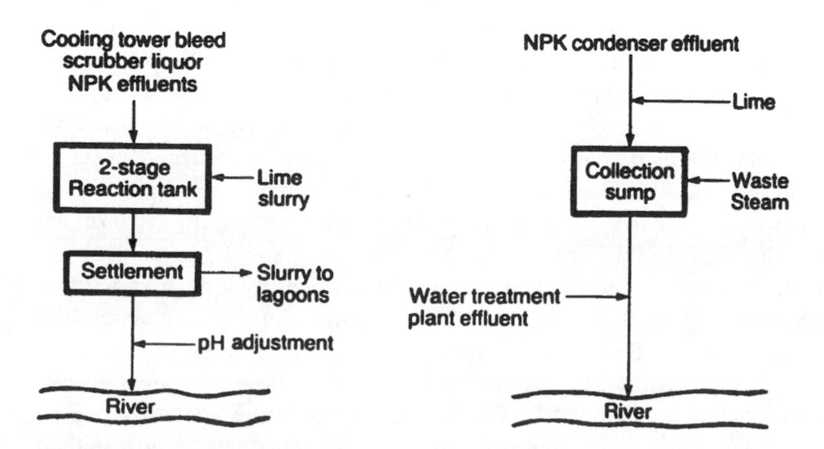

Figure 13 Phosphoric acid and N-P-K fertilizer production waste treatment (from Ref. 17).

producing raw materials for others within the complex. Examples of such major industries that can serve as the focus industry of an EBIC are fertilizer plants, steel mills, pulp and paper mills, and tanneries. Nemerow and Dasgupta [24] presented the example of a steel mill complex with a phosphate fertilizer and a building materials plant as the likely candidates for auxiliary or satellite industries (Fig. 14).

A second example presented was an EBIC centered phosphate fertilizer plant, with a cement production plant, a sulfuric acid plant, and a municipal solid wastes composting plant (its product to be mixed with phosphate fertilizer and sold as a combined product to the agricultural industry) as the satellite industries (Fig. 15). As previously mentioned, in the usual starting process of producing phosphoric acid and ammonium phosphate fertilizer by dissolving the phosphate rocks with sulfuric acid, a gypsumlike sludge is generated as a byproduct and some sulfur dioxide and fluorine are in the waste gases emitted at the high reaction temperatures. The large, relatively impure, quantities of phosphogypsum (5 vol. to 1 P_2O_5 fertilizer produced) are difficult to treat, and the fluorine present in the gas as hydrofluoric acid (concentrations from 1–10%, which is very low for commercial use) requires further costly and extensive treatment. Using such a fertilizer production facility as the focus industry of an EBIC would be a feasible solution to the environmental problems if combined, for example, with such satellite industries as: (a) a sulfuric acid plant to feed its products to the phosphoric acid plant and to use some of the hot water effluent from the cement plant and the effluent from the SO_2 scrubber of the phosphate fertilizer plant; (b) a municipal solid wastes composting plant utilizing hot water from the cement plant, serving as disposal facility for the garbage of a city, and producing composted organic solids to be used as fertilizer; and (c) a cement and plasterboard production plant utilizing the phosphogypsum waste sludge in the manufacture of products for the construction industry, producing hot water effluent to be used as mentioned above, and waste dust collected by a dust filter and used as a filler for the soil fertilizer produced by mixing composted garbage and phosphate fertilizer.

Development of an environmentally balanced industrial complex has been discussed extensively [49] by the United Nations Industrial Development Organization (UNIDO), Vienna, Austria.

9.6.9 Fluoride and Phosphorus Removal from a Fertilizer Complex Wastewater

A laboratory-scale treatability study was conducted for the Mississippi Chemical Corporation to develop a physicochemical wastewater treatment process for a fertilizer complex wastewater to control nitrogen, phosphorus, and fluoride and to recover ammonia [2,33]. The removal technique investigated consisted of precipitation of fluorides, phosphorus, and silica by lime addition, a second stage required for the precipitation of ammonia by the use of phosphoric acid and magnesium, and a third stage for further polishing of the wastewater necessary to remove residual phosphate. The wastewater quality parameters included the following concentrations: fluoride 2000 mg/L, ammonia 600 mg/L, phosphorus as P_2O_5 (P) 145 mg/L [50], and an acidic pH level of 3.5. Ammonia removals of 96% were achieved and the insoluble struvite complex produced by the ammonia removal stage is a potentially commercial-grade fertilizer product, whereas the fluoride and phosphorus in the effluent fell below 25 and 2 mg/L, respectively.

The multistage treatment approach was to first remove the fluorides by lime precipitation, with the optimum removal (over 99%) occurring with a two-step pH adjustment to about 10.4 (removal was more a function of lime dosage, rather than a pH solubility controlled phenomenon). Each step was followed by clarification, and the required lime equivalent dosage was 180% of the calcium required stoichiometrically. The effluent from the first stage had an average fluoride content of 135 mg/L (93% removal) and a phosphorus content of less than 5 mg/L, whereas the hydraulic design parameters were a 15 mm per stage reaction time

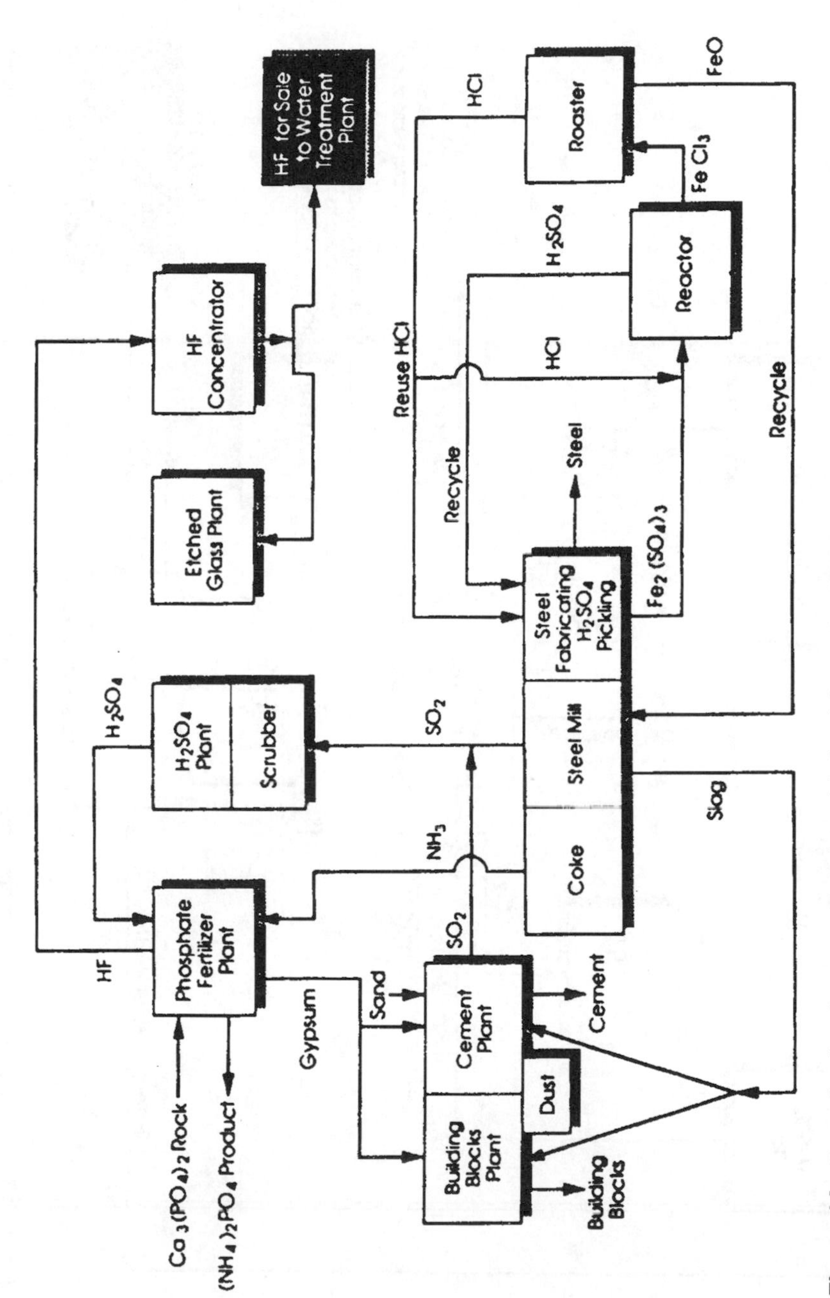

Figure 14 Example of environmentally balanced industrial complex (EBIC) centered about a steel mill plant (from Ref. 24).

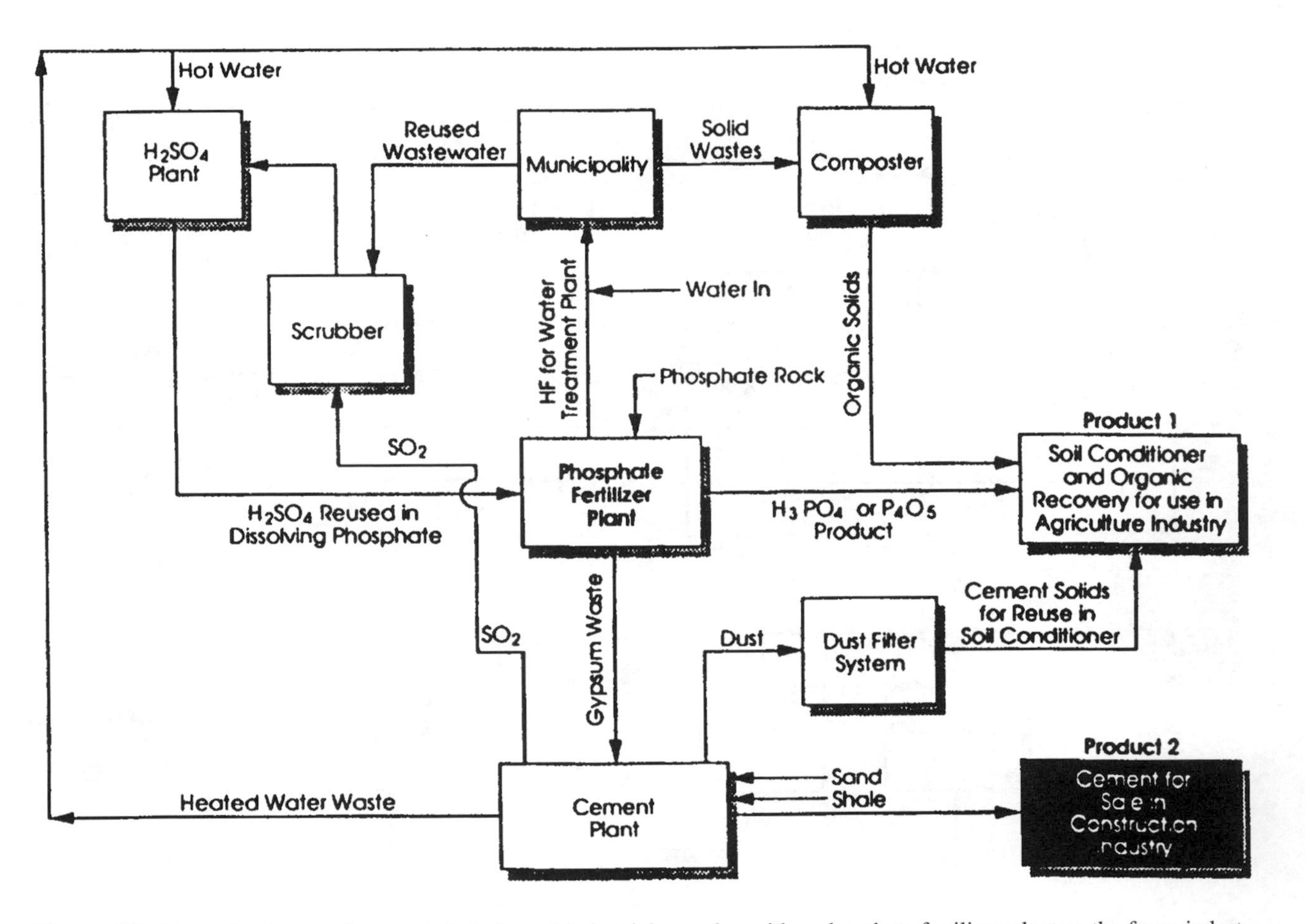

Figure 15 Example of an environmentally balanced industrial complex with a phosphate fertilizer plant as the focus industry (from Ref. 24).

and a 990 gpd/ft^2 (40.3 m^3/m^2/day) clarifier overflow rate. The resulting precipitated solids underflow concentration was 7.7% by weight.

The second-stage (ammonia removal) effluent contained unacceptable levels of F and P and had to be subjected to third-stage lime treatment. This raises the pH from 8.5 to 11.4 and produces an effluent with concentrations of F and P equal to 25 and 2 mg/L, respectively. The hydraulic design parameters were a 15 min reaction time and a 265 gpd/ft^2 (10.8 m^3/m^2/day) clarifier overflow rate. The resulting precipitated solids underflow concentration was 0.6% by wt. In all three stages, an anionic polymer was used to aid coagulation, solids settling, and effluent clarification.

The solids resulting from the first- and third-stage treatment consisted of calcium fluoride, calcium phosphate, and fluorapatite-type compounds. Typically, in the fertilizer industry, such sludges are disposed of by lagooning and subsequent landfilling. Other studies have investigated the recovery of the fluoride compounds, such as hydrofluoric and fluorosilisic acids, for use in the glass industry and in the fluoridation of drinking water supplies.

9.6.10 Phosphate Recovery by Crystallization Process

In municipal water applications and many industrial unit operations, phosphate-polluted wastewater is generated and, in general, conventional phosphate removal techniques are applied for treatment. These conventional processes are based on phosphate precipitation as calcium or iron salt or fixation in activated sludge. These processes, however, generate huge amounts of a water-rich sludge, which has to be disposed of at increasing costs. Typically, even after dewatering the sludge, the water content still is 60–85% and a relatively large part of the disposal costs comes from the expensive disposal of water. Owing to the high water content and the low quality of the sludge, reuse of phosphate is not an economically attractive option. Also, the area requirements for conventional phosphate precipitation is relatively high because the four process steps (coagulation, flocculation, sludge/water separation, dewatering) are performed serially.

An advanced alternative is to apply crystallization instead of precipitation [51,52]. The Crystalactor®, a fluid-bed type of crystallizer, has been developed by DHV (a multinational group headquartered in the Netherlands), to generate high-purity phosphate crystal pellets that can be reused in many applications and is more and more important because it is a sustainable solution to the environmental problems related to the mining and processing of natural phosphate resources. The chemistry of the process is comparable to conventional precipitation. By dosing of calcium or magnesium salt to the water (e.g., lime, calcium chloride, magnesium hydroxide, magnesium chloride), the solubility of calcium phosphate, magnesium phosphate, magnesium ammonium phosphate, or potassium magnesium phosphate is exceeded and subsequently phosphate is transformed from the aqueous solution into solid crystal material. The primary difference with conventional precipitation is that in the crystallization process the transformation is controlled accurately and that pellets with a typical size of approximately 1 mm are produced instead of fine dispersed, microscopic sludge particles. After atmospheric drying, readily handled and virtually water free highly pure phosphate pellets are obtained. Several reuse options for the pellets are: (a) raw material for the production of phosphoric acid in either the wet or thermo production processes; (b) intermediate products for fertilizer formulation; (c) raw material for kettle food; and (d) direct use as (slow-release) fertilizer.

In the food industry, wastewaters with a high organic load are released. The wastewater of a Dutch potato processing plant is treated in an anaerobic biological reactor because of the low sludge production, the low energy consumption, and the biogas production. The effluent is polished in an aerobic biological treatment plant. Cost-effective phosphate removal by struvite crystallization in the Crystalactor was implemented by dosing magnesium chloride and sodium hydroxide solutions into a part of the anaerobic stage effluent [52]. A flow of a maximum of 150 m^3/hour with 120 mg/L P was successfully treated in a reactor with a diameter of 1.8 m to an effluent containing about 10 mg/L P at a pH of 8–8.5. The struvite (magnesium ammonium phosphate) was accepted by a fertilizer producer to be used directly in the granulation process for fertilizers.

For a chemical industry, a pilot plant was operated on a waste stream that contained very high P and K concentrations (several grams per liter). It seemed that the production of potassium magnesium phosphate was not successful for this specific wastewater; however, magnesium phosphate was easily produced in order to reduce P to the desired concentration. In a pilot test for a food industry on wastewater from a biological anaerobic pretreatment unit, containing around 150 mg/L P and 10 mol/L of inorganic carbon (high concentration of carbonates), magnesium phosphate was successfully produced without disturbance by the presence of inorganic carbon. The effluent soluble P concentration was about 10 mg/L. Finally, this technology has also found applications in water softening, fluoride removal, and heavy metal recovery [53].

9.6.11 Phosphate Acid Plant Wastewater Treatment

The effluent wastewater from a chemical plant manufacturing technical grade phosphoric acid, due to low BOD and COD ratio and very low pH, was not amenable to biological treatment. So, treatability studies were carried out using the Fenton reaction and physicochemical (coagulation-settling) treatment using lime, alum, Fe salts, and polyaluminum chloride (PAC). The treatability studies indicated [54] that it is possible to remove 75–80% COD using Fenton's reagent at optimum doses of 1.0 g/L ferrous sulfate and 2 mL of 30% hydrogen peroxide. Simultaneously, significant quantities of suspended solids (SS), phosphate and fluoride were also removed. Polyaluminum chloride was found to be more effective towards SS, COD, phosphate, and fluoride removal, in comparison with other coagulants used in the many studies [50,54,55]. The addition of an anionic polyelectrolyte (Magnafloc 156) and PAC improved the performance further. A treatment scheme that consisted of neutralization (pH 4) + Fenton's reagent + neutralization (pH 7.5) + PAC/Magnafloc 156 was found to be effective in treating phosphoric acid plant wastewater to meet marine discharge standards [54].

9.6.12 Improvement and Phosphorus Removal in Wetlands and Sand Filters

Removal of phosphorus and nitrogen by a wetland process has been discussed by Hung *et al.* [56]. The efficiency of P removal in ecologically engineered wastewater treatment facilities, such as artificial wetlands and sand filters, was found to be enhanced by using a reactive sorbent. The sorbent must have a high P-sorption capacity and an adequate hydraulic conductivity. Several filter materials were tested in this study with regard to their P-sorption capacity [57]. The P-sorption experiments showed that, of the materials tested, the coarse (0.25–4 mm) blast furnace crystalline slag had the highest P-sorption capacity followed by the fine (0–0.125 mm) blast furnace crystalline slag, the fine amorphous slag, and the B horizon of a forest soil. The isotherm tests showed that the B horizon was the most efficient P retainer, followed in order by the coarse and the fine crystalline slag. It was also concluded that coarse crystalline slag possessed the highest hydraulic conductivity, suggesting that this material is the most suitable for use in ecological wastewater treatment systems.

9.6.13 Removal of Phosphate Using Feldspar

Feldspar, among many natural substances such as termite mount-clay, saw dust, kaolinite, and dolomite, offers significant removal ability for phosphate, sulfate, and color colloids. Optimization laboratory tests of parameters such as solution pH and flow rate, resulted in a maximum efficiency for removal of phosphate (42%), sulfate (52%), and color colloids (73%), x-ray diffraction, adsorption isotherms test, and recovery studies suggest that the removal process of anions occurs via ion exchange in conjunction with surface adsorption. Furthermore, reaction rate studies indicated that the removal of these pollutants by feldspar follows first-order kinetics. Percent removal efficiencies, even under optimized conditions, will be expected to be somewhat less for industrial effluents in actual operations due to the effects of interfering substances [58].

9.6.14 Phosphate Biological Uptake at Acid pH

Some microorganisms store phosphate as polyphosphate, thereby removing it from solution, but current methods to induce this are unreliable. However, it was recently discovered that polyphosphate production is increased by acid shock [59]. These studies have led to the identification of a significant, yet previously recognized, microbial stress response at acidic pH levels, which may be a novel strategy for the "one-step" removal of phosphate from wastewater effluents. It was possible to increase the level of phosphate removal by the microflora of a conventional activated sludge plant – under fully aerobic conditions – by more than 50% if the operational pH was adjusted within the range 5.5–6.5, as opposed to within 7.2–7.7, which are typical in current practice. Similar results were obtained in four other activated sludge plants of varying influent characteristics; enhancement of phosphate removal at pH 5.5 varied between 56 and 142% and involved a considerable fraction of the microbial population – bacterial, yeast, and fungal. Further research to assess the economic viability of a low-pH phosphate removal system is being carried out in Northern Ireland and Britain.

9.6.15 Emerging Phosphorus Removal Technologies

More recent research on treatment or recovery of phosphates and phosphorus from wastewater can be found from the literature [57,58,60–69,73].

The use of combined biological and physicochemical treatment processes for phosphorus removal was originally conceived by Beer and Wang at Coxsackie Sewage Treatment Plant, NY [58], and by Krofta and Wang at the Lenox Institute of Water Technology, MA [65,66]. They successfully used ferric chloride, lime, and alum for precipitation of phosphate from the activated sludge aeration basin effluent. Wang and Aulenbach [67] have discussed the theory and principles of biological phosphorus uptake under aerobic conditions, biological phosphorus release under anoxic/anaerobic conditions, and physicochemical precipitation of released phosphorus (in phosphate form) by an innovative A/O process. Wang [68] has adopted a commercially available dissolved air flotation (DAF) clarifier [70] in a combined biological physicochemical process system for high-rate phosphorus removal.

Essentially, Wang's innovative technology [68] is a combined biological-chemical precipitation process involving the use of the following process steps:

1. Treatment of incoming wastewater (usually primary effluent) in an aeration basin to transfer phosphorus from the incoming wastewater to the "P-stripped activated sludge microorganisms" under aerobic conditions.
2. Separation of the spent activated sludge microorganisms from the aeration basin effluent by either a conventional sedimentation clarifier (with 2 hours definition time

or DT), or by a high-rate dissolved air flotation (DAF) clarifier (with 15 minutes DT) [65,66,68].

3. Discharge of the almost P-free clarifier effluent into a receiving water, while discharge of the P-concentrated clarifier sludge (either settled sedimentation clarifier sludge, or the floated DAF clarifier sludge) to a phosphate stripper (i.e., a thickener-type holding tank with 5–15 hours DT) where the clarifier sludge is subjected to anoxic/anaerobic conditions to induce phosphorus release from the clarifier sludge into the aqueous phase [68].
4. Chemical precipitation of the newly released phosphate in the phosphate stripper effluent (i.e., a P-rich, low-volume sidestream containing 40–80 mg/L P; amounting to about 10–15% of the total wastewater flow) using lime, ferric chloride, or alum, and subsequently flotation of the P-rich precipitated chemical sludge for reuse as a fertilizer using a high-rate DAF clarifier (15 min DT).
5. Collection of the phosphate-stripped activated sludge, which has extremely high phosphorus uptake capacity from the stripper.
6. Return of the phosphate-stripped activated sludge to the aeration basin for reuse in a new cycle, where the phosphate-stripped activated sludge microorganisms are again induced to take up dissolved phosphorus in excess of the amount required for growth under aerobic conditions.

The above P-removal process system reduces the volume of the wastewater to be treated (10–15% of total wastewater flow), thereby reducing the chemical dosage required, the amount of chemical sludge produced, and associated costs. Lime can be used to remove phosphorus from the stripper supernatant at lower pH levels (8.5–9.0) than normally required, although alum and ferric chloride are equally effective. The cycling of sludge through an anoxic phase may also assist in the control of hulking by the destruction of filamentous organisms to which hulking is generally attributed. The process is capable of reducing the total phosphorus concentration of typical municipal wastewaters to 1 mg/L or less. Adoption of a DAF clarifier instead of sedimentation clarifier for both secondary clarification and P-rich precipitated chemical sludge separation significantly reduces process time, and, in turn, saves overall capital and O&M costs [68].

Wang [68] has also successfully used conventional biological sequencing batch reactors (SBR) instead of conventional activated sludge aeration for steps 1 and 6 above, and has used physicochemical sequencing batch reactors (PC-SBR) instead of the stripper and DAF for steps 3 and 4 above. The readers are encouraged to further improve upon this emerging phosphorus removal technology. Descriptions and discussions of PC-SBR can be found from the literature [3,49,69,71].

REFERENCES

1. Anon. Phosphate, the servant of mankind. *Oil Power*, **1951**, *26* (*3*), 61–62.
2. Arnold, D.W.; Wolfram, W.E. Ammonia removal and recovery from fertilizer complex wastewaters. In *Proceedings of 30th Industrial Waste Conference*, Purdue University, Lafayette, IN, 1975; Vol. 30, 760–767.
3. Beg, S.A. *et al.* Effect of toxicants on biological nitrification for treatment of fertilizer industry wastewater. In *Proceedings of 35th Industrial Waste Conference*, Purdue University, Lafayette, IN, 1980; Vol. 35, 826–834.
4. Beg, S.A. *et al.* Inhibition of nitrification by arsenic, chromium and fluoride. *J. Water Poll. Control Fed.*, **1982**, *54*, 482–488.

5. USEPA. *Process Design Manual for Phosphorus Removal;* Office of Technology Transfer, U.S. Environmental Protection Agency: Washington, DC, 1971.
6. USEPA. *Pretreatment of Pollutants Introduced into Publicly Owned Treatment Works;* Federal Guidelines, Office of Water Program Operations, U.S. Environmental Protection Agency: Washington, DC, 1973.
7. USEPA. *Basic Fertilizer Chemicals*, EPA-440/1-74-Olla; Effluent Guidelines Division, U.S. Environmental Protection Agency: Washington, DC, 1974.
8. USEPA. *Federal Guidelines on State and Local Pretreatment Programs*, EPA-430/9-76-017c; Construction Grants Program, U.S. Environmental Protection Agency: Washington, DC, 1977.
9. Federal Register. *Code of Federal Regulations*, CFR 40; U.S. Government Printing Office: Washington, DC, 1987; 412–430, 729–739.
10. Federal Register. *Notices, Appendix A, Master Chart of Industrial Categories and Regulations;* U.S. Government Printing Office: Washington, DC, 1990; Vol. 55, No. 1, Jan. 2, 102, 103.
11. Fertilizer Association of India. Liquid effluents. In *Pollution Control in Fertilizer Industry*, Tech. Rep. 4, Part I, 1979.
12. Fuller, R.B. The position of the pebble phosphate industry in stream sanitation. *Sewage Wks. J.* **1949**, *9* (*5*), 944.
13. Ghokas, S.I. *Treatment of Effluents from a Fertilizer Complex*, M.S. thesis; University of Manchester, UK, 1983.
14. Griffith, E.J. Modem mankind's influence on the natural cycles of phosphorus. In *Phosphorus and the Environment;* Ciba Foundation, New Series, 57, 1978.
15. Horton, J.P. *et al.* Processing of phosphorus furnace wastes. *J. Water Poll. Control Fed.*, **1956**, *28 (1), 70–77.*
16. Jones, W.E.; Olmsted, R.L. Waste disposal at a phosphoric acid and ammonium phosphate fertilizer plant. In *Proceedings of the 17th Industrial Waste Conference*, Purdue University, Lafayette, IN, 1962; Vol. 17, 198–202.
17. Kiff, R.I. Water pollution control in the fertilizer manufacturing industry. In *Manufacturing and Chemical Industries*; Barnes, D. *et al.*, Eds.; Longman Scientific & Technical: Essex, UK, 1987.
18. Koziorowski, B.; Kucharski, J. *Industrial Waste Disposal;* Pergamon Press: Oxford, UK, 1972; 142–151.
19. Lanquist, E. *Peace and Alafia River Stream Sanitation Studies;* Florida State Board of Health, June, Suppl. II to Vol. II, 1955.
20. Layer, W.; Wang, L.K. *Water Purification and Wastewater Treatment with Sodium Aluminate*, Report PB 85-214-492/AS; U.S. Department of Commerce, National Technical Information Service: Springfield, VA, 1984.
21. Lund, H.F. (Ed.) *Industrial Pollution Control Handbook;* McGraw-Hill: New York, 1971; 14-6–14-8.
22. Markham, J.H. *Effluent Control on a Fertilizer Manufacturing Site;* Fertilizer Soc. of London, London, 1983; Proc. 213.
23. Nemerow, N.L. *Industrial Water Pollution;* Addison-Wesley: Reading, MA, 1978; 583–588.
24. Nemerow, N.L.; Dasgupta, A. Environmentally balanced industrial complexes. In *Proceedings of the 36th Industrial Waste Conference*, Purdue University, Lafayette, IN, 1981; Vol. 36, 982–989.
25. Overcash, M.R. *Techniques for Industrial Pollution Prevention;* Lewis Publishers: Michigan, 1986; 87–89.
26. Robasky, J.G.; Koraido, D.L. Gauging and sampling industrial wastewater. *J. Chem. Engrs.* **1973**, *80* (*1*), 111–120.
27. Search, W.J. *et al. Source Assessment, Nitrogen Fertilizer Industry Water Effluents*, Report PB 292 937; U.S. Department of Commerce, National Technical Information Service: Springfield, VA, 1979.
28. Shahalam, A.B.M. *et al.* Wastes from processing of phosphate industry. In *Proceedings of 40th Industrial Waste Conference*, Purdue University, Lafayette, IN, 1985; Vol. 40, 99–110.
29. Specht, R.C. *Phosphate Waste Studies*, Bull. 42; Florida Eng. and Ind. Exp. Sta., University of Florida: Gainesville, 1950.
30. Specht, R.C. Disposal of wastes from the phosphate industry. *J. Water Poll. Control Fed.* **1960**, *32* (*9*), 963–974.

31. Wakefield, J.W. Semi-tropical industrial waste problems. In *Proceedings of the 7th Industrial Waste Conference*, Purdue University, Lafayette, IN, 1952; Vol. 7, 495–508.
32. Krofta, M.; Wang, L.K. *Flotation and Related Adsorptive Bubble Separation Processes*, 4th ed.; Lenox Institute of Water Technology: Lenox, MA 2001; 185 p.
33. Yapijakis, C. Treatment of phosphate industry wastes. In *Handbook of Industrial Waste Treatment*, Wang, L.K., Wang, M.H.S., Eds.; Marcel Dekker: New York, NY, 1992; Chapter 8, 323–383.
34. UNIDO. *Minimizing Pollution from Fertilizer Plants*, Report, Expert Group Meeting, Helsinki, ID/WG 175/19, 1974.
35. Wang, L.K. *Pretreatment and Ozonation of Cooling Tower Water, Part I*, PB84-192053; U.S. Department of Commerce, National Technical Information Service: Springfield, VA, 1983; 34 p.
36. Wang, L.K. *Pretreatment and Ozonation of Cooling Tower Water, Part II*, PB84-192046; U.S. Department of Commerce, National Technical Information Service: Springfield, VA, 1983; 29 p.
37. Caswell, C.A. The fallacy of the closed-cycle cooling concept. *Ind. Water Engng.* **1990**, *17* (*3*), 27–31.
38. Hallett, G.F. Area-view of proposed methods for evaluating cooling equipment. *Ind. Water Engng.* **1990**, *17* (*3*), 12–26.
39. Wang, L.K.; Wang, M.H.S. Decontamination of groundwater and hazardous industrial effluents by high-rate air flotation process. Presented at Proc. Great Lakes 90 Conf., Hazardous Materials Control Res. Inst., Silver Springs, MD, September 1990.
40. Wang, L.K. *Preliminary Design Report of a 10-MGD Deep Shaft-Flotation Plant for the City of Bangor*, PB88-200597/AS; U.S. Department of Commerce, National Technical Information Service: Springfield, VA, 1987; 42 p.
41. Wang, L.K.; Wang, M.H.S. Treatment of storm run-off by oil–water separation, flotation, filtration and adsorption, part A: wastewater treatment. In *Proceedings of the 44th Industrial Waste Conference*, Purdue University, Lafayette, IN, 1990; 655–666.
42. Wang, L.K. *et al.* Treatment of storm run-off by oil-water separation, flotation, filtration and adsorption, part B: waste sludge management. In *Proceedings of the 44th Industrial Waste Conference*, Purdue University, Lafayette, IN, 1990; 667–673.
43. Wang, L.K.; Wang, P.; Clesceri, N. Groundwater decontamination using sequencing batch processes. *Water Treatment*, **1995**, *10*, 121–134.
44. Krofta, M.; Wang L.K. Development of innovative Sandfloat systems for water purification and pollution control. *ASPE Journal of Eng. Plumbing*, **1984**, *0* (*1*), 1–16.
45. Krofta, M.; Wang, L.K. Tertiary treatment of secondary effluent by dissolved air flotation and filtration. *Civil Engng. for Practicing and Design Engrs.*, **1984**, *3*, 253–272.
46. Krofta, M.; Wang, L.K. *Development of an Innovative and Cost-Effective Municipal-Industrial Waste Treatment System*, PB88-168109/AS; U.S. Department of Commerce, National Technical Information Service: Springfield, VA, 1985; 27 p.
47. Krofta, M.; Wang, L.K. Wastewater treatment by biological-physicochemical two-stage process system. *Proceedings of the 41st Industrial Waste Conference*; Lewis Publishers, Inc.: Chelsea, MI, 1987; 67–72.
48. Krofta, M.; Wang, L.K. Development of low-cost flotation technology and systems for wastewater treatment. In *Proceedings of the 42nd Industrial Waste Conference*, Purdue University, Lafayette, IN, 1988; p. 185.
49. Wang, L.K.; Krouzek, J.V.; Kounitson, U. *Case Studies of Cleaner Production and Site Remediation*; United Nations Industrial Development Organization (UNIDO): Vienna, Austria, 1995, Training Manual No. DTT-5-4-95, 136.
50. Krofta, M.; Wang, L.K. Winter operation of nation's first two potable flotation plants. In *Proceedings of 1987 Joint Conference of AWWA and WPCF*, Cheyenne, Wyoming, USA, Sept. 20–23, 1987.
51. Piekema, P.G.; Gaastra, S.B. Upgrading of a wastewater treatment plant: Several nutrients removal. *Eur. Water Pollut. Control* **1993**, *3* (*3*), 21–26.
52. Piekema, P.G.; Giesen, A. *Phosphate Recovery by the Crystallization Process: Experience and Developments*; DHV Water BV: P.O. Box 484, 3800 Al Amersfoort, The Netherlands, 2001.
53. Bouwman, J.G.M.A; Luisman, A.H.C. Aiming at zero discharge and total reuse. *Eur. Semicond.* **2000**, *Sept.* 25–27.

54. Nawghare, P.; Rao, N.N.; Bejankiwar, R.; Szyprkowicz. Treatment of phosphoric acid plant wastewater using Fenton's reagent and coagulants. *J. Environ. Sci. and Health* **2001**, *A 36 (10)*, 2011.
55. Wang, L.K. *Poly Iron Chloride and Poly Aluminum Chloride*, Technical Report LIWT/1-2002/252; Lenox Institute of Water Technology: Lenox, MA, 2002, 26 p.
56. Hung, Y.; Gubba, S.; Lo, H.H.; Wang, L.K.; Yapijakis, C.; Shammas, N.K. Application of wetland for wastewater treatment. *OCEESA J.* **2003**, *20* (*1*), 41–46.
57. Johansson, L. Industrial by-products and natural substrata as phosphorus sorbents. *Environ. Technol.* **1999**, *20* (*3*), 309.
58. Beer, C.; Wang, L.K. Full-scale operations of plug flow activated sludge systems. *J. New Engl. Water Poll. Control Assoc.* **1975**, *9*(*2*), 145–173.
59. McGrath, J.W.; Quinn, J.P. *Phosphate Removal: a Novel Approach*; School of Biology and Biochemistry, Queens' University Belfast, Ireland, **2002**; http://www.qub.ac/uk/envres/EarthAir-water/phosphate-removal.htm.
60. Akin, B.S.; Ugurlu, A. Enhanced phosphorus removal by glucose fed sequencing batch reactor. *J. Environ. Sci. and Health* **2001**, *A 36* (*9*), p 1757.
61. CEEP Phosphates, a sustainable future in recycling. Centre Europeen d' Polyphosphates, 1999.
62. Giesen, A. Eliminate sludge. *Ind. Wastewater* **1998**, *6*.
63. Martin, D.F.; Dooris, P.M.; Sumpter, D. Environmental impacts of phosphogypsum vs. borrow pits in roadfill construction. *J. Environ. Sci. and Health* **2001**, *A 36* (*10*), 1975.
64. Priyantha, N.; Pereira, S. Removal of phosphate, sulfate, and colored substances in wastewater effluents using Feldspar. *Water Res. Mgnt.* **2000**, *14* (*6*), 417.
65. Krofta, M.; Wang, L.K. Improved biological treatment with a secondary flotation clarifier. *Civil Engng. for Practicing and Design Engrs.* **1983**, *2*, 307–324.
66. Krofta, M.; Wang, L.K. Wastewater treatment by biological-physicochemical two-stage process systems. In *Proceedings of the 41st Industrial Waste Conference*, May, 1987; 67–72.
67. Wang, L.K.; Aulenbach, D. *BOD and Nutrient Removal by Biological A/O Process Systems*, PB88-168430/AS; U.S. Department of Commerce, National Technical Information Service: Springfield, VA, 1986; 12 p.
68. Wang, L.K. *An Emerging Technology for Phosphorus Removal from Wastewaters*, Technical Report LIWT/1-2002/253; Lenox Institute of Water Technology: Lenox, MA, 2002, 18 p.
69. Wang, L.K. Laboratory simulation of water and wastewater treatment processes. *Water Treatment*, **1995**, *10*, 261–282.
70. Anonymous. Flotation equipment. *Environ. Protection* **2003**, *14* (*2*), 137–138.
71. Wang, L.K.; Kurylko, L.; Wang, M.H.S. *Sequencing Batch Liquid Treatment*, U.S. Patent No. 5,354,458, October 11, 1994.
72. Wang, L.K. *Preliminary Design Report of a 10-MGD Deep Shaft-Flotation Plant for the City of Bangor, Appendix*, PB88-200605/AS; U.S. Department of Commerce, National Technical Information Service: Springfield, VA, 1987; 171 p.
73. State of Florida. Industrial Wastewater Program—Phosphate Industry. State of Florida, Dept. of Environmental Protection, Miami, FL, Feb. 2004. www.dep.state.fl.us.

10
Treatment of Pulp and Paper Mill Wastes

Suresh Sumathi
Indian Institute of Technology, Bombay, India

Yung-Tse Hung
Cleveland State University, Cleveland, Ohio, U.S.A.

10.1 POLLUTION PROBLEMS OF PULP AND PAPER INDUSTRIES

Pulp and paper mills are a major source of industrial pollution worldwide. The pulping and bleaching steps generate most of the liquid, solid, and gaseous wastes (Table 1) [1]. Pulping is a process in which the raw material is treated mechanically or chemically to remove lignin in order to facilitate cellulose and hemicellulose fiber separation and to improve the papermaking properties of fibers. Bleaching is a multistage process to whiten and brighten the pulp through removal of residual lignin. Pulping and bleaching operations are energy intensive and typically consume huge volumes of fresh water and large quantities of chemicals such as sodium hydroxide, sodium carbonate, sodium sulfide, bisulfites, elemental chlorine or chlorine dioxide, calcium oxide, hydrochloric acid, and so on. A partial list of the various types of compounds found in spent liquors generated from pulping and bleaching steps is shown in Table 2 [2–4]. The effluents generated by the mills are associated with the following major problems:

- Dark brown coloration of the receiving water bodies result in reduced penetration of light, thereby affecting benthic growth and habitat. The color responsible for causing aesthetic problems is attributable to lignin and its degradation products.
- High content of organic matter, which contributes to the biological oxygen demand (BOD) and depletion of dissolved oxygen in the receiving ecosystems.
- Presence of persistent, bio-accumulative, and toxic pollutants.
- Contribution to adsorbable organic halide (AOX) load in the receiving ecosystems.
- Measurable long-distance transport (>100 km) of organic halides (such as chloroguaiacols), thereby contaminating remote parts of seas and lakes [5].
- Cross-media pollutant transfer through volatilization of compounds and absorption of chlorinated organics to wastewater particulates and sludge.

Significant solid wastes from pulp and paper mills include bark, reject fibers, wastewater treatment plant sludge, scrubber sludge, lime mud, green liquor dregs, and boiler and furnace ash. The bulk of the solid wastes is generated during wastewater treatment. Sludge disposal is a serious environmental problem due to the partitioning of chlorinated organics from effluents to

Table 1 Types of Pollutants Generated During Chemical (Kraft) Pulping and Bleaching Steps

Pollution generating step	Pollution output phase	Nature of pollution
Wood debarking and chipping, chip washing	Solid	Bark, wood processing residues
	Water	SS, BOD, color, resin acids
Chemical (Kraft) pulping, black liquor evaporation, and chemical recovery steps	Air	Total reduced sulfur (hydrogen sulfide, methyl mercaptan, dimethyl sulfide, dimethyl disulfide), VOC
Wood chip digestion, spent pulping liquor evaporator condensates	Water	High BOD, color, may contain reduced sulfur compounds, resin acids
Pulp screening, thickening, and cleaning operations	Water	Large volume of waters with SS, BOD, color
Smelt dissolution, clarification to generate green liquor	Solid	Green liquor dregs
Recausticizing of green liquor, clarification to generate white liquor	Solid	Lime slaker grits
Chlorine bleaching of pulp	Water	BOD, color, chlorinated organics, resin acids
	Air	VOC
Wastewater treatment	Solid	Primary and secondary sludge, chemical sludge
	Air	VOC
Scrubbing for flue gases	Solid	Scrubber sludge
Recovery furnaces and boilers	Air	Fine and coarse particulates, nitrogen oxides, SO_2
	Solid	Ash

SS, suspended solids; VOC, volatile organics; BOD, biochemical oxygen demand.
Source: Ref. 1.

solids. The major air emissions are fine and coarse particulates from recovery furnaces and burners, sulfur oxides (SOx) from sulfite mills, reduced sulfur gases and associated odor problems from Kraft pulping and chemical recovery operations, volatile organic compounds (VOC) from wood chip digestion, spent liquor evaporation and bleaching, nitrogen oxides (NOx), and SOx from combustion processes. Volatile organics include carbon disulfide, methanol, methyl ethyl ketone, phenols, terpenes, acetone, alcohols, chloroform, chloromethane, and trichloroethane [1].

The extent of pollution and toxicity depends upon the raw material used, pulping method, and pulp bleaching process adapted by the pulp and paper mills. For example, the pollution load from hardwood is lower than softwood. On the other hand, the spent liquor generated from pulping of nonwood fiber has a high silica content. Volumes of wastewater discharged may vary from near zero to 400 m^3 per ton of pulp depending on the raw material used, manufacturing process, and size of the mill [6]. Thus, the variability of effluent characteristics and volume from one mill to another emphasizes the requirement for a variety of pollution prevention and treatment technologies, tailored for a specific industry.

Table 2 Low-Molecular-Weight Organic Compounds Found in the Spent Liquors from Pulping and Bleaching Processes

Class of compounds				
Acidic				
Wood extractives	Lignin/carbohydrate derived	Phenolic	Neutral	Miscellaneous
Category: Fatty acid	*Category: Hydroxy*	*Category: Phenolic*	Hemicelluloses	*Category: Dioxins*
Formic acid (S)	Glyceric acid	Monochlorophenols	Methanol	2,3,7,8-tetrachloro-dibenzodioxin (2,3,7,8-TCDD)
Acetic acid (S)	*Category: Dibasic*	Dichlorophenols	Chlorinated acetones	2,3,7,8-tetrachloro-dibenzofuran (2,3,7,8-TCDF)
Palmitic acid (S)	Oxalic acid	Trichlorophenols	Chloroform	*Wood derivatives*
Heptadecanoic acid (S)	Malonic acid	Tetrachlorophenol	Dichloromethane	Monoterpenes
Stearic acid (S)	Succinic acid	Pentachlorophenol	Trichloroethene	Sesquiterpenes
Arachidic acid (S)	Malic acid	*Category: Guaiacolic*	Chloropropenal	Diterpenes: Pimarol
Tricosanoic acid (S)	*Category: Phenolic acid*	Dichloroguaiacols	Chlorofuranone	Abienol
Lignoceric (S)	Monohydroxy benzoic acid	Trichloroguaiacols	1,1-dichloro-methylsulfone	*Juvabiones*
Oleic (US)	Dihydroxy benzoic acid	Tetrachloroguaiacol	Aldehydes	Juvabiol
Linolenic acid (US)	Guaiacolic acid	*Category: Catecholic*	Ketones	Juvabione
Behenic acid (S)	Syringic acid	Dichlorocatechols	Chlorinated sulfur	*Lignin derivatives*
Category: Resin acid		Trichlorocatechols	Reduced sulfur compounds	Eugenol
Abietic acid		*Category: Syringic*		Isoeugenol
Dehydroabietic acid		Trichlorosyringol		Stilbene
Mono and dichloro dehydrabietic acids		Chlorosyringaldehyde		Tannins (monomeric, condensed and hydrolysable)
Hydroxylated-dehydroabietic acid				Flavonoids
Levopimaric acid				
Pimaric acid				
Sandracopimaric acid				

S, saturated; US, unsaturated
Source: Refs 2–4.

The focus of this chapter is to trace the origin and nature of the major pollution (especially water) problems within the pulp and paper industries and to present an overview of the pollution mitigation strategies and technologies that are currently in practice or being developed (emerging technologies).

10.2 NATURE AND COMPOSITION OF RAW MATERIALS USED BY PULP AND PAPER INDUSTRIES

The pulp and paper industries use three types of raw materials, namely, hard wood, soft wood, and nonwood fiber sources (straw, bagasse, bamboo, kenaf, and so on). Hard woods (oaks, maples, and birches) are derived from deciduous trees. Soft woods (spruces, firs, hemlocks, pines, cedar) are obtained from evergreen coniferous trees.

10.2.1 Composition of Wood and Nonwood Fibers

Soft and hard woods contain cellulose (40–45%), hemicellulose (20–30%), lignin (20–30%), and extractives (2–5%) [7]. Cellulose is a linear polymer composed of β-*D*-glucose units linked by 1–4 glucosidic bonds. Hemicelluloses are branched and varying types of this polymer are found in soft and hard woods and nonwood species. In soft woods, galactoglucomannans (15–20% by weight), arabinoglucurono-xylan, (5–10% by weight), and arabinogalactan (2–3% by weight) are the common hemicelluloses, while in hard woods, glucuronoxylan (20–30% by weight) and glucomannan (1–5% by weight) are found [2,3]. Lignin is a complex heterogeneous phenylpropanoid biopolymer containing a diverse array of stable carbon–carbon bonds with aryl/alkyl ether linkages and may be cross-linked to hemicelluloses [8]. Lignins are amorphous, stereo irregular, water-insoluble, nonhydrolyzable, and highly resistant to degradation by most organisms and must be so in order to impart resistance to plants against many physical and environmental stresses. This recalcitrant biopolymer is formed in plant cell walls by the enzyme-catalyzed coupling of p-hydroxycinnamyl alcohols, namely, p-coumaryl, coniferyl, and sinapyl alcohols that make up significant proportion of the biomass in terrestrial higher plants. In hardwoods, lignin is composed of coniferyl and sinapyl alcohols and in softwoods is largely a polymer of coniferyl alcohol. The solvent extractable compounds of wood termed as "extractives" include aliphatics such as fats, waxes, and phenolics that include tannins, flavonoids, stilbenes, and terpenoids. Extractives comprise 1–5% of wood depending upon the species and age of the tree. Terpenoids that include resin acids are found only in soft wood and are derived from the "pitch" component of wood. Compared to wood, the structures of nonwood species are not well studied. Grasses usually contain higher amounts of hemicelluloses, proteins, silica, and waxes [9]. On the other hand, grasses contain lower lignin content compared to wood and the bonding of lignin to cellulose is weaker and therefore easier to access.

10.3 PULPING PROCESSES

The steps involved in pulping are debarking, wood chipping, chip washing, chip crushing/digestion, pulp screening, thickening, and washing (Fig. 1). The two major pulping processes that are in operation worldwide are mechanical and chemical processes. Mechanical pulping methods use mechanical pressure, disc refiners, heating, and mild chemical treatment to yield pulps. Chemical pulping involves cooking of wood chips in pulping liquors containing chemicals under high temperature and pressure. Other pulping operations combine thermal, mechanical, and/or chemical methods. Characteristic features of various pulping processes are summarized in Table 3 and are further described shortly in the following subsections [3,10–12].

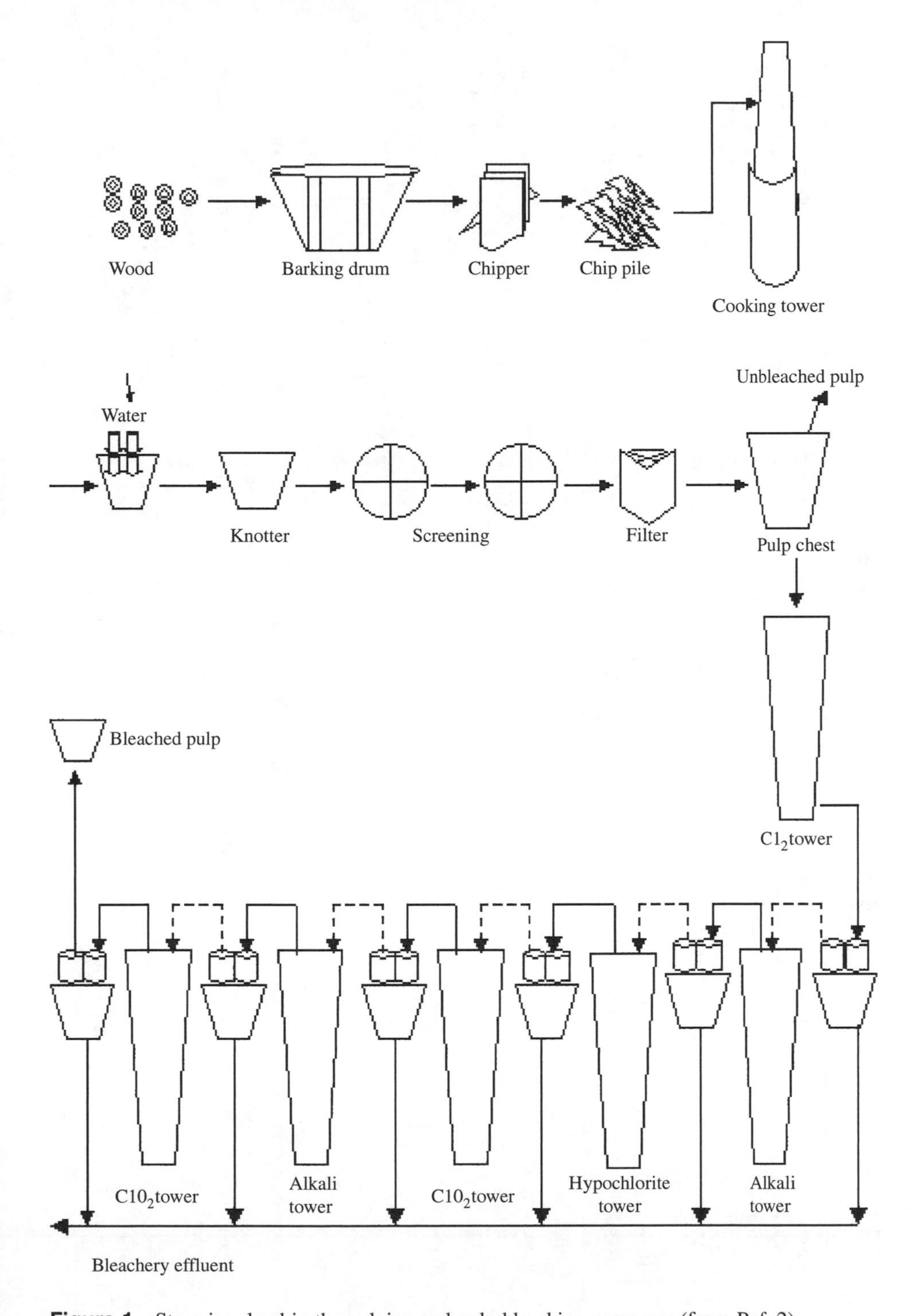

Figure 1 Steps involved in the pulping and pulp bleaching processes (from Ref. 2).

Table 3 Comparison of Various Pulping Processes

	Name of the pulping process				
Process features	Mechanical	CTMP	NSSC	Kraft	Sulfite
Pulping mechanism	Grinding stone, double disc refiners, steaming, followed by refining in TMP process	Chemical treatment using NaOH or $NaHSO_3$ + steaming followed by mechanical refining	Continuous digestion in $Na_2SO_3 + Na_2CO_3$ liquor using steam followed by mechanical refining	Cooking at 340–350°F, 100–135 psi for 2–5 hours in NaOH, Na_2S, and Na_2CO3; efficient recovery of chemicals	Sulfonation at 255–350°F, 90–110 psi for 6–12 hours in H_2SO_3 and Ca, Na, NH_4, $Mg(HSO_3)_2$
Cellulosic raw material	Hard woods like poplar and soft woods like balsam, fir, hemlock	Hard and soft woods	Hard woods like aspen, oak, alder, birch, and soft wood sawdust and chips	Any type of hard and soft wood, nonwood fiber sources	Any hard wood and nonresinous soft woods
Pulp properties	Low-strength soft pulp, low brightness	Moderate strength	Good stiffness and moldability	High-strength brown pulps, difficult to bleach	Dull white-light brown pulp, easily bleached, lower strength than Kraft pulp
Typical yields of pulp	92–96%	88–95%	70–80%	65–70% for brown pulps, 47–50% for bleachable pulps, 43–45% after bleaching	48–51% for bleachable pulp, 46–48% after bleaching
Paper products	Newspaper, magazines, inexpensive writing papers, molded products	Newspaper, magazines, inexpensive writing papers, molded products	Corrugating medium	Bags, wrappings, gumming paper, white papers from bleached Kraft pulp, cartons, containers, corrugated board	Fine paper, sanitary tissue, wraps, glassine strength reinforcement in newsprint

TMP, thermomechanical pump; CTMP, chemi-thermomechanical pump; NSSC, neutral sulfite semichemical pulp.
Source: Refs. 3, 10, and 12.

Nonconventional pulping methods such as solvent pulping, acid pulping, and biopulping are discussed in subsection 10.9.1.

10.3.1 Mechanical Pulps

Stone-Ground Wood Pulp

Wood logs are pushed under the revolving grindstone and crushed by mechanical pressure to yield low-grade pulps. Lignin is not removed during this process and therefore imparts a dark color to the pulp and paper product.

Refiner Mechanical Pulp

Wood chips are passed through a narrow gap of a double-disc steel refiner consisting of stationary and rotating plates having serrated surfaces. This process results in the mechanical separation of fibers that are subsequently frayed for bonding. The strength of the refiner pulp is better than that of ground-wood pulps.

Thermomechanical Pulp (TMP)

Wood chips are preheated in steam before passage through disc refiners. Heating is meant for softening the lignin portion of wood and to promote fiber separation. This pulp is stronger than that produced by the ground-wood process.

10.3.2 Semichemical Pulp

Wood chips are processed in mild chemical liquor and subjected to mechanical refining using disc refiners. Semichemical pulping liquors have variable composition ranging from sodium hydroxide alone, alkaline sulfite (sodium sulfite + sodium carbonate), mixtures of sodium hydroxide and sodium carbonate, to Kraft green or white liquors [3]. Sodium sulfite/ sodium carbonate liquor is most commonly used and the pulp product obtained thereafter is referred to as neutral sulfite semichemical (NSSC) pulp.

10.3.3 Chemithermo Mechanical Pulp (CTMP)

This process involves a mild chemical treatment of wood chips in sodium hydroxide or sodium bisulfite before or during steaming. Chemically treated chips are passed through mechanical disc refiners.

10.3.4 Chemical Pulps

Chemical pulping of wood is commonly carried out according to the Kraft (sulfate) or sulfite processes [13]. These methods are described in the following subsections.

Kraft Pulping

Kraft pulping involves the cooking of wood chips at 340–350°F and 100–135 psi in liquor that contains sodium hydroxide, sodium sulfide, and sodium carbonate. This process promotes cleavage of the various ether bonds in lignin and the degradative products so formed dissolve in alkaline pulping liquor. The Kraft process normally incorporates several steps to recover chemicals from the spent black liquor [3].

Sulfite Pulping

The sulfite process solubilizes lignin through sulfonation at 255–350°F under 90–110 psi. The pulping liquors are composed of mixture of sulfurous acid (H_2SO_3) and bisulfites (HSO_3^{2-}) of ammonium, sodium, magnesium, or calcium, and lignin is separated from the cellulose as lignosulfonates [3]. Bisulfite pulping is performed in the pH range of 3–5 while acid sulfite pulping is carried out with free sulfurous acid at pH 1–2. Sulfite pulping mills frequently adapt methods for the recovery of SO_2, magnesium, sodium, or ammonium base liquors [3].

10.4 COMPOSITION OF SPENT PULPING LIQUORS

10.4.1 Kraft Pulping Liquors (Black Liquors)

During Kraft pulping, about 90–95% of the reactive biopolymer, namely lignin, becomes solubilized to form a mixture of lignin oligomers that contribute to the dark brown color and pollution load of pulping liquors. Lignin oligomers that are released into the spent liquors undergo cleavage to low-molecular-weight phenylpropanoic acids, methoxylated and/or hydroxylated aromatic acids. In addition, cellulose and hemicelluloses that are sensitive to alkali also dissolve during the pulping processes [13]. Black liquors generated from the Kraft pulping process are known to have an adverse impact on biological treatment facilities and aquatic life. Emissions of total reduced sulfur (TRS) and hazardous air pollutants (HAP) are also generated. Black liquors typically consist of the following four categories of compounds derived from dissolution of wood [3]:

- ligninolytic compounds that are polyaromatic in nature;
- saccharic acids derived from the degradation of carbohydrates;

Table 4 Components of Kraft Black Liquor and Characteristics of Kraft Evaporator Condensate

Kraft black liquor characteristics	
Component	**Weight %, dry solids basis**
Lignin	30–45
Hemicellulose and sugars	1
Hydroxy acids	25–35
Extractives	3–5
Acetic acid	2–5
Formic acid	3–5
Methanol	1
Sulfur	3–5
Sodium	17–20
Kraft liquor evaporator condensate characteristics	
COD	1000–33,600 mg/L
Major organic component	Methanol, 60–90% of COD
Anaerobic degradability	80–90% of COD
Compounds that inhibit anaerobic metabolism	Reduced sulfur, resin acids, fatty acids, volatile terpenes

COD, chemical oxygen demand.
Source: Refs 3 and 6.

- solvent extractives that include fatty acids and resin acids;
- low-molecular-weight organic acids.

Table 4 shows the typical ranges of black liquor constituents and characteristics of Kraft evaporator condensates. The composition of liquors may vary significantly, depending upon the type of raw material used. Inorganic constituents in black liquor are sodium hydroxide, sodium sulfate, sodium thiosulfate, sodium sulfide, sodium carbonate, and sodium chloride [11].

10.4.2 Sulfite Pulping Liquors (Red Liquors)

Table 5 summarizes the composition of ammonia, sodium, magnesium, and calcium base sulfite pulping liquors. In general, spent ammonia base liquors have higher BOD_5, COD, and dissolved organics and exhibit more toxicity as compared to sodium, calcium, or magnesium base liquors. Higher toxicity is attributed to ammoniacal compounds in the spent liquors. The sulfite-spent liquors contain COD values typically ranging from 120–220 g/L and 50–60% of these are lignosulfonates [6]. The sulfite-spent liquor evaporator condensates have COD values in the range of 7500–50,000 mg/L. The major organic components in the condensates are acetic acid (30–60% of COD) and methanol (10–25% of COD). Anaerobic biodegradability of the condensates is typically 50–90% of COD and sulfur compounds are the major inhibitors of methanogenic activity [6].

Table 5 Composition of Ammonia, Sodium, Magnesium, and Calcium Base Sulfite Pulping Liquors

Parameter	Ammonia base mill[a]	Sodium base mill[b]	Magnesium base mill[c]	Calcium base mill[d]
Pulp liquor volume (m^3/ODT)	9.46	7.10	6.08	9.28
pH range	1.5–3.3	2.1–4.8	~3.4	5.3
BOD (kg/ODT)	413	235	222	357
COD (kg/ODT)	1728	938	975	1533
Dissolved organics (kg/ODT)	1223	595	782	1043
Dissolved inorganics (kg/ODT)	12.5	226	126	250
Lignin as determined by UV absorption (kg/ODT)	892	410	501	800
Total sugars (kg/ODT)	288	137	129	264
Reduced sugars (kg/ODT)	212	74	106	238
Toxicity emission factor[e] (TEF)	3663	714	–	422

[a]Average data based on 4 mills; [b]Average data based on 12 mills; [c] Average data based on 2 mills; [d]Composition of one mill; [e]Toxicity emission factors are based on static 96 hour bioassays and factored to the volume of liquor production. ODT = Oven dried ton of pulp.
Source: Refs. 3 and 10.

10.4.3 Thermomechanical Pulp (TMP), CTMP, and Semichemical Pulping Liquors

Thermomechanical pulp (TMP) and CTMP pulping liquors exhibit COD values in the ranges of 1000–5600 mg/L and 2500–13,000 mg/L, respectively [6]. Lignin derivatives can constitute anywhere from 15 to 50% of the soluble COD values in these spent liquors. The composition of spent NSSC pulping liquors and evaporator condensates are shown in Table 6. In general, anaerobic biodegradability of semichemical pulping and CTMP effluents are low as well as inhibitory to methanogenic metabolism [6].

10.4.4 Spent Liquors from Agro-Residue Based Mills

Agro-residue mills typically employ a soda or alkaline sulfite pulping process [14]. Typical compositions of the spent liquors generated from the small-scale, agro-residue utilizing pulp and paper mills are shown in Table 7. It is evident from the table that 45–50% of the total solids is represented by lignin. Most of the lignin present in the black liquor is the high-molecular-weight fraction, a key factor contributing to low BOD/COD ratio.

10.5 TOXICITY OF PULPING LIQUORS

A number of studies have evaluated the toxicity of pulping liquors, in particular the black liquors generated from Kraft mills. Table 8 shows a partial representation of toxicity data compiled by the NCASI (National Council of the Paper Industry for Air and Stream Improvement) and McKee and Wolf for Kraft mill pulping wastewaters [15,16]. The table indicates that hydrogen sulfide, methyl mercaptan, crude sulfate soap, and salts of fatty and resin acids are particularly

Table 6 Composition of Spent NSSC Pulping Liquor

Spent NSSC pulping liquor characteristics	
Parameter	**Average value**
Total solids (%)	12
Volatile solids (% of total solids)	48
COD (mg/L)	40,000
BOD_5 (mg/L)	25,000
Wood sugars (mg/L)	7000
Lignin (mg/L)	45,000
Acetate (mg/L)	18,000
pH range	6.5–8.5
Anaerobic degradability	NR
Compounds that have the potential to inhibit anaerobic process	Tannins, sulfur compounds
NSSC pulping liquor condensate characteristics	
COD	7000 mg/L
Major organic component	Acetic acid, 70% of COD
Anaerobic degradability	NR
Inhibitors of anaerobic degradation process	Sulfur compounds

NR, not reported; COD, chemical oxygen demand; BOD, biochemical oxygen demand.
Source: Refs. 3 and 6.

Table 7 Characteristics of Agro-Residue Based Spent Black Liquors

Parameter	Mill 1 (bagasse, wheat straw, and lake reed used as raw material)	Mill 2 (wheat straw used as the raw material)	Mill 3 (rice straw used as the raw material)
pH	9.7	10.2	8.8
Total solids (g/L)	44	42	38
Silica % (w/w) as SiO_2	2.4	3.2	12.0
Total organics % (w/w)	74.4	74.0	76.7
Lignin (g/L)	16.0	13.2	14.4
COD (mg/L)	48,700	45,600	40,000
BOD (mg/L)	15,500	13,800	16,500
COD/BOD	3.4	3.3	2.4

BOD, biochemical oxygen demand; COD, chemical oxygen demand.
Courtesy of MNES and UNDP India websites, Ref. 14.

toxic to Daphnia and fish populations. Among the toxic pollutants, compounds such as sodium hydroxide, hydrogen sulfide, and methyl mercaptan fall under the EPA's list of hazardous substances. Extractive compounds such as resin acids are known to contribute up to 70% of the total toxicity of effluents generated from chemical and mechanical pulping processes [17]. The concentrations of resin acids in the pulp mill discharges are two to four times higher than their LC_{50} values (0.5–1.7 mg/L) [17]. Some reports suggest that the transformation products of resin acids such as retene, dehydroabietin, and tetrahydroretene induce mixed function monooxygenases (MFO) in fish populations [17,18]. Hickey and Martin in 1995 found a correlation between the extent of resin acid contamination in sediments and behavior modification in benthic invertebrate species [19]. Johnsen et al. in 1995 reported that TMP mill effluents containing resin acids were lethal to rainbow trout following 2–4 weeks exposure at 200-fold dilution [20]. McCarthy et al. in 1990 demonstrated that resin acids are toxic to methanogens, thereby inhibiting the performance of these bacteria in anaerobic reactors [21].

Table 8 Toxicity of the Components of Kraft Pulp Mill Wastewaters

	Minimum lethal dose (ppm)	
Compound	Daphnia	Fish
Sodium hydroxide	100	100
Sodium sulfide	10	3.0
Methyl mercaptan	1.0	0.5
Hydrogen sulfide	1.0	1.0
Crude sulfate soap	5–10	5.0
Sodium salts of fatty acids	1.0	5.0
Sodium salts of resin acids	3.0	1.0
Sodium oleates	–	5.0
Sodium linoleate	–	10.0
Sodium salts of abietic acids	–	3.0

Source: Refs. 15 and 16.

10.6 PULP BLEACHING PROCESSES

About 5–10% of the original lignin cannot be removed from the pulp without substantial damage to the cellulosic fraction. Removal of the residual lignin, which is responsible for imparting dark color to the pulp, and the production of white pulp, requires a series of steps employing bleach chemicals (Fig. 1). Pulp bleaching is normally accomplished by sequential treatments with elemental chlorine (C_1), alkali (E_1), chlorine dioxide (D_1), alkali (E_2), and chlorine dioxide (D_2). The C stage consists of charging a slurry of the pulp (at 3–4% consistency) with elemental chlorine (60–70 kg/ton of pulp) at 15–30°C at pH 1.5–2.0 [2]. The chlorinated pulp slurry (at 10% consistency) is treated with alkali (35–40 kg/ton of pulp) at 55–70°C and pH 10–11. An optional hypochlorite (H) stage is introduced between the E_1 and D_1 stages for increasing the brightness of pulp. During the conventional bleaching, approximately 70 kg of each ton of pulp is expected to dissolve into the bleaching liquors [2]. The largest quantity of pulp is dissolved during the C_1 and E_1 stages. Alternate pulp bleaching techniques such as the elemental chlorine free (ECF), total chlorine free (TCF), and biobleaching are described in subsection 10.9.2.

10.6.1 Compounds Formed during Chlorine Bleaching Process

During pulp bleaching, lignin is extensively modified by chlorination (C stage) and dissolved by alkali (E stage) into the bleaching liquor. The E stage is intended for dissolving the fragmented chloro-lignin compounds and removal of noncellulosic carbohydrates. The most important reactions are oxidation and substitution by chlorine, which lead to the formation of chlorinated organic compounds or the AOX (Table 2). Chlorine bleaching liquors exhibit COD values ranging from 900–2000 mg/L and 65–75% of this is from chlorinated lignin polymers [6]. The types of chlorinated compounds found in the spent bleach liquors and their concentrations depend upon the quantity of residual lignin (Kappa number) in the pulp, nature of lignin, and bleaching conditions such as chlorine dosage, pH, temperatures, and pulp consistencies. The spent liquors generated from the conventional pulping and bleaching processes contain approximately 80% of the organically bound chlorine as high-molecular-mass material (MW above 1000) and 20% as the low-molecular-mass (MW of less than 1000) fraction [22].

The high-molecular-mass compounds, referred to as chlorolignins, cannot be transported across the cell membranes of living organisms and are likely to be biologically inactive. Nevertheless, these compounds are of environmental importance because they carry chromophoric structures that impart light-absorbing qualities to receiving waters. Long-term and low rates of biodegradation may generate low-molecular-weight compounds, causing detrimental effects on biological systems.

Efforts have been made to characterize the nature and content of individual components that are present in the low-molecular-mass fraction of the total mill effluents, which include the spent chlorination and alkali extraction stage liquors [2,4]. Approximately 456 types of compounds have been detected in the conventional bleach effluents, of which 330 are chlorinated organic compounds [22]. The compounds may be lumped into three main groups, namely, acidic, phenolic, and neutral (Table 2). Acidic compounds are further divided into the five categories of acids: fatty, resin, hydroxy, dibasic, and aromatic acids. The most important fatty acids are formic and acetic acids. The dominant resin acids are abietic and dehydroabietic acids. Among the hydroxy acids identified, glyceric acid predominates. Dibasic acids such as oxalic, malonic, succinic, and malic acids are derived from the lignin and carbohydrate fraction of wood and are present in significant amounts in the mill bleach effluents. Aromatic acids are formed from residual lignin through the oxidation of phenylpropanoid units and comprise four major categories: monohydroxy (phenolic), ortho-dihydroxy (catecholic), methoxy-hydroxy (guaiacolic), and dimethoxy-hydroxy (syringic) acids. The principal phenolics are chlorinated phenols, chlorinated catechols, chlorinated guaiacols, and chlorinated vanillin, derived from the chlorination and oxidative cleavage of lignin. The major neutral compounds are methanol, hemicellulose, and trace concentrations of aldehydes, ketones, chlorinated acetones, dichloromethane, trichloroethene, chloropropenal, chlorofuranone, chloroform, chlorinated sulfur derivatives, and 1,1-dichloromethylsulfone. In addition to the abovementioned compounds, the spent bleaching liquors have been reported to contain about 210 different chlorinated dioxins that belong to the two families: polychlorinated dibenzodioxins (PCDDs) and polychlorinated dibenzofurans (PCDFs) [22].

10.7 TOXICITY OF SPENT BLEACH LIQUORS

Compounds responsible for imparting toxicity to the spent bleach effluents originate during the chlorination (C) stage and caustic extraction (E) stages. The major classes of toxic compounds are resin acids, fatty acids, and AOX. Fatty and resin acids in bleach liquors often originate from the washing of unbleached pulps. They are recalcitrant to biodegradation as well as inhibitory to the anaerobic process. Adsorbable organic halides are the products of lignin degradation formed exclusively during the C stage of pulp bleaching and dissolved into the bleaching liquors during the E stage. About 1–3% of the AOX fraction is extractable into nonpolar organic solvents and is referred to as extractable organic halide (EOX). This extractable fraction poses greater environmental risks than the remaining 99% of the AOX and comprises compounds that are lipophilic with the ability to penetrate cell membranes and potential to bioaccumulate in the fatty tissues of higher organisms. Dioxins, in particular 2,3,7,8-tetrachlorodibenzodioxin (2,3,7,8-TCDD) and 2,3,7,8 tetrachlorodibenzofuran (2,3,7,8-TCDF) are highly toxic, bioaccumulable, carcinogenic, and cause an adverse impact on almost all types of tested species [2,22,23]. Additionally, the abovementioned dioxins and the other unidentified components of bleach liquors are also endocrine disrupting chemicals (EDC) that decrease the levels and activity of the estrogen hormone, thereby reducing reproductive efficiency in higher organisms [24]. However, limited information is available regarding these undesirable, genetically active, and endocrine-disrupting pollutants in receiving waters; further research is essential in this direction. Table 9 summarizes some findings related to the toxicity and impact of bleach mill discharges on selected aquatic organisms [25–31].

10.8 STRATEGIES FOR POLLUTION CONTROL IN PULP AND PAPER INDUSTRIES

Traditionally, discharge limits have been set for lumped environmental parameters such as BOD_5, COD, TSS, and so on. However, on account of the adverse biological effects of chlorinated organics coupled to the introduction of stricter environmental legislation, pulp and paper mills are faced with the challenges of not only reducing the BOD and suspended solids, but also controlling the total color as well as AOX in the effluents prior to discharge. In recent years,

Table 9 Summary of Selected Toxicology Studies to Assess the Ecological Impacts of Bleach Mill Effluents

Bleach process adopted by the mill	Organism studied	Physiological/biochemical effect(s)/ levels of toxicity	Research group
Kraft mill using 100% chlorine dioxide	Coastal fish community	High levels of mortality and low embryo quality	Sandstrom, 1994 [25]
New and old wood pulp bleaching employing various bleach sequences	Mesocosm and fish biomarker tests	Elemental chlorine containing bleach sequence, CEHDED was the most toxic	Tana et al., 1994 [26]
Kraft mill using 100% chlorine dioxide	Baltic sea amphipod (crustacean)	Reduced swimming activity	Kankaanpaa et al., 1995 [27]
Kraft bleach mill effluent produced by oxygen delignification or 100% chlorine dioxide	Freshwater fish	Induction of mixed function oxidase (MFO) enzymes following exposure to 4% and 12% effluent in artificial streams	Bankey et al., 1995 [28]
Kraft mill using 100% ClO_2	Aquatic organisms	Overall toxicity pattern of effluents in the bioassay was: Untreated ECF > untreated TCF > secondary treated ECF > secondary treated TCF	Kovacs et al., 1995 [29]
Kraft mill using 100% ClO_2	Fish populations	Changes in the reproductive development – reduction in gonad size, depression of sex hormones following exposure to bleach effluents subjected to secondary treatment	Munkittrick et al., 1997 [30]
Kraft mill that had used elemental chlorine historically	Microbial community and diatom species in lake sediments sampled from 2–8 cm depths	Drop in the ATP content, depressed butyrate-esterase activity indicating toxicity to microorganisms, and reduction in diatom species richness	Mika et al., 1999 [31]

the pulp and paper industry has taken great strides forward in recognizing and solving many of the environmental problems by adopting two strategies:

1. Pollution reduction measures within plants that include minimization of spills and modifications in the process through adaptation of cleaner technologies as alternatives to conventional technologies.
2. End-of-pipe pollution treatment technologies, which are essential either as a supplement or as backup measures to pollution reduction techniques in order to meet the effluent regulation standards.

These two approaches are equally important in meeting environmental regulations and are addressed in separate sections.

10.9 POLLUTION REDUCTION THROUGH PLANT PROCESS MODIFICATIONS

10.9.1 Nonconventional Pulping Technologies

Industries have developed alternate pulping techniques that do not use the conventional cooking methods. Some of these techniques are described briefly in the following subsections. Readers may note that some of these processes have not yet reached commercial stages.

Organic Solvent Pulping

Organic solvents such as methanol, ethanol, and other alcohols are used for pulping. This process is economical for small- to medium-scale mills with significant recovery of chemicals for reuse. However, pulping must be conducted in enclosed containers to prevent the loss of volatile solvents and for workers' safety. Additionally, some of these processes are more energy intensive than traditional methods. Major benefits include the elimination of odorous sulfur-containing compounds in the effluents and air.

Acid Pulping

Wood chips are treated with acetic acid at pressures that are significantly lower than those used for Kraft pulping. Drawbacks include loss of acid, although recovery is possible through the energy-intensive distillation process.

Biopulping

This method utilizes whole cells of microorganisms and microbial enzymes such as xylanases, pectinases, cellulases, hemicellulases, and ligninases, or their combinations, for pulping herbaceous fibers and improving the properties of pulp derived from wood [32]. Pretreatment of wood chips with lipases is known to reduce the problematic oily exudates during the pulping process as well as improving the texture of paper through the specific degradative action of these enzymes on pitch-derived extractives such as fatty acids and waxes. The innovative approach of using microorganisms or microbial enzymes to reduce the consumption of chemicals in the pulp and paper industry is known as biopulping. Biopulping has generated much interest among the pulp and paper industries because of the following advantages:

- Reduction in the chemical and energy requirements per unit of pulp produced. Thus, the process is expected to be cost effective and more affordable for medium- and small-scale mills.
- Reduction in the pollution load due to reduced application of chemicals.
- The yield and strength properties of the pulp are comparable (sometimes even better) to those obtained through conventional pulping techniques.

Nonwood fibers are more responsive to the action of pulping enzymes compared to wood, presumably due to lower lignin content and weaker hemicellulose–lignin bonds. This is clearly advantageous for developing countries, which are faced with the problem of shrinking forest wood resources. However, further research is required to optimize the conditions required for enzymatic pulping of herbaceous fibers and commercialization of the process.

10.9.2 Cleaner Pulp Bleaching Technologies

The use of elemental chlorine in pulp bleaching has been gradually discontinued in several countries to prevent the toxic effects of chlorinated organics in receiving waters and to meet regulatory requirements. Most nations have imposed stringent regulatory limits on AOX, ranging from 0.3 to 2.0 kg/ton of pulp [22]. Cleaner bleaching methods have been developed by industries based on elemental chlorine free (ECF), total chlorine free (TCF), microbial systems (bio-bleaching), extended delignification, and methods for monitoring and improved control of bleaching operations. Each of these approaches is discussed in the following subsections.

Elemental Chlorine Free (ECF) and Total Chlorine Free (TCF) Bleaching

Elemental chlorine has been replaced by chlorine dioxide and hypochlorite in the ECF bleach sequence, while oxygen, ozone, caustic soda, and hydrogen peroxide have been advocated for TCF bleaching of softwood and hardwood Kraft pulps. Benefits include significant reduction in the formation of chlorinated organics or their elimination and lower ecological impacts. Two Finnish mills eliminated elemental chlorine from the bleach sequence and substituted chlorine dioxide, thereby sharply reducing the concentration of chlorinated cymenes [33]. In another Finnish example, levels of chlorinated polyaromatic hydrocarbons in mill wastes were substantially reduced during production of bleached birch Kraft pulp without the use of elemental chlorine as compared to pine pulp bleached with elemental chlorine [34,35]. Research has also been conducted on the optimal usage of agents such as ozone and hydrogen peroxide [36,37]. However, alternatives such as ozonation, oxygenation, and peroxidation are not economically viable for medium- or small-capacity mills due to higher capital investments and plant operation costs.

Biobleaching

Biobleaching processes based on the pretreatment of pulp with microbial whole cells or enzymes have emerged as viable options. A number of studies examined the direct application of white rot fungi such as *Phanerochaete chrysosporium* and *Coriolus versicolor* for biobleaching of softwood and hardwood Kraft pulps [38–44]. It has been found that fungal treatment reduced the chemical dosage significantly as compared to the conventional chemical bleach sequence and enhanced the brightness of the pulp. Specific features of the fungal-mediated biobleaching processes are:

- Action through delignification that commences at the onset of the secondary metabolic (nitrogen starvation) phase in most fungi.
- Delignification is an enzymatic process mediated through the action of extracellular enzymes.
- The growth phase of the fungus has an obligate requirement for a primary substrate such as glucose.

The major drawbacks of the fungal bleaching process are that it is extremely slow for industrial application and requires expensive substrates for growth. To overcome these problems, enzyme preparations derived from selected strains of bacteria or fungi are recommended. The enzymatic method of pulp bleaching is being increasingly preferred by a number of pulp and paper industries, especially in the West, because it is a cost-effective and environmentally sound technology [32]. The distinct advantages of enzyme-mediated pulp bleaching are:

- minimal energy input;
- specificity in reactivity, unlike that of chemicals;

- reduced dosage of bleach chemicals in the downstream steps;
- improved quality of pulp through bleach boosting;
- reduced load of AOX in the effluents.

Two categories of enzymes, namely xylanases and peroxidases (lignin degrading), have been identified in the pulp bleaching processes. Of the two classes, the use of xylanases has achieved enormous success in aiding pulp bleaching [45]. Xylanase enzymes apparently cause hydrolytic breakdown of xylan chains (hemicellulose) as well as the cleavage of the lignin–carbohydrate bonds, thereby exposing lignin to the action of subsequent chemical bleaching steps [46]. However, most biobleaching studies using xylanases have been carried out with either hardwood or softwood, while nonwood resources are being increasingly used as the chief agricultural raw material for pulp production. Therefore, further research with regard to enzyme applications for nonwood pulp bleaching is warranted.

It is unlikely that xylanase treatment alone will completely replace the existing chemical bleaching technology, because this enzyme does not act directly on lignin, a crucial color-imparting polymer of the pulp. Nonspecific oxido-reductive enzymes such as lignin peroxidase, manganese peroxidase and in particular, laccases, which are lignolytic, are likely to be more effective in biobleaching [47]. The abovementioned enzymes can also act on a wide variety of substrates and therefore have significant potential for applications to pulp and paper effluent treatment [48,49]. The applicability of the laccase mediator system for lignolytic bleaching of pulps derived from hard wood, soft wood, and bagasse has been reviewed and compared by Call and Mucke [47]. The major advantage of enzymatic bleaching is that the process may be employed by the mills over and beyond the existing technologies with limited investment. Furthermore, there is ample scope for the improvement of the process in terms of cost and performance.

Extended Delignification

The key focus of this process is on the enhanced removal of lignin before subjecting the pulp to bleaching steps [50,51]. Such internal process measures also imply cost savings during the subsequent chemical bleaching steps and have a positive impact on the bleach effluent quality parameters such as COD, BOD, color, and AOX. Extended delignification may be achieved through:

- *Extended cooking*. This can be done by enhancing cooking time or temperature or by multiple dosing of the cooking liquors.
- *Oxygenation*. The pulp is mixed with elemental oxygen, sodium, and magnesium hydroxides under high pressure. An example is the PRENOX process [50]. According to Reeve, about 50% of the world capacity for Kraft pulp production incorporated oxygen delignification by the year 1994 [52].
- *Ozonation*. Ozone and sulfuric acid are mixed with the pulp in a pressurized reactor.
- *Addition of chemical catalysts*. Compounds such as anthraquinone or polysulfide or a mixture of the two are introduced into the Kraft cooking liquor.

Improved Control of Bleaching Operations

Installation of online monitoring systems at appropriate locations and controlled dosing of bleach chemicals can aid in the reduction of chlorinated organics in effluents.

10.10 TREATMENT OF PULP AND PAPER WASTEWATERS

Plant process modifications and cleaner technologies have the potential to reduce the pollution load in effluents. However, this approach cannot eliminate waste generation. End-of-pipe pollution treatment technologies are essential for meeting the prescribed limits for discharged pollutant concentrations such as color, AOX, BOD, COD, and so on. Assessment of the water quality of receiving ecosystems and periodic ecological risk assessments are required to validate the effectiveness of various treatment methods [53]. The most common unit processes employed by the pulp and paper mills during preliminary, primary, secondary, and tertiary (optional) stages of effluent treatment are listed in the flow sheet shown in Figure 2. Process technologies that are currently applied can be broadly classified as the physico-chemical and biological treatment methods. These technologies are discussed in the following subsections.

10.10.1 Physico-Chemical Processes

Several physico-chemical methods are available for the treatment of pulping and pulp bleaching effluents. The most prominent methods are membrane separation, chemical coagulation, and precipitation using metal salts and advanced oxidation processes.

Membrane Separation Techniques

Membrane processes operate on the basis of the following mechanisms:

- pressure driven, which includes reverse osmosis (RO), ultrafiltration (UF), and nanofiltration (NF);
- concentration driven, which includes diffusion dialysis, vapor permeation, and gas separation;
- electrically driven, which includes electrodialysis;
- temperature difference driven, including membrane distillation.

Membrane filtration (UF, RO, and NF) is a potential technology for simultaneously removing color, COD, AOX, salts, heavy metals, and total dissolved solids (TDS) from pulp mill effluents, resulting in the generation of high-quality effluent for water recycling and final discharges. The possibility of obtaining solid free effluents is a very attractive feature of this process. Ultrafiltration was used by Jonsson et al. [54] for the treatment of bleach plant effluents.

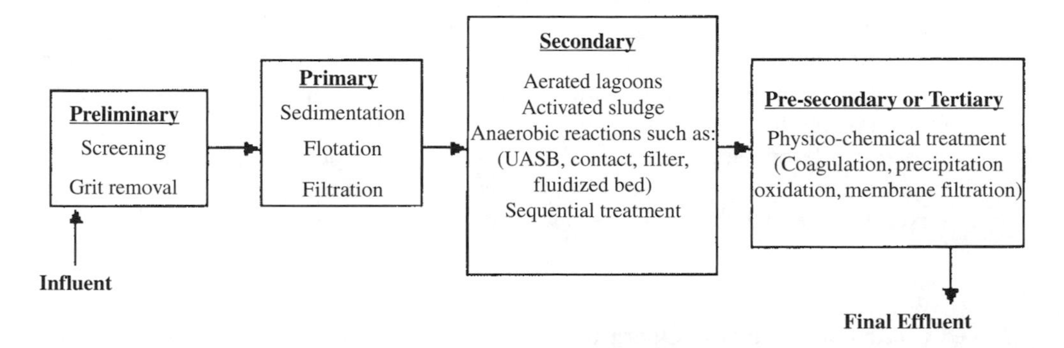

Figure 2 Flow sheet showing the unit processes employed by pulp and paper mills for effluent treatment.

Sierka et al. [55] described a study that compared the efficiencies of UF alone and UF in combination with RO for the removal of color and total organic carbon (TOC) in the D_o (acid stage) wastewaters discharged from the Weyerhaeuser Grande Priare pulp mill, which produces 300,000 tons of bleached Kraft bleached pulp per year. The bleach plant of this Kraft mill typically employs five stages ($ECD_oE_{op}DED$ sequence) for the production of tissue and specialty grade paper. D_o stage wastewaters were sterilized using 0.45 μm filters and subsequently passed through Amicon-stirred UF cells fitted with membranes having cutoff values of 500 Daltons (D) (YCO5), 1000 D (YM1), 3000 D (YM3), or 10,000 D (YM10). Table 10 summarizes the characteristics of the permeate and concentrates obtained following the ultrafiltration of D_o wastewater using various membranes. Based on these results, Sierka et al. concluded that most of the color (50%) is due to organic compounds with molecular size above 3000 D. Table 11 presents the results of additional studies conducted on D_o stage effluents that involved pretreatment by UF followed by RO. Clearly, the combination of the UF and RO steps gave excellent results by removing 99% of the color and more than 80% of the TOC from the D_o stage effluent.

Koyuncu et al. [56] presented pilot-scale studies on the treatment of pulp and paper mill effluents using two-stage membrane filtrations, ultrafiltration and reverse osmosis [56]. The combination of UF and RO resulted in very high removals of COD, color, and conductivity from the effluents. At the end of a single pass with seawater membrane, the initial COD, color and conductivity values were reduced to 10–20 mg/L, 0–100 PCCU (platinum cobalt color units) and 200–300 μs/cm, respectively. Nearly complete color removals were achieved in the RO experiments with seawater membranes.

A distinct advantage of the membrane technology is that it can be utilized at the primary, secondary, or tertiary phases of water treatment. Some membranes can withstand high concentrations of suspended solids, which presents a possible direct application for separating mixed liquor suspended solids (MLSS) in an activated sludge plant (membrane bioreactor) to replace the conventional sedimentation tank. Key variable parameters of membrane technology include variation in membrane pore size, transmembrane pressure, cross-flow velocity, temperature, and back flushing. The major disadvantages are high capital and maintenance costs, accumulation of reject solutes and decrease in the membrane performance, membrane fouling, and requirement for the pretreatment of discharges.

Table 10 Ultrafiltration Characteristics of D_o Wastewater Using Different Membranes

	Membrane			
	YCO5	YM1	YM3	YM10
UF input/output	500 D	1000 D	3000 D	10,000 D
TOC (mg/L)				
Feed	792.5	792.5	792.5	792.5
Permeate	282	465	546	634
Concentrate	1739	1548	1158	1033
Color (PCCU[a])				
Feed	1700	1700	1700	1700
Permeate	107	334	835	1145
Concentrate	3066	2972	2221	1876

[a] Platinum cobalt color units.
Source: Ref. 55.

Table 11 Characteristics of D_o Wastewater Subjected to Ultrafiltration and Reverse Osmosis

Input/output of UF/RO unit	pH	TOC (mg/L)	Color (PCCU[a])
Feedwater to UF unit	7.00	825	1750
Composite permeate from UF unit	–	555	1231
Feedwater to RO unit	–	555	1231
Composite permeate from RO unit	6.44	70.68	13.79
Concentrate from RO unit	6.47	3219	2741

[a] Platinum cobalt color units.
Experimental conditions: Pressure = 1104 kPa; temperature = 40°C; batch volume = 4 L; cutoff value of the UF membrane = 8000 D.
Source: Ref. 55.

Chemical Coagulation and Precipitation

This method relies on the addition of metal salts to cause agglomeration of small particles into larger flocs that can be easily removed by settling. The effectiveness of this process is dependent upon the nature of coagulating agent, coagulant dosage, pH, ionic strength, and the nature and concentration of compounds present in wastewaters. The not-so-easily biodegradable fraction of pulping and bleaching effluents consists of polar and hydrophobic compounds, notably resin acids, long-chain fatty acids, aromatic acids and phenols, lignin, and terpenes. Almost all of these toxic compounds can be effectively removed through coagulation using chloride and sulfate salts of Fe^{3+} and Al^{3+}. Typically, these trivalent cations remain in solution at acidic pH and form metal hydroxides that aggregate rapidly at higher pH conditions. Hydrogen bonding, electrostatic and surface interactions (adsorption) between the metal hydroxides and organic anions (containing hydroxyl and carboxyl groups) lead to the formation of metal hydroxide–organic compound precipitates [57,58]. Dissolved organics are also removed by physical adsorption to flocs.

Chemical precipitation of mill effluents from CTMP, BKME (bleached Kraft mill effluent), NSSC, and E & C bleach discharges have been extensively studied by Stephenson and Duff [59] using alum, lime, ferric chloride, ferrous sulfate, magnesium hydroxide, polyimine, polymers, and alum in combination with lime. They observed removal of 88% of total carbon and 90–98% of color and turbidity from mechanical pulping effluents using Fe^{3+}/Al^{3+} salts. In another publication, Stephenson and Duff reported significant reduction in the toxicity of wastewaters subsequent to the chemical coagulation process [60]. Ganjidoust et al. [61] compared the effects of a natural polymer, chitosan, and synthetic polymers, namely hexamethylene diamine epichlorohydrin polycondensate (HE), polyethyleneimine (PEI), polyacrylamide (PAM), and a chemical alum coagulant, alum on the removal of lignin (black liquor color and total organic carbon) from alkaline pulp and paper industrial wastewater. They observed that PAM, a nonionic polymer, had a poor effect, whereas HE and PEI, which are cationic polymers, coagulated 80% of the color and 30% of the total organic carbon from alkaline black liquor wastewater by gravity settling in 30 min. Alum precipitation removed 80% of the color and 40% of total organic carbon. By comparison, the natural coagulant chitosan was the most effective; it eliminated up to 90% of the color and 70% of the total organic carbon, respectively.

The major disadvantages of coagulation and precipitation are the generation of chemical sludge and the need for subsequent treatment of the sludge to eliminate the adsorbed toxic pollutants prior to disposal.

Advanced Oxidation Processes

Destruction of chromophoric and nonchromophoric pollutants in pulp and paper effluents may be achieved by advanced oxidation methods such as photocatalysis, photo-oxidation using hydrogen peroxide (H_2O_2)/UV or ozone (O_3)/UV systems, Fenton-type reactions, wet oxidation, and by employing strong oxidants such as ozone.

Photocatalysis has gained attention for its application to aqueous phase and wastewaters for near total oxidation and elimination of organic compounds [62]. The process involves mixing wastewater with aeration in a reactor at 20–25°C and the introduction of titanium dioxide (TiO_2) followed by irradiation using a UV lamp. Irradiation by UV light generates an electron hole on TiO_2 surface, which reacts with the adsorbed organic compounds or water molecules. TiO_2 can be provided as a suspension or as covered supports (immobilized on beads, inside tubes of glass/teflon, fiberglass, woven fibers, etc.). Various research groups have shown that photocatalysis is nonselective and that there is a nearly parallel reduction in the color, lignosulfonic acids, and other organic compounds in the treated pulp and paper mill effluents. Balcioglu et al. [63] observed enhanced biodegradability (increase in BOD_5/COD ratio) of raw Kraft pulp bleaching effluents and improved quality of the biologically pretreated effluents following TiO_2 photocatalytic oxidation. Yeber et al. [64] described the photocatalytic (TiO_2 and ZnO) treatment of bleaching effluents from two pulp mills. Photocatalysis resulted in the enhanced biodegradability of effluents with concomitant reduction in the toxicity.

Photo-oxidation systems using H_2O_2/UV or O_3/UV combinations generate hydroxyl radicals that are short lived but extremely powerful oxidizing organics through hydrogen abstraction. The result is the onsite total destruction of refractory organics without generation of sludges or residues. Wastewater is injected with H_2O_2 or saturated with O_3 and irradiated with UV light at 254 nm in a suitable reactor with no additional requirement for chemicals. The rate of oxidative degradation is generally much higher than systems employing UV or O_3 alone. Legrini et al. [62] have extensively reviewed the experimental conditions used by various researchers for conducting the photo-oxidation process as well as their application for removal of various types of organic compounds.

Fenton's reactions involving hydrogen peroxide (H_2O_2) and ferrous ion as the solution catalyst are an effective option for effluent treatment. Fenton's reaction as described by Winterbourn [65] requires a slightly acidic pH and results in the formation of highly reactive hydroxyl radicals ($^{\cdot}OH$), which are capable of degrading many organic pollutants. Rodriguez et al. [66] evaluated Fenton-type reactions facilitated by catecholic compounds such as 2,3-dihydroxybenzoic acid, 3,4-dihydroxybenzoic acid, and 1,2-dihydroxybenzene for treating pulp bleaching effluent. Their research indicated that 2,3-dihydrobenzoic acid was the most effective compound in enhancing hydroxyl radical formation in the iron–hydrogen peroxide reaction system at pH 4.0 with the concomitant reduction in the AOX concentration and toxicity of the bleach effluents.

Wet oxidation is a process where organic contaminants in liquids or soils are extracted into an aqueous phase and reacted with an oxidant at high temperature (220–290°C) and pressures (100–250 bar) to promote rapid destruction. Laari et al. [67] evaluated the efficiency of wet oxidation for the treatment of TMP processing waters. The major objective of this research was to reduce the concentration of lipophilic wood extractives (LWE) and to treat concentrated residues from evaporation and membrane filtration by low-pressure catalytic wet oxidation.

The wet oxidation of membrane and evaporation concentrates was effective in reducing 50% of the COD at 150°C and enhancing the biodegradability of wastewater.

Oxidants such as chlorine, oxygen, ozone, and peroxide have been proposed for the treatment of pulp bleach effluents. Ozonation has been reported to reduce the toxicity of CTMP and bleached Kraft mill (BKM) effluents at low dosages [22]. Hostachy et al. [68] reported detoxification and an increase in the biodegradability of bleach effluents by ozonation at low dosages [0.5–1 kg/ADMT (air-dried metric ton)] of pulp. The researchers observed significant elimination of the residual COD by catalyzed ozone treatment of hardwood and softwood pulp and paper mill final discharges. Such a treatment method may allow for reutilization of treated process waters and reduce consumption of freshwater during pulping steps. Helbe et al. [69] described a tertiary treatment process involving ozonation in combination with a fixed-bed biofilm reactor for the reuse of treated effluent in a pulp and paper industry. Sequential ozonation and bioreactor treatment gave maximum elimination of COD, color, and AOX from biologically treated effluent with minimum dosage of ozone. Further, the authors suggested that two-stage ozonation with intermediate biodegradation is more effective in terms of achieving higher removal of persistent COD.

The advantages of the various oxidation processes include nonselective and rapid destruction of pollutants, absence of residues, and improved biodegradability of the effluents. Some of the disadvantages are extremely short half-life of the oxidants and high expense of their generation.

10.10.2 Biological Processes

The most commonly used biological treatment systems for the pulp and paper mill discharges are activated sludge plants, aerated lagoons, and anaerobic reactors. Sequential aerobic-anaerobic systems (and vice versa) are a recent trend for handling complex wastewaters of pulp and paper mills that contain a multitude of pollutants. The application of various types of biological reactor systems for treating pulp and paper mill effluents are discussed in the following subsections.

Activated Sludge Process

This conventional aerobic biological treatment train consists of an aeration tank with complete mixing (for industrial discharges) followed by a secondary clarifier and has been typically used for the reduction of COD, BOD, TSS, and AOX in pulp and paper mill waste effluents. Oxygen is provided to the aerobic microorganisms through aeration or by using pure oxygen as in the deep shaft systems. Bajpai [22] has reviewed the efficiencies of activated sludge plants and reported that the overall removal of AOX can range from 15 to 65%, while the extents of removal of individual chlorinated organics such as chlorinated phenols, guaiacols, catechols, and vanillins can vary from 20 to 100%. Biotransformation and biodegradation seem to be the important mechanisms for reduction in the overall AOX concentrations and hydraulic retention time (HRT) is the key operating parameter.

There are a number of full-scale activated sludge plants that are in operation in countries such as the United States, Canada, and Finland, which treat effluents from Kraft, sulfite, TMP, CTMP, and newsprint mills [22]. Schnell et al. [70] reported the effectiveness of a conventional activated sludge process operating at an alkaline-peroxide mechanical pulping (APMP) plant at Malette Quebec, Canada. The full-scale plant achieved 74% reduction in filterable COD and nearly complete elimination of BOD_5, resin acids, and fatty acids in the whole mill effluent. The treated effluent tested nontoxic as measured by a Microtox assay. Saunamaki [71] reported excellent performances of activated sludge plants in Finland that were designed according to the low loading or extended aeration principle. Control of nutrients, aeration, low loading rates, introduction of equalization, and buffer basins seemed to be the key process control parameters for successful treatment. BOD_7 and COD removal averaged 94 and 82%, respectively, at paper mills while at pulp mills, the values were 82 and 60%, respectively. All paper mill activated sludge plants required dosing of nitrogen and phosphorus. Narbaitz et al. [72] evaluated the impacts of adding powdered activated carbon to a bench-scale activated sludge process ($PACT^{TM}$) fed with low-strength Kraft pulp mill wastewaters. Enhanced removal of AOX and marginal improvement in the levels of COD and toxicity reduction as compared to the conventional activated sludge process was observed.

Two common operational problems encountered during the treatment of pulp and paper wastewaters in activated sludge plants are:

- Limiting concentrations of nitrogen and phosphorus (N and P) that are vital for maintenance of active microbial population in an activated sludge plant.
- Growth of filamentous organisms or formation of pinpoint flocs that negatively impact the sludge settling rates, thereby reducing the effluent quality.

The problem of nutrient deficiency is frequently overcome through the external addition of nutrients with optimization of their dosage. A major drawback of supplementation is the requirement for extensive monitoring of treated effluents for N and P prior to discharge to avoid adverse environmental impacts such as the eutrophication of receiving waters. Alternate approaches have been investigated, such as the selection and incorporation of bacteria capable of fixing atmospheric nitrogen (nitrogen fixers) in the biological reactors or addition of solid N and P sources with low solubility to prevent excess loadings in the final effluents. As an example, Rantala and Wirola [73] have demonstrated the success of using a solid source of phosphorus with low solubility in activated sludge plants fed with CTMP mill wastewater. They observed that the total phosphorus concentration in the effluents was more than 2 mg/L in the activated sludge reactor fed with liquid phosphoric acid and less than 0.5 mg/L if fed with solids such as apatite or raw phosphate. Based on a full-scale trial study at a CTMP mill, the authors concluded that the addition of nutrient in the form of apatite is a viable alternative for reducing phosphorus load in the treated effluents.

Conventional practices for controlling sludge bulking are through chlorination or peroxidation of sludge or addition of talc powder. Clauss et al. [74] discussed two case studies on the application of fine Aquatal (product designed by Luzenac–Europe), a mineral talc-based powder, to activated sludge plants for counteracting the floc settlability problems. In the first case, Aquatal was added to aeration tanks to control sludge volume index (SVI) and reduce the concentration of suspended solids in the effluents. In a second case study, Aquatal was introduced to prevent sludge blanket bulking. In both cases, the mineral powder additive resulted in the formation of compact, well-structured heavier flocs that displayed increased settling velocities, and good thickening and dewatering properties. However, a major drawback of this method is that it addresses the symptoms of the problem rather than the root cause. A permanent solution based on the comparison of physiology, substrate requirement and degradation kinetics of floc forming and filamentous bacteria is needed.

A number of case studies have reported on the improvement of existing activated sludge plants in the pulp and paper industries through modifications. Two case studies are presented below.

(a) A Case Study on the Up-Gradation of an Activated Sludge Plant in Poland. Hansen et al. [75] described the up-gradation of an existing activated sludge plant of 400,000 ADMT pulp capacity mill in Poland that produces unbleached Kraft pulp. The discharge limits as set by

the Polish authorities were 150 mg/L of COD, 15 mg/L of BOD, 60 mg/L of SS and 88,000 m^3/day of water to the receiving river from the year 2000. To meet the new demands, two additional FlooBed reactors with a total volume of 50% of the existing activated sludge plant were installed. The three biological reactors were operated in series, with the activated sludge plant as the third stage, as shown in Figure 3. FlooBed is an activated sludge tank with microorganisms supported as thin films over floating carrier materials. The carrier is made of polyethylene with an area of 200 m^2/m^3 for biofilm growth. All three stages of the biological reactor were amended with the required concentrations of urea and phosphate. The up-graded plant operation was commissioned in 1998 and the efficiency of COD reduction was reported to have increased from 51 to 90%. The first-stage FlooBed reactor removed most of the easily biodegradable fraction, while the second FlooBed reactor mainly degraded the not-so-easily biodegradable fraction with continuing action on the easily biodegradable compounds. The third bioreactor (existing activated sludge plant) acted as a polisher and handled the residual biodegradable contaminants. The color of the untreated mill discharge was dark brown, while the effluent from the third-stage bioreactor was reported to be clear. According to the authors, the prescribed discharge limits were successfully met by the up-graded activated sludge plant. Since February 1999, the discharge from the plant is reported to have stabilized at 4 kg COD/ton of pulp produced.

(b) A Case Study on the Up-Gradation of an Activated Sludge Plant in Denmark. Andreasen et al. [76] presented a case study on the successful up-gradation of a Danish pulp industry activated sludge plant with an anoxic selector to reduce bulking sludge problems (Fig. 4). The wastewater of the pulp mill contained large amounts of biodegradable organics with insufficient concentrations of N and P. This condition led to excessive growth of filamentous microbes and poor settling properties of the sludge. The DSVI (sludge volume index) often exceeded 400 mL/g of suspended solids and, as a result, the sludge escaped from the settlers and caused a 70% reduction in the plant capacity. A selector dosed with nitrate was installed ahead of the activated sludge plant to remove a large fraction of easily degradable COD under denitrification conditions. Installation of this anoxic selector significantly improved the DSVI to less than 50 mL/g and enhanced the performance of settlers. Sludge loading of 20–30 kg COD/kg VSS corresponding to a removal rate of 16 kg filterable COD/kg NO_3-N and retention time of 17–22 min were chosen for the optimal performance of the selector. The dosing of nitrate was maintained above 1 mg/L in the selector to avoid anaerobic conditions. Phosphorus was not added due to stringent effluent discharge standards.

Aerated Lagoons (Stabilization) Basins

Aerated lagoons are simple, low-cost biological treatment systems that have been explored in laboratory-scale, pilot-scale, and full-scale studies for the treatment of pulp and paper industrial effluents. Distinct advantages of stabilization basins are lower energy requirement for operation and production of lower quantities of prestabilized sludge. In developed countries like Canada and the United States, the earliest secondary treatment plants for the treatment of pulp and paper effluents were aerated stabilization basins, while in developing countries such as India and China these simple, easy to operate, systems continue to be the most popular choice. Aerated lagoons have masonry or earthen basins that are typically 2.0–6.0 m deep with sloping sidewalls and use mechanical or diffused aeration (rather than algal photosynthesis) for the supply of oxygen [77]. Mixing of biomass suspension and lower hydraulic retention time (HRT) values prevent the growth of algae. Aerated lagoons are classified on the basis of extent of mixing. A completely mixed lagoon (also known as aerated stabilization basin, ASB) is similar to an activated sludge process where efficient mixing is provided to supply adequate concentrations of oxygen and to

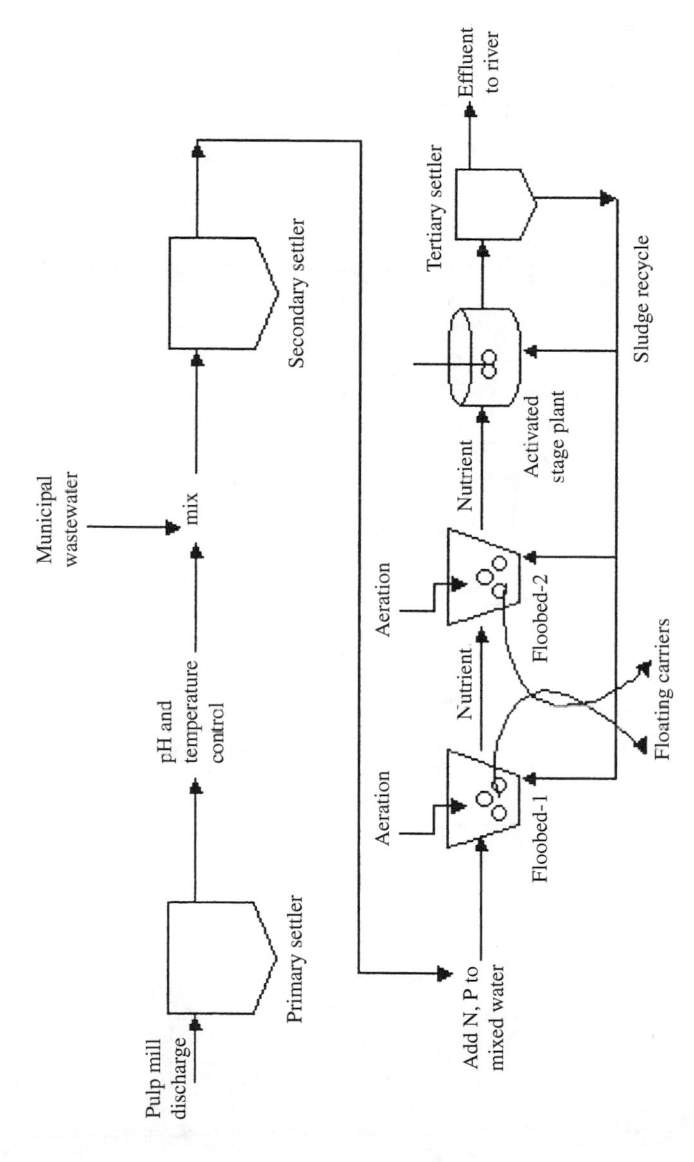

Figure 3 Up-gradation of an existing activated sludge plant in Poland by installation of FlooBed reactors (from Ref. 75).

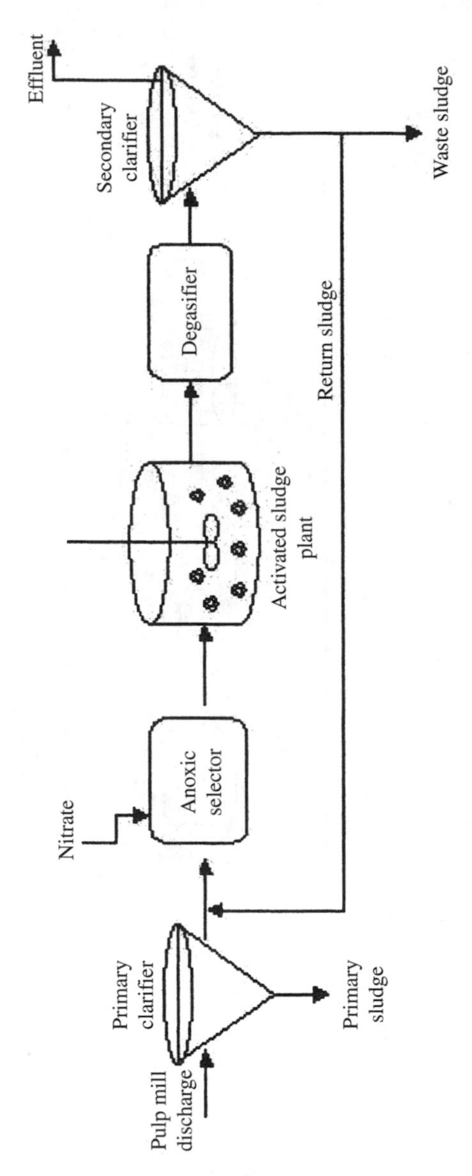

Figure 4 Up-gradation of a Danish pulp and paper mill activated sludge plant through installation of an anoxic selector (from Ref. 76).

keep all of the biomass in suspension (Fig. 5a). However, the system does not include a mechanism for recycling biomass or solids; consequently, the HRT approaches the SRT value. Aerobic bacteria oxidize a portion of the biodegradable organics into carbon dioxide and water, and the rest is utilized to generate biomass components. Several completely mixed aerated lagoons may be linked in series to increase the HRT/SRT value, thereby facilitating further stabilization of synthesized biomass and organic solids under aerobic conditions. In a partially mixed aerated lagoon (also known as facultative stabilization basin, FSB), the power input adequately satisfies the system's oxygen requirements but is insufficient for keeping the solids in suspension. This allows for settlement of biosolids by gravity sedimentation and subsequent benthal stabilization through anaerobic processes. Thus, the biological activity in facultative lagoons is partially aerobic and anaerobic. Partial mix lagoons are generally designed to include two to three cells in series (Fig. 5b). Typically the first cell is completely mixed with intense aeration while the final cell may have very low mixing in which the biomass is allowed to settle down to form benthal deposit. Growth of algae in the settling lagoon is prevented by minimum aeration and limiting HRT value of the overlying clear water zone.

Aerated lagoons have been employed as full-scale treatment systems or as polishing units in Kraft, TMP, and CTMP mills for the removal of BOD, low-molecular-weight AOX, resin and fatty acids [22]. Typical HRT values range from 5 to 10 days. Bajpai [22] compared the reduction efficiencies of individual chlorophenols across aerated lagoons and noted that the values ranged from 30 to 90%. Overall reduction of AOX in bleached kraft mill effluents typically vary from 15 to 60%. Removal of resin and fatty acids in CTMP effluents occur through aerobic oxygenation and degradation with efficiencies exceeding 95%. Welander et al. [78] observed significant improvement in the efficiency of aerated lagoons by installing a support matrix for microbial growth in 20 m^3 pilot-scale plants at two Swedish pulp and paper mills. The two plants were operated for nearly a year and exhibited 60–70% reduction in COD and phosphorus levels. However, efficiencies were much lower for full-scale plants. Kantardjieff and Jones [79] conducted pilot-scale studies on a Canadian integrated sulfite pulp and paper mill effluent using an aerobic biofilter (1 m^2, 3 m depth) as the main unit and aerated stabilization basins (3 m^3) as the polishing stage. The biofilter unit treated the most concentrated sulfite mill effluent and the resulting effluent was mixed with remaining mill wastewaters to be treated in the polishing ASB unit. Characteristics of the raw wastewater, biofilter, and ASB treated mill discharges are summarized in Table 12. In the final design, the ASB had two sections and was operated as a completely mixed system with a total HRT of 2.5 days. The final effluents met the prescribed discharge permit limits and were reported to be nontoxic.

Laboratory-scale treatability studies were conducted by Hall and Randle [80] to monitor and compare the performance of an activated sludge system, ASB, and FSB, operated in parallel to treat Kraft mill wastewaters. Results indicated that FSB and ASB achieved higher removal efficiencies of total and filterable AOX as compared to the activated sludge process under varying temperatures and SRT values. Higher removal rates of chlorinated organics were observed in FSB when the SRT value was increased from 15 to 30 days. The principal removal mechanism seemed to be sorption of AOX to biomass, settling, and anaerobic benthic dechlorination and degradation of the sorbed AOX. Slade et al. [81] evaluated three aerated stabilization basins in New Zealand, which treated elemental chlorine free (ECF) integrated bleached Kraft mill effluents. All three treatment systems achieved 90% removal of BOD without nutrient supplementation. Aerated basin receiving wastewater with a higher BOD : N ratio (100 : 0.8) exhibited nitrogen fixation capability. For phosphorus limited or lower BOD : N (100 : 2) ratio waste streams, benthic recycling seemed to be a crucial mechanism for nutrient supply in aerated basins. Bailey and Young [82] conducted toxicity tests using *Ceriodaphnia*

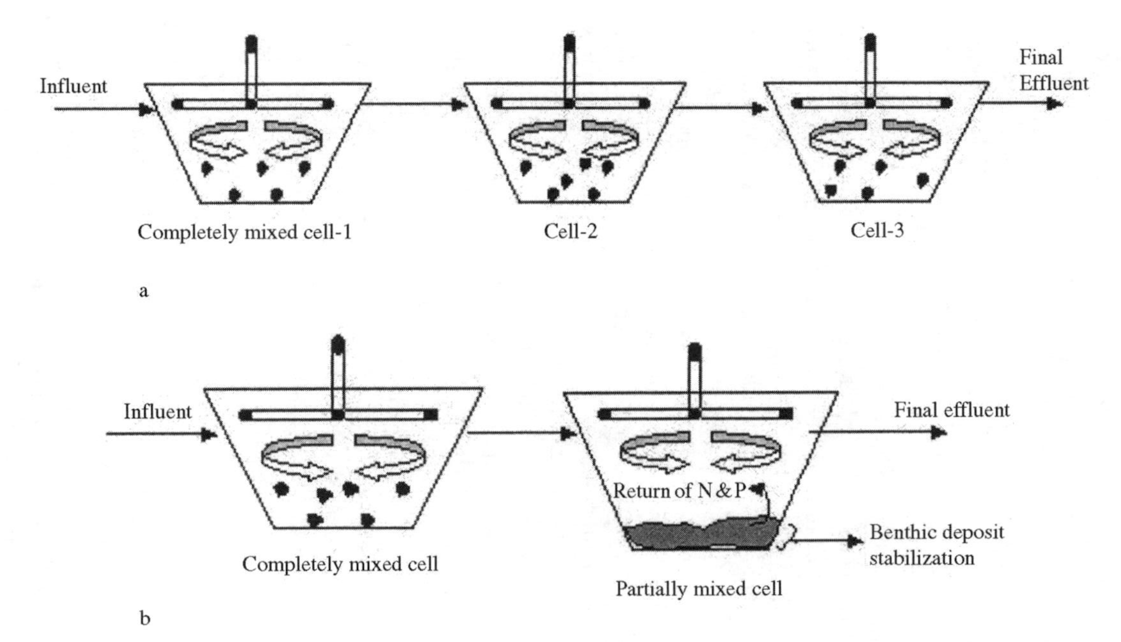

Figure 5 Types of aerated lagoons: (a) Biotransformation of organics and stabilization of biomass under aerobic conditions; (b) Biotransformation of organics under aerobic conditions followed by benthal stabilization of biosolids under anaerobic conditions (from Ref. 77).

Table 12 Characteristics of the Untreated and Biologically Treated Sulfite Mill Effluent

Parameters	Influent (average value) to aerobic filter	Aerobic filter effluent (average value)	Influent (average value) to aerated lagoon[a]	Aerated lagoon effluent (average value)
COD_{total} (mg/L)	2920	1493	1508	795
$COD_{soluble}$ (mg/L)	2737	1383	1305	721
BOD_{total} (mg/L)	795	144	320	50
SS (mg/L)	56	53	102	30
VSS (mg/L)	49	43	96	28
COD load ($kg/m^3/d$)	12	–	–	–
BOD load ($kg/m^3/d$)	34	–	–	–
COD removal (%)	–	49	–	44
BOD removal (%)	–	82	–	85

[a] Influent composed of a portion of aerobic filter effluent mixed with the remaining mill effluents.
Source: Ref. 79.

dubia, *Selenastrum capricornutum*, and rainbow trout, and suggested that mill effluents treated by ASBs exhibit less toxicity than other treatment methods.

Anaerobic Treatment Processes

Anaerobic processes have been employed to stabilize sewage sludge for more than a century. The application of this process for high-strength industrial wastewater treatment began with the development of high rate anaerobic reactors [83,84]. A spectrum of innovative reactors ranging from suspended to attached growth systems or a combination of both (hybrid) operate currently with a range of HRT and SRT values. Retention of biomass is accomplished through the sedimentation of microbial flocs or granules, use of reactor configuration that retains sludge, or immobilization on fixed surfaces or carrier materials [77]. High-rate reactors typically achieve 80–90% reduction in BOD_5, with biogas and methane production of 0.5 m^3/kg COD and 0.35 m^3/kg COD, respectively [83]. Biomass generation ranges from 0.05–0.1 kg VSS/kg COD removed. The various types of bioreactor configurations that are employed to treat industrial wastewaters include: (a) upflow anaerobic sludge blanket (UASB), (b) anaerobic contact (AC), (c) anaerobic filter (AF), (d) hybrid UASB with filter (UASB/AF), (e) expanded granular sludge blanket (EGSB), (f) fluidized bed (FB), (g) down-flow stationary fixed film (DSFF), and (h) anaerobic lagoons. Hulshoff Pol et al. [85] reported that in 1997, about 61% of the full-scale industrial anaerobic plants were designed as UASB-type reactors, while the rest employed contact processes (12%), lagoons (7%), filters (6%), hybrid reactors (4%), EGSB reactors (3%), fluidized bed reactors (2%), and fixed film reactors (2%).

Application of Anaerobic Bioreactors in the Pulp and Paper Industries. A number of factors govern the choice of a treatment process and reactor. The preferred choice for treatment of pulp and paper mill effluents is anaerobic degradation because these industries typically generate high-strength wastewaters with the potential to recover energy in the form of biogas. Moreover, anaerobic microorganisms are reported to be more efficient in dehalogenating and degrading chlorinated organics compared to aerobic microorganisms. Additional factors such as lower capital investment and limitation of land area often translate into a reactor that can accommodate high organic and hydraulic loadings with the least maintenance and operation problems. However, assessing the suitability of an anaerobic process and systematic evaluation of reactor configurations are essential prior to the full-scale implementation, in view of the heterogeneous nature of pulp and paper mill effluents. Laboratory-scale and pilot-scale studies on specific mill effluents must address the following key issues:

- Toxicity of the wastewater, especially to the methanogenic population. In general, wastewaters from chemical, NSSC pulping spent liquor condensates, and TMP are nontoxic. On the other hand, unstable anaerobic operations have been noticed with untreated NSSC spent liquors, effluents from debarking, CTMP, and chemical bleaching. Resin acids, fatty acids, terpenes, condensed and hydrolyzable tannins, sulfate, sulfite, reduced sulfur compounds, alkylguaiacols, and chlorinated phenols have been reported to be highly inhibitory to the methanogenic population [86]. Inhibitory waste streams must be diluted or treated by methods such as precipitation, aerobic biodegradation, autooxidation, and polymerization for the selective removal of toxic compounds before anaerobic treatment.
- Anaerobic biodegradability of the components in various effluents (lignin derivatives, resin and fatty acids are known to be highly resistant to anaerobic degradation).
- Maximum loading capacity and reliability of the process under fluctuating loads and shock loading conditions.
- Ease of start-up following interruption of the process.
- Cost of construction, operation and maintenance of reactors.
- Recovery of chemicals and energy.

Anaerobic reactor configurations that have found application in the treatment of pulp and paper mill effluent include anaerobic contact, UASB, anaerobic filter, UASB/AF hybrid, and fluidized bed reactors. Specific features of these reactors are described in the following sections.

Anaerobic Contact Reactor. The anaerobic contact system as illustrated in Figure 6 consists of a completely mixed anaerobic reactor with suspended growth of biomass, a degasifier unit, and a sedimentation unit intended for the separation of clarified effluent from biosolids. Part of the biomass is recycled to the bioreactor through a recycle line. The purpose of the degasifier is to remove gases such as carbon dioxide and methane and to facilitate efficient settling of solids. The volumetric loading rate (VLR) varies from 0.5 to 10 kg COD/m^3 day with HRT and SRT values in the range 0.5–5 days and 10–20 days, respectively [77]. Volatile suspended solids (VSS) in the bioreactor typically range from 4–6 g/L to 25–30 g/L. The process is applicable to wastewaters containing high concentrations (>40%) of suspended solids. Major disadvantages of contact systems are poor settleability of sludge and susceptibility to shock loadings and toxicity.

Up-Flow Anaerobic Sludge Blanket (UASB) Reactor. The UASB reactor is a suspended growth reactor in which the microorganisms are encouraged to develop into dense, compact, and readily settling granules (Fig. 7). Granulation is dependent upon the environmental conditions of the reactor and facilitates the maintenance of high concentrations (20–30 g/L) of VSS in the reactor. The flow of influent starts at the bottom of the reactor, passes through the blanket of dense granules in the bottom half portion of the reactor and reaches into the gas–liquid–solid separator located in the top portion of the reactor. The gas–liquid–solid separating device consists of a gas collection hood and a settler section. Most of the treatment occurs within the blanket of granules. The gas collected in the hood area of the bioreactor is continually removed while the liquid flows into the settler section for liquid–solid separation and settlement of solids back into the reactor. The combined effects of wastewater flow (upflow liquid velocity of 1 m/hour) and biogas production facilitate mixing and contact between the wastewater and microorganisms in the granules. The volumetric loading rates vary from 10 to 30 kg COD/m^3 day

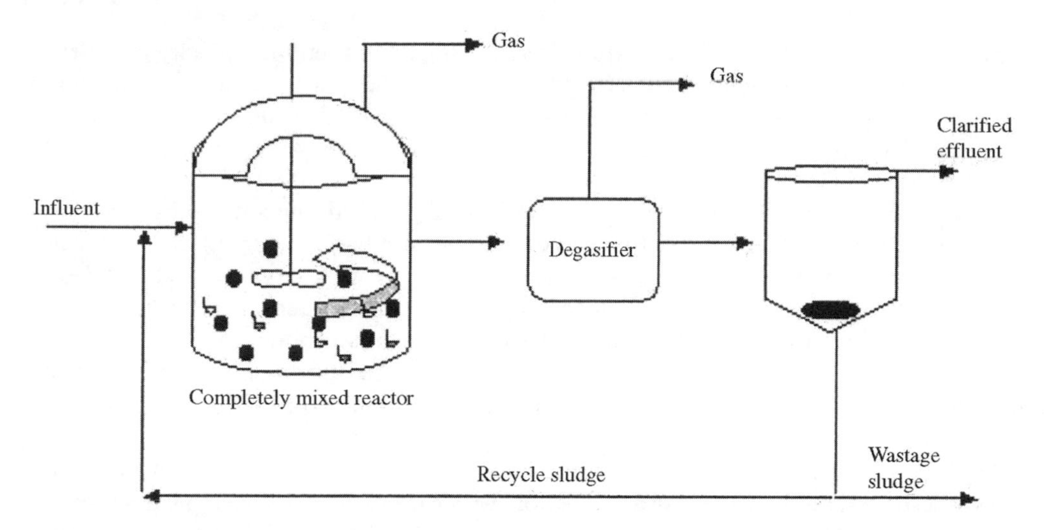

Figure 6 Diagrammatic representation of anaerobic contact process (from Ref. 77).

and typical HRT values range from 4 to 12 hours. The UASB reactors can handle effluents with a high content of solids. However, the quality of granules and hence the performance of the reactor is highly dependent upon the toxicity and other characteristics of wastewater. A modified version of the UASB reactor is the expanded granular sludge blanket (EGSB) reactor, which is

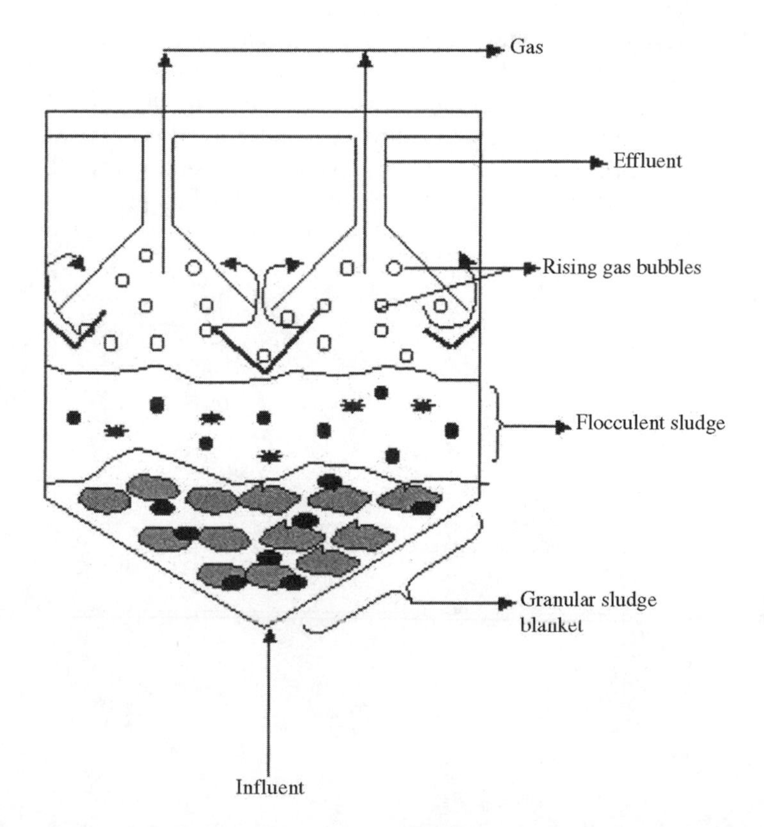

Figure 7 Diagrammatic representation of upflow anaerobic sludge blanket reactor process (from Ref. 77).

designed for higher up-flow velocity (3–10 m/hour) of liquid. Higher flow velocity is achieved by using tall reactors or recycling of treated effluent or both. The VLR in EGSB reactors ranges from 20 to 40 kg COD/m^3 day. Another modification to the UASB is an internal circulation reactor (IC) that has two UASB compartments on top of each other with biogas separation in each stage [6].

Anaerobic Filter. An anaerobic filter consists of packed support media that traps biomass as well as facilitates attached growth of biomass as a biofilm (Fig. 8). Such a reactor configuration helps in the retention of suspended biomass as well as gas–liquid–solid separation. The flow of liquid can be upward or downward, and treatment occurs due to attached and suspended biomass. Treated effluent is collected at the bottom or top of the reactor for discharge and recycling. Gas produced in the media is collected underneath the bioreactor cover and transported for storage or use. Volumetric loading rates vary from 5 to 20 kg COD/m^3 day with HRT values of 0.5–4 days.

Hybrid UASB/Filter Reactor. This hybrid reactor is a suspended growth reactor primarily designed as a UASB reactor at the bottom with packing media (anaerobic filter) on the top of the reactor. The influent is uniformly distributed at the bottom of the reactor and flows upwards sequentially through the granular sludge blanket and filter media where gas–liquid–solid separation takes place. Treated effluent is collected at the top for discharge or recycling. Gas collected under the cover of the bioreactor is withdrawn for use or storage. Process loadings for this system are similar to those of UASB.

Fluidized Bed Reactor. Fluidized bed systems are upflow attached growth systems in which biomass is immobilized as a thin biofilm on light carrier particles such as sand (Fig. 9). A high specific surface area of carrier particles facilitates accumulation of VSS concentrations ranging from 15 to 35 g/L in the bioreactor. The upflow velocities are much higher compared to UASB, AF, or hybrid reactors, preventing the growth of suspended biomass. Carrier particles with biomass are fluidized to an extent of 25–300% of the resting bed volume by the upward flowing influent and the recirculating effluent. Such a system allows for high mass transfer rates with minimal clogging and reduces the risk of toxic effects of the incoming wastewater. HRT values ranging from 0.2 to 2 days and VLR above 20 kg COD/m^3 day are common [77].

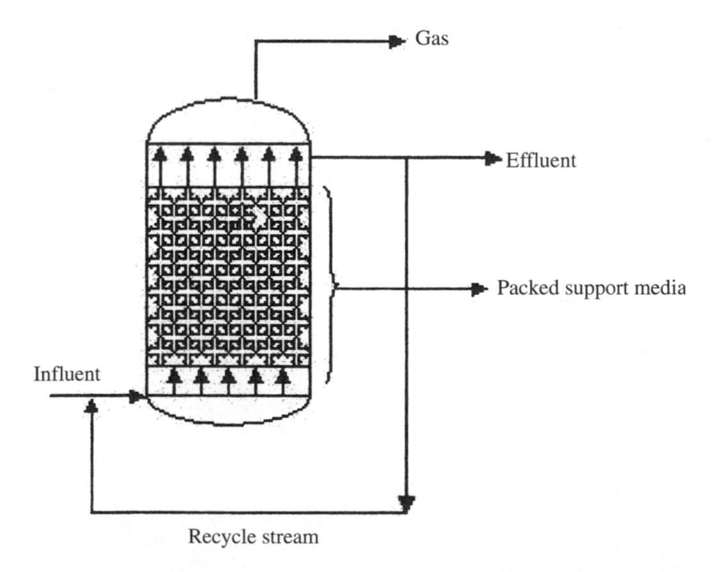

Figure 8 Diagrammatic representation of anaerobic filter process (from Ref. 77).

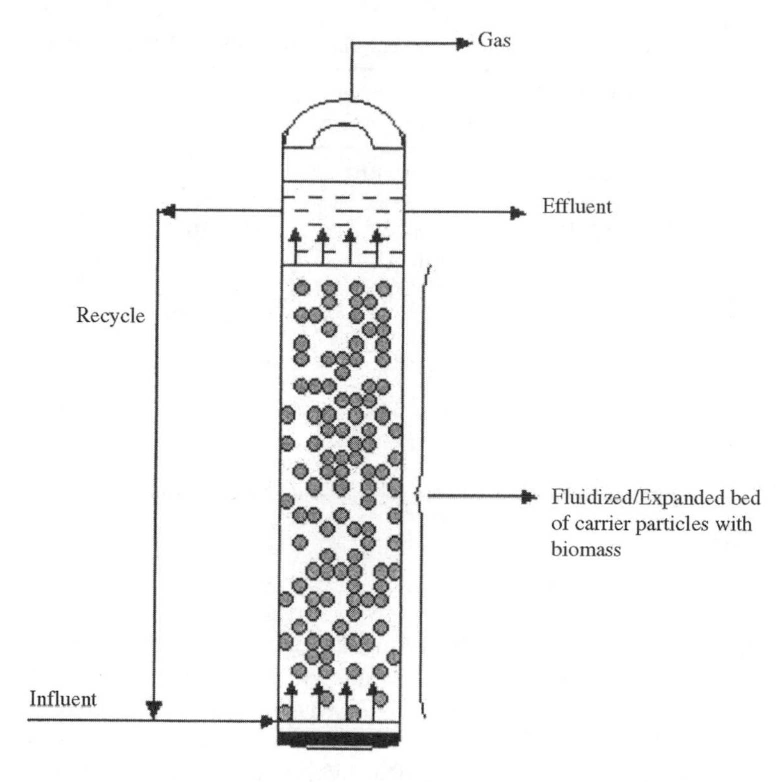

Figure 9 Diagrammatic representation of fluidized and expanded bed reactor process (from Ref. 77).

Anaerobic Technologies Suppliers and Anaerobic Plants in the Pulp and Paper Industry. Table 13 shows the major suppliers of full-scale anaerobic treatment plants for industrial wastewater and the corresponding technologies. It may be noted from this table that a significant number of installations for pulp and paper mill wastewater treatment are performed by Paques BV, The Netherlands, and Bioethane Systems International, Canada. In total, 89 installations were performed in the year 1998, of which 69 UASB, 15 anaerobic contact, and 3 fluidized bed reactors were chosen [6]. Table 14 summarizes the performance of selected full-scale applications of anaerobic technologies in the pulp and paper industry. The UASB reactor configuration is distinctively the best choice, followed by the anaerobic contact process. There are limited applications of the other anaerobic reactor configurations such as fluidized bed and anaerobic filter. Agro-pulping black liquors have a high content of nonbiodegradable lignin that contributes to 50% of the COD; subsequently, there is a need to develop viable solutions to handle such wastewaters.

Thermophilic Anaerobic Reactor Applications. Pulp and paper industries typically discharge warm (50°C) effluents, and conventional reactors operating under mesophilic conditions require cooling of such wastewaters. Attempts have been made periodically by various groups to investigate the possibility of applying thermophilic anaerobic processes to pulp and paper discharges, but to date there is no conclusive evidence to prove the superior performances of thermophilic reactors as compared to their mesophilic counterparts.

Lepisto and Rintala [87] used four different types of thermophilic (55°C) anaerobic processes, namely an upflow anaerobic sludge blanket (UASB) digester, UASB enriched with sulfate, UASB with recirculation, and fixed-bed digester with recirculation for investigating the

Table 13 Commercial Suppliers of Full-Scale Anaerobic Treatment Plants for Industrial Wastewaters

Process configuration	Technology supplier/Trade name. Number of plants: pulp and paper/total number of industries				
	ADI[a]	Biothane[b]	Degremont[c]	Paques[d]	Purac[e]
UASB	–	Biothane UASB 17/189	Anapulse 3/44	Biopaq 49/266	–
Contact process	ADI-BVF 1/67	Biobulk 0/4	Analift 1/22	–	Anamet 13/63
IC reactor	–	–	–	Biopaq-IC 1/40	
Anaerobic filter	–	–	Anafiz 1/8	–	–
Fluidized bed reactor	–	–	Anaflux 3/13	–	–

[a]ADI Systems Inc., USA; [b]Biothane Systems International, Canada; [c]Degremont, France; [d]Paques BV, The Netherlands; [e]Purac AB, Sweden.
Courtesy of UNDP India website, Ref. 6.

removal of chlorinated phenolics from soft wood bleaching effluents. All four processes eliminated/reduced chlorinated phenols, catechols, guiacols, and hydroquinones from bleach mill effluents. The ranges of COD and AOX removal were 30–70% and 25–41%, respectively, in the four reactors. Jahren et al. [88] treated TMP whitewaters using three different types of thermophilic anaerobic reactor configurations. The anaerobic hybrid reactor consisted of a UASB and a filter that degraded 10 kg COD/m^3/day. The anaerobic multistage reactor, composed of granular sludge and carrier elements, gave a degradation rate of 9 kg COD/m^3/day at loading rates of 15–16 kg COD/m^3/day and HRT of 2.6 hours. The anaerobic moving bed biofilm digester handled loading rates of up to 1.4 kg COD/m^3/day.

Sequential Anaerobic-Aerobic Treatment Systems

Biological reactors employing combination anaerobic and aerobic environments can be more effective for the detoxification of pulp and paper mill discharges through the following processes:

- Reduction of biodegradable organics under anaerobic and aerobic conditions;
- Transformation and degradation of chlorinated compounds presumably via reductive dehalogenation and subsequent aerobic metabolism;
- Aerobic metabolism of extractable compounds such as resin acids via hydroxylation reactions.

Haggblom and Salkinoja-Salonen [89] treated Kraft pulp bleaching in an anaerobic fluidized bed reactor, followed by an aerobic trickling filter. The sequential treatment process reduced 65% of AOX and 75% of the chlorinated phenolic compounds. The anaerobic reactor was efficient in dechlorination, thereby eliminating most of the toxicity and improving biodegradability of the subsequent aerobic reactor at shorter retention times. The researchers identified two species of *Rhodococcus* bacteria that were capable of degrading polychlorinated phenols, guaiacols, and syringols in the bleaching effluents. Wang et al. [90] examined continuous-flow sequential reactors operated in anaerobic-aerobic and aerobic-aerobic modes. The objective of this research was to enhance reductive dehalogenation and degradation of

Table 14 Commercial Suppliers of Full-Scale Anaerobic Treatment Plants for Industrial Wastewaters

	Paper mill	Influent type and characteristics	Supplier	Reactor details	Start-up year	Plant performance
1	Macmillan Bloedel Ltd., Canada	Corrugated Cardboard/NSSC Flow: 6300 m^3/day COD load: 107 T/day	Biothane	UASB V = 7000 m^3 (2 × 3500 m^3 R) VLR = 15.4 kg	1989	Gas production: 1140 m^3/hour COD reduction: 55% BOD reduction: 85%
2	Stone Container, Canada	CTMP/NSSC recycle Flow: 656 m^3/day COD load: 7.7 T/day	Biothane	UASB V = 15,600 m^3 (2 × 7800 m^3 R) VLR = 12 kg	1988	BOD reduction: 85%
3	Industriewater, The Netherlands	Total Flow from 3 mills = 12,400 m^3/day COD load: 22 T/day	Paques BV	UASB V = 2184 m^3 VLR = 6–7 kg	1990	Gas production: 2000–3600 m^3/hour COD reduction: 70% BOD reduction: 80%
4	Satia Paper Mills Ltd., India	Chemical pulping using agricultural residues Flow: 4500 m^3/day COD load: 53 T/day	Paques BV	UASB V = 5200 m^3 (2 × 2600 m^3 R) VLR = 10 kg	1997	Gas production: 10,000–12,000 m^3/hour COD reduction: 50–60% BOD reduction: 60–70%

(continues)

Table 14 Continued

	Paper mill	Influent type and characteristics	Supplier	Reactor details	Start-up year	Plant performance
5	Papetries Lecoursonnois Paper Mill, France	Corrugated medium and coated paper Flow: 600 m^3/day COD load: 3.6 T/day	Degremont	Fluidized bed capacity of the reactor = 20–30 m^3/hour	1994	COD reduction: 75% BOD reduction: 85%
6	Modo Paper AB, Sweden	Sulfite condensate Flow: 6000 m^3/day COD load: 72 T/day	Purac AB	V = 30,000 m^3 (2 × 15,000 m^3 R) Reactor loading rate = 125 T COD/day	1984	Gas production: 23,000 m^3/hour COD reduction: 81% BOD reduction: 98%
7	Pudumjee Pulp and Paper Mills, India	40 tpd bagasse bleached pulp Flow: 2200 m^3/day COD load: 20–22 T/day	Sulzer Brothers Ltd., Switzerland	Continuous stirred tank (An-OPUR) V = 13,000 m^3 (2 × 6500 m^3 R) VLR = 3–5 kg	1989	Gas production: 5000–6000 m^3/hour COD reduction: 60–70% BOD reduction: 85–90%

R = Reactor; VLR (volumetric loading rate) expressed as COD/m^3/day; T = Ton.
Courtesy of UNDP India website, Ref. 6.

organics. The researchers noted that the anaerobic-aerobic reactor improved the biodegradability of bleach wastewaters and performed better compared to the aerobic-aerobic reactor. Rintala and Lepisto [91] performed experiments on a mixture comprised of 20% Kraft mill chlorination (C) stage effluent, 30% alkaline extraction stage (E) effluent, and 50% tap water in anaerobic-aerobic and aerobic post-treatment reactors (250 mL volume) operated at 55°C and partially packed with polyurethane. They observed that the second stage of the anaerobic-aerobic reactor removed negligible COD if the anaerobic stage performed well. However, the aerobic reactor exhibited excellent removal efficiencies during periods of anaerobic reactor upsets. Lee et al. [92] evaluated a continuous-flow sequential anaerobic-aerobic lagoon process for the removal of AOX from Kraft effluents on bench and pilot scales. Bench-scale studies demonstrated AOX removal efficiencies of approximately 70% at HRT values of 2, 5, and 10 days. In comparison, aerobic lagoons removed only 20, 35, and 36% of AOX at the above HRT values.

Mycotic Systems for the Removal of Color and AOX

Fungal systems, particularly white-rot fungi such as *Phanerochaete chrysosporium*, *Ganoderma lacidum*, and *Coriolus versicolor* have been investigated for their abilities to decolorize and degrade Kraft pulp bleaching effluents [93–95]. For instance, Huynh et al. [96] used the MyCoR (mycelial color removal) process for the treatment of E_1 stage effluents. The MyCoR process utilizes a fixed film reactor such as a rotating biological contactor (RBC) for the surface growth of *Phanerochaete chrysosporium.* A related process known as MyCoPOR uses polyurethane foam cubes (surface area of 1 cm^2) as porous carrier material in trickling filter reactors for supporting the growth of *Phanerochaete chrysosporium.* The immobilized fungus is capable of removing color and AOX, and has the capacity to degrade several chlorinated phenols such as 2,4,6-trichlorophenol, polychlorinated guaiacols, and polychlorinated vanillins, all of which are toxic compounds that are present in the pulp bleaching effluents. Prasad and Joyce [97] employed *Phanerochaete chrysosporium* in a rotating biological contactor for treatment of E_1 stage effluent from a Kraft mill bleach plant effluent containing approximately 190 mg/L of AOX. The extent of reduction in AOX, color, COD, and BOD levels were 42, 65, 45, and 55%, respectively, in the aerobic fungal reactor system through biotransformation mechanisms. Efficiency of the treatment process remained constant for up to 20 days with replacement of effluent every 2 days. A subsequent anaerobic treatment removed additional 40% AOX, 45% of soluble COD, and 65% of total BOD at a loading rate of 0.16 kg COD/m^3 day. Thus, the fungal-anaerobic system proved to be more effective than the fungal process alone. Taseli and Gokcay [98] observed 70% removal of AOX by a fungal culture immobilized on glass wool packed in an upflow column reactor. Optimum dechlorination by the fungus was observed at pH 5.5 and 25°C with HRT of 7–8 hour and required very low levels of carbon and dissolved oxygen. *Coriolus versicolor* has shown promising results in terms of color removal, COD reduction, and degradation of chlorolignins in bleach effluents and Kraft mill liquors [22]. Various researchers have applied the fungus in the form of mycelial pellets or as mycelium, entrapped within calcium alginate beads. The fungus has an obligate requirement for simple growth substrates such as glucose, sucrose, and starch for effective performance.

Mechanisms of White-Rot Fungal Mediated Degradation and Decolorization Processes. The ligninolytic enzyme systems of white-rot fungi exhibit catalytic activities that are beneficial for the transformation and mineralization of a wide range of organopollutants (including halogenated organics) with structural similarities to lignin [48,49]. Currently, three types of extracellular lignin modifying enzymes (LMEs), namely lignin peroxidase (LiP), Mn dependent peroxidase (MnP), and laccase (Lac) are known to aid in the catabolism of lignin [99].

LiP catalyzes the oxidation of a low-molecular-weight redox mediator, veratryl alcohol, which in turn mediates one-electron oxidation of lignin to generate aryl cation radicals [100]. The radicals facilitate a wide variety of reactions such as carbon–carbon cleavage, hydroxylation, demethylation, and so on. Dezotti et al. [101] reported enzymatic removal of color from extraction stage effluents using lignin and horseradish peroxidases immobilized on an activated silica gel support.

MnP catalyzes hydrogen-peroxide-dependent oxidation of Mn^{2+} to Mn^{3+}, which in turn oxidizes phenolic components of lignin [102]. Oxidative demethylation, dechlorination, and decolorization of bleach plant effluent by the MnP of *Phanerochaete chrysosporium* has been demonstrated [103,104].

Laccase generates free radicals from low-molecular-weight redox mediators such as 3-hydroxanthranilate and initiates condensation of phenolic compounds [105]. The Lac-catalyzed polymerization reaction has the potential for enhancing the subsequent lime-induced precipitation of high-molecular-weight chromophoric pollutants. Archibald et al. [106] and Limura et al. [107] demonstrated efficient decolorization and up to 85% removal of chloroguaiacol using laccase elaborated by *Trametes versicolor*. Davis and Burns [108] demonstrated decolorization of pulp mill bleach effluents by employing laccase that was covalently immobilized on activated carbon. They reported color removal at the rate of 115 units/unit enzyme/hour.

Treatment of bleach mill effluents using the white-rot fungi is promising and offers the option to expand the range of pollutants that cannot be biodegraded by the prokaryotes (bacteria). White-rot fungal remediation may be particularly suited for those recalcitrant compounds for which bioavailability and toxicity are the key issues.

10.11 TREATMENT OF WASTE GAS EMISSIONS FROM PULP AND PAPER INDUSTRIES

A spectrum of gaseous phase pollutants is emitted from pulp and paper mills, which include VOCs, total reduced sulfur compounds (TRS), NOx, and SOx. Air pollution control, particularly nuisance odor abatement, has gained importance in recent years. VOCs formed during pulping, bleaching, and liquor evaporation are conventionally treated by physico-chemical methods such as adsorption to activated coal filters, absorption, thermal oxidation (incineration), catalytic oxidation, and condensation [109]. However, the major limitations of these air pollution control technologies include energy costs, transfer of pollutants from one phase to another, or generation of secondary pollutants. A more recent trend is the development of a low-cost and effective biological treatment of air through usage of biofilters and bioscrubbers that can remove trace concentrations of pollutants. The biological vapor phase remediation process involves three steps: the transfer of pollutants from gaseous phase to liquid phase; transfer from liquid to microbial surface and uptake of pollutants by the microorganisms; and finally transformation and/or degradation of the pollutants by microbial enzymes. Most of the biodegradative enzymes concerned are intracellular. A brief description of the biofilters and bioscrubbers is provided in the following subsections.

10.11.1 Biofilters

A typical biofilter setup consists of a blower, humidification chamber and a biofilter unit, and an additional polishing unit (optional) such as granular activated carbon backup. The biofilter is composed of microbial communities supported on a packing surface material such as wet peat, wood chips, charcoal, compost, diatomaceous earth pellets, and so on. The microorganisms derive all their nutrients from the surrounding liquid film; therefore, an additional device for direct sprinkling of water on the packing surface is usually included. The operational steps in biofiltration involve contacting the waste air with a stream of water in a humidifier and passing the moisture loaded air at a slow rate through microbial film supported on packing surfaces. The superficial gas flow varies from 1 to 15 cm/s for a bed height of 1–3 m. Removal efficiencies of 90% can be achieved at volumetric loading rates of 0.1–0.25 kg organics/m^3/day. The major drawbacks of biofilters are the lack of control over pH and space requirements.

Characteristics of Contaminants Treated by Biofilters and Their Removal Mechanisms

Biofilter units, which have been used conventionally for odor control in wastewater treatment plants, have gained acceptance over the last decade for control of VOCs in industries such as the pulp and paper mills. The technology is best suited for contaminated air streams that have trace concentrations of readily soluble and biodegradable VOCs. Eweis et al. [109] reviewed research findings related to the application of biofiltration for removal of VOCs ranging in concentration from 0.01 to 3000 ppm. Types of contaminants that can be removed through biofiltration include reduced sulfur compounds, ammonia, NOx, and chlorinated organic compounds. Most full-scale biofilter applications have been for odor control through removal of sulfide ranging in concentrations from less than a ppm to several hundred ppm. H_2S is highly soluble in water and oxidized by the aerobic, chemoautotophic bacteria to H_2SO_4, which results in highly acidic conditions during biofiltration. Although sulfur-oxidizing bacteria function well at acidic pH, there is a need to control pH effectively for maintaining the optimal performance of other groups of bacteria. The most effective method for removal of NOx from air streams is based on denitrification, which requires an electron donor. Chlorinated compounds may be dechlorinated to produce HCl with the concomitant drop in pH.

10.11.2 Bioscrubbers

The bioscrubber is a modification of biofilter technology. Contaminated air is drawn into a chamber and sprayed with a fine mist of liquid stream containing a suspension of microorganisms. The liquid is continually circulated between the spray chamber and an activated sludge process unit where biodegradation occurs.

10.11.3 Biotreatment of Flue Gases

Nitrogen and sulfur oxides released in flue gases are removed by scrubbing, which involves dissolution of these gases in a solution of $NaHCO_3$ and Fe(II)-EDTA. A new system based on the regeneration of scrubbing solution via biological denitrification and desulfurization is shown in the following flow chart [110]:

Influent gas (SOx and NOx) $\longrightarrow$ Scrubber $\longrightarrow$ Denitrification tank

$\longrightarrow$ UASB reactor ($S^{4+} \longrightarrow S^{2-}$) $\longrightarrow$ Aerobic reactor ($S^{2-} \longrightarrow S^0$)
$\longrightarrow$ Filter $\longrightarrow$ Recovery of elemental sulfur

10.12 CONCLUSIONS

Over the last decade, much interest has been generated in monitoring environmental problems and associated risks of wastes, in particular, wastewaters generated by the pulp and paper industries. A major goal is to reassess the target pollutant levels and consider the use of risk-based discharge permit values rather than the absolute endpoint values. This risk-based approach requires analytical tools that can quantify the ecotoxic characteristics of discharges rather than the absolute concentration of specific pollutants or the values of lumped pollution parameters such as BOD, COD, and so on.

The development of pollution treatment strategies, technologies, and their implementation in pulp and paper industries requires an integrated holistic approach that requires a detailed understanding of the manufacturing processes as well as the physical, chemical, and biological properties of the multitude of pollutants generated. The nature and extent of pollution varies significantly from one mill to another. Therefore, the selection of treatment technology(ies) and optimization of the process calls for laboratory- and pilot-scale studies to be conducted with actual wastewater and processes under consideration. Finally, the success of a specific process such as biological treatment is highly dependent upon the preceding operations such as segregation of mill process streams, primary treatment such as filtration, chemical precipitation, and oxidation. Physico-chemical treatment methods such as the oxidation processes can effectively increase the biodegradability of recalcitrant pollutants in the subsequent bioreactor. Thus, choosing the right combination and sequence of treatment methods is the key to the successful handling of pollution problems in the pulp and paper industries.

REFERENCES

1. Smook, G.A. (Ed.) *Handbook for Pulp and Paper Technologists*, 2nd Ed.; Angus Wilde: Vancouver, Canada, 1992.
2. Kringstad, K.P.; Lindstrom, K. Spent liquors from pulp bleaching (critical review). *Environ. Sci. Technol.* **1984**, *18*, 236A–247A.
3. U.S. EPA. *Technical support document for best management practises for spent liquor management, spill prevention and control*; USEPA: Washington, DC, 1997.
4. Sharma, C.; Mohanty, S.; Kumar, S.; Rao, N.J. Gas chromatographic determination of pollutants in Kraft bleachery effluent from Eucalyptus pulp. *Anal. Sci.* **1999**, *15*, 1115–1121.
5. Grimvall, A.; Boren, H.; Jonsson, S.; Karlsson, S.; Savenhed, R. Organohalogens of natural and industrial origin in large recipients of bleach-plant effluents. *Water. Sci. Technol.* **1991**, *24* (*3/4*), 373–383.
6. Rintala, J.A.; Jain, V.K.; Kettunen, R.H. Comparative status of the world-wide commercially available anaerobic technologies adopted for biomethanation of pulp and paper mill effluents. *4th International Exhibition and Conference on Pulp and Paper Industry, PAPEREX-99*, New Delhi, India, 14–16 December, 1999.
7. Sjostrom, E. (Ed.) *Wood Chemistry, Fundamentals and Applications*; Academic Press: New York, 1981.
8. Sjostrom, E. (Ed.) *Wood Chemistry, Fundamentals and Applications*, 2nd ed.; Academic Press: San Diego, CA, 1993.
9. Folke, J. Environmental aspects of bagasse and cereal straw for bleached pulp and paper. *Conference on Environmental Aspects of Pulping Operations and Their Wastewater Implications*, Edmonton, Canada, 27–28 July 1989.
10. Kocurek, M.J.; Ingruber, O.V.; Wong, A. (Eds.) *Pulp and Paper Manufacture, Volume 4, Sulfite Science and Technology*; Joint Textbook Committee of the Paper Industry: Tappi, Atlanta, GA, 1985.

11. Green, R.P.; Hough, G. (Eds.) *Chemical Recovery in the Alkaline Pulping Process*, Revised Edition; Tappi, Atlanta, GA, 1992.
12. Biermann, C.J. (Ed.) *Essentials of Pulping and Paper Making*; Academic Press: New York, 1993.
13. Rydholm, S.A. (Ed.) *Pulping Processes*; Interscience: New York, 1965.
14. Jain, R.K.; Gupta, A.; Dixit, A.K.; Mathur, R.M.; Kulkarni, A.G. Enhanced biomethanation efficiency of black liquor through lignin removal process in small agro based paper mills. *Bioenergy News* **2001**, *5* (*3*), 2–6.
15. National Council of the Paper Industry for Air and Stream Improvement, Inc. (NCASI). *The toxicity of Kraft pulping wastes to typical fish food organisms*, Technical Bulletin No. 10; NCASI: New York, 1947.
16. Mckee, J.E.; Wolf, H.W. (Eds.) *Water Quality Criteria*, 2nd Ed., Publication 3-A; State Water Resources Control Board, The Resources Agency of California: Pasadena, CA, 1963.
17. Liss, S.N.; Bicho, P.A.; McFarlane, P.N.; Saddler, J.N. Microbiology and degradation of resin acids in pulp mill effluents: a mini review. *Can. J. Microbiol.* **1997**, *75*, 599–611.
18. Lepp, H.J.T.; Oikari, A.O.J. The occurrence and bioavailability of retene and resin acids in sediments of a lake receiving BKME (bleached Kraft mill effluent). *Water. Sci. Technol.* **1999**, *40* (*11–12*), 131–138.
19. Hickey, C.W.; Martin, M.L. Relative sensitivity of five benthic invertebrate species to reference toxicants and resin acid contaminated sediments. *Environ. Toxicol. Chem.* **1995**, *14*, 1401–1409.
20. Johnsen, K.; Mattsson, K.; Tana, J.; Stuthridge, T.R.; Hemming, J.; Lehtinen, K.-J. Uptake and elimination of resin acids and physiological responses in rainbow trout exposed to total mill effluent from an integrated newsprint mill. *Environ. Toxicol. Chem.* **1995**, *14*, 1561–1568.
21. McCarthy et al. Role of resin acids in the anaerobic toxicity of chemi-thermomechanical pulp wastewaters. *Water. Res.* **1990**, *24*, 1401–1405.
22. Bajpai, P. Microbial degradation of pollutants in pulp mill effluents. *Adv. Appl. Microbiol.* **2001**, *48*, 79–134.
23. Kringstad, K.P.; Ljungquist, P.O.; de Sousa, F.; Stromberg, L.M. Identification and mutagenic properties of some chlorinated aliphatic compounds in the spent liquor from Kraft pulp chlorination. *Environ. Sci. Technol.* **1981**, *15*, 562–566.
24. McMaster, M.E.; Van Der Kraak, G.J.; Munkittrick, K.R. An evaluation of the biochemical basis for steroidal hormonal depression in fish exposed to industrial wastes. *J. Great Lakes Res.* **1996**, *22*, 153–171.
25. Sandstrom, O. Incomplete recovery in a fish coastal community exposed to effluent from a Swedish bleached Kraft mill. *Can. J. Fish. Aqua. Sci.* **1994**, *51*, 2195–2202.
26. Tana, J.; Rosemarin, A.; Lehtinee, K.-J.; Hardig, J.; Grahn, O.; Landner, L. Assessing impacts on Baltic coastal ecosystems with mesocosm and fish biomarker tests: a comparison of new and old pulp bleaching technologies. *Sci. Total. Environ.* **1994**, *145*, 213–214.
27. Kankaanpaa, H.; Lauren, M.; Mattson, M.; Lindstrom, M. Effects of bleached Kraft mill effluents on the swimming activity of Monoporeia affinis (Crustacea, Amphipoda). *Chemosphere* **1995**, *31*, 4455–4473.
28. Bankey, L.A.; Van Vels, P.A.; Borton, D.L.; LaFleur, L.; Stegeman, J.J. Responses of cytochrome P4501A1 in freshwater fish exposed to bleached Kraft mill effluent in experimental stream channels. *Can. J. Fish. Aqua. Sci.* **1995**, *52*, 439–447.
29. Kovacs, T.G.; Gibbons, J.S.; Trembaly, L.A.; O'Connor, B.I.; Martel, P.H.; Vos, R.H. The effects of a secondary treated bleached Kraft mill effluent on aquatic organisms as assessed by short term and long term laboratory tests. *Ecotox. Environ. Safe.* **1995**, *31*, 7–22.
30. Munkittrick, K.R.; Servos, M.R.; Carey, J.H.; Van Der Kraak, G.J. Environmental impacts of pulp and paper wastewater: evidence for a reduction in environmental effect at north American pulp mills since 1992. *Water. Sci. Technol.* **1997**, *35* (*2–3*), 329–338.
31. Mika, A.K.; Liukkonen, M.; Wittmann, C.; Suominen, K.P.; Salkinoja-Salonen, M.S. Integrative assessment of sediment quality history in pulp mill recipient area in Finland. *Water. Sci. Technol.* **1999**, *40* (*11–12*), 139–146.

32. Kirk, T.K.; Jeffries, T.W. Roles for microbial enzymes in pulp and paper processing. In *Enzymes for Pulp and Paper Processing*; Jeffries, T.W., Viikari, L., Eds.; American Chemical Society Symposium, **1996**, Series 655, 2–14.
33. Rantio, T. Chlorinated cymenes in the effluents of two Finnish pulp mills in 1990–1993. *Chemosphere* **1995**, *31*, 3413–3423.
34. Koistinen, J.; Paasivirta, J.; Nevalainen, T.; Lahtipera, M. Chlorophenanthrenes, alkylchlorophenanthrenes and alkylchloronapthalenes in Kraft mill products and discharges. *Chemosphere* **1994**, *28*, 1261–1277.
35. Koistinen, J.; Paasivirta, J.; Nevalainen, T.; Lahtipera, M. Chlorinated fluorenes and alkylfluorenes in bleached Kraft pulp and pulp mill discharges. *Chemosphere* **1994**, *28*, 2139–2150
36. Mielisch, H.J.; Odermatt, J.; Kordsachia, O.; Patt, R. TCF bleaching of Kraft pulp: Investigation of the mixing conditions in an MC ozone stage. *Holzforshung* **1995**, *49*, 445–452.
37. Axegard, P.; Bergnor, E.; Elk, M.; Ekholm, U. Bleaching of softwood Kraft pulps with H_2O_2, O_3 and ClO_2. *Tappi J.* **1996**, *79*, 113–119.
38. Mehta, V.; Gupta, J.K. Biobleaching eucalyptus Kraft pulp with *Phanerochaete chrysosporium* and its effect on paper properties. *Tappi J.* **1992**, *75*, 151–152.
39. Reid, I.D.; Paice, M.G.; Ho, C.; Jurasek, L. Biological bleaching of softwood Kraft pulp with the fungus, *Trametes (Coriolus) versicolor. Tappi J.* **1990**, *73*, 149–153.
40. Kirkpatrick, N.; Reid, I.D.; Ziomek, F.; Paice, M.G. Biological bleaching of hardwood Kraft pulp using *Trametes (Coriolus) versicolor* immobilized in polyurethane foam. *Appl. Microbiol. Biotechnol.* **1990**, *33*, 105–108.
41. Paice, M.G.; Jurasek, L.; Ho, C.; Bourbonnais, R.; Archibald, F. Direct biological bleaching of hardwood kraft pulp with the fungus, *Coriolus versicolor. Tappi J.* **1989**, *72*, 217–221.
42. Tran, A.V.; Chambers, R.P. Delignification of an unbleached hardwood Kraft pulp by *Phanerochaete chrysosporium. Appl. Microbiol. Technol.* **1987**, *25*, 484–490.
43. Ziomek, E.; Kirkpatrick, N.; Reid, I.D. Effect of polymethylsioloxane oxygen carriers on the biological bleaching of hardwood Kraft pulp by *Trametes versicolor. Appl. Microbiol. Biotechnol.* **1991**, *25*, 669–673.
44. Murata, S.; Kondo, R.; Sakai, K.; Kashino, Y.; Nishida, T.; Takahra, Y. Chlorine free bleaching process of Kraft pulp using treatment with the fungus *IZU-154. Tappi J.* **1992**, *75*, 91–94.
45. Viikari, L.; Pauna, M.; Kantelinen, A.; Sandquist, J.; Linko, M. Bleaching with enzymes. In *Proceedings of the Third International Conference on Biotechnology in the Pulp and Paper Industry*, Stockholm, **1986**, 67–69.
46. Daneault, C.; Leduce, C.; Valade, J.L. The use of xylanases in Kraft pulp bleaching: a review. *Tappi J.* **1994**, *77*, 125–131.
47. Call, H.P.; Mucke, I. History, overview and applications of mediated lignolytic systems, especially laccase-mediator systems (Lignozym-process). *J. Biotechnol.* **1997**, *53*, 163–202.
48. Hammel, K.E. Organopollutant degradation by ligninolytic fungi. *Enzyme Microb. Technol.* **1989**, *11*, 776–777.
49. Hammel, K.E. Organopollutant degradation by ligninolytic fungi. In *Microbial Transformation and Degradation of Toxic Organic Chemicals*; Young, L.Y., Cerniglia, C.E., Eds.; Wiley-Liss: New York, 1995; 331–346.
50. Gullichsen, J. Process internal measures to reduce pulp mill pollution load. *Water Sci. Technol.* **1991**, *24* (3/4), 45–53.
51. McDonough, T. Bleaching agents (pulp and paper). In *Kirk-Othmer Encyclopedia of Chemical Technology*; Grayson, M., Ed., 4th ed.; John Wiley and Sons: New York, 1992; 301–311.
52. Reeve, D.W. Introduction to the principles and practice of pulp bleaching. In *Pulp Bleaching: Principles and Practice*; Dence, C.W., Reeve, D.W., Eds., Tappi,; Atlanta, GA, 1996; 2–24.
53. Sprague, J.B. Environmentally desirable approaches for regulating effluents from pulp mills. *Water Sci. Technol.* **1991**, *24 (3/4)*, 361–371.
54. Jonsson, A.-S.; Jonsson, C.; Teppler, M.; Tomani, P.; Wannstrom, S. Treatment of paper coating color effluents by membrane filtration. *Desalination* **1996**, *105*, 263–276.

55. Sierka, R.A.; Cooper, S.P.; Pagoria, P.S. Ultrafiltration and reverse osmosis treatment of an acid stage wastewater. *Water Sci. Technol.* **1997**, *35 (2–3)*, 155–161.
56. Koyuncu, I.; Yalcin, F.; Ozturk, I. Color removal of high strength paper and fermentation industry effluents with membrane technology. *Water Sci. Technol.* **1999**, *40 (11–12)*, 241–248.
57. Randtke, S.J. Organic contaminant removal by coagulation and related process combination. *J. Am. Wat. Wks. Assoc.* **1988**, *80*, 40–56.
58. Licsko, I. Dissolved organics removal by solid–liquid phase separation (adsorption and coagulation). *Water Sci. Technol.* **1993**, *27* (11), 245–248.
59. Stephenson, R.J.; Duff, S.J.B. Coagulation and precipitation of a mechanical pulping effluent–I. Removal of carbon, color and turbidity. *Water Res.* **1996**, *30*, 781–792.
60. Stephenson, R.J.; Duff, S.J.B. Coagulation and precipitation of a mechanical pulping effluent–II. Toxicity removal and metal recovery. *Water Res.* **1996**, *30*, 793–798.
61. Ganjidoust, H.; Tatsumi, K.; Yamagishi, T.; Gholian, R.N. Effect of synthetic and natural coagulant on lignin removal from pulp and paper wastewater. *Water Sci. Technol.* **1997**, *35 (2–3)*, 291–296.
62. Legrini, O.; Oliverous, E.; Braun, E.M. Photochemical process for water treatment. *Chem. Rev.* **1993**, *93*, 671–698.
63. Balcloglu, A.I.; Cecen, F. T. reatability of Kraft pulp bleaching wastewater by biochemical and photocatalytic oxidation. *Water Sci. Technol.* **1999**, *40* (1), 281–288.
64. Yeber, M.C.; Rodriguez, J.; Baeza, J.; Freer, J.; Zaror, C.; Duran, N.; Mansilla, H.D. Toxicity abatment and biodegradability enhancement of pulp mill bleaching effluent by advanced chemical oxidation. *Water Sci. Technol.* **1999**, *40 (11–12)*, 337–342.
65. Winterbourn, C. Toxicity of iron and hydrogen peroxide: the Fenton reaction. *Toxic. Lett.* **1995**, *82/83*, 969–974.
66. Rodriguez, J.; Contreras, D.; Parra, C.; Freer, J.; Baeza, J.; Duran, N. Pulp mill effluent treatment by Fenton-type reaction catalyzed by iron complexes. *Water Sci. Technol.* **1999**, *40 (11–12)*, 351–355.
67. Laari, A.; Korhonen, S.; Tuhkanen, T.; Verenich, S.; Kallas, J. Ozonation and wet oxidation in the treatment of thermomechanical pulp (TMP) circulation waters. *Water Sci. Technol.* **1999**, *40 (11–12)*, 51–58.
68. Hostachy, J.-C.; Lenon, G.; Pisicchio, J.-L.; Coste, C.; Legay, C. Reduction of pulp and paper mill pollution by ozone treatment. *Water Sci. Technol.* **1997**, *35 (2–3)*, 261–268.
69. Helbe, A.; Schlayer, W.; Liechti, P.-A.; Jenny, R.; Mobius, C.H. Advanced effluent treatment in the pulp and paper industries with a combined process of ozonation and fixed bed biofilm reactors. *Water Sci. Technol.* **1999**, *40 (11–12)*, 345–350.
70. Schnell, A.; Sabourin, M.J.; Skog, S.; Garvie, M. Chemical characterization and biotreatability of effluents from an integrated alkaline-peroxide mechanical pulping/machine finish coated (APMP/MFC) paper mill. *Water Sci. Technol.* **1997**, *35 (2–3)*, 7–14.
71. Saunamaki, R. Activated sludge plants in Finland. *Water Sci. Technol.* **1997**, *35 (2–3)*, 235–243.
72. Narbaitz, R.M.; Droste, R.L.; Fernandes, L.; Kennedy, K.J.; Ball, D. PACT™ process for treatment of Kraft mill effluent. *Water Sci. Technol.* **1997**, *35 (2–3)*, 283–290.
73. Rantala, P.-R.; Wirola, H. Solid, slightly soluble phosphorus compounds as nutrient source in activated sludge treatment of forest industry wastewaters. *Water Sci. Technol.* **1997**, *35 (2–3)*, 131–138.
74. Clauss, F.; Balavoine, C.; Helaine, D.; Martin, G. Controlling the settling of activated sludge in pulp and paper wastewater treatment plants. *Water Sci. Technol.* **1999**, *40 (11–12)*, 223–229.
75. Hansen, E.; Zadura, L.; Frankowski, S.; Wachocvicz, M. Upgrading of an activated sludge plant with floating biofilm carrier at frantschach Swwiecie S.A. to meet the new demands of year 2000. *Water Sci. Technol.* **1999**, *40 (11–12)*, 207–214.
76. Andreasen, K.; Agertved, J.; Petersen, J.-O.; Skaarup, H. Improvement of sludge settleability in activated sludge plants treating effluent from pulp and paper industries. *Water Sci. Technol.* **1999**, *40 (11–12)*, 215–221.
77. Grady, Jr. C.P.L.; Daigger, G.T.; Lim, H.C. (Eds.) *Biological Wastewater Treatment*; Marcel Dekker: New York, 1999.

78. Welander, T.; Lofqvist, A.; Selmer, A. Upgrading aerated lagoons at pulp and paper mills. *Water Sci. Technol.* **1997**, *35* (*2–3*), 117–122.
79. Kantardjieff, A.; Jones, J.P. Practical experiences with aerobic biofilters in TMP (thermomechanical pulping) sulfite and fine paper mills in Canada. *Water Sci. Technol.* **1997**, *35* (*2–3*), 227–234.
80. Hall, E.R.; Randle, W.G. AOX removal from bleached Kraft mill wastewater. A comparison of three biological treatment processes. *Water Sci. Technol.* **1992**, *26* (*1–2*), 387–396.
81. Slade, A.H.; Nicol, C.M.; Grigsby, J. Nutrients within integrated bleached Kraft mills: sources and behaviour in aerated stabilization basins. *Water Sci. Technol.* **1999**, *40* (*11–12*), 77–84.
82. Bailey, H.C.; Young, L. A comparison of the results of freshwater aquatic toxicity testing of pulp and paper mill effluents. *Water Sci. Technol.* **1997**, *35* (*2–3*), 305–313.
83. Hall, E.R. Anaerobic treatment of wastewaters in suspended growth and fixed film processes. In *Design of Anaerobic Processes for the Treatment of Industrial and Municipal Wastes*; Malin, J.F.; Pohland, F.G., Eds.; Technomics: Lancaster, PA, 1992, 41–118.
84. Iza, J.; Colleran, E.; Paris, J.M.; Wu, W.M. International workshop on anaerobic treatment technology for municipal and industrial wastewaters – summary paper. *Water Sci. Technol.* **1991**, *24* (*8*), 1–16.
85. Hulshoff Pol, L.; Euler, H.; Eitner, A.; Grohganz, D. GTZ sectoral project "Promotion of anaerobic technology for the treatment of municipal and industrial sewage and wastes." *Proceedings of the 8th International Conference on Anaerobic Digestion*, Sendai, Japan, May 25–29, 1997; Vol. II, 285–292.
86. Sierra-Alvarez, R.; Field, J.A.; Kortekaas, S.; Lettinga, G. Overview of the anaerobic toxicity caused by organic forest industry wastewater pollutants. *Water Sci. Technol.* **1994**, *29* (*5–6*), 353–363.
87. Lepisto, R.; Rintala, J.A. The removal of chlorinated phenolic compounds from chlorine bleaching effluents using thermophilic anaerobic processes. *Water Sci. Technol.* **1994**, *29* (*5–6*), 373–380.
88. Jahren, S.J.; Rintala, J.A.; Odegaard, H. Anaerobic thermophilic (55°C) treatment of TMP whitewater in reactors based on biomass attachment and entrapment. *Water Sci. Technol.* **1999**; *40* (*11–12*), 67–76.
89. Haggblom, M.; Salkinoja-Salonen, M. Biodegradability of chlorinated organic compounds in pulp bleaching effluents. *Water Sci. Technol.* **1991**, *24* (*3/4*), 161–170.
90. Wang, X.; Mize, T.H.; Saunders, F.M.; Baker, S.A. Biotreatability test of bleach wastewaters from pulp and paper mills. *Water Sci. Technol.* **1997**, *35* (*2–3*), 101–108.
91. Rintala, J.A.; Lepisto, R. Thermophilic anaerobic-aerobic and aerobic treatment of Kraft bleaching effluents. *Water Sci. Technol.* **1993**, *28* (*2*), 11–16.
92. Lee, E.G.-H.; Crowe, M.F.; Stutz, H. Anaerobic-aerobic lagoon treatment of Kraft mill effluent for enhanced removal of AOX. *Water Pollut. Res. J. Can.* **1993**, *28*, 549–569.
93. Eaton, D.; Chang, H.-M.; Kirk, T.K. Fungal decolorization of Kraft bleach effluents. *Tappi J.* **1980**, *63*, 103–106.
94. Wang, S.-H.; Ferguson, J.F.; McCarthy, J.L. The decolorization and dechlorination of Kraft bleach plant effluent solutes by use of three fungi: *Ganderma lacidum, Coriolus versicolor* and *Hericium erinaceum. Holzforschung*, **1992**, *46*, 219–233.
95. Livernoche, D.; Jurasek L.; Desrochers, M.; Dorica J.; Veliky, I.A. Removal of color from Kraft mill wastewaters with the cultures of white-rot fungi and with immobilized mycelium of *Coriolus versicolor. Biotechnol. Bioengg.* **1983**, *25*, 2055–2065.
96. Huynh, V.-B.; Chang, H.-M.; Joyce, T.W.; Kirk, T.K. Dechlorination of chloro-organics by a white-rot fungus. *Tappi J.* **1985**, *68*, 98–102.
97. Prasad, D.Y.; Joyce, T.W. Sequential treatment of E_1 stage Kraft bleach plant effluent. *Biores. Technol.* **1993**, *44*, 141–147.
98. Taseli, B.K.; Gokcay, C.F. Biological treatment of paper pulping effluents by using a fungal reactor. *Water Sci. Technol.* **1999**, *40* (*11–12*), 93–100.
99. Pointing, S.B. Feasibility of bioremediation by white-rot fungi. *Appl. Microbiol. Biotechnol.* **2001**, *57*, 20–33.
100. Reddy, C.A.; D'Souza, T.M. Physiology and molecular biology of the lignin peroxidases of *Phanerochaete chrysosporium. FEMS Microbiol. Rev.* **1994**, *13*, 137–152.

101. Dezotti, M.; Innocentini-Mei, L.H.; Duran, N. Silica immobilized enzyme catalyzed removal of chlorolignins from eucalyptus Kraft effluent. *J. Biotechnol.* **1995**, *43*, 161–167.
102. Wariishi, H.; Valli, K.; Gold, M.M. H. Manganese (II) oxidation by manganese peroxidase from the basidiomycete *Phanerochaete chrysosporium. J. Biol. Chem.* **1992**, *267*, 23688–23695.
103. Michel, F.C.J.; Dass, S.B.; Grulke, E.A.; Reddy, C.A. Role of manganese peroxidases (MNP) and lignin peroxidases (LiP) of *Phanerochaete chrysosporium* in the decolorization of Kraft bleach plant effluent. *Appl. Environ. Microbiol.* **1991**, *57*, 2368–2375.
104. Jaspers, C.J.; Jiminez, G.; Penninck, M.J. Evidence for a role of manganese peroxidase in the decolorization of Kraft pulp bleach plant effluent *Phanerochaete chrysosporium*: effects of initial culture conditions on enzyme production. *J. Biotechnol.* **1994**, *37*, 229–234.
105. Bourbonnais, R.; Paice, M.G.; Freiermuth, B.; Bodie, E.; Borneman, S. Reactivities of various mediators and laccases with Kraft pulp and lignin model compounds. *Appl. Environ. Microbiol.* **1997**, *63*, 4627–4632.
106. Archibald, F.S.; Paice, M.G.; Jurasek, L. Decolorization of Kraft bleachery effluent chromophores by *Coriolus* (*Trametes*) *versicolor. Enzyme Microb. Technol.* **1990**, *12*, 846–853.
107. Limura, Y.; Hartkainen, P.; Tatsumi. K. Dechlorination of tetrachloroguaiacol by laccase of white-rot basidiomycete Coriolus versicolor. *Appl. Microbiol. Biotechnol.* **1996**, *45*, 434–439.
108. Davis, S.; Burns, R.G. Covalent immobilization of laccase on activated carbon for phenolic treatment. *Appl. Microbiol. Biotechnol.* **1992**, *37*, 474–479.
109. Eweis, J.B.; Ergas, S.J.; Chang, D.P.Y.; Schroeder, E.D. (Eds.) *Bioremediation Principles*; McGraw-Hill, Singapore, 1998.
110. Vandevivere, P.; Verstraete, W. Environmental applications, In Basic Biotechnology; Ratledge, C., Kristiansen, B., Eds., 2nd ed.; Cambridge University Press, Cambridge, UK, 2001; 531–557.

11
Treatment of Pesticide Industry Wastes

Joseph M. Wong
Black & Veatch, Concord, California, U.S.A.

11.1 INTRODUCTION

Pesticides are chemical or biological substances intended to control weeds, insects, fungi, rodents, bacteria, and other pests. They protect food crops and livestock, control household pests, promote agricultural productivity, and protect public health. The importance of pesticides to modern society can be summarized by a statement made by Norman E. Borlaug, the 1970 Nobel Peace Prize winner: "Let's get our priorities in perspective. We must feed ourselves and protect ourselves against the health hazards of the world. To do that, we must have agricultural chemicals. Without them, the world population will starve" [1].

However, the widespread use of pesticides has also caused significant environmental pollution problems. Examples of these include the biological concentration of persistent pesticides (e.g., DDT) in food chains and contamination of surface and groundwater used for drinking sources. Because they can affect living organisms, pesticides are highly regulated in the United States to ensure that their use will be safe for humans and the environment. Recently, the National Research Council's Committee on the Future Role of Pesticides in U.S. Agriculture conducted a comprehensive study and concluded that although they can cause environmental problems, chemical pesticides will continue to play a role in pest management for the foreseeable future. In many situations, the benefits of pesticide use are high relative to risks or there are no practical alternatives [2].

This chapter deals with the characterization, environmental regulations, and treatment and disposal of liquid wastes generated from the pesticide industry.

11.2 THE PESTICIDE INDUSTRY

The pesticide industry is an important part of the economy. Worldwide and U.S. pesticide sales in 1990 were expected to reach more than $20 billion and $6 billion, respectively (*Chemical Week*, January 3, 1990). Usually the highest usage of pesticides is in agriculture, accounting for about 80% of production [3]. Agricultural pesticide use in the United States averaged 1.2 billion pounds of ingredient in 1997, and was associated with expenditures exceeding $11.9 billion. This use involved over 20,700 products and more than 890 active ingredients [2]. Household and garden pesticide uses are other significant markets. The United States constituted about 40% of the world market for household pesticides, with annual sales exceeding $1 billion in 2002 [4]. China is the second largest national market with over $580 million of household insecticides purchased each year [5]. The United States also dominates the world market for garden pesticides with sales of over $1.5 billion per year. The United Kingdom is a distant second with sales of $155 million [5].

Pesticides are classified according to the pests they control. Table 1 lists the various pesticides and other classes of chemical compounds not commonly considered pesticides but included among the pesticides as defined by U.S. federal and state laws [1]. The four most widely used types of pesticides are: (a) insecticides, (b) herbicides, (c) fungicides, and (d) rodenticides [6].

The major components of the pesticide industry include manufacturing and formulation/packaging [7]. During manufacture, specific technical grade chemicals are made. Formulating/packaging plants blend these chemicals with other active or inactive ingredients to achieve the endproducts' desired effects, and then package the finished pesticides into marketable containers. A brief overview of these sectors of the industry follows.

Table 1 Pesticide Classes and Their Uses

Pesticide class	Function
Acaricide	Kills mites
Algicide	Kills algae
Avicide	Kills or repels birds
Bactericide	Kills bacteria
Fungicide	Kills fungi
Herbicide	Kills weeds
Insecticide	Kills insects
Larvicide	Kills larvae (usually mosquito)
Miticide	Kills mites
Molluscicide	Kills snails and slugs (may include oysters, clams, mussels)
Nematicide	Kills nematodes
Ovicide	Destroys eggs
Pediculicide	Kills lice (head, body, crab)
Piscicide	Kills fish
Predicide	Kills predators (coyotes, usually)
Rodenticide	Kills rodents
Silvicide	Kills trees and brush
Slimicide	Kills slimes
Termiticide	Kills termites
Chemicals classed as pesticides not bearing the -cide suffix:	
Attractant	Attracts insects
Chemosterilant	Sterilizes insects or pest vertebrates (birds, rodents)
Defoliant	Removes leaves
Desiccant	Speeds drying of plants
Disinfectant	Destroys or inactivates harmful microorganisms
Growth regulator	Stimulates or retards growth of plants or insects
Pheromone	Attracts insects or vertebrates
Repellent	Repels insects, mites and ticks, or pest vertebrates (dogs, rabbits, deer, birds)

Source: Ref. 1

11.2.1 Pesticide Manufacturing

There are more than 100 major pesticide manufacturing plants in the United States. Figure 1 presents the geographical locations of these plants [7]. Specific pesticide manufacturing operations are usually unique and are characteristic only of a given facility.

Almost all pesticides are organic compounds that contain active ingredients for specific applications. Based on 500 individual pesticides of commercial importance and perhaps as many as 34,000 distinct major formulated products, pesticide products can be divided into six major groups [8]:

1. Halogenated organic.
2. Organophosphorus.
3. Organonitrogen.
4. Metallo-organic.
5. Botanical and microbiological.
6. Miscellaneous (not covered in the preceding groups).

Plants that manufacture pesticides with active ingredients use diverse manufacturing processes, including synthesis, separation, recovery, purification, and product finishing such as drying [9].

Chemical synthesis can include chlorination, alkylation, nitration, and many other substitution reactions. Separation processes include filtration, decantation, extraction, and centrifugation. Recovery and purification are used to reclaim solvents or excess reactants as well as to purify intermediates and final products. Evaporation and distillation are common recovery and purification processes. Product finishing may involve blending, dilution, pelletizing, packaging, and canning. Examples of production facilities for three groups of pesticides follow.

Halogenated Aliphatic Acids

Figure 2 shows a simplified process flow diagram for halogenated aliphatic acid production facilities [8]. Halogenated aliphatic acids include chlorinated aliphatic acids and their salts, for example, TCA, Dalapon, and Fenac herbicides. Chlorinated aliphatic acids can be prepared by nitric acid oxidation of chloral (TCA) or by direct chlorination of the acid. The acids can be sold as mono- or dichloro acids, or neutralized to an aqueous solution with caustic soda. The neutralized solution is generally fed to a dryer from which the powdered product is packaged.

As shown on Figure 2, wastewaters potentially produced during the manufacture of halogenated aliphatic acids include the following:

- vent gas scrubber water from the caustic soda scrubber;
- wastewater from the chlorinator (reactor);
- excess mother liquor from the centrifuges;
- process area cleanup wastes;
- scrubber water from dryer units;
- washwater from equipment cleanout.

Nitro Compounds

This family of organonitrogen pesticides includes the nitrophenols and their salts, for example, Dinoseb and the substituted dinitroanilines, trifluralin, and nitralin. Figure 3 shows a typical commercial process for the production of a dinitroaniline herbicide [8]. In this example, a chloroaromatic is charged to a nitrator with cyclic acid and fuming nitric acid. The crude product is then cooled to settle out spent acid, which can be recovered and recycled. Oxides of nitrogen

Figure 1 Geographical distribution of major pesticide manufacturers in the United States. Most of the plants are located in the eastern half of the continent (from Ref. 7).

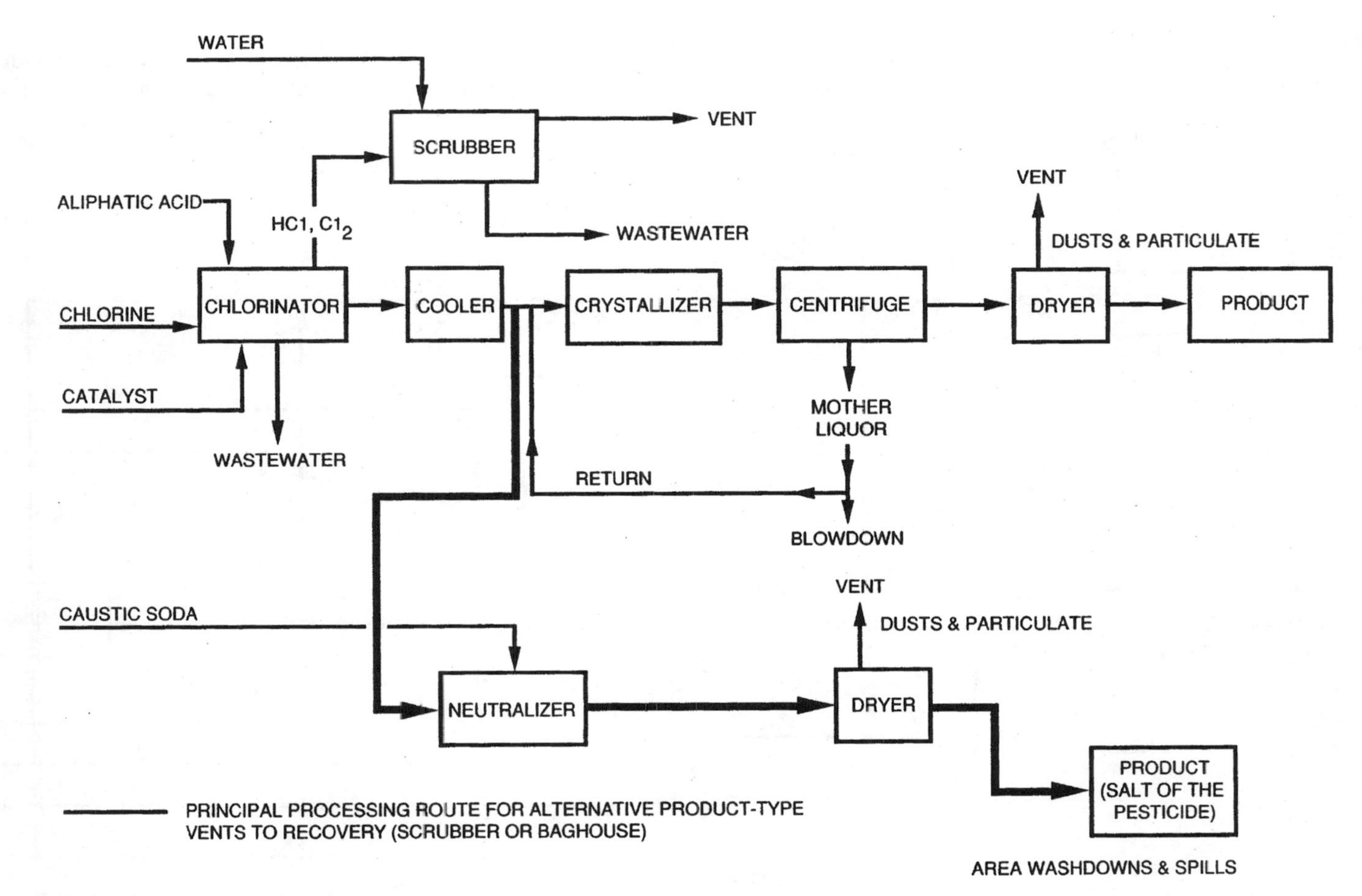

Figure 2 General process flow diagram for halogerated aliphatic acid production facilities. Major processes for pesticide production, including chlorination, cooling, crystallization, centrifying, and drying. The salt of the pesticide is produced by another route (from Ref. 8).

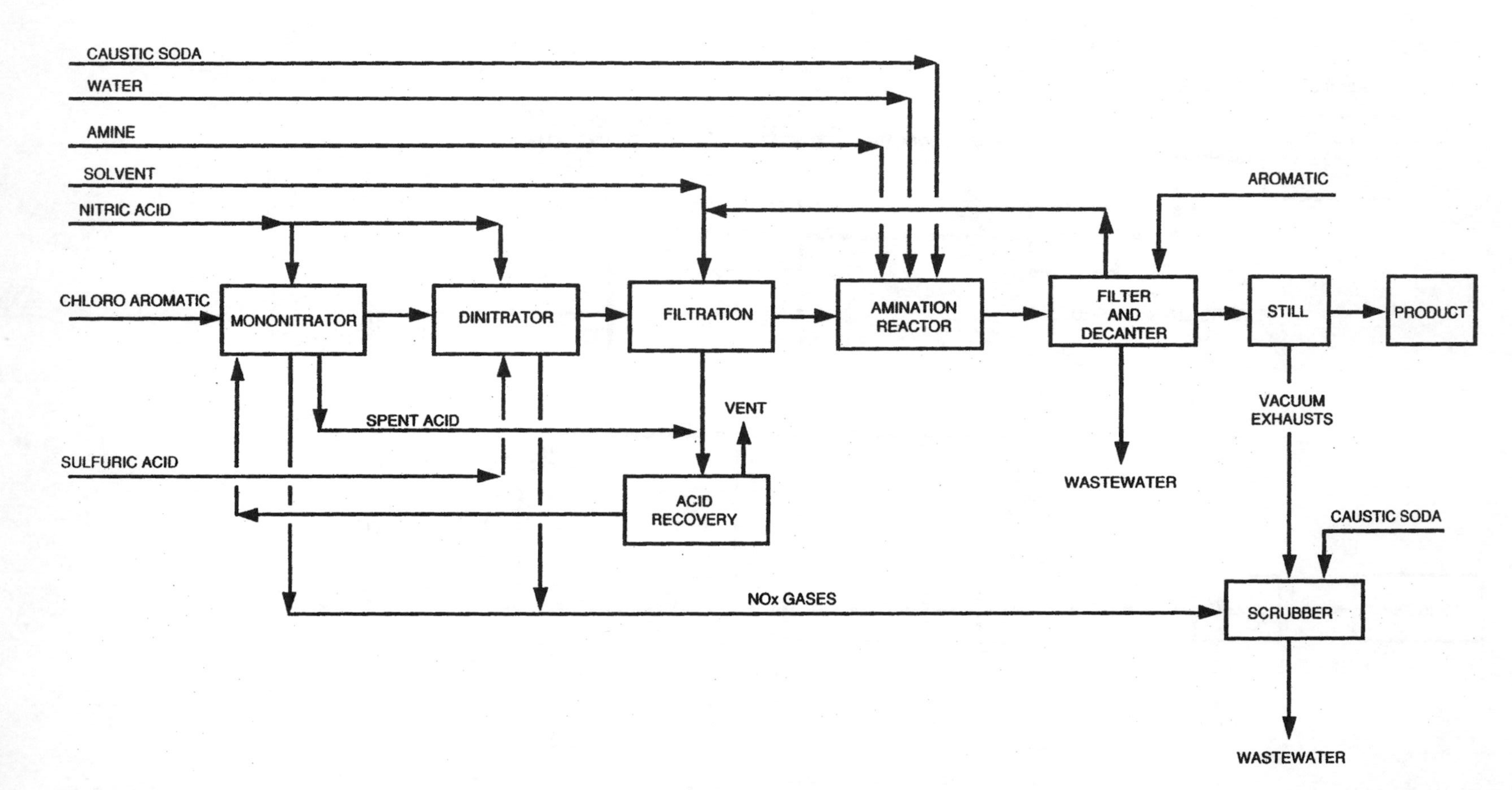

Figure 3 General process flow diagram for nitro-type pesticides. Major processes for pesticide production are mononitration, dinitration, filtration, amination, filtering, and vacuum distillation (from Ref. 8).

are vented and caustic scrubbed. The mononitrated product is then charged continuously to another nitrator containing 100% sulfuric acid and fuming nitric acid at an elevated temperature.

The dinitro product is then cooled and filtered (the spent acid liquor is recoverable), the cake is washed with water, and the resulting washwater is sent to the wastewater treatment plant. The dinitro compound is then dissolved in an appropriate solvent and added to the amination reactor with water and soda ash. An amine is then reacted with the dinitro compound. The crude product is passed through a filter press and decanter and finally vacuum distilled. The saltwater layer from the decanter is discharged for treatment. The solvent fraction can be recycled to the reactor, and vacuum exhausts are caustic scrubbed. Still bottoms are generally incinerated.

Wastewaters potentially generated during the manufacture of the nitro family of pesticides include the following:

- aqueous wastes from the filter and the decanter;
- distillation vacuum exhaust scrubber wastes;
- caustic scrubber wastewaters;
- periodic kettle cleanout wastes;
- production area washdowns.

Metallo-Organic Pesticides

Metallo-organic active ingredients mean organic active ingredients containing one or more metallic atoms, such as arsenic, mercury, copper, and cadmium, in the structure. Figure 4 shows a general process flow diagram for arsenic-type metallo-organic pesticide production [8]. Monosodium acid methanearsenate (MSMA) is the most widely produced organoarsenic herbicide in this group.

The first step of the process is performed in a separate, dedicated building. The drums of arsenic trioxide are opened in an air-evacuated chamber and automatically dumped into 50% caustic soda. A dust collection system is used. The drums are carefully washed with water, the washwater is added to the reaction mixture, and the drums are crushed and sold as scrap metal. The intermediate sodium arsenite is obtained as a 25% solution and is stored in large tanks prior to further reaction. In the next step, the 25% sodium arsenite is treated with methyl chloride to produce the disodium salt DSMA (disodium methanearsenate, hexahydrate). This DSMA can be sold as a herbicide; however, it is more generally converted to MSMA, which has more favorable application properties [8].

To obtain MSMA, the DSMA solution is partially acidified with sulfuric acid and the resulting solution concentrated by evaporation. As the aqueous solution is being concentrated, a mixture of sodium sulfate and sodium chloride precipitates out (about 0.5 kg per 100 kg of active ingredient). These salts are a troublesome disposal problem because they are contaminated with arsenic. The salts are removed by centrifugation, washed in a multistage, countercurrent washing cycle, and then disposed of in an approved landfill.

Methanol, a side product of methyl chloride hydrolysis, can be recovered and reused. In addition, recovered water is recycled. The products are formulated on site as solutions and are shipped in 1 to 30 gallon containers.

Wastewaters that can be generated from the production of these pesticides include the following:

- spillage from drum washing operations;
- washwater from product purification steps;
- scrub water from the vent gas scrubber unit;
- process wastewater;

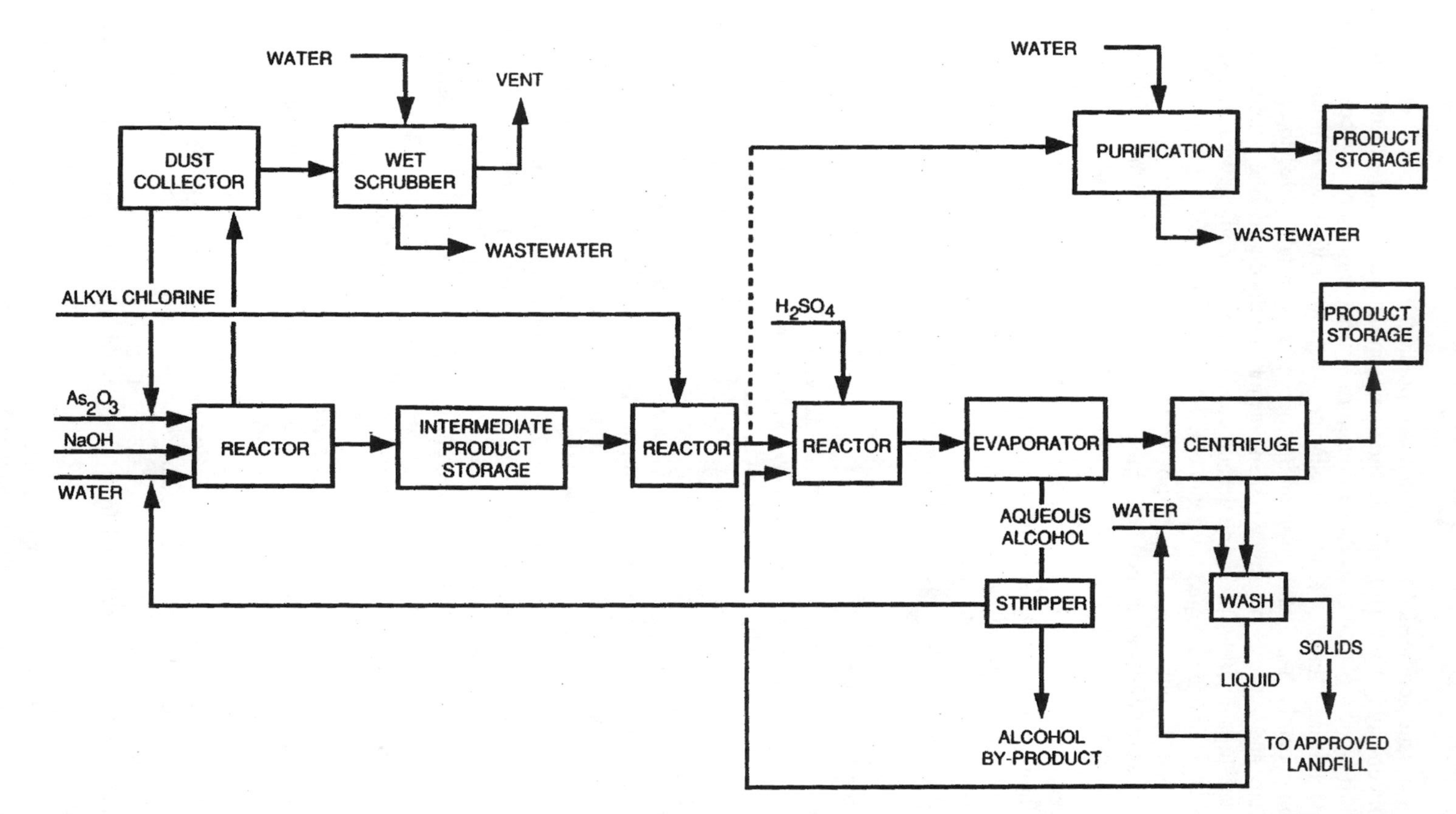

Figure 4 General process flow diagram for arsenic-type metallo-organic production. Sodium arsenate is formed in the first reactor, disodium methanearsenate (DSMA) in the second reactor; DSMA is purified as a product or further changed to monosodium methanearsenate (MSMA) by acidification and purified (from Ref. 8).

- area washdowns;
- equipment cleanout wastes.

11.2.2 Pesticide Formulating/Packaging

After a pesticide is manufactured in its relatively pure form (the technical grade material) the next step is formulation – processing a pesticide compound into liquids, granules, dusts, and powders to improve its properties of storage, handling, application, effectiveness, or safety [9]. The technical grade material may be formulated by its manufacturer or sold to a formulator/packager.

In the United States, there are more than a thousand pesticide formulating/packaging plants covering a broad range of formulations [7]. Many small firms have only one product registration, and produce only a few hundred pounds of formulated pesticides each year. However, USEPA [7] identified one plant operating in the range of 100 million pounds of formulated product per year. The approximate production distribution of formulators/packagers is presented in Table 2 [7].

The most important unit operations involved in formulation are dry mixing and grinding of solids, dissolving solids, and blending [8]. Formulation systems are virtually all batch-mixing operations. The units may be completely enclosed within a building or may be in the open, depending primarily on the geographical location of the plant. Production units representative of the liquid and solid formulation/packaging equipment in use as well as wastewater generation are described in the following.

Liquid Formulation Units

A typical liquid formulation unit is depicted in Fig. 5 [8]. Until it is needed, technical grade pesticide is usually stored in its original shipping container in the warehouse section of the plant. When this material is received in bulk, however, it is transferred to holding tanks for storage.

Batch-mixing tanks are frequently open-top vessels with a standard agitator and may or may not be equipped with a heating/cooling system. When solid technical grade material is used, a melt tank is used before this solid material is added to the mix tank. Solvents are normally stored in bulk tanks and are either metered into the mix tank or are determined by measuring the tank level. Necessary blending agents (emulsifiers, synergists, etc.) are added directly. From the mix tank, the formulated material is frequently pumped to a holding tank before being put into containers for shipment. Before packaging, many liquid formulations must be filtered by conventional cartridge filters or equivalent polishing filters.

Air pollution control equipment used on liquid formulation units typically involves exhaust systems at all potential sources of emission. Storage and holding tanks, mix tanks, and

Table 2 Formulator/Packager Production Distribution

Production (million lb/year)	Formulator/ Packagers (%)
<0.5	24
>0.5 to <5.0	41
>5.0 to <50	35
Total	100

Source: Ref. 7.

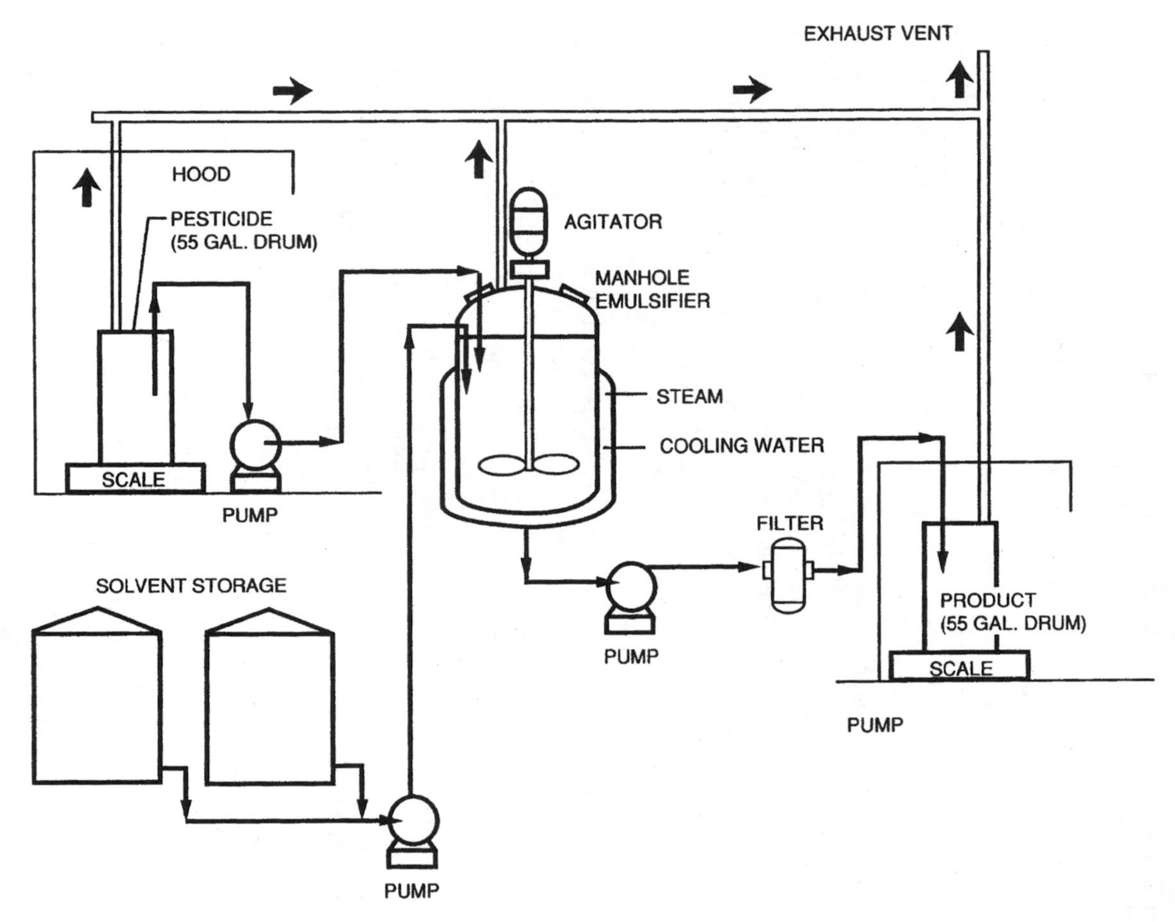

Figure 5 Liquid formulation unit. Technical grade pesticide products are blended with solvents and emulsifiers or other agents in a mix tank. Formulated products are filtered before packaging (from Ref. 8).

container-filling lines are normally provided with an exhaust connection or hood to remove any vapors. The exhaust from the system normally discharges to a scrubber system or to the atmosphere.

Dry Formulation Units

Dry products can include dusts, powders, and granules. Dusts and powders are manufactured by mixing technical grade material with the appropriate inert carrier and grinding the mixture to obtain the correct particle size. Several rotary or ribbon blender-type mixers mix the product. Figure 6 shows a typical dry formulation unit for pesticides [8].

Baghouse systems efficiently control particulate emissions from grinding and blending processes. Vents from feed hoppers, crushers, pulverizers, blenders, mills, and cyclones are typically routed to baghouses for product recovery. This method is preferable to using wet scrubbers. However, even scrubber effluent can be largely eliminated by recirculation.

Granules are formulated in systems similar to the mixing sections of dust plants. The active ingredient is adsorbed onto a sized, granular carrier such as clay or a botanical material. This is accomplished in various capacity mixers that generally resemble cement mixers. If the technical grade material is a liquid, it can be sprayed directly onto the granules. Solid material is usually melted or dissolved in a solvent to provide adequate dispersion on the granules. Screening to remove fines is the last step prior to intermediate storage before packaging.

Packaging and Storage

Packaging the finished pesticide into a marketable container is the last operation conducted at a formulation plant. This operation is usually carried out in conventional filling and packaging units. By moving from one unit to another, the same liquid filling line is frequently used to fill products from several formulation units. Packages of almost every size and type are used, including 1, 2, and 5 gallon cans, 30 and 55 gallon drums, glass bottles, bags, cartons, and plastic jugs.

Aerosol products (for home use) undergo leak testing in a heated water bath to comply with U.S. Department of Transportation regulations. This water bath also serves as a quality control checkpoint for leaks. Bath water must be kept clean for inspection.

Generally, onsite storage is minimized. The storage facility is often a building completely separate from the formulation and filling operation or is at least located in the same building but separate from the formulation units to avoid contamination and other problems. Technical grade material, except for bulk shipments, is usually stored in a special section of the product storage area.

Wastewater Sources

In pesticide formulating/packaging plants, wastewaters can be generated at several sources, including the following [8]:

- formulation equipment cleanup;
- spill washdown;
- drum washing;
- air pollution control devices;
- area runoff;
- laboratory drains.

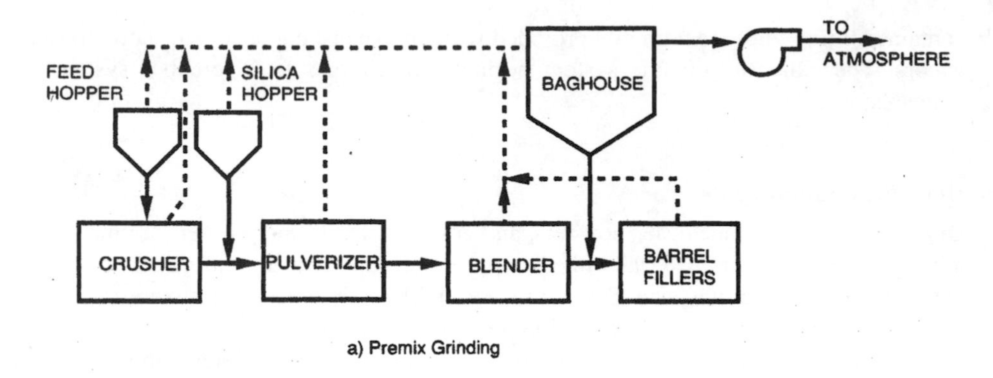

a) Premix Grinding

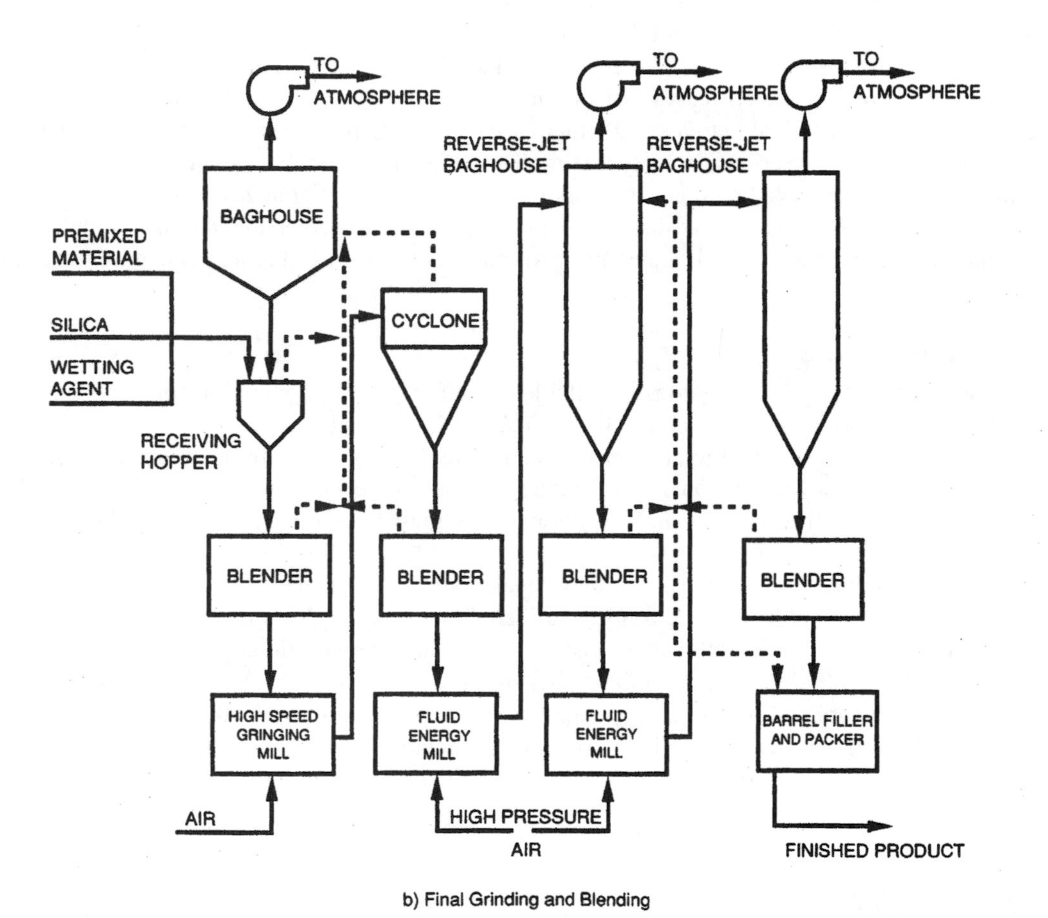

b) Final Grinding and Blending

Figure 6 Dry formulation unit. Technical grade products are ground and mixed with appropriate inert materials; the premixed material is further blended with more inert materials and wetting agents in several steps to obtain the correct particle size (from Ref. 8).

The major source of contaminated wastewater from formulation plants is equipment cleanup. Formulation lines, including filling equipment, must be cleaned periodically to prevent cross-contamination of product. Sometimes equipment is washed with formula solvent and rinsed with water. Hence, this waste may contain pesticide ingredients as well as solvents.

For housekeeping purposes, most formulators clean buildings that house formulation units on a routine basis. Prior to washdown, as much dust, dirt, and so on as possible is swept and vacuumed. The washdown wastewater, which generally contains pesticide ingredients, is normally contained within the building and is disposed of in whatever manner is used for other contaminated wastewater.

A few formulation plants process used pesticide drums so they can be sold to a drum reconditioner or reused by the formulator for appropriate products, or simply to decontaminate the drums before they are disposed of. Drum-washing procedures range from a single rinse with a small volume of caustic solution or water to complete decontamination and reconditioning processes. Hence, drum-washing wastewater usually contains caustic solution as well as washed pesticide ingredients in the drums.

Water-scrubbing devices are often used to control emission to the air. Most of these devices generate wastewater streams that are potentially contaminated with pesticide ingredients. Although the quantity of water in the system is high – about 20 gallons per 1000 cfm – water consumption is kept low by a recycle–sludge removal system.

Natural runoff at formulating/packaging plants, if not properly handled, can become a major factor in the operation of wastewater systems simply because of the relatively high flow and because normal plant wastewater volumes are generally extremely low. Isolation of runoff from contaminated process areas or wastewaters, however, eliminates its potential for becoming significantly contaminated with pesticide ingredients. Hence, the content of area runoff depends on the degree of weather protection and area isolation. Modern stormwater pollution prevention regulations in the United States have virtually eliminated this pollution source.

Most of the larger formulation plants have some type of control laboratory on site. Wastewater from the control laboratories relative to the production operations can range from an insignificantly small, slightly contaminated stream to a rather concentrated source of contamination. In many cases, this stream can be discharged into the sanitary sewer. Larger, more highly contaminated streams, however, must be treated along with other contaminated wastewaters.

11.3 WASTE CHARACTERISTICS

Wastewater sources from pesticide manufacturing and formulating/packaging facilities have been described in the previous section. This section discusses wastewater quality and quantity.

11.3.1 Pesticide Manufacturing

Because of the variety and uniqueness of pesticide manufacturing processes and operations, the flow and characteristics of wastewater generated from production plants vary broadly. In 1978, 1979, 1980, 1982, and 1984, the USEPA conducted surveys to obtain basic data concerning manufacturing, disposal, and treatment as well as to identify potential sources of priority pollutants in pesticide manufacturers [7]. The results of these surveys and USEPA's interpretations and evaluations are summarized in the following.

Wastewater Flows

Based on survey results from individual plants, USEPA determined the amount of flow per unit of pesticide production (gal/1000 lb) and the amount of flow (million gallons per day, or MGD) at these plants.

Figure 7 presents a probability plot of the flow ratio (gal/1000 lb) for 269 of the 327 pesticide process areas for which data were available [7]. Significant information in this figure shows that 11% of all pesticide processes have no flow, 50% of all pesticide processes have flows equal to or less than 1000 gal/1000 lb, and 84% have flows equal to or less than 4500 gal/1000 lb.

Figure 8 presents a probability plot of pesticide wastewater flows (MGD) at individual plants [7]. This figure shows that 50% of all plants have flows less than 0.01 MGD, and that virtually all plants (98%) have flows less than 1.0 MGD.

Wastewater Constituents

Because of the nature of pesticides and their components, wastewaters generated from manufacturing plants usually contain toxic (e.g., toxic priority pollutants as defined by USEPA) and conventional pollutants. Based on the results of the surveys and process evaluations, USEPA determined the pollutants or groups of pollutants likely to be present in raw wastewater from these facilities. The agency also selected raw waste loads for these pollutants in order to design treatment and control technologies. The approach taken was to design for the removal of maximum priority pollutant raw waste concentrations as reported in the surveys. Table 3 presents the summary of these raw waste load design levels [7].

The pollutants or groups of pollutants likely to be present in raw wastewater include volatile aromatics, halomethanes, cyanides, haloethers, phenols, polynuclear aromatics, heavy

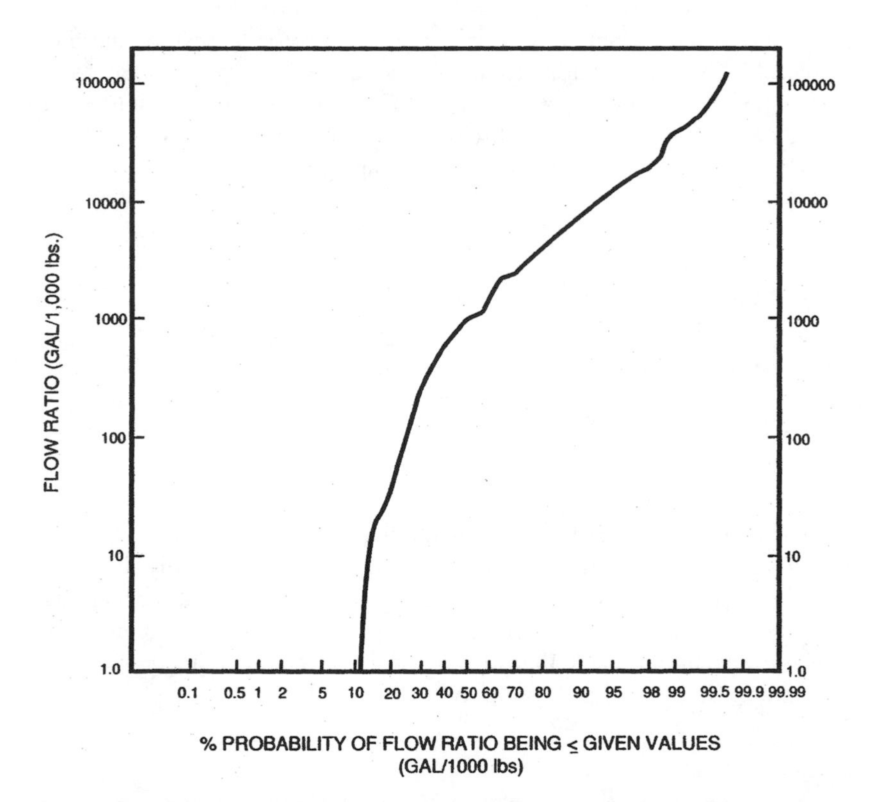

Figure 7 Probability plot of pesticide product flow ratios. Of pesticide production processes, 11% have no flow, 50% have flows less than 1000 gal/1000 lb; 84% have flows less than 4500 gal/1000 lb (from Ref. 7).

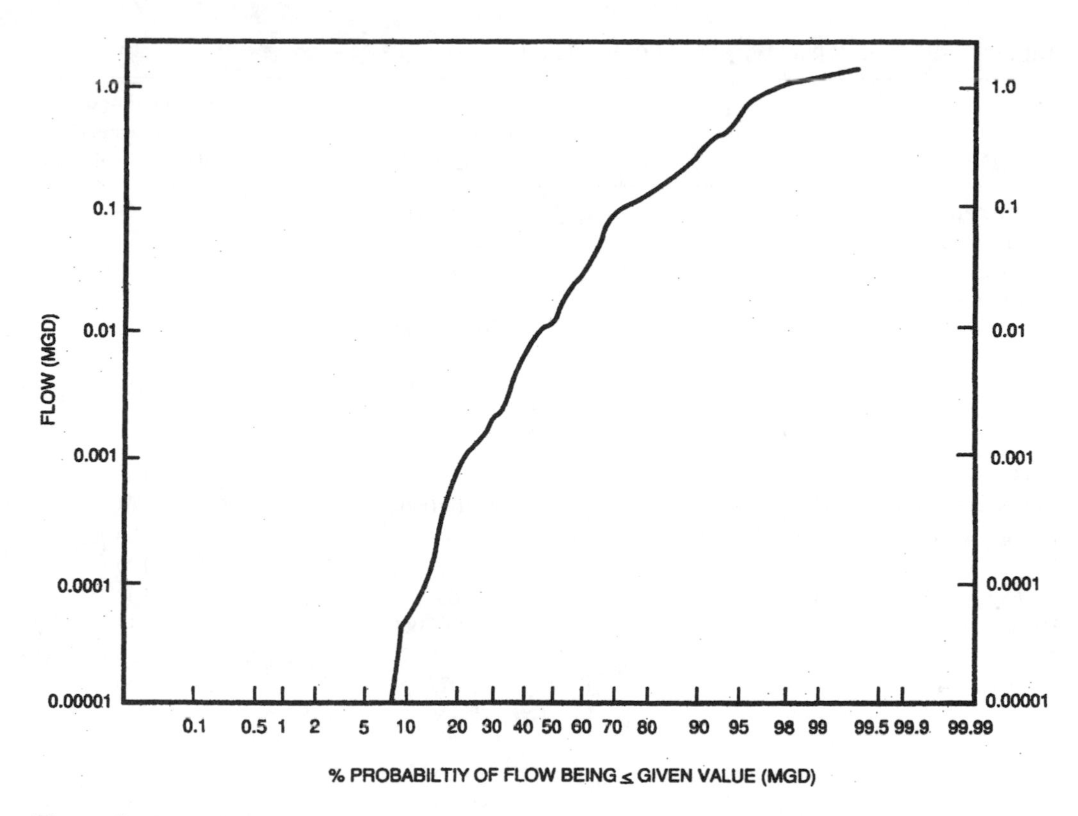

Figure 8 Probability plot of pesticide product wastewater flows. Of pesticide manufacturing plants, 50% have flows less than 0.01 MGD; 98% have flows less than 1.0 MGD (from Ref. 7).

metals, chlorinated ethanes and ethylenes, nitrosamines, phthalates, dichloropropane and dichloropropene, pesticides, dienes, TCDD, and other common constituents such as BOD, COD, and TSS. The sources and significance of these pollutants are briefly discussed [7].

Volatile Aromatics

Benzene and its derivatives are used widely throughout the chemical industry as solvents and raw materials. Mono-, di-, and trichlorobenzenes are used directly as pesticides for their insecticidal and fungicidal properties. Benzene, toluene, and chlorobenzene are used as raw materials in the synthesis of at least 15 pesticides, although their main use is as a carrier solvent in 76 processes. Additional priority pollutant aromatics and chlorinated aromatics exist as impurities or as reaction byproducts because of the reactions of the basic raw materials and solvent compounds.

Halomethanes

Halomethanes, including methylene chloride, chloroform, and carbon tetrachloride (di-, tri-, and tetrachloromethane, respectively), are used mainly as raw materials and solvents in approximately 28 pesticide processes. Bromomethanes can be expected in at least five pesticides as raw materials, byproducts, or impurities and in the case of methyl bromide, can function as a fumigant.

Table 3 Summary of Raw Waste Load Design Levels.

Pollutant group	Design level (mg/L)	Detected pesticide wastewaters at design level[a] (%)
Volatile aromatics	127–293,000	24
Halomethanes	122–2,600	23
Cyanides	5,503	6.0
Haloethers	0.582	17
Phenols	100–42,000	45
Nitro-substituted aromatics	ND[b]	100
Polynuclear aromatics	1.06–1.2	25
Metals		
Copper	4,500	17
Zinc	247	100
Chlorinated ethanes and ethylenes	98–10,000	18
Nitrosamines	1.96	100
Phthalates	ND	100
Dichloropropane and dichloropropene	ND	100
Pesticides	10–11,200	45
Dienes	2,500–15,000	50
TCDD	0.022	100
Miscellaneous	N/A[c]	N/A
PCBs	N/A	N/A
Benzidine	N/A	N/A
BOD	1,470	33
COD	3,886	45
TSS	266	14

[a]Remainder of known pesticide wastewaters are below design level prior to biological oxidation.
[b]ND = not detected.
[c]NA = not applicable.
Source: Ref. 7.

Cyanide

Cyanide is a known or suspected pollutant in approximately 24 pesticide processes. The primary raw materials that favor the generation of cyanides as either byproducts or impurities are cyanamides, cyanates, thiocyanates, and cyanuric chloride. Cyanuric chloride is used exclusively in the manufacture of triazine pesticides.

Haloethers

Five compounds classified as priority pollutants contain an ether moiety and halogen atoms attached to the aryl and alkyl groups. Five pesticides are suspected to contain at least one compound from this class. *Bis*(2-chloroethyl) ether (BCEE) is used as a raw material in two pesticides; BCEE itself functions as a fungicide or bactericide in certain applications. In the other three pesticides, the ethers are suspected to be present as raw material impurities.

Phenols

Phenols are compounds having the hydroxyl (OH) group attached directly to an aromatic ring. Phenols commonly found in pesticide wastewaters include chlorophenols, nitrophenols, and methylphenols (cresols). These compounds may be found throughout the pesticide industry as raw materials, impurities in raw materials, or as byproducts of reactions using related compounds such as chlorobenzenes. The presence of nitrated phenols is expected in six pesticides. Methylated phenols are not expected to be significant because they are not used as raw materials, but they may appear as impurities of reaction from one pesticide because of using 4-methylthio-m-cresol as a raw material.

Polynuclear Aromatics

Seventeen priority pollutant compounds can be classified as polynuclear aromatics (PNA). These compounds consist of two or more benzene rings that share a pair of carbon atoms. They are all derived from coal tar, with naphthalene being the largest constituent. Naphthalene derivatives such as alpha-naphthylamine and alpha-naphthol are used in some pesticide processes; therefore, naphthalene is by far the most prevalent PNA priority pollutant in the industry. Acenaphthene, anthracene, fluorene, fluoranthene, and phenathrene are found as raw material impurities. Acenaphthene is found in one pesticide process as a raw material. The remaining ten PNAs are not suspected to be present in pesticide processes.

Heavy Metals

In the pesticide industry, metals are used principally as catalysts or as raw materials that are incorporated into the active ingredients, for example, metallo-organic pesticides. Priority pollutant metals commonly incorporated into metallo-organic pesticides include arsenic, cadmium, copper, and mercury. For metals not incorporated into the active ingredients, copper is found or suspected in wastewaters from at least eight pesticides, where it is used as a raw material or catalyst; zinc becomes part of the technical grade pesticide in seven processes; and mercury is used as a catalyst in one pesticide process. Nonpriority pollutant metals such as manganese and tin are also used in pesticide processes.

Chlorinated Ethanes and Ethylenes

The chlorinated ethanes and ethylenes are used as solvents, cleaning agents, and intermediates. Vinyl chloride (chloroethylene) is used in the production of plastic polyvinyl chloride (PVC). In the pesticide industry, approximately 23 products are suspected to contain a member of this group of priority pollutants. The main pollutants include 1,2-dichloroethane, which is used as a solvent in seven pesticides and tetrachloroethylene, which is used as a solvent in two pesticides.

Nitrosamines

N-nitrosamines are a group of compounds characterized by a nitroso group (N=O) attached to the nitrogen of an aromatic or aliphatic secondary amine. N-nitrosodi-N-propylamine is a suspected reaction byproduct from the nitrosation of di-N-propylamine. Two pesticides are suspected to contain some form of nitrosamines.

Phthalates

Phthalate esters are used widely as plasticizers in commercial polymers and plastic endproducts such as PVC. One phthalate classified as a priority pollutant is suspected to be present in three pesticide processes. Dimethyl phthalate is known to be a raw material in two products.

Dichloropropane and Dichloropropene

1,3-Dichloropropene is a raw material in one pesticide. 1,3-Dichloropropene and the combined pollutants 1,2-dichloropropane-1, 3-dichloropropene are pesticide products as well as priority pollutants and function as insecticidal fumigants.

Priority Pollutant Pesticides

There are only 18 priority pollutants commonly classified as pesticides. Only two are still in production: heptachlor and chlordane. Aldrin, dieldrin, and endrin aldehyde are suspected as reaction byproducts in the endrin process. Heptachlor epoxide occurs as a reaction byproduct in both chlordane and heptachlor manufacturing. DDD, DDE, and DDT can occur as a reaction byproduct in the manufacture of endosulfan.

Dienes

Four manufactured pesticides and two pesticides currently not manufactured use a priority pollutant diene as a raw material. The basic material for all six pesticides is hexachlorocyclopentadiene (HCCPD). The priority pollutant hexachlorobutadiene is suspected to be present in the pesticide wastewater because it is a byproduct of HCCPD synthesis and is used as a solvent in manufacturing mirex.

TCDD

2,3,7,8-Tetrachlorodibenzo-p-dioxin (TCDD) is believed to be a byproduct in chemical processing generated by a halophenol or chlorobenzene starting material. An intermediate reaction will occur at an elevated temperature (equal to or greater than 160°C), an alkaline condition, or in the presence of a free halogen. The end reaction results in either direct dioxin, intermediate dioxin, or predioxin formation that will ultimately form dibenzo-p-dioxins [10]. TCDD is suspected in wastewaters from pesticide manufacture that uses such raw materials as 2,4,5-trichlorophenol (2,4,5-T) and 1,2,4,5-tetra-chlorobenzene, which are characteristic of TCDD precursors. A TCDD level as high as 111 mg/L has been found in drums of waste from the production of the pesticide 2,4,5-T.

Other Pollutants

The pesticide industry routinely monitors conventional and nonconventional pollutants in manufacturing wastewaters. According to the USEPA surveys [7], chemical oxygen demand (COD) concentrations ranged from 14.0 mg/L to 1,220,000 mg/L; Total organic carbon (TOC) ranged from 53.2 mg/L to 79,800 mg/L; biochemical oxygen demand (BOD) ranged from nondetected to 60,000 mg/L; and total suspended solids (TSS) ranged from 2.0 mg/L to 4090 mg/L. Many other pollutants can be present in pesticide wastewaters that are not unique to this industry, including pollutants such as ammonia, oil and grease, fluoride, and inorganic salts. Nonpriority pollutant pesticides would naturally occur in their manufacturing wastewaters due to imperfect separations.

11.3.2 Pesticide Formulating/Packaging

Washing and cleaning operations provide the principal sources of wastewater in formulating and packaging operations. Because these primary sources are associated with cleanup of spills, leaks, area washdowns, and stormwater runoff, there is apparently no basis from which to correlate the pollutants generated to the product made.

According to USEPA's survey [8] of 71 pesticide formulating/packaging plants, 59 reported no generation of wastewater. For the plants that generated wastewater, neither the rate of production nor the type of product formulated had a direct bearing on the quality or quantity of wastewater generated. The three largest plants of a major pesticide formulator each generated less than 5800 gal/day. Other plants generated from 5 to 1000 gal/day. The average flows generated in formulating/packaging plants were between 50 and 1000 gal/day [11].

The pollutants contained in the wastewaters are expected to be similar to those from manufacturing facilities. Pesticides and solvents are the principal pollutants of concern. Although their volumes are small, the wastewaters from pesticide formulating/packaging plants could be highly contaminated and toxic.

11.4 ENVIRONMENTAL REGULATIONS

Many federal and state regulations govern the registration, manufacture, transportation, sale, use, and disposal of pesticides in the United States. Pesticides are regulated by the USEPA primarily under the Federal Insecticide, Fungicide, and Rodenticide Act (FIFRA) and the Federal Food, Drug, and Cosmetic Act (FDCA). The FIFRA requires pesticides to be registered by USEPA and authorizes the agency to prescribe conditions for their use. The FDCA requires the agency to establish maximum acceptable levels of pesticide residues in foods. The transportation of hazardous pesticides is regulated by the Hazardous Materials Transportation Act (HMTA). In addition, certain states such as California and Florida aggressively enforce their own pesticide laws.

The disposal of pesticides and pesticide wastes is regulated by the Clean Air Act (CAA), the Clean Water Act (CWA), the Resource Conservation and Recovery Act (RCRA), and the Comprehensive Environmental Response, Compensation, and Liability Act (CERCLA). This section deals with the regulations for liquid waste disposal, which is mainly under the CWA. However, when the waste is disposed of as a hazardous waste, it is regulated by the RCRA.

11.4.1 Clean Water Act

The U.S. Congress enacted the Federal Water Pollution Control Act (FWPCA) in 1972. The act was significantly amended in 1977 and has since become known as the CWA. It was again amended by the Water Quality Act of 1987. The CWA applies to all industries that generate wastewater discharges. Some of its provisions are particularly applicable to the pesticide industry.

Effluent Guidelines for Pesticides

Under Section 304 of the CWA, USEPA was required to establish "effluent guidelines" for a number of different industrial categories by specifying the effluent limits that must be met by dischargers in each category. Two types of standards were required for each industry: (a) effluent limitations that require the application of the best practicable control technology (BPT) currently available, and (b) effluent limitations that require application of the best available technology (BAT).

Effluent limitations reflecting BPT currently available for the pesticide manufacturing and formulating industrial category were promulgated by USEPA on April 25, 1978 (43 Federal

Regulation 17,785, 1978). The pesticide industry was divided into three subcategories under the BPT regulations: (a) organic pesticide chemicals manufacturing, (b) metallo-organic pesticide chemical manufacturing, and (c) pesticide chemicals formulating and packaging.

For the first subcategory, the rules limit the number of pounds or kilograms of COD, BOD, TSS, and pesticide chemicals that a plant may discharge during any 1 day or any 30 consecutive days. Table 4 presents the BPT effluent limitations for the organic pesticide chemicals manufacturing subcategory (40 CFR pt. 455). For the second and third subcategories, the regulations permit "no discharges of process wastewater pollutants into navigable waters." The BPT regulations are based on pesticide removal by hydrolysis or adsorption followed by biological treatment [3].

The USEPA issued BAT regulations for the pesticide industry in October 1985 (50 Federal Regulation 40701, 1985). However, four chemical companies and three chemical trade organizations challenged these regulations in *Chemical Specialties Manufacturers Association vs. EPA*, No. 86-8024 (11th Cir. July 25, 1986), modified (11th Cir. August 29, 1986). As a result, the agency voluntarily withdrew its regulations and, on remand by the Eleventh Circuit, agreed to initiate a new round of rule making on the pesticide industry standards (51 Federal Regulation 44,911, 1986). The new regulations were later proposed by USEPA in 1992 [12] and finalized in 1996 (61 FR 57551, No. 6, 1996). All of the updated effluent guidelines and standards for the pesticide manufacturing and formulation industries are included in 40 CFR Part 455 – Pesticide Chemicals.

Pretreatment Standards for Pesticides

Section 306(b) of the CWA requires USEPA to promulgate pretreatment standards applicable to the introduction of wastes from industry and other nondomestic sources into publicly owned treatment works (POTWs). USEPA issued the General Pretreatment Regulations on June 26, 1978, and amended these regulations several times in the following years (40 CFR pt. 403).

The pretreatment standards for existing and new sources for the organic pesticide chemicals manufacturing subcategory were promulgated on September 28, 1993 (58 Federal

Table 4 BPT[b] Effluent Limitations for Organic Pesticide Chemicals Manufacturing Subcategory

Effluent characteristics	Maximum for any 1 day	Average of daily values for 30 consecutive days shall not exceed
COD	13.000	9.0000
BOD_5	7.400	1.6000
TSS	6.100	1.8000
Organic pesticide chemicals	0.010	0.0018
pH	(′)[a]	(′)

Source: 40 CFR 455.22.
[a] (′) Within the range 6.0–9.0.
[b]BPT, best practicable control technology currently available.
Note: For COD, BOD_5, and TSS, metric units: Kilogram/1,000 kg of total organic active ingredients. English units: Pound/1,000 lb of total organic active ingredients. For organic pesticide – metric units: Kilogram/1,000 kg of organic pesticide chemicals. English units: Pound/1,000 lb of organic pesticide chemicals.

Register 50690). The main concern for this subcategory is the discharge of priority pollutants into POTWs. Table 6 in 40 CFR Part 455 listed 24 priority pollutants with maximum daily and maximum monthly discharge limitations. With the exception of cyanide and lead, all the priority pollutants are organic compounds. Presently there are no pretreatment standards for the metallo-organic pesticide chemicals manufacturing subcategory. The pretreatment standard for the pesticide chemicals formulating and packaging subcategory is no discharge of process wastewater pollutants to POTWs (40 CFR Part 455.46).

The general pretreatment regulations prohibit an industry or nondomestic source from introducing pollutants that will pass through or interfere with the operation or performance of POTWs [40 CFR Section 403.5(a)]. In addition, the CWA requires USEPA to establish "categorical" pretreatment standards, which apply to existing or new industrial users in specific categories (40 CFR Section 403.6). The discharge of wastewater from the pesticide industry to POTWs will also be subject to the general discharge prohibitions against "pass through" and "interference" with the POTWs. These pretreatment requirements are usually enforced by POTWs, with approved pretreatment programs. As an example, Table 5 shows the general

Table 5 Industrial Waste Pretreatment Limits for a Publicly Owned Treatment Works

Toxic substance	Maximum allowable concentration (mg/L)
Aldehyde	5.0
Antimony	5.0
Arsenic	1.0
Barium	5.0
Beryllium	1.0
Boron	1.0
Cadmium	0.7
Chlorinated hydrocarbons including, but not limited to, pesticides, herbicides, algaecides	Trace
Chromium, total	1.0
Copper	2.7
Cyanides	1.0
Fluorides	10.0
Formaldehydes	5.0
Lead	0.4
Manganese	0.5
Mercury	0.010
Methyl ethyl ketone and other water-insoluble ketones	5.0
Nickel	2.6
Phenol and derivatives	30.0
Selenium	2.0
Silver	0.7
Sulfides	1.0
Toluene	5.0
Xylene	5.0
Zinc	2.6
pH, su	5.0 to 10.5

Source: City of San Jose (California) Municipal Code, 1988.

industrial effluent limits established by the City of San Jose, CA (San Jose Municipal Code, 1988).

Toxic Pollutant Effluent Standards

Section 307 of the CWA requires USEPA to maintain and publish a list of toxic (priority) pollutants, to establish effluent limitations for the BAT economically achievable for control of such pollutants, and to designate the category or categories of sources to which the effluent standards shall apply [3]. Effluent standards have been promulgated for the following toxic pollutants: aldrin/dieldrin; DDT, DDD, and DDE; endrin; toxaphene; benzidine; and PCBs (40 CFR 129.4). These standards, which may be incorporated into National Pollutant Discharge Elimination System (NPDES) permits, limit or prohibit the discharge of process wastes or other discharge from manufacturing processes into navigable waters. For example, any discharge of aldrin or dieldrin is prohibited for all manufacturers (40 CFR 129.100(b)(3)).

Water Quality-Based Limitations

In the United States, as control of conventional pollutants has been significantly achieved, increased emphasis is being placed on reduction of toxic pollutants. The USEPA has developed a water quality-based approach to achieve water quality where treatment control-based discharge limits have proved to be insufficient [13].

The procedures for establishing effluent limitations for point sources discharging to a water quality-based segment generally involves the use of some type of mathematical model or allocation procedure to apportion the allowable loading of a particular toxicant to each discharge in the segment. These allocations are generally made by the state regulatory agency and reviewed, revised, and approved by the USEPA in accordance with Section 303 of the CWA.

To control the discharge of toxic pollutants in accordance with Section 304(1) of the CWA, state and regional regulatory agencies may also establish general effluent limitations for a particular water body. For example, Table 6 shows the discharge limits for toxic pollutants

Table 6 Effluent Limitations for Selected Toxic Pollutants for Discharge to Surface Waters (All Values in μg/L)

	Daily average	
	Shallow water	Deep water
Arsenic	20	200
Cadmium	10	30
Chromium(VI)	11	110
Copper	20	200
Cyanide	25	25
Lead	5.6	56
Mercury	1	1
Nickel	7.1	71
Silver	2.3	23
Zinc	58	580
Phenols	500	500
PAHs	15	150

Source: Water Quality Control Plan, San Francisco Bay Basin, 1986.

established by the San Francisco Bay Regional Water Quality Control Board in 1986. This regional agency has also adopted biomonitoring and toxicity requirements for municipal and industrial dischargers. Biomonitoring, or whole-effluent toxicity testing, has become a requirement for most discharges in the United States. As of 1988, more than 6000 discharge permits incorporated toxicity limits to protect against acute and chronic toxicity [13] and practically all discharge permits in the United States have toxicity limits as of 2003.

When a discharge exceeds the toxicity limits, the discharger must conduct a toxicity identification/reduction evaluation (TI/RE). A TI/RE is a site-specific investigation of the effluent to identify the causative toxicants that may be eliminated or reduced, or treatment methods that can reduce effluent toxicity.

11.4.2 Resource Conservation and Recovery Act

The Resource Conservation and Recovery Act (RCRA) was enacted in 1976 and was revised substantially by the Hazardous and Solid Waste Amendment (HSWA) of 1984 (40 CFR pts. 260–280). The RCRA regulates the management of "solid wastes" that are "hazardous." The definition of "solid wastes" in these regulations generally encompasses all "discarded" materials (including solid, liquid, semisolid, and contained gaseous materials) and many "secondary materials" (e.g., spent solvents, byproducts) that are recycled or reused rather than discarded [3]. Products such as commercial pesticides are not ordinarily "solid wastes," but they become solid wastes if and when they are discarded or stored, treated, or transported prior to such disposal.

The "solid wastes" that are RCRA hazardous wastes are those either listed in 40 CFR pt. 261, or exhibit one of the four "characteristics" [ignitability, corrosivity, reactivity, and "extraction procedure" (EP) toxicity] identified in Part 261 [a more stringent Toxicity Characteristic Leaching Procedure (TCLP) replaced EP in 1986 (51 Federal Regulation 21,648 1986)]. Both the characteristics and the lists sweep many pesticides and pesticide wastes into the RCRA regulatory program.

The USEPA has developed extensive lists of waste streams (40 CFR Sections 261.31, 261.32) and chemical products (40 CFR Section 261.33) that are considered hazardous wastes if and when disposed of or intended for disposal. The waste streams listed in Sections 261.31 and 261.32 include numerous pesticide manufacturing and formulating process wastes. The lists of commercial chemical products in Section 261.33 include two sublists; both include numerous insecticides, herbicides, and other pesticides. The E List (Table 7) identifies pesticides and other commercial chemicals regulated as "acutely hazardous wastes" when discarded. The F List (Table 8) identifies pesticides that are regulated as toxic (hazardous) wastes when discarded.

Listed pesticides (formulated, manufacturing-use, and off-specification) are regulated as hazardous wastes under the RCRA if they are discarded rather than used for their intended purposes. State listings are often more extensive. Both onsite and offsite disposal options are regulated under the RCRA. Onsite facilities that generate more than 1 kg/month of acutely hazardous wastes in the RCRA E List or 1000 kg/month of any waste as defined in 40 CFR 261.31, 261.32, or 261.33 will require an RCRA hazardous waste permit for treatment or for storage for more than 90 days. Offsite disposal must be handled by an RCRA-permitted facility.

11.5 CONTROL AND TREATMENT FOR PESTICIDE MANUFACTURING WASTES

The management of wastes from pesticide manufacturing plants includes source control, in-plant control/treatment, end-of-pipe treatment, and other control methods for concentrated

Table 7 Pesticide Active Ingredients That Appear on the RCRA Acutely Hazardous Commercial Products List (RCRA E List)

Acrolein	Endrin
Aldicarb	Famphur
Aldrin	Fluoroacetamide
Allyl alcohol	Heptachlor
Aluminum phosphide	Hydrocyanic acid
4-Aminopyridine	Hydrogen cyanide
Arsenic acid	Methomyl
Arsenic pentoxide	Alpha-naphthylthiourea (ANTU)
Arsenic trioxide	Nicotine and salts
Calcium cyanide	Octamethylpyrophosphoramide (OMPA, schradan)
Carbon disulfide	Parathion
p-Chloroaniline	Phenylmercuric acetate (PMA)
Cyanides (soluble cyanide salts)	Phorate
Cyanogen	Potassium cyanide
2-Cyclohexyl-4,6-dinitrophenol	Propargyl alcohol
Dieldrin	Sodium azide
0,0-Diethyl S-[2-ethylthio)ethyl] phosphorodithioate (disulfoton, Di-Syston®)	Sodium cyanide
0,0-Diethyl 0-pyrazinyl phosphorothioate (Zinophos®)	Sodium fluoroacetate
Dimethoate	Strychnine and salts
0,0-Dimethyl 0-p-nitrophenyl phosphorothioate (Methyl parathion)	0,0,0,0-tetraethyl dithiopyrophosphate (sulfotepp)
4,6-Dinitro-o-cresol and salts	Tetraethyl pyrophosphate
4,6-Dinitro-o-cyclohexylphenol	Thallium sulfate
2,4-Dinitrophenol	Thiofanox
Dinoseb	Toxaphene
Endosulfan	Warfarin
Endothall	Zinc phosphide

Note: There are currently no inert pesticide ingredients on the RCRA E List.
Source: 40 CFR 261.33(e).

wastes such as incineration. Source control can reduce the overall pollutant load that must be treated in an end-of-pipe treatment system. In-plant control/treatment reduces or eliminates a particular pollutant before it is diluted in the main wastewater stream, and may provide an opportunity for material recovery. End-of-pipe treatment is the final stage for meeting regulatory discharge requirements and protection of stream water quality. These and other control techniques are discussed in more detail in the following sections.

11.5.1 Source Control

Source control and waste minimization can be extremely effective in reducing the costs for in-plant controls and end-of-pipe treatment, and in some cases can eliminate the need for some treatment units entirely. The first step is to prepare an inventory of the waste sources and continuously monitor those sources for flow rates and contaminants. The next step is to develop in-plant operating and equipment changes to reduce the amount of wastes. The following are some of the techniques available for the pesticides manufacturing facilities.

Table 8 Pesticides and Inert Pesticide Ingredients Contained on the RCRA Toxic Commercial Products List (RCRA F List)

Active ingredients

Acetone
Acrylonitrile
Amitrole
Benzene
Bis(2-ethylhexyl) phthalate
Cacodylic acid
Carbon tetrachloride
Chloral (hydrate)
Chlorodane, technical
Chlorobenzene
4-Chloro-m-cresol
Chloroform
o-Chlorophenol
4-Chloro-o-toluidine hydrachloride
Creosote
Cresylic acid (cresols)
Cyclohexane
Cyclohexanone
Decachlorooctahydro-1,3,4-metheno-2H-cyclobuta[c,d]-pentalen-2-one (Kepone, chlordecone)
1,2-dibromo-3-chloropropane (DBCP)
Dimbutyl phthalate
S-2,3-(Dichloroallyl diisopropylthiocarbamate) (diallate, Avadex)
o-Dichlorobenzene
p-Dichlorobenzene
Dichlorodifluoromethane (Freon 12®)
3,5-Dichloro-N-(1,1-dimethyl-2-propyny1) benzamide (pronamide, Kerb®)
Dichloro diphenyl dichloroethane (DDD)
Dichloro diphenyl trichloroethane (DDT)
Dichloroethyl ether
2,4-Dichlorophenoxyacetic, salts and esters (2,4-D)
1,2-Dichloropropane
1,3-Dichloropropene (Telone)
Diethyl phthalate
Epichlorohydrin (1-chloro-2,3-epoxypropane)
Ethyl acetate
Ethyl 4,4′-dichlorobenzilate (chlorobenzilate)
Ethylene dibromide (EDB)
Ethylene dichloride
Ethylene oxide
Formaldehyde
Furfural
Hexachlorobenzene
Hexachlorocyclopentadiene
Hydrofluoric acid
Isobutyl alcohol
Lead acetate
Lindane
Maleic hydrazide
Mercury
Methyl alcohol (methanol)
Methyl bromide
Methyl chloride
2,2′-Methylenebis (3,4,6-trichlorophenol) (hexachlorophene)
Methylene chloride
Methyl ethyl ketone
4-Methyl-2-pentanone (methyl isobutyl ketone)
Naphthalene
Nitrobenzene
p-Nitrophenol
Pentachloronitrobenzene (PCNB)
Pentachlorophenol
Phenol
Phosphorodithionic acid, 0,0-diethyl, methyl ester
Propylene dichloride
Pyridine
Resorcinol
Safrole
Selenium disulfide
1,2,4,5-Tetrachlorobenzene
1,1,2,2-Tetrachloroethane
2,3,4,6-Tetrachlorophenol
Thiram
Toluene
1,1,1-Trichloroethane
Trichloroethylene
Trichloromonofluoromethane (Freon 11®)
2,4,5-Trichlorophenol
2,4,6-Trichlorophenol
2,4,5-Trichlorophenoxyacetic acid (2,4,5-T)
2,4,5-Trichlorophenoxypropionic acid (Silvex)
Xylene

(*continues*)

Table 8 Continued

Inert ingredients	
Acetone	Formaldehyde
Acetonitrile	Formic acid
Acetophenone	Isobutyl alcohol
Acrylic acid	Maleic anahydride
Aniline	Methyl alcohol (methanol)
Benzene	Methyl ethyl ketone
Chlorobenzene	Methyl methacrylate
Chloroform	Naphthalene
Cyclohexane	Saccharin and salts
Cyclohexanone	Thiourea
Dichlorodifluoromethane (Freon 12®)	Toluene
Diethyl phthalate	1,1,1-Trichloroethane
Dimethylamine	1,1,2-Trichloroethane
Dimethyl phthalate	Trichloromonofluoromethane (Freon 11^R)
1,4-Dioxane	Vinyl chloride
Ethylene oxide	Xylene

Source: 40 CFR 261.33(f).

Waste segregation is an important step in waste reduction. Process wastewaters containing specific pollutants can often be isolated and disposed of or treated separately in a more technically efficient and economical manner. Highly acidic and caustic wastewaters are usually more effectively adjusted for pH prior to being mixed with other wastes. Separate equalization for streams of highly variable characteristics is used by many plants to improve overall treatment efficiency [7].

Wastewater generation can be reduced by general good housekeeping procedures such as substituting dry cleanup methods for water washdowns of equipment and floors. This is especially applicable for situations where liquid or solid materials have been spilled. Flow measuring devices and pH sensors with automatic alarms to detect process upsets are two of many ways to effect reductions in water use. Prompt repair and replacement of faulty equipment can also reduce wastewater losses.

Barometric condenser systems can be a major source of contamination in plant effluents and can cause a particularly difficult problem by producing a high-volume, dilute waste stream [8]. Water reduction can be achieved by replacing barometric condensers with surface condensers. Vacuum pumps can replace steam jet eductors. Reboilers can be used instead of live steam; reactor and floor washwater, surface runoff, scrubber effluents, and vacuum seal water can be reused.

In some cases, wastewater can be substantially reduced by substituting an organic solvent for water in the synthesis and separation steps of the production process, with subsequent solvent recovery. Specific pollutants can be eliminated by requesting specification changes from raw material suppliers in cases when impurities are present and known to be discharged in process wastewaters [7].

Raw material recovery can be achieved through solvent extraction, steam-stripping, and distillation operations. Dilute streams can be concentrated in evaporators and then recovered. Recently, with the advent of membrane technology, reverse osmosis (RO) and ultrafiltration (UF) can be used to recover and concentrate active ingredients [14].

11.5.2 In-Plant Control/Treatment

There are six primary in-plant control methods for removal of priority pollutants and pesticides in pesticide manufacturing plants. These methods include steam-stripping, activated carbon adsorption, chemical oxidation, resin adsorption, hydrolysis, and heavy metals separation. Steam-stripping can remove volatile organic compounds (VOCs); activated carbon can remove semivolatile organic compounds and many pesticides; and resin adsorption, chemical oxidation, and hydrolysis can treat selected pesticides [7]. Heavy metals separation can reduce toxicity to downstream biological treatment systems. Discussion of each of these methods follows.

Steam-Stripping

Steam-stripping is similar to distillation. Steam contacts the wastewater to remove the soluble or sparingly soluble VOCs by driving them into the vapor phase. The steam, which behaves both as a heating medium and a carrier gas, can be supplied as live or reboiled steam. As shown in Fig. 9 [11], a steam-stripping system generally includes an influent storage drum, feed/bottom heat exchangers, pumps, a stripping column (packed column or tray tower), an overhead condenser, an effluent storage drum, and sometimes a reflux drum. Reflux is used to enrich or concentrate the VOCs in the condensate. Enrichment of the condensate could provide higher energy content so that it can be burned for energy recovery [9].

In the pesticide industry, steam-stripping has proven effective for removing groups of priority pollutants such as volatile aromatics, halomethanes, and chloroethanes as well as a variety of nonpriority pollutant compounds such as xylene, hexane, methanol, ethylamine, and ammonia [11]. Thus, this process is used to reduce or remove organic solvents from waste

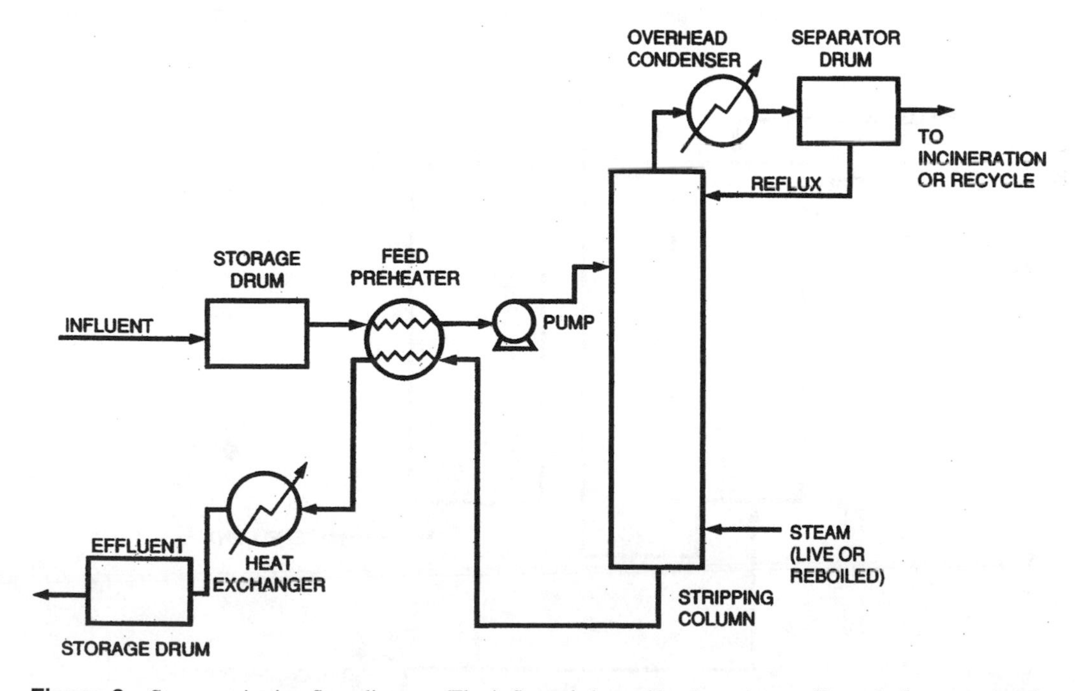

Figure 9 Steam-stripping flow diagram. The influent is heated by the stripper effluent before entering the stripping column near the top; the liquid stream flows downward through the packing, and steam flows upward, carrying volatile compounds; the overhead is condensed and liquid returned to the column; volatile compounds are either recycled or incinerated (from Ref. 11).

streams. A comprehensive study on steam-stripping of organic priority pollutants indicated that effluent concentrations of these pollutants can be reduced to as low as 0.05 mg/L from influent concentrations at their solubility [15]. Pesticides usually have high molecular weights and low volatility and are not effectively removed by steam-stripping.

One variation of steam-stripping is vacuum-stripping, which uses vacuum to create the driving force for pollutant separation. Vacuum strippers normally operate at an absolute pressure of 2 in. of mercury. At least eight pesticide manufacturing plants in the United States use steam-stripping or vacuum-stripping for VOCs removal [7]. The flow rates vary from 0.01 to 0.09 MGD. For example, one pesticide plant uses a steam-stripper to remove methylene chloride from a segregated stream with a flow rate of 0.0165 MGD. The stripper contains 15 ft of packing consisting of 1 in. polypropylene saddles. The steam feed rate is about 1860 lb/hour. Stripped compounds are recycled to the process, thus realizing a net economic savings.

Activated Carbon Adsorption

Activated carbon adsorption is a well-established process for adsorption of organics in wastewater, water, and air streams. Granular activated carbon (GAC) packed in a filter bed or of powdered activated carbon (PAC) added to clarifiers or aeration basins is used for wastewater treatment. In the pesticide industry, GAC is much more widely used than PAC. Figure 10 shows the process flow diagram of a GAC system with two columns in series, which is common in the pesticide industry [11].

Activated carbon studies on widely used herbicides and pesticides have shown that it is successful in reducing the concentration of these toxic compounds to very low levels in wastewater [16]. Some examples of these include BHC, DDT, 2,4-D, toxaphene, dieldrin, aldrin, chlordane, malathion, and parathion. Adsorption is affected by many factors, including

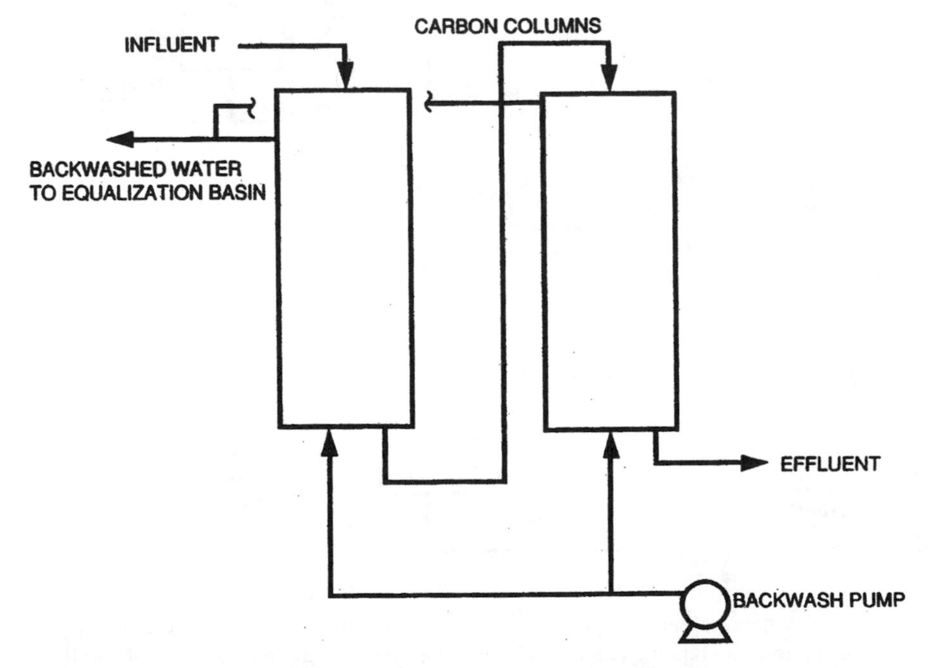

Figure 10 Carbon adsorption flow diagram. The carbon columns are operated in series; backwash water is provided by a pump (from Ref. 11).

molecular size of the adsorbate, solubility of the adsorbate, and pore structure of the carbon. A summary of the characteristics of activated carbon treatment that apply to the pesticide industry follows [11]:

1. Increasing molecular weight is conducive to better adsorption.
2. The degree of adsorption increases as adsorbate solubility decreases.
3. Aromatic compounds tend to be more readily absorbed than aliphatics.
4. Adsorption is pH-dependent; dissolved organics are generally adsorbed more readily at a pH that imparts the least polarity to the molecule.

According to the USEPA surveys, at least 17 pesticide plants in the United States use GAC treatment [7]. Flow rates vary from a low of 0.0004 MGD to a high of 1.26 MGD (combined pesticide flow). Empty bed contact times of the GAC systems vary from a low of 18 minutes to a high of 1000 minutes. The majority of these plants use long contact times and high carbon usage rate systems that are applied as a pretreatment for removing organics from concentrated waste streams. Three plants operate tertiary GAC systems that use shorter contact times and have lower carbon usage rates. Most of the full-scale operating data from the GAC plants indicate a 99% removal of pesticides from the waste streams. The common surface loading rate for primary treatment is 0.5 gallon per minute per square foot (gpm/ft^2) and for tertiary treatment, 4 gpm/ft^2.

Activated carbon adsorption is mainly a waste concentration method. The exhausted carbon must be regenerated or disposed of as hazardous waste. For GAC consumptions larger than 2000 lb/day, onsite regeneration may be economically justified [7]. Thermal regeneration is the most common method for GAC reactivation, although other methods such as washing the exhausted GAC with acid, alkaline, solvent, or steam are sometimes practiced for specific applications [17].

Figure 11 shows a typical flow diagram for a thermal regeneration system [11]. Thermal regeneration is conventionally carried out in a multiple hearth furnace or a rotary kiln at

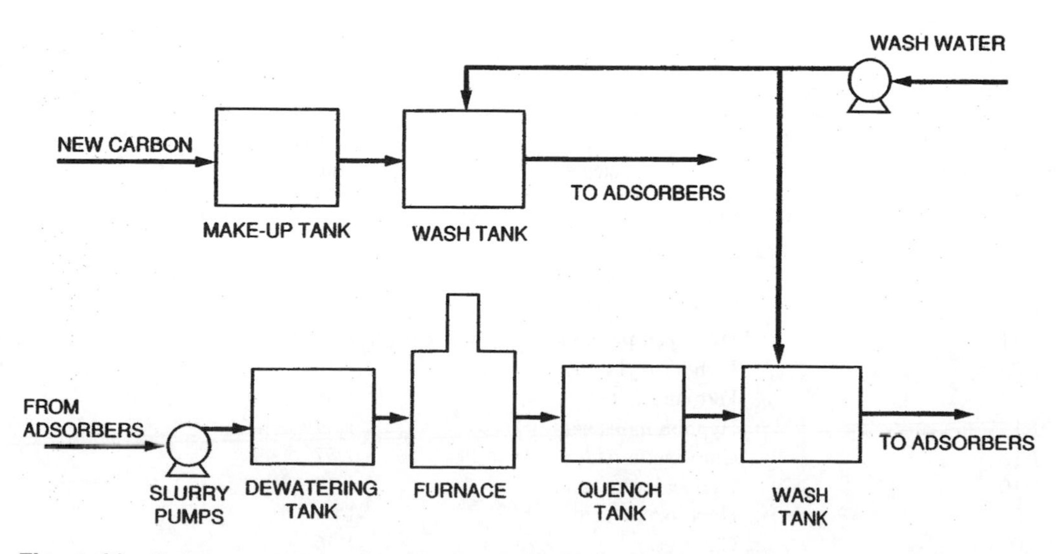

Figure 11 Carbon regeneration flow diagram. Exhausted carbon is sluiced from adsorbers, dewatered, and regenerated in a thermal furnace (multiple hearth, rotary kiln, infrared, or fluidized bed); the regenerated carbon is quenched and washed before returning to the adsorbers; new carbon is washed and added to make up for the loss during regeneration (from Ref. 11).

temperatures from 870 to 980°C. The infrared furnace is a newer type and was installed in a pesticide plant for a GAC system treating mainly aqueous discharge from vacuum filtration of the mother liquor [7]. Infrared furnace manufacturers have claimed ease of operation with quick startup and shutdown capabilities [18]. Another newer type of reactivation process is the fluidized bed process where the GAC progresses downward through the reactivator counterflow to rising hot gases, which carry off volatiles as they dry the spent GAC and pyrolyze the adsorbate. Both the infrared furnace and fluidized bed reactivation processes have been pilot-tested by USEPA in drinking water treatment plants [18].

Other adsorbing materials besides GAC have also been investigated for treating pesticide-containing wastewaters [19]. Kuo and Regan [20] investigated the feasibility of using spent mushroom compost as an adsorption medium for the removal of pesticides including carbaryl, carbofuran, and aldicarb from rinsate. The adsorption of carbamate pesticides on the sorbent exhibited nonlinear behavior that could be characterized by the Freundlich isotherm. Competitive adsorption was observed for pesticide mixtures with adsorption in the order: carbaryl > carbofuran > aldicarb. In another study, Celis and coworkers [21] studied montmorillonites and hydrotalcite as sorbent materials for the ionizable pesticide imazamox. At the pH of the sorbent [6–7], the calcined product of hydrotalcite was found to be the best sorbent for imazamox anion. Sudhakar and Dikshit [22] found that wood charcoal removed up to 95% of endosulfan, an organochlorine insecticide, from water. The sorption followed second-order kinetics with an equilibrium time of 5 hours. In a separate study, pine bark, a wood industry byproduct, was evaluated as an economical adsorbent for tertiary treatment of water contaminated with various organochlorine pesticides [23].

Chemical Oxidation

Oxidizing agents have been shown to be extremely effective for removing many complex organics from wastewater, including phenols, cyanide, selected pesticides such as ureas and uracils, COD, and organo-metallic complexes [11]. Many oxidants can be used in wastewater treatment. Table 9 shows the oxidation potentials for common oxidants [24]. The most widely used oxidants in the

Table 9 Oxidation Potential of Oxidants

Relative oxidation power ($Cl_2 = 1.0$)	Species	Oxidative potential (V)
2.23	Fluorine	3.03
2.06	Hydroxyl radical	2.80
1.78	Atomic oxygen (singlet)	2.42
1.52	Ozone	2.07
1.31	Hydrogen peroxide	1.78
1.25	Perhydroxyl radical	1.70
1.24	Permanganate	1.68
1.17	Hypobromous acid	1.59
1.15	Chlorine dioxide	1.57
1.10	Hypochlorous acid	1.49
1.07	Hypoiodous acid	1.45
1.00	Chlorine	1.36
0.80	Bromine	1.09
0.39	Iodine	0.54

Source: Ref. 24.

pesticide industry are chlorine and hydrogen peroxide (H_2O_2). However, the use of chlorine may create objectionable chlororganics such as chloromethanes and chlorophenols in the wastewater. When organic pollutant concentrations are very high, the use of chemical oxidation may be too expensive because of the high chemical dosages and long retention time required.

At least nine United States pesticide manufacturers use chemical oxidation to treat wastewater [7]. In these systems, more than 98% of cyanide, phenol, and pesticides are removed; COD and other organics are reduced considerably. Some plants use chemical oxidation to reduce toxic compounds from the wastewater to make the streams more suitable for subsequent biological treatment.

Reynolds *et al.* [25] conducted a comprehensive review of aqueous ozonation of five groups of pesticides: chlorinated hydrocarbons, organophosphorus compounds, phenoxyalkyl acid derivatives, organonitrogen compounds, and phenolic compounds. Generally, chlorinated compounds were more resistant to ozonation than the other groups. With the exception of a few pesticides, most of the compounds in the four other groups could achieve complete destruction upon ozonation. The presence of bicarbonate ions could decrease reaction rates by acting as free radical scavengers. Contact times and pH were important parameters. Atrazine destruction by ozonation was evaluated in a bench-scale study in the presence of manganese [26]. Mn-catalyzed ozonation was enhanced in the presence of a small amounts of humic substances (1 mg/L as DOC).

A newer development in chemical oxidation is the combination of ultraviolet (UV) irradiation with H_2O_2 and/or ozone (O_3) oxidation. This combination generates hydroxyl radical, which is a stronger oxidant than ozone or H_2O_2. The UV light also increases the reactivities of the compounds to be oxidized by exciting the electrons of the molecules to higher energy levels [27]. As a result, lower chemical dosages and much higher reaction rates than other oxidation methods can be realized. When adequate chemical dosages and reaction times are provided, pesticides and other organic compounds can be oxidized to carbon dioxide, inorganic salts, and water [28]. Beltran *et al.* [29] evaluated atrazine removal in bubble reactors by treating three surface waters with ozone, ozone in combination with H_2O_2 or UV radiation. Surface water with low alkalinity and high pH resulted in the highest atrazine removal, and ozonation combined with H_2O_2 or UV radiation led to higher atrazine removal and higher intermediates formation as compared to single ozonation or UV radiation.

The UV/O_3 process has been shown to be effective in destroying many pesticides in water [30]. Pilot tests conducted in California on synthetic pesticide wastewaters demonstrated that 15 mg/L each of organic phosphorous, organic chlorine, and carbamate pesticides can be UV-oxidized to nondetectable concentrations [31]. Figure 12 shows a UV/oxidation process flow diagram with the option of feeding both O_3 and H_2O_2. The combination of O_3 and H_2O_2 without UV can also generate the powerful hydroxyl radicals and can result in catalyzed oxidation of organics [32].

The UV/O_3 process was investigated as a pretreatment step to biological treatment by measuring biodegradability (BOD_5/COD), toxicity (ED_{50}), and mineralization efficiency of treated pesticide-containing wastewater [33]. The investigator found that after treatment of an industrial pesticide wastewater by the UV/O_3 process for one hour, COD was reduced by only 6.2% and TOC by merely 2.4%. However, the value of BOD_5/COD increased significantly so that the wastewater was easily biodegradable ($BOD_5/COD > 0.4$) and the toxicity obviously declined (EC_{50} reduction $> 50\%$). The UV light intensity used was 3.0 mW/cm^2 and O_3 supply rate was 400 g/m^3/hour. The investigator concluded that using UV/O_3 as pretreatment for a biological unit is an economical approach to treating industrial wastewaters containing xenobiotic organics as most part of the mineralization work is done by the biological unit rather than photolytic ozonation.

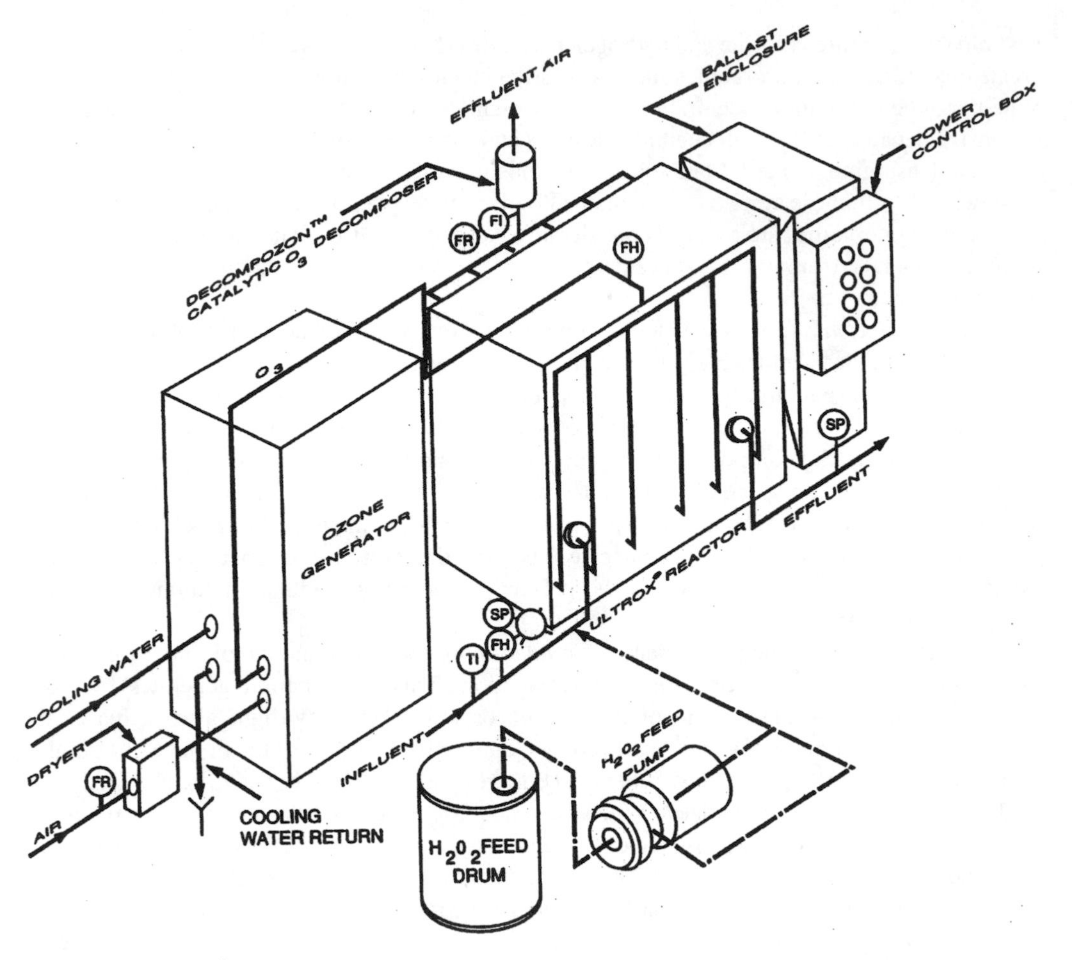

Figure 12 Ultrox® ultraviolet/oxidation process flow schematic. Equipment includes an O_3 generation and feed system and an oxidation reactor mounted with UV lamps inside; H_2O_2 feed is optional. (Courtesy of Ultrox International.)

Balmer and Sulzberger [34] found that the kinetics of atrazine degradation by hydroxyl radicals in photo-Fenton systems were controlled by iron speciation, which further depended upon pH and oxalate concentration. Nguyen and Zahir [35] found that the photodecomposition of the herbicide methyl viologen with UV light was a hemolytic process leading to the formation of methyl pyridinium radicals, which then underwent photolysis at a much faster rate, producing environmentally benign byproducts. In a separate study, Lu [36] investigated the photocatalytic oxidation of the insecticide propoxur, in the presence of TiO_2 supported on activated carbon. Photodegradation of the insecticide followed a pseudo-first-order kinetics described by the Langmuir–Hinshelwood equation. Photocatalytic oxidation of the fungicide metalaxyl in aqueous suspensions containing TiO_2 was explained in terms of the Langmuir–Hinshelwood kinetic model [37].

Resin Adsorption

Adsorption by synthetic polymeric resins is an effective means for removing and recovering specific chemical compounds from wastewater. The operation is similar to that of GAC adsorption. Polymeric adsorption can remove phenols, amines, caprolactam, benzene, chlorobenzenes, and chlorinated pesticides [11]. The adsorption capacity depends on the type and concentration of specific organics in the wastewater as well as pH, temperature, viscosity, polarity, surface tension, and background concentration of other organics and salts. For example, a high salt background will enhance phenol adsorption; increasing the pH will cause the adsorptive capacity to change sharply because the phenolic molecule goes from a neutral, poorly dissociated form at low and neutral pH to an anionic charged dissociated form at high pH [7].

The binding energies of the resin are normally lower than those of activated carbon for the same organic molecules, which permits solvent and chemical regeneration and recovery. Regeneration can be conducted with caustic or formaldehyde or in solvents such as methanol, isopropanol, and acetone. Batch distillation of regenerant solutions can be used to separate and return products to the process.

The USEPA surveys identified four resin adsorption systems in the pesticide industry [7]. Phenol, pesticide, and diene compounds are all effectively removed by these systems. At least one system realized a significant product recovery via regeneration and distillation. The design surface loading rates vary from 1.0 to 4.0 gpm/ft^2 with empty bed contact times of 7.5 to 30 minutes.

Amberlite XAD-4 resin, a synthetic, polymeric adsorbant, was used in one pesticide plant to treat an influent with 1000 mg/L of para-nitrophenol (PNP). With an effluent PNP concentration of 1.0 mg/L, the capacity of the resin was 3.3 lb PNP/cu ft of resin. Kennedy [38] conducted a study regarding the treatment of effluent from a manufacturer of chlorinated pesticides with Amberlite XAD-4 and GAC. Results indicated that the leakage of unadsorbed pesticides from the XAD-4 column was significantly lower than that from the GAC column. An economic analysis indicated that pesticide waste treatment via XAD-4 resin and chemical regeneration would be more economical than GAC adsorption using external thermal regeneration. Chemical regeneration becomes more advantageous because of its feasibility for regenerant recovery and reuse and recycle of adsorbed materials.

Hydrolysis

Hydrolysis is mainly an organic detoxification process. In hydrolysis, a hydroxyl or hydrogen ion attaches itself to some part of the pesticide chemical molecule, either displacing part of the group or breaking a bond, thus forming two or more new compounds. The agents for acid hydrolysis most commonly used are hydrochloric acid and sulfuric acid [11]. Alkaline hydrolysis uses sodium hydroxide most frequently, but the alkaline carbonates are also used. Sometimes high temperature and pressure or catalytic enzymes are required to attain a reasonable reaction time.

Hydrolysis can detoxify a wide range of aliphatic and aromatic organics such as esters, ethers, carbohydrates, sulfonic acids, halogen compounds, phosphates, and nitriles. It can be conducted in simple equipment (in batches in open tanks) or in more complicated equipment (continuous flow in large towers). However, a potential disadvantage is the possibility of forming undesirable reaction products. This possibility must be evaluated in bench- and pilot-scale tests before hydrolysis is implemented.

The primary design parameter to be considered in hydrolysis is the half-life of the original molecule, which is the time required to react 50% of the original compound. The half-life is generally a function of the type of molecule hydrolyzed and the temperature and pH of the reaction. Figure 13 shows the effect of pH and temperature for the degradation of malathion by hydrolysis [11].

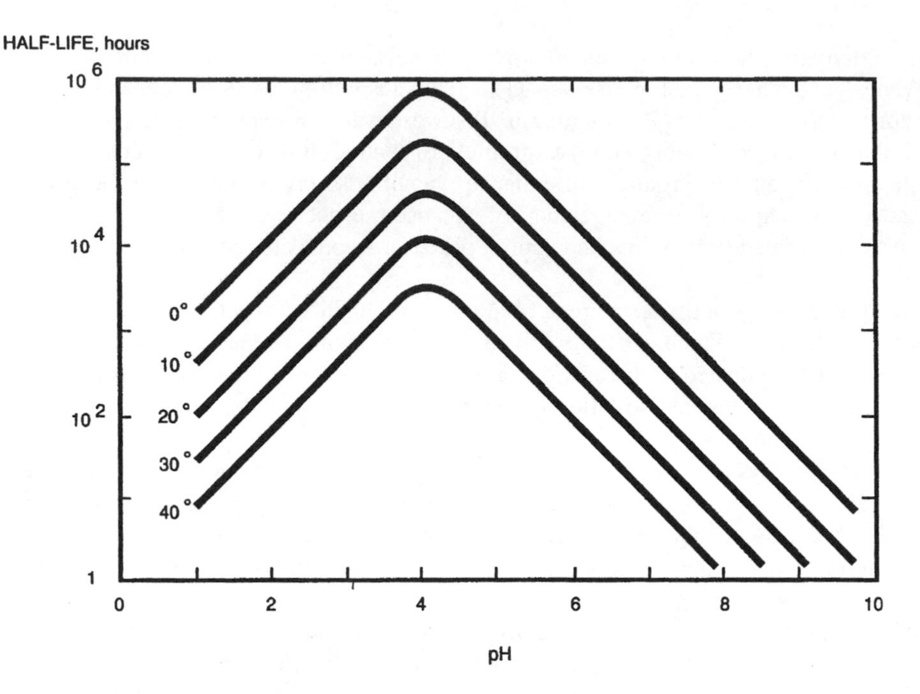

Figure 13 Effect of pH and temperature on malathion degradation by hydrolysis (temperature in degrees C); degradation is faster at higher temperatures and pH values further away from 4.0 to 4.2 (from Ref. 11).

In a study the insecticide carbofuran was hydrolyzed to carbofuran-phenol and monomethyl amine in an anaerobic system. Carbofuran-phenol was resistant to further degradation, while monomethyl amine was further mineralized in the methanogenic culture. Huang and Stone [39] found that the hydrolysis of the secondary amide naptalam, which has a carboxylate side group, was inhibited by dissolved metal ions such as Cu^{2+} and Zn^{2+} and by Al_2O_3 and FeOOH surfaces. In contrast, the hydrolysis of secondary amide propanil and tertiary amide furalaxyl, which lack carboxylate side groups, was unaffected by the presence of Cu^{2+} during the 45-day reaction period. In a separate study, Skadberg *et al.* [40] investigated the stimulation of 2,6-DCP transformation using electric current under varying pH, current, and Cu concentrations. Formation of H_2 at the cathode was found to induce dechlorination with simultaneous removal of Cu.

The USEPA surveys identified nine pesticide plants using full-scale hydrolysis treatment systems [7]. In the industry, a detention time of up to 10 days is used to reduce pesticide levels by more than 99.8%, resulting in typical effluent less than 1 mg/L. The effluents are treated further in biological treatment systems, GAC systems, or chemical oxidation systems, or are discharged to POTWs, if permitted.

Heavy Metals Separation

Metallic ions in soluble form are commonly removed from wastewater by conversion to an insoluble form followed by separation processes such as flocculation, sedimentation, and filtration. Chemicals such as lime, caustic soda, sulfides, and ferrous or ferric compounds have been used for metals separation. Polymer is usually added to aid in flocculation and sedimentation.

For removing low levels of priority metal pollutants from wastewater, using ferric chloride has been shown to be an effective and economical method [41]. The ferric salt forms iron oxyhydroxide, an amorphous precipitate in the wastewater. Pollutants are adsorbed onto and trapped within this precipitate, which is then settled out, leaving a clear effluent. The equipment is identical to that for metal hydroxide precipitation. Trace elements such as arsenic, selenium, chromium, cadmium, and lead can be removed by this method at varying pH values. Alternative methods of metals removal include ion exchange, oxidation or reduction, reverse osmosis, and activated carbon.

At least three pesticide plants use priority pollutant metals separation systems in the United States [7]. One plant uses hydrogen sulfide precipitation to remove copper from its pesticide wastewater. The operating system consists of an agitated precipitator to which the H_2S is added, a soak vessel to which sulfur dioxide is added, a neutralization step using ammonia, and a gravity separation and centrifuging process. Copper is removed from an influent level of 4500 mg/L to 2.2 mg/L.

A second plant uses sodium sulfide for the precipitation of copper from pesticide wastewater. Effluent copper concentration can be lowered to 23 μg/L in this wastewater.

A third plant uses a chemical precipitation step for removing arsenic and zinc from contaminated surface water runoff. Ferric sulfate and lime are alternately added while the wastewater is vacuum-filtered and sludge is contract-hauled. The entire treatment system consists of dual-media filtration, carbon adsorption, ion exchange, chemical precipitation, and vacuum filtration. Sampling results across the entire treatment system indicated that arsenic was reduced from 6.9 to 0.2 mg/L and zinc from 0.34 to 0.11 mg/L.

One caution about metals removal for wastewater with complex organics is that precipitation may be hindered by the formation of soluble metal complexes. Bench- and pilot-scale tests are required for new applications of technology on a particular wastewater stream. Porras-Rodriguez and Talens-Alesson [42] found that flocs resulting from the adsorption of Al^{3+} to lauryl sulfate micelles possessed pollutant-sequestering properties. In studies conducted by these researchers, the pesticide 2,4-D appeared to associate with the micelle-bound Al^{3+} following a Guoy–Chapman–Stern isotherm.

11.5.3 End-of-Pipe Treatment Methods

End-of-pipe treatment methods commonly used in the pesticide industry include equalization, neutralization, biological treatment, and filtration. These methods are discussed as follows.

Equalization

Equalization consists of a wastewater holding vessel or a pond large enough to dampen flow and/or pollutant concentration variation that provides a nearly constant discharge rate and wastewater quality. Capacity is determined by wastewater volume and composition variability. The equalization basin may be agitated or may use a baffle system to prevent short circuiting. Aeration is sometimes needed to prevent septicity. Equalization is used prior to wastewater treatment processes that are sensitive to fluctuation in waste composition or flow, such as biological treatment processes. The recommended detention time for equalization in the pesticide industry is 12 hours prior to pretreatment and 24 hours prior to biological treatment [7].

Neutralization

Neutralization is practiced in the pesticide industry to raise or lower the pH of a wastewater stream to meet discharge requirements or to facilitate downstream treatment. Alkaline wastewater may be neutralized with hydrochloric acid, carbon dioxide, sulfur dioxide, and most commonly, sulfuric acid. Acidic wastewater may be neutralized with limestone or lime slurries, soda ash, caustic soda, or anhydrous ammonia. Often a suitable pH can be achieved through mixing acidic and alkaline process wastewaters. Selection of neutralizing agents is based on cost, availability, safety, ease of use, reaction byproducts, reaction rates, and quantities of sludge formed.

In the pesticide industry, neutralization is provided prior to GAC and resin adsorption, pesticide hydrolysis, and biological treatment. The neutralization basin is sized on the basis of an average retention time of 6 minutes and 70 horsepower per million gallons for mixing requirements [7].

Biological Treatment

Biological treatment processes are widely used throughout the pesticide industry to remove organic pollutants measured by parameters such as BOD and COD. Biological treatment involves using microorganisms (bacteria) under controlled conditions to consume organic matter in the wastewater as their food, making this a useful process for removing certain organic materials from the wastewater. Because the process deals with living organisms, every factor influencing the growth and health of the culture must be considered, including an adequate food supply (organic materials), the availability of proper nutrients (phosphorus and nitrogen), a temperate climate, and a nontoxic and relatively uniform environment free of temperature shocks and similar disturbances.

Biological treatment processes are probably the most cost-effective techniques for treating aqueous waste streams containing organic contaminants [43]. Because of the potential presence of toxic materials in pesticide wastewater that may inhibit biological treatment, physical/chemical processes are usually used before biological treatment. Some materials that may inhibit or interfere with biological treatment are heavy metals, cyanides, chlorinated organic compounds, and high salt content (>20,000 to 30,000 mg/L).

Many pesticides are complex compounds that may not be easily biodegraded. Some factors that affect biodegradability include [44]:

- *Solubility and availability.* Compounds in emulsified or chelated forms are not readily available to microorganisms and are removed slowly. A prime example is DDT and many of its isomers, which are extremely insoluble in water.
- *Molecular size.* The physical size of complex molecules often limits the approach of enzymes and reduces the rate at which organisms can break down the compound. Many pesticide compounds and their isomers are of large and complex structure, making them resistant to degradation. Examples are some carbamates and carboxylic acid-based compounds.
- *Molecular structure.* In general, aliphatic (straight and cyclic) compounds are more degradable than aromatic compounds. Thus some pesticide compounds and parts of some molecules can be degraded easily while other parts cannot. In some cases, partial degradation will occur, but the pesticidal activity of the waste stream may not be reduced significantly if the toxic components of the compound are bioresistant.
- *Substitutions.* The substitution of elements other than carbon in the molecular chain often make the compound more resistant. Esters and epoxides, salts, and so on are more resistant than the base pesticidal compound.
- *Functional groups.* Halogen substitution to an aromatic compound renders it less degradable. The number of substitutions and the location are important. Chlorophenols are an excellent example of increasing resistance with increasing substitution. Amino and hydroxyl substitutions often increase degradability.

Under proper conditions, biological treatment effectively removes priority pollutants, nonconventional pollutants (TOC), and conventional pollutants (e.g., BOD). The mechanism of pollutant removal may be one or more of the following: (a) biological degradation of the pollutant, (b) adsorption of the pollutant onto sludge, which is separately disposed, or (c) volatilization of the pollutant into the air.

The USEPA [7] surveys identified 31 pesticide plants using biological treatment processes to treat wastewater, including: (a) 14 aerated lagoon systems with detention times ranging from approximately 2 days to 95 days, (b) 13 activated sludge systems with detention times from 7.15 hours to 79 hours, and (c) four trickling filter systems. Biochemical oxygen demand (BOD) removals ranging from 87.4 to 98.8% were achieved at major industry biological treatment systems. Chemical oxygen demand (COD) removals at these plants ranged from 60.5 to 89.7%.

Removal of priority pollutant phenols reached more than 90% when influent levels were high (i.e., 60–1000 mg/L). However, if influent levels were at approximately 1 mg/L or less, the removal ranged from 4.5% or less to 97.6%. Cyanide removals reached more than 50% when influent levels were more than 1 mg/L, and less than 50% for raw wasteloads less than 1 mg/L. Volatile priority pollutants were removed from more than approximately 1 mg/L down to their detection limits of 0.005 to 0.01 mg/L.

Approximately 50% of priority pollutant metals, including copper and zinc, were removed at influent concentrations of 1 mg/L or less. These metals were adsorbed onto sludge because they were not volatile or biodegradable. Priority pollutant dienes were not expected to be biodegraded or volatilized due to their relatively low solubility, but like metals, will adsorb on sludge. Pesticides were removed in biological systems to varying degrees, based on the characteristics of the individual compound. Some plants were achieving removals in excess of 50% at influent concentrations of approximately 1 mg/L.

One concern about biological treatment is the potential toxicity of pesticides, which could inhibit microorganism growth. Results from bench-scale treatability studies performed by one pesticide plant showed that a pesticide in concentrations up to 3000 mg/L did not inhibit aerobic degradation of sewage at typical aerator food-to-microorganism ratios [7].

Chemagro Corporation used an activated sludge system to degrade wastewater with a mixture of organophosphates such as guthion, meta-systox, coumaphos, and fenthion [45]. The system used a first-stage activated sludge process to absorb some of the shock this waste had on the microbes and a second-stage activated sludge process for the ultimate degradation. Atkins [44] reported that using a trickling filter ahead of an activated sludge was successful for the treatment of 2,4-D waste. The trickling filter can handle large pulse loads of toxics because the contact time is short, and thus it is a good pretreatment device for biologically degradable toxic wastes.

A sequential batch reactor (SBR) is a variation of activated sludge using the sequential steps of fill, aerate, settle, and withdrawal. It has become more popular recently because of the advances in instrumentation and automatic control technology. Mangat and Elefsiniotis [46] examined the biodegradation of herbicide 2,4-D by SBRs. After four months of acclimation, they obtained more than 99% removal of 2,4-D at steady-state operation. In addition, they demonstrated that the removal rates of 2,4-D were influenced by the types of supplemental substrates (phenol and dextrose). A sequential batch biofilm reactor (SBBR) incoculated with *Agrobacterium radiobacter* strain J14a was used to treat formulated atrazine rinsate from an agricultural chemical formulation facility [47]. The SBBR reduced 30 mg/L of atrazine to less than 1 mg/L within 12 hours at 22°C. Galluzo and co-workers [48] found that aerobic cometabolic biodegradation of atrazine was responsible for the removal of 30–35% of the pesticide in continuous flow packed columns with humic acid and peptone–tryptone–yeast extract–glucose as the primary substrate.

The use of PAC in combination with activated sludge (PACT®) is a biophysical process that has high potential to improve biological treatment for pesticide wastewater due to the adsorption of toxic compounds by activated carbon. Although the PACT® process has not been widely used on a full-scale basis in the pesticide industry, this trend is expected to change when regulatory agencies adopt more stringent discharge requirements. One particular requirement is whole effluent toxicity reduction, which has become a criterion for many direct discharges in the United States since the late 1980s [13].

Wong and Maroney [49] reported on a pilot plant comparison of PACT® and extended aeration (activated sludge) for treating petroleum refinery wastewater. Results indicated that although both processes performed similarly in COD removal, only the PACT® system yielded an effluent meeting the discharge requirements for whole effluent toxicity reduction. Similar results in toxicity reduction have been reported for wastewaters from other industries [50].

Another important advance in biological treatment is the use of membrane bioreactors (MBR). The MBR uses a UF membrane system immersed in an aeration tank to accomplish both biochemical oxidation and excellent solid/liquid separation, without the need for a separate clarification step [51]. The MBR can offer reliability, compactness, and excellent treated water quality as the MBR effluent is of UF permeate quality. It is particularly attractive in situations when long solids retention times are required, such as for biological degradation of pesticide ingredients. Successful treatment of herbicide-contaminated wastewater in an MBR by sulfate-reducing consortia was reported by Gonzalez-Gonzalez *et al.* [52].

Filtration

Multimedia filtration has been used in the pesticide industry for two purposes: (a) suspended solids removal prior to activated carbon or resin adsorption applied as a pesticide removal pretreatment, and (b) tertiary polishing after biological oxidation and before tertiary activated carbon treatment. With more stringent discharge requirements adopted by regulatory agencies, the increasing use of filtration for tertiary polishing is widespread in every industry, including the pesticide industry. The common design criteria for dual- or multimedia filtration for this industry include a surface loading rate of 4 gpm/ft^2 with a run length of 12 hours [7]. With the advent of membrane technologies, low-pressure membrane processes such as UF and microfiltration (MF) are replacing granular media in certain applications, especially when water reuse is practised. The effluent turbidity from UF/MF system is usually below 0.1 nephelometric turbidity unit (NTU) as these membranes exclude particles larger than 0.01–0.1 micron [53].

11.5.4 Control Methods for Concentrated Wastes

The pesticide industry generates many concentrated wastes that are considered hazardous wastes. These wastes must be detoxified, pretreated, or disposed of safely in approved facilities. Incineration is a common waste destruction method. Deep well injection is a common disposal method. Other technologies such as wet air oxidation, solvent extraction, molten-salt combustion, and microwave plasma destruction have been investigated for pesticide waste applications.

Incineration

Incineration is an established process for virtually complete destruction of organic compounds. It can oxidize solid, liquid, or gaseous combustible wastes to carbon dioxide, water, and ash. In the pesticide industry, thermal incinerators are used to destroy wastes containing compounds such as hydrocarbons (e.g., toluene), chlorinated hydrocarbons (e.g., carbon tetrachloride), sulfonated solvents (e.g., carbon disulfide), and pesticides [7]. More than 99.9% pesticide removal, as well as more than 95% BOD, COD, and TOC removal, can be achieved if sufficient temperature, time, and turbulence are used.

Sulfur- and nitrogen-containing compounds will produce their corresponding oxides and should not be incinerated without considering their effects on air quality. Halogenated hydrocarbons not only may affect air quality but also may corrode the incinerator. Also, organometallic compounds containing cadmium, mercury, and so on, are not recommended for incineration because of the potential for air and solid waste contamination.

Many types of incinerators may be used for thermal destruction of hazardous wastes, including the following basic types [54]:

- multiple hearth;
- fluidized bed;
- liquid injection;
- fume;
- rotary kiln;
- multiple chamber;
- cyclonic;
- auger combustor;
- ship-mounted.

Each of these incinerators has advantages and disadvantages that must be evaluated before final process selection. Figure 14 shows a typical flow diagram of an incineration system incorporating any of these incinerators [11]. Residence times and operating temperature ranges for the various types of incinerators are listed in Table 10 [54]. A matrix matching waste types against incineration equipment is presented in Table 11 [54]. This matrix offers a general guideline for using different types of incinerators for different wastes (e.g., solid, liquid, and fume).

In addition to using the proper type of incinerator and operating conditions to destroy the pesticide wastes, the incineration system must be equipped with the proper emission controls to ensure that toxic gases and particulates do not escape into the environment [55]. The ash (which may contain hazardous substances) must be properly disposed. Many wet collection systems (scrubbers) can be used for removing gaseous pollutants. The various types of scrubbers available include venturi, plate, packed tower, fiber bed, spray tower, centrifugal, moving bed, wet cyclone, self-induced spray, and jet. Dry collection equipment is available for the removal of particulate pollutants and includes settling chambers, baffle chambers, skimming chambers, dry cyclones, impingement collectors, electrostatic precipitators, and fabric filters. The incinerator ash, scrubber water, and particulate collection can then be landfilled, chemically treated, or otherwise processed for disposal.

The USEPA surveys identified at least 14 pesticide plants using incineration for flows ranging up to 39,000 gal/day and heat capacities up to 77 million Btu/hour [7]. Many incinerators are devoted entirely for the destruction of pesticide wastes, but in some cases, only a small part of the capacity is devoted for this purpose.

As an example of incinerator use in the pesticide industry, one plant operates two incinerators to dispose of wastewater from six pesticide products [7]. They are rated at heat release capacities of 35 and 70 million Btu/hour and were designed to dispose of two different wastes. The first primary feed stream consists of approximately 95% organics and 5% water. The second stream consists of approximately 5% organics and 95% water. The energy generated in burning the primary stream is anticipated to vaporize all water in the secondary stream and to oxidize all the organics present. Wastes from two of the six pesticide processes use 0.55% and 4.68% of the incinerator capacity, respectively. The volume of the combined pesticide

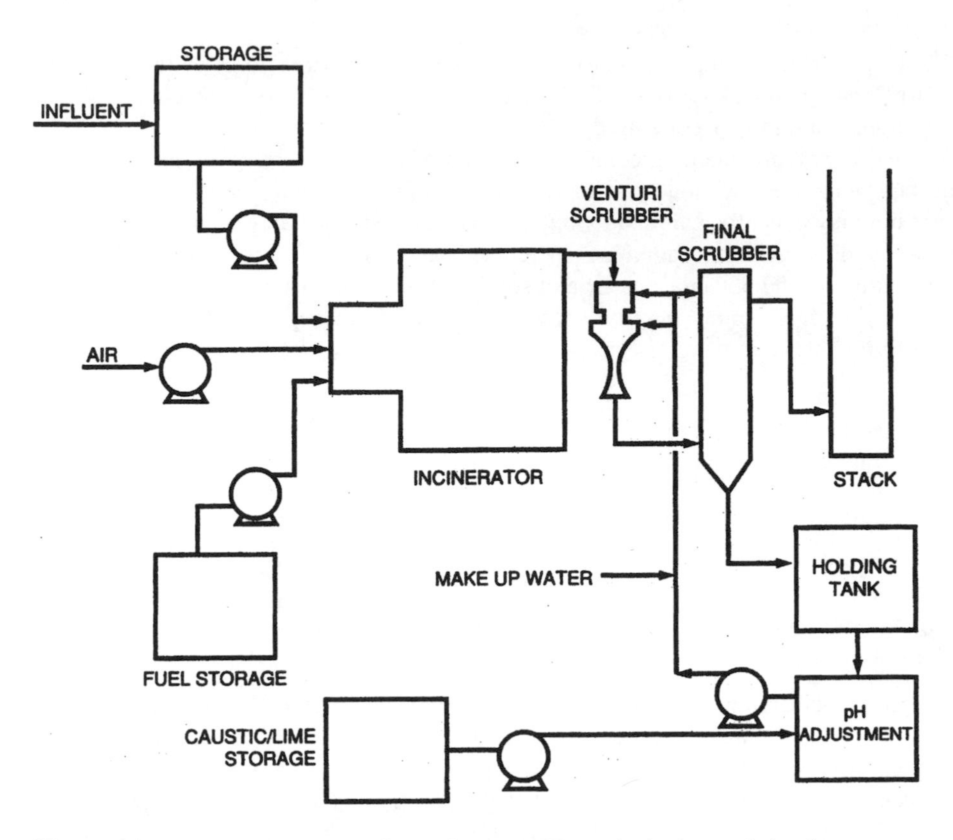

Figure 14 Incineration system flow diagram. Waste is incinerated in the presence of air and supplemental fuel; the incinerator can be multiple hearth, fluidized bed, liquid injection, rotary kiln, or other types; caustic or lime scrubbers are used to remove gaseous pollutants from exhaust gases (from Ref. 11).

wastewater incinerated is 0.0074 MGD. The scrubber effluent is discharged from the tertiary treatment system at a rate of 0.992 MGD.

Since 1974, the USEPA has conducted many incineration tests for pesticide destruction. Most pesticides tested were capable of being destroyed to an efficiency of more than 99.99%. The only exception was Mirex, with 98–99% destruction. However, investigators felt that destruction could be improved to the 99.99% level with a somewhat more effective incinerator design. Incineration has become very controversial in recent years because of the potential to generate dioxin under high temperature conditions.

Deep Well Disposal

Deep well disposal involves injecting liquid wastes into a porous subsurface stratum that contains noncommercial brines [57]. The wastewaters are stored in sealed subsurface strata isolated from groundwater or mineral resources. Disposal wells may vary in depth from a few hundred feet (100 m) to 15,000 ft (4570 m), with capacities ranging from less than 10 to more than 2000 gpm. The disposal system consists of the well with high-pressure injection pumps and pretreatment equipment necessary to prepare the waste for suitable disposal into the well.

Table 10 Operating Parameters for Incinerators

Incinerator type	Temperature range (°F)	Residence time
Multiple hearth	1400–1800	0.25–1.5 hour
Fluidized bed	1400–1800	Seconds–hours
Liquid injection	1800–3000	0.1–2 s
Fume	1400–3000	0.1–2 s
Rotary kiln	1500–3000	Liquids and gases: seconds Solids: hours
Multiple chamber	1000–1800	Liquids and gases: seconds Solids: minutes
Cyclonic	1800–3000	0.1–2 s
Auger combustor	1400–1800	Seconds–hours
Ship-mounted	1800–3000	0.1–2 s

Source: Ref. 54.

In the United States, injection wells are classified into three categories: Class 1 wells are used to inject hazardous wastes; Class 2 wells are used to inject fluids brought to the surface in connection with the production of oil and gas or for disposal of salt water; and Class 3 covers solution mining wells [58]. Class 1 wells are heavily regulated by the USEPA and state agencies because of the potential for groundwater contamination.

The USEPA surveys identified 17 pesticide plants using deep well injection for the disposal of wastewater [7]. One plant used incinerators to remove pesticides as well as benzene and toluene from the wastewater before disposal by deep well injection. Using deep well injection to dispose of hazardous wastes is expected to decrease in the future because of more stringent regulatory requirements and increased concerns about the long-term fate of these wastes in the injection zone.

Other Technologies

Other potential technologies that can be applied to the treatment of concentrated wastes from pesticide manufacturing include wet air oxidation, solvent extraction, molten salt combustion, and microwave plasma destruction.

Table 11 Matrix of Incinerator Application

	Waste					
Incinerator	Solid	Sludge	Slurry	Liquid	Fume	Containerized
Multiple hearth	✓	✓	✓	✓		
Fluidized bed	✓	✓	✓	✓	✓	
Liquid injection			✓	✓	✓	
Fume					✓	
Rotary kiln	✓	✓	✓	✓	✓	✓
Multiple chamber	✓	✓	✓	✓	✓	✓
Cyclonic		✓	✓	✓	✓	
Auger combustor	✓	✓	✓	✓	✓	
Ship-mounted		✓	✓	✓		

Source: Ref. 54.

Wet air oxidation (WAO) is a liquid phase oxidation and/or hydrolysis process at elevated temperature (175–345°C) and pressure (300–3000 psig) in the presence of oxygen. The WAO process can be used as a pretreatment step to destroy toxics or substantially reduce organics before using other conventional treatment processes. When raw waste loads reach a level of 20,000–30,000 mg/L COD, the process becomes thermally self-sustaining [7]. Phenols, cyanides, nitrosoamines, dienes, and pesticides have been shown to be effectively removed by WAO. Zimpro [59] reported that in pilot-plant tests using WAO, a wastewater composite of about 40 pesticides showed a 99+% pesticide destruction and 85% COD reduction. Another investigation indicated that the pesticide Amiben was degraded by 88% to 99.5% and atrazine by 100% [60].

The use of solvent extraction as a unit process operation is common in the pesticide industry; however, it is not widely practised for removing pollutants from waste effluents. Solvent extraction is most effectively applied to segregated process streams as a roughing treatment for removing priority pollutants such as phenols, cyanide, and volatile aromatics [7]. One pesticide plant used a full-scale solvent extraction process for removing 2,4-D from pesticide process wastewaters. As a result, 2,4-D was reduced by 98.9%, from 6710 mg/L to 74.3 mg/L.

Molten-salt combustion is a process by which hazardous materials can be oxidized below the surface of a salt or salt mixture in the molten state [27]. Molten sodium carbonate and a molten mixture of sodium carbonate and sodium sulfate (90 : 10, w/w) have been used. Operating temperatures range from 800 to 1000°C. Hazardous materials and air are fed into the combustion chamber below the surface of the melt. Generally, the heat produced during oxidation is adequate for maintaining the salt or salts in the molten state. Oxidation products include carbon dioxide, steam, and elemental gases such as nitrogen. Hydrochloric acid and sulfur dioxide, which form during the oxidation of chlorine- and sulfur-bearing compounds, react with sodium carbonate. In bench-scale tests, pesticides such as DDT, chlordane, and 2,4-D had more than 99.9% destructions [60].

In microwave plasma destruction, organic material is channeled through a plasma detector tube where destruction is initiated by microwave radiation-producing electrons. The electrons react with the organic molecules to form free radicals and final simple reaction products such as SO_2, CO_2, CO, H_2O, HPO_3, $COCl_2$, and Br_2 [60]. In bench-scale tests, the plasma method resulted in extensive detoxification (>99% destruction) for several pesticides, including malathion, phenylmercuric acetate (PMA), and Kepone® [55].

11.6 CONTROL AND TREATMENT FOR PESTICIDE FORMULATING WASTES

Management of wastes in pesticide formulating/packaging plants is simpler than in manufacturing plants because the volumes are much smaller. In the past, evaporation was the predominant disposal technique for wastewater generated in formulating plants [8]. However, due to concerns over air pollution and other nonwater quality environmental impacts, evaporation is currently not favored. Contract hauling for incineration is the recommended disposal method for small volumes of concentrated wastes, which are handled and transported as hazardous wastes in compliance with RCRA regulations.

For high-flow plants, treatment and recycle/reuse of the wastewater is recommended. The physical/chemical treatment processes used in pesticide manufacturing plants can be applied to formulating plants. The USEPA evaluated treatment and recycle technology for four plants that discharge high volumes of formulating/packaging wastewater [7]. These plants confirmed that treatment and recycle technology are feasible in their facilities but identified selected production processes that are not amenable to reuse. Such processes demand high purity source water to guarantee product integrity. The water volume requirements are low; therefore, wastewater from these processes is contract hauled and incinerated.

One of the four plants presently treats and reuses 75% of its waste stream as vent scrubber washwater. A second plant incinerates formulating/packaging process waste and discharges incinerator blowdown that contains levels of pesticides measured as not detected.

Waste reduction/minimization have also been evaluated and practised by the pesticide formulating/packaging industry [61]. Some techniques include using high-pressure spray nozzles to wash tanks and to clean production floors and other equipment, which can reduce wastewater volumes by at least 50%. Some plants use storage tanks to hold wash liquids (water or solvents) to be used for makeup purposes when the same product is formulated again [62]. This procedure reduces the total quantity of washwater discharged and minimizes product loss. Other techniques used in the pesticide manufacturing plants can also be applied in formulating/packaging plants and are not repeated here.

REFERENCES

1. Ware, G.W. *Pesticides: Theory and Application*; W. H. Freeman and Company; San Francisco, 1983.
2. National Research Council. The *Future Role of Pesticides in U.S. Agriculture*; The National Academies Press: Washington, DC, 2000.
3. McKenna, Conner & Cuneo. *Pesticide Regulation Handbook*, Revised Edition; Executive Enterprises Publications Co., Inc.: New York, 1987.
4. Worldwatch Institute. *Vital Signs 2002*; Worldwatch Institute: Washington, DC, 2003.
5. Agrow Reports. *World Non-Agricultural Pesticide Market*; PJB Publications: London, 2000.
6. Coble, H.D. *Pesticide, The World Book Encyclopedia*; World Book, Inc.: Chicago, 1989; 317–318.
7. USEPA. *Development Document for Effluent Limitations Guidelines, and Standards for the Pesticide Point Source Category*, Report EPA 440/1-85-079; USEPA, 1985.
8. USEPA. *Development Document for Interim Final Effluent Limitations Guidelines for the Pesticide Chemical Manufacturing Point Source Category*, Report EPA 440/1-75/060-d; USEPA, 1976.
9. Wong, J.M. Pesticides Wastewater Management. Paper presented at the *3rd Annual Hazardous Materials Management Conference and Exhibition/West*, Long Beach, California, 1987.
10. Dryden, F.E. *et al. Assessment of Dioxin-Forming Chemical Processes*, prepared for USEPA IERL, Cincinnati, OH; Walk, Haydel and Associates, Inc.: New Orleans, Louisiana, 1979.
11. USEPA. *Development Document for Effluent Limitations Guidelines and Standards for the Pesticides Point Source Category*, Report EPA 440/1-82/079-b; USEPA, 1982.
12. USEPA. *Development Document for Best Available Technology and New Source Performance for the Pesticide Chemical Industry, Proposed*, Report EPA/821-R-92.005; USEPA, 1992.
13. Thomas, N.A. Use of biomonitoring to control toxics in the United States. Wat. Sci. Technol. **1988**, *20*, 10.
14. World Bank Group. *Pollution Prevention and Abatement Handbook – Pesticides Manufacturing*; World Bank Group, July 1998.
15. Hwang, S.T.; Fahrenthold, P. Treatability of the organic priority pollutants by steam stripping. Am. Inst. Chem. Engrs. Symposium Series, Water–1979 **1980**, *197* (*76*): 37–60.
16. Becker, D.L.; Wilson, S.C. The use of activated carbon for the treatment of pesticides and pesticide waste. In *Carbon Adsorption Handbook*; Cheremisinoff, P.N., Ellenworth, F., Eds., Ann Arbor Science: Ann Arbor, Michigan, 1978.
17. Lyman, W.J. Applicability of carbon adsorption to the treatment of hazardous industrial wastes. In *Carbon Adsorption Handbook*; Cheremisinoff, P.N., Ellenworth, F., Eds.; Ann Arbor Science: Ann Arbor, Michigan, 1978.
18. Clark, R.M.; Lykins, B.W., Jr. *Granular Activated Carbon Design, Operation and Cost*; Lewis Publishers: Chelsea, MI, 1989.

19. Bhandari, A.; Xia, K.; Starrett, S.K. Pesticides and herbicides. Wat. Environ. Res. Literature Review, **2000**, *72*, 5.
20. Kuo, W.S.; Regan, R.W., Sr. Removal of pesticides from rinsate by adsorption using agricultural residuals as medium. J. Environ. Sci. Health, **1999**, *B34*, 431.
21. Celis, R.; Koskinen, W.C.; Cecchi, A.M.; Bresnahan, G.A.; Carrisoza, M.J.; Ulibarri, M.A.; Pavlovic, I.; Hermosin, M.C. Sorption of the ionizable pesticide Imazamox by organo-clays and organohydrotalcites. J. Environ. Sci. Health, **1999**, *B34*, 929.
22. Sudhakar, Y.; Dikshit, A.K. Kinetics of endosulfan sorption on two wood charcoal. J. Environ. Sci. Health, **1999**, *B34*, 587.
23. Bras, I.P.; Santos, L.; Alves, A. Organochlorine pesticides removal by pinus bark sorption. Environ. Sci. Technol., **1999**, *33*, 631.
24. Hager, D.G.; Smith, C.E. The UV–hydrogen peroxide process: an emerging technology for groundwater treatment. Paper presented at *HazMat West 85*, Long Beach, California, 1985.
25. Reynolds, G.; Graham, N.; Perry, R.; Rice, R.B. Aqueous ozonation of pesticides: A review. Ozone Sci. Engrg. **1989**, *11* (*4*), 339–382.
26. Ma, J.; Graham, N.J.D. Degradation of atrazine by manganese-catalyzed ozonation: influence of humic substances. Wat. Res. **1999**, *33*, 785.
27. Tucker, S.P.; Carson, G.A. Deactivation of hazardous chemical wastes. Environ. Sci. Technol. **1985**, *19* (*3*), 215–220.
28. Zeff, J.D. New developments in equipment for detoxifying halogenated hydrocarbons in water and air. Paper presented at the *Halogenated Solvents Alliance Meeting*, San Francisco, September, 1985.
29. Beltran, F.J.; Rivas, J.; Acedo, B. Atrazine removal by ozonation processes in surface waters. J. Environ. Sci. Health, **1999**, *B34*, 229.
30. Mauk, C.E.; Prengle, H.W.; Payne, N.E. *Oxidation of Pesticides by Ozone and Ultraviolet Light*; Houston Research Inc.: Houston, TX, 1976.
31. Ultrox International. UV/ozone treatment of pesticide and groundwater, prepared for Department of Health Services, California, Grant No. 85-00169, 1988.
32. Glaze, W.H.; Kang, J.W.; Chapin, D.H. The chemistry of water treatment processes involving ozone, hydrogen peroxide and ultraviolet radiation. Ozone Sci. Engrg. **1987**, *9* (*4*), 335.
33. Kuo, W.S. Effects of photolytic ozonation on biodegradability and toxicity of industrial wastewater. Environ. Sci. Health **1999**, *A34* (*4*), 919–933.
34. Balmer, M.E.; Sulzberger, B. Atrazine degradation in irradiated iron/oxalate systems: Effect of pH and oxalate. Environ. Sci. Technol. **1999**, *33*, 2418.
35. Nguyen, C.; Zahir, K.O. UV induced degradation of herbicide methyl viologen: kinetics and mechanism and effect of ionic media on degradation rates. J. Environ. Sci. Health, **1999**, *B34*, 1.
36. Lu, M. Photocatalytic oxidation of Propoxur insecticide with titanium dioxide supported on activated carbon. J. Environ. Sci. Health **1999**, *B34*, 207.
37. Topalov, A.; Molnar-Gabor, D.; Csnadi, J. Photocatalytic oxidation of the fungicide Metalaxyl dissolved in water over TiO_2. Wat. Res. **1999**, *33*, 1372.
38. Kennedy, D.C. Treatment of effluent from manufacture of chlorinated pesticides with a synthetic, polymeric adsorbent, Amberlite XAD-4. Environ. Sci. Technol. **1973**, *7* (*2*), 138–141.
39. Huang, C.; Stone, A.T. Hydrolysis of naptalam and structurally related amides: inhibition by dissolved metal ions and metal hydroxide surfaces. J. Agric. Food Chem. **1999**, *47*, 4425.
40. Skadberg, B.; Geoly-Horn, S.L.; Sangamalli, V.; Flora, J.R.V. Influence of pH, current, and copper on the biological dechlorination of 2,6-dichlorophenol in an electrochemical cell. Wat. Res. **1999**, *33*, 1997.
41. Merrill, D.T.; Manzione, M.A.; Peterson, J.J.; Parker, D.S.; Chow, W.; Hobbs, A.D. Field evaluation of arsenic and selenium removal by iron coprecipitation. J. WPCF **1986**, *58*, 18–26.
42. Porras-Rodriguez, M.; Talens-Alesson, F.I. Removal of 2,4-dichlorophenoxyacetic acid from water from adsorptive micellar flocculation. Environ. Sci. Technol. **1999**, *33*, 3206.
43. Kiang, Y.H.; Metry, A.A. *Hazardous Waste Processing Technology*; Ann Arbor Science: Ann Arbor, MI, 1982.

44. Atkins, P.R. *The Pesticide Manufacturing Industry – Current Waste Treatment and Disposal Practices*, USEPA Project No. 12020 FYE, 1972.
45. Lue-Hing, C.; Brady, S.D. Biological treatment of organic phosphorus pesticide wastewaters. Purdue Univ. Eng. Ext. Servo 132 (pts. 1/2), 1968;1166–1177.
46. Mangat, S.S.; Elefsiniotis, P. Biodegradation of the herbicide 2,4-dichlorophenoxyacetic acid (2,4-D) in sequencing batch reactor. Wat. Res. **1999**, *33*, 861.
47. Portzman, R.S.; Lee, P.H.; Ong, S.K.; Moorman, T.B. Treatment of formulated Atrazine rinsate by agrobacterium radiobactor strain J14a in a sequencing batch biofilm reactor. Wat. Res. **1999**, *33*, 1399.
48. Galluzo, M.J.; Banrji, S.K.; Bajpai, R.; Surampalli, R.Y. Atrazine removal through biofiltration. Practice Periodical of Hazardous, Toxic, and Radioactive Waste Manage. **1999**, *3*, 163.
49. Wong, J.M.; Maroney, P.M. Pilot plant comparison of extended aeration and PACT® for toxicity reduction in refinery wastewater. *Proceedings, 44th Purdue Industrial Waste Conference*, West Lafayette, IN, 1989.
50. Zimpro, Inc. CIBA-GEIGY Meeting Tough Bioassay Test. Reactor, June 1986; 13–14.
51. Stephenson, T.; Judd, S.; Jeferson, B.; Brindle, K. *Membrane Bioreactors for Wastewater Treatment*; IWA Publishing: London, 2000.
52. Gonzalez-Gonzalez, L.R.; Buenrostro-Zagal, J.F.; Luna-Martinez, A.D.; Sandoval-Gomez, Y.G.; Schettino-Bermudez, B.S. Treatment of an herbicide-contaminated wastewater in a membrane bioreactor by sulfate-reducing consortia. *Proceedings, 7th International Symposium on In Situ and On-Site* Bioremediation, Orlando, Florida, June 2–5, 2003.
53. Wong, J.M. Technologies and case histories for industrial wastewater recovery and reuse. *Proceedings, Seminar on Application of Biotechnology in Industrial Wastewater Treatment and Reuse*, Kaohsiung, Taiwan, ROC, October 30, 2002.
54. Kiang, Y.H.; Metry, A.A. *Hazardous Waste Processing Technology*; Ann Arbor Science: Ann Arbor, MI, 1982.
55. Dillon, A.P. *Pesticide Disposal and Detoxification Processes and Techniques*; Noyes Data Corporation: Park Ridge, NJ, 1981.
56. Oberacker, D.A. Incineration options for disposal of waste pesticides. In *Pesticide Waste Disposal Technology*; Bridges, J.S., Dempsey, C.R., Eds.; Noyes Data Corporation: Park Ridge, NJ, 1988.
57. Eckenfelder, W.W., Jr. *Industrial Water Pollution Control*, 2nd ed.; McGraw-Hill: New York, 1989.
58. McNally, R. Tougher rules challenge future for injection wells. Petrol. Eng. Int. **1987**, *July*, 28–30.
59. Zimpro, Inc. *Report on Wet Air Oxidation for Pesticide Chemical Manufacturing Wastes*, prepared for G. M. Jett, USEPA; Rothchild: Wisconsin, 1980.
60. Honeycutt, R.; Paulson, D.; LeBaron, H.; Rolofson, G. Chemical treatment options for pesticide wastes disposal. In *Pesticide Waste Disposal Technology*; Bridges, J.S., Dempsey, C.R., Eds.; Noyes Data Corporation: Park Ridge, NJ, 1988.
61. Lewis, D.A. Waste minimization in the pesticide formulation industry. J. APCA **1988**, *38*; (*10*), 1293–1296.
62. World Bank Group. *Pollution Prevention and Abatement Handbook – Pesticides Formulation*; World Bank Group, July 1998.

12

Treatment of Rubber Industry Wastes

Jerry R. Taricska
Hole Montes, Inc., Naples, Florida, U.S.A.

Lawrence K. Wang
Lenox Institute of Water Technology and Krofta Engineering Corporation, Lenox, Massachusetts and Zorex Corporation, Newtonville, New York, U.S.A.

Yung-Tse Hung
Cleveland State University, Cleveland, Ohio, U.S.A.

Joo-Hwa Tay
Nanyang Technological University, Singapore

Kathleen Hung Li
NEC Business Network Solutions, Irving, Texas, U.S.A.

12.1 INDUSTRY DESCRIPTION

12.1.1 General Description

The U.S. rubber processing industry encompasses a wide variety of production activities ranging from polymerization reactions closely aligned with the chemical processing industry to the extrusion of automotive window sealing strips. The industry is regulated by seven Standard Industrial Classification (SIC) codes [1]:

- SIC 2822: Synthetic Rubber Manufacturing (vulcanizable elastomers);
- SIC 3011: Tire and Inner Tube Manufacturing;
- SIC 3021: Rubber Footwear;
- SIC 3031: Reclaimed Rubber;
- SIC 3041: Rubber Hose and Belting;
- SIC 3069: Fabricated Rubber Products, Not Elsewhere Classified; and
- SIC 3293: Rubber Gaskets, Packing, and Sealing Devices.

Approximately 1650 plants exist in the United States and have production ranges from 1.6×10^3 kkg/year (3.5×10^6 lb/year) to 3.7×10^8 kkg/year (8.2×10^8 lb/year). Table 1 presents a summary of the rubber processing industry regarding the number of subcategories and the number and types of dischargers. Table 2 presents a subcategory profile of best practical control technology currently available (BPT) regulations (daily maximum and 30-day averages) [2]. The effluent limitations are shown as kilogram of pollutants per 1000 kg of raw material processed (kg/kkg).

Table 1 Industry Summary

Industry: Rubber processing
Total number of subcategories: 11
Number of subcategories studied: 3[a]
Number of dischargers in industry:
- Direct: 1054
- Indirect: 504
- Zero: 100

[a]Wet digestion, although not a paragraph 8 exclusion, was not studied because of the lack of plant-specific data. Emulsion and solution crumb rubber, although candidates for exclusion, were studied, because data were available.
Source: USEPA.

The rubber processing industry is divided into 11 subcategories based on raw waste loads as a function of production levels, presence of the same or similar toxic pollutants resulting from similar manufacturing operations, the nature of the wastewater discharges, frequency and volume of discharges, and whether the discharge is composed of contact or noncontact wastewater. Other primary considerations are treatment facilities and plant size, age, and location. The 11 subcategories are listed below. A brief description of each subcategory follows.

- Subcategory 1: Tire and Inner Tube Manufacturing;
- Subcategory 2: Emulsion Crumb Rubber Production;
- Subcategory 3: Solution Crumb Rubber Production;
- Subcategory 4: Latex Rubber Production;
- Subcategory 5: Small-Sized General Molding, Extruding, and Fabricating Rubber Plants;
- Subcategory 6: Medium-Sized General Molding, Extruding, and Fabricating Rubber Plants;
- Subcategory 7: Large-Sized General Molding, Extruding, and Fabricating Rubber Plants;
- Subcategory 8: Wet Digestion Reclaimed Rubber;
- Subcategory 9: Pan, Dry Digestion, and Mechanical Reclaimed Rubber;
- Subcategory 10: Latex-Dipped, Latex-Extruded, and Latex Molded Goods;
- Subcategory 11: Latex Foam.

Subcategory 1. Tire and Inner Tube Manufacturing

The production of tires and inner tubes involves three general steps: mixing and preliminary forming of the raw materials, formation of individual parts of the product, and constructing and curing the final product. In total, 73 plants use these general steps to produce tires in the United States.

The initial step in tire construction is the preparation or compounding of the raw materials. The basic raw materials for the tire industry include synthetic and natural rubber, reinforcing agents, fillers, extenders, antitack agents, curing and accelerator agents, antioxidants, and pigments. The fillers, extenders, reinforcing agents, pigments, and antioxidant agents are added and mixed into the raw rubber stock. This stock is nonreactive and can be stored for later use. When curing and accelerator agents are added, the mixer becomes reactive, which means it has a short shelf-life and must be used immediately.

Table 2 BPT Limitations for Subcategories of Rubber Processing Industry (kg/kkg of raw material)

	Tire and inner tube plants[b]		Emulsion crumb rubber		Solution crumb rubber		Latex rubber		Small GMEF[c]		Medium GMEF[c]	
Pollutant	Daily max	30-day avg.[a]	Daily max	30-day avg.[a]	Daily max	30-day avg.[a]	Daily max	30-day avg.[a]	Daily max	30-day avg.[a]	Daily max	30-day avg.[a]
COD			12.0	8.0	5.9	3.9	10.0	6.8				
BOD_5			0.60	0.40	0.60	0.40	0.51	0.34				
TSS	0.096	0.064	0.98	0.65	0.98	0.65	0.82	0.55	1.3	0.64	0.80	0.40
Oil and grease	0.024	0.016	0.24	0.16	0.24	0.16	0.21	0.14	0.70	0.25	0.42	0.15
Lead									0.0017	0.0007	0.0017	0.0007
Zinc												
pH[d]												

	Large GMEF[c]		Wet digestion reclaimed		Pan, dry digestion, mechanical reclaimed		LDEM[e]		Latex foam	
Pollutant	Daily max	30-day avg.[a]	Daily max	30-day avg.[a]	Daily max	30-day avg.[a]	Daily max	30-day avg.[a]	Daily max	30-day avg.[a]
COD			15	6.1	6.2[f]	2.8				
BOD_5							3.7	2.2	2.4	1.4
TSS	0.50	0.25	1.0	0.52	0.38	0.19	7.0	2.9	2.3	0.94
Oil and grease	0.26	0.093	0.40	0.14	0.40	0.14	2.0	0.73		
Lead	0.00017	0.0007								
Zinc									0.058	0.024
Chromium							0.0086[g]	0.0036		

[a]Computed from average daily value taken over 30 consecutive days.
[b]Oil and grease limitations for nonprocess wastewater from plants placed in operation before 1959: daily max = 10 mg/L; 30-day avg. = 5 mg/L.
[c]General molded, extruded, and fabricated rubber.
[d]Limitation is 6–9 pH units for all subcategories.
[e]Latex-dipped, latex-extruded, and latex-molded goods.
[f]Allowable when the pan, dry digestion, mechanical reclaimed processes are integrated with a wet digestion reclaimed rubber process.
[g]Allowable when plants employ chromic acid for cleaning operations.
Source: USEPA.

After compounding, the stock is sheeted out in a roller mill and extruded into sheets or pelletized. This new rubber stock is tacky and must be coated with an antitack solution, usually a soapstone solution or clay slurry, to prevent the sheets or pellets from sticking together during storage.

The rubber stock, once compounded and mixed, must be molded or transformed into the form of one of the final parts of the tire. This consists of several parallel processes by which the sheeted rubber and other raw materials, such as cord and fabric, are made into the following basic tire components: tire beads, tire treads, tire cords, and the tire belts (fabric). Tire beads are coated wires inserted in the pneumatic tire at the point where the tire meets the wheel rim (on which it is mounted); they ensure a seal between the rim and the tire. The tire treads are the part of the tire that meets the road surface; their design and composition depend on the use of the tire. Tire cords are woven synthetic fabrics (rayon, nylon, polyester) impregnated with rubber; they are the body of the tire and supply it with most of its strength. Tire belts stabilize the tires and prevent the lateral scrubbing or wiping action that causes tread wear.

The processes used to produce the individual tire components usually involve similar steps. First, the raw stock is heated and subjected to a final mixing stage before going to a roller mill. The material is then peeled off rollers and continuously extruded into the final component shape. Tire beads are directly extruded onto the reinforcing wire used for the seal, and tire belt is produced by calendering rubber sheet onto the belt fabric.

The various components of the tire are fitted together in a mold to build green, or uncured, tires which are then cured in an automatic press. Curing times range from less than one hour for passenger car tires to 24 hours for large, off-the-road tires. After curing, the excess rubber on the tire is ground off (deflashed) to produce the final product.

This subcategory is often subdivided into two groups of plants: (a) those starting operations prior to 1959 (applies to 39 plants) and (b) those starting operations after 1959. This subdivision must be recognized in applying limitations on plant effluents of oil and grease because BPT limitations are different for the two groups of plants. For plants placed in operation after 1959, the 30-day average oil and grease limitation is 0.016 kg/kkg of product. For plants placed in operation prior to 1959, the limitation is the same (0.016 kg/kkg) but only for process wastewater. Process wastewater for these pre-1959 plants comes from soapstone solution applications, steam cleaning operations, air pollution control equipment, unroofed process oil unloading areas, mold cleaning operations, latex applications, and air compressor receivers. Water used only for tread cooling and discharges from other areas of such plants is classified as nonprocess wastewater, in which oil and grease levels are limited to 5 mg/L as a 30-day average and 10 mg/L as a daily maximum.

Emulsion polymerization, the traditional process for synthetic rubber production, is the bulk polymerization of droplets of monomers suspended in water. Emulsion polymerization is operated with sufficient emulsifier to maintain a stable emulsion and is usually initiated by agents that produce free radicals. This process is used because of the high conversion and the high molecular weights that are possible. Other advantages include a high rate of heat transfer through the aqueous phase, easy removal of unreacted monomers, and high fluidity at high concentrations of product polymer. Over 90% of styrene butadiene rubber (SBR) is produced by this method. Approximately 17 plants use the emulsion crumb rubber process.

Raw materials for this process include styrene, butadiene, catalyst, activator, modifier, and soap solution.

Polymerization proceeds stepwise through a train of reactors. This reactor system contributes significantly to the high degree of flexibility of the overall plant in producing different grades of rubber. The reactor train is capable of producing either "cold" (277–280 K, 103–206 kPa) or "hot" (323 K, 380–517 kPa) rubber. The cold SBR polymers, produced at the lower temperature and stopped at 60% conversion, have improved properties when compared to hot SBRs. The hot process is the older of the two. For cold polymerization, the monomer–additive emulsion is cooled prior to entering the reactors. Each reactor has its own set of cooling coils and is agitated by a mixer. The residence time in each reactor is approximately one hour. Any reactor in the train can be bypassed. The overall polymerization reaction is ordinarily carried to no greater than 60% conversion of monomer to rubber since the rate of reaction falls off beyond this point and product quality begins to deteriorate. The product rubber is formed in the milky white emulsion phase of the reaction mixture called latex. Short stop solution is added to the latex exiting the reactors to quench the polymerization at the desired conversion. The quench latex is held in blowdown tanks prior to the stripping operation.

The stripping operation removes the excess butadiene by vacuum stripping, and then removes the excess styrene and water in a perforated plate stripping column. The water and styrene from the styrene stripper are separated by decanting and the water is discharged to the treatment facility. The recovered monomers are recycled to the monomer feed stage. The latex is now stabilized and is precipitated by an electrolyte and a dilute acid. This coagulation imparts different physical characteristics to the rubber depending on the type of coagulants used. Carbon black and oil can be added during this coagulation/precipitation step to improve the properties of the rubber. This coagulated crumb is separated from the liquor, resuspended and washed with water, then dewatered, dried, and pressed into bales for shipment. The underflow from the washing is sent to the wastewater treatment facility.

Subcategory 3: Solution Crumb Rubber Production

Solution polymerization is bulk polymerization in which excess monomer serves as the solvent. Solution polymerization, used at approximately 13 plants, is a newer, less conventional process than emulsion polymerization for the commercial production of crumb rubber. Polymerization generally proceeds by ionic mechanisms. This system permits the use of stereospecific catalysts of the Ziegler–Natta or alkyl lithium types which make it possible to polymerize monomers into a *cis* structure characteristic that is very similar to that of natural rubber. This *cis* structure yields a rubbery product, as opposed to a *trans* structure which produces a rigid product similar to plastics.

The production of synthetic rubbers by solution polymerization processes is a stepwise operation very similar in many aspects to production by emulsion polymerization. There are distinct differences in the two technologies, however. For solution polymerization, the monomers must be extremely pure and the solvent should be completely anhydrous. In contrast to emulsion polymerization, where the monomer conversion is taken to approximately 60%, solution polymerization systems are polymerized to conversion levels typically in excess of 90%. The polymerization reaction is also more rapid, usually being completed in 1 to 2 hours.

Fresh monomers often have inhibitors added to them while in storage to prevent premature polymerization. These inhibitors and any water that is present in the raw materials must be removed by caustic scrubbers and fractionating drying columns to provide the solution process with the high purity and anhydrous materials needed. The purified solvent and monomers are then blended into what is termed the "mixed feed," which may be further dried in a desiccant column.

The dried mixed feed is now ready for the polymerization step, and catalysts can be added to the solution (solvent plus monomers) just prior to the polymerization stage or in the lead polymerization reactor.

The blend of solution and catalysts is polymerized in a series of reactors. The reaction is highly exothermic and heat is removed continuously by either an ammonia refrigerant or by chilled brine or glycol solutions. The reactors are similar in both design and operation to those used in emulsion polymerization. The mixture leaves the reactor train as a rubber cement, that is, polymeric rubber solids dissolved in solvent. A short stop solution is added to the cement after the desired conversion is reached.

The rubber cement is then sent to storage tanks where antioxidants and extenders are mixed in. The rubber cement is pumped from the storage tank to the coagulator where the rubber is precipitated with hot water under violent agitation. The solvent and unreacted monomer are first steam stripped overhead and then condensed, decanted, and recycled to the feed stage. The bottom water layer is discharged to the wastewater treatment facility.

The stripped crumb slurry is further washed with water, then dewatered, dried, and baled as final product. Part of the water from this final washing is recycled to the coagulation stage, and the remainder is discharged for treatment.

Subcategory 4: Latex Rubber Production

The emulsion polymerization process is used by 17 production facilities to produce latex rubber products as well as solid crumb rubber. Latex production follows the same processing steps as emulsion crumb rubber production up to the finishing process. Between 5 and 10% of emulsion polymerized SBR and nearly 30% of nitrile rubber production (NBR) are sold as latex. Latex rubber is used to manufacture dipped goods, paper coatings, paints, carpet backing, and many other commodities.

Monomer conversion efficiencies for latex production range from 60% for low-temperature polymerization to 98% for high-temperature conversion.

The monomers are piped from the tank farm to the caustic soda scrubbers where the inhibitors are removed. Soap solution, catalysts, and modifiers are added to produce a feed emulsion which is fed to the reactor train. Fewer reactors are normally used than the number required for a crumb product line. When polymerization is complete, the latex is sent to a holding tank where stabilizers are added.

A vacuum stripper removes any unwanted butadiene, and the steam stripper following it removes the excess styrene. Neither the styrene nor butadiene is recycled. Solids are removed from the latex by filters, and the latex may be concentrated to a higher solids level.

Subcategories 5, 6, 7: Small-, Medium-, and Large-Sized General Molding, Extruding, and Fabricating Plants

These three closely related subcategories are divided based on the volume of wastewater emanating from each. These subcategories include a variety of processes such as compression molding, transfer molding, injection molding, extrusion, and calendering. An estimated 1385 plants participate in these subcategories.

A common step for all of the above processes is the compounding and mixing of the elastomers and compounding ingredients. The mixing operation is required to obtain a thorough and uniform dispersion of the rubber and other ingredients. Wastewater sources from the mixing operation generally derive from leakage of oil and grease from the mixers.

Compression molding is one of the oldest and most commonly used manufacturing processes in the rubber fabrication industry. General steps for the processes include warming the raw materials, preforming the warm stock into the approximate shape, cooling and treating with antitack solution, molding by heat and pressure, and finally deflashing. Major products from this process include automotive parts, medical supplies, and rubber heels and soles.

Transfer molding involves the forced shifting of the uncured rubber stock from one part of the mold to another. The prepared rubber stock is placed in a transfer cavity where a ram forces the material into a heated mold. The applied force combined with the heat from the mold softens the rubber and allows it to flow freely into the entire mold. The molded item is cured, then removed and deflashed. Final products include V-belts, tool handles, and bushings with metal inserts.

Injection molding is a sophisticated, continuous, and essentially automatic process that uses molds mounted on a revolving turret. The turret moves the molds through a cyclic process that includes rubber injection, curing, release agent treatment, and removal. Deflashing occurs after the product has been removed. A wide range of products is made by this process, including automotive parts, diaphragms, hot-water bottles, and wheelbarrow tires.

The extrusion process takes unvulcanized rubber and forces it trough a die, which results in long lengths of rubber of a definite cross-section. There are two general subdivisions of this technique; one extrudes simple products and the other builds products by extruding the rubber onto metal or fabric reinforcement. Products from these techniques include tire tread, cable coating, and rubber hose.

Calendering involves passing unformed or extruded rubber through a set or sets of rolls to form sheets or rolls of rubber product. The thickness of the material is controlled by the space between the rolls. The calender may also produce patterns, double the product thickness by combining sheets, or add a sheet of rubber to a textile material. The temperature of the calender rolls is controlled by water and steam. Products produced by this process include hospital sheeting and sheet stock for other product fabrication.

This subcategory represents a process that is used to recover rubber from fiber-bearing scrap. Scrap rubber, water, reclaiming and defibering agents, and plasticizers are placed in a steam-jacketed, agitator-equipped autoclave. Reclaiming agents used to speed up depolymerization include petroleum and coal tar-base oils and resins as well as various chemical softeners such as phenol alkyl sulfides and disulfides, thiols, and amino acids. Defibering agents chemically do the work of the hammer mill by hydrolyzing the fiber; they include caustic soda, zinc chloride, and calcium chloride.

A scrap rubber batch is cooked for up to 24 hours and then discharged into a blowdown tank where water is added to facilitate subsequent washing operations. Digester liquor is removed by a series of screen washings. The washed rubber is dewatered by a press and then dried in an oven. Two major sources of wastewater are the digester liquor and the washwater from the screen washings.

Two rubber reclaiming plants use the wet digestion method for reclamation of rubber.

Subcategory 9: Pan, Dry Digestion, and Mechanical Reclaimed Rubber

This subcategory combines processes that involve scrap size reduction before continuing the reclaiming process. The pan digestion process involves scrap rubber size reduction on steel rolls, followed by the addition of reclaiming oils in an open mixer. The mixture is discharged into open pans, which are stacked on cars and rolled into a single-cell pressure vessel where live steam is used to heat the mixture. Depolymerization occurs in 2 to 18 hours. The pans are then discharged and the cakes of rubber are sent on for further processing. The steam condensate is highly contaminated and is not recycled.

The mechanical rubber reclaiming process, unlike pan digestion, is continuous and involves fiber-free scrap being fed into a horizontal cylinder containing a screw that works the scrap against the heated chamber wall. Reclaiming agents and catalysts are used for depolymerization. As the depolymerized rubber is extruded through an adjustable orifice, it is quenched. The quench vaporizes and is captured by air pollution control equipment. The captured liquid cannot be reused and is discharged for treatment.

Subcategory 10: Latex-Dipped, Latex-Extruded, and Latex-Molded Goods

These three processes involve the use of latex in its liquid form to manufacture products. Latex dipping consists of immersing an impervious male mold or article into the latex compound, withdrawing it, cleaning it, and allowing the adhering film to air dry. The straight dip process is replaced by a coagulant dip process when heavier films are desired. Fabric or other items may be dipped in latex to produce gloves and other articles. When it has the required coating, the mold is leached in pure water to improve physical and electrical properties. After air drying, the items are talc-dusted or treated with chlorine to reduce tackiness. Water is often used in several processes, for makeup, cooling, and stripping. Products from dipping include gloves, footwear, transparent goods, and unsupported mechanical goods.

Latex molding employs casts made of unglazed porcelain or plaster of paris. The molds are dusted with talc to prevent sticking. The latex compound is then poured into the mold and allowed to develop the required thickness. The mold is emptied of excess rubber and then oven dried. The mold is removed and the product is again dried in an oven. Casting is used to manufacture dolls, prosthetics, printing matrices, and relief maps.

Subcategory 11: Latex Foam

No latex foam facilities are known to be in operation at this time.

12.1.2 Wastewater Characterization

The raw wastewater emanating from rubber manufacturing plants contains toxic pollutants that are present due to impurities in the monomers, solvents, or the actual raw materials, or are associated with wastewater treatment steps. Both inorganic and organic pollutants are found in the raw wastewater, and classical pollutants may be present in significant concentrations. Wastewater from reclaimed rubber manufacturing had 16,800–63,400 mg/L total solids, 1000–24,000 mg/L suspended solids, 3500–12,500 mg/L BOD (biochemical oxygen demand), 130–2000 mg/L chlorides, pH of 10.9–12,2, while wastewaters from synthetic rubber manufacturing had 1900–9600 mg/L total solids, 60–3700 mg/L suspended solids, 75–1600 mg/L BOD, and pH of 3.2–7.9 [3].

Table 3 presents an industry-wide profile of the concentration of toxic pollutants found at facilities in each subcategory (no data are available for Subcategories 9, 10, and 11). Table 4 gives a subcategory profile of the pollutant loadings (no data are available for Subcategories 8, 10, and 11). These tables were prepared from available screening and verification sampling data. The minimum detection limit for toxic pollutants is 10 μg/L and any value below 10 μg/L is presented in the following tables as BDL, below detection limit.

In-plant management practices may often control the volume and quality of the treatment system influent. Volume reduction can be attained by process wastewater segregation from noncontact water, by recycling or reuse of noncontact water, and by the modification of plant processes. Control of spills, leakage, washdown, and storm runoff can also reduce the treatment system load. Modifications may include the use of vacuum pumps instead of steam ejectors, recycling caustic soda solution rather than discharging it to the treatment system, and incorporation of a more efficient solvent recovery system.

12.1.3 Tire and Inner Tube Manufacturing

The tire and inner tube manufacturing industry has several potential areas for wastewater production, but water recycle is used extensively. The major area for water use is in processes

Table 3 Concentrations of Toxic Pollutants Found in the Rubber Processing Industry by Subcategory, Verification, and Screening

	Tire and inner tube manufacturing							
	Treatment influent				Treatment effluent			
Toxic pollutants (μg/L)	Number of samples	Average	Median	Maximum	Number of samples	Average	Median	Maximum
Metals and Inorganics								
Chromium	1	10			1	BDL		
Copper	1	BDL			0			
Lead	2	25		50	0			
Zinc	5	260	150	770	1	330		
Phenols								
2,4,6-Trichlorophenol	0				1	<14		
Aromatics								
Toluene	0				1	<10,000		
Halogenated aliphatics								
1,2-*Trans*-dichloroethylene	0				1	16		
Methylene chloride	0				2	<5,000		<10,000
Trichloroethylene	0				1[a]	40		
Pesticides and metabolites								
Isophorone	0				1	BDL		
	Emulsion crumb rubber manufacturing							
Metals and Inorganics								
Cadmium	2	46		90	1			BDL
Chromium	5	230	250	720	2	140		220
Copper	1			200	0			
Lead	1			390	0			

(continues)

Table 3 *Continued*

Toxic pollutants (μg/L)	Emulsion crumb rubber manufacturing							
	Treatment influent				Treatment effluent			
	Number of samples	Average	Median	Maximum	Number of samples	Average	Median	Maximum
Mercury	3	BDL	BDL	BDL	3	BDL	BDL	BDL
Nickel	2	380		590	1			400
Selenium	1			20	1			<24
Zinc	3	100	BDL	290	2	BDL		BDL
Phthalates								
Bis(2-ethylhexyl)phthalates	3	310	260	530	3	250	200	430
Dimethyl phthalate	1			11	2	BDL		14
Nitrogen compounds								
Acrylonitrile[b]	4	BDL		BDL	4			BDL
Phenols								
2-Nirophenol	1			BDL	1			BDL
Phenol	3	180	57	440	3	30	19	37
Aromatics								
Acenapthene[c]	1			BDL	1			BDL
Acenapthylene[c]	1			BDL	1			BDL
Benzene								
Benzopyrene[c]								
Ethylbenzene								
Napthalene[c]								
Toluene								
Halogenated aliphatics								
Dichlorobromoethane	1			>3,100	1			BDL

Carbon tetrachloride	1			BDL	1			BDL
Chloroform	3	130	100(c)	270	2	BDL		BDL
1,1-Dichloroethane	1			BDL	1			BDL
1,2-Dichloroethane	1			93	0			
1,2-*Trans*-dichlorcethylene	1			16	0			
Methylene chloride	3	29	15	73	3	220	150	520
1,1,2,2-Tetrachloroethane	1			BDL	1			BDL
			Solution crumb rubber manufacturing					
Metals and Inorganics								
Metals and inorganics								
Cadmium	3	31	BDL	90	2	BDL		BDL
Chromium	4	350	310	720	3	170	67	410
Copper	3	72	BDL	200	2	BDL		14
Lead	1			390				
Mercury	3	BDL	BDL	BDL	2	BDL		BDL
Nickel	1	160			0			
Zinc	2	8,100		16,000	1			190,000
Phthalates								
Bis(2-ethylhexyl)phthalate	3	260	140	530	3	190	120	430
Dimethyl phthalate	1			BDL	1			BDL
Phenols								
Phenol	3	210	180	440	3	15	BDL	37
Aromatics								
Acenapthene	1			BDL	1			BDL
Acenapthylene	1			BDL	1			BDL
Benzopyrene	1			BDL	1			BDL
Benzene	3	1,200	50	3,400	3	BDL	BDL	10
Ethylbenzene	2	BDL		10	2	BDL		10
Tolunene	4	BDL	BDL	10	5	88	BDL	420

(*continues*)

Table 3 *Continued*

Toxic pollutants (µg/L)	Solution crumb rubber manufacturing							
	Treatment influent				Treatment effluent			
	Number of samples	Average	Median	Maximum	Number of samples	Average	Median	Maximum
Halogenated aliphatics								
Carbon tetrachloride	1	35			1			1,400
Chloromethane	1	4,900			1			2,200
Chloroform	2	BDL		BDL	2	BDL		BDL
1,2-*Trans*-dichloroethylene								
Methylene chloride	2	BDL		15	2	<260		520
1,1,2,2-Tetrachloroethane	1			BDL	1			BDL
1,1,2,2-Trichloroethane	1			BDL	1	BDL		BDL
Trichloroethylene	1			BDL	1			BDL
Pesticides and metabolites								
Acrolein	1			BDL	1			BDL
	General molding, extruding, and fabricating							
Metals and Inorganics								
Lead	1			20	0			
Mercury	0				1			BDL
Zinc	0				1			970
Phthalates								
Bis(2-ethylhexyl)phthalate	1			17	2	BDL		16
Di-*n*-butyl phthalate	0				1			36
Nitrogen compounds								
N-nitrosodiphenylamine	2	35		53	0			

Phenols						
Pentachlorophenol	1		BDL	0		12,000
Phenol	0			1		
Aromatics						
Benzene	0			1		BDL
Halogenated aliphatics						
Chloroform	1		25	2	BDL	10
1,1-Dichloroethane	0			1		110
1,2-Dichloroethane	0			1		BDL
1,2-*Trans*-dichloroethylene	0			1		290
1,1,2,2-Tetrachloroethane	0			1		BDL
1,1,1-Trichloroethane	0			1		7,100
1,1,2-Trichloroethane	0			1		BDL
Trichloroethylane	0			1		1,600
			Wet digestion reclaimed rubber			
Metals and Inorganics						
Cadmium[d]	1		10	0		
Lead	1		50	0		
Zinc[d,e]	2	250	350	0		
Nitrogen compounds						
Phenol[e]	1		BDL	0		
Pesticides and metabolites						
Isophorone[e]	1		BDL	0		

Analytic methods: V.7.3.29, Data sets 1,2.
BDL, below detection limit.
[a]40 μg/L of trichloroethylene also measured in city water.
[b]Detection limit of acrylonitrile by direct aqueous injection was 2300 μg/L.
[c]This value believed to be a glassware contaminant.
[d]These pollutants appear to be attributed to tire operation.
[e]Wastewater is from both tire and reclaiming processes.
Source: USEPA.

Table 4 Industry Profile of Toxic and Classical Pollutant Loadings, Verification, and Screening Data (Toxic Pollutants Kg/kkg)

	Tire and inner tube manufacturing							
	Treatment influent				Treatment effluent			
Toxic Pollutants (Kg/Mg)	Number of samples	Average	Median	Maximum	Number of samples	Average	Median	Maximum
Toxic metals								
Chromium	0				1			0.000005
Copper	1			0.001	0			
Lead	1			0.001	0			
Zinc	3	0.003	0.004	0.006	1			0.0007
Toxic organics								
Phenol	1			BDL	0			
Methylene chloride	0				1			BDL
2,4,6-Trichlorophenol	0				1			BDL
Isophorone	1			BDL	1			BDL
Classical pollutants (kg/day)								
TSS	4	590	200	2,000	47	270	32	2,400
Oil and grease	8	17	7.2	120	35	7.3	2.1	42
pH, pH units	10	7.6	2.4	9.4	44	7.5	7.5	10.3
Toxic pollutants (kg/Mg)	Emulsion crumb rubber manufacturing							
Toxic metals								
Cadmium	2	0.0004		0.0006	1			0.00001
Chromium	5	2.6	0.003	13	4	3.0	0.0005	12
Copper	1			0.003	0			
Lead	1			0.006	0			
Mercury	2	0.00003		0.00003	2	0.00002		0.00002
Nickel	2	0.006		0.008	1			0.005
Selenium	1			<1.0	1			1.3
Zinc	3	0.002	BDL	0.005	2	BDL		BDL

Toxic organics								
Bis(2-ethylhexyl)phthalate	3	<2.4	0.0017	<7.3	3	<2.3	0.0016	<7.0
Dimethyl phthalate	2	0.0002		0.0002	2	0.0001		0.0002
Acrylonitrile	4	BDL	BDL	BDL	5	<240	BDL	<1,200
N-nitrosodiphenylamine	1			BDL	1			BDL
2-Nitrophenol	1			<0.5	1			0.26
Phenol	4	0.75	0.003	3.0	4	<0.25	0.0004	<0.98
Benzene	3	0.01	0.0007	0.01	2	0.0003		0.0005
Ethylbenzene	5	<0.01	BDL	<0.05	4	<0.001	BDL	<0.005
Nitrobenzene	1			<0.0004	1			<0.0004
Toluene	6	<0.009	0.002	<0.05	5	<0.001	0.000001	
Carbon tetrachloride	1			0.00001	1			<0.000002
Chloroform	3	0.14	0.0004	0.40	2	0.04		0.09
1,1-Dichloroethane	1			0.00002	1			<0.00002
1,1-*Trans*-dichloroethylene	1			ND	0			
1,2-Dichloroethane	1			ND	0			
Methylene chloride	3	<1.3	0.0002	<3.8	3	<1.9	0.00007	<5.7
1,1,2,2-Tetrachloroethane	1			0.00002	1			0.000001
Acenapthene	1			BDL	1			BDL
Acenapthylene	1			BDL	1			BDL
Napthalene	1			BDL	1			BDL
Benzo-pyrene	1			BDL	1			BDL
Dichlorobromomethane	1			<1.6	1			7.0
Acrolein	1			BDL	1			BDL
				Solution crumb rubber manufacturing				
Metals and Inorganics								
Toxic metals								
Cadmium	3	0.01	0.0007	0.04	2	0.5		0.09
Chromium	4	<4.3	0.0006	<17	3	<0.42	0.0004	<1.3
Copper	3	<0.09	0.00007	<0.28	2	<0.17		<0.34
Lead	1			0.006	0			
Mercury	1			0.00003	1			0.00001
Nickel	1			0.003	0			
Zinc	2	0.07		0.14	1			2.0

(continues)

Table 4 *Continued*

	Solution crumb rubber manufacturing							
	Treatment influent				Treatment effluent			
Toxic Pollutants (Kg/Mg)	Number of samples	Average	Median	Maximum	Number of samples	Average	Median	Maximum
Toxic organics								
Bis(2-ethylhexyl)phthalate	3	<1.8	0.007	<5.4	3	<2.7	0.006	<8.1
Dimethyl phthalate	1			0.0001	1			0.00008
Phenol	3	2.4		7.1	3	<0.25	0.0005	<0.76
Benzene	4	43	0.0004	130	4	<0.002	<0.0001	<0.007
Ethylbenzene	2	0.00005		0.00005	2	<0.00001		<0.00002
Toluene	4	0.001	0.00006	0.004	5	0.003	0.0001	0.007
Carbon tetrachloride	1			0.0003	1			0.0001
Chloroform	2	0.06		0.12	2	0.03		0.06
Methylene chloride	2	0.0001		0.0002	2	0.004		0.007
1,1,2,2-Tetrachloroethane	1			0.004	1			<0.007
1,1,2-Trichloroethane	1			<0.0000008	1			<0.000001
Trichloroethylene	1			<0.0000008	1			<0.000001
Acenapthene								
Acenapthylene								
Benzo-pyrene								
Chloromethane	1			0.04	2	0.02		0.02
Acrolein	1			BDL	1			BDL
Classical pollutants (kg/day)								
BOD_5	6	2,900	900	15,000	9	200	86	1,100
COD	4	2,300	2,500	4,400	8	310	320	1,200
TSS	4	540	920	1,200	8	270	85	1,100
Oil and grease	3	96	110	130	7	<25	11	<92
pH (pH units)	1			9.5	4	6.8	7.5	8.2

			Latex rubber manufacturing					
Metals and Inorganics								
Toxic metals								
Chromium	2	BDL		BDL	2	BDL		BDL
Zinc	2	BDL		BDL	2	BDL		BDL
Toxic organics								
Bis(2-ethylhexyl)phthalate	1			0.0004	1			<0.00004
Di-*n*-butyl phthalate	1			BDL	1			BDL
Acrylonitrile	2	BDL		BDL	2	BDL		BDL
Pentachlorophenol	1			0.0001	1			<0.00004
Phenol	1		0.0001		1			<0.00004
Benzene	1			BDL	1			BDL
Ethylbenzene	3	0.002	BDL	0.006	3	<0.000007	BDL	<0.00002
Toluene	2	BDL		BDL	2	BDL		BDL
Methylene chloride	1			BDL	1			BDL
Butylbenzyl phthalate	1			BDL	1			BDL
Napthalane	1			BDL	1			BDL
Classical pollutants (kg/day)								
BOD_5	0				5	86	15	340
COD	0				3	120	150	160
TSS	1			640	5	130	12	590
Oil and grease	0				3	2.8	3.2	4.0
pH (pH units)	0				3	10	10	10.0
Toxic pollutants (kg/Mg)			General molding, extruding, and fabricating rubber manufacturing					
Metals and Inorganics								
Toxic metals								
Lead	1			0.003	1			0.0001
Zinc	0				1			0.14
Toxic organics								
Bis(2-ethylhexyl)phthalate	1			0.002	0			0.005

(continues)

Table 4 *Continued*

Toxic Pollutants (Kg/Mg)	General molding, extruding, and fabricating rubber							
	Treatment influent				Treatment effluent			
	Number of samples	Average	Median	Maximum	Number of samples	Average	Median	Maximum
Di-*n*-butyl phthalate	0				1			
N-nitrosodiphenylamine	1			0.0007	0			
Pentachlorophenol	1			0.00003	0			
Phenol	0				1			1.7
Benzene	0				1			0.001
Chloroform	0				1			0.0003
1,1-Dichloroethane	0				1			0.02
1,1-*Trans*-dichloroethylene	0				1			0.04
1,2-Dichloroethane	0				1			0.0006
Tetrachloroethylene	0				1			0.0006
1,1,1-Trichloroethane	0				1			1.0
1,1,2-Trichloroethane	0				1			0.0002
Trichloroethylene	0				1			0.23

Analytic methods: V.7.3.29, Data sets 1,2.
BDL, below detection limit.
ND, not detected.
Source: USEPA.

requiring noncontact cooling. The general practice of the industry is to recirculate the majority of this water with a minimal blowdown to maintain acceptable concentrations of dissolved solids. Another water use area is contact water used in cooling tire components and in air pollution control devices. This water is also recirculated. Steam condensate and hot and cold water are used in the molding and curing areas. The majority of the water is recycled back to the boiler or hot water tank for use in the next recycle. Soapstone areas and plant and equipment cleanup are the final water use areas. Most facilities try to recycle soapstone solution because of its high solids content. Plant and equipment cleanup water is generally sent to the treatment system. Table 5 presents a summary of the potential wastewater sources for this subcategory.

Grease, oils, and suspended solids are the major pollutants within this industry. Organic pollutants, pH, and temperature may also require treatment. The organics present are due generally to poor housekeeping procedures.

12.1.4 Emulsion Crumb Rubber Production

In-process controls for the reduction of wastewater flows and loads for emulsion crumb rubber plants include recycling of finishing line wastewaters and steam stripping of heavy monomer decanter wastewater. Recycling of finishing line wastewater occurs at nearly all emulsion crumb plants with the percent recycle depending primarily upon the desired final properties of the crumb. Approximately 75% recycle is an achievable rate, with recycle for white masterbatch crumb below this level and that for black masterbatch crumb exceeding it.

Organic toxic pollutants found at emulsion crumb rubber plants come from the raw materials, impurities in the raw materials, and additives to noncontact cooling water. BOD, COD, and TSS levels may also reach high loadings.

Table 5 Summary of Potential Process-Associated Wastewater Sources from the Tire and Inner Tube Industry

Plant area	Source	Nature and origin of wastewater contaminants
Oil storage	Runoff	Oil
Compounding	Washdown, spills, leaks, discharges from wet air pollution equipment	Solids from soapstone dip tanks; oil from seals in roller mills; oil from solids from Banbury seals; solids from air pollution equipment discharge
Bead, tread, tube formation	Washdown, spills, leaks	Oil and solvent-based cements from the cementing operation; oil from seals in roller mills
Cord and belt formation	Washdown, spills, leaks	Organics and solids from dipping operation; oil from seals, in roller mills, calenders, etc.
Green tire painting	Washdown, spills, air pollution equipment	Organics and solids from spray-painting operation; soluble organics and solids from air pollution equipment discharge
Molding and curing	Washdown, leaks	Oil from hydraulic system; oil from presses
Tire finishing	Washdown, spills, air pollution equipment	Solids and soluble organics from painting operations; solids from air pollution equipment discharge

Source: USEPA.

Table 6 lists potential wastewater sources and general wastewater contaminants for the emulsion crumb rubber industry.

12.1.5 Solution Crumb Rubber Production

Solution crumb rubber production plants have lower raw wastewater loads than emulsion crumb plants because of the thorough steam stripping of product cement to remove solvent and permit effective coagulation. Recycling in this industry is comparable to that in the emulsion crumb industry, with about 75% of the wastewater being recirculated.

Toxic pollutants found in the wastewater streams are normally related to solvents and solvent impurities, product additives, and cooling water treatment chemicals. Table 7 presents a listing of the potential wastewater sources and the associated contaminants for this industry.

12.1.6 Latex Rubber Production

No in-process contact water is currently used by the latex rubber industry. No raw material recycling is practised because of poor control of monomer feeds and the buildup of impurities in the water.

Organic toxic pollutants and chromium are present in the raw wastewater and normally consist of raw materials, impurities, and metals used as cooling water corrosion inhibitors.

Table 8 presents potential wastewater sources and general contaminants for this industry.

12.1.7 General Molding, Extruding, and Fabricating Rubber Plants

Toxic pollutants resulting from production processes within this industry are generally the result of leaks, spills, and poor housekeeping procedures. Pollutants include organics associated with the raw materials and lead from the rubber curing process.

Table 6 Summary of Wastewater Sources from Emulsion Crumb Rubber Production Facilities

Processing unit	Source	Nature of wastewater contaminants
Caustic soda scrubber	Spent caustic solution	High pH, alkalinity, and color. Extremely low average flow rate
Monomer recovery	Decant water layer	Dissolved and separable organics. Source of high BOD and COD discharges
Coagulation	Coagulation liquor overflow	Acidity, dissolved organics, suspended and high dissolved solids, and color. High wastewater flow rates relative to other sources
Crumb dewatering	Crumb rinse water overflow	Dissolved organics, and suspended and dissolved solids. Source of highest wastewater volume from emulsion crumb rubber production
Monomer strippers	Stripper cleanout rinse water	Dissolved organics, and suspended and dissolved solids. High quantities of uncoagulated latex
Tanks and reactors	Cleanout rinse water	Dissolved organics, and suspended and dissolved solids. High quantities of uncoagulated latex
All plant areas	Area washdowns	Dissolved and separable organics, and suspended and dissolved solids

Source: USEPA.

Table 7 Summary of Wastewater Sources from Solution Crumb Rubber Production

Processing unit	Source	Nature of wastewater contaminants
Caustic soda scrubber	Spent caustic solution	High pH, alkalinity, and color. Extremely low average flow rate
Monomer and solvent drying columns	Water removed from monomers and solvent	Dissolved and separable organics. Very low flow
Solvent purification	Fractionator bottoms	Dissolved and separable organics
Monomer recovery	Decant water layer	Dissolved and separable organics
Crumb dewatering	Crumb rinse water overflow	Dissolved organics, and suspended and dissolved solids. Source of highest volume wastewater flow
All plant areas	Area washdowns	Dissolved and separable organics, and suspended and dissolved solids

Source: USEPA.

12.1.8 Rubber Reclamation

Wastewater effluents from this subcategory contain high levels of toxic organic and inorganic pollutants. These pollutants generally result from impurities in the tires and tubes used in the reclamation process. The wastewater from the pan process is of low volume [0.46 m^3/kkg (56 gal/1000 lb)], but is highly contaminated, requiring treatment before discharge. The mechanical reclaiming process uses water only to quench the reclaimed rubber, but it uses a much higher quantity (1.1 m^3/kkg). Steam generated from the quenching process is captured in a scrubber and sent to the treatment system. Wet digestion uses 5.1 m^3 of water per kkg (610 gal/1000 lb) of product in processing, of which 3.4 m^3/kkg (407 gal/1000 lb) of product is used in air pollution control.

12.1.9 Latex-Dipped, Latex-Extruded, and Latex-Molded Goods

Wastewater sources in this subcategory are the leaching process, makeup water, cooling water, and stripping water. Toxic pollutants are present at insignificant levels in the wastewater discharges.

Table 8 Summary of Wastewater Sources from Latex Rubber Production

Processing unit	Source	Nature of wastewater contaminants
Caustic soda scrubber	Spent caustic solution	High pH, alkalinity, and color. Extremely low average flow rate
Excess monomer stripping	Dacent water layer	Dissolved and separable organics
Latex evaporators	Water removed during latex concentration	Dissolved organics, suspended and dissolved solids. Relatively high wastewater flow rates
Tanks, reactors, and strippers	Cleanout rinse water	Dissolved organics, suspended and dissolved solids. High quantities of uncoagulated latex
Tank cars and tank trucks	Cleanout rinse water	Dissolved organics, suspended and dissolved solids. High quantities of uncoagulated latex
All plant areas	Area washdowns	Dissolved and separable organics, and suspended and dissolved solids

Source: USEPA.

12.1.10 Latex Foam

No information is available on the wastewater characteristics of this subcategory.

12.2 PLANT-SPECIFIC DESCRIPTION

Only two subcategories of the rubber industry have not been recommended as Paragraph 8 exclusions of the NRDC Consent Decree: Wet Digestion Reclaimed Rubber, and Pan, Mechanical, and Dry Digestion Reclaimed Rubber. Of these two, plant specific data are available only for the latter. Of the nine remaining subcategories, plant-specific information is available only for Emulsion Crumb Rubber and Solution Crumb Rubber, and is presented below. Two plants in each subcategory are described. They were chosen as representative of their subcategories based on available data.

Plant 000012 produces 3.9×10^4 kkg/year (8.7×10^7 lb/year) of emulsion crumb rubber, primarily neoprene. The contact wastewater flow rate is approximately 8.45 m^3/day (2.25×10^3 gpd) and includes all air pollution control equipment, sanitary waste, maintenance and equipment cleanup, and direct contact wastewater. The treatment process consists of activated sludge, secondary clarification, sludge thickening, and aerobic sludge digestion. Noncontact wastewater, with a flow rate of approximately 1.31×10^5 m^3/day (3.46×10^7 gpd), is used on a once-through basis and is returned directly to the river source. Contact wastewater is also returned to the surface stream after treatment.

Plant 000033 produces three types of emulsion crumb rubber in varying quantities. Styrene butadiene rubber (SBR) forms the bulk of production, at nearly 3.7×10^5 kkg/year (8.2×10^8 lb/year), with nitrile butadiene rubber (NBR) and polybutadiene rubber (PBR) making up the remainder of production [4.5×10^3 kkg/year (1.0×10^7 lb/year) and 4.5×10^3 kkg/year, respectively]. Wastewater consists of direct contact process water, noncontact blowdown, and noncontact ancillary water. The total flow of contact water is approximately 1.27×10^4 m^3/day (3.355×10^6 gpd), and the total flow of noncontact water is 340.4 m^3/day (9×10^4 gpd). Treatment of the wastewater consists of coagulation, sedimentation, and biological treatment with extended aeration. Treated wastewater is discharged to a surface stream.

Tables 9 and 10 present plant-specific toxic pollutant data for the selected plants. Table 11 gives plant-specific classical pollutant data, including BPT regulations set for each specific plant.

12.2.1 Solution Crumb Rubber Production

Plant 000005 produces approximately 3.2×10^4 kkg/year (7.0×10^7 lb/year) of isobutene–isopropene rubber. Wastewater generally consists of direct processes and MEC water. Contact wastewater flow rate is approximately 1040 m^3/day (2.75×10^5 gpd), and noncontact water flows at about 327 m^3/day (8.64×10^4 gpd). Treatment consists of coagulation, flocculation, and dissolved air flotation, and the treated effluent becomes part of the noncontact cooling stream of the onsite refinery.

Plant 000027 produces polyisoprene crumb rubber [4.5×10^4 kkg/year (1.0×10^8 lb/year)] polybutadiene crumb rubber, and ethylene-propylene-diene-terpolymer rubber [EPDM; 4.5×10^4 kkg/year (1.0×10^8 lb/year)]. Wastewater consists of contact process water, MEC, cooling tower blowdown, boiler blowdown, and air pollution control. Wastewater is produced at about 12,100 m^3/day (3.2×10^6 gpd). Treatment consists of API separators, sedimentation, stabilization, and lagooning, followed by discharge to a surface stream.

Table 9 Plant-Specific Verification Data for Emulsion Production Plant 000012

	Local in process line				
Pollutant	Stripper decant	Spray wash water	Treatment influent	Treatment effluent	Raw intake water
Toxic pollutant (μg/L)					
Cadmium	<1	<1	<2	<1	<1.0
Mercury	1.5	1.7	2.5	1.6	1.5
Nickel	60	690	610	400	<10
Bis(2-ethylhexyl)phthalate	290	490	260	<230	260
Dimethyl phthalate	<14	<14	<14	<14	<16
N-nitrosodiphenylamine	1.5	<1.0	5.2	<2.0	<1.0
Phenol	19	29	41	19	<2
Nitrobenzene	<30	<30	<30	<30	<30
Toluene	70	<0.5	250	<0.5	<0.5
Carbon tetrachloride	41	0.1	4.7	<0.2	0.3
Chloroform	110	14	27	4.1	8.5
1,1-Dichloroethylene	51	<1.7	<1.7	<1.7	<1.7
Methylene chloride	4.8	1.0	<0.1	1.0	<0.1
Tetrachloroethylene	<0.1	<0.1	1.4	<0.1	<0.1
1,1,1-Trichloroethane	<1.6	0.3	<1.1	0.3	0.2

Analytic methods: V.7.3.29, Data set 2.
Flow rate (cu. m/day): contact = 8.45; noncontact = 131,000.
Source: USEPA.

Tables 12 and 13 show plant specific toxic pollutant data for the above plants. Classical pollutant data and BPT regulations are presented in Table 14.

12.2.2 Dry Digestion Reclaimed Rubber

A data summary for plant 000134 is given in Table 15. Production, wastewater flow, and treatment data are currently not available for a plant within this subcategory.

12.3 POLLUTANT REMOVABILITY

In this industry, numerous organic compounds, BOD, and COD are typically found in plant wastewater effluent. Industrywide flow and production data show that these pollutants can be reduced by biological treatment. In emulsion crumb and latex plants, uncoagulated latex contributes to high suspended solids. Suspended solids are produced by rubber crumb fines and include both organic and inorganic materials. Removal of such solids is possible using a combination of coagulation/flocculation and dissolved air flotation.

Solvents, extender oils, and insoluble monomers are used throughout the rubber industry. In addition, miscellaneous oils are used to lubricate machinery. Laboratory analysis indicates the presence of oil and grease in the raw wastewater of these plants. Oil and grease entering the wastewater streams are removed by chemical coagulation, dissolved air flotation, and, to some extent, biological oxidation.

Table 10 Plant-Specific Verification Data for Emulsion Crumb Rubber Production Plant 000033

Pollutant (μg/L)	Location in process line							
	SBR stripper	Finishing comp.	NBR finishing	Treatment influent	Treatment effluent	NBR decant	Raw intake, well	Raw intake, river
Cadmium	<1	80	<1	40	40	<2	<1	<1
Chromium	6	400	20	250	220	10	6	5
Copper	71	80	<1	1,400	410	<1	1	<1
Mercury	0.8	63	2.2	3.2	3.1		0.7	0.6
Selenium	<4	<30	<6	<20	<25	<6	<4	<4
Bis(2-ethylhexyl)phthalate	<350	<210	<170	<130	<130	<120	<110	<110
Acrylonitrile	<26	<23	94	32	<23	48,000	<23	<23
2-Nitrophenol	<4	<4	<17	<10	<5	<5	<4	<4
Phenol	41	67	32	61	<20	<16	10	<3
Ethylbenzene	<38	<0.1	<0.1	<0.1	<0.1	<23	<0.1	<0.1
Toluene	<0.1	<0.1	<0.1	<0.1	<0.1	<25	<0.1	<0.1
Chloroform	1.5	2.5	5.2	8.3	1.8	37	1.2	41
Dichlorobromomethane	<0.3	<0.1	<0.5	<0.3	<0.1	5.2	<0.1	6.2
Methylene chloride	<110	<0.1	<80	<67	<110	180	<2	<2

Analytic methods: V.7.3.29, Data set 2.
Flow rate (cu.m/day): SBR–contact = 10,200, noncontact = 190; NBR–contact = 1,250, noncontact = 75.7; PBR–contact = 1,250, noncontact = 75.7; Total–contact = 12,700, noncontact = 340.
Source: USEPA.

Table 11 Plant-Specific Classical Pollutant Data for Selected Emulsion Crumb Rubber Production Plants, Verification Data

	Waste load, plant 000012					
Parameter	Influent		Effluent		BPT regulation	
BOD_5	1,200	(2,600)	5.0	(11)	44	(97)
COD	2,100	(4,600)	130	(280)	880	(1,900)
TSS	8	(18)	35	(77)	71	(160)
Oil and grease	<8	(<18)	8	(18)	18	(39)
pH (pH units)					6 to 9	
Phenol	0.014	(0.03)	30	(67)		
	Waste load, plant 000033					
BOD_5	2,700	(5,900)	140	(320)	460	(1,000)
COD	8,600	(19,000)	2,700	(5,900)	9,200	(20,000)
TSS	2,100	(4,700)	240	(540)	750	(1,700)
Oil and grease	240	(530)	140	(310)	180	(410)
pH (pH units)					6 to 9	
Phenol	4.8	(10.5)	0.35	(0.75)		

Analytic methods: V.7.3.29, Data set 2.
Blanks indicate data not available.
Source: USEPA.

Wastewater sampling indicates that toxic pollutants found in the raw wastewater can be removed. Biological oxidation (activated sludge) adequately treats all of the organic toxic pollutants identified in rubber industry wastewater streams. Significant removal of metals was also observed across biological treatment. The metals are probably absorbed by the sludge mass and removed with the settled sludge. Treatment technologies currently in use are described in the following subcategory descriptions.

12.3.1 Emulsion Crumb Rubber Plants

There are a total of 17 plants in the United States producing emulsion-polymerized crumb rubber. Five of these plants discharge to POTWs; 10 discharge to surface streams; one plant discharges to an evaporation pond; and one plant employs land application with hauling of settled solids. Of the five plants discharging to POTWs, four pretreat using coagulation and primary treatment and one employs equalization with pH adjustment. All 10 of the plants discharging to surface streams employ biological waste treatment ranging from conventional activated sludge to nonaerated wastewater stabilization lagoons.

Organic pollutants are generally found to be reduced to insignificant levels (<10 μg/L) by biological treatment. Most metals are also found to be reduced across biological treatment; they are generally at very low levels in the treated effluent. However, significant metal concentrations may be found in some treated effluent.

At emulsion crumb rubber facilities, a well-operated biological treatment facility permits compliance with BPT limitations and reduces organic toxic pollutant levels. Toxic metals that may not be reduced include chromium, cadmium, copper, selenium, and mercury. Tables 16 and 17 show pollutant removal efficiencies at two emulsion crumb plants.

Table 12 Plant-Specific Verification Data for Solution Crumb Rubber Production Plant 000005

	Location in process line		
Pollutant (μg/L)	Screen tank 1 and 2 comp.	Expeller 1 and 2 comp.	DAF influent
Cadmium	<1[a]	<1[a]	<1[a]
Chromium	3	6	75
Copper	6	7	9
Zinc	14,000	12,000	14,000
Bis(2-ethylhexyl)phthalate	<60	100	200
Phenol	9	<5	7
Benzene	<22	<13	<22
Ethylbenzene	<38	<2	<12
Toluene	<26	<3	<26
Carbon tetrachloride	0.06	0.06	35
Chloroform	0.90	0.88	2.2
Methyl chloride	14,000	2,600	4,900
Methylene chloride	<1	<1	<1
1,1,2-Trichloroethane	<0.1	<0.1	<0.1
Trichloroethylene	<0.1	<0.1	<0.1

	Location in process line			
	DAF effluent	Well water (a)	Boiler feedwater (a)	Boiler blowdown
Cadmium	<1[a]	<1	<1	
Chromium	410	3	3	5,700
Copper	14	3	7	
Zinc	13,000	30	13,000	3,700
Bis(2-ethylhexyl)phthalate	24	98	50	
Phenol	5	<2	6	
Benzene	<110	<43	<43	
Ethylbenzene	<38	<110	<110	
Toluene	<26	<72	<72	
Carbon tetrachloride	14	0.10	0.06	
Chloroform	1.3	1.0	0.98	
Methyl chloride	2,000	190	31	
Methylene chloride	<1	<1	35	
1,1,2-Trichloroethane	<0.1	<0.1	<0.1	
Tetrachloroethylene	<0.1	<0.1	<0.1	

Analytic methods: V.7.3.29, Data set 2.
[a]Screening data.
Flow rate (cu.m/day): contact = 1000; noncontact = 327.
Source: USEPA.

12.3.2 Solution Crumb Rubber Plants

There are 13 solution crumb rubber plants in the United States. Twelve of these plants discharge treated wastewater to surface streams; the other plant discharges its treated wastewater into a neighboring oil refinery's noncontact cooling water system.

Table 13 Plant-Specific Verification Data for Solution Crumb Rubber Production Plant 000027

Pollutant (μg/L)	SN/CB process	EPDM process	Treatment influent	Treatment effluent	Well water	Boiler blowdown
Cadmium	<1	<1	<1	<1	<1	<1
Chromium	450	820	440	19	2	2,600
Copper	4	<2	<7	<5	2	<1
Mercury	1.8	2.3	1.1	2.0	4.0	1.4
Bis-(2-ethylhexyl) phthalate	77	120	<140	<120	170	<46
Phenol	13	670	180	<12	<2	7
Benzene	<0.1	39,000	3,300	<0.1	<0.1	<0.1
Ethylbenzene	<0.1	<0.1	<0.1	<0.1	<0.1	<0.1
Toluene	<0.1	<43	<0.1	<0.1	<0.1	<0.1
Chloroform	1.0	22	3.2	0.9	1.1	1.0
1,1,2,2-Tetrachloroethene	<0.1	<0.1	<0.1	<0.1	<0.1	<0.1

Analytic methods: V.7.3.29, Data set 2.
Total flow rate: 12,100 cu. m/day.
Source: USEPA.

Ten of the plants discharging to surface streams employ some form of biological treatment for waste load reduction. Two of the plants discharging to surface streams use in-process controls, oil removal, and primary treatment prior to discharge. In-process control employed at one plant consists of steam stripping of wastewaters, while in-process control at the second plant was not disclosed. The plant discharging to the oil refinery noncontact cooling water system used coagulation, flocculation, and dissolved air flotation prior to discharge.

The results of the verification program showed that all organic toxic pollutants were reduced across biological treatment. Chloromethane, used as a solvent at plant 000005, was present at significant levels in treated effluent.

Tables 18 and 19 show pollutant removal efficiencies at two selected solution crumb rubber plants.

12.3.3 Latex Rubber Plants

There are 17 latex rubber production facilities in the United States. Of these, nine plants discharge to POTWs; seven discharge to surface streams; and one employs land application with contractor disposal of solids. All seven plants discharging to surface streams employ biological treatment before discharge. Pretreatment for the POTW dischargers consists of coagulation, flocculation, and primary treatment for seven of the nine dischargers, equalization for one discharger, and biological treatment for the other plant.

12.3.4 Tire and Inner Tube Manufacturing

There are a total of 73 tire and inner tube manufacturing facilities in the United States, of which 39 were placed in operation prior to 1959. Twenty-three of the pre-1959 plants do not treat their wastewaters, and six of these plants discharge to POTWs. A total of 17 plants placed in operation since 1959 provide no treatment of their wastewaters, and 10 of these plants discharge into POTWs.

Table 14 Plant-Specific Classical Pollutant Data for Solution Crumb Rubber Production Plant

Parameter	Waste load											
	Plant 00005						Plant 000027					
	Influent		Treated effluent		BPT regulation		Influent		Treated effluent		BPT regulation	
BOD_5	95	(210)	68	(150)	51	(110)	1,200	(2,700)	<90	(<200)	160	(360)
COD	250	(550)	140	(300)	500	(1,100)	2,700	(5,900)	450	(1,000)	1,600	(3,600)
TSS	19	(41)	11	(25)	83	(180)	1,300	(2,800)	11	(25)	270	(590)
Oil and grease	100	(230)	14	(30)	20	(45)	45	(100)	<90	(<200)	66	(150)
pH (pH units)					6 to 9						6 to 9	
Phenol	<0.01	(0.01)	<0.01	(0.01)			1.0	(2.3)	0.16	(0.35)		

Analytic methods: V.7.3.29, Data set 2.
Parameter data: kg/day (lb/day).
Source: USEPA.

Table 15 Plant-Specific Verification Data for Pan, Dry Rubber Digestion, and Mechanical Reclaiming Plant 000134

Pollutant (μg/L)	Location in process line			
	Treatment influent, automatic sampler	Treatment effluent, grab composite		Treatment effluent, automatic samples
Cadmium	<1	<1		3
Chromium	6	4		21
Copper	31	<1		12
Lead	70	290		670
Mercury		1.9		2.3
Zinc	100	2,700		2,500
Bis(2-ethylhexyl)phthalate	16,000	<80		4,300
Di-*n*-butyl phthalate				
2,4-Dimethylphenol	58,000	56,000		15,000
Phenol	26,000	21,000		4,900
Benzene	<60	<10		<0.1
Chlorobenzene				
Ethylbenzene	8,600	<0.1		<0.1
Toluene	2,700	<0.1		<0.1
Acenaphthylene	<33	<8		<8
Anthracene				
Phenanthrene	1,400[c]	<49		<300
Fluorene	2,000[c]	<40		<12
Naphthalene	100,000[c]	<12		<44
Pyrene	6,700[c]	<10		<14
Chloroform	1.9	1.3		1.4
Methylene chloride				
	Cooling tower blowdown, grab composite	Steam condensate, grab composite	Boiler blowdown	Intake water
Cadmium	<1	35	1	<1
Chromium	<2	33	2	<1

(*continues*)

Table 15 *Continued*

Pollutant (μg/L)	Location in process line			
	Treatment influent, automatic sampler	Treatment effluent, grab composite		Treatment effluent, automatic samples
Copper	3	20	6	<1
Lead	29	330	22	10
Mercury	0.5	1.0	0.9	0.8
Zinc	100		220	30
Bis(2-ethylhexyl)phthalate	120	2,800	940	1,300
Di-*n*-butyl phthalate		1,900		
2,4-Dimethylphenol	<510	730	11	<6
Phenol	<130	950	12	<4
Benzene	<0.4	27[b]	<0.1	<0.1
Chlorobenzene		25,000[b]		
Ethylbenzene	0.1	<0.1[b]	<0.1	<0.1
Toluene	<0.1	<0.1[b]	<0.1	<0.1
Acenaphthylene	<8	<16	<8	<8
Anthracene		<190		
Phenanthrene	<110	<140	340	<4
Fluorene	<12	<12	<12	<12
naphthalene	<17	<1,400[c]	<12	<12
Pyrene	<32	<29	<8	15
Chloroform	4.9	3.3[b]	1.0[a]	36[a]
Methylene chloride		1,300[b]		

Analytic methods: V.7.3.29, Data set 2.
Blanks indicate data not available.
[a]Based on first 24-hour composite sample.
[b]Based on second and third 24-hour composite samples.
[c]Interference may have caused this value to be too high.
Source: USEPA.

Table 16 Toxic Pollutant Removal Efficiency at Emulsion Crumb Rubber Plant 000012, Verification Data (Treatment Technology: Activated Sludge, Discharge Point: Surface Stream)

	Concentration (μg/L)		
Pollutant	Influent	Effluent	Percent removal
Cadmium	1	<1	NM
Mercury[a]	2.5	1.6	36
Nickel	610	400	34
Bis(2-ethylhexyl)phthalate[b]	260	220	15
Dimethyl phthalate	<14	<14	NM
N-nitrosodiphenylamine	5.2	1.6	69
Phenol	41	19	54
Nitrobenzene	<30	<30	NM
Toluene	250	<0.1	>99
Carbon tetrachloride	4.7	0.1	98
Chloroform	27	4.1	85
1,1-Dichloroethylene	<1.7	<1.7	NM
Methylene chloride	<0.1	0.9	NM
Tetrachloroethene	1.4	<0.1	>93
1,1,1-Trichloroethane	1.0	3.3	NM

Analytic methods: V.7.3.29, Data set 2.
NM, not meaningful.
[a]Intake measured at 1.5 μg/L, making plant's contribution minimal.
[b]Analytical methodology for phthalates is questionable. Therefore, significance of values reported is unknown.
Source: USEPA.

The toxic pollutants present in raw wastewaters from tire and inner tube manufacturing operations are volatile organic pollutants that are used as degreasing agents in tire production. These toxic pollutants (methylene chloride, toluene, trichloroethylene) were found to be reduced to insignificant levels across sedimentation ponds.

12.3.5 Rubber Reclamation Plants

There are nine rubber reclaiming plants in the United States. Two of these use wet digestion, and all nine use pan, mechanical, and dry digestion. Eight of the plants discharge to POTWs. The other plant employs cartridge filtration and activated carbon for oil removal, followed by activated sludge. Table 20 shows the pollutant removal efficiency at a dry digestion reclaiming plant.

12.3.6 Rubber Fabricating Operations

Rubber fabricating operations include latex-dipped, extruded, and molded goods (LDEM), and general molded, extruded, and fabricated rubber (GMEF). There are an estimated 1385 rubber fabricating plants in the United States.

No treatment method descriptions are currently available for this subcategory. Wastewater treatment technology consistent with equalization and sedimentation may permit compliance with BPT regulations.

Table 17 Toxic Pollutant Removal Efficiency at Emulsion Crumb Rubber Plant 000033, Verification Data

Pollutant	Concentration ($\mu g/L$) Influent	Effluent	Percent removal
Cadmium[a]	40	40	0
Chromium[a]	250	220	12
Copper[a]	1,400	410	71
Mercury[a]	3.2	4.9	NM
Selenium[a]	<20	20	NM
Bis(2-ethylhexyl) phthalate[b] (range)	100 (65–140)	94 (59–130)	6
Acrylonitrile	32,000	<23,000	>28
2-Nitrophenol	9	3	67
Phenol	60	19	68
Ethylbenzene	<0.1	<0.1	NM
Toluene	<0.1	<0.1	NM
Chloroform	8.2	1.8	78
Dichlorobromomethane	0.3	0.1	67
Methylene chloride[c]	66	110	NM

Analytic methods: V.7.3.29, Data set 2.
NM, not meaningful.
[a]Found at potentially significant levels in treatment effluent although generally higher than during screening.
[b]Analytical methodology for phthalates is questionable; therefore, significance of values reported is unknown.
[c]Suspected contaminant from glassware cleaning procedures or analytical methods.
Treatment technology: Primary flocculation/separation, aerated lagoons.
Discharge point: Surface stream.
Source: USEPA.

12.4 TREATMENT METHODS

The treatment methods for ruber wastewaters consist of various biological processes, and physico-chemical processes including coagulation, ozonation, activated carbon adsorption, aeration, sulfonation, chlorination, and aeration, and biological nutrient removal processes. The purpose of the treatment is to meet USEPA effluent limitations [4].

12.4.1 Coagulation

Municipal wastewaters containing synthetic latexes have been treated with coagulants with 84% BOD removal efficiency using $Al_2(SO_4)_3$, $Fe_2(SO_4)_3$, and $FeCl_3$ [5].

12.4.2 Odor Control

For butadiene wastewaters with initial odor concentration of 1200, it was reduced to 100 and 250 after bubble aeration and spray aeration, respectively. For styrene wastewaters with initial odor concentration of 1000, it was reduced to 50 with bubble aeration and there was no odor reduction using spray aeration [6]. Sulfonation treatment for butadiene wastewater with initial odor

Table 18 Toxic Pollutant Removal Efficiency at Solution Crumb Rubber Plant 000005

Pollutant	Concentration (μg/L)		Percent removal
	Influent	Effluent	
Cadmium	<1	<1	NM
Copper	9	14	NM
Chromium	75	410	NM
Zinc	14,000	13,000	7
Bis(2-ethylhexyl)phthalate[a]	180	24	87
Phenol	7	5	29
Benzene	<22	110[b]	NM
Ethylbenzene	<46	<39	NM
Toluene	<26	<26	NM
Carbon tetrachloride	35	14	60
Chloroform	2.2	1.3	41
Methyl chloride	4,900	2,200[c]	55
Methylene chloride	<1.0	<1.0	NM
1,1,2-Trichloroethane	<0.1	<0.1	NM
Trichloroethylene	<0.1	<0.1	NM

Analytic methods: V.7.3.29, Data set 2; NM, not meaningful.
[a]Analytical methodology for phthalates is questionable; therefore, significance of values reported is unknown.
[b]Average of 320 μg/L, <11 μg/L.
[c]Found at significant levels in treatment effluent.
Treatment technology: Primary flocculation/clarification (DAF).
Discharge point: Treated effluent is discharged to a nearly oil refinery's cooling water system.
Source: USEPA.

Table 19 Toxic Pollutant Removal Efficiency at Solution Crumb Rubber Plant 000027, Verification Data

Pollutant	Concentration (μg/L)		Percent removal
	Influent	Effluent	
Cadmium	<1	1	NM
Chromium	440	20	95
Copper	7	5	29
Mercury[a]	1.1	2.0	NM
Bis(2-ethylhexyl)phthalate[b]	120	110	8
Phenol	180	11	94
Benzene	3,200	<0.1	>99
Ethylbenzene	<0.1	<0.1	NM
Toluene	<0.1	<0.1	NM
Chloroform	3.0	0.9	70
1,1,2,2-Tetrachloroethane	<0.1	<0.1	NM

Analytic methods: V.7.3.29, Data set 2; NM, not meaningful.
[a]Intake measured at 4 μg/L, making plant's contribution zero.
[b]Analytical methodology for phthalates is questionable; therefore, significance of values reported is unknown.
Treatment technology: Sedimentation, waste stabilization lagoons.
Discharge point: Surface stream.
Source: USEPA.

Table 20 Toxic Pollutant Removal Efficiency at Dry Digestion Reclaiming Plant 000134, Verification Data

Pollutant (μg/L)	Concentration (μg/L) Influent	Concentration (μg/L) Effluent	Percent removal	Cooling tower blowdown (μg/L)[a]	Percent removal
Cadmium	1	3	NM	<1	>67
Chromium	6	21	NM	2	90
Copper	28	12	57	2	83
Lead[b]	70	670	NM	29	96
Mercury		2.3		0.5	78
Zinc	100	2,500	NM	100	96
Bis(2-ethylhexyl)phthalate[c]	16,000	4,200	74	100	98
2,4-Dimethylphenol	58,000	25,000	57	120	>99
Phenol	26,000	4,900	81	27	>99
Benzene	60	<0.1	>99	<0.1	NM
Ethylbenzene	8,600	<0.1	>99	<0.1	NM
Toluene	2,700	<0.1	>99	<0.1	NM
Acenaphthylene	<33	<8	NM	<8	NM
Fluorene	2,000	<12	>99	<12	NM
Naphthalene[d]	100,000	42	>99	13	69
Phenanthrene	1,300	300	77	<4	>99
Pyrene	6,800	11	>99	4	64
Chloroform	1.9	1.4	26	4.9	NM

Analytic methods: V.7.3.29, Data set 2.
NM, not meaningful.
[a]Effluent from treatment goes into cooling tower and is discharged with noncontact cooling water as cooling tower blowdown.
[b]Potentially significant levels observed in cooling tower blowdown.
[c]Analytical methodology for phthalates is questionable; therefore, significance of values raported is unknown.
[d]Significance of blowdown value questionable due to high detection limit, low values observed in the carbon column effluent (treatment influent), and the fact that compound is not a metabolic byproduct of activated sludge treatment.
Treatment technology: Cartridge filtration, activated carbon (oil removal), activated sludge sedimentation.
Discharge point: Noncontact cooling water system, blowdown of this system to surface stress.
Source: USEPA.

concentration of 4100 reduced odor to 250 and 500 using 100 ppm Na_2SO_3 and Na_2S respectively after 17 days of treatment. For styrene wastewaters of initial odor concentration of 128, it was reduced to 4 after 9 days treatment with 100 ppm Na_2SO_3 and for the same wastewaters with initial odor concentration of 65 ppm, it was reduced to 4 ppm using Na_2S using the same dosage and same duration of treatment [6].

12.4.3 Biological Treatment

A trickling filter has been used to treat neutralized rubber wastewaters with initial BOD of 445 mg/L. It removed 92.1% BOD with a 24-hour detention time [5]. The activated sludge process has removed 85% BOD from combined rubber and domestic wastewater wastewaters [7].

12.4.4 Nutrient Removal

Attached-growth waste stabilization ponds have been used to remove 65–70% TKN (total Kjeldahl nitrogen), and 70–83% NH_3-N from concentrated latex and rubber sheet plant wastewaters [8]. A combined algae and water hyacinth system has been used to remove 96.41% COD, 98.93% TKN, 99.28% NH_3-N, 100% NO_2-N, and 100% NO_3-N [9].

12.5 TREATMENT TECHNOLOGY COSTS

The investment cost, operating, and maintenance costs, and energy costs for the application of control technologies to the wastewaters of the rubber processing industry have been analyzed. These costs were developed to reflect the conventional use of technologies in this industry. Several unit operation/unit process configurations have been analyzed for the cost of application of technologies and to select BPT and BAT level of treatment. The applicable treatment technologies, cost methodology, and cost data are available in a detailed presentation [10].

REFERENCES

1. U.S. Department of Labor – Occupational Safety and Health Administration. *SIC Division Structure*, http://www.osha.gov/cgi-bin/sic/sicser5, 2003.
2. USEPA. *Subchapter N – Effluent Guidelines and Standards in CFR Title 40*, Protection of Environment; http://www.epa.gov/docs/epacfr40/chapt-I.info/subch-N.htm, 2003.
3. Sechrist, W.D.; Chamberlain, N.S. Chlorination of phenol bearing rubber wastes. In *Proceedings of 6th Industrial Waste Conference*, Purdue University, Lafasette, IN, November 1951; 396.
4. USEPA. *Development Document for Effluent Limitation Guidelines and New Source Performance Standards for the Tire and Synthetic Segment of the Rubber Processing Point Source Category*, U.S. Environmental Protection Agency: Washington, DC. http://www.epa.gov/cgi-bin/, 1974.
5. Morzycki, J. Effluents from sewage contaminated with latex. Chem. Abs. **1966**, *64*, 3192.
6. Black, O.R. Study of wastes from rubber industry. Sewage Works J. **1946**, *18*, 1169.
7. Mills, R.E. Progress report on the bio-oxidation of phenolic and 2,4-D waste waters. In *Proceedings of 4th Ontario Industrial Waste Conference*, June 1957; 30.

8. Rakkoed, A.; Danteravanich, S.; Puetpaiboon, U. Nitrogen removal in attached growth waste stabilization ponds of wastewater from a rubber factory. Water Sci. Technol. **1999**, *40(1)*, 45–52.
9. Bich, N.N.; Yaziz, M.I.; Bakti, N.A.K. Combination of *Chlorella vulgaris* and *Eichhornia crassipes* for wastewater nitrogen removal. Water Res. **1999**, *33* (*10*), 2357–2362.
10. USEPA and Envirodyne Engineers, Inc. *Review of the Best Available Technology for the Rubber Processing Point Source Category*, Technical Report No. 68-01-4673; US Environmental Protection Agency: Washington, DC, 1978.

13

Treatment of Power Industry Wastes

Lawrence K. Wang
Lenox Institute of Water Technology and Krofta Engineering Corporation, Lenox, Massachusetts and Zorex Corporation, Newtonville, New York, U.S.A.

13.1 INTRODUCTION

13.1.1 Steam Electric Power Generation Industry

The steam electric power generation industry is defined as those establishments primarily engaged in the steam generation of electrical energy for distribution and sale. Those establishments produce electricity primarily from a process utilizing fossil-type fuel (coal, oil, or gas) or nuclear fuel in conjunction with a thermal cycle employing the steam–water system as the thermodynamic medium. The industry does not include steam electric power plants in industrial, commercial, or other facilities. The industry in the United States falls under two Standard Industrial Classification (SIC) Codes: SIC 4911 and SIC 4931.

There are about 1000 steam electric power generating plants in operation in the United States. Of these plants, approximately 35% generate in excess of 500 megawatts (MW) and approximately 12% generate 25 MW or less. These steam electric power generating plants represent about 79% of the entire electric utility generating capacity, and they generate about 85% of electricity produced by the entire electric utility industry. Within the steam electric power generation industry, plants built after 1970 represent 44% of the total capacity, and plants built before 1960 represent 26% of capacity.

"Small units" are defined by the U.S. Environmental Protection Agency (USEPA) as generating units of less than 25 MW capacity. "Old units" are defined as generating units of 500 MW or greater rated net generating capacity that were first placed into service on or before January 1, 1970, as well as any generating unit of less than 500 MW capacity first placed in service on or before January 1, 1974.

The term "10-year, 24-hour rainfall event" refers to a rainfall event with a probable recurrence interval of once in 10 years as defined by the National Weather Service.

13.1.2 Power Generation, Waste Production, and Effluent Discharge

In the operation of a power plant, combustion of fossil fuels – coal, oil, or gas – supplies heat to produce steam, which is used to generate mechanical energy in a turbine. This energy is subsequently converted by a generator to electricity. Nuclear fuels, currently uranium, are used in a similar cycle except that the heat is supplied by nuclear fusion wastewater discharge. A number of different operations by steam electric power plants discharge chemical wastes. Many wastes are discharged more or less continuously as long as the plant is operating. These include wastewaters from the following sources: cooling water systems, ash handling systems, wet-scrubber air pollution control systems, and boiler blowdown. Some wastes are produced at regular intervals, as in water treatment operations, which include a cleaning or regenerative step as part of their cycle (ion exchange, filtration, clarification, evaporation). Other wastes are also produced intermittently but are generally associated with either the shutdown or startup of a boiler or generating unit, such as during boiler cleaning (water side), boiler cleaning (fire side), air preheater cleaning, cooling tower basin cleaning, and cleaning of miscellaneous small equipment.

The discharge frequency for these varies from plant to plant. Some or all of the various types of wastewater streams occur at almost all of the plant sites in the industry. However, most plants do not have distinct and separate discharge points for each source of wastewater; rather, they combine certain streams prior to final discharge.

Additional wastes exist that are essentially unrelated to production. These depend on meteorological or other factors. Rainfall runoff, for example, causes drainage from coal piles, ash piles, floor and yard drains, and from construction activity.

The summary for the steam electric power generating (utility) point source category in terms of the number of dischargers in industry is as follows:

- direct dischargers in industry: 1050;
- indirect dischargers in industry: 100;
- zero dischargers in industry: 10.

Current BPT regulations for the steam electric power industry for generating, small and old units can be found elsewhere [1].

13.2 INDUSTRY SUBCATEGORY AND SUBDIVISIONS

Subcategories for the steam electric utility point source category are developed according to chemical waste stream origin within a plant. This approach is a departure from the usual method of subcategorizing an industry according to different types of plants, products, or production processes. Categorization by waste source provides the best mechanism for evaluating and controlling waste loads since the steam electric power plant waste stream source has the strongest influence on the presence and quantity of various pollutants as well as on flow. The breakdown of the stream electric power generation industry into subcategories and subdivisions is based on similarities in wastewater characteristics throughout the industry. The eight broad subcategories and their subdivisions are presented below:

1. Once-through cooling water.
2. Recirculating cooling system blowdown.
3. Ash transport water:
   - fly ash transport;
   - bottom ash transport.
4. Low volume wastes:
   - clarifier blowdown;
   - makeup water filter backwash;
   - ion exchange softener regeneration;
   - evaporator blowdown;
   - lime softener blowdown;
   - reverse osmosis brine;

- demineralizer regenerant;
- powdered resin demineralizer;
- floor drains;
- laboratory drains;
- sanitary wastes; and
- diesel engine cooling system discharge.

5. Metal cleaning wastes:
   - boiler tube cleaning;
   - cleaning rinses;
   - fireside wash; and
   - air preheater wash.
6. Ash pile, chemical handling, and construction area runoff coal pile runoffs.
7. Coal pile runoff.
8. Wet flue gas cleaning blowdown.

13.2.1 Once-Through Cooling Water Subcategory

In a steam electric power plant, cooling water is utilized to absorb heat that is liberated from the steam when it is condensed to water in the condensers. The cooling water is withdrawn from a water source, passed through the system, and returned directly to the water source. Shock (intermittent) chlorination is employed in many cases to minimize the biofouling of heat transfer surfaces. Continuous chlorination is used only in special situations. Based on 308 data, approximately 65% of the existing steam electric power plants have once-through cooling water systems.

13.2.2 Recirculating Cooling Water Subcategory

In a recirculating cooling water system, the cooling water is withdrawn from the water source and passed through condensers several times before being discharged to the receiving water. After each pass through the condenser, heat is removed from the water through evaporation. Evaporation is carried out in cooling ponds or canals, in mechanical draft evaporative cooling towers, and in natural draft evaporative cooling towers. In order to maintain a sufficient quantity of water for cooling, additional makeup water must be withdrawn from the water source to replace the water that evaporates.

When water evaporates from the recirculating cooling water system, the dissolved solids content of the water remains in the system, and the dissolved solids concentration tends to increase over time. If left unattended, the formation of scale deposits will result. Scaling due to dissolved solids buildup is usually controlled through the use of a bleed system called cooling tower blowdown. A portion of the cooling water in the system is discharged via blowdown, and since the discharged water has a higher dissolved solids content than the intake water used to replace it, the dissolved solids content of the water in the system is reduced.

Chemicals such as sulfuric acid are used to control scaling in the system. Biofoulants such as chlorine and hypochlorite are widely used by the industry. These additives are discharged in the cooling tower blowdown.

13.2.3 Ash Transport Water Subcategory

Steam electric power plants using oil or coal as a fuel produce ash as a waste product of combustion. The total ash product is a combination of bottom ash and fly ash. Because the ash composition of oil is much less than that of coal, the presence of ash is an extremely important consideration in the design of a coal-fired boiler. Improper design leads to the accumulation of ash deposits on furnace walls and tubes, leading to reduced heat transfer, increased pressure drop, and corrosion. Accumulated ash deposits are removed and transported to a disposal system.

The method of transport may be either wet (sluicing) or dry (pneumatic). Dry handling systems are more common for fly ash than bottom ash. The dry ash is usually disposed of in a landfill, but the ash is also sold as an ingredient for other products. Wet ash handling systems produce wastewaters that are either discharged as blowdown from recycle systems or are discharged to ash ponds and then to receiving streams in recycle and once-through systems.

Ash From Oil-Fired Plants

Fly ash is a light material, which is carried out of the combustion chamber in the flue gas stream. The ash from fuel oil combustion is usually in the form of fly ash. The many elements that may appear in oil ash deposits include vanadium, sodium, and sulfur.

Ash From Coal-Fired Plants

More than 90% of the coal used by electric utilities is burned in pulverized coal boilers. In these boilers, 65–80% of the ash produced is in the form of fly ash. This fly ash is carried out of the combustion chamber in the flue gases and is separated from these gases by electrostatic precipitators and/or mechanical collectors. The remainder of the ash drops to the bottom of the furnace as bottom ash. While most of the fly ash is collected, a small quantity may pass through the collectors and be discharged to the atmosphere. The vapor is that part of the coal material that is volatilized during combustion. Some of these vapors are discharged into the atmosphere; others are condensed onto the surface of fly ash particles and may be collected in one of the fly ash collectors.

13.2.4 Low-Volume Wastes Subcategory

Low-Volume Blowdowns

Low-volume wastes include wastewaters from all sources except those for which specific limitations are otherwise established in 40 CFR 423. Waste sources include, but are not limited to, wastewaters from wet scrubber air pollution control systems [2], ion exchange water treatment systems, water treatment evaporator blowdown, laboratory and sampling streams, floor drainage, cooling tower basin cleaning wastes, and blowdown from recirculating house service water systems. Sanitary wastes and air conditioning wastes are specifically excluded from the low-volume waste subcategory.

Boiler Blowdown

Power plant boilers are either of the once-through or drum-type design. Once-through boilers operate under supercritical conditions and have no wastewater streams directly associated with their operation. Drum-type boilers operate under subcritical conditions where steam generated in the drum-type units is in equilibrium with the boiler water. Boiler water impurities are concentrated in the liquid phase. Boiler blowdown serves to maintain concentrations of dissolved and suspended solids at acceptable levels for boiler operation. The sources of impurities in the blowdown are the intake water, internal corrosion of the boiler, and chemicals added to the boiler. Phosphate is added to the boiler to control solids deposition.

In modern high-pressure systems, blowdown water is normally of better quality than the water supply. This is because plant intake water is treated using clarification, filtration, lime/lime soda softening, ion exchange, evaporation, and in a few cases reverse osmosis to produce makeup for the boiler feedwater. The high-quality blowdown water is often reused within the plant for cooling water makeup or it is recycled through the water treatment and used as boiler feedwater.

13.2.5 Metal Cleaning Wastes Subcategory

Metal cleaning wastes result from cleaning compounds, rinse waters, or any other waterborne residues derived from cleaning any metal process equipment, including, but not limited to, boiler tube cleaning, boiler fireside cleaning, and air preheater cleaning.

Boiler Tube Cleaning

Chemical cleaning is designed to remove scale and corrosion products that accumulate in the steam side of the boiler. Hydrochloric acid, which forms soluble chlorides with the scale and corrosion products in the boiler tubes, is the most frequently used boiler tube cleaning chemical. In boilers containing copper, a copper complexer is used with hydrochloric acid to prevent the replating of dissolved copper onto steel surfaces during chemical cleaning operations. If a complexer is not used, copper chlorides, formed during the cleaning reaction, react with boiler tube iron to form soluble iron chlorides while the copper is replated onto the tube surface.

Alkaline cleaning (flush/boil-out) is commonly employed prior to boiler cleaning to remove oil-based compounds from tube surfaces. These solutions are composed of trisodium phosphate and a surfactant and act to clear away the materials that may interfere with reactions between the boiler cleaning chemicals and deposits.

Citric acid cleaning solutions are used by a number of utilities in boiler cleaning operations. The acid is usually diluted and ammoniated to a pH of 3.5 and then used for cleaning in a two-stage process. The first stage involves the dissolution of iron oxides. In the second stage, anhydrous ammonia is added to raise the pH of the cleaning solution to between 9 and 10 and air is bubbled through the solution to dissolve copper deposits.

Ammoniated EDTA has been used in a wide variety of boiler cleaning operations. The cleaning involves a one-solution, two-stage process. During the first stage, the solution solubilizes iron deposits and chelates the iron solution. In the second stage, the solution is oxidized with air to induce iron chelates from ferric to ferrous and to oxidize copper deposits into solution where the copper is chelated. The most prominent use of this agent is in circulating boilers that contain copper alloys.

When large amounts of copper deposits in boiler tubes cannot be removed with hydrochloric acid due to the relative insolubility of copper, ammonia-based oxidizing compounds have been effective. Used in a single separate stage, the ammonia sodium bromate step includes the introduction into the boiler system of solutions containing ammonium bromate to rapidly oxidize and dissolve the copper.

The use of hydroxyacetic/formic acid in the chemical cleaning of utility boilers is common. It is used in boilers containing austenitic steels because its low chloride content prevents possible chloride stress corrosion cracking of the austenitic-type alloys. It has also found extensive use in the cleaning operations for once-through supercritical boilers. Hydroxyacetic/formic acid has chelation properties and a high iron pick-up capability; thus it is used on high iron content systems. It is not effective on hardness scales.

Sulfuric acid has found limited use in boiler cleaning operations. It is not feasible for removal of hardness scales due to the formation of highly insoluble calcium sulfate. It has found some use in cases where a high-strength, low-chloride solvent is necessary. Use of sulfuric acid requires high water usage in order to rinse the boiler sufficiently.

Boiler Fireside Washing

Boiler firesides are commonly washed by spraying high-pressure water against boiler tubes while they are still hot.

Air Preheater Washing

Air preheaters employed in power generating plants are either the tubular or regenerative types. Both are periodically washed to remove deposits that accumulate. The frequency of washing is typically once per month; however, frequency variations ranging from 5 to 180 washings per year are reported. Many preheaters are sectionalized so that heat transfer areas may be isolated and washed without shutdown of the entire unit.

13.2.6 Ash Pile, Chemical Handling, and Construction Area Runoff Subcategory

Fly ash and bottom ash stored in open piles, chemicals spilled in handling, and soil distributed by construction activities will be carried in the runoff caused by precipitation events.

13.2.7 Coal Pile Runoff Subcategory

In order to ensure a consistent supply of coal for steam generation, plants typically maintain an outdoor 90-day reserve supply. The piles are usually not enclosed, so the coal comes in contact with moisture and air, which can oxidize metal sulfides to sulfuric acid. Precipitation then results in coal pile runoff with minerals, metals, and low pH (occasionally) in the stream.

13.2.8 Wet Flue Gas Cleaning Blowdown Subcategory

Depending on the fossil fuel sulfur content, an SO_2-scrubber may be required to remove sulfur emissions in the flue gases. These scrubbing systems result in a variety of liquid waste streams depending on the type of process used. In all of the existing FGD (flue gas desulfurization) systems, the main task of absorbing SO_2 from the stack gases is accomplished by scrubbing the existing gases with an alkaline slurry. This may be preceded by partial removal of fly ash from the stack gases. Existing FGD processes may be divided into two categories: nonregenerable FGD processes including lime, limestone, and lime/limestone combination, and double alkali systems.

In the lime or limestone FGD process, SO_2 is removed from the flue gas by wet scrubbing with a slurry of calcium oxide or calcium carbonate [3]. The waste solid product is disposed by ponding or landfill. The clear liquid product can be recycled. Many of the lime or limestone systems discharge scrubber waters to control dissolved solids levels.

A number of processes can be considered double alkali processes, but most developmental work has emphasized sodium-based systems, which use lime for regeneration. This system pretreats the flue gas in a prescrubber to cool and humidify the gas and to reduce fly ash and chlorides. The gas passes through an absorption tower where SO_2 is removed into a scrubbing solution, which is subsequently regenerated with lime or limestone in a reaction tank.

The disadvantage of all nonregenerable systems is the production of large amounts of throwaway sludges. Onsite disposal is usually performed by sending the waste solids to a settling pond. The supernatant from the ponds may be recycled; however, according to 308 data, 82% of the plants with FGD systems discharged the supernatant into surface waters.

13.3 WASTEWATER

13.3.1 Characterization

Wastewater produced by a steam electric power plant can result from a number of operations at the site. Many wastewaters are discharged more or less continuously as long as the plant is operating. These include wastewaters from the following sources: cooling water systems, ash handling systems, wet-scrubber air pollution control systems, and boiler blowdown. Some wastes are produced at regular intervals, as in water treatment operations that include a cleaning or regenerative step as part of their cycle (ion exchange, filtration, clarification, evaporation). Other wastes are also produced intermittently but are generally associated with either the shutdown or startup of a boiler or generating unit such as during boiler cleaning (water side), boiler cleaning (fire side), air preheater cleaning, cooling tower basin cleaning, and cleaning of miscellaneous small equipment. Additional wastes exist that are essentially unrelated to production. These depend on meteorological or other factors. Rainfall runoff, for example, causes drainage from coal piles, ash piles, floor and yard drains, and from construction activity. A diagram indicating potential sources of wastewaters containing chemical pollutants in a coal-fueled steam electric power plant is shown in Figure 1.

Data on wastestream characteristics presented in this section are based on the results of screening sampling carried out at eight plants, verification sampling carried out at 18 plants, and periodic surveillance and analysis sampling carried out as part of compliance monitoring at eight plants. These data were stored on a computerized data file [1]. All waste streams discussed in this chapter were analyzed during the screening program, while the verification program focused on the following waste streams: once-through cooling water, cooling tower blowdown, and ash handling waters. The wastewater characteristics of the various waste streams are discussed in the following sections. Where they are available, only verification data are presented. Where verification data are limited or not available, screening and/or surveillance and analysis data are presented. The data source is clearly indicated in each table and in the text.

The following is a summary of all priority pollutants detected in any of the waste streams from steam electric power plants:

- Benzene
- Chlorobenzene
- 1,2-Dichloroethane
- 1,1,1-Trichloroethane
- 1,1,2-Trichloroethane
- 2-Chloronaphthalene
- Chloroform
- 2-Chlorophenol
- 1,2-Dichlorobenzene
- 1,4-Dichlorobenzene
- 1,1-Dichloroethylene
- 1,2-*trans*-Dichloroethylene
- 2,4-Dichiorophenol

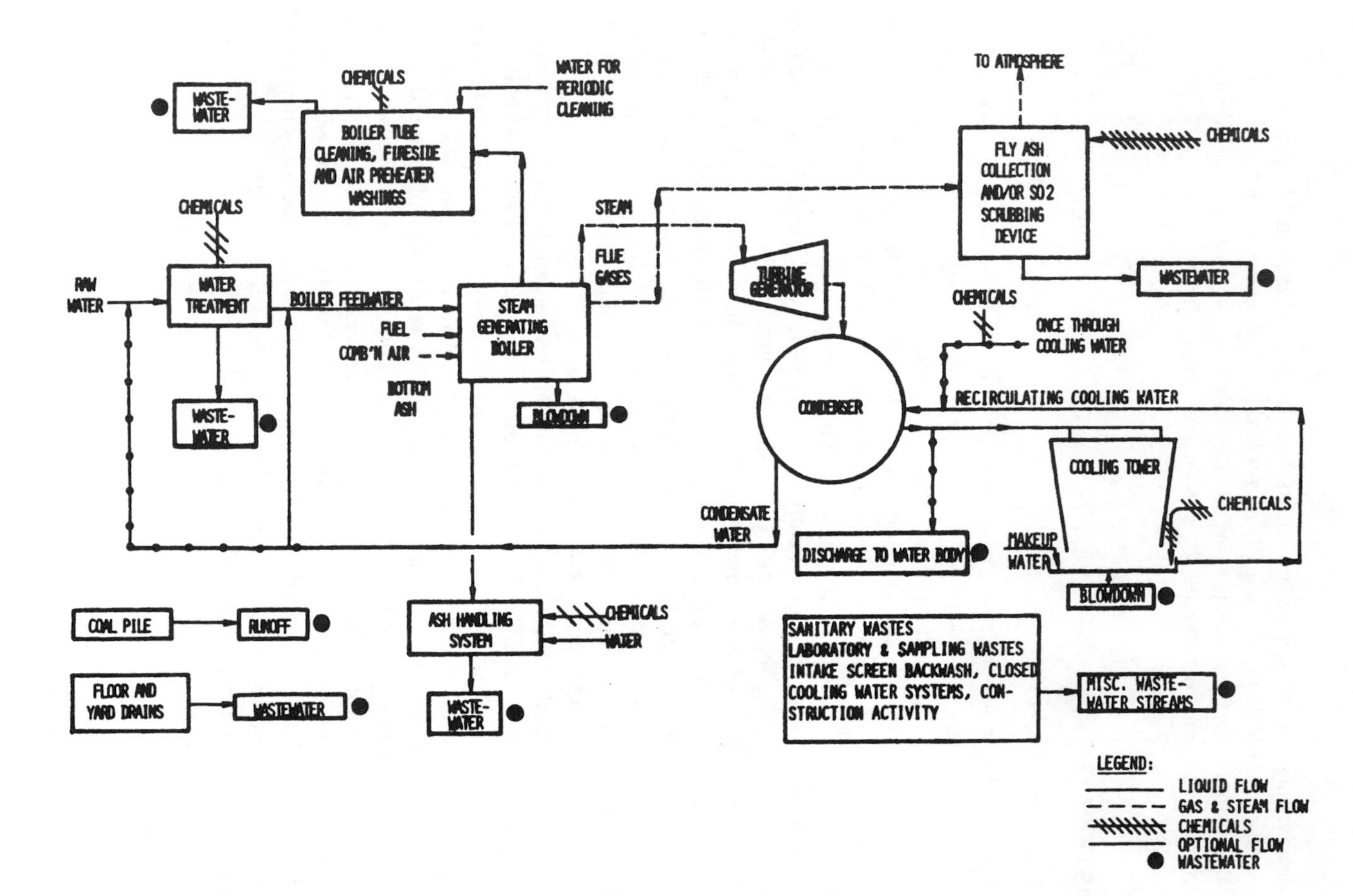

Figure 1 Potential sources of wastewater in a stream electric power generation plant. (Courtesy of USEPA.)

- Ethylbenzene
- Methylene chloride
- Bromoform
- Dichlorobromomethane
- Trichlorofluoromethane
- Chlorodibromomethane
- Nitrobenzene
- Pentachlorophenol
- Phenol
- *Bis*(2-ethylhexyl) phthalate
- Butyl benzyl phthalate
- Di-*n*-butyl phthalate
- Di-*n*-octyl phthalate
- Diethyl phthalate
- Dimethyl phthalate
- Tetrachloroethylene
- Toluene
- Trichloroethylene
- 4,4-DDD
- Antimony (total)
- Arsenic (total)
- Asbestos (total-fibers/L)
- Beryllium (total)
- Cadmium (total)
- Chromium (total)
- Copper (total)
- Cyanide (total)
- Lead (total)
- Mercury (total)
- Nickel (total)
- Selenium (total)
- Silver (total)
- Thallium (total)
- Zinc (total)

13.3.2 Cooling Water

In general, wastewater characteristics of once-through cooling water and recirculating cooling water systems are similar. Pollutants discharged from both systems are caused by the erosion or corrosion of construction materials plus the chemical additives used to control erosion, scaling, and biological growth (biofouling). The wastewater generated from a recirculating cooling water system also depends on the design limits for dissolved solids in the system.

Erosion

The fill material in natural draft cooling towers is frequently asbestos cement. Erosion of this fill material may result in the discharge of asbestos in cooling water blowdown. In a testing program for detection of asbestos fibers in the waters of 18 cooling systems, seven of the 18 sites contained detectable concentrations of chrysotile asbestos in the cooling tower waters at the time of sampling.

Corrosion

Corrosion is an electrochemical process that occurs when metal is immersed in water and a difference in electrical potential between different parts of the metal causes a current to pass through the metal between the region of lower potential (anode) and the region of higher potential (cathode). The migration of electrons from anode to cathode results in the oxidation of the metal at the anode and the dissolution of metal ions into the water.

Copper alloys are used extensively in power plant condensers, and as a result, copper can usually go into a corrosion product film or directly into solution as an ion or as a precipitate in the initial stages of condensation by tube corrosion. As corrosion products form and increase in thickness, the corrosion rate decreases until a steady state is achieved. Studies indicate that copper release is a function of flow rate more so than of the salt content of the makeup water.

Data on copper concentrations in both once-through cooling and recirculatory cooling systems indicate that corrosion products are more of a problem in cooling tower blowdown than in once-through systems discharge. The concentration of pollutants (via evaporation) in recirculating systems probably accounts for most of the difference in the level of metals observed between once-through discharge and cooling tower blowdown.

Chemical Treatment

Chemical additives are needed at some plants with recirculating cooling water systems in order to prevent corrosion and scaling. Chemical additives are also occasionally used at plants with once-through cooling water systems for corrosion controls.

Scaling occurs when the concentration of dissolved materials, usually calcium- and magnesium-containing species, exceeds their solubility levels. The addition of scaling control chemicals allows a higher dissolved solids concentration to be achieved before scaling occurs.

Therefore, the amount of blowdown required to control scaling can be reduced. Chemicals added to once-through cooling water to control corrosion or to recirculating cooling water to control corrosion and scaling is usually present in the discharges. Chromium and zinc are the active components of most of the popular corrosion inhibitors.

The solvent and carrier components that may be used in conjunction with scaling and corrosion control agents are as follows:

- Dimethyl formamide
- Methanol
- Ethylene glycol monomethyl ether
- Ethylene glycol monobutyl ether
- Methyl ethyl ketone
- Glycols to hexylene glycol
- Heavy aromatic naphthalene
- Cocoa diamine
- Sodium chloride
- Sodium sulfate
- Polyoxyethylene glycol
- Talc
- Sodium aluminate
- Monochlorotoulene
- Alkylene oxide–alchohol glycol ethers

Chlorine and hypochlorite are used to control biofouling in both once-through and recirculating cooling water systems. The addition of chlorine to the water causes the formation of toxic compounds and chlorinated organics, which may be priority pollutants.

Eleven plants with once-through cooling water systems were sampled as part of the verification program and the surveillance and analysis sampling efforts. Four of these plants have estuarine or salt water intakes, and the remaining seven plants have fresh water intakes. Sampling was carried out only during the period of chlorination. Samples were analyzed for all organic priority pollutants except the pesticides, and for total organic carbon and total residual chlorine (nine plants). Table 1 is a summary of the data collected in the verification and surveillance and analysis sampling efforts. Only the priority pollutants that were detected are shown.

The following is a summary of once-through cooling system flow rates based on responses to 308 USEPA questionnaires.

1. The once-through cooling system flow rate range of 239 steam electric power generation plants using coal is 0.189–6,280,000 m^3/plant (mean 1,130,000 m^3/plant), which is equal to 0.001–209,000 m^3/MW (mean 4,310,000 m^3/MW).
2. The once-through cooling system flow rate range of 104 steam electric power generation plants using gas is 0.29–7,230,000 m^3/plant (mean 783,000 m^3/plant), which is equal to 0.006–13,800,000 m^3/MW (mean 2,410,000 m^3/MW).
3. The once-through cooling system flow rate range of 138 steam electric power generation plants using oil is 0.007–26,700,000 m^3/plant (mean 1,490,000 m^3/plant), which is equal to 0.00004–219,000 m^3/MW (mean 5260 m^3/MW).

The fuel designations in the above surveyed data were determined by the fuel that contributes the most Btu for power generation during the survey.

The surveyed data also indicate that there were net increases in all of the following compounds: total dissolved solids, total suspended solids, total organic carbon, total residual chlorine, free available chlorine 2,4-dichlorophenol, 1,2-dichlorobenzene, phenolics, chromium, lead, copper, mercury, silver, iron, arsenic, zinc, barium, calcium, manganese, sodium, methyl chloride, aluminum, boron, and titanium.

Eight power plants with cooling towers were sampled at intake and discharge points during the verification sampling program. The results of the verification sampling program for cooling tower blowdown (recirculating cooling water) are presented in Table 2A and 2B for intake information and discharge information, respectively. Only the priority pollutants that were detected are shown.

The following is a summary of cooling tower blowdown flow rates based on responses to 308 questionnaires:

1. The cooling tower blowdown flow rate range of 82 steam electric power generation plants using coal is 0–152,000 m^3/plant (mean 8440 m^3/plant), which is equal to 0–239 m^3/MW (mean 11.2 m^3/MW).
2. The cooling tower blowdown flow rate range of 120 steam electric power generation plants using gas is 0–10,900 m^3/plant (mean 1190 m^3/plant), which is equal to 0–99 m^3/MW (mean 11.6 m^3/MW).
3. The cooling tower blowdown flow rate range of 47 steam electric power generation plants using oil is 0–12,100 m^3/plant (mean 1040 m^3/plant), which is equal to 0–63.2 m^3/MW (mean 7.04 m^3/MW).

The data also indicate that there was a net increase from the influent concentration to the effluent concentration for the following compounds: trichlorofluoromethane, bromoform,

Table 1 Summary of Priority Pollutants in the Steam Electric Industry Once-Through Cooling Water

Pollutant	Intake					Discharge				
	Number of samples	Number of detections	Range of detections	Median of detections	Mean of detections	Number of samples	Number of detections	Range of detections	Median of detections	Mean of detections
Classical pollutants (mg/L)										
Total residual chlorine	11	6	<0.01–3	<0.01	<0.5	7	6	0.02–5.5	0.4	0.82
COD	11	1	35			7	0			
TDS	11	11	7–35,000	340	7,600	7	7	360–33,000	12,000	12,000
TSS	11	8	3–100	13	25	7	7	4–90	14	28
TOC	11	11	6–34	12	14	7	7	7.9–30	17	19
Free residual chlorine	10	1	500			7	1	190		
Phenolics	11	6	<0.005–0.015	0.010	0.009	7	5	0.007–0.26	0.02	0.06
Barium	11	5	0.01–0.06	0.024	0.029	6	0			
Calcium	11	5	0.084–51	45	37	6	0			
Manganese	11	5	0.053–0.2	0.066	0.096	6	0			
Magnesium	11	4	13–33	19	21	6	0			
Sodium	11	5	<15–49	21	24	6	0			
Iron	11	5	0.25–4	0.84	1.7	6	1	0.76		
Aluminum	11	4	0.28–2.4	1.2	1.3	6	0			
Boron	11	3	0.07–0.09	0.083	0.081	6	0			
Tin	11	2	0.03–0.036		0.033	6	0			
Titanium	11	3	0.018–0.051	0.040	0.036	6	0			
Molybdenum	11	1	0.009			6	0			
Cobalt	11	2	<0.005–0.01		0.0075	6	0			
Vanadium	11	0				6	0			
Toxic pollutants (μg/L)										
Toxic metals and inorganics										
Antimony	11	3	7–16	10	11	7	1	14		
Aresenic	11	3	3–5	4	4	7	0			
Chromium	11	6	7–24	13	14	6	1	14		
Copper	11	6	7–22	18	16	6	1	24		
Cyanide	11	1	10			7	0			
Lead	11	2	9–10		9.5	6	1	11		
Mercury	11	3	0.2–1.2	1.0	0.8	7	1	1		

Selenium	11	2	3.0–3.8		3.4	7	0	
Silver	11	1	30			6	1	36
Nickel	11	3	8–120	17	48	6	1	120
Zinc	11	4	32–340	79	130	5	1	24
Cadmium	11	2	13–40		26	6	1	16
Toxic organics								
Bromoform	9	0				11	1	31
Chlorodibromomethane	10	0				11	1	2.6
Bis(2-ethylhexyl) phthalate	11	4	<10–420	<10	110	10	0	
Gamma-BHC	10	0				11	1	<0.1
2,4-Dichlorophenol	11	0				11	1	6
1,2-Dichlorobenzene	11	0				11	1	30
1,4-Dichlorobenzene	11	1	18			10	0	
Benzene	10	3	<10–40	20	22	0		
2-Chloronapthalene	11	0				0		
Chloroform	10	1	<10			0		
1,1-Dichloroethylene	10	1	10			0		
Ethyl benzene	10	1	<10			0		
Methylene chloride	10	1	<10			0		
Phenol	11	3	9–26	<10	13	0		
Butyl benzyl phthalate	11	1	<10			0		
Di-*n*-butyl phthalate	11	4	<10– < 10	<10	<10	0		
Toluene	10	1	<10			0		
Trichloroethylene	10	1	<10			0		
1,1,1-Trichloroethane	11	0				0		
Pentachlorophenol	11	0				0		
Diethyl phthalate	11	1	50		50	0		
Tetrachloroethylene	11	1	<10			0		
Benzidene	11	0				0		
Methyl chloride	11	1	50			0		

Source: USEPA.

Table 2A Summary of Priority Pollutants in the Steam Electric Industry Recirculating Cooling Water

	Intake				
Pollutant	Number of samples	Number of detections	Range of detections	Median of detections	Mean of detections
Classical pollutant (mg/L)					
TDS	8	6	190–26,000	300	4,600
TSS	7	7	0.005–110	2	20
TOC	8	8	<1–34	15	16
Phenolics	7	5	0.002–0.02	0.007	0.008
Total residual chlorine	7	4	0.005–14	0.59	3.8
Sedium	8	5	17–6,000	95	1,300
Tin	8	2	0.06–0.3		0.18
Titanium	8	3	0.02–0.2	0.03	0.083
Iron	8	4	1.0–4.0	2	2.2
Vanadium	8	4	0.011–0.2	0.016	0.061
Barium	8	5	0.02–0.5	0.1	0.19
Boron	8	5	0.06–5.0	0.3	1.9
Calcium	8	6	6.9–340	56	120
Cobalt	8	3	0.008–0.01	0.01	0.009
Manganese	8	6	0.008–7.6	0.14	1.4
Magnesium	8	5	4.5–100	22	33
Molybdenum	8	4	0.02–0.08	0.02	0.035
Aluminum	8	2	0.7–2		1.4
Toxic pollutants (μg/L)					
Toxic metals and inorganics					
Antimony	8	2	4–7		5.5
Arsenic	8	0			
Cadmium	8	5	1.4–100	8	24
Chromium	8	6	7–440	71	140
Copper	8	8	10–700	30	110

Cyanide	8	2	4–15,000		7,500
Lead	8	5	6–500	11	120
Nickel	8	6	1.7–200	82	82
Silver	8	4	1.3–40	1.8	11
Selenium	8	2	2–2		2
Zinc	8	5	15–340	53	100
Thallium	8	1	20		
Mercury	8	1	0.5		
Toxic organics					
Benzene	8	2	1.2–2.4		1.8
Carbon tetrachloride	8	1	<1		
Chloroform	7	1	1.4		
1,2-Dichlorobenzene	8	2	5.3–18		12
Dichlorobromomethane	6	0			
Chlorodibromomethane	6	0			
Toluene	8	2	2–9.1		5.5
Trichloroethylene	8	1	4		
1,4-Dichlorobenzene	8	1	2.4		
2,4,6-Trichlorophenol	8	0			
2,4-Dichlorophenol	8	4	3–12	7	7.2
Pentachlorophenol	7	2	4–12		8
Bromoform	7	0			
			Discharge		
Classical pollutant (mg/L)					
TDS	8	6	430–34,000	1,600	11,000
TSS	8	8	2–460	30	84
TOC	8	8	8–76	10	28
Phenolics	7	4	2.5–20	8	9.6
Total residual chlorine	7	6	0.005–110	1.3	19
Sodium	8	5	33–7,000	210	1,500

Table 2B Summary of Priority Pollutants in the Steam Electric Industry Recirculating Cooling Water–Discharge Information

Pollutant	Discharge				
	Number of samples	Number of detections	Range of detections	Median of detections	Mean of detections
Tin	8	4	0.03–0.5	0.045	0.16
Titanium	8	2	0.02–0.2		0.11
Iron	8	4	0.3–4	1.5	1.8
Vanadium	8	5	0.01–0.2	0.02	0.054
Barium	7	3	0.02–0.2	0.1	0.11
Boron	8	5	0.06–4	0.07	1.2
Calcium	8	6	6.9–460	82	130
Cobalt	8	4	0.008–0.06	0.01	0.022
Manganese	8	5	0.05–0.1	0.1	0.076
Magnesium	8	5	4.9–57	20	28
Molybdenum	8	4	0.02–0.1	0.05	0.055
Aluminum	8	2	0.2–1		0.6
Toxic pollutant (μg/L)					
Toxic metals and inorganics					
Antimony	8	2	5–7		6
Arsenic	8	2	2–4		3
Cadmium	8	6	1–200	4.5	36
Chromium	8	8	2–550	42	150
Copper	8	8	42–3,800	55	560
Cyanide	8	2	3–5		4
Lead	8	4	3–800	85	240

Nickel	8	8	4–140	9	56
Silver	8	5	0.7–80	3	20
Selenium	8	0			
Zinc	8	5	38–780	200	270
Thallium	8	2	8–20		14
Mercury	8	2	0.2–1		0.6
Toxic organics					
Benzene	8	2	<1–1.5		1.0
Carbon tetrachloride	7	0			
Chloroform	8	2	<1–2.4		1.4
1,2-Dichlorobenzene	7	1	26		
Dichlorobromomethane	8	2	2.6–8.2		5.4
Chlorodibromomethane	8	2	<1–58		30
Toluene	7	1	24		
Trichloroethylene	8	1	4		
1,4-Dichlorobenzene	7	0			
2,4,6-Trichlorophenol	8	1	35		
2,4-Dichlorophenol	8	2	8–8		
Pentachlorophenol	8	2	4–4		8
Bromoform	8	1	150		4

Source: USEPA.

chlorodibromomethane, *bis*(2-ethylhexyl) phthalate, antimony, arsenic, cadmium, chromium, mercury, nickel, selenium, silver, thallium, benzene, tetrachloroethylene, toluene, copper, cyanide, lead, zinc, chloroform, phenol, asbestos, total dissolved solids, total suspended solids, total organic carbon, total residual chlorine, 1,2-dichlorobenzene, 2,4-dichlorophenol, boron, calcium, magnesium, molybdenum, total phenolics, sodium, tin, vanadium, cobalt, iron, chloride, 2,4,6-trichlorophenol, and pentachlorophenol. It must be recognized, however, that recirculating cooling systems tend to concentrate the dissolved solids present in the make-up water and, thus, a blowdown stream with many different compounds showing concentration increases is to be expected. Of the priority pollutants detected as net discharges, the concentration increase was greater than 10 ppb only for *bis*(2-ethylhexyl) phthalate, cadmium, chromium, nickel, selenium, silver, toluene, copper, cyanide, lead, zinc, phenol, 1,2-dichlorobenzene, total phenolics, and 2,4,6-trichlorophenol.

13.3.3 Ash Transport

The chemical compositions of both types of bottom ash, dry or slag, are quite similar. The major species present in bottom ash are silica (20–60 wt.% as SiO_2), alumina (10–35 wt.% as Al_2O_3), ferric oxides (5–35 wt.% as Fe_2O_3), calcium oxide (1–20 wt.% as CaO), magnesium oxide (0.3–0.4 wt.% as MgO), and minor amounts of sodium and potassium oxides (1–4 wt.%). In most instances, the combustion of coal produces more fly ash than bottom ash. Fly ash generally consists of very fine spherical particles, ranging in diameter from 0.5 to 500 microns. The major species present in fly ash are silica (30–50 wt.% as SiO_2), alumina (20–30 wt.% Al_2O_3), and titanium dioxide (0.4–1.3 wt.% as TiO_2). Other species that may be present include sulfur trioxide, carbon, boron, phosphorus, uranium, and thorium.

In addition to these major components, a number of trace elements are also found in bottom ash and fly ash. The trace elemental concentrations can vary considerably within a particular ash or between ashes. Generally, higher trace element concentrations are found in the fly ash than bottom ash; however, there are several cases where bottom ash exceeds fly ash concentration. Fly ash demonstrates an increased concentration trend with decreasing particle sizes.

During the verification sampling effort, the ash pond overflows of nine facilities were sampled to further quantify those effluent pollutants identified in the screening program. The data are presented in Table 3A and 3B for information of intake and discharge, respectively.

13.3.4 Low-Volume Wastes

Low-volume waste sources include water treatment processes that prevent scale formation such as clarification, filtration, lime/lime soda softening, ion exchange, reverse osmosis, and evaporation. Also included are drains and spills from floor and yard drains and laboratory streams.

Clarification Wastes

Clarification is the process of agglomerating the solids in a stream and separating them by settling. Chemicals that are commonly added to the clarification process do not contain any of the listed priority pollutants.

Table 3A Summary of Priority Pollutants in the Steam Electric Industry Ash Pond Overflow–Intake Information

	Intake				
Pollutant	Number of samples	Number of detections	Range of detections	Median of detections	Mean of detections
Classical pollutants (mg/L)					
Oil and grease	10	1	25		
TDS	11	8	130–530	260	300
TSS	9	8	0.005–170	13	41
TOC	10	7	5–21	10	9.4
Phenolics	10	4	0.006–0.04	0.01	0.02
Chloride	10	1	14		
Aluminium	11	5	0.2–2	0.5	0.78
Barium	11	7	0.017–0.06	0.03	0.032
Boron	11	6	0.06–0.1	0.075	0.077
Calcium	11	8	6.9–57	33	33
Cobalt	11	3	0.007–0.04	0.01	0.019
Manganese	11	8	0.04–0.8	0.082	0.19
Magnesium	11	8	4.5–23	6.9	13
Molybdenum	11	2	0.009–0.06		0.034
Sodium	11	8	15–57	26	28
Tin	11	3	0.01–0.036	0.03	0.025
Titanium	11	4	0.018–0.04	0.025	0.027
Iron	11	10	0.2–20	0.5	3
Vanadium	10	3	0.01–0.04	0.013	0.021
Yttrium	11	0			
Toxic pollutants (μg/L)					
Toxic metals and inorganics					
Antimony	11	3	3–7	4	4.7
Arsenic	11	2	2–7		2.5
Beryllium	11	0			
Cadmium	11	5	2.1–40	6.5	11
Chromium	11	9	3–4,000	10	460
Copper	11	11	8–700	20	84
Cyanide	11	2	4–15,000		7,500
Lead	11	9	1.7–20	10	11
Mercury	11	2	0.2–0.5		0.35
Nickel	11	10	1.7–2,000	18	210
Selenium	11	2	2–3		2.5
Silver	11	2	1.5–1.6		1.6
Thallium	11	1	1		
Zinc	11	7	15–88	32	4.3
Toxic organics					
Benzene	11	3	1.2–10	2.4	4.5
Carbon tetrachloride	11	1	1		
Chloroform	11	3	0.17–10	1.4	3.9
1,2-Dichlorobenzene	11	1	5.3		
Ethylbenzene	9	0			

(continues)

Table 3A *Continued*

Pollutant	Intake				
	Number of samples	Number of detections	Range of detections	Median of detections	Mean of detections
Toluene	11	2	2–9.1		5.5
Trichloroethylene	11	2	0.57–4		1.3
1,1-Dichloroethylene	11	0			
1,4-Dichlorobenzene	11	1	2.4		
Methylene chloride	10	1	10		
Phenol	11	1	9		
Bis(2-ethylhexyl) phthalate	11	1	10		
Butyl benzyl phthalate	11	1	10		
Di-*n*-butyl phthalate	11	1	10		
Diethyl phthalate	11	1	50		
Dimethyl phthalate	11	0	2.7		
Tetrachloroethylene	11	2	0.4–10		5.2
1,1,2,2-Tetrachloroethane	11	1	24		
1,1,1-Trichloroethane	11	1	0.68		
Pentachlorophenol	11	1	3.8		
Pesticides					
4,4′-DDD	11	1	1		

Table 3B Summary of Priority Pollutants in the Steam Electric Industry Ash Pond Overflow–Discharge Information

Pollutant	Discharge				
	Number of samples	Number of detections	Range of detections	Median of detections	Mean of detections
Classical pollutants (mg/L)					
Oil and grease	12	2	1–24		12
TDS	12	9	4–2,400	490	700
TSS	12	11	5–160	15	33
TOC	12	8	3–150	16	32
Phenolics	12	7	0.006–0.04	0.01	0.02
Chloride	12	2	37–37		37
Aluminium	12	6	0.06–5	0.55	1.4
Barium	12	8	0.04–0.2	0.06	0.083
Boron	12	8	0.08–2	0.6	0.95
Calcium	12	9	21–140	64	71
Cobalt	12	4	0.007–0.05	0.015	0.022
Manganese	12	8	0.01–1	0.1	0.24
Magnesium	12	9	5.6–20	9.5	11
Molybdenum	12	9	0.005–0.3	0.05	0.094
Sodium	12	9	15–70	32	34
Tin	12	6	0.007–0.036	0.025	0.022

(*continues*)

Table 3B *Continued*

Pollutant	Discharge				
	Number of samples	Number of detections	Range of detections	Median of detections	Mean of detections
Titanium	12	2	0.02–0.05		0.035
Iron	12	9	0.17–8	0.9	1.7
Vanadium	12	5	0.02–0.14	0.022	0.028
Yttrium	12	1	0.02		
Toxic pollutants (μg/L)					
Toxic metals and Inorganics					
Antimony	12	4	6–10	6.5	7.2
Arsenic	11	2	9–300		150
Beryllium	12	3	2–2.5	2.5	2.3
Cadmium	12	8	1–90	3.5	15
Chromium	12	12	4–1,000	12	120
Copper	12	11	8–80	25	43
Cyanide	12	2	22–22		22
Lead	12	8	1.2–120	8.5	25
Mercury	11	3	0.2–1.5	1.0	0.9
Nickel	12	12	5.2–470	21	65
Selenium	12	3	3–13	8	8
Silver	12	6	0.5–24	4.8	7.8
Thallium	12	0			
Zinc	12	8	15–1,200	100	260
Toxic organics					
Benzene	12	3	2–10	2	4.0
Carbon tetrachloride	11	0			
Chloroform	11	2	0.25–10		5.1
1,2-Dichlorobenzene	11	0			
Ethylbenzene	12	3	1–10		4.0
Toluene	12	2	3.5–3.5	1	3.5
Trichloroethylene	10	0			
1,1-Dichloroethylene	12	1	10		
1,4-Dichlorobenzene	12	1	2.4		
Methylene chloride	12	2	10–32		21
Phenol	12	1	4		
Bis(2-ethylhexyl)phthalate	12	1	10		
Butyl benzyl phthalate	12	0			
Di-*n*-butyl phthalate	12	1			
Diethyl phthalate	12	1	10		
Dimethyl phthalate	12	1	10		
Tetrachloroethylene	11	0	10		
1,1,2,2-Tetrachloroethane	11	0			
1,1,1-Trichloroethane	11	0			
Pentachlorophenol	12	1	6.5		
Pesticides					
4,4′-DDD					

Source: USEPA.

Ion Exchange Wastes

Ion exchange processes can be designed to remove all mineral salts in a one-unit operation and, as such, is the most common means of treating supply water. The process uses an organic resin that must be regenerated periodically by backwashing and releasing the solids. A regenerant solution is passed over the bed and it is subsequently washed.

The resulting exchange wastes are generally acidic or alkaline with the exception of sodium chloride solutions, which are neutral. While these wastes do not have significant amounts of suspended solids, certain chemicals such as calcium sulfate and calcium carbonate have extremely low solubilities and are often precipitated because of common ion effects.

Spent regenerant solutions, constituting a significant part of the total flow of wastewater from ion exchange regeneration, contains ions that are eluted from the ion exchange material plus the excess regenerant that is not consumed during regeneration. The eluted ions represent the chemical species that were removed from water during the service cycle of the process. Table 4A and 4B present a summary of ion exchange demineralizer regenerant wastes characterized in the surveillance and analysis study, for the information of intake and discharge, respectively.

Lime Softener Wastewater

Softening removes hardness using chemical precipitation. The two major chemicals used are calcium hydroxide and sodium carbonate, thus no priority pollutants will be introduced into the system.

Reverse Osmosis Wastewater

Reverse osmosis is a process used by some plants to remove dissolved salts. The waste stream from this process consists of reverse osmosis brine. In water treatment schemes reported by the industry, reverse osmosis was always used in conjunction with demineralizers, and sometimes with clarification, filtration, and ion exchange softening.

Floor and Yard Drain Wastewater

As a result of the numerous potential sources of wastewater from equipment drainage and leakage throughout a steam electric facility, the pollutants encountered in such wastewaters may be diverse. There have been little data reported for these waste streams; however, the pollutant parameters that may be of concern are oil and grease, pH, and suspended solids.

Laboratory Drain Wastewater

The wastes from the laboratories vary in quantity and constituents, depending on the use of the facilities and the type of power plant. The chemicals are usually present in extremely small quantities. It has been common practice to combine laboratory drains with other plant plumbing.

Boiler Blowdown

Boiler blowdown is generally of fairly high quality because the boiler feedwater must be maintained at high quality. Boiler blowdown having a high pH may contain a high dissolved solids concentration depending on boiler pressure. The sources of impurities in the blowdown are the intake water, internal corrosion of the boiler, and chemicals added to the boiler system. Impurities contributed by the intake water are usually soluble inorganic species (Na^+, K^+, Cl^-,

Table 4A Summary of Priority Pollutants in the Steam Electric Industry Demineralizer Regenerant–Intake Information

	Intake				
Pollutant	Number of samples	Number of detections	Range of detections	Median of detections	Mean of detections
Classical pollutants (mg/L)					
TDS	3	2	210–290		250
TSS	2	1	2.8		
TOC	3	2	2.3–9		5.6
Aluminium	3	1	0.50		
Barium	3	1	0.017		
Boron	3	0			
Calcium	3	1	49		
Manganese	3	1	0.065		
Magnesium	3	1	15		
Molybdenum	3	0			
Sodium	2	0			
Titanium	3	1	0.018		
Iron	3	2	0.001–0.84		0.42
Toxic pollutants (μg/L)					
Toxic metals and inorganics					
Antimony	3	0			
Arsenic	3	2	2–3		2.5
Cadmium	3	1	4		
Copper	3	3	9–22	22	18
Chromium	3	1	10		
Cyanide	3	0			
Lead	3	0			
Mercury	3	3	0.2–1.5	1	0.9
Nickel	3	1	8		
Selenium	3	1	1		
Silver	3	0			
Zinc	3	3	10–100	88	66
Thallium	3	0			
Toxic organics					
Benzene	3	1	<10		
Chloroform	3	2	4.4–68		36
1,1-Dichloroethylene	3	0			
Methylene chloride	3	1	<10		
Bromoform	3	1	23		
Dichlorobromomethane	3	1	0.87		
Chlorodibromomethane	3	1	0.17		
Phenol	3	2	4.2–9		6.6
Bis(2-ethylhexyl) phthalate	3	1	<10		
Butyl benzyl phthalate	3	1	<10		
Di-*n*-butyl phthalate	3	1	<10		
Diethyl phthalate	3	1	50		
Tetrachloroethylene	3	1	<10		

(*continues*)

Table 4A *Continued*

Pollutant	Intake				
	Number of samples	Number of detections	Range of detections	Median of detections	Mean of detections
Trichloroethylene	3	2	0.13–<10		<5.1
Chlorobenzene	3	0			
1,1,2-Trichloroethane	3	1	0.23		
1,2-Dichlorobenzene	3	0			
1,3-Dichlorobenzene	3	0			
1,4-Dichlrorobenzene	3	0			
Nitrobenzene	3	0			

Table 4B Summary of Priority Pollutants in the Steam Electric Industry Demineralizer Regenerant–Discharge Information

Pollutant	Discharge				
	Number of samples	Number of detections	Range of detections	Median of detections	Mean of detections
Classical pollutant (mg/L)					
TDS	3	2	3,000–4,600		3,800
TSS	3	2	9.2–17		13
TOC	3	2	4.8–8		6.4
Aluminium	3	1	0.28		
Barium	3	0			
Boron	3	1	0.063		
Calcium	3	1	170		
Manganese	3	1	0.009		
Magnesium	3	1	17		
Molybdenum	3	1	0.015		
Sodium	3	1	160		
Iron	3	2	0.790–5		2.9
Toxic pollutant (μg/L)					
Toxic metals and inorganics					
Antimony	3	1	20		
Arsenic	1	0			
Cadmium	3	2	5–35		20
Copper	2	2	14–65		40
Chromium	3	2	14–26		20
Cyanide	3	2	0.04–47		24
Lead	3	1	24		
Mercury	2	2	1.6–6		3.8
Nickel	3	2	200–230		220
Selenium	2	1	4		
Silver	3	1	58		
Zinc	2	1	54		
Thallium	3	1	180		

(*continues*)

Table 4B *Continued*

Pollutant	Discharge				
	Number of samples	Number of detections	Range of detections	Median of detections	Mean of detections
Toxic organics					
Benzene	3	0			
Chloroform	3	3	1.8–140	38	60
1,1-Dichloroethylene	3	1	<10		
Methylene chloride	3	2	60–>220		>140
Bromoform	2	1	>10		
Dichlorobromomethane	3	1	70		
Chlorodibromomethane	3	1	30		
Phenol	3	2	3.8–4		3.9
Bis(2-ethylhexyl) phthalate	3	1	<10		
Butyl benzyl phthalate	2	0			
Di-*n*-butyl phthalate	3	1	<10		
Diethyl phthalate	3	1	<10		
Tetrachloroethylene	3	1	<10		
Trichloroethylene	3	1	0.38		
Chlorobenzene	3	1	0.67		
1,1,2-Trichloroethane	3	1	0.68		
1,2-Dichlorobenzene	3	1	39		
1,3-Dichlorobenzene	3	1	0.3		
1,4-Dichlrorobenzene	3	1	5.2		
Nitrobenzene	3	1	81		

Source: USEPA.

SO_4^{2-}, etc.) and precipitates containing calcium/magnesium cations. Products of boiler corrosion are soluble and insoluble species of iron, copper, and other metals. A number of chemicals are added to the boiler feedwater to control scale formation, corrosion, pH, and solids deposition. Table 5 presents a summary of toxic and classical pollutants detected in verification analyses of boiler blowdown.

13.3.5 Metal Cleaning Wastes

Metal cleaning wastes include wastewater from chemical cleaning of boiler tubes, air preheater washwater, and boiler fireside washwater.

Chemical Cleaning of Boiler Tubes

The characteristics of waste streams emanating from the chemical cleaning of utility boilers are similar in many respects. The major constituents consists of boiler metals; that is, alloy metals used for boiler tubes, hot wells, pumps, and so on. Although waste streams from certain cleaning operations that are used to remove certain deposits, such as alkaline degreaser to remove oils and organics, do not contain heavy concentrations of metals, the primary purpose of the total boiler cleaning operation (all stages combined) is removal of heat

Table 5 Summary of Priority Pollutants in the Steam Electric Industry Boiler Blowdown

	Intake					Discharge				
Pollutant	Number of samples	Number of detections	Range of detections	Medium of detections	Mean of detections	Number of samples	Number of detections	Range of detections	Medium or detection	Mean or detections
Classical pollutants (mg/L)										
TDS	4	2	210–290		250	4	3	7–100	11	39
TSS	3	1	2.8			4	3	0.8–<5	<5	1.9
TOC	3	2	2.3–9		5.6	4	3	1.2–3	<3	1.9
Oil and grease	4	0				4	1	5		
Phenolics	4	2	4.2–<20		7.1	4	2	6.4–<20		8.2
Aluminum	4	0				4	1	0.21		
Calcium	4	1	49			4	2	<5–<5		<5
Manganese	4	1	0.065			4	0			
Magnesium	4	1	15			4	0			
Molybdenum	4	0				4	2	0.055–0.061		0.058
Sodium	3	0				4	1	<15		
Iron	4	2	0.01–0.84		0.43	3	1	0.06		
Titanium	4	1	0.018			4	0			
Toxic pollutants (μg/L)	4									
Toxic metals	4									
Antimony	4	0				4	3	6–20	10	12
Arsenic	4	2	2–3		2.5	4	2	2–2		2
Cadmium	4	1	4			4	1	5		
Chromium	4	1	10			4	1	6		
Copper	4	3	9–22	22	18	4	4	8–520	17	140
Lead	4	0				4	2	36–40		38
Mercury	4	3	0.2–1.5	1.0	0.9	3	1	1.7		
Nickel	4	1	8			4	1	1.3		
Selenium	4	0				4	1	5.7		
Zinc	4	3	10–100	88	66	4	3	10–72	68	50
Toxic organics	4									

Benzene	4	1	<10			4	2	30–290	150
1,1,1-Trichloroethane	4	0				4	1	<10	
1,1,2,2-Tetrachloroethane	4	0				4	1	<10	
Chloroform	4	3	4.4–6.8	<5	4.6	4	2	0.12–<10	2.6
1,1-Dichloroethylene	4	0				4	2	<10–<60	<35
Ethyl benzene	4	0				4	2	<10–<10	<10
Methylene chloride	4	1	<10			4	1	910	
Bis(2-thylhexyl)phthalate	4	1	<10			4	1	<10	
Butyl benzyl phthalate	4	1	<10			4	0		
Di-*n*-butyl phthalate	4	1	<10			4	2	<10–<10	<10
Diethyl phthalate	4	1	50			4	2	<10–<10	<10
Tetrachloroethylene	4	1	<10			4	2	<10–<10	<10
Toluene	4	0					2		
Trichloroethylene	4	2	0.13–<10			4	0		
Dichlorobromomethane	4	2	0.87–23		2.6	4	0		
Chlorodibromomethane	4	2	0.17–3.8		12	4	0		
1,1,2-Trichloroethane	4	1	0.23		2	4	0		
Bromoform		1	0.07			4	1	<10	
1,3-Dichloropropene	4	0				4	1	<10	
Phenol	4	0				4	1	10	

Source: USEPA.

transfer-retarding deposits, which consist mainly of iron oxides resulting from corrosion. This removal of iron is evident in all total boiler cleaning operations through its presence in boiler cleaning wastes.

Cleaning mixtures used include alkaline chelating rinses, proprietary chelating rinses, organic solvents, acid cleaning mixtures, and alkaline mixtures with oxidizing agents for copper removal. Wastes from these cleaning operations will contain iron, copper, zinc, nickel, chromium, hardness, and phosphates. In addition to these constituents, wastes from alkaline cleaning mixtures will contain ammonium ions, oxidizing agents, and high alkalinity; wastes from acid cleaning mixtures will contain fluorides, high acidity, and organic compounds; wastes from alkaline chelating rinses will contain high alkalinity and organic compounds; and wastes from most proprietary processes will be alkaline and will contain organic and ammonium compounds. Other waste constituents present in spent chemical cleaning solutions include wide ranges of pH, high dissolved solids concentrations, and significant oxygen demands (BOD and/or COD). The pH of spent solutions ranges from 2.5 to 11.0 depending on whether acidic or alkaline cleaning agents are employed.

Table 6 presents a summary of toxic and classical pollutants detected in three common cleansing solutions: ammoniacal sodium bromate, hydrochloric acid without copper complexer, and hydrochloric acid with copper complexer.

Boiler Fireside Wastewater

When boiler firesides are washed, the waste effluents produced contain an assortment of dissolved and suspended solids. Acid wastes are common for boilers fired with high-sulfur fuels. Sulfur oxides absorb onto fireside deposits, causing low pH and a high sulfate content in the waste effluent.

Air Preheater Wastewater

Fossil fuels with significant sulfur content will produce sulfur oxides that absorb on air preheater deposits. Water washing of these deposits produces an acidic effluent. Alkaline reagents are often added to washwater to neutralize acidity, prevent corrosion of metallic surfaces, and maintain an alkaline pH. Alkaline reagents might include soda ash (Na_2CO_3), caustic soda (NaOH), phosphates, and/or detergent. Preheater washwater contains suspended and dissolved solids, which include sulfates, hardness, and heavy metals including copper, iron, nickel, and chromium.

13.3.6 Ash Pile, Chemical Handling, and Construction Area Runoff

Runoff wastewater characteristics change all the time. No reliable data have been gathered. The readers are referred to the literature for similar technical information [4,5].

13.3.7 Coal Pile Runoff

No reliable data have been gathered for the wastewater characteristics of coal pile runoff. Example 3 (Section 13.6.3) presents the technical information on the characteristics of a combined wastewater (consisting of coal pile runoff, regeneration wastewater, and fly ash), and its treated effluent.

Table 6 Summary of Priority Pollutants in the Steam Electric Industry Metal Cleaning Wastes

	Ammoniated EDTA Solutions					Ammonia Sodium Bromide Solutions				
Pollutant	Number of samples	Number of detections	Range of detections	Median of detections	Mean of detections	Number of samples	Number of detections	Range of detections	Median of detections	Mean of detections
Classical pollutants (mg/L)										
TDS	2	2	60,000–74,000		67,000	3	3	340–1,400	1,000	920
TSS	1	1	24			3	3	8–77	71	52
COD						2	2	24–120		72
Oil and grease	1	1	41			2	2	<5–<5		<5
pH, pH Units	7	7	8.8–10	9.2	9.3	2	2	10–10		10
Phosphorous	1	1	260			2	2	10–30		20
Bromide						2	2	<5–52		28
Chloride						1	1	60		
Fluoride						2	2	1.5–6.1		3.0
Aluminum	1	1	31			2	2	<0.2–<0.2		<0.2
Calcium	2	2	21–45		33	3	2	0.4–3		1.7
Barium						2	2	<0.1–<0.1		<0.1
Sodium	1	1	370			3	3	3.7–59	15	26
Potassium						2	2	70–220		140
Tin						2	2	<1–<1		<1
Iron	7	7	2,200–8,300	6,900	6,300	5	4	0.15–4.9	1.8	2.2
Manganese	2	2	50–73		61	3	3	0.01–0.04	0.03	0.03
Magnesium	2	2	11–21		16	3	2	0.67–2.9		1.8

(*continues*)

Table 6 *Continued*

Pollutant	Ammoniated EDTA Solutions					Ammonia Sodium Bromide Solutions				
	Number of samples	Number of detections	Range of detections	Median of detections	Mean of detections	Number of samples	Number of detections	Range of detections	Median of detections	Mean of detections
Toxic pollutants (μg/L)										
Toxic metals										
Arsenic						3	3	2.5–310,000	40	100,000
Cadmium						2	2	<10–<10		<10
Chromium	3	3	10,000–26,000	12,000	16,000	2	2	<1–<20		<10
Copper	7	7	17–12,000,000	120,000	1,900,000	4	3	<5–<50	<5	<20
Lead						6	6	100,000–790,000	370,000	420,000
Mercury						3	3	<10–100	<10	37
Nickel	3	3	12,000–140,000	68,000	73,000	3	3	<0.2–15,000	<0.2	5,000
Selenium						5	4	80–260,000	1,500	57,000
Silver						3	3	<2–24,000	<2	8,000
Zinc	3	3	79,000–140,000	120,000	110,000	2	2	<10–<20		<15
						5	5	60–1,000	500	510

Source: USEPA

13.3.8 Wet Flue Gas Cleaning Blowdown

The readers are referred to another source for more detailed information regarding wet flue gas cleaning blowdown characteristics and treatment [2,3].

13.4 WASTE TREATMENT

13.4.1 End-of-Pipe Treatment Technologies

Wastewater effluents discharged to publicly owned treatment facilities are sometimes treated by physical or chemical systems to remove pollutants potentially hazardous to the POTW or which may be treated inadequately in the POTW. Such treatment methods are numerous, but they generally fall into one of three broad categories in accordance with their process objectives. These include pH control, removal of dissolved materials, and separation of phases.

The following is a summary of end-of-pipe treatment technologies commonly employed in the steam electric power generation industry, their objectives, equipment and processes required, and efficiency [12–22].

Neutralization

This is a process for pH adjustment, usually to within the range 6–9. Acid or base is used as required; this is usually in the form of sulfuric acid or lime.

Chemical Reduction

This is a process mainly used in power plants for reduction of hexavalent chromium to trivalent chromium. Sulfur dioxide, sodium bisulfite, sodium metabisulfite, and ferrous salts are common reducing agents to be used in the process. A pH range of 2–3 should be controlled. The process efficiency of removal is about 99.7%.

Precipitation

This is a process mainly used in power plants for removal of ions by forming insoluble salts. Common precipitating agents are lime, hydrogen sulfide, organic precipitants, and sods ash. Optimum pH depends on the ions to be removed. The removal efficiency for inorganic pollutants is as follows:

- Copper, 96.6%;
- Nickel, 91.7%;
- Chromium, 98.8 %;
- Zinc, 99.7%;
- Phosphate, 93.6%.

Ion Exchange

This is a process mainly used in power plants for removal of ions by sorption on the surface of a solid matrix. Synthetic cation and/or anion exchange resins are required depending on the pollutants to be removed. It may require pH adjustments. The removal efficiency for inorganic pollutants is as follows:

- Cyanide, 99%;
- Chromium, 98%;

- Copper, 95%;
- Iron, 100%;
- Cadmium, 92%;
- Nickel, 100%;
- Zinc, 75%;
- Phosphate, 90%;
- Sulfate, 97%;
- Aluminum, 98%.

Liquid/Liquid Extraction

This is a process mainly used in power plants for removal of soluble organics or chemically charged pollutants. The required chemicals are immiscible solvents that may contain chelating agents. It may require pH adjustments. The removal efficiency for inorganic pollutants is as follows:

- Phenol, 99%;
- Chromium, 99%;
- Nickel, 99%;
- Zinc, 99%;
- Fluoride, 68%;
- Iron, 99%;
- Molybdenum, 90%;

Disinfection

This is a process for destruction of microorganisms. Chlorine, hypochlorite salts, phenol, phenol derivatives, ozone, salts of heavy metals, chlorine dioxide, and so on are effective disinfectants. It may require pH adjustments.

Adsorption

This is a process mainly used in power plants for removal of sorbable contaminants. Activated carbon, synthetic sorbents are the common adsorbents to be used in the process. It may require pH adjustments. The process removal efficiency depends on the nature of the pollutants and the composition of the waste.

Chemical Oxidation

This is a process mainly used in power plants for destruction of cyanides using chlorine, hypochlorite salts, or ozone. The process removal efficiency is about 99.6% [12–19].

Distillation

This is a process mainly used in power plants for separation of dissolved matters by evaporation of the water. Multistage flash distillation, multiple-effect vertical long-tube vertical evaporation, submerged tube evaporation, and vapor compression are effective process equipment. It may require pH adjustment. The process removal efficiency is about 100%.

Reverse Osmosis (RO)

This is a process mainly used in power plants for separation of dissolved matter by filtration through a semipermeable membrane. Tubular membrane, hollow filter modules, or spiral-wound flat sheet membrane can be adopted for the RO process. Total dissolved solids (TDS) removal efficiency is about 93% [20].

Electrodialysis (ED)

This is a process mainly used in power plants for removal of dissolved polar compounds. Solute is exchanged between two liquids through a selective semipermeable membrane in response to differences in chemical potential between the two liquids. The process removal efficiency for TDS is about 62–96%.

Freezing

This is a process mainly used in power plants for separation of solute from liquid by crystallizing the solvent. Either direct refrigeration, or indirect refrigeration can be used. The process removal efficiency is over 99.5%.

13.4.2 Solid–Liquid Separation Technologies

The solid/liquid separation technologies commonly employed in the steam electric power generation industry include the following.

Skimming

This is a process for removal of floating solids from liquid wastes. It requires between about 1 and 15 minutes of retention time, and has a removal efficiency of 70–90%.

Clarification (Conventional)

This is a process for removal of suspended solids by settling. Typical examples are settling ponds and settling clarifiers. It requires 45 minutes to 2 hours retention time (RT), and can reduce TSS to 15 mg/L or below.

Flotation

This is an innovative separation process for removal of suspended solids and oil and grease by flotation followed by skimming. It requires very short RT (less than 30 minutes), and can achieve 90–99% removal efficiency [15,18].

Microstraining

This is mechanical separation process for removal of suspended solids by passing the wastewater through a microscreen. A removal efficiency for TSS is 50–80% depending on the pore size of the microscreen to be used.

Filtration

This is physical operation for removal of suspended solids by filtration through a bed of sand and gravel. TSS removal efficiency is 50–90% depending on the type of filter media used and the filtration rate.

Screening

This is a unit operation for removal of large solid matter by passing through screens. The efficiency for large solid removal is 50–99% depending on the type of coarse screen or bar screen to be used.

Thickening

This is a process for concentration of sludge by removing water. Either gravity thickening or dissolved air flotation thickening can be used. The thickening efficiency depends on the nature of sludge to be processed [15].

Pressure Filtration

This is a unit operation for separation of solid from liquid by passing through a semipermeable membrane or filter media under pressure. It requires 1 to 3 hours of RT, and reduces 50% of moisture content.

Heat Drying

This is a process for reducing the water content of sludge by heating. Flash drying, spray drying, rotary kiln drying, or multiple hearth drying can be used.

Ultrafiltration

This is a separation process for removal of macromolecules of suspended matter from the waste by filtration through a semipermeable membrane under pressure. Total solids removal of 95% and above can be achieved.

Sandbed Drying

This is a process for removal of moisture from sludge by evaporation and drainage through sands. The RT is as long as 1 to 2 days. It is practiced extensively by industry due its low cost.

Vacuum Filtration

This is a process for solid–liquid separation by vacuum. It requires about 1 to 5 minutes RT, and can produce 30% solid in filter cake.

Centrifugation

This is a liquid/solid separation process using centrifugal force. The moisture of the sludge can be reduced to 65–70%.

Emulsion Breaking

This process is effective for separation of emulsified oil and water. It requires 2 to 8 hours of RT. Over 99% removal efficiency can be achieved if aluminum salts, iron salts, and other demulsifiers are used at optimum pH conditions. It is practised extensively by the industry [21,22].

13.4.3 Cleaner Production, Industrial Ecology, and Other Issues

Traditional industries operated in a one-way, linear fashion; natural resources from the environment are used for producing products for our society, and the generated wastes are dumped back into our environment. However, natural resources such as minerals and fossil fuels are present in finite amounts, and the environment has a limited capacity to absorb waste. The field of industrial ecology has emerged to address these issues in the power industry. The thermal energy generated from the power industry, for instance, may be reused for many domestic and industrial applications, for cost saving as well as thermal pollution control. The greenhouse gas, carbon dioxide, in flue gas is a pollutant, but can also be reused as a chemical agent in wastewater treatment [8]. An extension of the concept of sustainable manufacturing – industrial ecology – seeks to use resources efficiently and regards "wastes" as potential products [11,23].

All electric power plants should practice cleaner production and industrial ecology strategies.

Additional issues of air and thermal pollution for electric power generation are addressed in detail by Wisconsin Public Service Commission [9].

The World Nuclear Association [10] provides detailed technical information on nuclear electricity and related environmental, health and safety issues.

13.5 TREATMENT TECHNOLOGY COSTS

The investment cost, operating and maintenance costs, and energy costs for the application of control technologies to the wastewaters of the steam electric power generating industry have been analyzed. These costs were developed to reflect the conventional use of technologies in this industry. Several unit operation/unit process configurations have been analyzed for the cost of application of technologies and to select BPT and BAT levels of treatment. A detailed presentation of the applicable treatment technologies, cost methodology, and cost data are available in the literature [6,7].

13.6 PLANT-SPECIFIC EXAMPLES

13.6.1 Example 1

Plant 1226 is a bituminous coal-, oil-, and gas-fired electricity plant [1]. The recirculator cooling water system influent was sampled from a stream taken from the river and the effluent from the cooling tower blowdown stream. The effluent stream is used again in the ash sluice stream. Table 7 presents the data. The following additives are combined with the cooling tower influent:

- chlorine (biocide);
- calgon Cl-5 (corrosion inhibitor);
- sulfuric acid (scale prevention).

The addition is necessary for the control of pipe corrosion.

Table 7 Plant-Specific Treatment Data for Plant 1226 Recirculating Cooling Water

Pollutant	Influent	Effluent[a]
Classical pollutants (mg/L)		
TDS	190	1,000
TSS	14	8
TOC	10	11
Phenolics	0.01	0.008
TRC[b]	ND	<0.01
Aluminum	0.7	0.4
Barium	0.02	0.02
Boron	ND	0.06
Calcium	6.9	6.9
Cobalt	0.007	0.008
Manganese	0.2	0.1
Magnesium	4.5	4.9
Sodium	33	210
Titanium	0.02	0.02
Iron	2.0	3.0
Vanadium	ND	0.03
Flow (L/s)	745	630
Toxic pollutants (μg/L)		
Toxic metals		
Antimony	7	7
Arsenic	3	4
Cadmium	2.1	1.8
Chromium	7	20
Copper	10	48
Lead	11	3
Mercury	0.5	0.2
Nickel	14	6
Silver	1.3	0.7
Zinc	40	38
Toxic organics		
Chloroform	NA	<1
Bromoform	NA	150
Dichlorobromomethane	NA	8.2
Chlorodibromomethane	NA	58

[a]Percent removal not meaningful because water does not undergo any treatment.
[b]Total residual chorine.
Source: USEPA.

13.6.2 Example 2

Plant 1245 is an oil- and gas-fired electric generating facility. The samples chosen are the influent and effluent from a once-through cooling tower stream. The influent sample was taken from the makeup stream comprised of river water, with the effluent stream being a direct discharge from the condensers to the river. The cooling water does not undergo any treatment to remove pollutants. The data reflect the changes that may occur to such a stream due to evaporation and pipe corrosion. Table 8 presents plant-specific data for plant 1245.

Table 8 Plant-Specific Treatment Data for Plant 1245 Once-Through Cooling Water

Pollutant	Influent	Effluent[a]
Classical pollutants (mg/L)		
TDS	35,000	33,000
TSS	6	14
TOC	14	25
Phenolics	<5	<5
TRC[b]	<10	120
Flow (L/s)	4,380	4,380

[a]Percent removal is not meaningful due to the fact that the water does not undergo any treatment.
[b]Total residual chlorine.
Source: USEPA.

Table 9 Plant-Specific Treatment Data for Plant 3920 Fly Ash Pond Water

Pollutant	Influent	Effluent	Percent removal
Classical pollutant (mg/L)			
TDS	220	880	NM
TSS	12	73	NM
TOC	5	3	40
Phenolics	0.04	0.04	0
Barium	0.03	0.06	NM
Boron	0.08	1	NM
Calcium	28	120	NM
Cobalt	ND	0.007	NM
Manganese	0.05	0.3	NM
Magnesium	7.2	6.7	7
Molybdenum	ND	10	NM
Sodium	18	35	NM
Tin	ND	ND	NM
Aluminum	ND	5	NM
Iron	0.5	2	NM
Flow (L/s)	61.3	61.3	
Toxic pollutants (μg/L)			
Toxic metals			
Cadmium	ND	ND	NM
Chromium	11	30	NM
Copper	8	30	NM
Lead	20	8	60
Nickel	25	18	28
Zinc	ND	140	NM
Beryllium	ND	2	NM
Silver	ND	ND	NM

ND, not detected.
NM, not meaningful.
Source: USEPA.

Table 10 Plant-Specific Treatment Data for Plant 1742 Ash Pond Water and Once-Through Cooling Water

	Ash Pond			Once-Through Cooling Water		
Pollutant	Influent	Effluent	Percent removal	Influent	Effluent	Percent removal
Classical pollutant (mg/L)						
TDS	340	370	NM	340	1,200	NM
TSS	100	15	85	100	90	10
TOC	10	150	NM	10	9	10
Phenolics	0.006	0.01	NM	0.006	0.260	NM
TRC				NA	830	NM
Aluminum	2	ND	>99	2	NA	
Barium	0.06	0.05	17	0.06	NA	
Boron	0.09	0.2	NM	0.09	NA	
Calcium	51	51	NM	51	NA	
Cobalt	0.01	0.05	NM	0.01	NA	
Manganese	0.2	0.3	NM	0.2	NA	
Magnesium	23	20	13	23	NA	
Molybdenum	0.009	0.05	NM	0.009	NA	
Sodium	21	26	NM	21	NA	
Tin	0.03	0.03	0	0.03	NA	
Titanium	0.04	ND	>99	0.04	NA	
Iron	4	8	NM	4	NA	
Vanadium	ND	20	NM	ND	NA	
Flow (L/s)	4.38	4.38		1,440	1,440	
Toxic pollutants (μg/L)						
Toxic metals						
Cadmium	40	10	75	40	NA	
Chromium	22	1,000	NM	22	NA	
Copper	20	78	NM	20		
Lead	9	9	0	9	NA	
Nickel	17	470	NM	17	NA	
Silver	ND	ND	0	ND	NA	
Zinc	70	ND	>99	70	NA	
Mercury	ND	1.5	NM	NA	NA	

ND, not detected; NM, not meaningful; NA, not analyzed.
Source: USEPA.

Table 11 Plant-Specific Treatment Data for Plant 3001 Multiple Ash Ponds Water

Pollutant	Influent	Effluent	Percent removal
Classical pollutant (mg/L)			
TDS	530	490	8
TSS	170	30	82
Oil and grease	25	24	4
Phenolics	NA	0.01	NM
Aluminum	0.5	2	NM
Barium	0.04	0.2	NM
Boron	0.06	2	NM
Calcium	38	64	NM
Manganese	0.04	ND	>99
Cadmium	ND	0.008	NM
Magnesium	23	11	52
Molybdenum	ND	0.03	NM
Sodium	57	70	NM
Tin	ND	0.007	NM
Iron	0.2	ND	>99
Vanadium	ND	0.02	NM
Flow (L/s)	23.3	unknown	
Toxic pollutants (μg/L)			
Toxic metals			
Chromium	10	190	NM
Copper	10	ND	>99
Lead	ND	3	NM
Nickel	6	35	NM
Toxic organics			
1, 1, 2, 2-Tetrachloroethane	24	ND	>99

Analytic methods: V.7.3.31, Data set 2; ND, not detected; NM, not meaningful; NA, not analyzed.
Source: USEPA.

13.6.3 Example 3

Plant 3920 is a bituminous coal- and oil-fired plant with a generating capacity of 557 MW. This plant uses 1,220,000 Mg/year of coal. An ash settling pond was used to remove wastes from coal pile runoff, regeneration wastes, and fly ash. The influent data were obtained from the pond inlet whereas the effluent data were from the discharge stream to the river. The results of this treatment are shown in Table 9.

13.6.4 Example 4

Plant 1742 is a bituminous coal- and oil-fired plant producing 22 MW of electricity. Table 10 represents data that are from both the ash pond and the once-through cooling tower.

13.6.5 Example 5

Plant 3001 is a coal- and gas-fired facility with a generating capacity of 50 MW. The plant uses approximately 277,000 Mg/year of coal. The fly ash and bottom ash from the boiler are combined and put through a series of three settling ponds. The effluent from the ponds is discharged to the river. Table 11 shows the effectiveness of this treatment technology.

REFERENCES

1. USEPA. *Development Document for Effluent Limitations Guidelines and Standards for the Steam Electric Point Source Category*, EPA-440/1-80/029-b; U.S. Environmental Protection Agency: Washington, DC, 1980; 597p.
2. Wang, L.K.; Taricska, J.; Hung, Y.T.; Eldridge, J.; Li, K. Wet and dry scrubbing. In *Air Pollution Control Engineering*; Wang, L.K., Pereira, N.C., Hung, Y.T., Eds.; Humana Press, Inc.: Totowa, NJ, 2004.
3. Wang, L.K.; Williford, C.; Chen, W.Y. Desulfurization and emission control. In *Advanced Air and Noise Pollution Control*; Wang, L.K.; Pereira, N.C., Hung, Y.T., Eds.; Humana Press, Inc.: Totowa, NJ, 2004.
4. Wang, L.K. Treatment of storm run-off by oil–water separation, flotation, filtration and adsorption, part A: wastewater treatment. In *Proceedings of the 44th Industrial Waste Conference*, Purdue University, Lafayette, IN, 1990; 655–666.
5. Wang, L.K. Treatment of storm run-off by oil–water separation, flotation, filtration and adsorption, part B: waste sludge management. In *Proceedings of the 44th Industrial Waste Conference*, Purdue University, Lafayette, IN, 1990; 667–673.
6. Wang, J.C.; Aulenbach, D.B.; Wang, L.K. Energy models and cost models for water pollution controls. In *Clean Production*; Misra, K.B., Ed.; Springer-Verlag: Berlin, Germany, 1996; 685–720.
7. Wang, L.K.; Chen, J.L.; Hung, Y.T. Performance and costs of air pollution control technologies. In *Advanced Air and Noise Pollution Control*; Wang, L.K., Pereira, N.C., Hung, Y.T., Eds.; Humana Press, Inc.: Totowa, NJ, 2004.
8. Wang, L.K.; Krouzek, J.V.; Kounitson, U. *Case Studies of Cleaner Production and Site Remediation*; United Nations Industrial Development Organization (UNIDO): Vienna, Austria, 1995; Training Manual No. DTT-5-4-95, 136.
9. Wisconsin Public Service Commission. *Air Quality Issues for Electric Power Generation*, Publication No. 6015B; Wisconsin Public Service Commission: State of Wisconsin, PO Box 7854, Madison, WI, 1998.
10. World Nuclear Association. Nuclear electricity. In *Nuclear Energy Made Simple*, Chapter 3; World Nuclear Association, 2003; www.world-nuclear.org/education/ne/ne3/htm.
11. Wang, L.K. Industrial ecology. In *Encyclopedia of Life Support Systems: Hazardous Waste Management*, Chapter 15; Grasso, D., Vogel, T., Smets, B., Eds.; Eolss Publishers Co., Ltd.: London, 2003; www.eolss.net/E-1-08-toc.aspx.
12. Wang, L.K. *Pretreatment and Ozonation of Cooling Tower Water, Part I*; U.S. Department of Commerce, National Technical Information Service: Springfield, VA, 1983; Technical Report No. PB84-192053, 34 p., April.
13. Wang, L.K. *Pretreatment and Ozonation of Cooling Tower Water, Part II*; U.S. Department of Commerce, National Technical Information Service: Springfield, VA, 1983; Technical Report No. PB84-192046, 29 p., Aug.
14. Wang, L.K. *Prevention of Airborne Legionairs' Disease by Formulation of a New Cooling Water for Use in Central Air Conditioning Systems*; U.S. Department of Commerce, National Technical Information Service, 1984; Technical Report No. PB85-215317/AS, 97 p., Aug.
15. Wang, L.K.; Krofta, M. Treatment of cooling tower water by dissolved air–ozone flotation. In *Proceedings of the Seventh Mid-Atlantic Industrial Waste Conference*, 1985; p. 207–216, June 1985.
16. Wang, L.K. *Recent Development in Cooling Water Treatment with Ozone*; Lenox Institute of Water Technology: Lenox, MA, 1988; Technical Report No. LIR/03-88/285, 237 p., March.
17. Wang, L.K. *Treatment of Cooling Tower Water with Ozone*; Lenox Institute of Water Technology: Lenox, MA, 1988; Technical Report No. LIR/05-88/303, 55 p., May.
18. Wang, L.K. *Analysis of Sludges Generated from Flotation Treatment of Storm Runoff Water*; U.S. Department of Commerce, National Technical Information Service: Springfield, VA, 1988; Technical Report No. PB88-20062I/AS, 20 p.

19. Wang, L.K.; and Krofta, M. *Treatment of Cooling Tower Water with Ozone*. Lenox Institute of Water Technology: Lenox, MA, 1988; Report No. LIR/05-88/303, 55 p., May.
20. Wang, L.K.; Kopko, S.P. *City of Cape Coral Reverse Osmosis Water Treatment Facility*; U.S. Department of Commerce, National Technical Information Service: Springfield, VA, 1997; 15p., www.afssociety.org/publications, Association of Filtration Society.
21. Wang, L.K. *Evaluation and Development of Physical–Chemical Techniques for the Separation of Emulsified Oil from Water*, Project Report No. 189; Veridian Engineering (formerly Arvin Calspan Corp.): Buffalo, NY, 1973; 31 p., May.
22. Wang, L.K. Separation of emulsified oil from water. *Chem. Indust.* **1975**, 562–564.
23. Van Berkel, C.W.M. Cleaner production: a profitable road for sustainable development of Australian industry. *Clean Air* **1999**, *33* (*4*), 33–38.

Index